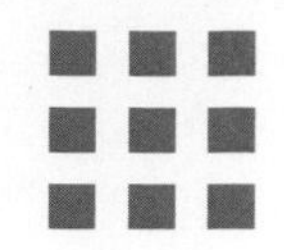

Security and Loss Prevention

Security and Loss Prevention: An Introduction

Sixth Edition

Philip P. Purpura

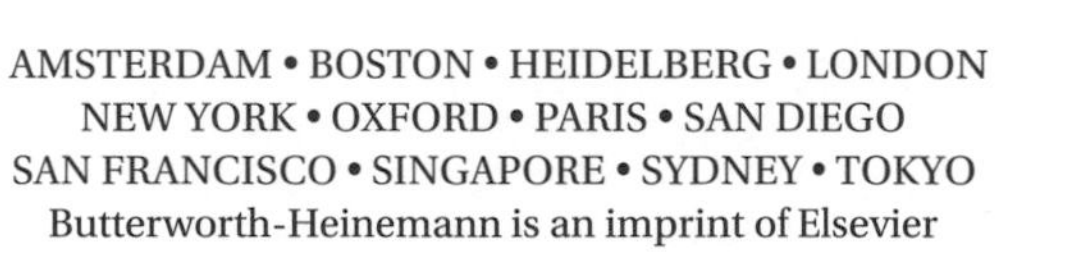
AMSTERDAM • BOSTON • HEIDELBERG • LONDON
NEW YORK • OXFORD • PARIS • SAN DIEGO
SAN FRANCISCO • SINGAPORE • SYDNEY • TOKYO
Butterworth-Heinemann is an imprint of Elsevier

Acquisition Editor: Mary Jane Peluso
Editorial Project Manager: Amber Hodge
Production Project Manager: Mohanambal Natarajan
Marketing Manager: Cindy Minor
Cover Designer: Russell Purdy

Butterworth-Heinemann is an imprint of Elsevier
The Boulevard, Langford Lane, Kidlington, Oxford OX5 1GB, UK
225 Wyman Street, Waltham, MA 02451, USA

Sixth edition 2013

Library of Congress Cataloging-in-Publication Data
Purpura, Philip P., 1950-
Security and loss prevention : an introduction / Philip Purpura. — Sixth edition.
pages cm
Includes bibliographical references and index.
ISBN 978-0-12-387846-5 (alk. paper)
1. Private security services. 2. Burglary protection. 3. Employee theft—Prevention. 4. Fire prevention.
5. Shoplifting—Prevention. 6. Security systems. I. Title.
HV8290.P87 2013
658.4′73—dc23 2012044092

British Library Cataloguing in Publication Data
A catalogue record for this book is available from the British Library

For information on all **Butterworth-Heinemann** publications
visit our web site at store.elsevier.com

Printed and bound in USA
13 14 15 16 17 10 9 8 7 6 5 4 3 2 1

Dedication

To my family.

To the millions of military, public safety, security and loss prevention, and other professionals who seek global security and safety.

Contents

About the Author xv

Preface xvii

Acknowledgments xxi

Part I Introduction to Security and Loss Prevention 1

1. The History of Security and Loss Prevention: A Critical Perspective 3

Why Critical Thinking? 4

How can we Think Critically? 5

Why Think Critically about the History of Security and Loss Prevention? 6

Security and Loss Prevention Defined 7

History 8

Early Civilizations 9

More Contemporary Times 12

Case Problems 20

References 21

2. The Business, Careers, and Challenges of Security and Loss Prevention 23

Introduction: Why is Security a Huge Industry? 24

Terms and Definitions of the Security Industry 24

Metrics 27

The Security Industry 31

Careers: Loss Prevention Services and Specialists 35

Occupational Outlook 38

Limitations of the Criminal Justice System: Implications for Loss Prevention Practitioners 38
Challenges of the Security Industry 40
Case Problems 49
References 50

Part II Reducing the Problem of Loss 53

3. Foundations of Security and Loss Prevention 55

The Security and Loss Prevention Profession 56
Methods for Protection Programs 64
Standards and Regulations 75
Evaluation of Loss Prevention Programs 80
Proprietary Security 83
Case Problems 86
References 87

4. Law 89

Introduction 90
Judicial Systems 90
Origins of Law 93
Tort Law and Controls over Private Security 94
Legal Theory of Premises Security Claims 97
Contract Law 101
Civil Justice Procedures 102
Administrative Law 103
Criminal Justice Procedures 106
Case Problems 112
The Decision for "You Be the Judge" 113
References 113

5. Internal and External Relations 115
Internal and External Relations 115
Internal Relations 118
External Relations 124
Special Problems 128
Case Problems 132
References 134

6. Applicant Screening and Employee Socialization 135
Introduction 136
Employment Law 137
Social Media, Sexual Harassment, and Employee Venting 144
Screening Methods 145
Employee Socialization 154
Case Problems 160
The Decision for "You Be the Judge" 163
References 163

7. Internal Threats and Countermeasures 165
Introduction 166
Internal Theft 168
Management Countermeasures 175
Physical Security Countermeasures 180
Case Problems 213
References 214

8. External Threats and Countermeasures 217
Introduction 218
Methods of Unauthorized Entry 218
Countermeasures 219

Case Problems 256
References 258

9. Services and Systems: Methods for Making Wise Purchasing Decisions 261
Introduction 261
Pitfalls when Purchasing Security Services and Systems 262
Purchasing Security Services 263
Purchasing Security Systems 268
Outsourcing 271
Case Problem 273
The Decision for "You Be the Judge" 273
References 273

10. Investigations 275
Introduction 275
Types of Investigations 277
Law 282
Evidence 283
Interviewing and Interrogation 285
Information Sources 289
Surveillance 296
Information Accuracy 297
Report Writing 297
Testimony 299
Case Problems 301
References 302

11. Accounting, Accountability, and Auditing 305
Introduction 305
Accounting 306

Accountability 307
Auditing 312
Fraud 315
Governance, Risk Management, and Compliance 317
Case Problems 319
References 319

12. Resilience, Risk Management, Business Continuity, and Emergency Management 321
Introduction 322
Resilience: A Critical Thinking Perspective 322
Risk Management 325
Insurance 332
Claims 339
Business Continuity 342
Emergency Management 345
The Military 353
Case Problems 358
The Decision for "You Be the Judge #1" 359
The Decision for "You Be the Judge #2" 359
References 360

13. Life Safety, Fire Protection, and Emergencies 363
Life Safety 363
Fire Protection 368
Fire Prevention and Fire Suppression Strategies 372
Public Safety Agencies 386
Emergencies 389
Case Problems 398
References 398

14. Safety in the Workplace 401

Introduction 401

Accident Statistics and Costs 403

History of Safety Legislation 403

Occupational Safety and Health Administration 406

Safety Strategies 415

Case Problems 420

References 420

Part III Special Problems and Countermeasures 423

15. Terrorism and Homeland Security 425

Terrorism 426

Homeland Security 454

Private Sector 466

Case Problems 470

References 470

16. Protecting Critical Infrastructure 475

Critical Infrastructure 476

Critical Infrastructure Sectors 484

Transportation Systems 509

National Monuments and Icons 520

Border and Transportation Security 521

Case Problems 524

References 525

17. Protecting Commercial and Institutional Critical Infrastructure 529

Introduction 530

Commercial Facilities 530

Robbery and Burglary 556

Banking and Finance 559
Educational Institutions 565
Healthcare and Public Health 574
Case Problems 584
References 584

18. Topics of Concern 589
Workplace Violence 590
Human Resources Protection 596
Substance Abuse in the Workplace 606
Information Security 611
Communications Security 624
Case Problems 632
References 632

19. Your Future in Security and Loss Prevention 637
Introduction 637
Security and Loss Prevention in the Future 638
Education 647
Research 651
Training 654
The Concept of the Security Institute 655
Employment 655
Case Problems 658
References 659

Index 661

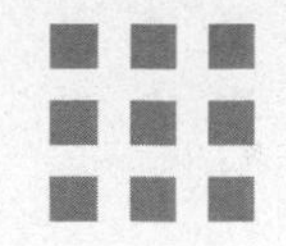

About the Author

Philip P. Purpura, Certified Protection Professional, is a college educator, consultant, expert witness, and writer. He is Director of the Security and Justice Institute and Coordinator of the Security for Houses of Worship Project in South Carolina. Purpura began his security career in New York City and held management and proprietary and contract investigative positions. He has also worked with a public police agency. Purpura is the author of seven other books: *Security: An Introduction* (Boca Raton, FL: Taylor & Francis/CRC Press, 2011); *Terrorism and Homeland Security: An Introduction with Applications* (Burlington, MA: Elsevier Butterworth-Heinemann, 2007); *Security Handbook, 2nd ed.* (Boston, MA: Butterworth-Heinemann, 2003; Albany, NY: Delmar, 1991); *Police & Community: Concepts & Cases* (Needham, MA: Allyn & Bacon, 2001); *Criminal Justice: An Introduction* (Boston, MA: Butterworth-Heinemann, 1997); *Retail Security & Shrinkage Protection* (Boston, MA: Butterworth-Heinemann, 1993); and *Modern Security & Loss Prevention Management* (Boston, MA: Butterworth, 1989).

Purpura has been a contributing editor to three security periodicals; has written numerous articles published in journals, magazines, and newsletters; and has been involved in a variety of editorial projects for publishers. He holds an associate's degree in police science from the State University of New York at Farmingdale, and bachelor's and master's degrees in criminal justice from the University of Dayton and Eastern Kentucky University, respectively. He also studied in Europe, Asia, and the former Soviet Union. He serves as Chairperson of the ASIS International Council on Academic and Training Programs.

Preface

The sixth edition of *Security and Loss Prevention: An Introduction* continues to draw on many disciplines for answers to protection challenges facing practitioners while helping the reader understand the security and loss prevention profession. As in previous editions, a major focus is on loss problems and countermeasures. This book shows an awareness that, beyond the attention terrorism has received since the 9/11 attacks, security and loss prevention practitioners continue to face the same basic challenges they faced prior to the attacks, such as violence in the workplace, theft, and cybercrime. At the same time, this edition contains an enhanced effort to include relevant and practical research from journals to assist decision-makers when choosing security strategies that are evidence-based.

Terminology, concepts, and theories, at the foundation of this profession, are emphasized. The book has been updated with new research, statistics, laws, standards, guidelines, strategies of protection, technology, events, and issues. The contents retain basic information on the body of knowledge of security and loss prevention.

The many disciplines within this book include law, criminal justice, intelligence, business, accounting, risk management, business continuity, emergency management, fire protection, safety, sociology, and psychology. The publications supporting this book include a variety of periodicals, books, and research reports from numerous organizations.

Security and Loss Prevention: An Introduction is aligned with ASIS International research on security tasks, knowledge, and skills of practitioners. The foundation of this book also rests on the annual ASIS International academic/practitioner symposiums that produced core competencies, course outlines, undergraduate and graduate curriculum models, accreditation criteria, and directions for research, among other products. Another source of input is the U.S. Department of Homeland Security publication titled "Security Specialist Competencies: An Interagency Security Committee Guideline." This Guideline seeks uniformity and consistency of core competencies among Federal agencies in the training and professional development of security specialists.

This book helps applicants prepare for the Certified Protection Professional examination, which is sponsored by ASIS International. Numerous topics included in the examination are covered in this book.

To provide the reader with a view of some of the updates in this sixth edition, here is a list of new terms and topics:

threats to brand
social disorganization and anomie of private military/private security companies
occupational outlook
an expansion of theoretical foundations of security
controllers and super controllers

peer reviewed
budgets
evidence-based research
black collar crime
Twitter, YouTube, and Facebook
infinity screening
sexting
employees venting in social media
megapixel and HD cameras
video analytics
sustainability
green security
anticlimb fence
smart fence
Ponzi scheme
counter social media techniques
resilience
apps
enterprise risk management
enterprise security risk management
PS-Prep
supply chain
IPDACT
homegrown and lone wolf terrorists
"new terrorism"
intelligence
Fukushima nuclear disaster
British Petroleum Oil Spill
Chemical Facility Anti-Terrorism Standards
hacktivism
cloud computing
active shooter
near field communication
zero-day exploits
tradecraft

Because we are in the "information age" and the protection of both information technology (IT) and information is so important today, a major theme of this book is to connect the traditional security manager, physical security specialist, and investigator to IT security. This is not a claim to make the reader an IT security expert. Rather, the reader will learn about similarities and differences of physical and IT security, internal and external IT risks and countermeasures, and the mindset of the IT security specialist. In addition, the convergence of physical security and IT security, as covered in the contents, makes an understanding of IT especially important to the security and loss prevention practitioner.

Our information age has brought with it an explosion of data, information, and hype that challenge us to probe and shape knowledge for our personal and professional life. To assist the reader, critical thinking skills are presented as a tool to go beyond collecting "facts" as we seek to understand causes, motives, and change. Because all security strategies either protect people and assets, accomplish nothing, or help offenders (see Chapter 3), practitioners will benefit by thinking critically as they plan, select vendors, and implement and manage security strategies that will make or break their careers.

The sixth edition includes learning objectives and key terms at the beginning of each chapter, key terms in bold within each chapter, definitions, examples, illustrations, photos, boxed scenarios, boxed international topics for global perspectives, and career boxes that explain various specializations in security.

The student or practitioner will find this book to be user-friendly and interactive, as in previous editions. Several features will assist the reader in understanding not only the basics but also the reality of the field. The reader is placed in the role of the practitioner through various exercises. Within each chapter, the loss problems are described and are followed by a discussion of the nuts-and-bolts countermeasures. Sidebars in each chapter emphasize significant points and facilitate critical thinking about security issues. Cases titled "You Be the Judge" appear in the text. These fictional accounts of actual cases deal with security-related legal problems. The reader is asked for a verdict based on the material at hand and is then directed to the end of the chapter for the court's ruling. This sixth edition contains a new learning aid titled "You Decide! What Is Wrong With This Facility?" With the assistance of chapter content, the reader exposes vulnerabilities at a facility and offers suggestions for improved protection. Additional boxed cases appear in chapters and offer bits of interesting information or analyze a loss problem relevant to the subject matter of the chapter. The case problems at the ends of chapters also bridge theory to practice and ask the reader to apply the general concepts of the chapter to real-world situations. These exercises enable the student to improve analytical and decision-making skills, consider alternative strategies, stimulate controversy in group discussions, make mistakes and receive feedback, and understand corporate culture and ethical guidelines.

This new edition also serves as a helpful directory. Professional organizations and sources of information that enhance protection programs are included with website addresses at the ends of chapters.

The first few chapters provide an introduction and foundation for security and loss prevention programs and strategies. Chapter 1 defines security and loss prevention and presents a critical perspective on the history of this profession. The second chapter concentrates on the growth of the security industry and profession and related challenges. The next three chapters provide a foundation from which protection programs can become more efficient and effective. Chapter 3 focuses on why security is a profession, theory, planning, budgets, risk analysis, standards and regulations, evaluation, research, and the basics of organization. Chapter 4 provides an overview of the judicial system of the United States, civil and criminal law, premise security claims, administrative law, and labor law. This chapter also covers arrest law, searches, use of force, and questioning subjects. Chapter 5 explains the "why" and "how" of working with people and organizations to assist loss prevention efforts. Topics include internal and external relations, marketing, social media, and the news media. Chapters 6, 7, and 8 emphasize strategies for curbing internal and external crime threats. These strategies include job applicant screening, management countermeasures, and physical security. Chapter 9, on purchasing security services and systems, is vital because not all security specialists are wise consumers, and the best plans are useless when followed by poor purchasing decisions. Chapter 10 provides practical information on investigations including types of investigations, legal issues, technology, and evidence in our digital age. The strategies of accountability, accounting, and auditing are described in Chapter 11, with an explanation as to why these tools are essential for survival. Chapters 12 through 16 explain the threats of terrorism, natural disasters, and accidents, while emphasizing an "all hazards" preparation and protection approach. Chapter 12 focuses on resilience, risk management, business continuity, and emergency management as a foundation for Chapter 13 on life safety, fire protection, emergencies, and disasters. Chapter 14 emphasizes workplace safety and OSHA. The topics of terrorism and homeland security are explained in Chapter 15. Chapter 16 concentrates on protecting critical infrastructure, key resources, and borders. Chapter 17 describes security and loss prevention at retail, financial, educational, and healthcare organizations. The topics of workplace violence, human resources protection, substance abuse, and information security are in Chapter 18. The final chapter explains trends, education, research, training, and employment.

The traditional focus of security—security officers, fences, and alarms—is too narrow to deal with an increasingly complex world. Practitioners are being asked to do more with fewer resources and prove that the money spent on protection has a return on investment. In a world of rapid change, senior management expects security and loss prevention practitioners to produce answers and results quickly. The true professional maintains a positive attitude and sees problems as challenges that have solutions.

The tremendous growth of the security and loss prevention profession provides fertile ground to advance in a rewarding career. In such a competitive world, the survival and protection of businesses and institutions, technological innovations, and the national interest depend greatly on security and loss prevention programs. This book should inspire and motivate students and practitioners to fulfill these vital protection needs.

Acknowledgments

I would like to thank the many people who contributed to this sixth edition. Gratitude goes to my family for their support and patience. I am thankful to security practitioners, researchers, educators, librarians, and others who helped to provide a wealth of information to support the contents of this book. The hardworking team at Elsevier, including Amber Hodge and Mohanambal Natarajan, are to be recognized for their talents and skills in publishing this book. I am grateful for the team effort, among so many people, for without it this book could not be published.

Introduction to Security and Loss Prevention

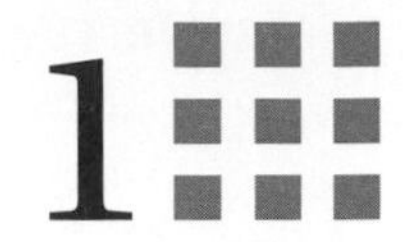

The History of Security and Loss Prevention

A Critical Perspective

OBJECTIVES

After studying this chapter, the reader will be able to:

1. Explain the purpose of critical thinking and how to think critically
2. Define security and loss prevention
3. List the benefits of studying the history of security and loss prevention
4. Trace the early development of security and policing
5. Describe the growth of security companies in the United States
6. Explain security as it relates to railroads, labor unions, the Great Wars, and the Third Wave
7. Describe 21st century / post-9/11 security challenges

KEY TERMS

- critical thinking
- security
- loss prevention
- Chief Security Officer (CSO)
- layered security
- redundant security
- circumvent
- Great Wall of China
- Hammurabi, King of Babylon
- polis
- Praetorian Guard
- vigiles
- feudalism
- comitatus
- posse comitatus
- Posse Comitatus Act
- frankpledge system
- tithing
- Magna Carta
- Statute of Westminster
- watch and ward
- Henry Fielding
- Bow Street Runners
- Sir Robert Peel
- Metropolitan Police Act
- Allan Pinkerton
- Henry Wells
- William Fargo
- William Burns
- Washington Perry Brink
- George Wackenhut
- Edwin Holmes
- first wave societies
- second wave societies
- third wave societies

Why Critical Thinking?

Security and loss prevention practitioners face enormous challenges. Dealing with crimes, accidents, and natural disasters requires sound planning and action and the ability to adapt to a changing environment. Critical thinking offers an avenue for practitioners to enhance positive results in their work. The techniques of critical thinking are also helpful to students in any discipline to improve their thinking. We all think, but when we apply critical thinking, we produce a clearer, more accurate picture of our world.

The events of September 11, 2001, marked a turning point in the history of security, in particular, an increase in the importance of critical thinking. In a devastating terrorist onslaught, knife-wielding hijackers crashed two airliners into the World Trade Center in New York City, creating an inferno that caused the 110-story twin skyscrapers to collapse. Nearly 3,000 people were killed, including responding firefighters and police. On the same morning another hijacked airliner crashed into the Pentagon, causing additional deaths and destruction. A fourth hijacked airliner failed to reach its target; it crashed when heroic passengers learned of the other attacks and struggled with hijackers to control the airliner. The attacks were immensely successful and cost-effective for the terrorists. With a loss of 19 terrorists and expenses between $400,000 and $500,000, the attackers were able to kill thousands, cause hundreds of billions of dollars in economic damage and spending on counterterrorism, and significantly affect global history. With such a huge kill ratio and investment payoff for the terrorists, governments and the private sector have no option but to succeed in controlling terrorism.

Because of these devastating attacks, not only have military strategies, homeland security, public safety, and private sector security changed, but also our way of thinking has changed. We cannot afford to have failures in our planning and in our imagination of what criminals can do. To improve security, we must seek new tools to assist us in our thinking processes.

Critical thinking counters "business as usual." It helps us to become active learners: to not only absorb information, but to probe and shape knowledge. The critical thinker cuts through "hype" and emotion and goes beyond collecting "facts" and memorizing information in an effort to understand causes, motives, and changes. Critical thinking skills provide a foundation for creative planning while helping us to anticipate future events.

The critical thinker asks many questions, and the questions are often easier to formulate than the answers. Critical thinking requires us to "jump out of our own skin" to see the world from the perspective of others. Although this is not an easy process, by doing it we are much better informed before we make our conclusions and decisions.

At the same time, critical thinking is not to be used as a tool to open up the floodgates of criticism in the workplace. It is to be applied discreetly in order to understand the world and to meet challenges.

A professional's success depends on his or her thinking process applied to everyday duties and long-range planning. Critical thinking adds an extra edge to the repertoire of tools available to security and loss prevention practitioners.

Kiltz (2009: 4) notes that there is a variety of definitions of **critical thinking** in the literature. Terms used to define it include the ability to compare and evaluate viewpoints, study evidence, draw inferences, and defend opinions. Critical thinking can result in a belief, action, and the solution to a problem.

Safi and Burrell (2007: 54) write:

> *Theorists have hypothesized that critical thinking is correlated with internal motivation to think. Cognitive skills of analysis, interpretation, explanation, evaluation, and correcting one's own reasoning are at the heart of critical thinking.*
>
> *Critical thinking can be learned with practice and guidance by changing the actions involved in making decisions so that they become part of permanent behavior in homeland security intelligence analysis, threat protection and security planning.*

To prime the reader's mind for the explanation of critical thinking, Chapter 3 applies critical thinking to security planning by suggesting that all security strategies be placed under one of the following three models: it protects people and assets, it accomplishes nothing, or it helps offenders.

How can we Think Critically?

In today's world, there are many agencies that seek to influence our thinking, for example, the media, advertisers, politicians, educators, and writers, including the writer of this book. Although an effort has been made to write an objective book here, biases naturally surface. For example this book presents a North American interpretation of security. Objectivity is fostered in this book through an introduction to critical thinking skills, a multidisciplinary approach, international perspectives, boxed topics and questions, a variety of references, web exercises, and case problems at the end of chapters that bridge theory to practice and ask the reader to make decisions as a practitioner.

With so much information competing to influence us, choices become difficult and confusing. And, as we think through complex challenges, we need a method of sorting conflicting claims, differentiating between fact and opinion, weighing "evidence" or "proof," being perceptive to our biases and those of others, and drawing logical conclusions. Ellis (1991: 184–185) suggests a four-step strategy for critical thinking:

Step 1: Understand the point of view.

- Listen/read without early judgment.
- Seek to understand the source's background (e.g., culture, education, experience, and values).
- Try to "live in their shoes."
- Summarize their viewpoint.

Step 2: Seek other views.

- Seek viewpoints, questions, answers, ideas, and solutions from others.

Step 3: Evaluate the various viewpoints.

- Look for assumptions (i.e., an opinion that something is true without evidence), exceptions, gaps in logic, oversimplification, selective perception, either/or thinking, and personal attacks.

Step 4: Construct a reasonable view.

- Study multiple viewpoints, combine perspectives, and produce an original viewpoint that is a creative act and the essence of critical thinking.

Why Think Critically about the History of Security and Loss Prevention?

Critical thinking can be applied even while reading this chapter on the history of security and loss prevention. The intent here is to stimulate the reader to go beyond memorizing historical events, names, and dates. A person who has read several books in this field, may find that the history chapters sound very similar. Did the writers, including this one, become complacent and repeat what was already written about the history of this field? How does the reader know that the history of security and loss prevention as presented in this book and in others is objective?

Recorded history is filled with bias. Historians and scholars decide what subjects, events, innovations, countries, ethnic groups, religions, men, and women should be included in or excluded from recorded history. In reference to the history of security and loss prevention, what has been missed? What subjects have been overemphasized? (A case problem at the end of this chapter asks the reader to critically think about the history of security and loss prevention.) In the policing field, for example, Weisheit, Baker, and Falcone (1995: 1) note that history and research reflect a bias toward urban police at the expense of rural police. Do security researchers and writers overemphasize large proprietary security programs and large security service firms? What about the thousands of proprietary security programs at small companies and the thousands of small security service firms?

Another question is: what role have women and minorities played in the history of this field, early on and since the civil rights movement of the 1960s? Calder (2010: 79 and 99) writes that "before World War II industrial guards were white males, typically 35–65 years old, and veterans of World War I, military, or public police services." He adds: "Blacks, other ethnic minorities, and women were simply not hired in major resource and manufacturing industries, and anyone serving in a private policing position was always a white male."

Still another question might be: what other country has had the most impact on police and security in the United States? Our language, government, public and private protection, law, and many other aspects of our lives have deep roots in England. However, what about the roles of other countries in the development of police and private security methods? Stead (1983: 14–15) writes of the French as innovators in crime prevention as early as the 1600s under King Louis XIV. During that time, crime prevention was emphasized through preventive patrol and street lighting. Germann, Day, and Gallati (1974: 45–46) write of early Asian investigative methods that used psychology to elicit confessions.

A critical thinking approach "opens our eyes" to a more objective perspective of historical events. Encouraging greater objectivity does not mean the author is seeking to rewrite history or to change the basic strategies of security and loss prevention. Rather, the aim is to expand the reader's perception and knowledge skills, which are foundational for smarter protection in a complex world.

Just as critical thinking skills are applied to a critical perspective of history in this chapter, students and practitioners are urged to continue this thinking process throughout this book.

Security and Loss Prevention Defined

Within our organized society, security has traditionally been provided primarily by our armed forces, law enforcement agencies, and private security. During the last decades of the 20th century, the methods of private security became more specialized and diverse. Methods not previously associated with security emerged as important components of the total security effort. Security officers, fences, and alarms have been the hallmarks of traditional security functions. Today, with society becoming increasingly complex, various specializations—auditing, safety, fire protection, cyber security, crisis management, and intelligence, to name a few—are continually being added to the security function. For this reason, many practitioners group all of these functions under the single term "loss prevention."

Astor (1978: 27) argues that various organizations have switched from the term "security" to "loss prevention" because of the negative connotations of security. He points out:

> *In the minds of many, the very word "security" is its own impediment.*
>
> *Security carries a stigma; the very word suggests police, badges, alarms, thieves, burglars, and some generally negative and even repellent mental images. … Simply using the term "loss prevention" instead of the word "security" can be a giant step toward improving the security image, broadening the scope of the security function, and attracting able people.*

In many organizations, traditional security functions and other specialized fields (auditing, safety, fire protection, etc.) are subsumed under loss prevention. **Security** is narrowly defined as traditional methods (security officers, fences, and alarms) used to increase the likelihood of a crime-controlled, tranquil, and uninterrupted environment for an individual or organization who is in pursuit of an objective. **Loss prevention** is broadly defined as almost any method (e.g., security officers, safety, auditing) used by an individual or organization to increase the likelihood of preventing and controlling loss (e.g., people, money, productivity, materials) resulting from a host of adverse occurrences (e.g., crime, fire, accident, natural disaster, error, poor supervision or management, bad investment). This broad definition provides a foundation for the loss prevention practitioner, whose innovations are limited only by his or her imagination. It is hoped that these concepts will not only guide the reader through this book but also reinforce a trend in the use of these definitions.

Various employment titles are applied to individuals who perform security and loss prevention duties within organizations. The titles include Vice President, Director, or Manager of any of the following: Security, Corporate Security, Loss Prevention, or Assets Protection.

Another title is **Chief Security Officer (CSO)**. The *Chief Security Officer Organizational Standard* (ASIS International, 2008: 1), approved by the American National Standards Institute, is designed as "a model for organizations to use when developing a leadership function to provide a comprehensive, integrated security risk strategy to contribute to the viability and success of the organization." This standard, developed in response to an increasingly serious threat environment, recommends that the CSO report to the most senior level executive of the organization. The standard includes education, experience and competency requirements, responsibilities, and a model position description. The CSO designation and the standard supporting it provide an excellent reference that the security profession and senior management can draw on to improve the protection of people and assets and help organizations survive in a world filled with risks.

Research conducted by Booz Allen Hamilton (2005) for ASIS International, the Information Systems Security Association, and the Information Systems Audit and Control Association found that placing all security functions under one individual ("the strongest or most powerful of the various security elements") may not be beneficial ("an obvious and flawed option") for all organizations because it can reduce valuable input from key managers with regard to enterprise-wide security. The study recommended a "business-focused council of leaders" consisting of representatives from various specializations—such as risk management, law, safety, and business continuity—who "come together using the corporate strategy as a common element on which to focus."

***Security* is narrowly defined; *loss prevention* is broadly defined.**

History

We should study the history of security and loss prevention because

- We learn about the origins of the profession and how it developed.
- We can see how gaps in security and safety within society were filled by the private sector.
- We can learn about noted practitioners and theorists and their challenges, failures, and successes.
- We can compare security in the past to security in the present to note areas of improvement and areas requiring improvement.
- We can learn how security services and systems have been controlled and regulated.
- We can learn about the interaction of private security and public police over time.
- History repeats itself. We should strive to avoid the mistakes of the past and continue with its successes.
- We can learn how social, economic, political, and technological forces have affected security over time.
- The past assists us in understanding the present, and it offers us a foundation from which to anticipate future events.

Early Civilizations

Prehistoric human beings depended on nature for protection because they had not learned how to build strong houses and fortifications. In cold climates, caves provided protection and shelter, whereas in the tropics, trees and thickets were used. Caves were particularly secure because rocky walls guarded those inside on all sides except at the cave mouth. To protect the entrance, **layered security** (i.e., diverse methods of security) was employed: large rocks acted as barriers when they were rolled in front of entrances; dogs, with their keen sense of smell, served to alarm and attack; and fires added an additional safeguard. **Redundant security** (i.e., two or more of the same type of security, such as two or more dogs) was also applied. By living on the side of a mountain with access via a narrow, rocky ledge, cave dwellers were relatively safe from enemies and wild animals. Early Pueblo Indians, living in what is now New Mexico and Arizona, ensured greater protection for themselves in their dwellings by constructing ladders that could be pulled in, and this defense proved useful until enemies attacked with their own ladders. *In fact, from the earliest civilizations until today, security measures have never been foolproof, and adversaries have typically attempted to* ***circumvent*** *(i.e., to go around) defenses.*

Throughout history, layered and redundant security have been used to block attempts by adversaries to circumvent defenses.

The **Great Wall of China** is the longest structure ever built. It was constructed over hundreds of years beginning in the 400s BC. Hundreds of thousands of workers lived near the wall and participated in the huge project, which stretched 4,000 miles and reached heights of 25 feet. Unfortunately, the wall provided protection only from minor attacks; when a major invasion force struck, the defense could not withstand the onslaught. The army of Mongol leader Genghis Khan swept across the wall during the AD 1200s and conquered much of China. Since 1949, the Chinese government has restored some sections of the mostly collapsed wall, which is a major tourist attraction (Feuerwerker, 1989: 373–374).

It is interesting to note the changing character of security through history. In earlier years, huge fortifications could be built with cheap labor, and a king could secure a perimeter with many inexpensive guards. Today, physical barriers such as fences and walls are expensive, as is the posting of security forces at physical barriers; often, technological solutions are less expensive than hiring personnel.

As societies became more complex, the concepts of leadership, authority, and organization began to evolve. Mutual association created social and economic advantages but also inequities, so people and assets required increased protection. Intergroup and intragroup conflicts created problems whose "solutions" often took the form of gruesome punishments, including stoning, flaying, burning, and crucifying. A person's criminal record was carried right on his or her body through branding and mutilation. By 1750 BC, the laws of **Hammurabi, King of Babylon**, not only codified the responsibilities of the individual to the group and the rules for private dealings between individuals, but also discussed retributive penalties (Germann, Day, and Gallati, 1974: 43).

Ancient Greece

Between the ninth and third centuries BC, ancient Greece blossomed as an advanced commercial and culturally rich civilization. The Greeks protected their advancing civilization with the **polis**, or city-state, which consisted of a city and the surrounding land protected by a centrally built fortress overlooking the countryside. The Greeks' stratified society caused the ruling class to be in constant fear of revolution from below. Spartans, for example, kept their secret agents planted among the lower classes and subversives. *During the time of the Greek city-states, the first police force evolved to protect local communities, although citizens were responsible for this function.* The Greek rulers did not view local policing as a state responsibility, and when internal conflicts arose, they used the army. During this era, the Greek philosopher Plato introduced an advanced concept of justice, in which an offender would be forced to not only pay a sort of retribution but also undergo some kind of reform or rehabilitation.

Ancient Egyptians sealed the master locksmith in the tomb to prevent security leaks.

Ancient Rome

The civilization of ancient Rome was fully developed both commercially and culturally before the birth of Christ. Rome was located only 15 miles from the sea and could easily share in the trade of the Mediterranean. This city sat on seven hills overlooking the Tiber River, which permitted ease in fortification and defense. A primitive but effective alarm system was created by placing geese, who have very sensitive hearing, at strategic locations so that the sound of an approaching army would trigger squawking.

The Roman regime was well designed to carry on the chief business of the Roman state, which was war. A phalanx of 8,000 foot-soldiers equipped with helmets, shields, lances, and swords became the basic unit of a Roman army. Later, a more maneuverable legion of 3,600 men, additionally armed with iron-tipped javelins, was used. These legions were also employed to maintain law and order. The first emperor of Rome, Augustus (63 BC–AD 14), created the **Praetorian Guard** to provide security for his life and property. These urban cohorts of 500 to 600 men were deployed to keep the peace in the city. Some believe that, after about AD 6, this was the most effective police force until recent developments in law enforcement. After AD 6, modern-day coordinated patrolling and preventive security began with the Roman nonmilitary **vigiles**, night watchmen who were active in both policing and firefighting (Post and Kingsbury, 1977; Ursic and Pagano, 1974).

The Romans have an interesting history in fire protection. During the 300s BC, slaves were assigned firefighting duties. Later, improved organization established divisions involving hundreds of people, who carried water in jars to fires or brought large pillows so victims trapped in taller structures could jump with improved chances for survival. The completion of the aqueducts to Rome aided firefighting by making water easier to obtain. Hand pumps and leather hoses were other innovations.

The Middle Ages in Europe

During the Dark Ages, the period in history after the destruction of the ancient Greek and Roman empires, **feudalism** gradually developed in Europe. Overlords supplied food and security to those who farmed and provided protection around castles (Figure 1-1) fortified by walls, towers, and a drawbridge that could be raised from its position across a moat. Even then, security required registration, licensing, and a fee—Henry II of England (reigned AD 1154–1189) destroyed more than 1,100 unlicensed castles that had been constructed during a civil war (Brinton et al., 1973: 167).

Another feudal arrangement was the war band of the early Germans, the **comitatus**, by which a leader commanded the loyalty of his followers, who banded together to fight and win booty. To defend against these bands of German barbarians, many landowners throughout Europe built their own private armies. (The term **posse comitatus** denotes a body of citizens that authority can call on for assistance against offenders. The **Posse Comitatus Act** is a Civil War–era act that generally prohibits the military from engaging in civilian law enforcement. This law has been labeled as archaic because it limits the military from responding to disasters.)

Much of the United States' customs, language, laws, and police and security methods can be traced to the nation's English heritage. For this reason, England's history of protection is examined here.

FIGURE 1-1 Castles provided protection for local residents during earlier centuries.

Between the 7th and 10th centuries, the frankpledge system and the concept of tithing fostered increased protection. The **frankpledge system**, which originated in France and spread to England, emphasized communal responsibility for justice and protection. The **tithing** was a group of 10 families who shared the duties of maintaining the peace and protecting the community.

In 1066, William, Duke of Normandy (in present-day France), crossed the English Channel and defeated the Anglo-Saxons at Hastings. Under his rule, a highly repressive police system developed under martial law as the state appropriated responsibility for peace and protection. Community authority and the tithing system were weakened. William divided England into 55 districts, or *shires*. A *reeve*, drawn from the military, was assigned to each district. (Today, we use the word *sheriff*, derived from *shire-reeve*.) William is credited with changing the law to make a crime an offense against the state rather than against the individual and was instrumental in separating police from judicial functions. A traveling judge tried the cases of those arrested by the shire-reeves.

In 1215, King John signed the **Magna Carta**, which guaranteed civil and political liberties. Local government power increased at the expense of the national government, and community protection increased at the local level.

Another security milestone was the **Statute of Westminster**, issued by King Edward I in 1285 to organize a police and justice system. A **watch and ward** was established to keep the peace. Every town was required to deploy men all night, to close the gates of walled towns at night, and to enforce a curfew.

What are some of the similarities between security strategies of earlier civilizations and those of today?

More Contemporary Times

England

For the next 500 years, repeated attempts were made to improve protection and justice in England. Each king was confronted with increasingly serious crime problems and cries from the citizenry for solutions. As England colonized many parts of the world and as trade and commercial pursuits brought many people into the cities, urban problems and high crime rates persisted. Merchants, dissatisfied with the protection afforded by the government, hired private security forces to protect their businesses.

By the 18th century, the Industrial Revolution had compounded urban problems. Many citizens were forced to carry arms for their own protection because a strong government policing system was absent. Various police and private security organizations did strive to reduce crime; **Henry Fielding**, in 1748, was appointed a magistrate (judge), and he devised the strategy of preventing crime through police action by helping to form the famous **Bow Street Runners**, the first detective unit. This unit ran to the crime scene immediately upon being notified in an effort to catch the offenders. The merchant police were formed to protect businesses, and the Thames River police provided protection at the docks. During this period, more than 160 crimes, including stealing food, were punishable by death. As pickpockets were being hanged others moved among the spectators, picking pockets.

Do you think policing and justice were impotent during the early Industrial Revolution in England? Do you think we have a similar problem today in the United States?

Peel's Reforms

In 1829, **Sir Robert Peel** worked to produce the **Metropolitan Police Act**, which resulted in a revolution in law enforcement. Modern policing was born. Peel's innovative ideas were accepted by Parliament, and he was selected to implement the act, which established a full-time, unarmed police force with the major purpose of patrolling London. Peel is also credited with reforming criminal law by limiting its scope and abolishing the death penalty for more than 100 offenses. It was hoped that such a strategy would gain public support and respect for the police. Peel was very selective in hiring his personnel, and training was an essential part of developing a professional police force. Peel's reforms are applicable today and include crime prevention, the strategic deployment of police according to time and location, a command of temper rather than violent action, record keeping, and crime news distribution (Dempsey and Forst, 2010: 8–9).

Although Sir Robert Peel produced a revolution in law enforcement in 1829, both crime and the private security industry continued to grow.

Early America

The Europeans who colonized North America brought with them the heritage of their mother countries, including various customs of protection. The watchman system and collective responses to safety and security challenges remained popular. A central fortification in populated areas provided increased security from hostile threats. As communities expanded in size, the office of sheriff took hold in the South, whereas the functions of constable and watchman were the norm in the Northeast. The sheriff's duties involved apprehending offenders, serving subpoenas, and collecting taxes. Because a sheriff was paid a higher fee for collecting taxes, policing became a lower priority. Constables performed a variety of tasks such as keeping the peace, bringing suspects and witnesses to court, and eliminating health hazards. As in England, the watch system had its share of inefficiency, and to make matters worse, those convicted of minor crimes were sentenced to serve time on the watch.

The watch also warned citizens of fire. In colonial towns, each home had to have two fire buckets, and homeowners were subject to a fine if they did not respond to a fire, buckets in hand. A large fire in Boston in 1679 prompted the establishment of the first paid fire department in North America (Bugbee, 1978: 5).

The Growth of Policing

The middle of the 1800s was a turning point for both law enforcement and private security in America, as it was in England. Several major cities (e.g., New York, Philadelphia, and San Francisco) organized police forces, often modeled after the London Metropolitan Police. However, corruption was widespread. Numerous urban police agencies in the Northeast received large boosts in personnel and resources to combat the growing militancy of the labor unions in the late 1800s and early 1900s. Many of the large urban police departments were originally formed as strikebreakers (Holden, 1986: 23). Federal policing also experienced growth during this period. The U.S. Treasury established an investigative unit in 1864. As in England, an increase in public police did not quell the need for private security.

The Growth of Security Companies

In 1850, **Allan Pinkerton**, who had been a cooper and also the Chicago Police Department's first detective, opened a private detective agency with a Chicago lawyer named Edward A. Rucker. McCrie (2010: 543 and 548) refers to Allan Pinkerton as the "founder of the security services industry" and as the provider of "the first substantive executive protection evaluation of a U.S. president's [President Lincoln] vulnerability while traveling in public." McCrie (2010: 550) writes, "Pinkerton popularized his life and embellished his agency's reputation with essays and a series of books." In its logo the company used the image of a wide-awake human eye and the slogan "we never sleep," which is the foundation for the term "private eye" (PI) often used today. Such private security businesses thrived because public police were limited by geographic jurisdiction, which handicapped them when investigating and apprehending fleeing offenders. Pinkerton (Figure 1-2) and others became famous as they pursued criminals across state boundaries throughout the country.

FIGURE 1-2 Major Allan Pinkerton, President Lincoln, and Major General John McClernand, Antietam, MD, October 1862. *Courtesy: National Archives.*

The History of Loss Prevention in a Nutshell

Loss prevention has its origin in the insurance industry. Before the Civil War, insurers gave minimal attention to the benefits of loss prevention. For instance, in the fire insurance business, executives generally viewed fires as good for business. Insurance rates were based on past loss experience, premiums were paid by customers, losses were paid to customers affected by a fire, and a profit was expected by the insurer. When excessive fire losses resulted in spiraling premiums, the changing nature of the fire insurance business created a hardship for both the insurer and the insured. Insurance executives were forced to raise premiums to cover losses, and customers complained about high rates. The predominance of wooden construction (even wooden chimneys) in dense urban areas made fire insurance unaffordable for many. A serious fire peril persisted.

After the Civil War, loss prevention gained momentum as a way to reduce losses and premiums. Fire insurance companies formed the National Board of Fire Underwriters, which, through engineering, investigation, research, and education, was able to prevent losses. In 1965, the board was merged into the American Insurance Association (AIA). AIA activities brought about the development of the National Building Code, a model code adopted by many municipalities to reduce fire losses.

Today, executives throughout the insurance industry view loss prevention as essential. Many insurers have loss prevention departments to aid themselves and customers. Furthermore, customers (i.e., the insured), to reduce premiums, have become increasingly concerned about preventing losses. The security function in many businesses includes loss prevention duties involving fire protection and safety.

During the 1800s, because public police were limited by geographic jurisdiction, private security filled the need for an agency that could chase fleeing offenders across city, county, or state lines and became a growth industry.

To accompany Americans' expansion westward during the 19th century and to ensure the safe transportation of valuables, **Henry Wells** and **William Fargo** supplied a wide-open market by forming Wells, Fargo & Company in 1852, opening the era of bandits accosting stagecoaches and their shotgun riders. Burns International Services Corporation acquired Wells Fargo. Today the name Wells Fargo is exclusive to Wells Fargo & Company, a large financial services business (Figure 1-3).

Another security entrepreneur, **William Burns**, was a Secret Service agent who directed the Bureau of Investigation, an organization that preceded the FBI. In 1910, this experienced investigator opened the William J. Burns Detective Agency (Figure 1-4), which became the investigative arm of the American Bankers Association.

In 1999, Securitas acquired Pinkerton, and in 2000, it acquired Burns and other security companies (Securitas, 2011). Securitas is headquartered in Stockholm, Sweden.

In 1859, **Washington Perry Brink** also took advantage of the need for the safe transportation of valuables. From freight and package delivery to the transportation of payrolls, his

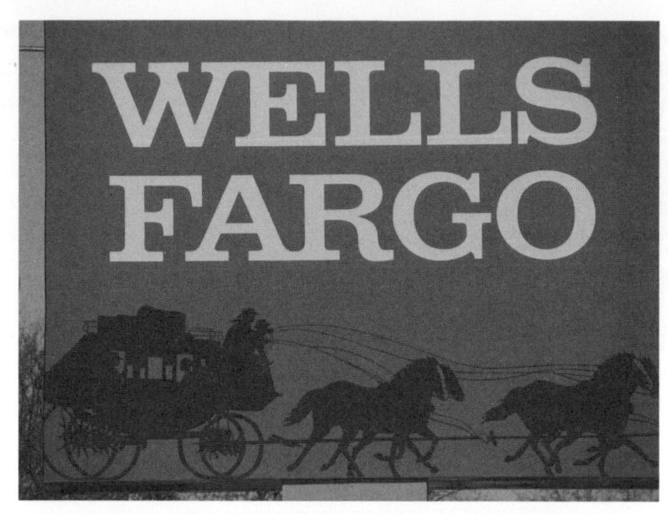

FIGURE 1-3 The name Wells Fargo has a long history dating back to the westward expansion during the 19th century.

FIGURE 1-4 In 1910, William J. Burns, the foremost American investigator of his day and the first director of the government agency that became the FBI, formed the William J. Burns Detective Agency.

service required increased protection through the years as cargo became more valuable and more vulnerable. Following the killing of two Brink's guards during a robbery, the armored truck was initiated in 1917. Today, Brink's, headquartered in Richmond, Virginia, is a leading global security services provider for banks, retailers, and other business and government customers (Brink's, 2011).

The Wackenhut Corporation was another leader in the security industry. Founded in 1954 by **George Wackenhut** and other former FBI agents, the corporation extended its services to government agencies, which resulted in the company receiving numerous contracts from its inception. The Wackenhut Corporation was the U.S.-based division of Group 4 Securicor. In 2004, G4S was created from the merger between Securicor and Group 4 Falck, with headquarters in the United Kingdom (G4S, 2011).

During the mid-1800s, an English inventor named Tildesley devised the first burglar alarm. The mechanical device included a door lock linked to chimes that sounded when unauthorized access was attempted. In 1852, Augustus Pope, a Boston inventor, secured a patent for the first electric burglar alarm system that could also signal the fire department. **Edwin Holmes** of Boston purchased the patent in 1857 and pioneered the electronic security alarm business (Lee, 2011). He had a difficult time convincing people that an alarm would sound on the second floor of a home when a door or window was opened on the first floor. His sales strategy was to carry door-to-door a small model of a home containing his electric alarm system. Soon sales soared, and the first central office burglar alarm monitoring operation began. Holmes Protection Group, Inc. was acquired by ADT Security Services, Inc. at the end of the 20th century.

Since 1874, ADT Security Services, Inc. has been a leader in electronic security services. Originally known as American District Telegraph, ADT has acquired numerous security companies since its inception. ADT is a provider of electronic security services (i.e., intrusion, fire protection, closed-circuit television, access control) to millions of commercial, federal, and residential customers.

Railroads and Labor Unions

The history of private security businesses in the United States must include two important events of the 19th century: the growth of railroads and labor unions.

Although railroads were valuable in providing the vital east-west link that enabled the settling of the American frontier, these powerful businesses used their domination of transportation to control several industries, such as coal and kerosene. Farmers were especially hurt in economic terms because they had no alternative but to pay high fees to transport their products via the railroads. The monopolistic practices of railroads created considerable hostility; when Jesse James and other criminals robbed trains, citizens applauded. Railroads could not rely on public police protection because of jurisdictional boundaries. Consequently, numerous states passed laws enabling railroads to organize proprietary security forces with full arrest powers and the authority to apprehend criminals across multiple jurisdictions. Railroad police numbered 14,000 by 1914. During World War I, they were deputized by the federal government to ensure protection of this vital transportation network.

The growth of labor unions at the end of the 19th century resulted in increased business for security firms who acted as strikebreakers for large corporations. However, this venture proved costly. For example, a bloody confrontation between Pinkerton men and the workers at the Carnegie steel plant in Homestead, Pennsylvania, resulted in eight deaths (three security men and five workers). Pinkerton's security force surrendered. The plant was then occupied by federal troops. Senate hearings followed the Homestead disaster and "anti-Pinkertonism" laws were enacted to restrict private security. However, local and state police forces became the ones to deal with the strikers (Shelden, 2001: 84).

Later, the Ford Motor Company and other businesses were involved in bloody confrontations. Henry Ford had a force of about 3,500 security personnel, spies, and "sluggers" (i.e., private detectives) who were augmented by various community groups such as the Knights of Dearborn and the Legionnaires. The negative image brought to the public eye by newspaper coverage of union organizers being beaten tarnished many businesses and security firms. Prior to World War II, pressure from Congress, the Roosevelt Administration, labor unions, and the ACLU caused corporate management to shift its philosophy to a softer "public relations" approach (Shelden, 2001: 92). Interestingly, Calder (2010: 94) writes: "Unionization of security guards in the years just before and extending through World War II secured the full transformation of company plant protection personnel from despised and distrusted surrogates of management power into employees who performed functions valued by society, by company managements, and by other employees."

The Great Wars

World War I brought about an increased need for protection in the United States. Sabotage and espionage were serious threats. Key industries and transportation systems required expanded and improved security. The social and political climate in the early 20th century reflected urban problems and worldwide nationalism. The combination of the "war to end all wars" (i.e., World War I), Prohibition, intense labor unrest, and the Great Depression overtaxed public police. Private security companies helped fill the void.

By the late 1930s, Europe was at war again, and the Japanese were expanding their economic influence in the Far East. A surprise Japanese bombing of the Pacific Fleet at Pearl Harbor in 1941 jolted the United States into World War II, and intense security concerns appeared again. The United States went into full production, and protection of vital industries became crucial, leading the federal government to bring plant security personnel into the army as an auxiliary to military police. During the war, more than 200,000 of these security workers were sworn in (Calder, 2010: 88).

The "great wars" showed the need to protect arms industries and their secrets, so the federal government established security standards that evolved into the National Industrial Security Program (see Chapter 16). Another event that facilitated the growth and professionalization of the security industry occurred in 1955 with the formation of the American Society for Industrial Security, today known as ASIS International (see the end of Chapter 2).

The Third Wave

In the decades following World War II, private security expanded even more; during the 1950s, the Korean War and the unrelenting "cold war" created worldwide tension and competition between the democracies and communist regimes. In 1952, the Department of Defense strengthened the security requirements of defense industries to protect classified information and materials. When the Soviets successfully launched the first earth satellite (Sputnik, in 1957) and first reached the moon with an unmanned rocket (1959), Americans were stunned. The technological race became more intense, and information protection became more important.

The turbulent 1960s created massive social and political upheaval in the United States. Public police forces were overwhelmed by responses to the unpopular Vietnam war; protests over the denial of civil rights to minority groups; the assassinations of President John F. Kennedy, Senator Robert Kennedy, and the Reverend Martin Luther King, Jr.; and rising crime and drug problems. Private security boomed.

Protests, crime, terrorism, and limited public police resources marked the 1970s, 1980s, and 1990s. By this time, the advanced nations of the world had developed into what Alvin Toffler's *The Third Wave* (1980) and John Naisbitt's *Megatrends* (1982) call **third wave societies:** societies based on information and technology. (**First wave societies** had agriculture as a foundation; these dominated the world for thousands of years, deriving energy from human and animal power. Offenders stole cattle, gold, and other valuables. **Second wave societies** occurred during the Industrial Revolution, when production was powered by irreplaceable energy sources such as coal and oil. Criminals focused on money and booming economic conditions.) With the depletion of world resources, the world is becoming more dependent on technology and information. "Third wave" criminals exploit technology to commit their crimes, the extent of which is limited only by technological innovation and the offenders' imaginations. Today, security requires broad applications because IT systems can be accessed remotely. *An offender no longer has to physically trespass to steal and do harm to an organization.* We can only guess at the number of times the traditional security manager has done an excellent job of ensuring that security officers are patrolling, physical security is operational, and the facility is protected, except that a hacker has penetrated the corporate IT system and stolen proprietary information or caused other harm to the business.

Twenty-First Century/Post-9/11 Security Challenges

The last decade of the 20th century offered warnings of what was to come in the next century. The 1990s brought the first bombing of the World Trade Center, the bombing of the Murrah Federal Building in Oklahoma City, the first war with Iraq, criminals exploiting the Internet, the increased value of proprietary information, and attention to violence in the workplace.

As we know, not long into the 21st century, on September 11, 2001, terrorists attacked the World Trade Center and the Pentagon. Following the attacks, a crisis in confidence in government occurred. Citizens asked: How could the most powerful nation on earth be subject to such a devastating attack? What went wrong? Who is to blame? In response to the crisis, President George W. Bush declared war on terrorism, Afghanistan became a war zone, the Department of Homeland Security was created, and a massive global effort ensued to curb terrorism. The attacks also led to greater police powers for search and seizure and electronic surveillance in the United States under the Patriot Act, which raised the age-old question of how to balance police powers and constitutional rights.

The 9/11 attacks, subsequent bioterrorism (i.e., anthrax attacks through the U.S. Postal Service), the war in Afghanistan, and the second war in Iraq show the difficult challenges facing our world in this new century. The United States and its allies are faced with not only conflict in Iraq, Afghanistan, and other regions, but also old and emerging state competitors and the proliferation of weapons of mass destruction.

The 21st century has also recorded huge natural disasters that, along with the problem of terrorism, necessitate a rethinking of emergency management and business continuity. In 2005, Hurricanes Katrina and Rita devastated Gulf Coast states. Katrina flooded New Orleans. The December 2004 Sumatran tsunami killed almost 300,000 people and affected 18 countries around the Indian Ocean. In 2011, Japan experienced three devastating disasters simultaneously: an earthquake, a tsunami, and the resulting destruction of nuclear power plants that released radiation. Loss of life, injuries, future health problems, and monetary losses will continue for many years. The human and financial strain on nations in preparing for and responding to natural and accidental disasters is overwhelming. These challenges require global cooperation.

Security and loss prevention practitioners are faced with serious challenges and questions as they assist their employers with surviving in a constantly changing world filled with risks. How can businesses and institutions protect employees, assets, and operations from terrorism, natural disasters, and other risks? What does the future hold? Who will pay for protection? This book offers some insights into the answers to these difficult questions.

Search the Internet

Access the Internet and seek an international perspective by visiting the New Scotland Yard website, which includes links to history: www.met.police.uk.

Use search engines to check the sites of major security companies. Examples: G4S, www.g4s.com; Securitas, www.securitas.com; and Brink's, www.brinks.com.

What can be learned from these sites?

Case Problems

1A. As a security manager you are asked to speak to a local college class on the history and development of the security and loss prevention profession. What five significant points in the history of this profession do you emphasize?

1B. As a part-time security officer and a full-time college student, you are working on a course assignment to think critically about the history of security and loss prevention. The assignment requires you to focus on some aspect of the history of security and loss prevention that you believe is biased or inaccurate and to explain your interpretation of historical events. Prepare a typed report.

1C. Suppose you are at a security conference and you decide to attend an educational session entitled "The Most Serious Challenges Facing Security Professionals in Their Jobs." What topics do you think the speaker will focus on?

References

ASIS International. (2008). *Chief security officer organizational standard.* <www.asisonline.org> retrieved February 25, 2011.

Astor, S. (1978). *Loss prevention: Controls and concepts.* Stoneham, MA: Butterworth.

Booz Allen Hamilton. (2005). Convergence of Enterprise Security Organizations. (November 8). <www.securitymanagement.com> retrieved January 27, 2006.

Brink's. (2011). *Secure logistics worldwide.* <www.brinks.com> retrieved February 26, 2011.

Brinton, C., et al. (1973). *Civilization in the west.* Englewood Cliffs, NJ: Prentice-Hall.

Bugbee, P. (1978). *Principles of fire protection.* Boston: National Fire Protection Association.

Calder, J. (2010). Law, politics, and occupational consciousness: Industrial guard unions in the United States, 1933–1945. *Journal of Applied Security Research, 5.*

Dempsey, J., & Forst, L. (2010). *An introduction to policing* (5th ed.). Clifton Park, NY: Delmar, Cengage.

Ellis, D. (1991). *Becoming a master student* (6th ed.). Rapid City, SD: College Survival, Inc.

Feuerwerker, A. (1989). *Great wall of China.* Chicago, IL: *World Book Encyclopedia.*

G4S. (2011). *Facts and figures.* <www.g4s.com> retrieved February 26, 2011.

Germann, A., Day, F., & Gallati, R. (1974). *Introduction to law enforcement and criminal justice.* Springfield, IL: Thomas Pub.

Holden, R. (1986). *Modern police management.* Englewood Cliffs, NJ: Prentice-Hall.

Kiltz, L. (2009). Developing critical thinking skills in Homeland security and emergency management courses. *Journal of Homeland Security and Emergency Management, 6*(1).

Lee, S. (2011). Spatial analyses of installation patterns and characteristics of residential burglar alarms. *Journal of Applied Security Research, 6* (January–March).

McCrie, R. (2010). Allan Pinkerton (August 25, 1819–July 1, 1884): Founder of the security services industry. *Journal of Applied Security Research, 5.*

Naisbitt, J. (1982). *Megatrends.* New York: Warren Books.

Post, R., & Kingsbury, A. (1977). *Security administration: An introduction* (3rd ed.). Springfield, IL: Charles C. Thomas.

Safi, A., & Burrell, D. (2007). Developing critical thinking leadership skills in Homeland security professionals, law enforcement agents and intelligence analysts. *Homeland Defense Journal, 5* (June).

Securitas. (2011). *Securitas history.* <www.securitas.com> retrieved February 26, 2011.

Shelden, R. (2001). *Controlling the dangerous classes.* Boston, MA: Allyn & Bacon.

Stead, P. (1983). *The police of France.* New York: Macmillan.

Toffler, A. (1980). *The third wave.* New York: Morrow.

Ursic, H., & Pagano, L. (1974). *Security management systems.* Springfield, IL: Charles C. Thomas.

Weisheit, R., Baker, L., & Falcone, D. (1995). *Crime and policing in rural and small town America: An overview of the issues.* Washington, DC: National Institute of Justice.

2 The Business, Careers, and Challenges of Security and Loss Prevention

OBJECTIVES

After studying this chapter, the reader will be able to:

1. Explain the risks and losses facing our society
2. Define and illustrate metrics and explain why it is important
3. Describe the security industry
4. Define and explain privatization
5. Describe the types of employment available in the security and loss prevention vocation
6. Explain the limitations of the criminal justice system
7. List and discuss the challenges of the security industry

KEY TERMS

- private sector
- public sector
- human resources
- asset
- operations of enterprises
- risk
- threat
- hazard
- all-hazards
- all-hazards preparedness concept
- threats to brand
- metrics
- internal metrics
- crime pattern analysis
- methodological problems
- direct losses
- indirect losses
- external metrics
- contract security
- proprietary security
- privatization
- mercenary
- deterrence
- reactive strategies
- proactive strategies
- Private Security Officer Employment Authorization Act of 2004
- ethics
- ASIS International
- International Foundation for Protection Officers

The millions of people who have chosen security and loss prevention as their vocation work to protect people, assets, and the operations of enterprises.

Introduction: Why is Security a Huge Industry?

Security is "big business" globally, and there are many career opportunities and challenges in this profession. Several drivers are affecting the growth of this industry:

- Crimes, fires, accidents, and natural and other disasters pose threats to the general public and the public and private sectors worldwide.
- Offenders from throughout the world are constantly seeking to exploit vulnerabilities that they can pinpoint in government, businesses, institutions, nonprofits, and individuals.
- Nations, businesses, and individuals seek advantage over rivals by whatever means are available.
- The exploitation of technology and the theft of information are enormous problems.
- The 9/11 attacks, other terrorist threats, and the conflicts in Iraq, Afghanistan, and other locations illustrate the need for global security and safety.
- Government laws and mandates to improve homeland security and protect critical infrastructure and key resources are placing increasing responsibilities on protection programs.
- The challenges of risk management, such as the increasing costs of insurance, are forcing businesses, institutions, and organizations to enhance their security and loss prevention programs to reduce risks.
- There is increased pressure on the security and loss prevention vocation and industry to improve professionalism, performance, and value to enterprises.

Terms and Definitions of the Security Industry

This chapter begins with basic terms and definitions as a foundation for explaining what this business is about and what it seeks to accomplish. Subsequent paragraphs explain risks and the importance of metrics as a way to show the value and performance of security and loss prevention.

Another important topic in this business is customer service (explained in Chapter 5). The practitioners of this vocation must be mindful of value, performance, and customer service as vital strategies to survive in this business. These strategies are important for both proprietary and contract security organizations.

Businesses, often referred to as the **private sector**, exist to generate profit. Government, often referred to as the **public sector**, serves the general public and is supported by tax dollars. Both sectors contain security and loss prevention programs to protect *people, assets,* and the *operations of enterprises* from a broad variety of *threats* and *hazards* that can result in harm and losses. Here we emphasize protecting people connected in some way to organizations, as opposed to protecting the general public through the efforts of our armed forces and law enforcement agencies and other first responders. The people in organizations are referred to as employees or **human resources**. They must be protected if an enterprise is to continue its operations and pursue its objectives. However, other groups of people connected to organizations also require protection. Examples are customers, vendors, and visitors. In fact, people are not required to be on the premises to receive protection, as in the case of an organization protecting customer information as required by law. From a business perspective, an **asset** is a resource controlled by an organization that can produce economic benefit. Examples are many and include cash, stocks, inventory, property, equipment, patents, copyrights, and goodwill. The **operations of enterprises** are the utilization of human resources and assets to pursue organizational objectives.

Practitioners in the security and loss prevention vocation must be mindful of value, performance, and customer service as vital strategies to survive in this business.

Risk is an especially important term in the security profession. It must be *measured* and *managed*. (Risk analysis and risk management are explained in subsequent chapters.) Several disciplines helpful to the security profession offer definitions of risk as illustrated next. Quinley and Schmidt (2002: 4), from the insurance and risk management discipline, define risk as "a measure of the frequency or probability of a negative event and the severity or consequences of that negative event." Miller and Jentz (2011: 241), business law specialists, explain risk as "a prediction concerning potential loss based on known and unknown factors." Haddow, Bullock and Coppola (2011: 380), from the emergency management discipline, define risk as follows: "A measure of the likelihood that a hazard will manifest into an actual emergency or disaster event and the consequences of that event should it occur." ASIS International (2009: 48), from the security discipline, sees risk as "characterized by reference to potential events, consequences, or a combination of these and how they can affect the achievement of objectives." **Risk** is defined here as the measurement of the frequency, probability, and severity of losses from exposure to threats or hazards.

A **threat** is a serious, impending, or recurring event that can result in loss, and it *must be dealt with immediately*. For instance, management in a corporation is informed that an employee stated that he intends to kill other employees.

The U.S. Department of Homeland Security (2004: 66) defines hazard as "something that is potentially dangerous or harmful, often the root cause of an unwanted outcome." **Hazard** is defined here as a source of danger that has the potential to cause unwanted outcomes such as injury, death, property damage, economic loss, and environmental damage that adversely impact our society.

Table 2-1 depicts several threats and hazards that can cause losses. The frequency and cost of each loss vary. Each type of threat or hazard has its own specialist to work toward a solution. For instance, a rash of robberies at a liquor store may require assistance from public law

Table 2-1 Threats/Hazards

Criminal acts	Natural disasters	Miscellaneous
Arson	Earthquake	Accident
Assault	Excessive Snow/Ice	Bad Investment
Burglary	Floods/Excessive Rain	Business Interruption
Counterfeiting	Hurricane	Equipment Failure
Cybercrime	Landslide	Error
Embezzlement	Lightning	Explosion
Espionage	Pandemic	Fire
Extortion	Pestilence	Harm to Brand
Fraud	Tidal Wave	Litigation
Identity Theft	Tornado	Mine Disaster
Kidnapping	Tsunami	Nuclear Accident
Larceny/Theft	Volcanic Eruption	Oil Spill
Murder		Pollution
Product Tampering		Poor Safety
Riot		Poor Supervision
Robbery		Power Outage
Sabotage		Sexual Harassment
Sexual Assault		Sonic Boom
Shoplifting		Strike
Substance Abuse		Unethical Conduct
Terrorism		War
Transnational Crime		Waste
Vandalism		

enforcement officers and the installation of a more sophisticated alarm system from the private sector. Numerous injuries at a manufacturing plant may require assistance from a safety specialist. The loss prevention manager—a specialist in his or her own right or else with the assistance of a specialist—must plan, implement, and monitor programs to anticipate, prevent, and reduce loss.

Another important term is **all-hazards**. It refers to multiple types of hazards, including natural disasters (e.g., hurricane) and human-made events (e.g., inadvertent events or accidents, such as an aircraft crash, and deliberate events, such as the terrorist bombing of an aircraft). Another type is technological events. An example is an electric service blackout resulting from a variety of possible causes. The term *all-hazards* is important because it relates to the **all-hazards preparedness concept**. This means that different hazards contain similarities, and organizations can benefit, to a certain degree, from generic approaches to emergency management and business continuity. This provides an opportunity for efficient and cost-effective planning. In other words, how one hazard is prepared for may be similar to the preparation for other hazards. This topic is covered in greater depth in a later chapter.

Threats to Brand

Threats to brand are a major concern of businesses. They can bring harm to an entity's reputation and good name, resulting in lost business. The 1982 Tylenol scare, when seven people died after purchasing and consuming this product laced with cyanide, is a famous example of threat to brand. It caused enormous financial problems for the manufacturer. The 2010 British Petroleum oil spill in the Gulf of Mexico caused the oil company to incur huge financial losses following this environmental disaster that also devastated Gulf Coast businesses. In 2009, *ABC News* reported that a Domino's Pizza employee at a Conover, North Carolina, store was filmed by another employee as he put cheese in his nose, blew mucous on a sandwich, and put a sponge used to wash dishes between his buttocks. The video of these actions was viewed by millions through *Twitter*, blogs, and *YouTube*. The offending employees were arrested for food tampering. This case illustrates how social networks can be used to harm a brand. Businesses must maintain awareness of what is being communicated. Complaints and issues should be investigated and followed by a well-planned response (Goldman, 2009). Horovitz (2009) offers suggestions for businesses: monitor social media; respond quickly; respond at the flashpoint (Domino's responded on a blog and on the Twitter site where chatter was mounting); establish strong policies, procedures, and training; and foster a positive culture while satisfying employee needs as well as possible.

In 2010, Domino's was subject to media attention again when a former employee, Jamal Thomas, was arrested for arson against two stores in New York City. The *Daily News* reported that the former employee claimed, "Domino's is a terrible place to work." He kept his uniform and key and "visited several stores in uniform, claiming to be part of a secret Domino's team measuring employee satisfaction." The *Daily News* reported that he was really casing Domino's stores to commit arson. Thomas was seen on video footage at one of the stores. When apprehended, he was charged with arson, reckless endangerment, and other crimes. Besides the negative media attention, the losses incurred by Domino's were estimated to be over $1 million dollars (Paddock et al., 2010).

Metrics

Since *the business* of security and loss prevention is to protect people, assets, and the operations of enterprises, it is imperative that management in this vocation develop methods to *measure* security and loss prevention, loss events, and the cost of losses. Senior executives are sure to ask for evidence of the program's successes, effectiveness, and return on investment, and when losses do occur, the cost, why it occurred, and what corrective action followed. Methods of measurement can also be used to identify and analyze risks, brief senior executives, help support a budget increase and additional resources, and justify human resources (e.g., proving an employee's value to the organization). Senior executives are also interested in what is driving the security and loss prevention program. Is it risks, legal requirements, regulations, or other factors?

Metrics are measurements that quantify results. These measurements show the *value* and *performance* that security and loss prevention bring to the business enterprise. Using metrics is a vital component of quality management and is essential when communicating with senior executives. Examples of metrics related to security are numerous and depend on business objectives and need. What may be a high priority metric for one business may not be important in another business.

Modern technology provides the opportunity to remotely monitor global operations, collect data, analyze it, and report to superiors. In addition, the Internet provides easy access to a variety of information and data.

Internal Metrics

Internal metrics focus on measurements within an organization. Examples of internal metrics include the annual number of theft incidents, monetary losses, and the value of company assets recovered; the number of employees hired and then released because it was later learned that they had falsified their backgrounds; the number of malfunctions of an access control system each month; and the turnover of security officers.

Management at one security program has over 200 metrics that record items such as badges issued to employees, door alarms that activate, CCTV cameras that malfunction, and security violations. This company links metrics to corporate financial programs to extract costs for analysis, reporting, and planning (Treece and Freadman, 2010: 90–94).

Kovacich and Halibozek (2006: xxvii) define a security metric as "the application of quantitative, statistical, and/or mathematical analyses to measuring security functional costs, benefits, successes, failures, trends, and workload—in other words, tracking the status of each security function in those terms." They describe two basic methods of tracking costs and benefits. The first is through recurring costs from day-to-day operations, such as security officer duties and investigations of loss of assets. Metrics can depict trends, such as whether the cost of protection is going up or down. The second method is through formal project plans that have a set schedule of beginning and ending dates, with specific objectives and costs that serve to track (i.e., use metrics to assess) time, expenses, and accomplishments.

Wailgum (2005) writes that metrics vary by security executive, organization, and industry. He offers an example of how metrics are used by a retail chain of thousands of stores. One metric used is the number of robberies per 1,000 stores. This figure is compared to the rates of other, similar retail chains. Another metric is cash loss as a percentage of sales for every retail unit. Another illustration offered by Wailgum is of a utility company that uses metrics to gauge its compliance with federal regulations. It compares performance on "readiness reviews" among different facilities. Readiness reviews assess whether employees understand threat plans and what to do when the threat level is raised or lowered. Metrics at this utility company also focus on penetration testing of possible security breaches. How far can someone reach without a badge? Can someone talk his or her way around delivery procedures? These evaluations can be quantified and compared over time.

Metrics can assist **crime pattern analysis** (CPA). This generic term includes a variety of approaches and techniques that search for crime patterns and trends. Ballantyne (2011: 98) explains three major types of CPA: (1) spatial (i.e., a study of the relationship between crime and place) and temporal (i.e., time, day, week, month, and year of event), (2) based on the offender, and (3) a fusion of both. He adds that CPA includes three stages: (1) collecting data on crimes and searching for patterns, (2) studying why the patterns occurred and applying prevention strategies, and (3) evaluating the impact the prevention strategies had on crime. Another tool of CPA is crime cluster analysis that studies a group of crimes with common characteristics. Ballantyne (2011: 98) writes: "The clusters can be put into such categories as location of the crime, type of crime, way the crime had been committed, type of items stolen and even type of victim." Security practitioners and public police apply a variety of software to study trends and investigate crimes.

Metrics are subject to **methodological problems**. This means that measurements are not perfect; they are subject to a host of factors that may result in inaccuracies. Factors that can distort metrics include variations in the use of definitions by those who report, inconsistencies in reporting (e.g., not reporting certain incidents), exclusion of indirect losses, bias, miscalculations, and the possibility of fraud. In addition, comparisons among organizations are difficult. Furthermore, *metrics may produce more evidence about the way the enterprise is managed than about actual events and losses.* For instance, one company that maintains a zero tolerance for violence in the workplace shows data indicating numerous incidents of threats and violence, while another similar company, with a corporate culture that almost ignores the problem, shows few incidents.

Why Emphasize Both Direct and Indirect Losses?

The methodology of metrics should include direct and indirect losses. Businesses, institutions, and organizations can suffer extensive direct losses from threats and hazards. Additionally, indirect losses can be devastating and often surpass direct losses. **Direct losses** are immediate, obvious losses, whereas **indirect losses** are prolonged and often hidden. A burglary at a business, for example, may show the direct loss of $1,000 from a safe. However, on close inspection of indirect losses, the total losses may include the following: damage to the door or window where the break-in occurred; replacement of the destroyed safe; raise in insurance policy deductible; increase in insurance premium; loss of sales from a delay in re-opening the business; customer dissatisfaction; and employee time diverted to speaking with police, insurance representatives, and repair people. If the event was serious, additional indirect losses could include adverse media attention, permanent loss of customers, employee morale problems, turnover, and negative reactions from shareholders. *Security professionals can help justify their position and their value to the business by demonstrating total losses resulting from each incident.*

Do you think a security professional who emphasizes total losses (i.e., direct and indirect) helps or harms his or her job security?

External Metrics

External metrics focus on society-wide measurements, such as those pertaining to crimes and natural disasters in a geographic area. These measurements help enterprises make business decisions on the geographic location for business investment and the types of insurance to purchase, among other decisions. External metrics provide input for security and loss prevention planning and budgeting.

Public and private sector organizations collect data and publish reports that may or may not be available to the general public. The U.S. Department of Justice publishes crime statistics that are easily accessible on the Internet. The U.S. Department of State publishes information on terrorism and safety in other countries that is likewise easily accessible on the Internet. Risk management and insurance firms offer services for a fee that estimate risks and costs from all-hazards. A government agency may withhold data and reports due to reasons of national security, whereas a corporation may see its data and reports as proprietary information and/or subject to sale. As with internal metrics, methodological problems exist with external metrics.

In the following we look at some external metrics in public sector organizations in order to gauge the risks facing not only society but also business enterprises, institutions, and other organizations. An emphasis is placed on the risk of crimes, fires, and accidents because, traditionally, security and loss prevention practitioners have expended considerable time and resources on these problems. However, as noted earlier in this book, our world is becoming increasingly complex and practitioners are facing a broader variety of risks.

Crimes

In 1930, Congress passed the first legislation mandating the collection of crime data. The task was assigned to the Federal Bureau of Investigation (FBI) within the U.S. Department of Justice, and each year the *Uniform Crime Report* (UCR) has provided data on crime trends. This helps to calculate quality of life in communities throughout the country, set government policy, plan funding, and gauge the effectiveness of anticrime strategies. The data are collected by police agencies that then report them to the FBI. There are many weaknesses of the UCR, and the data have been called notoriously inaccurate. Examples of UCR problems are as follows: it represents *reported* crimes, while many crimes are not reported to police; when crimes are reported to police, the crimes may not be recorded; only local and state crimes are reported, not federal crimes or crimes at institutions (e.g., jails and prisons); definitions of crimes vary among states; and the data have been subject to political manipulation. Fuller (2012: 25) writes: "The misreporting of crime statistics by police is a serious problem in the collection of crime statistics." To improve the UCR, the FBI is refining a newer system—the National Incident-Based Reporting System—that collects data that are more specific. In 1972, because of the shortcomings of the UCR, the National Crime Victim Survey (NCVS) began to collect crime data from households. It was found that there was a significantly higher rate of crime than what was reported in the UCR. Forcible rape, robbery, aggravated assault, burglary, and larceny were reported by the NCVS at rates two to three times greater than rates reported in UCR data. NCVS data are gathered by the Bureau of Justice Statistics (U.S. Department of Justice) and the U.S. Census Bureau (Fuller, 2012: 29).

Although the methodology of collecting crime data by official sources is problematic (even though the United States has one of the best systems in the world), crime is a serious problem, and millions of crimes are committed each year in the United States. In 2009, the UCR reported

1,318,398 violent crimes and 9,320,971 property crimes that were reported and recorded (FBI, 2010). Other sources of crime data are available from individual states, educational institutions, and self-report surveys conducted by researchers who use questionnaires to gather anonymous input on crimes committed by offenders.

Barkan (2006: 75) writes that crime data gathered outside the United States is highly inconsistent. Some nations gather crime data in ways similar to the United States, while other nations do not collect such data. Three major sources of international crime data are the International Criminal Police Organization (INTERPOL), the World Health Organization, and the United Nations.

Chamard (2006: 2) notes that there are few studies on the victimization of businesses in the United States, yet businesses suffer disproportionately from crime in comparison to households and individuals. She argues that the UCR underreports crimes against businesses. Chamard refers to a British study of small businesses that found that nearly two-thirds had suffered some form of victimization over a five-year span. She reports data from cross-national surveys suggesting that "retail businesses may have an overall burglary risk that is ten-times greater than the risk faced by households."

Fires

Fire represents another serious hazard facing organizations. Two major sources that measure the fire problem are the National Fire Protection Association (NFPA) and the U.S. Fire Administration (USFA). The National Fire Data Center (NFDC) of the USFA periodically publishes Fire *in the United States*, a 10-year overview of fires in the United States. It is designed to motivate corrective action, set priorities, serve as a model for state and local analysis of fire data, and serve as a baseline for evaluating programs. *Because of the time it takes states to submit data, the publication lags the date of data collection.*

The U.S. Fire Administration (2010) reported that in 2009 there were 89,200 nonresidential building fires, 90 deaths, 1,500 injuries, and 2.7 billion in dollar losses. During the same year, there were 356,000 residential building fires, 2,480 deaths, 12,600 injuries, and 7.2 billion in dollar losses.

Accidents

Accidents pose another serious hazard facing organizations and their employees when on and off the job. The U.S. Department of Labor (2011) reported 4,547 fatal occupational injuries in the United States during 2010. The number of nonfatal workplace injuries and illnesses reported from private industry during the same year was nearly 3.1 million. Security professionals must work with a variety of specialists (e.g., safety, insurance, medical, and legal) to prevent and respond to accidents.

The Security Industry

The security industry is a multi-billion-dollar business. Every decade seems to bring an increased need for security. Internal theft, cybercrime, street crime, terrorism, natural disasters, and so on bring greater demands for protection. In the United States, thousands of companies provide contract security services and investigations. In addition, there are manufacturers of products and systems and thousands of alarm installation firms.

The Freedonia Group, a market research firm, reported the following on the security industry:

- Sales of private contract security services in the United States will increase nearly five percent annually to $63 billion in 2014. Guarding is the largest security service segment (40 percent of industry revenue and 67 percent of employment). The second largest service segment is alarm monitoring (30 percent of industry revenue and 10 percent of employment). Pre-employment screening is the fastest growing segment of this industry (Freedonia, 2010a).
- Sales of electronic security products and systems in the United States are expected to increase 9.3 percent annually to $17 billion in 2014. The main driver of this demand is increased perceived risk of crime despite an actual downtrend in crime. Access controls make up the largest and fastest-growing product segment, forecast to grow 12.8 percent per year through 2014. Alarms accounted for the next-largest share of the electronic security equipment market, forecast to grow 7.2 percent per year through 2014 (Freedonia, 2010b).
- The global market for private contract security services increased nearly 7.5 percent annually through 2012 to almost $200 billion. This market will be driven by an improving global economy, rising urbanization, and increased fears of domestic crime and terrorism in many countries. Guarding is the largest segment of this market at 48 percent of total revenue (Freedonia, 2008).
- Global demand for security equipment is expected to rise 7.4 percent per year through 2014 to about $100 billion. In 2009, the electronic security products market accounted for over 60 percent of total security equipment demand with stronger sales than mechanical security equipment through 2014. This demand will be driven by improvements in design that make products more user-friendly and cost-effective (Freedonia, 2010c).

Contract versus Proprietary Security

Contract security refers to businesses that seek a profit by offering a host of security services to businesses, institutions, and organizations. Examples of services include security officers, investigations, consulting, and monitoring alarm systems. Depending on its unique needs and after weighing several factors, an entity requiring security may prefer to establish its own security, known as **proprietary** (in-house) **security**, of which there are thousands. In addition, an organization may use both contract and proprietary security (Figure 2-1).

The *Report of the Task Force on Private Security* (U.S. Department of Justice, 1976b: 146–147 and 249–257) lists several factors to consider concerning contract versus proprietary security officers. Contract security generally is less expensive, although there are exceptions. The service company typically handles recruitment, selection, training, and supervision. Hiring unqualified security officers and turnover are two primary disadvantages of contract services. Many contract officers are "moonlighting" and subject to fatigue. Questions concerning insurance and liability between the security company and the client may be unclear.

FIGURE 2-1 Businesses often employ both proprietary and contract security personnel.

A major advantage of a proprietary force is that greater control is maintained over personnel, including selection, training, and supervision, and, of course, such a force is more familiar with the unique needs of the company. Salaries and benefits, however, often make establishment of a proprietary force more expensive.

Privatization

Privatization is the *contracting out* of government programs, either wholly or in part, to for-profit and not-for-profit organizations. There is a growing interdependence between the public and private sectors. A broad array of services is provided to government agencies by the private sector today, from consulting services to janitorial services. Both government and the private sector are operating hospitals, schools, and other institutions formerly dominated by government. For crime control efforts, we see private security patrols in residential areas, private security officers in courts, and private prisons. Businesses make themselves attractive to governments when they claim that they can perform services more efficiently and at a lower cost than the public sector.

Privatization is not a new concept. In the late 1600s, many security and incarceration services for early urban areas were supplied by the private sector. By the 1700s, government dominated these services. Today, privatization can be viewed as a movement to demonopolize and decentralize services dominated by government. The movement to privatize criminal justice services encourages shared responsibility for public safety (Bowman, 1992: 15–57).

Another factor fueling the privatization movement is victim and citizen dissatisfaction with the way government is handling crime. Increasing numbers of citizens are confronting crime through neighborhood watches, citizen patrols, crime stoppers, hiring private attorneys to assist prosecutors, and dispute resolution.

Critics of privatization argue that crime control by government is rooted in constitutional safeguards and crime control should not be contracted to the private sector. Use of force and searches by the private sector, punishment in private prisons, and liabilities of governments and contractors are examples of the thorny issues that face privatization.

For many years, private security companies have guarded military installations in the United States. Supporters note that these private security officers are supervised and trained by Department of Defense (DOD) specifications, and they provide an invaluable service. Critics argue that there is apprehension about the caliber and training of these security officers and that some work for American subsidiaries of foreign-owned companies. This concern led to a U.S. Government Accountability Office (GAO) investigation, which found that some contractors had hired felons and that training records had been falsified. The GAO discovered that the contracts cost 25 percent more than contracts later put out for bid. In addition, the Army relied on what the contractors said they were doing and provided very little monitoring. The DOD pledged to improve management and oversight of private security officers (Marks, 2006; U.S. Government Accountability Office, 2006).

Private military and private security companies (PMPSCs) are providing a wide variety of services for a profit in geographic areas facing conflict. These firms are growing rapidly worldwide. Governments, corporations, organizations, the United Nations, and individuals are relying on these companies for many reasons. A national government, for example, may need additional personnel because human resources and expertise are stretched to the limit and a rebel group is gaining strength. A corporation, organization (e.g., nonprofit aid group), or individual may require protection for operations in a foreign country containing an insurgency and weak police and military forces. In addition, a country may be dangerous because of a hodgepodge of self-defense forces, mercenary units, militias, and vigilante squads that maintain their own agendas. The uncontrolled proliferation of conventional arms has hastened the growth of armed groups (Purpura, 2007).

Zarate (1998) writes that the term *mercenary* is difficult to define and United Nations members disagree on its definition. He refers to a **mercenary** as "a soldier-for-hire, primarily motivated by pecuniary interests, who has no national or territorial stake in a conflict and is paid a salary above the average for others of his rank." Zarate sees the legal issues involving mercenaries as unclear. He offers these issues for debate: (1) PMPSCs could be hired by insurgents or foreign governments to destabilize an established regime, or a government could hire a PMPSC to suppress a national liberation movement (e.g., group fighting colonialism or racism); and (2) PMPSCs that assist multinational corporations will act solely for the benefit of the corporation in foreign countries and create semi-sovereign entities supported by the government. He adds that PMPSCs are regulated in most countries. In the United States, regulations are stringent, and *registration with the U.S. government is required.*

The hostilities in Iraq and Afghanistan offered excellent opportunities for PMPSCs, especially those firms with employees possessing military backgrounds who can provide support to the U.S. military and train the armies of weak countries. The U.S. Army (2006: 6–4 and 8–17) notes, "training support from contractors enables commanders to use soldiers and marines more efficiently." The support includes institutional training, developing security ministries and headquarters, and establishing administrative and logistic systems. In addition, theater support contractors (i.e., local vendors in hostile areas) supply a host of products and services. Examples are concrete security barriers, fencing, construction, sanitation, and trucking.

Nordland (2012) reported that whereas the U.S. military announces the names of war dead, private companies usually only notify family members. Other issues are the limited compensation, if any, for survivors of contractors killed and medical care for injured contractors. In 2011, in Afghanistan, 430 American contractors and 418 American soldiers were killed.

Schwartz (2011) offers several benefits of PMPSCs. He writes that using contractors can allow the federal government to adapt more quickly to changing global events. They can be hired and deployed faster in many cases, and they save the government money because contractors can be let go when their services are no longer needed. As the United States pursues a global war on terrorism with limited armed forces private contractors are in demand.

Rothe and Ross (2010: 593–617) argue about the social disorganization and anomie (i.e., normlessness) associated with PMPSCs. They note that PMPSCs often operate in a disorganized environment (e.g., a chaotic, war-torn area) and they have a propensity to make up their own rules without accountability, especially because "back-up" is typically unavailable. For example, they may run vehicles off the road and fire rounds into anything that gets too close to them, like something out of the *Mad Max* movie. As one contractor stated, "Your car can be a 3,000-pound weapon when you need it. Hit and run… The police aren't coming to your home because you left the scene of an accident." Rothe and Ross also note that a disorganized environment is created from high turnover in a PMPSC. They quote a Blackwater employee: "Blackwater is like a fucking restaurant. You've got hundreds of people coming through." Rothe and Ross are critical of PMPSCs because of "minimal oversight, no transparency, and no standing international criminal laws to regulate them."

One source of standards for PMPSCs is ASIS International. It published *Management System for Quality of Private Security Company Operations—Requirements with Guidance, ANSI/ASIS PSC.1-2012 and Conformity Assessment and Auditing Management Systems for Quality of Private Security Company Operations, ANSI/ASIS PSC.2-2012.* These standards build upon international human rights and humanitarian laws and apply to PMPSCs operating in locations subject to conflict and/or disaster.

What are your views on privatization?
Do you believe that the war on terrorism is heavily privatized?

Careers: Loss Prevention Services and Specialists

Many services and specialists can be involved in developing an effective loss prevention program. These services and specialists can be proprietary or outsourced. Table 2-2 lists various facilities requiring security and loss prevention programs. Since the 9/11 attacks, many of these facilities have been referred to as "critical infrastructure." Table 2-3 lists security and loss prevention services and products from the security industry. Table 2-4 lists specialists and consultants who can aid loss prevention efforts. These tables are not conclusive because protection programs are becoming increasingly specialized and diversified. Therefore, additional specialization will evolve to aid these programs.

Most loss prevention managers are generalists, which means that they have a broad knowledge of the field plus specialized knowledge of the risks facing their employer. When feasible, these managers should develop a multidisciplinary staff to assist in protection objectives. The staff should represent various specializations as appropriate, such as security, fire protection, information technology, and safety.

Table 2-2 Locations Requiring Security and Loss Prevention Programs

• Agriculture/Food Facilities	• Industrial
• Airport/Airline	• Library
• Campus/School	• Medical
• Commercial Buildings	• National Monument
• Cultural Properties	• Nuclear Plant
• Defense Industries and Facilities	• Park/Recreation
• Drinking Water	• Port
• Emergency Services	• Postal/Shipping
• Energy Facilities	• Retail Store/Mall
• Financial Institution	• Sports/Entertainment Facility
• Gaming/Wagering	• Telecommunications
• Government Building	• Transportation
• Hotel/Motel/Restaurant	• Wholesale/Warehouse
• Housing/Residential	• Others

Table 2-3 Security Services and Products from the Private Sector

Services	Products
• Armored Transportation (Figure 2-2)	• Access Control Systems
• Business Continuity/Emergency Management	• Barriers
• Canine	• Closed-Circuit Television
• Central Alarm Station	• Doors
• Consulting	• Fire Alarm Systems
• Counterterrorism	• Glazing
• Detection of Deception (e.g., polygraph)	• Intrusion Detection Systems
• Executive Protection	• Lighting
• Honesty Shopping	• Locks/Keys
• Information Security	• Safes/Vaults
• Investigations	• Vehicles
• Risk Analysis/Security Survey	• Weapons
• Security Officers	• Others
• Technical Surveillance Countermeasures	
• Undercover Investigations	
• Others	

Table 2-4 Specialists and Consultants Who Can Assist Loss Prevention Efforts

• Accountant	• Forensic Scientist
• Auditor	• Landscape Architect
• Architect	• Life Safety/Fire Consultant
• Business Consultant	• Lighting Consultant
• Computer/IT Consultant	• Locksmith
• Criminologist	• Marketing Consultant
• Critical Infrastructure Consultant	• Risk Management Consultant
• Education/Training Consultant	• Others
• Engineering Consultant	

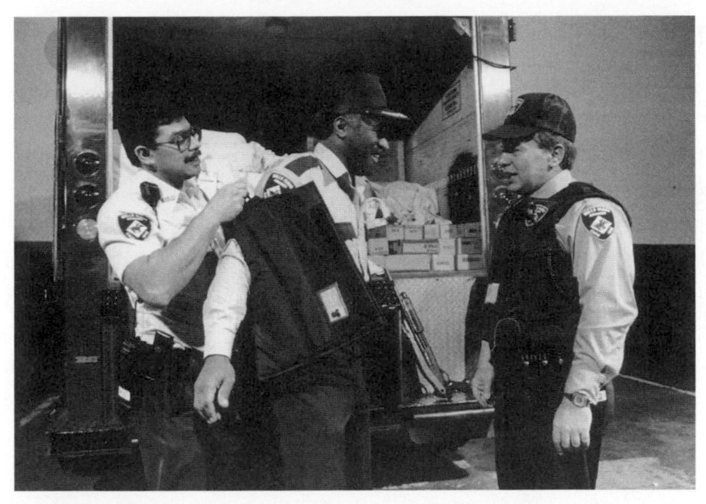

FIGURE 2-2 Employees wear bullet-resistant vests and personal hold-up alarms to help protect them as they pick up and deliver valuable shipments.

Career: Manufacturing Security

Manufacturers make products, which, in turn, are sold either to wholesalers, distributors, or directly to consumers. Professionals within the manufacturing security specialty are responsible for issues involving not only sales transactions, but also transport issues, ordering and purchasing of raw materials, and the protection of resources against loss or theft. Prevention of loss can be accomplished only through employing competent security directors and managers who can help integrate the security function into the total operation instead of allowing it to remain isolated. Depending on the products being manufactured, individuals in this specialty may work in a variety of environments, including exposure to varying weather conditions and involvement with chemical processing areas.

Entry-level management positions generally require at least an associate's degree and often a bachelor's degree. Most positions require two or more years of general security experience and may require some specialty-specific training or experience. A Certified Protection Professional (CPP) designation is preferred for many positions.

Mid-level management positions generally require a bachelor's degree and five or more years of general security experience and may require some specialty-specific training or experience. A CPP designation is required or desired for many positions. To enhance the background and marketability of the security professional, additional certifications can be pursued, for example, the Certified Business Continuity Professional (DRI International, 2012).

Modified from source: Courtesy of ASIS International (2005). "Career Opportunities in Security." www.asisonline.org.

Occupational Outlook

Except as noted, the following information is from the *Occupational Outlook Handbook*, 2010–2011 Edition (U.S. Bureau of Labor Statistics, 2009).

Security officer job responsibilities differ from one employer to another. Security officers generally *observe* and *report* while working at stationary posts or while on patrol. They work full-time or part-time and may be armed or unarmed. States usually require contract security businesses and security officers to be licensed and the training varies. Employment is expected to grow by 14 percent between 2008 (1,086,000 officers) and 2018 (1,239,500 officers), which is faster than the average for all occupations. Demand will grow as these practitioners perform some duties formerly handled by public police officers, such as security at public events and in residential and commercial areas.

Interestingly, a study of police and private security collaborations by the U.S. Department of Justice, Office of Community Oriented Policing Services (2009: 37), argued that the Bureau of Labor Statistics category of "security guards and gaming surveillance officers" employed over 1 million persons, however, "that category likely includes no more than half of those employed in private security overall." The rest of security workers include those in the alarm, access control, CCTV, and other industries, as well as managers of contract security firms and security departments of various organizations. The study estimated that the total number of U.S. security workers is "2 million or more—that is, more than twice the number of law enforcement officers."

Private detectives and investigators collect information through several methods such as interviewing people, searching online, conducting surveillance, traveling to locations (e.g., courthouses to check public records), and studying crime scenes. Private investigators (PIs) are typically required to be licensed by state government. They work for corporations, financial institutions, insurance companies, attorneys, private individuals, and others while performing a variety of types of investigations that focus on applicant background checks, asset searches, fraud, litigation, and domestic disputes, among other areas. About 21 percent are self-employed. These private sector practitioners may work irregular hours, are generally unarmed, and often have previous law enforcement experience. Employment is expected to grow by 22 percent between 2008 (45,500) and 2018 (55,500), which is faster than the average for all occupations.

Police and detectives work for government and serve the public. Police officers patrol a specific area and respond to calls for service. Detectives perform investigative work, gather facts, and collect evidence. Police and detectives spend a lot of time writing reports. Following the 9/11 terrorist attacks, public police have increased their involvement in security and intelligence duties. The applicant screening, training, and educational requirements are generally more extensive than in the private sector. Employment is expected to grow by 10 percent between 2008 (883,600) and 2018 (968,400).

Limitations of the Criminal Justice System: Implications for Loss Prevention Practitioners

In 2006, federal, state, and local governments spent over $214 billion for criminal and civil justice and employed 2,427,452 personnel (U.S. Department of Justice, Bureau of Justice Statistics,

2009). Despite these expenditures, many theories underpinning crime prevention and law enforcement measures have come under increased scrutiny. For example, the strategy of **deterrence** has been questioned. This criminal justice system strategy seeks to prevent crime through state-imposed punishment. This model presumes that offenders rationally consider the risks and rewards of crime prior to acting. In fact, numerous offenders simply find that "crime pays," because the certainty of punishment is a myth and the odds are in the offender's favor. John E. Conklin (2001: 466), a criminologist, writes: "People have imperfect knowledge of the maximum penalties for various crimes, but their perceptions of the severity of sanctions, the certainty with which they will be administered, and the promptness of punishment influence their choice of behavior." Since the impact of deterrence is questionable, crime prevention is a better strategy for reducing losses.

Other limitations of the criminal justice system are tight budgets and too few resources and personnel, which are the main reasons why public safety agencies play a minor role in private loss prevention programs. Public police agencies cannot afford to assign personnel to patrol inside business establishments or watch for employee theft. They are too busy with street crime, traffic problems, and other issues of public safety. An occasional (public) police patrol and a response to a crime are the primary forms of assistance that public police can provide to businesses. The Dallas Police Department, for example, no longer responds to calls from retailers who report a shoplifting incident with a value of less than $50; this is a trend because of limited budgets (Richardson, 2012). Prosecutors are unwilling to prosecute certain crimes against businesses because of heavy caseloads. Consequently, public protection is being supplemented or replaced by private security and volunteer efforts in many locales.

Besides public police agencies, other first-responding agencies—fire departments, emergency medical services, and emergency management agencies—also have limited resources. The private sector must develop security and loss prevention programs and hire specialists to ameliorate all-hazards.

The United States employs an average of 2.3 full-time law enforcement officers for every 1,000 inhabitants in communities. This force ratio has remained steady for decades, although large cities have higher ratios (FBI, 2010**).**

The Basic Differences between Public Police and Private Security

The primary differences between public police and private security personnel pertain to the *employer*, the *interests served, basic strategies,* and *legal authority.* Public police are employed by governments and serve the general public. Tax dollars support public police activities. On the other hand, private security personnel are employed by and serve private concerns (e.g., businesses) that provide the funds for this type of protection. There are exceptions to these general statements. For instance, government agencies sometimes contract protection needs to private security companies to cut costs. Also, public police may be involved in efforts to assist business owners in preventing crimes through public education.

Another difference involves basic strategies. Public police devote considerable resources to **reactive strategies** to crimes. This entails rapid response to serious crimes, investigation, and apprehension of offenders. Law enforcement is a key objective. In contrast, private security personnel stress the *prevention* of crimes; arrests are often de-emphasized. These generalizations contain exceptions. For instance, public police have introduced **proactive strategies** to reduce crime through such methods as aggressive patrol tactics and enhanced information systems to pinpoint "hot spots" of crime and better identify suspects they encounter on the street.

The legal authority of public police and private security personnel is another distinguishing characteristic. Public police derive their authority from statutes and ordinances, whereas private security personnel function commonly as private citizens. Public police have greater arrest, search, and interrogation powers. Depending on the jurisdiction and state laws, private security personnel may be deputized or given special commissions that increase powers.

In your opinion, what are the most serious problems of the criminal justice system, and what solutions would you recommend?

Challenges of the Security Industry

Calder (2010: 72–73) writes that the 1929 Wickersham Commission, appointed by President Herbert Hoover to study the justice system and crime control, published one of many reports that "was most likely the first administrative review of American private security." The report included the cost of crime to communities and businesses and the fact that businesses had increasingly hired private security guards and investigators for protection against robberies and theft. Calder credits a sociologist named Jeremiah Shalloo with "the first academic study acknowledging the expansion and diversification of private policing" in 1933. Shalloo's writings included "critical revelations concerning the status of company police organizations … reclusive and largely unregulated by state and local governments."

During the 1970s, business and government leaders viewed the growing private security industry as an ally of the criminal justice system. Both crime-fighting sectors were seen as having mutual and overlapping functions in controlling crime. With this thinking in mind, the U.S. Department of Justice provided financial support for the production of important research reports. *The Rand Report* (U.S. Department of Justice, 1972: 30) focused national attention on the problems and needs of the private security industry. This report stated that "the typical private guard is an aging white male, poorly educated, usually untrained, and very poorly paid." This conclusion has met with criticism because the research sample was small and thus did not represent the entire security industry. However, with the assistance of this report and its recommendations, the professionalism of private security improved.

The Rand Report was a great aid (because of limited literature in the field) to the *Report of the Task Force on Private Security* published a few years later (U.S. Department of Justice, 1976b). This report of the National Advisory Committee on Criminal Justice Standards and Goals represented the first national effort to set realistic and viable standards and goals designed to maximize the ability, competency, and effectiveness of the private security industry for its role in the prevention and reduction of crime. A major recommendation of the report was that all businesses that sell security services should be licensed and all personnel in private security work should be registered.

The task force's urging of stricter standards for the security industry reflected the need to reduce ineptitude and industry abuses while striving toward professionalism. The task force focused on the problem of minimal hiring, training, and salary standards that are still challenges today. These minimal standards enable companies to reduce costs and provide potential clients with low bids for contract service. Thus, professionalism is sacrificed to keep up with the competition. The task force recommended improved hiring criteria, higher salaries (especially to reduce turnover), and better training, among other improvements. Both studies recommended state-level regulation of the security industry in general as a means of creating more uniformity.

Another major study of the security industry, funded by the U.S. Department of Justice and published in the mid-1980s, was titled *Private Security and Police in America: The Hallcrest Report* (Cunningham and Taylor, 1985). It focused research on three major areas: (1) the contributions of both public police and private security to crime control, (2) the interaction of these two forces and their level of cooperation, and (3) the characteristics of the private security industry. Several industry problems and preferred solutions discussed in this report are covered in subsequent pages here.

The U.S. Department of Justice funded a second Hallcrest Report a few years later titled *Private Security Trends: 1970–2000, The Hallcrest Report II* (Cunningham et al., 1990). This report provided a study of security trends leading up to the 21st century. Some of its predictions are listed next. This list can assist us in thinking critically about security today.

- Since the mid-1980s, companies have been less inclined to hire security managers with police and military backgrounds and more inclined to hire those with a business background.
- During the 1990s, in-house security staffs would diminish, with an increased reliance on contract services and equipment.
- The negative stereotypical security personnel are being replaced with younger, better-educated officers and greater numbers of women and minority group members. However, the problems of quality, training, and compensation remain.
- The false-alarm problem is continuing. There is a massive waste of public funds when police and fire agencies must respond to current levels of false alarms. Between 97 percent and 99 percent of all alarms are false.

Cooperation between Public Police and Private Security

Partnering between public police and private security has had a history of challenges. Early and more recent research shows continuing problems between these sectors, however, cooperation is improving. An early study titled *Law Enforcement and Private Security: Sources and Areas of Conflict* (U.S. Department of Justice, 1976a: 6) stated the following:

> *A persistent problem noted by several research reports involves disrespect and even conflict between public and private police.*

> *Some law enforcement officers believe that being a "public servant" is of a higher moral order than serving private interests.... They then relegate private security to an inferior status.... This perceived status differential by law enforcement personnel manifests itself in lack of respect and communication, which precludes effective cooperation.*

These problems persist in the 21st century, even though many public police work part-time in the private security industry and, upon retiring, secure positions in this vocation. To reduce conflict, the *Task Force Report* and the *Hallcrest Reports* recommend that a liaison be implemented between public and private police. During the 1980s, the International Association of Chiefs of Police, the National Sheriffs' Association, and the American Society for Industrial Security began joint meetings to foster better cooperation between the public and private sectors (Cunningham et al., 1991). Suggested forms of increased cooperation included appointing high-ranking practitioners from both sectors to increase communication, instituting short training lessons in established training programs, and sharing expertise.

According to a 2004 national policy summit on public police and private security cooperation (U.S. Department of Justice, Office of Community Oriented Policing Services, 2004), the two groups have a lot to offer each other; however, they are not always comfortable working together. Police continue to criticize security for having lower standards for screening and training, and they may see security as a threat to their domain. Police often fall short of understanding the important role of private security. At the same time, some security personnel see police as elitists and claim that police are not concerned about security until they seek a position in the security field. Here are observations by participants of the 2004 national policy summit:

- Issues include respect (that is, law enforcement's lack of respect for security), trust, training differentials, and competition.
- Community policing calls on law enforcement to develop partnerships and relationships with various sectors of the community.
- Information sharing is difficult. Corporations do not feel they receive timely information from police, and they fear that information they give to the police may end up in the newspaper. Police fear that the corporate sector may not treat law enforcement information discreetly.
- Each side should educate the other about its capabilities before a crisis erupts, so each will know when to call on the other and what help to expect (and to offer).

Increasingly, we are seeing cooperation between both groups that takes many forms and occurs at many levels of government. To varying degrees, both groups share information with each other, attend each other's conferences, and plan together for protection and emergencies. They even work side by side at certain sites, such as downtown districts, government buildings, and special events.

At the 2004 national policy summit, public police and private security leaders recommended that both groups should make a formal commitment to cooperate. Also, the U.S. Department of Homeland Security (DHS) and/or the U.S. Department of Justice should fund relevant research and training; create an advisory council to oversee partnerships to address tactical issues and intelligence sharing, improve selection and training for private security personnel, create a national partnership center, and organize periodic summits on relevant issues. On the local level, immediate action should be taken to improve joint response to critical incidents, coordinate infrastructure protection, improve communications and data interoperability, bolster information sharing, prevent and investigate high-tech crime, and plan responses to workplace crime (U.S. Department of Justice, Office of Community Oriented Policing Services, 2004: 3–4).

A more recent study, *Operation Partnership: Trends and Practices in Law Enforcement and Private Security Collaborations* (U.S. Department of Justice, Office of Community Oriented Policing Services, 2009), noted the increase in law enforcement–private security partnerships since 2000 and described successful practices. Interestingly, the study listed barriers and challenges that were written in earlier research reports: lack of awareness by public police of the capabilities of private security, trust issues between sectors, information sharing and privacy concerns, and personnel issues in the private security industry (e.g., minimal background checks and training).

The following are some examples of collaborations reviewed during the *Operation Partnership* study:

- The *BOMA (Building Owners and Managers Association) Chicago Security Committee* is a partnership of police and security directors of large buildings to monitor and report criminal activity. Following the 9/11 attacks, membership increased.
- *Dallas Law Enforcement and Private Security Program* works to enhance communications between both sectors and offers numerous training workshops for private security.
- The *Southeast Transportation Security Council* involves police and the transportation industry in preventing and recovering stolen cargo.
- *InfraGard*, an FBI-sponsored partnership with the private sector, concentrates on sharing information. It has chapters throughout the country.
- The *U.S. Secret Service Electronic Crimes Task Forces and Working Groups* is a nationwide network of law enforcement, private security, and academic specialists who work on preventing and investigating attacks on the nation's financial and other infrastructures.

Chapter 15 addresses public-private sector partnerships that focus on the problem of terrorism. These include Information Sharing and Analysis Centers, Fusion Centers, and the Overseas Security Advisory Council.

The 9/11 attacks caused the federal government to develop numerous programs and strategies to enhance homeland security. The federal government expanded its role in regulating certain aspects of private industries to protect against terrorism and other risks, especially since the private sector owns 85 percent of the critical infrastructure in the United States. Federal law enforcement agencies are working with private sector executives, including security executives, to control terrorism. The U.S. Treasury Department, for example, has directed not only finance companies but also brokerage firms, casinos, and other businesses to reduce terrorist financing by taking specific steps at their own expense, such as establishing or expanding anti-money-laundering programs. The U.S. Food and Drug Administration provides guidelines to the food industry to prevent terrorist attacks on the nation's food supply.

A variety of other public–private sector information-sharing programs are operating. For example, the Homeland Security Information Network-Critical Infrastructure (HSIN-CI) enables key executives in the public and private sectors to receive alerts and notifications from the DHS via phone (landline and wireless), email, fax, and pager. It is set up on a regional basis with regional websites; not every region receives the same alerts.

Regulation of the Industry

The security vocation has its share of charlatans who tarnish the industry, and, as with many types of services offered to the public, government intervention has taken the form of licensing and registration. The *Task Force Report* and *Hallcrest Reports* recommend regulation of the security industry by all states. To protect consumers, the majority of states have varied laws that regulate, through licensing and registration, contract security officer services, private investigators, polygraph and other detection-of-deception specialists, and security alarm businesses. Security consultants are generally not regulated, but consumers should verify professional memberships and certifications. Although government regulation does not guarantee that all security practitioners will perform in a satisfactory manner, it does prevent people who have criminal records from entering the profession. In the case of applicants who have lived in multiple states, their backgrounds should be checked in each jurisdiction.

Another way the industry is regulated is through local, state, and federal agencies that contract out private security services and mandate various contract requirements (e.g., education, training, and character) to enhance professionalism and competence. Certain industries are regulated by government and require stringent security standards. For example, the U.S. Nuclear Regulatory Commission issues rules and standards that must be followed by licensees of nuclear reactors to ensure safety and security. A variety of businesses, institutions, and other organizations also issue requirements. Notably, when a contract is awarded simply to the lowest bidder, professionalism may be sacrificed.

Attempts have been made through Congress to pass a national law to regulate the security industry. In 1995, Rep. Bob Barr (R-GA) introduced H.R. 2092, The Private Security Officers Quality Assurance Act. It languished in Congress in various forms until Rep. Matthew Martinez (D-CA) introduced a similar bill. Known as the Barr-Martinez Bill, if passed, it would have provided state regulators with expedited FBI criminal background checks of prospective and newly hired security officers. The minimal training standards were struck from the bill. These would have required eight hours of training and four hours of on-the-job training for unarmed officers and an additional 15 hours for armed officers. Critics argued that states are against such a national law and state regulations are sufficient.

Finally, Congress passed the **Private Security Officer Employment Authorization Act of 2004** (included in the National Intelligence Reform Act of 2004), which enables security service businesses and proprietary security organizations in all 50 states to check whether applicants have a criminal history record with the FBI, which may emanate from one or more states. (National training requirements were not included in the legislation.) This law is significant and adds strength to criminal records checks of security officer applicants. Traditional name-based searches can result in false positives (i.e., an applicant's name is incorrectly matched to an offender's name or similar name) and false negatives (i.e., an applicant's name does not result in a match because the applicant is using a different name or a database was omitted from investigation). Because applicants may use different names and other bogus identifying information, fingerprints provide increased accuracy in background investigations. Under the act, security firm access to the FBI's Integrated Automated Fingerprint Identification System (IAFIS) also includes access to the Violent Gangs and Terrorism Organization File. Although not all security organizations are required to check these fingerprint databases for a match with their applicants, the effort and expense of doing so prevents litigation and the possibility that an applicant is a terrorist or other criminal. Cost is one of several factors that influence the extent of background investigations (Friedrick, 2005: 11).

Which government agency (or agencies) in your state regulates the private security industry? What are the requirements for contract security ("guard") companies, private detectives, and security alarm installers? Are proprietary security organizations regulated in your state?

The Need for Training

Numerous research reports and other publications have pointed to the need for more training of personnel in the security industry. Training of all security officers should be required by law *prior to assignment.* Training topics of importance include customer service, ethics, criminal law and procedure, constitutional protections, interviewing, surveillance, arrest techniques, post assignments and patrols, report writing, safety, fire protection, and specialized topics for client needs. Firearms training is also important. State training requirements vary.

With liability a constant threat, many security firms are gambling by not preparing officers for the job. Security training is inconsistent and sometimes nonexistent. The *Task Force Report* and the *Hallcrest Reports* stress the need for improved recruitment, selection, pay, and training within the security industry.

The harsh realities of the contract security business hinder training. It is easy to forget about national reports and recommendations when a businessperson is under pressure to reduce expenses and turn a profit while dealing with employee turnover, must ensure that security officers are on client posts, and faces competition from low bidders. Furthermore, the security industry plays a strong role in influencing government laws and regulations that can result in added expenses for businesses. On the bright side, many security service companies are professional, set high standards for themselves, and have improved the industry.

ASIS International, in its effort to increase professionalism and develop guidelines for the security industry, published *Private Security Officer Selection and Training* (ASIS International, 2010). This guide offers minimum recommendations for the selection and training of private security officers. Proprietary and contract security and regulatory bodies can apply it. Its employment screening criteria include being at least 18 years of age for unarmed and 21 years of age for armed security officers; a high school diploma, GED, or equivalent; fingerprints; criminal history check; drug screening; and other background checks. The first edition guide, published in 2004, recommended training requirements of 48 hours within the first 100 days of employment. The second edition guide, published in 2010, offered increased flexibility by recommending three phases of training: pre-assignment training as required by applicable law, on-the-job training of 8 to 16 hours, and annual training of eight hours.

The U.S. Department of Homeland Security (2012) published *Security Specialist Competencies* that explains core competencies for federal security specialists to help them protect federal facilities. These competencies promote a baseline of knowledge that is applicable to the public as well as private sectors. Topics in this publication include risk management, personnel security, information security, physical security, safety, and many other topics.

Ethics

A code of ethics is a partial solution to strengthening the professionalism of security practitioners. Such a code helps to guide behavior by establishing standards of ethical conduct. Twomey, Jennings, and Fox (2001: 28–31) describe **ethics** as a branch of philosophy dealing with values that relate to the nature of human conduct. They write that "conduct and values within the context of business operations become more complex because individuals are working together to maximize profit. Balancing the goal of profits with the values of individuals and society is the focus of *business ethics.*" Twomey and colleagues note that capitalism succeeds because of trust; investors provide capital for a business because they believe the business will earn a profit.

Customers rely on business promises of quality and the commitment to stand behind a product or service. Having a code makes good business sense because consumers make purchasing decisions based on their own experience or the experiences of others. Reliance on promises, not litigation, nurtures good business relationships.

Twomey and colleagues write of studies that show that those businesses with the strongest value systems survive and do so successfully. Citing several companies, they argue that "bankruptcy and/or free falls in the worth of shares are the fates that await firms that make poor ethical choices."

A host of other problems can develop for a business and its employees when unethical decisions are made. Besides a loss of customers, unethical decisions can result in criminal and civil liabilities. Quality ethics must be initiated and supported by top management, who must set an example without hypocrisy. All employees must be a part of the ethical environment through a code of ethics and see it spelled out in policies, procedures, and training.

International business presents special challenges when promoting ethical decision making because cultures differ on codes of ethics. Business management must take the lead, research, and define guidelines for employees.

Another challenge develops when a security practitioner's employer or supervisor violates ethical standards or law. No business wants to become part of the problem and subject itself to a tarnished reputation or criminal and civil liabilities. When faced with such difficult dilemmas, the business should refer to its professional background and code of ethics.

The Internet is a rich source of information on ethics. The *Task Force Report* and ASIS International are sources for codes of ethics for the security profession. The former has one code for security management (as does ASIS International) and a code for security employees in general. A sample of the wording, similar in both codes, includes "to protect life and property," and "to be guided by a sense of integrity, honor, justice and morality" (U.S. Department of Justice, 1976b: 24). Management, supervision, policies, procedures, and training help to define what these words mean.

Following are guidelines for making ethical decisions:

- Does the decision violate law, a code of ethics, or company policy?
- What are the short-term and long-term consequences of your decision for your employer and yourself?
- Is there an alternative course of action that is less harmful?
- Are you making a levelheaded decision, rather than a decision based on emotions?
- Would your family support your decision?
- Would your supervisor and management support your decision?

The False Alarm Problem

A persistent problem that causes friction between public police and the private sector is false alarms. It is generally agreed that more than 95 percent of all alarm response calls received by public police are false alarms. However, the definition of "false alarm" is subject to debate. It is often assumed that if a burglar is not caught on the premises, the alarm was false. Police do not always consider that the alarm or the approaching police could have frightened away a burglar.

The *Task Force* and *Hallcrest Reports* discuss the problem of false alarms. Many police agencies nationwide spend millions of dollars each year in personnel and equipment to respond to these calls. For decades, municipalities and the alarm industry have tried various solutions. Police agencies have selectively responded or not responded to alarms. City and county governments have enacted false alarm control ordinances that require a permit for an alarm system and impose fines for excessive false alarms. End users cause most of the problems.

Police and the alarm industry have developed a model alarm ordinance that includes verification, a graduated fine system, withholding response to chronic abusers, and an appeal process. The model ordinance is at www.siacinc.org.

Martin (2005: 160–163) writes that jurisdictions across North America are adopting ordinances and policies that specify a nonresponse policy to unverified alarm activations. As an alarm industry advocate, he argues that this change places a financial burden on businesses and increases the risk of burglary. In Salt Lake City, a person must be physically present on the property to visually verify that a crime is in progress before the police accept a call for service; CCTV confirmation is not considered sufficient. Martin sees public police as more qualified to respond to alarms than private security officers. Martin refers to the strategy of Enhanced Call Verification to reduce false alarms. This strategy requires two calls from the alarm-monitoring center to the user to determine if an error has occurred before police are called.

Berube (2010: 353 and 371) argues that "false alarm reduction methods, such as verification, can result in delaying police response and decreasing the deterrence effect that we have come to expect from burglar alarm systems." "Deterrence of intrusion systems comes from the ability of proper responding authorities to apprehend the intruder." "For alarms to deter, burglars must equate their use to an increase in the risk of apprehension." Berube also discusses the topic of privatizing response, which brings us to a key question: "Who should pay for alarm response—the public or the users?"

What do you think are the most serious problems facing the private security industry and what solutions would you suggest?

Search the Internet

Several security associations exist to promote professionalism and improve the security field. Go to the website of **ASIS International** (www.asisonline.org), formerly the American Society for Industrial Security, founded in 1955. With a membership near 40,000, it is the leading general organization of protection executives and specialists. Its monthly magazine, *Security Management*, is an excellent source of information. This association offers courses and seminars. ASIS International offers three certifications: Certified Protection Professional (CPP); Physical Security Professional (PSP); and Professional Certified Investigator (PCI). For each certification the candidate must pay a fee for administration, meet eligibility requirements, and successfully pass an examination.

Whereas ASIS International is the leading professional association of security executives and specialists, the **International Foundation for Protection Officers** (IFPO), founded in 1988, is the leading professional association of security officers who are on the front lines of protecting businesses, institutions, other entities, and our infrastructure. It offers a variety of publications. The IFPO (www.ifpo.org) has global reach and serves to help professionalize officers through training and certification. It has developed several distance delivery courses and programs: Entry Level Protection Officer (ELPO), Basic Protection Officer (BPO), Certified Protection Officer (CPO), Security Supervisor (SSP), and the Certified Security Supervisor (CSS). All programs are designed for self-paced home study and some are available online. Many corporations and institutions have included these programs in their professional development programs for security personnel.

How does each group promote professionalism and improve the security field?

Here are additional websites related to this chapter:

Bureau of Justice Statistics: www.ojp.usdoj.gov
Federal Bureau of Investigation: www.fbi.gov/ucr/ucr.htm
International Association of Chiefs of Police: www.theiacp.org
International Association of Security and Investigative Regulators: www.iasir.org
Markkula Center for Applied Ethics: www.scu.edu/ethics
National Association of Security Companies: www.nasco.org
National Fire Protection Association: www.nfpa.org
Security Industry Association: www.siaonline.org
Sourcebook of Criminal Justice Statistics: www.albany.edu/sourcebook/pdf/t112006.pdf
U.S. Bureau of Labor Statistics: www.bls.gov
U.S. Department of Homeland Security: www.dhs.gov/dhspublic

Case Problems

2A. As Manager of Loss Prevention for five retail stores, you are seeking a promotion to Regional Director of Loss Prevention. The position involves responsibilities for several more stores and subordinates. You are scheduled for an interview for the position soon, and an interview topic will be metrics. Before studying metrics related specific to the company and your work, you want to review the basics of metrics to serve as a foundation. Prepare a list of five general questions on metrics and answer them. Examples of basic questions are: What are metrics? What are the purposes of metrics?

2B. As a city police detective specializing in white-collar crime, you enjoy the challenges of your work. You look forward to retirement in five years and plan to become an insurance fraud investigator. You often work with the private sector, and you are frequently in contact with corporate security investigators. As an active member of a local police-security council, you are assigned the task of developing a plan to improve police-security cooperation at all levels in the city. What are the specific plans that you will present to the council?

2C. As a uniformed security officer, how would you handle the following situations?

- Another security officer says that you can leave two hours early during second shift and she will "punch you out."
- You are assigned to a stationary post at a shipping and receiving dock, and a truck driver asks you to "look the other way" for $500 cash.
- A security officer you work with shows you how to make the required physical inspections around the plant without leaving your seat.
- During the holidays, some coworkers planning a party on the premises ask you if you want to contribute to a fund to hire a stripper/prostitute.
- You are testifying in criminal court in a shoplifting case, and the defense attorney asks you to state whether you ever lost sight of the defendant when the incident occurred. The case depends on your stating that you never lost sight of the defendant. You actually did lose sight of the defendant once. How do you respond?
- Your best friend wants you to provide a positive recommendation for him when he applies for a job where you work, even though he has an arrest record.
- You see your supervisor take company items and put them in the trunk of her vehicle.

2D. As a corporate security manager, how would you handle the following situations?

- Two contract security officers fail to show up for first shift. The contract manager says that screened and trained replacements are unavailable, but two new applicants are available. You are required to make an immediate decision. Do you accept the two applicants, who lack background investigations and state-mandated training?
- A vendor offers you a condominium at the beach for a week if you support his firm's bid for an access control system. You know that the system is not the best and that it will cost your company slightly more than the best system. The condominium will save you about $1,500 on your summer vacation. What is your decision?

- Your employer is violating environmental laws. You know that if the government learns of the violations, your company will be unable to survive financially and criminal and civil action may result. You will certainly lose your job. What do you do?
- You placed a pinhole-lens camera in an office supply closet to catch a thief. While reviewing the CCTV video footage, you see your boss inappropriately touching a coworker who recently filed a sexual harassment suit against the boss, who vehemently denied the allegations. No one knows of the placement of the camera and the video footage except you. Your boss, who is the vice president of finance, has been especially helpful to your career, your excellent raises and bonuses, and the corporate security budget. What do you do?
- You are testifying in a case of negligent security concerning a manufacturing plant in your region of responsibility. The plaintiff's attorney asks you if any security surveys have ever been conducted at the manufacturing plant where the murder occurred. You know that a survey conducted prior to the murder showed the need for increased security. Such information would help the plaintiff. What is your response to the question?

References

ASIS International. (2009). *Organizational resilience: Security, preparedness, and continuity management systems—requirements with guidance for use*. American National Standard. <www.asisonline.org> retrieved March 23, 2011.

ASIS International. (2010). *Private security officer selection and training*. <www.asisonline.org> retrieved February 25, 2011.

Ballantyne, S. (2011). Crime pattern analysis and the security director: A primer. *Security, 48* (March).

Barkan, S. (2006). *Criminology: A sociological understanding* (3rd ed.). Upper Saddle River, NJ: Pearson Prentice-Hall.

Berube, H. (2010). An examination of alarm system deterrence and rational choice theory: The need to increase risk. *Journal of Applied Security Research, 5*.

Bowman, G., et al. (1992). *Privatizing the United States justice system*. Jefferson, NC: McFarland Pub.

Calder, J. (2010). Law, politics, and occupational consciousness: Industrial guard unions in the united states, 1933–1945. *Journal of Applied Security Research, 5*.

Chamard, S. (2006). *Partnering with businesses to address public safety problems*. Washington, D.C.: U.S. Department of Justice, Office of Community Oriented Policing. (April).

Conklin, J. (2001). *Criminology* (7th ed.). Boston: Allyn & Bacon.

Cunningham, W., et al. (1991). *Private security: Patterns and trends*. Washington, D.C.: National Institute of Justice. (August).

Cunningham, W., et al. (1990). *Private security trends: 1970–2000, The Hallcrest report II*. Boston: Butterworth-Heinemann.

Cunningham, W., & Taylor, T. (1985). *Private security and police in America: The Hallcrest report*. Portland, OR: Chancellor Press.

DRI International. (2012). The Certified Business Continuity Professional (CBCP). <www.drii.org/certification/cbcp.php> retrieved April 11, 2012.

FBI. (2010). *Crime in the United States, 2009.* <www.fbi.gov> retrieved March 11, 2011.

Freedonia. (2008). *World security services: Industry study with forecasts for 2012 & 2017.* (September). <www.freedoniagroup.com> retrieved March 7, 2011.

Freedonia. (2010a). *Private security services: U.S. industry study with forecasts for 2014 & 2019.* (November). <www.freedoniagroup.com> retrieved March 7, 2011.

Freedonia. (2010b). *Electronic security systems: U.S. industry study with forecasts for 2014 & 2019.* (July). <www.freedoniagroup.com> retrieved March 7, 2011.

Freedonia. (2010c). *World security equipment: Industry study with forecasts for 2014 & 2019.* (December). <www.freedoniagroup.com> retrieved March 7, 2011.

Friedrick, J. (2005). *New law empowers security firms, private employers with information* (2). Security Director News. (September).

Fuller, J. (2012). *Think criminology.* New York, NY: McGraw-Hill.

Goldman, R. (2009). *Domino's employee video taints food and brand.* ABC News. (April 16, 2009). <http://abcnews.go.com/Business/story?id=7355967&page=1> retrieved March 12, 2011.

Haddow, G., Bullock, J., & Coppola, D. (2011). *Introduction to emergency management* (4th ed.). Burlington, MA: Butterworth-Heinemann.

Horovitz, B. (2009). *Domino's nightmare holds lessons for marketers.* USA Today. (April 16).

Kovacich, G., & Halibozek, E. (2006). *Security metrics management: How to manage the costs of an assets protection program.* Boston: Elsevier Butterworth-Heinemann.

Marks, A. (2006). Security at military bases: A job for private firms? *The Christian Science Monitor* (April 27) <www.csmonitor.com/2006/0427/p02s01-usmi.htm> retrieved April 28, 2006.

Martin, S. (2005). *What's best for alarm response policies?* (49). Security Management. (March).

Miller, R., & Jentz, G. (2011). *Business law today* (9th ed.). Mason, OH: South-Western Cengage.

Nordland, R. (2012). *Risks of Afghan war shift from soldiers to contractors.* The New York Times. (February 11).

Paddock, B., et al. (2010). *Jamal Thomas, fired Domino's employee, accused of burning two pizza joints, also torched church.* Daily News. (September 14).

Purpura, P. (2007). *Terrorism and Homeland security: An introduction with applications.* Burlington, MA: Elsevier Butterworth-Heinemann.

Quinley, K., & Schmidt, D. (2002). *Business at risk: How to assess, mitigate, and respond to terrorist threats.* Cincinnati, OH: The National Underwriter Co.

Richardson, W. (2012). *Dallas PD stops responding to retail thefts under $50.* Security Director News. (January 18).

Rothe, D., & Ross, J. (2010). *Private military contractors, crime, and the terrain of unaccountability* (27). Justice Quarterly. (August).

Schwartz, M. (2011). *The department of defense's use of private security contractors in Afghanistan and Iraq: Background, analysis, and options for congress.* (February 21). <www.crs.gov> retrieved March 11, 2011.

Treece, D., & Freadman, M. (2010). Metrics is not a four-letter word. *Security, 47* (November).

Twomey, D., Jennings, M., & Fox, I. (2001). *Anderson's business law and the regulatory environment* (14th ed.). Cincinnati, OH: West Legal Studies.

U.S. Army, (2006). *Counterinsurgency (FM-324).* Washington, DC: Headquarters, Department of the Army.

U.S. Bureau of Labor Statistics. (2009). Occupational Outlook Handbook, 2010–2011 Edition. <www.bls.gov> retrieved April 12, 2011.

U.S. Department of Homeland Security. (2004). *National response plan.* <www.dhs.gov> retrieved January 12, 2005.

U.S. Department of Homeland Security. (2012). *Security specialist competencies: An interagency security committee guideline.* (January). <www.dhs.gov> retrieved February 12, 2012.

U.S. Department of Justice. (1972). *Private police in the United States: Findings and recommendations 1. (The Rand Report).* Washington, D.C.: U.S. Government Printing Office.

U.S. Department of Justice. (1976a). *Law enforcement and private security: Sources and areas of conflict.* Washington, D.C.: U.S. Government Printing Office.

U.S. Department of Justice. (1976b). *Report of the task force on private security.* Washington, D.C.: U.S. Government Printing Office.

U.S. Department of Justice, Bureau of Justice Statistics. (2009). *2006 Justice expenditures and employment extracts.* <www.ojp.usdoj.gov/bjs> retrieved March 11, 2011.

U.S. Department of Justice, Office of Community Oriented Policing Services. (2004). *National policy summit: Building private security/public policing partnerships to prevent and respond to terrorism and public disorder.* <www.cops.usdoj.gov> retrieved January 4, 2005.

U.S. Department of Justice, Office of Community Oriented Policing Services. (2009). *Operation partnership: Trends and practices in law enforcement and private security collaborations.* <http://lepsc.org> retrieved March 3, 2011.

U.S. Department of Labor. (2011). *Census of fatal occupational injuries, 2010.* <www.bls.gov> December 13, 2011.

U.S. Fire Administration. (2010). *U.S. fire administration fire estimates.* <www.usfa.fema.gov> retrieved November 16, 2011.

U.S. Government Accountability Office. (2006). *Contract security guards: Army's guard program requires greater oversight and reassessment of acquisition approach.* (April) <http://www.gao.gov/cgi-bin/getrpt?GAO-06-284> retrieved April 5, 2006.

Wailgum, T. (2005). *Where the Metrics Are.* CSO (February) <www.csoonline.com> retrieved February 9, 2005.

Zarate, J. (1998). The emergence of a new dog of war: Private international security companies, international law, and the new world order. *Stanford Journal of International Law, 34* (January).

Reducing the Problem of Loss

Foundations of Security and Loss Prevention

OBJECTIVES

After studying this chapter, the reader will be able to:

1. Detail precisely how the security and loss prevention field has reached the status of a profession
2. Explain rational choice theory, routine activity theory and situational crime prevention
3. Discuss planning and budgeting and why each is important
4. List and explain the three-step risk analysis process
5. Discuss standards and regulations and how each is relevant to security
6. Explain and illustrate how to evaluate security and loss prevention programs
7. Describe the characteristics of proprietary security programs

KEY TERMS

- criminology
- criminal justice
- scientific method
- research validity
- research reliability
- incapacitation
- rehabilitation
- deterrence
- prevention
- rational choice theory
- routine activity theory
- Oscar Newman
- defensible space
- territoriality
- natural surveillance
- C. Ray Jeffery
- Crime Prevention Through Environmental Design (CPTED)
- situational crime prevention
- controllers
- super controllers
- peer-reviewed
- value added
- return on investment
- work product
- security business proposal
- security master plan
- planning
- budgeting
- risk analysis
- cost/benefit analysis
- annual loss exposure

- systems perspective
- National Technology Transfer and Advancement Act
- consensus standards
- standard of care
- code war
- regulations
- Code of Federal Regulations (CFR)
- codes
- recommended practices
- Interagency Security Committee
- Underwriters Laboratories (UL)
- National Fire Protection Association
- Security Industry Association
- American Society for Testing and Materials
- American National Standards Institute
- International Organization for Standardization
- pretest-posttest design
- experimental control group design
- evidence-based research
- division of work
- authority
- responsibility
- power
- delegation of authority
- chain of command
- span of control
- unity of command
- line personnel
- staff personnel
- formal organization
- information organization
- organization chart
- directive system
- policies
- procedures
- manual
- autocratic style
- democratic style

The Security and Loss Prevention Profession

The security and loss prevention field has reached the status of a profession. If we look to other professions as models to emulate, we see the following in each, just as we see in the security and loss prevention profession: a history and body of knowledge recorded in books and periodicals; a theoretical foundation; academic programs; and associations that promote advancement of knowledge, training, certification, and a code of ethics.

Theoretical Foundations

The challenges of security and loss prevention in a complex world have created an intense search among practitioners to seek answers to protection problems. Many fields of study offer answers for the practitioner, and thus, a multidisciplinary approach to the problem of loss is best, as illustrated in this book. Here we present theories and concepts that serve as part of the theoretical foundation for security and loss prevention.

The study of criminology and criminal justice provide answers to security and loss prevention challenges. **Criminology** focuses on the scientific study of the nature, extent, and causes of crime, and lawmaking and our society's response to crime. The study of **criminal justice** emphasizes the response to crime, especially police, courts, and correctional agencies (Vito and Maahs, 2012).

Although sociology has had a major influence on criminology, many other disciplines have made criminology interdisciplinary. These disciplines include biology, psychology, political science, law, criminal justice, and social work.

When students first begin to study criminology, they often struggle to develop a clear picture of the numerous theories of why individuals commit crime and society's response to crime. For the security practitioner who begins to study criminology, frequently asked questions are: "How do these theories relate to my job in security?" How can these theories help me to improve security?" At the same time, there are practitioners in security and in the criminal justice system, some with "college degrees," who state that the theories of criminology are "a bunch of bullshit that don't apply to the job." This view is dispelled next.

Scientists apply a variety of methods to evaluate theories. For instance, as covered early on in high school science and social science classes and again in college, the **scientific method** consists of a statement of the problem, hypothesis, testing, and conclusion. As in just about any discipline, the scientific method is applicable to security. (We see the scientific method applied to the problem of employee theft later in the chapter.) **Research validity** is applied in testing to ensure that the most appropriate questions are asked in the research design and that we measure what we intend to measure. **Research reliability** also improves the quality of research when it is repeated and the results are similar. Researchers must be cautious because research may show that a theory is unsupported or, in other cases, the research methodology applied to test a theory may be faulty.

Practitioners in security and criminal justice, whether they know it or not, are performing duties at their jobs that are based on theories. **Incapacitation** is a theory that seeks to prevent crime by taking action against an individual so they are unable to commit crimes in our communities. Examples are incarceration and the death penalty. Many violent offenders are imprisoned for long sentences and are unable to victimize those outside of prison. Hundreds of thousands of law enforcement, court, and correctional practitioners perform duties every day based upon the theory of incapacitation. Theories applied in the field are not without challenges. **Rehabilitation** has been subject to controversy for many years. It refers to an endless number of programs and strategies that seek to transform offenders into law-abiding citizens. Correctional employees, probation and parole officers, and employees of non-profit organizations expend massive resources in an attempt to rehabilitate offenders through counseling, education and training, employment assistance, substance abuse treatment, intensive supervision, and many other programs. Evaluative research results show successful and unsuccessful programs. Research is applied to refine programs and eliminate those that are unsuccessful.

Deterrence theory also has it problems, as explained in Chapter 2, as one of the limitations of the criminal justice system. This theory views humans as rational and hedonistic and that when punishment following a crime is certain, swift, and severe, crime by an individual is less likely to be repeated. At the same time, a message is sent to society showing the consequences of violating criminal laws. However, because many crimes are not discovered, reported, or successfully prosecuted, prevention is a key theory and strategy as emphasized in this book. **Prevention** refers to proactive methods implemented so that crime and other harmful incidents are less likely to occur, or if a harmful incident occurs, to minimize losses.

In reference to the usefulness of criminological theories and research to security students and practitioners, it is important to understand that *the history of criminology and the theorists behind it have concentrated on people and why they commit crime.* Biological theories focus on such topics as whether criminality is inherited and comparisons of the brain structure of criminals and non-criminals. Psychological theories look to factors inherent in an individual that are linked to crime, such as negative personality traits, psychopathic personality, and low intelligence. Sociological theories examine how society itself (the social structure) causes crime; theorists in this discipline study deteriorated neighborhoods and schools, unemployment, poverty, violent households, limited legitimate means to succeed, and other variables that may affect crime. Each theory is subject to controversy and criticism.

Besides explaining *why* people commit crimes, another group of criminological theories explains *where* and *how* crime occurs. These theories give attention to crime incidents, the environment or place where crime occurs, and the situations and opportunities that exist that may result in crime incidents. For the security practitioner, such theories help to pinpoint, for instance, why crime occurred on a premises and what changes can be made to enhance security. The following theories provide input for planning security and justifying budgets.

Rational Choice Theory

Rational choice theory, developed by Cornish and Clark (1986), is linked to deterrence theory in that individuals make rational decisions to avoid punishment and are deterred by criminal sanctions. Major elements of rational choice theory are that individuals (1) study the consequences of crime against the benefits of crime prior to committing a crime and (2) select criminal behavior when the rewards offset the costs. Factors that may be considered by an individual contemplating a criminal act are the possibility of being caught, arrested, imprisoned, and stigmatized. For an individual who is employed, a criminal act can also result in losing one's job and career. However, the potential rewards may appear more attractive, such as monetary gain, respect from others, and the thrill of "getting away with a crime." This theory provides a foundation for security strategies that increase the risks and consequences of crime. Examples are formidable security that will likely catch an offender and a strong cooperative relationship with police and prosecutors.

Let us look at a more difficult security challenge that may weaken both rational choice theory and its applicability to security planning. Suppose an individual is unemployed, in poverty, desperate because of a variety of reasons, and has no concern about being stigmatized or separated (because of incarceration) from family and friends. The rewards of committing a criminal act may be attractive and the act may be rationalized as, "I have nothing to lose." In this case, alternative theories as described next will enhance security.

Because of efforts to reduce the expensive problem of false alarms (see Chapter 2) through, for example, local government ordinances that specify nonresponse by police to unverified alarm activations, do you think such efforts are reducing apprehensions? In addition, do you think because of "nonresponse," the use of alarm systems does not deter, as rational choice theory would suggest?

Routine Activity Theory

Routine activity theory, from Cohen and Felson (1979), emphasizes that crime occurs when three elements converge: (1) a motivated offender, (2) a suitable target, and (3) the absence of a capable guardian. This theory includes the routine activities of both offender and victim. An offender may routinely walk through specific neighborhoods looking for homes that appear as easy targets for burglary or into buildings in a commercial area to seek opportunities for theft. Because in many families all adults work, homes are often unoccupied during the day, which can become suitable targets for burglary. "Neighborhood Watch" and alarm systems can prevent crime. Commercial buildings without access controls or other security methods, likewise, can become suitable targets. A capable guardian can be ordinary people who can intervene or serve as witnesses, or police or security personnel. From a corporate security perspective, for example, salespeople, truck drivers, and others who are "on the road" can become suitable targets when a capable guardian is unavailable and a motivated offender is encountered. Thus, security practitioners should establish preventive programs to protect employees through training, security and safety tips, policies, procedures, technology, and other methods.

Situational Crime Prevention

Vito and Maahs (2012: 68) write that rational choice theory and routine activity theory share a view that situational factors (e.g., suitable target) influence whether a crime occurs. These theories are sometimes grouped together as opportunity theories. The policies and strategies emanating from opportunity theories focus on reducing opportunities for crime by changing the environment, such as designing features in and around buildings to prevent crime. The terminology describing this approach varies among researchers (and causes confusion) as explained by Vito and Maahs (2012: 68):

> *Some scholars and policymakers continue to use the term CPTED [Crime Prevention Through Environmental Design] to describe this work, although others prefer terms such as environmental criminology or situational crime prevention. Regardless of the terminology, the literature provides numerous methods or principles for crime prevention.*

Here we describe early contributors to the concept of preventing crime through environmental design. The work of **Oscar Newman** (1972) is the bedrock of many security designs worldwide. He argued that informal control of criminal behavior could be enhanced through architectural design that creates **defensible space** (e.g., private areas that residents can show guardianship over) while changing residents' use of public places and reducing fear. His architectural designs of defensible space concentrated on public housing and included fences, landscape, street layout, and controls at entrances.

A security strategy that developed from Newman's designs, which also facilitates defensible space, is **territoriality**. Subtle and obvious features of a building and its curtilage send a message outward to offenders to avoid trespassing because residents or employees have "ownership" of the area. Examples are shrubbery, brickwork, gates, and signs. These features make a statement that an area is private. Since residents or employees know of these private areas and the behavior of legal occupants, an intruder is more likely to be spotted and reported.

Another security strategy that evolved from Newman's work is **natural surveillance**. It refers to features at buildings that enable people to observe areas inside and outside of buildings while affording less hiding spaces for offenders who have an increased chance of being observed if they commit a crime. Examples of this concept are windows that face pedestrian walkways, playgrounds, and parking lots so people and property can be observed, and limited and/or trimmed foliage for ease of observation and to limit hiding places for offenders. The design of lighting, walkways, and roads also aid natural surveillance.

At about the same time Newman was developing the concept of defensible space, C. Ray Jeffery (1977) was developing **Crime Prevention Through Environmental Design (CPTED)**. His concept includes elements of defensible space while being broader in scope. Jeffery favored a host of strategies to prevent crime, besides changing the physical environment, and he favored broad application of CPTED to urban design, businesses, institutions, and facilities (i.e., beyond Newman's emphasis on public housing). Jeffery even looked to improved policing in the community and the underlying causes of crime (e.g., poverty).

Since the early development of Newman's and Jeffery's work, many professionals in architecture, landscape architecture, urban planning, engineering, security, and other fields have expanded on their work. Today, CPTED includes many security strategies that are limited only by our imaginations.

Situational crime prevention techniques offer practical strategies to reduce *opportunities* for crime through not only environmental design, but also through people's behavior. It contains the physical design characteristics of CPTED and managerial and user behaviors that impact opportunities for criminal behavior. Lab (2004) writes that, instead of making changes in a community, situational crime prevention focuses on specific problems, places, people, or times. Ronald Clarke (1997) studied successful crime prevention programs and developed practical situational crime prevention techniques that have been applied successfully in many environments. Following is a list of major techniques of situational crime prevention that have wide applications:

- ***Increase the effort by the offender***. Examples: *harden the target* through tamper-resistant packaging and auto steering column locks. *Control access* with fences and strong locks. *Supervise exits* by implementing merchandise tags. *Deflect offenders* with street closures. *Control tools and weapons* by restricting spray-paint sales to youth and supplying plastic cups at bars.
- ***Increase the risks for offenders***. Examples: *promote guardianship* by traveling in groups at night. *Aid natural surveillance* with improved lighting. *Reduce anonymity* by requiring IDs. *Assign place managers*, such as two clerks at a convenience store. *Enhance surveillance* through CCTV, alarm systems, and security personnel.
- ***Reduce rewards for offenders***. Examples: *conceal targets* by not labeling valuables. *Remove targets* by hiding valuables. *Mark property* for easy identification if stolen. *Disrupt criminal markets* by monitoring pawnshops and websites. *Deny benefits* by quick graffiti cleaning and using ink security tags on retail merchandise.

- ***Reduce provocations.*** Examples: *minimize stress* through quality service. *Avoid disputes* by reducing crowds. *Control emotional arousal* through enforcement of appropriate behavior. *Neutralize peer pressure* by phrases such as "say no." *Discourage imitation* through rapid investigation and enforcement.
- ***Remove excuses.*** Examples: *set rules and policies. Post signs* such as "no loitering." *Alert conscience* by implementing roadside speed displays. *Aid compliance* by using trash bins. *Control substance abuse* by monitoring bar patron consumption.

The theories presented here are beginning points from which to build. The social sciences are by no means the only disciplines helpful to security and loss prevention. In subsequent chapters, theory and concepts are drawn from law, marketing, accounting, fire science, safety, and risk management.

You Decide!

What is Wrong with this Facility?

The Chandler Office Building is a six-story, forty-year-old structure located in an urban area. It houses numerous business tenants, some of whom work late into the night. During the day, one security officer is stationed at the front entrance and one is stationed at the rear entrance. A card access system is available after hours for access. Parking is available on all four sides of the building. The architects who designed the building gave little thought, if any, to defensible space, CPTED, rational choice theory, routine activity theory, and situational crime prevention. Low walls, berms, and valleys that accommodate a variety of overgrown shrubbery and trees surround the building. The outside lighting is poor and made worse by large trees that surround light fixtures; dark shadows are cast on walkways and parking lots. Motorists seeking to avoid a traffic light and intersection use the back parking lot as a short cut.

As a security specialist, you have been assigned the task of improving security at this office building following a robbery and murder that occurred between the rear entrance and the nearby back parking lot in the late evening. Furthermore, numerous larcenies occurred in the building and in the parking lots during the last five years. What is wrong with this facility and what are your security plans?

Controllers and Super Controllers

Sampson et al. (2010: 1–17) add their perspective on routine activity theory by writing that, although this theory provides an understanding of crime patterns by explaining the convergence of *offenders, targets,* and *places,* this theory lacks a full account of why controllers may be ineffective. (These theorists modified the third element of routine activity theory by referring to it as "places" rather than as "the absence of a capable guardian.") Sampson et al. explain that each of the three necessary conditions for crime has a **controller** (handler, guardian, or manager) who can manage one of the conditions. "Handlers" are what sociologists refer to as "agents of socialization," such as parents, teachers, mentors, and religious leaders with whom offenders may have emotional attachment. "Guardians" (e.g., citizens looking out for each other, police, and security officers) protect targets.

"Managers" are owners of places or owner representatives (e.g., site managers, clerks, and bartenders) who are involved in crime prevention through such methods as training, policies, and procedures. Sampson et al. argue that crime prevention efforts can reduce crime and a major "issue is not whether crime prevention can work, but getting people and organizations to take the necessary actions."

Sampson and colleagues show interest in researching "why failures to prevent crime exist and what can be done to prevent such failures." Thus, according to these researchers, to enhance crime prevention, controllers must be mobilized and become more effective through incentives facilitated by **super controllers**---people, organizations, and institutions that can provide incentives to prevent crime and control the controllers. Sampson et al. goes as far as writing, "Crime prevention efforts that fail to be implemented are highly likely to have failed because they did not succeed in getting super controllers behind their efforts." For instance, public police and state alcohol and beverage control agencies (multiple-super controllers) work with bars to reduce violence through partnering, training, and regulations. The U.S. Department of Justice Office of Community Oriented Policing Services (2006 and 2007), has assisted controllers and super controllers through publications such as *Assaults in and Around Bars*, 2nd Edition, and *Understanding Risky Facilities*. These publications offer strategies such as policies and procedures, access controls, and responsible beverage service. Many other super controllers exist in our society. Examples are the U.S. Nuclear Regulatory Commission that regulates the nuclear industry, including security and safety issues, and the U.S. Transportation Security Administration that regulates security and safety in all modes of transportation. Super controllers vary in terms of legal authority, regulations, and enforcement powers as they manage controllers. The concept of super controllers is broad and includes individuals, groups, the media, social networks (e.g., *Twitter*, *Facebook*, blogs, and *YouTube*), the insurance industry, and political institutions.

Super controllers also consist of professional organizations and government that promote industry standards. Sampson and colleagues view industry standards as a form of "group pressure" to remove excuses for not taking action to prevent crime. Through premises liability cases of alleged inadequate security, courts (also a super controller) have held defendants (e.g., a business) liable for a person's victimization resulting from the defendant not adopting industry standards.

Do you think that within our society (e.g., citizens, industries, and institutions), there is a failure to address certain crime problems because of such reasons as a lack of will or resources, ignorance about what to do, or a view that crime incidents are less expensive than crime prevention efforts and expenditures? Do you think laws and government regulations are the best options so super controllers can facilitate incentives (e.g., avoiding legal action and a fine for not adhering to security regulations) to prompt crime prevention action by controllers? Explain your answers.

Academic Programs

Adolf (2011: 126), reporting on ASIS International research of 2008, wrote that there are 94 American colleges offering security courses and 30 undergraduate and graduate security degree programs. Security degrees are offered on the associate, bachelor's, master's, and doctoral levels. The ASIS International website maintains a list of these programs and courses for the United States and a few other countries. (See www.asis.online.org.)

Adolf reminds us that "formal schooling serves as an integral part of a profession. Additionally, the development of a profession includes the predictable development of academic programs of study, research articulating the needs of the profession, and research to expand the knowledge base of the profession." He also emphasizes the importance of business education as an integral component of security studies. Adolf adds that security education is growing with the proliferation of online college programs and that there is a need for accreditation of security academic programs to enhance the quality of such curriculums. Read more on the topics of academic programs and research in Chapter 19.

Security Periodicals

There are many security periodicals published by a variety of associations, organizations, and businesses. Periodicals serve as a platform not only for the theoretical foundation of the security and loss prevention profession, but also to introduce readers to new developments, security strategies and technology, research, laws, issues, professional development, and a host of other topics.

A critical thinking approach is worthwhile when reading security periodicals. **Peer-reviewed** means that before a professional periodical publishes an article it is first studied by specialists in the field relevant to the periodical, and that the article meets specific criteria, such as quality writing and research methodology. Warner (2010: 23) sees it as a "pre-publication vetting process" that is "essential to the quality of an academic paper." Journals are often peer-reviewed. On the other hand, there are many security periodicals (e.g., trade magazines), as in other fields, that contain "sales pitches" woven into the content. A security practitioner should exercise caution and seek multiple sources of information before making an important decision or purchasing a service or system based on limited research.

What follows are noted periodicals in this discipline. The website for each periodical is located at the end of this chapter.

The *Journal of Applied Security Research* is published by Taylor & Francis. It is peer-reviewed and contains excellent articles on a variety of security topics. This periodical is the official journal of the Security & Crime Prevention Section of the Academy of Criminal Justice Sciences.

The *Security Journal* is published by Palgrave Macmillan Journals and supported by the ASIS International Foundation. The editorial staff is from the United States and the United Kingdom. This peer-reviewed journal publishes excellent articles on a variety of topics on the latest developments and techniques of security management. Articles include findings and recommendations of independent research. Two other journals from this publisher are *Risk Management: An International Journal* and *Crime Prevention and Community Safety: An International Journal.*

The *Journal of Physical Security* is affiliated with Argonne National Laboratory. It is the first scholarly peer-reviewed journal devoted to physical security. This periodical began because of the need for peer-reviewed journals in the physical security field and because papers on physical security are scattered among a wide variety of publications. The articles are technical and non-technical and offer the reader excellent perspectives and opportunities for critical thinking about security.

Security Management is a monthly magazine published by ASIS International. Each issue contains a wealth of informative articles on a broad range of topics written by experienced security professionals and editors.

Security Associations

Up to this point in this book, a number of professional associations have been identified and their objectives summarized. Others will follow. The private sector contains numerous associations dedicated to improving the world in which we live and promoting safety and security. These missions are accomplished by enhancing the knowledge, skills, and capabilities of members through training and education; offering certifications to demonstrate competence; conducting research; producing "best practices" and standards; disseminating information; communicating the group's goals; forming partnerships to reach common objectives; participating in community service; and advocating positions on key issues.

Professional associations typically require annual dues that pay for administrative and other expenses that are beneficial to members. Membership benefits include opportunities to network among peers; subscriptions to the group's periodicals and informative e-mails; members-only website privileges; discounts on national and regional educational programs and seminars; opportunities to serve on specialized committees; and career guidance and placement services.

Readers are urged to join one or more professional associations that fit their specializations and interests for a truly enriching experience for career development and service. What we invest in our membership activities affects what we receive in return.

By typing various security specializations into a search engine, such as "computer security," "school security," "healthcare security," and "transportation security," the reader can obtain a wealth of information on associations, professional development, training, certifications, and periodicals.

If you were to begin a career in the security and loss prevention vocation, would you join a professional association? Why or why not?

Methods for Protection Programs

Business Concepts

Suppose you are a security and loss prevention executive planning a comprehensive protection program for a business. Where do you begin? First, careful planning is essential. Your college education, experience, and affiliations with associations for professional development and networking all combine to enhance your chances of success in your job. Here we review business concepts to help you.

The concept of **value added** means that all corporate departments must demonstrate their value to the organization by translating expenditures into bottom-line impact. Corporate financial officers ask, "Is security contributing to our business and profit success and, if so, how?"

Historically, the security function has been challenged to show its value to an organization. This is still a challenge today. "Scare tactics" (i.e., adequate funding for security or else a host of crimes and losses will occur) and the "wonders of security technology" and how it will save personnel costs may or may not sway a superior into supporting a security program and budget. However, the value of security is enhanced by showing how security is aligned with business goals and success, customer needs, and the specific threats, hazards, metrics, risks, industry standards, regulations, and other factors that face the business. In addition, modern practitioners state their protection plans in financial terms that justify expenditures, save the organization money, and strive to bring in a return on investment that has a positive impact on the "bottom line" (i.e., a line in a financial statement that signifies net income or loss). A **return on investment** (ROI) yields a profit or benefit from an investment. It can be realized, for example, through a CCTV system that serves to eliminate the need for a security officer at one or more posts, is capable of providing 24/7 surveillance throughout a facility and reducing losses, verifies burglar alarm activations so the business does not have to pay fines for false alarms, and helps to locate production problems to improve efficiency and cut costs. The savings over time can pay for the investment in the CCTV system. Other examples of ROI include conducting an undercover investigation to pinpoint not only theft, but also substance abuse and safety problems; hiring a bad check specialist who recovers several times his or her salary; and purchasing an access control system that performs multiple roles, such as producing time and attendance data.

Modern security and loss prevention practitioners state their protection plans in financial terms that justify expenditures, save the organization money, and strive to bring in a return on investment.

Other, not so obvious, factors influence the success of security. These include communicating and marketing security initiatives to all customers (e.g., management, employees in general, those who purchase the company's products or services, and public safety agencies), speaking the language and understanding the background and culture of each customer group, and partnering with customers to augment resources.

Upon following the above guidelines as a prerequisite to security planning and budgeting, security practitioners must realize that businesses exist in a rapidly changing environment and adjustments are inevitable for both businesses and the security programs. Another challenge for the security practitioner is that security initiatives may depend on too few resources. Creativity and the acceptance of new challenges can provide opportunities. For instance, the security function can show value and ROI by protecting brand, becoming involved in business continuity, and vetting prospective business partners, suppliers and customers.

Once a security executive is ready to put his/her thoughts into action, a security and loss prevention plan and budget is prepared. This can be considered a **work product** (also called a "deliverable" or "outcome"), defined as the culmination of labor to produce a desired result. Examples of work products are endless (e.g., fundraising plan, training program, vehicle, building, or weapon).

For a security executive, a major type of work product is a **security business proposal**. It includes a plan, the purpose of security, its value and benefits to the organization, how security will be put into effect, and most of all, financials such as the budget and the return on investment. (Another type of business proposal is presented by a vendor of security services or systems to a potential customer; it contains such topics as what is being offered for sale, contract terms, and billing.)

An additional approach to producing a security program plan and budget is through a **security master plan**. This work product is broader than a security business proposal and covers multiple years. It is aligned with the business's goals and includes input from stakeholders (e.g., employees) who are affected by security. Other topics of the security master plan include an organizational chart of security personnel, security strategies and goals, physical security, and plans for investigations, information security, and emergencies. There is no standard security business proposal or security master plan; these work products vary in terms of content and depend on the security needs of the entity.

Planning and Budgeting

Planning results in a design used to reach objectives. It is better to know where one is going and how to get there than to adhere to a philosophy of "we'll cross that bridge when we get there." A serious consequence of poor planning is the panic atmosphere that develops when serious losses occur; emotional decisions are made when quickly acquiring needed services and systems. This sets up an organization to be a target for unscrupulous salespeople who prey on the panic.

Budgeting is closely related to planning because it pertains to the money required to fulfill plans. Businesses use many different types of budgets to plan, set goals, forecast revenues and expenditures, and evaluate performance. Because budgets are estimates, adjustments occur over time. Budgets common to businesses are explained next.

- *Cash flow budget*: serves as an estimate of future cash receipts and expenditures for a period. It aids management in deciding whether outside financing is needed.
- *Sales budget*: an estimate of future sales. It helps to set sales goals.
- *Production budget*: the number of units a business must manufacture to meet sales objectives. It includes the costs of manufacturing the units, such as raw materials and labor.
- *Project budget*: costs of a specific company project. The costs include materials and labor, among many other costs.
- *Capital budget*: planned outlays for such investments as a manufacturing plant, equipment, and product development.
- *Master budget*: financial plans for subunits of a company, showing such items as income and expenses.

Risk Analysis

A risk analysis helps to identify the threats, hazards, and risks facing an organization. It provides helpful input for planning security. There are many perspectives and methods of risk analysis from government, the private sector, and researchers. Instead of engaging in a debate about the topic, we present some perspectives here, beginning with a basic three-step process of risk analysis. This simplified approach will assist the reader and provide a foundation for those who decide to enter "the jungle of risk analysis" and join the fray by studying the literature and differences in terminology, definitions, and methodology. For instance, the term *risk analysis* is used interchangeably with *risk assessment, risk evaluation,* and other terms. The term *vulnerability assessment* refers to studying weaknesses that can lead to harm. Examples are an easily circumvented card access system at an office building or a poorly designed refund policy at a retail store. A *threat assessment* looks for sources of harm such as dishonest employees, terrorists, or flooding.

James F. Broder (2006: 4), author of *Risk Analysis and the Security Survey,* writes, "Risk assessment analysis is a rational and orderly approach, as well as a comprehensive solution, to problem identification and probability determination. It is also a method for estimating the expected loss from the occurrence of an adverse event. The key word here is *estimating* because risk analysis will never be an exact science—we are discussing probabilities."

This chapter defines **risk analysis** as a method to estimate the expected loss from specific risks using the following three-step process: (1) conducting a loss prevention survey; (2) identifying vulnerabilities; and (3) determining probability, frequency, and cost.

Conducting a Loss Prevention Survey

The purpose of a loss prevention survey is to pinpoint threats, hazards, and vulnerabilities and develop a foundation for improved protection. The survey should tailor its questions to the unique needs of the organization to be surveyed. Essentially, the survey involves a day-and-night, possibly multiple-day, physical examination of the location requiring a loss prevention program. The merging of information technology (IT) specialists into the risk analysis process is vital for comprehensive protection. For example, IT specialists can be involved in penetration testing of the IT system to identify protection needs.

The list that follows, expanded from Crawford (1995: 85–90), is a beginning point for topics for the survey:

1. Overall threats and hazards, geography, climate (possible natural disasters), and nearby hazards and potential targets that can impact the entity
2. Social, economic, and political climate surrounding the facility and in the region
3. Past events, litigation, and complaints
4. Condition of physical security, fire protection, safety measures, and business continuity plans
5. Hazardous substances and protection measures
6. Policies and procedures and their enforcement
7. Quality of security personnel (e.g., applicant screening, training, and supervision; properly registered and licensed)
8. Protection of people, assets, and operations on and off the premises
9. Protection of IT systems and information
10. Protection of communications systems (e.g., telephones, fax machines, and e-mail)
11. Protection of buildings, grounds, and utilities
12. Protection of parking lots
13. Protection of products, services, goodwill, and image

The survey document usually consists of a checklist in the form of questions that remind the practitioner or committee of what to examine. A list attached to the survey can contain the targets—for example, people, money, inventory, equipment, IT systems—that must be protected and the present strategies, if any, used to protect them. Also helpful are computer software generating three-dimensional views of the facility, geographic information systems, maps, and blueprints.

Identifying Vulnerabilities

Once the survey is completed, vulnerabilities can be isolated. For example, access controls may be weak for both the facility and IT; certain policies and procedures may be ignored; and specific people, assets, or enterprise operations may require improved protection. Vulnerabilities may also show that security, fire, and life-safety strategies are outdated and must be brought up to current codes and standards.

Determining Probability, Frequency, and Cost

The third step requires an analysis of the probability, frequency, and cost of each loss. Shoplifting and employee theft are common in retail stores, and numerous incidents can add up to serious losses. Fires and explosions are hazards at a chemical plant; even one incident can be financially devastating. The frequency of shoplifting and employee theft incidents at a retail store will likely be greater than the frequency of fires and explosions at a chemical plant. However, it is impossible to pinpoint accurately when, where, and how many times losses will occur. When the questions of probability, frequency, and cost of losses arise, practitioners can rely on their own experience, metrics, communication with fellow practitioners, information provided by trade publications, risk management or security consulting firms, and risk analysis software.

As an alternative to the preceding three-step process, the "General Security Risk Assessment Guideline" (ASIS International, 2003) offers a seven-step process to identify risks at specific locations and to begin planning to select and implement security and loss prevention strategies.

1. Understand the organization and identify the people and assets at risk.
2. Specify loss risk events/vulnerabilities. Each site often has a unique history of losses and unique weaknesses.
3. Establish the probability of loss risk and frequency of events.
4. Determine the impact of the events. This includes the cost of losses from tangible or intangible assets.
5. Develop options to mitigate risks.
6. Study the feasibility of implementation of options. This focuses on practical security options that are aligned with the objectives of the enterprise.
7. Perform a **cost/benefit analysis**. This process seeks to analyze the value of the benefits from an expenditure.

Annual Loss Exposure

It is argued that security directors of large, complex organizations should use quantitative, rather than qualitative, risk analysis when exposures cannot be evaluated intuitively, especially for the protection of IT and e-business (Jacobson, 2000: 142–144). The process begins with a mathematical model that can be simple or complex. A simple formula is ALE = I × F, where "ALE" is **annual loss exposure**, "I" is impact (i.e., dollar loss if the event occurs), and "F" is frequency (i.e., the number of times the event will occur each year). Software tools are available to organize and automate complex risk analyses. Debate continues over when to use quantitative risk analysis, its cost and value, how much guesswork goes into the process, and which formula and software are best. Two points are clear: (1) there are many opinions and styles of risk analysis, and (2) what works for one organization may not work for another.

The "General Security Risk Assessment Guideline" recognizes that, in certain cases, data may be lacking for a quantitative risk analysis. Consequently, a qualitative approach can be applied through multi-step processes as described previously. The "Guideline" offers both qualitative and quantitative approaches.

Why do you think the ASIS International's "General Security Risk Assessment Guideline" offers quantitative and qualitative approaches?

Department of Homeland Security's Approach to Risk Analysis

The National Research Council (2010), a non-profit institution chartered by Congress to conduct research and provide policy advice, prepared a report entitled *Department of Homeland Security's Approach to Risk Analysis.* The primary conclusion from this report is that the DHS has established a conceptual framework for risk analysis (Risk = Threat × Vulnerability × Consequences) and it has built models, data streams, and processes to conduct risk analyses for its missions. "However, with the exception of risk analysis for natural disaster preparedness, the committee did not find any DHS risk analysis capabilities and methods that are yet adequate for supporting DHS decision making, because their validity and reliability are untested." The formula was deemed inadequate for estimating the risk of terrorism. In addition, the committee was critical of DHS capabilities in improving the situation and recommended strengthening its scientific practices through external peer review.

International Perspective: Security Impact Assessments and Culture

Malampy and Piper (2010: 26) write that Asia-Pacific governments have favored the concept of a "security risk assessment" as part of environmental impact statements for large capital projects (e.g., a chemical plant). Some of the countries have codified this requirement into law. Without compliance, no permits for construction are issued. This concept has yet to be accepted in the United States; however, interest is developing in the public and private infrastructure sectors. Rather than the term "security risk assessment," the term used in the United States is "security impact assessment" (SIA). Malampy and Piper note benefits of an SIA, such as security planners being involved early on to ensure alignment with business objectives and avoiding expensive security retrofits if security planners are brought into the capital project during construction. Borrowing from the work of Stephen Gale and Lawrence Husick of the Center on Terrorism and Counter-Terrorism in Philadelphia, Malampy and Piper explain that a SIA includes the following:

- A standard for measuring the financial value of investments made in security
- The impacts on security of both the proposed action and the failure to act
- Any adverse security effects that would be avoided should the proposal be implemented, as well as those that are unavoidable
- Alternatives to proposed action
- The cost of a successful attack
- Expenditures of implementing the proposed action

Because of the diversity in our world, security issues are viewed differently in each country. This is the case with SIAs and many other issues. Berrong (2010: 48–50) writes that, because many organizations operate globally, finding a security fit in each country can be tricky. Upon interviewing a global security manager, Berrong notes that the main challenges are cultural; she explains three approaches to enhance security in global operations. First, know the overseas company's security culture, how management perceives threats, hazards and risks, and its risk tolerance. Second, understand that security priorities and what works in the United States may not apply in another country. For instance, the risk of terrorism is viewed differently in each country. Also, each country has its unique cultural traits that affect security. In the United States, dogs are applied to trace explosives detection; however, in Pakistan, dogs may not be used "because Muslims believe dogs to be ritually impure." Third, realize that employees who work in overseas operations and are citizens of another country are likely to have views and biases that may conflict with security methods originating in the United States. In addition, if such an employee is assigned to work in the United States, a similar challenge can arise. Likewise, a United States citizen working overseas brings views and biases on security issues with him or her.

Input for Planning

Besides a risk analysis providing input for planning protection, many other factors go into the planning process, as described in the following:

1. Has the problem been carefully and accurately identified?
2. How much will it cost to correct the problem, and what percent of the budget will be allocated to the particular strategy?
3. Is the strategy practical?
4. Is the strategy cost effective? For example, a loss prevention manager debates an increase in the staff of loss prevention officers or the purchase of a CCTV system. The staff increase will cost $40,000 per officer (three officers × $40,000 per year = $120,000 per year). CCTV will cost $150,000. After considering the costs and benefits of each, the manager decides on the CCTV system because by the second year the expense will be lower than will hiring three extra officers.
5. Does the cost of the strategy exceed the potential loss? For instance, it would not be cost effective to spend $5,000 to prevent theft from a $50 petty cash fund.
6. Does the strategy relate to unique needs? Often, strategies good for one location may not be appropriate for another location.
7. How will the strategy relate to the entire loss prevention program? CCTV, for example, can be applied in a retail store to prevent both shoplifting and employee theft. Security

personnel are needed to respond to incidents. CCTV can also save on personnel costs and management can remotely monitor multiple sites.

8. Does the strategy conform to the goals and objectives of the organization and their loss prevention program?
9. How does the strategy compare to contract loss prevention programs? Will the strategy interfere with any contract service?
10. How will insurance carriers react?
11. If a government contract is involved, what regulations must be considered?
12. Does the strategy create the potential for losses greater than what is being prevented? Applying a chain and lock to the inside handles of a double door hinders a burglar's entry; however, if a fire occurs, emergency escape is blocked.
13. Does the strategy reduce the effectiveness of other loss prevention strategies? A chain-link fence with colorful plastic woven through the links and high hedges will prevent people from seeing into the property and impede an offender's penetration, but this strategy also will cause observation problems for patrolling security and police.
14. Will the loss prevention strategy interfere with productivity or business operations? In a high-risk environment, for example, how much time is required to search employees who leave and return at lunchtime? What if 1,000 employees leave for lunch? As another example, if the loss prevention manager requires merchandise loaded into trucks to be counted by three separate individuals, will this strategy slow the shipping process significantly?
15. Will the strategy receive support from management, employees, customers, clients, and visitors? Can any type of adverse reaction be predicted?
16. Must the protection strategy conform to local codes, ordinances, or laws? For instance, in certain jurisdictions, perimeter fences must be under a specific height and the use of barbed wire is prohibited. Are there any guidelines, standards, or regulations pertaining to the strategy?
17. Will the strategy lower employee morale or lead to a distrust of management?
18. Are there any possible problems with civil liberties violations?
19. How will the union react to the strategy?
20. Was participatory management used to aid in planning the strategy?
21. Can the strategy be effectively implemented with the present number of loss prevention personnel?
22. Will the strategy cause a strain on personnel time?
23. What are the characteristics of the area surrounding the location that will receive the loss prevention strategy? To illustrate, if loss prevention strategies are planned for a manufacturing plant, what factors outside the plant impact protection? (Factors of consideration include crime, fire, and accident rates.) In addition, certain nearby sites may be subject to disaster: nuclear plants, airports, railroads (transportation of hazardous materials), educational institutions (student unrest), forests (fire), hazardous industries (chemicals), and military installations, among others. Weather conditions are important to consider as well. Storms can activate alarms. Excessive rainfall can cause losses due to flooding. Heavy snow can result in a variety of losses. Earthquake and volcanic actions are additional factors to consider.

24. When considering loss prevention strategies such as burglar or fire alarms, what is the response time of public services such as police, fire, and emergency medical service? Where is the nearest facility housing each service?
25. Are loss prevention strategies able to repel activity from local criminals, a gang, or organized crime?
26. Will the strategy shift crime (i.e., crime displacement) to another location, time, type of victim, or crime?
27. Does the strategy attempt to "loss-proof" or eliminate all losses? This often is an impossible objective to reach. Loss prevention practitioners sometimes are surprised by the failure of a strategy that was publicized as a panacea.
28. If the strategy is not implemented, what is the risk of loss?
29. Will a better, less expensive strategy accomplish the same objective?
30. Can any other present strategy be eliminated when the new strategy is implemented?
31. What other strategies are more important? Are priorities established?
32. Should a pilot program be implemented (say, at one manufacturing plant instead of at all plants) to study the strategy for problems and corrective action?
33. How will the strategy be evaluated?

Security strategies generally take the form of personnel, systems, and policies and procedures.

Incomplete Protection Plans

Andrew Smith, security manager at Tecsonics, Inc., a fast-growing electronics firm, was assigned the task of preparing a plan and one-year budget for information security. Two months later, Andrew was in front of senior executives who were eager to learn how proprietary information would be protected in such a fiercely competitive industry. At the beginning of the presentation, Andrew emphasized that the survival and growth of Tecsonics depended on its information security program. The four major strategies of protection for the first year included electronic soundproofing, also referred to as *shielding*, which would involve the use of a copper barrier throughout one conference room to cut off radio waves from a spy's bugging equipment. Cost: $250,000. The second strategy was to spend $85,000 for countermeasure "sweeps" to detect bugs in select locations on the premises. The third strategy was to upgrade the access control system at a cost of $250,000. The fourth strategy was to hire an IT security specialist for $110,000 annually.

Before Andrew was five minutes into his presentation, the rapid-fire questions began: "What is the return on investment?" "Are the plans cost effective?" "Did you perform a risk analysis?" "What are other similar businesses doing?" "Why should we spend so much money on shielding and sweeps when we can use cheaper methods, such as holding meetings in unexpected locations, preparing good policies and procedures, promoting employee education, and keeping certain sensitive information out of the IT system and locked up in a safe?" One sarcastic, hard-nosed executive quipped, "You would probably spend millions on shielding and sweeps and not realize that one of our male scientists could go out of town to a seminar, get cornered by a foxy broad, and get drunk as she pumps him for information!" Unfortunately for Andrew, everybody was laughing while he wished he had done a better job of preparing for his presentation (Purpura, 1989: 43–44).

Planning from a Systems Perspective

The **systems perspective** looks at interactions among subsystems. When actions take place in one subsystem, other subsystems are affected. For example, the criminal justice system is composed of three major subsystems: police, courts, and corrections. If, during one day, the police make 100 arrests for illegal drug offenses, then the court and corrections subsystems must react by accommodating these arrestees. There are many other examples of systems: a loss prevention department, a business, government, a school, an automobile, the human body, and so on. All these systems have subsystems that interact and affect the whole system. In each system, subsystems are established to attain overall system objectives and goals.

Similar to other systems, a loss prevention department can be analyzed in terms of *inputs, processes, outputs,* and *feedback.* As an example, look at a loss prevention department's immediate reaction and short-term planning concerning an employee theft incident (Figure 3-1). The loss prevention department receives a call from a supervisor who has observed an employee stealing: the *input.* The *process* is the analysis of the call, planning, and the action taken (i.e., dispatching of loss prevention personnel). The *outputs* (i.e., activity at the scene) are the arrival of the personnel, questioning, and note taking. *Feedback* involves communications from the on-site loss prevention personnel to the loss prevention department; this helps to determine if the output was proper or if corrections are necessary. For instance, suppose an arrest occurred (output). Then, this feedback would result in contacting public police so they will transport the offender.

The systems perspective described in Figure 3-1 is for planning a short-term, immediate action. Figure 3-2 illustrates long-term loss prevention planning from a systems perspective.

In Figure 3-2, the *inputs* of goals and objectives relate to senior management's expectations of the loss prevention function. The resources are explained in a budget and include human resources and security services and systems. Information, research, reports, risk analyses, and metrics are all inputs that aid in decision-making within the planning *process.* The *outputs* are the loss prevention programs and strategies evolving from the planning process that will

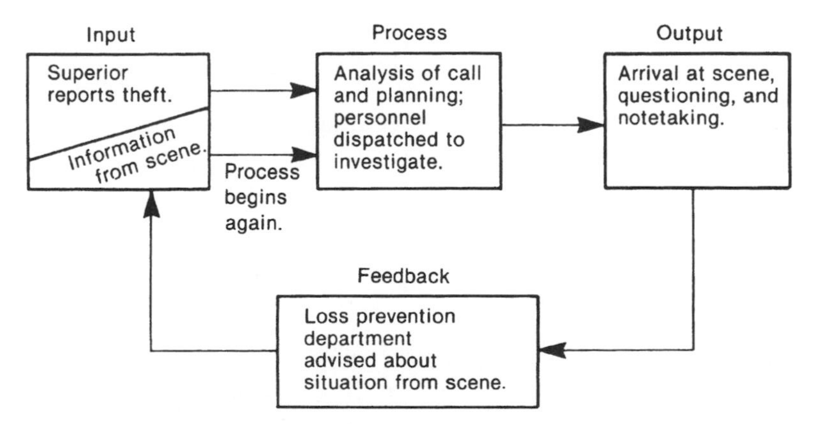

FIGURE 3-1 A systems perspective of a loss prevention department's immediate reaction and short-term planning relevant to an employee theft incident.

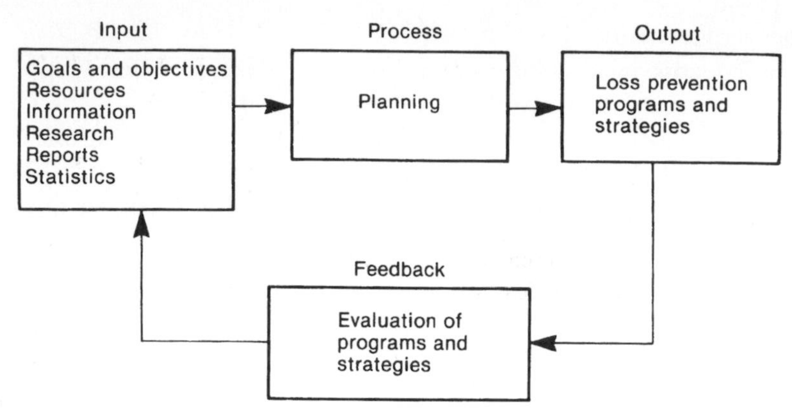

FIGURE 3-2 Long-term loss prevention planning from a systems perspective.

prevent and reduce losses while increasing profits. *Feedback*, an often-overlooked activity, is essential for effective planning, and evaluation is an integral part of feedback. Ineffective programs and strategies must be eliminated. Other programs and strategies may need modification. Research and evaluation help to justify programs and strategies.

Loss prevention practitioners must be prepared when asked by senior management, "How do you know the loss prevention plans and strategies are working?"

Critical Thinking for Security Planning

Maxims Of Security

1. Security is never foolproof. The term *foolproof* is a misnomer. Instead of *burglarproof, bulletproof,* or *fireproof,* replace *proof* with *resistant.*
2. State-of-the-art security has its vulnerabilities. History is filled with grand security strategies that failed.
3. Through history, security designers have been in constant competition with adversaries seeking to devise methods to circumvent defenses. This is a "cat and mouse" cycle where one side takes the lead, often temporarily, until the other side produces a superior technique. We see this competition with such areas as physical security, cyber security, and counterterrorism.
4. Related to #3, the design of security plays a role in an adversary's plan of penetration and attack. Terrorists and other criminals fine-tune their plans according to how a target is protected.
5. Security often is as good as the time it takes to get through it. The longer the time delay facing the offender, the greater the protection and chances that he or she will abort the offense, be apprehended, or seek another target (i.e., crime displacement).
6. The "harder" the target or defensive strategy, the more likely the offender will seek weaker defenses at the same target or seek a completely new target.
7. Security must focus on not only what is leaving a facility (e.g., company assets; proprietary information), but also what is entering (e.g., weapons, explosives, illegal drugs, anger, and computer viruses).

The above maxims apply to physical security and cyber security.

Three Models of Security

All security strategies fall under one of the following models:

1. It protects people, assets, and/or the operations of enterprises.
2. It accomplishes nothing.
3. It helps offenders.

Illustrations of security protecting people, assets, and/or the operations of enterprises are seen when a hospital security officer escorts nurses to their vehicles at night, when a safe proves too formidable for a burglar who then leaves the scene, or when a manufacturing process is protected through access controls and other security methods. Security accomplishes nothing when security officers sleep on the job or fail to make their rounds or when alarm systems remain inoperable. Sometimes unknown to security practitioners and those they serve are the security strategies that actually help offenders. This can occur when security officer applicants are poorly screened, hired, and then commit crimes against their employers. The ordinary padlock is an example of how physical security can assist offenders. An unlocked padlock hanging on an opened gate can invite *padlock substitution,* in which the offender replaces it with his or her identical padlock, returns at night to gain access, and then secures the gate with the original padlock. Such cases are difficult to investigate because signs of forced entry are likely to be absent and employees may be blamed for losses. Fences, another example, are often built with a top rail and supports for barbed wire that are strong enough to hold up an offender climbing, rather than only the fence and barbed wire. In addition, attractive-looking picket fences have been knocked down by offenders and used as ladders.

Security practitioners should identify and classify all security strategies under these models to expose useful, wasteful, and harmful methods. This endeavor should be a perpetual process within risk analysis, careful planning, critical thinking, testing, and research to facilitate cost-effective, results-oriented security. Although these challenging goals require time and effort, the net result is a superior security and loss prevention program.

Using a critical thinking approach, what are your views on any of the maxims or models of security? How can they be modified or enhanced?

Standards and Regulations

Standards and regulations serve as resources for employees in the public and private sectors who seek to optimize professionalism, competence, and quality within their organizations that are reflected in the products and services they generate. There is much confusion and controversy over standards and regulations. Here we begin with a federal government perspective, basic definitions, and then information specific to *security and loss prevention planning.*

Standards

The **National Technology Transfer and Advancement Act** (NTTAA), which became law in March 1996, directs federal agencies to adopt private sector standards, wherever possible, rather than create proprietary, non-consensus standards. The act also directs the National

Institute of Standards and Technology to bring together all levels of government for the same purpose.

There are many types of standards that serve a variety of purposes. They are classified in numerous ways. Standards based on purpose include terminology standards that standardize nomenclature, test and measurement standards that define methods to assess performance, and product and service standards that promote quality. Standards may also be classified by the intended user group. Examples are company standards that are meant for a single business, industry standards for a particular industry, international standards, and government standards. Standards may also specify requirements, such as performance standards that describe how a product is supposed to function and design standards that define how a product is to be built.

Hundreds of U.S., international, and regional organizations develop standards. Research on standards can be confusing and time-consuming. The end of this chapter offers federal government websites helpful for finding standards and regulations.

Consensus standards are accepted industry practices developed through a consensus process (i.e., open to review and modifications by experts who agree on how a specific task should be performed prior to the final standard). Consensus standards do not have the force of law unless a jurisdiction adopts them as law. Professional groups also publish "guidelines" or "guides" (also without the force of law) that offer organizations information and factors to consider when developing programs (e.g., security or fire protection).

Standards may help an industry prevent legislation that is burdensome and expensive to that industry. For instance, if an industry voluntarily produces standards that promote security and safety, then government security regulations for that industry may be averted.

Although standards may not be adopted as law by a jurisdiction, they may be used to establish a standard of care or used during litigation. Angle (2005: 22) defines **standard of care** as "the concept of what a reasonable person with similar training and equipment would do in a similar situation." In the emergency medical field, for example, a practitioner can prevent a claim of negligence if he or she performs in the same way as another reasonable person with the same training and equipment. In other words, everyone has certain expectations of performance.

An employer can face a claim of negligence by failing to adhere to policies and procedures, standards, or legal mandates (Fried, 2004: 30). Suppose employees are injured during an emergency evacuation and sue. Experts may be hired by opposite sides of the civil case to provide expertise and testimony on whether the employer provided a safe workplace and properly planned for emergencies. Comparisons may be made of what reasonable, prudent management would do. If the defense side can show that management did everything "reasonable management" would have done under the same circumstances, then the employer has a good chance of showing that it acted reasonably and should not be liable for injuries.

Fried makes interesting observations about standards and litigation. He notes that if industry experts seek to develop a standard to benefit themselves or others, or to sell products or services, then the standard would not have significant weight in court. Another question sure to surface is how the standard was developed. Consensus standards often do not meet the scientific rigor of, say, how the medical profession uses multiple blind tests to determine if a drug works. Fried (2004: 30) writes: "Thus, for any standard to pass legal muster under the Daubert challenge (requiring proof that a standard or conclusion is based on scientific or sound research), the standard needs to be tested and proven to be correct. Otherwise the standard is just a suggestion."

For example, although the National Fire Protection Association (NFPA) outlines ideal actions and what an ideal manager should achieve based on industry experts, failure to meet a standard does not mean law was violated or someone is negligent. A judge may declare a standard introduced in court as not applicable in that jurisdiction because no court or legal body authorized it, even though an expert witness might argue that an unofficial standard has become an industry standard of practice. Fried writes that if over 50 percent of employers follow a conduct—whether or not from a standard—that conduct could be considered the reasonable industry standard that should be followed.

In reference to settling lawsuits involving negligent security, courts have ruled inconsistently. Acceptable security in one jurisdiction may be unacceptable in another. However, security standards foster uniform security. Those against standards cite costs and argue that it is impossible to standardize security because each location and business is unique.

A government jurisdiction can adopt a standard as law and enforce it. This has been done by many jurisdictions with the NFPA 101 Life Safety Code. The NFPA has no enforcement authority.

Competition over Standards

There is competition among organizations over producing and publishing standards. A major issue is which organization is best suited to prepare standards in a particular field. Swope (2006: 20–23) uses the term **code war** to describe rival groups lobbying governments to accept their competing sets of standards. For instance, for many years the business community has urged governments to standardize building codes among cities and states to reduce construction costs. On one side of the competition is the International Code Council (ICC), a group supported by government building and code enforcement officials, architects, and building owners and managers. The National Fire Protection Association (NFPA) heads the other side. It is supported by fire chiefs and unions representing building specialists. Government and industry prefer one code, but they are faced with a major question: "Which code to choose?" Swope (2006: 22) writes that the code war also involves the ICC and NFPA acting as "two publishing houses engaged in a war for book sales." When a government adopts a model code or when codes change, officials, architects, engineers, construction specialists, and others must purchase up-to-date copies. Each code group earns tens of millions of dollars annually from publication sales.

A major difference between the ICC and the NFPA is how they update their codebooks. The ICC allows only building officials to vote on changes. The ICC sees this as a way to prevent special interests (e.g., trade unions, manufacturers) from influencing the codes. The NFPA permits all of its members to vote, although checks and balances are applied to prevent undue influence.

Other groups are also in competition. For instance, ASIS International and the NFPA each have committees involved in the subject of security. ASIS International produced several excellent security guidelines and standards (many discussed in this book) on such topics as risk assessment, chief security officer, private security officer selection and training, business continuity, and organizational resilience, among others. The group has several standards under development (ASIS International, 2011).

The NFPA produced "NFPA 730, Guide for Premises Security." It is based on risk assessment principles and includes information on physical security, security personnel, and security at certain occupancies (e.g., educational, healthcare). Another publication is "NFPA 731, Installation of Electronic Premises Security Systems." It is a standard with specific requirements for the installation of various security systems (Moore, 2005).

Is the competition over standards beneficial or detrimental to our society and the security profession? Justify your view.

Regulations

Regulations are rules or laws enacted at the federal, state, or local levels with the requirement to comply. These requirements may address health, product safety, environmental effects, or other matters in the public interest. A regulation may consist of agency-developed technical specifications or private sector standards. The NTTAA endorses the use of private sector standards.

Federal regulations are contained in the **Code of Federal Regulations (CFR)**. A widely known example of federal regulations is found in Title 29 CFR, Occupational Safety and Health Administration (OSHA). OSHA is a federal agency, under the U.S. Department of Labor, established to administer the law on safety and health in the workplace.

Angle (2005: 21) offers these definitions: "**Codes** are standards that cover broad subject areas, which can be adopted into law independently of other codes or standards." "**Recommended practices** are standards, which are similar in content to standards and codes, but are non-mandatory in compliance."

It is important to note that the terms *regulations*, *standards*, *codes*, and *recommended practices* are used interchangeably in the literature and by the public and private sectors. OSHA, for example, uses the term *standards* in the context of *regulations*.

Post-9/11 Standards and Regulations

Prior to the 9/11 attacks, many business enterprises set their own internal security standards and policies. This still occurs; however, certain industries have been following external regulations and standards for many years. Examples include businesses involved with government contracts, or the nuclear, financial, or aviation industries. Institutions such as healthcare facilities and colleges have also been influenced by regulations and standards. OSHA has played a major role in influencing businesses and institutions in promoting a safe environment, even through crime prevention methods. In addition, some locales have enacted ordinances to increase security at convenience stores.

The passage of the NTTAA resulted in a proliferation of standards and standards developers, and the 9/11 attacks resulted in even more growth of security standards and regulations.

The federal government has spearheaded efforts to protect infrastructure and key assets through many initiatives and regulations (e.g., chemical industry) as described in this book. Security and loss prevention practitioners must continue to study and apply guidelines, standards, and regulations pertaining to their specific industries. These resources help to formulate internal policies and procedures. Businesses involved in transportation, for instance, refer to the U.S. Department of Transportation and the U.S. Department of Homeland Security. The banking and finance industry refers to the U.S. Department of the Treasury. Drinking water and wastewater treatment systems are guided by the Environmental Protection Agency. Alongside government regulations, many organizations publish standards of a general nature as well as for specific industries.

The **Interagency Security Committee** (ISC), U.S. Department of Homeland Security, has a mandate to improve physical security at buildings and nonmilitary Federal facilities in the U.S. The ISC provides standards, best practices, and online training courses. The *Physical Security Criteria for Federal Facilities* establishes a baseline set of physical security measures for Federal facilities and a framework for customization of security for unique risks. The *Design-Basis Threat* focuses on profiles and capabilities of adversaries. Other publications are available from the ISC. The Federal Emergency Management Agency is another source; it offers physical security "guides" against terrorism.

Several benefits result from adherence to guidelines, standards, and regulations. These include greater safety and security, the prevention of losses and litigation, and enhanced risk management.

Standard-Setting Organizations

ASIS International is a standard-setting organization approved and accredited by the American National Standards Institute (ANSI), which means that ASIS conforms to strict ANSI-approved procedures as it prepares standards for the security profession. Here we look at standards from the perspective of manufacturing and installing security and loss prevention products and systems. Such standards serve as written and tested guidelines that promote uniformity and quality. Additional benefits of such standards include preventing people from installing unsafe systems, helping manufacturers define how their products converge with IT networks, and assisting the industry in the highly standards-oriented federal market.

Manufacturers have been producing their products in accordance with safety standards for many years. During the 1920s, for example, **Underwriters Laboratories (UL)**, an independent testing organization, worked with insurers to establish a rating system for alarm products and installations. This system assists insurers in setting premiums for customers. An alarm company may show customers that its service is of a higher standard than a competitor's is. UL has various listings, and it requires that an alarm company advertise its own listing specifically. What a company has to do to obtain a listing as a central station burglar alarm company differs widely from what it has to do to be listed as a residential monitoring station. For example, to obtain a listing as the former, the company must provide fire-resistant construction, backup power, access controls, and optimal response time following an alarm.

In general, consumers are more familiar with UL as an organization promoting the electrical safety of thousands of retail products. Companies pay a fee to have UL test their products for safety according to UL standards. The famous UL label is often seen attached to the product. Sometimes the UL label is attached to a product without testing and authorization.

The **National Fire Protection Association (NFPA)** establishes standards for fire protection equipment and construction that have been adopted by government agencies and companies in the private sector. Beginning in 1898, in cooperation with the insurance industry, the NFPA has produced standards covering sprinklers, fire hoses, fire doors, and other forms of fire protection.

The **Security Industry Association (SIA)** is an international trade association that promotes education, research, and technical standards. It represents manufacturers, distributors, service providers, and others in the security industry. SIA standards promote the interests of its membership by developing common, open, interoperability protocols and performance standards and through standards-related venues. SIA follows American National Standards Institute (ANSI) principles on developing standards like other standards writing bodies.

The **American Society for Testing and Materials (ASTM International),** organized in 1898, has grown into one of the largest voluntary standards development systems in the world. ASTM is a nonprofit organization providing a forum for producers, consumers, government, and academia to meet to write standards for materials, products, systems, and services. Among its standards-writing committees are committees that focus on security, safety, and fire protection.

The **American National Standards Institute (ANSI)**, organized in 1918, is a nonprofit organization that coordinates U.S. voluntary national standards and represents the United States in international standards bodies such as the International Organization for Standardization. ANSI approves and accredits standards for products and personnel certification that are developed by standards-setting organizations, government agencies, consumer groups, businesses, and others. It does not develop its own standards.

The **International Organization for Standardization** is a worldwide federation of national standards bodies from just about all countries. It is a nongovernmental group based in Geneva, Switzerland, established in 1947 with the purpose of promoting standardization globally to facilitate the international exchange of goods and services. Its agreements are published as International Standards, also known as ISO standards. According to Wikipedia (2011), ISO does not stand for "International Standards Organization." ISO is not an acronym; it comes from the Greek word *isos*, meaning "equal."

Between ASIS International and the NFPA, which group do you think is most appropriate to produce security guidelines and standards? Explain your answer.

Evaluation of Loss Prevention Programs

How can a loss prevention program be evaluated? First, a research design can assist in the evaluation. A look at a few simplified research designs demonstrates how loss prevention programs are determined to be successful or unsuccessful.

One design is the **pretest–posttest design.** A robbery prevention program can serve as an example. First, the robbery rate is measured (by compiling metrics) before the robbery prevention program is implemented. The program then is implemented and the rate measured again. Robbery rates before and after program implementation is compared. If the robbery rate declined, then the robbery prevention program *may* be the causative factor.

A loss prevention training program can serve as another example. Loss prevention personnel are tested prior to the training program. The training program is implemented. After the program is completed, another similar test is given the personnel. The pretest scores are compared to the posttest scores. Higher posttest scores *may* indicate that the training program was effective.

Another evaluation design or research method is called the **experimental control group design**. As an example, a crime prevention program within a corporation is subject to evaluation. Within the corporation, two plants that are characteristically similar are selected. One plant (the experimental plant) receives the crime prevention program, while the other plant (the control plant) does not. Before the program is implemented, the rate of crime at each plant is measured. After the program has been in effect for a predetermined period, the rate of crime is again measured at each plant. If the crime rate has declined at the "experimental" plant while remaining the same at the "control" plant, then the crime prevention program *may* be successful.

A good researcher should be cautious when formulating conclusions. In the crime prevention program at the plant, crime may have declined for reasons unknown to the researcher. For instance, offenders at the experimental plant may have refrained from crime because of the publicity surrounding the crime prevention program. However, after the initial impact of the program and the novelty expired, offenders could possibly continue to commit crimes. In other words, the program may have been successful in the beginning but soon became ineffective. Thus, continued evaluations are vital to strengthen research results.

Scientific Method

To assist planning and research, the scientific method can be applied. As written earlier in this chapter, the four steps are: statement of the problem, hypothesis, testing, and conclusion. As an example, *employee* theft will serve as the problem. The hypothesis is a statement whereby the problem and a possible solution (i.e., loss prevention strategy) are noted. Testing involves an attempt to learn whether the strategy reduces the problem. Several research designs are possible. The following is an example of the presentation of the problem according to the style of scientific methodology:

- *Problem*: Employee theft.
- *Hypothesis*: Employee theft can be reduced by applying CCTV.
- *Testing*: Control group (plant A—no CCTV); experimental group (plant B—CCTV).
- *Conclusion*: After several months of testing, plant A maintained previous levels of employee theft, whereas plant B showed a drop in employee theft. Therefore, CCTV *may* be an effective loss prevention strategy to reduce employee theft.

To strengthen research results, continued testing is necessary. In the example, CCTV can be tested at other, similar locations. Furthermore, other strategies can be combined with CCTV to see if the problem can be reduced even more.

Sources of Research Assistance

More research is needed to identify successful and unsuccessful loss prevention methods. Four potential sources of research assistance are in-house, university, private consulting, and insurance companies.

In-house research may be the best because proprietary personnel are familiar with the unique problems at hand. Salary costs could be a problem, but a loss prevention practitioner with a graduate degree can be an asset to loss prevention planning and programming. In-house research, however, can result in increased bias by the researcher because superiors may expect results that conform to their points of view.

University researchers usually have excellent credentials. Many educators are required to serve the community and are eager to do research that can lead to publication.

Private consulting firms often have qualified personnel to conduct research and make recommendations to enhance protection. This source can be the most expensive because these firms are in business for profit. Careful consideration and a scrutiny of the consulting staff are wise. The buyer should beware.

Insurance companies are active in studying threats, hazards, and risks. They also participate in varying degrees in research projects relevant to crime, fire, and safety. A major function of this industry is risk management, whereby strategies are recommended to reduce possible losses.

Performance Measures and Evidence-Based Loss Prevention

Traditionally, public police performance has been measured by reported crime rates, overall arrests, crimes cleared by arrest, and response time to incidents. These metrics have become institutionalized, and substantial investments have been made to develop IT systems to capture such data. However, this data may tell little about police effectiveness in reducing crime and the fear of crime (U.S. Department of Justice, 1993: ix). Furthermore, do these measures reflect outmoded policing, and do they fail to account for many important contributions police make to the quality of life, such as efforts at community cohesion and crime prevention? The following are examples of performance measures for police that have implications for security:

- *Goal*: Promoting secure communities
- *Methods and activities*: Promoting crime prevention and problem-solving initiatives
- *Performance indicators*: Programs and resources allocated to crime prevention, public trust and confidence in police, and reduced public fear of crime (DiIulio, 1993: 113–135)

What lessons can security practitioners learn from research on performance measures in the public sector? Do security programs contain systems of measuring performance that are outdated and reflect outmoded security strategies? Performance measures do exist in security programs today, but how can they be improved and how can security strategies be improved? One avenue to advance security programs is through evidence-based research. We can look to the Loss Prevention Retail Council—a group of major retailers, practitioners, and researchers—who promote quality evidence-based research on a host of retail security topics such as interviews of offenders, predictive models, and field trials of security strategies at stores. This work is shaping modern retail loss prevention. **Evidence-based research** is defined by Hayes (2010: 60) as, "... an approach that tries to specify the way in which decision-makers should make decisions by identifying credible evidence for why problems are occurring, how solutions actually work (that is, deter offenders), how well a solution works, and how cost-effective it appears to be." Hayes explains ways in which such research is applied. For instance, multiple loss prevention measures can be compared for effectiveness. Another approach is to research the types of places or incidents that benefit the most from the application of a specific loss prevention strategy.

Proprietary Security

Proprietary security programs and organizations in general are characterized by the organizational terms and practical management tools described next. These terms and topics also apply to contract security businesses.

Basics of Organization: The Vocabulary

- **Division of work:** Work is divided among employees according to such factors as function, clientele, time, and location.
- **Authority:** The right to act.
- **Responsibility:** An obligation to do an assigned job.
- **Power:** The ability to act.
- **Delegation of authority:** A superior delegates authority to subordinates to spread the workload. A superior can delegate authority, but a person must accept responsibility. Responsibility cannot be delegated. For example, a sergeant delegates to loss prevention officers the authority to check employee lunch boxes when employees leave the plant. The sergeant is still responsible for the lunch boxes being checked.
- **Chain of command:** Communications go upward and downward within an organized hierarchy for the purpose of efficiency and order.
- **Span of control:** The number of subordinates that one superior can adequately supervise. An example of a broad span of control would be one senior investigator supervising 20 investigators. An example of a narrow span of control would be one senior investigator supervising five investigators. An adequate span of control depends on factors such as the amount of close supervision necessary and the difficulty of the task.
- **Unity of command:** To prevent confusion during an organized effort, no subordinate should report to more than one superior.
- **Line personnel:** Those in the organized hierarchy who have authority and function within the chain of command. Line personnel can include uniformed loss prevention officers, sergeants, lieutenants, captains, and other superiors.
- **Staff personnel:** Specialists with limited authority who advise line personnel. For example, the loss prevention specialist advises the captain about a more efficient method of scheduling security officers for duty.

- **Formal organization:** An official organization designed by senior management whereby the "basics of organization" are applied to produce the most efficient organization possible.
- **Informal organization:** An unofficial organization produced by employees with specific interests. For example, several employees spend time together (during breaks and lunch) because they are active as community volunteers.
- **Organization chart:** A pictorial chart that visually represents the formal organization. Many of the "basics of organization" actually can be seen on organization charts (Figure 3-3).
- **Basics of organization:** The practical management tools.
- **Directive system:** A formal directive system is a management tool used to communicate information within an organized group. The communications can be both verbal and written (e.g., e-mail). A verbal directive can be a superior informing a subordinate of what work needs to be done. The formal verbal directive system also can include meetings in which verbal communications are exchanged.
- **Policies:** Policies are management tools that control employee decision making. Policies reflect the goals and objectives of management.
- **Procedures:** Procedures are management tools that point out a particular way of doing something; they guide action. Many procedures actually are plans that fulfill the requirements of policies. The loss prevention manager must maintain an open mind when feedback evolves from policies and procedures. For instance, suppose only four security officers are assigned to thoroughly search (according to policies and procedures) the belongings carried by 1,000 employees as they end their shift. Long lines of irritated employees are likely to develop. Consequently, changes must be made.

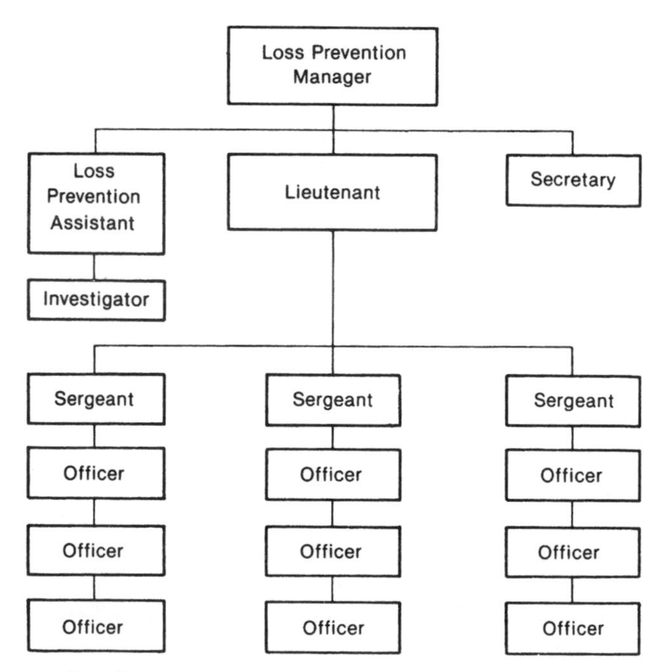

FIGURE 3-3 Small loss prevention department.

- **Manual:** A manual is like a "rule book" for an organized group; it contains policies and procedures.

  It is important to note that if an organization's structure is too rigid, employees may be hindered from fulfilling business objectives. Participation by subordinates in management decision-making and a team atmosphere of cooperation have been shown to increase morale and motivation. Leadership style also influences success. The **autocratic style** sees managers making all decisions, whereas the **democratic style** seeks opinions from employees as input for decisions. Following decades of research, psychologists have concluded that no one style is best. Effective managers use both. For example, a manager may be autocratic with a trainee and democratic with a seasoned employee (Levy and Weitz, 2001: 534).

What are the Duties of a Security Manager?

The duties of security managers vary widely. A generalization is presented here.

Security managers plan, prioritize tasks, concentrate on day-to-day challenges, and strive to stay within their budget. They delegate tasks and supervise subordinates and contractors who perform security services. Another duty is to ensure that security systems are functioning properly. Security managers may spend all day at their desks and computers sending and responding to e-mails, preparing reports, conducting research on the Internet, reading, or talking to peers on the telephone. *With modern technology, such tasks can be performed almost anywhere.* On other days, they may split their time in the office doing administrative work with visits to various locations on and off the premises for inspections or investigations. They may attend meetings, conduct or receive training, and attend a college course. Since they are interacting with people so much, they must have excellent human relations and communications skills.

Physical security and access controls are often part of their job; only authorized people should be on the premises. Security managers also realize that IT systems face internal and external threats; consequently, security managers and IT specialists are increasingly working together. If an alleged crime occurs or an incident requires investigation, the manager may conduct the investigation, delegate the task, or contract the work to an outside firm. Workplace safety is another important issue, and this includes fire safety, prevention of accidents, and emergency procedures. All security personnel and volunteers from the workforce may receive special training to prevent and suppress fires, render first aid, and prepare for disasters. As risks surface, security managers prepare policies and procedures, in cooperation with other employees, in an effort to prevent losses.

A company may require the manager to spend a great deal of time providing training programs to employees on a host of topics, from employee protection to information security. Security managers may be responsible for a certain geographic area and visit corporate locations over several weeks to conduct a variety of duties as described previously. To reduce costs, businesses may add additional duties—outside traditional security duties—to the security manager's position. Examples include business continuity planning, investigating employment applicants, and supervision of landscaping, parking, a fleet of vehicles, a mail system, and a cafeteria. Security managers must be flexible and available for emergencies at any hour, since the employer and employees depend on them for protection.

What is your opinion of the duties of a security manager?

Search the Internet

The Internet contains a wealth of information on the security and loss prevention profession. Here are websites relevant to this chapter:

Academy of Security Educators and Trainers: www.asetcse.org
American National Standards Institute: www.ansi.org
ASIS International, *Security Management:* www.asisonline.org
American Society for Testing and Materials: www.astm.org
finding regulations: standards.gov/regulations.cfm
finding standards: standards.gov/standards.cfm
International Foundation for Protection Officers: www.ifpo.org
International Organization for Standardization: www.iso.org
Journal of Applied Security Research: www.tandf.co.uk/journals/WASR
Journal of Physical Security: jps.anl.gov
National Fire Protection Association: www.nfpa.org
Security Industry Association: www.siaonline.org
Security Journal: www.palgrave-journals.com
Underwriters Laboratories (UL): www.ul.com/global/eng/pages

Case Problems

3A. As a security executive, prepare at least fifteen situational crime prevention strategies to curb the loss of laptop computers in a company office building. (Refer to the situational crime prevention techniques earlier in the chapter.)

3B. As the Chief Security Officer for a global manufacturer of a variety of products, you have in your budget only enough money to subscribe to three security periodicals. You search the Internet and select three. Which ones did you choose and why?

3C. As a loss prevention manager, you have been asked to prepare a speech to a group of security practitioners. The topic is "Planning Security and Loss Prevention Programs and Strategies." What topics do you include in your speech? For your audience, prepare information on each topic and why each is important.

3D. Refer to the box in this chapter titled "Incomplete Protection Plans." If you were an outside security consultant hired by Andrew Smith, what would you suggest to help him prior to his meeting with senior executives?

3E. As a security manager, you believe that the meeting today with the Vice President of Finance will not bring good news. Each business quarter seems to show poor profits and the need for cutbacks in all departments. When you enter her office for the meeting, the VP, Alaine Nell, gets right to the point: "Your security budget to protect the three access points at the plant must be cut by 50 percent." She draws a sketch of the huge square plant and notes that the three access points are costing $500,000 annually for 24-hour-a-day security officer protection, including overtime. She states that each post requires four officers (three on and one off during each 24-hour period) at $40,000 apiece. She requests a financial plan for the next five years. Prepare such a plan, providing hypothetical information if needed.

3F. You are a security specialist at a port. Your supervisor has assigned you the task of preparing a list of the laws, regulations, standards, government agencies, and private sector organizations pertaining to security at ports. Prepare the list by conducting research on the Internet.

3G. Design an organization chart for a loss prevention department of 35 people at an industrial plant. Write a one-page justification of your design to satisfy management. Provide hypothetical information about the plant if needed.

References

Adolf, D. (2011). Security studies and higher education. *Journal of Applied Security Research*, 6, January–February.

ASIS International. (2003). *General Security Risk Assessment Guideline*. <www.asisonline.org> retrieved April 15, 2011.

ASIS International. (2011). ASIS standards and guidelines update. *Security Management*, 55, April.

Angle, J. (2005). *Occupational safety and health in the emergency services* (2nd ed.). Clifton Park, NY: Thomas Delmar Pub.

Berrong, S. (2010). Taking security global. *Security Management*, 54, August.

Broder, J. (2006). *Risk analysis and the security survey* (3rd ed.). Burlington, MA: Elsevier Butterworth Heinemann.

Clarke, R. (1997). *Situational crime prevention: Successful case studies* (2nd ed.). New York, NY: Harrow and Heston Pub.

Cohen, L., & Felson, M. (1979). Social change and crime rate trends: A routine activity approach. *American Sociological Review*, 44, August.

Cornish, D., & Clarke, R. (1986). *The reasoning criminal: Rational choice perspectives on offending*. New York: Springer-Verlag.

Crawford, J. (1995). Security, heal thyself. *Security Management*, May.

DiIulio, J. (1993). *Performance measures for the criminal justice system (October)*. Washington, D.C.: Bureau of Justice Statistics.

Felson, M. (1998). *Crime and everyday life: Insights and implications for society*. Thousand Oaks, CA: Pine Forge Press.

Fried, G. (2004). Ask the expert. *Public Venue Security*, May/June.

Hayes, R. (2010). Developing EBLP professionals. *Loss Prevention*, July–August.

Jacobson, R. (2000). What is a rational goal for security? *Security Management*, 44, December.

Jeffery, C. R. (1977). *Crime prevention through environmental design*. Beverly Hills, CA: Sage Pub.

Lab, S. (2004). Crime prevention: Approaches, practices and evaluations, 5th ed. <www.lexisnexis.com/anderson/criminaljustice>.

Levy, M., & Weitz, B. (2001). *Retailing management*. New York: McGraw-Hill.

Malampy, W., & Piper, J. (2010). Security risk assessments: Integrating the concept. *Security Technology Executive*, 20, August.

Moore, W. (2005). NFPA 730 and NFPA 731: The new NFPA security guide and standard increase the quality and reliability of security system installations. *NFPA Journal*, January/February.

National Research Council. (2010). *Department of Homeland Security's Approach to Risk Analysis.* <www.nap.edu>, retrieved April 10, 2011.

Newman, O. (1972). *Defensible space.* New York: Macmillan.

Purpura, P. (1989). *Modern security and loss prevention management.* Boston: Butterworth-Heinemann.

Sampson, R., et al. (2010). Super controllers and crime prevention: A routine activity explanation of crime prevention success and failure. *Security Journal*, 23. <www.palgrave-journals.com> retrieved January 19, 2011.

Swope, C. (2006). The code war. *Governing*, 19, January.

U.S. Department of Justice. (1993). *A police guide to surveying citizens and their environment.* Washington, D.C.: U.S. Government Printing Office. (October).

U.S. Department of Justice, Office of Community Oriented Policing Services. (2006). Assaults in and Around Bars, 2nd ed. <www.cops.usdoj.gov/files/ric/Publications/e08064507.pdf> retrieved April 5, 2011.

U.S. Department of Justice, Office of Community Oriented Policing Services. (2007). Understanding Risky Facilities. <www.cops.usdoj.gov/files/ric/Publications/e02071462.pdf> retrieved April 5, 2011.

Vito, G., & Maahs, J. (2012). *Criminology: Theory, research, and policy.* Sudbury, MA: Jones & Bartlett.

Warner, J. (2010). What's with all this peer-review stuff anyway? *Journal of Physical Security*, 4.

Wikipedia. (2011). International Organization for Standardization. en.<wikipedia.org/wiki/International_Organization_for_Standardization> retrieved April 16, 2011.

4 Law

OBJECTIVES

After studying this chapter, the reader will be able to:

1. Summarize the judicial systems of the United States
2. Explain the origins of law
3. List and define at least five torts
4. Discuss the theory of premises security claims and negligence
5. Explain contract law
6. Outline civil justice procedures
7. Explain administrative law
8. Summarize labor law and its relationship to private security
9. Outline criminal justice procedures
10. Explain the legal guidelines for arrest, use of force, searches, and questioning

KEY TERMS

- dual court system
- limited jurisdiction courts
- courts of general jurisdiction
- appellate jurisdiction
- intermediate state appellate courts
- state supreme court
- U.S. Magistrate's Courts
- U.S. District Courts
- U.S. Courts of Appeals
- U.S. Supreme Court
- federal and state constitutions
- common law
- case law
- legislative law
- felonies
- misdemeanors
- civil law
- substantive law
- procedural law
- tort law
- false imprisonment
- malicious prosecution
- battery
- assault
- trespass to land
- trespass to personal property
- infliction of emotional distress
- defamation (libel and slander)
- invasion of privacy
- negligence
- legal duty
- proximate cause

- foreseeability
- prior similar incidents rule
- totality of the circumstances test
- conscious disregard
- contract
- remedies for breach of contract
- nondelegable duty
- respondeat superior
- vicarious liability
- plaintiff
- defendant
- discovery
- motion
- pretrial conference
- verdict
- administrative agencies
- administrative inspections
- administrative searches
- Health Insurance Portability and Accountability Act of 1996 (HIPAA)
- Sarbanes-Oxley (SOX) Act of 2002
- Financial Modernization Act of 1999
- National Labor Relations Act
- Weingarten rights
- arrest
- probable cause
- booking
- initial appearance
- preliminary hearing
- arraignment
- plea bargaining
- trial
- reasonably force
- deadly force
- exclusionary rule
- Miranda warnings

Introduction

A good foundation in law is an essential prerequisite to loss prevention programming. Many crucial decisions by practitioners are circumscribed by legal parameters, and the consequences of these decisions can be serious. An arrest without the proper legal authority and evidence can result in civil and criminal action against security personnel. Negligence is a serious concern that results from the failure to exercise due care in the use of force, for example. This is why training is so important; it becomes a major issue in a lawsuit against security. Numerous lawsuits have also been directed at those responsible for security (e.g., business owners), who are claimed to be negligent for not providing a safe environment, which caused a person to become a crime victim. Consequently, security and loss prevention decisions must take into consideration the legal environment. We begin with the structure of our judicial systems.

Judicial Systems

Our nation has a **dual court system** of federal and state courts. This means that, in addition to a federal court system, each of the 50 states, the District of Columbia, and territories such as Guam has its own court system (Purpura, 1997: 200–212).

State Court Systems

States commonly have a three- or four-tier court system. Keeping in mind state variation, the tiers, from lowest to highest, are:

- Courts of limited or special jurisdiction
- Courts of general jurisdiction (major trial courts)
- Intermediate appellate courts
- State supreme court (the court of last resort)

Limited jurisdiction courts handle minor cases—the bulk of the judicial caseload, traffic violations, and misdemeanors. An example of this court is the justice of the peace, which in many states has been reformed or eliminated. When found, it exists more frequently in rural than urban areas. The magistrate's court resembles the justice of the peace, although there may be differences in some states. Small, medium, and large cities have established municipal courts to handle a variety of justice duties. Courts of special jurisdiction hear cases only in a single area of law, such as criminal, drug, family, juvenile, or probate (wills and transfers of assets).

State laws or constitutions define jurisdiction between lower and higher courts that may hold concurrent jurisdiction over some misdemeanors. **Courts of general jurisdiction** have jurisdiction over all cases involving civil law and criminal law. A person accused of a felony is prosecuted in this type of court. Courts of general jurisdiction have a variety of names, which can create confusion in understanding court systems among the states. Courts of general jurisdiction may have appellate jurisdiction over lower court decisions (such as a misdemeanor tried in a court of limited jurisdiction). **Appellate jurisdiction** is the authority of a court to review and revise the decision of an inferior court.

Appeals to **intermediate state appellate courts** offer litigants a chance to change an unfavorable trial court decision by arguing that the lower court judgment was based on a reversible error. The party bringing the appeal might contend, for instance, that the trial court erred when it allowed inadmissible testimony, or that the jury was given improper instructions.

All states have a court of last resort, often called the **state supreme court**. It is the highest court in the state and it interprets the law, applies it to the case at hand, and renders a decision. This court receives much of its caseload from the intermediate appellate court, if one exists.

About 95 percent of all legal cases initiated in the United States are filed in state courts. In 2008, lower courts and courts of general jurisdiction handled approximately 106 million incoming trial court cases—the most ever reported. This number included about 57.5 million traffic and ordinance violations, handled primarily in lower courts. During the same year, about 300,000 incoming appellate cases were reported, with two-thirds processed in intermediate appellate courts (Conference of State Court Administrators, 2010).

Federal Court System

The federal system is divided into two major categories: the legislative courts and the constitutional courts. Congress established legislative courts under Article I of the U.S. Constitution. An example of these courts of special jurisdiction is the tax court, which handles disputes between taxpayers and the Internal Revenue Service. Congress also created district and appellate courts in the District of Columbia and territorial courts. Article III of the U.S. Constitution provided for the U.S. Supreme Court and authorized Congress to create the lower federal judiciary. Four kinds of constitutional courts are: U.S. Magistrate's Courts, U.S. District Courts, U.S. Courts of Appeal, and the U.S. Supreme Court.

The **U.S. Magistrate's Courts** are the lowest level of jurisdiction of the federal court system. U.S. magistrate judges conduct initial appearances and preliminary hearings, determine bail and conditions of release, conduct preliminary stages of felony cases, issue search and arrest warrants to federal law enforcement officers such as FBI agents, review cases involving prisoners, and deal with numerous civil matters. Their authority covers all tasks performed by federal district judges except trying and sentencing felony defendants. Decisions by magistrate judges can be appealed to the federal district courts.

U.S. District Courts are similar to the courts of general jurisdiction in state court systems. These courts hear criminal and civil cases.

The **U.S. Courts of Appeals**, often referred to as circuit courts, have jurisdiction over specific geographic areas and are the intermediate appellate courts of the federal system. The Courts of Appeals review cases from the district courts of each region, the U.S. Tax Court, and from certain federal administrative agencies.

The **U.S. Supreme Court** is the highest court of our nation. Few cases are heard by this court. It handles cases of national significance, such as those involving the U.S. Constitution, treaties, a citizen and a state, or two states. In the 2009 term, the total number of cases filed in this court was 8,159, an increase of 5.4 percent over the previous term. During the 2009 term, 82 cases were argued and 77 were disposed of (Supreme Court of the United States, 2011).

The approximate caseload of other federal courts is as follows for the fiscal year ending in 2010: U.S. Magistrate's Courts—1,027,191; U.S. District Courts—361,323; and U.S. Courts of Appeals—55,992 (Administrative Office of the United States Courts, 2010). The majority of all these cases involved criminal and civil matters and prisoner petitions.

International Perspective: Black-Collar Crime

An American sociologist name Edwin H. Sutherland defined the term "white-collar crime" in 1939. It refers to violations of law committed by business executives who use deceit for financial gain. Sutherland emphasized that this crime has as serious an impact on our world as does street crime.

The term "black-collar crime" is used in the United Kingdom (UK) to identify judicial misconduct of a criminal nature. The UK's Office for Judicial Complaints received 1,339 complaints for the years 2008–2009. Action was taken in 89 cases, 25 judges and magistrates were removed, and 20 others resigned during the investigative process. Complaints focused on judicial decisions, case management, and the failure to fulfill judicial duty. One group of judges was taking bribes from ex-husbands to render favorable decisions in child custody cases (Wrennall, 2010: 10-12).

In the United States, the process of handling complaints of judicial misconduct varies among the states. Here we focus only on the federal system, where 1,448 complaints were filed in 2010, down seven percent from 2009. Most allegations were in the categories of erroneous decision, other misconduct, personal bias against the litigant or attorney, and delayed decision. Two cases in 2010 reached the stage of the appointment of a Special Investigation Committee (Administrative Office of the United States Courts, 2010: 38).

Origins of Law

Five major sources of law are federal and state constitutions, common law, case law, legislative law, and administrative law. **Federal and state constitutions** are at the foundation of our government and legal system. The U.S. Constitution specifies powers of the federal government, powers reserved to the states, and the rights of citizens as contained in the Bill of Rights. Federal courts interpret the Constitution and render decisions on issues, such as whether a federal or state law is unconstitutional. State constitutions, although subordinate to the U.S. Constitution, are the supreme law within the respective state and superior to any local government constitution or charter.

Common law is based on principles of justice determined by reasoning according to custom and universal consent handed down from one generation to another through history. As civilization advanced, standards of behavior (i.e., common law) became formalized through judicial decisions, rather than from written statutes from legislative bodies. Rassas, L. (2011: 6) writes: "Common law, also referred to as case law, refers to the principles that are established by courts through the issuance of judicial decisions." Specific acts were, and still are, deemed criminal. These acts, even today, are referred to as common law crimes: treason, murder, robbery, battery, larceny, arson, kidnapping, and rape, among others. Common law also covers personal conflicts (i.e., civil law) such as payments of claims for personal injury and contracts.

English common law has a long history. Settlers brought common law from the English courts to the American colonies, where it served as a major source of law as the legal system evolved in the United States. After our nation gained independence from England, the common law influence remained. Nineteen states have perpetuated common law through case law (i.e., judicial precedent). Eighteen states have abolished common law and written it into statutes. The remaining states have either adopted common law via ratification or are unclear about how it is reflected in the state system.

The **case law** of today, sometimes referred to as *judge-made law*, evolved from English common law. It includes the interpretation of statutes or constitutional concepts by federal and state appellate courts. Previous case decisions or *precedent cases* have a strong influence on court decisions. Precedents clarify both statutes and court views to limit ambiguity. When a new case comes into existence, earlier case decisions (i.e., precedents) are used as a reference for decision-making. Because the justice system is adversarial in nature, opposing attorneys refer to past cases (i.e., precedents) that support their individual contentions. The court makes a decision between the opposing parties. Societal changes are often reflected in decisions. Because the meanings of legal issues evolve from case law, these court decisions are the law. Of course, later court review of previous decisions can alter legal precedent.

Legislative law from the federal government is passed by Congress under the authority of the U.S. Constitution. Likewise, individual state constitutions empower state legislatures to pass laws. Legislative laws permit both the establishment of criminal laws and civil laws and a justice system to preside over criminal and civil matters. A later court decision may clarify a legislative law or decide that it is unconstitutional; this illustrates the system of "checks and balances" that enables one government body to check on another.

Criminal law pertains to crimes against society (Tables 4-1 and 4-2). Each state and the federal government maintain a criminal code that classifies and defines offenses. **Felonies** are considered to be more serious crimes, such as burglary and robbery. **Misdemeanors** are less serious crimes, such as trespassing and disorderly conduct.

Civil law adjusts conflicts and differences between individuals (Tables 4-1 and 4-2). Examples of civil law cases include accidental injuries, marital disputes, breaches of contract, dissatisfied sales customers, and disputes with government agencies.

Law is also classified as substantive and procedural (Table 4-2). **Substantive law** with respect to criminal law defines criminal offenses (e.g., burglary and robbery) and specifies punishments. With respect to civil law, substantive law defines the rights and duties among people. **Procedural law** (Table 4-2) covers the formal rules for enforcing substantive law and the steps required to process a case, whether criminal or civil.

Legal Quiz

Think about the following questions while reading. The answers are provided throughout this chapter.

1. Can a person be arrested and sued for the same act?
2. Do most employers monitor Internet access by employees?
3. Do employers customarily place covert closed-circuit television (CCTV) cameras in restrooms?
4. Are employers prohibited by law from installing Global Positioning System (GPS) technology in company vehicles?
5. Is it true that the vast majority of civil and criminal cases never make it to trial?
6. Do administrative searches by government regulatory agencies require a search warrant?
7. Does the U.S. Supreme Court require private security personnel to read Miranda warnings to suspects prior to questioning?

Tort Law and Controls over Private Security

Public police officers have greater police powers than private security officers, who typically possess citizen's arrest powers. Even though they have greater police powers, public officers are limited in their actions by the Bill of Rights of the U.S. Constitution. On the other hand, private officers, possessing fewer powers, for the most part, are not as heavily restricted by constitutional limitations. Authority and limitations on private officers result from **tort law**, the body of state legislative statutes and court decisions that govern citizens' actions toward each other and allow lawsuits to recover damages for injury (Table 4-1). Tort law is the foundation for civil actions in which an injured party may litigate to prevent an activity or recover damages from

Table 4-1 Criminal Law and Tort Law

	Criminal law	Tort law
Definition	Crime as a public wrong	A civil or private wrong
Punishment/Sanction	Fine, probation, incarceration or death	Money damages
Who brings the action	The State	The plaintiff
Who can appeal	Defendant found guilty can appeal. The State usually does not appeal after losing a case.	Both parties can appeal
Standard of proof required	Proof beyond a reasonable doubt (a higher standard of proof than is required in a civil case)	A preponderance of evidence (greater weight of facts proven)

Source: Reprinted from ***Criminal Justice: An Introduction***, Philip. P. Purpura, Copyright Reed Elsevier Inc, 1997, with Permission from Elsevier.

Table 4-2 Substantive Law and Procedural Law

Criminal law		Civil law	
Substantive Law (Legal definitions of crimes, penalties)	Procedural Law (Law guiding the criminal justice process)	Substantive Law (Law defining rights and duties among people)	Procedural Law (Law guiding the civil justice process)
Capital crimes	Jurisdiction of cases	Tort law	Jurisdiction of cases
Felonies	Rules of evidence	Contract law	Rules of evidence
Misdemeanors	Other rules	Other civil law	Other rules
Violations			

Source: Reprinted from ***Criminal Justice: An Introduction***, Philip. P. Purpura, Copyright Reed Elsevier Inc, 1997, with Permission from Elsevier.

someone who has violated his or her person or property. Most civil actions are based not on a claim of intended harm, but on a claim that the defendant was negligent. This is especially so in cases involving private security officers. Tort law refers to actions that affect the safety and rights of others. When these are disregarded, negligence results. *The essence of the tort law limitations on private security officers is fear of a lawsuit and the payment of damages.* Businesses usually cover the risks and costs of lawsuits through insurance.

The primary torts relevant to private security are as follows:

1. **False imprisonment:** The intentional and forceful confinement or restriction of the freedom of movement of another person, also called *false arrest.* The elements necessary to create liability are detention and its unlawfulness.
2. **Malicious prosecution:** Groundless initiation of criminal proceedings against another.
3. **Battery:** Intentionally harmful or offensive touching of another.
4. **Assault:** Intentional causing of fear of harmful or offensive touching.
5. **Trespass to land:** Unauthorized entering upon another person's property.
6. **Trespass to personal property:** Taking or damaging another person's possessions.
7. **Infliction of emotional distress:** Intentionally causing emotional or mental distress in another.

8. **Defamation (libel and slander):** Injury to the reputation of another by publicly making untrue statements. *Libel* refers to the written word; *slander,* to the spoken word.
9. **Invasion of privacy:** Intruding on another's physical solitude, the disclosure of private information about another, public misrepresentation of another's actions.
10. **Negligence:** Causing injury to persons or property by failing to use reasonable care or by taking unreasonable risk.

Civil action is only one factor hindering abuses by the private sector. State laws regulate the private security industry. A local government, especially in a large city, may also regulate the industry. This usually pertains to licensing and registration requirements. Improper or illegal action is likely to result in suspension or revocation of a license. Criminal law presents a further deterrent against criminal action by private sector personnel. Examples include laws prohibiting impersonation of a public official, using electronic surveillance, breaking and entering, and assault. *Conduct can result in both a crime and a tort: in other words, a security practitioner can be arrested and sued.*

Some court cases have applied select constitutional limitations, from the Bill of Rights, to private sector action, especially if private security personnel are working with public police or if off-duty police are working part-time to assist private sector security efforts. Public police and private security are also subject to federal and state laws that impose civil liability. For example, 42 USC Section 1983 is a federal law that imposes civil liability for intentional violations of constitutional, civil, or statutory rights of individuals by persons "acting under color of state law" (i.e., the misuse of power by a person possessing government authority). This action is more likely against public police than private security. Nemeth (2005: 159) asks whether a private security officer who is granted authority by regulatory bodies and licensure agencies, is empowered by legislation, and who detains a suspect is "acting under color of state law". His answer to this question is that some claimants have persuaded courts that private security action falls "under color of state law," but such decisions are rare, and he notes that it is difficult to define Section 1983 as a liability of private security officers.

Union contracts also can limit private security. These contracts might stipulate, for instance, that employee lockers cannot be searched and that certain investigative guidelines must be followed.

Match the Tort

You are a security manager. Identify the tort(s) related to each scenario.

- A retail customer was frightened by one of your security officers who hurt the customer's arm and, in front of other customers, loudly accused the customer of shoplifting.
- You learn that an employee of a company department that is missing cash has been locked in an office by the department manager and ordered not to leave.
- An administrative assistant in the human resources department told people in the community that a specific employee is going to be fired for sexual harassment, even though you are only investigating allegations of such behavior.

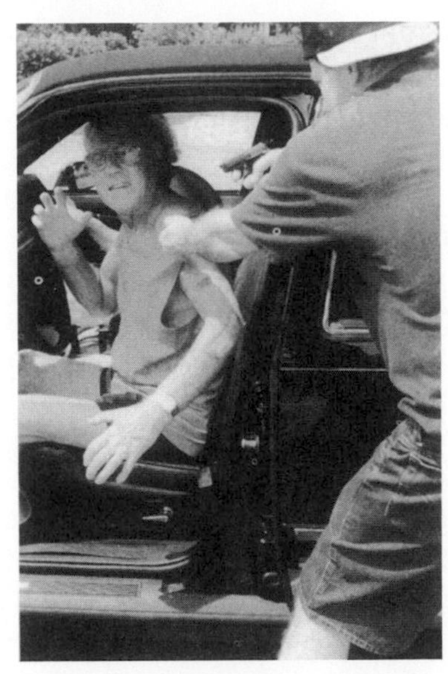

FIGURE 4-1 Assaults in parking lots have led to lawsuits alleging negligent security. *Source: Philip P. Purpura,* Criminal Justice: An Introduction *(Boston: Butterworth-Heinemann, 1997), p. 4.*

Legal Theory of Premises Security Claims

Negligence results when failure to exercise a reasonable amount of care in a situation causes harm to another (Figure 4-1). For instance, management should take steps to ensure the safety of individuals on the premises. Numerous premises security claims have been directed at management for failing to provide adequate protection for employees, customers, residents, or students who were injured by a third-party criminal act (e.g., sexual assault, robbery).

The legal theory of premises security claims is presented here. States allow monetary damages to a plaintiff who is injured as a proximate cause of the defendant's (e.g., management's) breach of a legal duty to provide protection. The definitions of *legal duty, proximate cause,* and the related concept of *foreseeability* distinguish the law among the states. **Legal duty** refers to management's duty to maintain the premises in a reasonably safe condition for invitees (e.g., customers on the property of a retail store). **Proximate cause** means that the breach of the legal duty is the actual cause of the harm. **Foreseeability** refers to whether the harm was likely to occur based on the safety history (e.g., crime metrics) of the premises and nearby property. If harm was likely, there is a duty to protect invitees. Nemeth (2005: 136–137) writes:

> *The whole theory of negligence operates from the measure of the average man or woman—the 'reasonable person' standard. What should we expect from the average person in his or her dealings with others?. In the law of negligence, the unreasonable person is needlessly careless and even reckless and fails to take those precautions necessary to prevent injury to others.*

To succeed in a negligence case, the plaintiff must prove that the defendant (e.g., a business) owed a duty to protect the victim; the defendant breached the duty; and the breach of the duty was the proximate cause of the crime and harm to the victim. To defend the case, the defendant seeks to show that a duty was not owed to the plaintiff; that there was no breach of duty; or by refuting the argument that any breach of a duty was a proximate cause of the victimization (Kaminsky, 1995: 7–11).

The legal theory of premises security claims has evolved over the years. The **prior similar incidents rule**, for instance, means that a plaintiff must demonstrate that there had been a prior similar incident on the premises. However, courts became critical of this rule in the 1990s because it had the effect of being a "one free rape rule." At that time, the majority of jurisdictions shifted to a **totality of the circumstances test**: prior similar incidents and other factors, such as the nature of the business, its surrounding locale, security training, and whether customary security measures for the particular industry were applied (Gordon and Brill, 1996: 2–3). A third test that courts place on security-related cases is **conscious disregard**: the determination of whether or not management or the security program knew of a problem or vulnerability but did nothing to correct it.

A lawsuit means that security faces a very big test. Management either pays for adequate protection initially to prevent a serious incident or possibly pays later, following a lawsuit. Most of these cases never make it to the trial stage and are settled out of court, which complicates research. In addition to providing basic protection for people and assets, a good security and loss prevention program becomes an investment in litigation prevention, even though premises security claims continue within our society. Apartment and condominium complexes and parking lots at a variety of locations are particularly prone to claims of inadequate security. However, many other types of sites are subject to lawsuits, such as shopping malls, retail stores, educational institutions, hotels and motels, restaurants and bars, office buildings, and healthcare facilities.

Negligence involves many types of situations pertaining to protection programs. Security personnel have been held liable for negligent use of force and firearms. Managers and supervisors have been found negligent in the areas of applicant screening and the training and supervision of employees. Such negligence is not restricted to the security industry. Any organization and employee can be subject to negligence charges. A hospital, for example, may find itself in a lawsuit after an orderly sexually assaults a patient. The human resources department may be found liable because it did not check on the orderly's background.

Do businesses and institutions in your locale provide reasonable protection for people on the premises? Should premises security lawsuits be permitted? Why or why not?

You Be the Judge*

On August 1, Ken Yates and his family arrived at their motel about 6:30 P.M. While he was unloading the car, Yates was approached, threatened at gunpoint, and told to hand over his money. He resisted and was shot in the ensuing scuffle. Even though there had been a security officer on duty, no one intervened to help Yates.

Yates sued the motel, claiming that his injuries were caused by management's failure to provide an adequate number of trained security officers. Yates's expert witness supported this claim. However, he conceded that, even if the motel had two officers on duty, the incident might have occurred outside the officers' direct observation. In addition, under cross-examination, the expert witness conceded that he had never developed or studied a security plan for a hotel or motel whose security needs required two security officers, so his expertise in the matter was limited.

The trial court dismissed the case, saying there was no evidence that having two officers on duty would have prevented the assault. Yates appealed, insisting that the inn could have done more to ensure his safety.

How did the appeals court rule?

Make your decision; then turn to the end of the chapter for the court's decision.

---

*Reprinted with permission from *Security Watch* (January 15, 2000), Aspen Pub.

Social Network Tools: We Learn from the Mistakes of Others

Social network tools such as Facebook and Twitter can assist public police and security practitioners by alerting others about dangers, helping with background investigations of employment applicants, and securing information and evidence in criminal investigations. At the same time, when these tools are used inappropriately, enormous problems can result. People can tarnish their reputations and, if they are employed, their employers' reputations. Employers are developing policies to reduce this problem. In addition, attorneys and their support staff research these sites to gather information and evidence helpful to their side of a case, with the possibility that adverse information and photos will be introduced in the courtroom. Here, we learn from the mistakes of others.

Goode (2011) reported the following events:

- A police officer listed his occupation on Facebook as "human waste disposal." After the officer was involved in a fatal on-duty shooting, a local reporter discovered the comment on the officer's page.
- An officer involved in a high-speed chase posted that it ended up in "a little bit of a tussle" and that he had a good time during the "dust-up."
- In an Arkansas case, a federal appeals court cited as evidence of a police officer's character photos he posted on MySpace showing him pointing a gun at the camera, flanked by a skull and the legend, "the Punisher."
- During one police department's check of an applicant's social networking pages, a candidate posted the following: "Just returned from the interview with the Southfield Police Department and I can't wait to get a gun and kick some ass." The candidate was rejected.

Business Justification versus Employees' Expectations of Privacy

Technical advances have the potential to invade privacy, and there is the potential for abuse. The issues include employer monitoring of computer, Internet, e-mail, and telephone use by employees and the application of closed-circuit television (CCTV) and Global Positioning System (GPS) technology. Courts attempt to balance business justification with employees' expectations of privacy.

Employers should communicate with employees about monitoring and why it is being used (e.g., as a supervisory tool). Written policies should clearly state guidelines and the consequences of violations regarding employee use of the Internet. Employers may choose to state that employees should assume there is no privacy, that they should reserve personal Internet access and e-mail for their home computers, and that the employer reserves the right to access employee e-mail.

Telephone monitoring is another thorny issue. Businesses are capable of generating reports on the number of calls made by telemarketers, and to improve quality, supervisors may listen to telephone conversations. Legal risks may surface. Essentially, an employer may listen to a telephone conversation when one party consents. It is best for an employer to notify employees of telephone monitoring and obtain their consent in writing. If a conversation becomes personal during monitoring, the listening must stop. Since many states require "two-party" consent to monitor, businesses may provide customers with the following message: "This call may be monitored for quality purposes."

CCTV presents additional concerns. It is a widely used method of preventing crime and conducting investigations. Cameras may be placed overtly or covertly to observe people in public areas. Employers must avoid placing cameras in restrooms, dressing rooms, locker rooms, and areas containing an expectation of privacy.

Employers are increasingly installing GPS technology in company vehicles and other items to track movements to improve efficiency and protection (McShane, 2005). GPS technology consists of satellites and receivers that allow people and devices to pinpoint a vehicle's precise location on the earth. The application of this technology is broad, and as the cost drops, the number of uses is increasing. Portable GPS devices are used by people to help with navigation (e.g., new cars are being equipped with GPS systems). The military uses GPS to guide cruise missiles to prespecified targets.

Employers will strengthen their positions when they inform employees of the types of monitoring conducted in the workplace, post signs, and ask employees to sign consent forms containing policies. The enforcement of policies must be done in a uniform manner to avoid charges of unfair employment practices.

Zachary (2006) writes that employee use of e-mail and the Internet can expose employers to potential liability. Examples include confidential information inadvertently misdirected, libelous communications, and workplace jokes that were later used against an employer in hostile environment sexual harassment cases. Zachary notes that employers have a legally protected interest in controlling employee computer usage. She emphasizes the importance of policies on employee computer use and refers to the following New Jersey appellate court decision to support her contention. The following case summary is from Zachary. In *Jane Doe, individually and as guardian ad litem for Jill Doe, a minor, v. XYC Corp.*, 887 A.2d 1156 (Super. Ct. N.J. 2005), the court held that an employer had a legal duty to report an employee's viewing of child pornography to authorities and take effective action within the company to stop the activities. The employee was an accountant for the defendant company and worked in a cubicle that opened onto a hallway. People walking past the cubicle could see the employee's computer screen. An IT manager reviewing computer log reports learned that the employee was visiting pornographic sites. Two IT managers told the employee to stop, but they did not inform their superiors. About a year later, the employee's immediate supervisor told the senior network administrator about the recurring problem, and log reports again showed that the employee was visiting pornographic sites. A senior IT manager informed the senior network administrator that he was not to access any employee logs because it was against company policy. The IT manager was wrong. The company actually had a policy providing it with the right to review, audit, access, and disclose all e-mails over the company system. Also, employees were to visit websites of a business nature only. Violators were to be reported to the human resources department for possible disciplinary action. The policy was to be signed by each employee.

In October 2000, the employee married the plaintiff, who had a 10-year-old daughter. A few months later, a coworker of the employee complained to management about the employee visiting pornographic sites at work, but no action was taken. At home, the employee began to secretly videotape and photograph the stepdaughter in seminude positions. At work, the problem continued, the employee was confronted, and the employee agreed to stop viewing pornography on the company's computer. Later, the employee transmitted three photos of the stepdaughter over the Internet from his workplace. When photos of the stepdaughter were found in the company dumpster, a search warrant was executed at the employee's workspace. E-mails showed contact with pornographic sites and others with a similar interest. The employee was terminated and later arrested. Thereafter, the employee's wife filed a suit for negligence against the employer for permitting the employee to use, view, and download child pornography at work without reporting him.

The trial court ruled in favor of the company, noting that the company did not have a duty to investigate the employee's private communications and did not have control over the employee's conduct at home. The appellate court disagreed. It found the following: there was no expectation of privacy on the part of the employee; the company was on notice of the activities (i.e., internal reports from coworkers); and it had a duty to investigate, report the activities to the authorities, and stop the activities. Under tort law, a company has a duty to control an employee acting outside the scope of employment in order to prevent that employee from intentionally harming others. The case was remanded to the lower court to answer the question of whether the employer's breach of duty was the proximate cause of harm to the stepdaughter.

What is your opinion of privacy issues, such as employer monitoring in the workplace?

Contract Law

A **contract** is an agreement between parties to do or to abstain from doing some act. The law may enforce the agreement by requiring that a party perform its obligation or pay money equivalent to the performance. These court requirements are known as **remedies for breach of contract**. Specific circumstances may create defenses for failure to perform contract stipulations. Contracts may be express or implied. In an express contract, written or oral, the terms are stated in words. An implied contract is presumed by law to have been made from the circumstances and relations of the parties involved.

Several areas in the security and loss prevention field are relevant to the law of contracts. A company that provides a service or system to a client company may be liable for breach of contract following a dispute. A client company may also be subject to a lawsuit for breach of contract. A contract usually states liabilities for each party. For instance, if a third party is harmed (e.g., a person is illegally arrested on the premises by a private security officer from a contract service), the contract will commonly establish who is responsible and who is to have insurance for each risk. However, courts have held a specific party liable in third-party suits even though the contract stipulated that another party was to be responsible in the matter. This principle is known as **nondelegable duty**.

In the common law principle of **respondeat superior** (i.e., let the master respond), an employer (master) is liable for injuries caused by an employee (servant). This is also called

vicarious liability. Typically, the injured party will look beyond the employee, to the employer, for compensation for damages. Proper supervision and training can prevent litigation. Businesses typically pay for liability insurance to cover this risk.

Civil Justice Procedures

Civil procedures require considerable time, and the process can be expensive. Over 95 percent of civil (and criminal) cases are settled before reaching the trial stage. Many of the strategies and procedures used in civil trials are also used in criminal trials, although the rules vary. These procedures include discovery, motions, jury selection, questioning of witnesses in court by opposing attorneys, judge's instructions to a jury, verdict, posttrial motions and many more.

The following list describes civil procedures both before and during a trial (Purpura, 2003: 162–165). The descriptions are generalizations.

1. **Plaintiff** initiates a lawsuit resulting from the commission of a tort, breach of contract, or other event.
2. **Defendant** prepares a defense after receiving a summons (i.e., order to appear in court) and complaint (i.e., a pleading of facts and claims filed in court). Failure to appear in court and defend a complaint can result in losing the lawsuit.
3. Opposing attorneys confer and communicate issues, evidence, and settlement options. This may take place any time throughout the proceedings.
4. **Discovery** involves opposing attorneys obtaining all factual information (e.g., documents and evidence) in possession of the other. This stage helps to narrow the issues and saves time in dispensing justice.
5. Motions filed in court. A **motion** is a request to the judge for a decision on an issue or to take action on an issue—for example, a motion may be filed to dismiss the case.
6. Answer is the response to the lawsuit. The defendant files an answer with the court and includes denials and counterclaims.
7. **Pretrial conference** is the stage in which opposing attorneys and the judge meet to work toward a settlement or to face trial.
8. If the case proceeds to trial, opposing attorneys question potential jurors and a jury is selected.
9. Opening statements are comments from opposing attorneys who explain facts to the judge and jury once the trial begins.
10. Presentation of the plaintiff's case is done by the plaintiff's attorney. The plaintiff's attorney presents evidence, such as witnesses and documents, to support allegations in the complaint.
11. Defendant's motion for dismissal is the stage in which the defendant's attorney moves for a dismissal if he or she believes that the plaintiff's case failed to prove the allegations. If the judge denies the motion for dismissal, the trial continues.
12. Presentation of the defendant's case occurs when the defendant's attorney presents evidence to disprove (rebut) the plaintiff's case.

13. Plaintiff's rebuttal occurs when the plaintiff's attorney attempts to disprove the presentation by the defendant's attorney (from the preceding stage).
14. Defendant's rebuttal occurs when the defendant's attorney attempts to disprove newer issues argued by the plaintiff's attorney (from the preceding stage).
15. Motion for directed verdict means that either or both parties move for a directed verdict. In other words, the judge takes the decision away from the jury and informs them of what to decide. This situation results from a failure of evidence, overwhelming evidence, or the law that applies to the facts favors one of the parties. If the motion is denied, the trial continues.
16. Closing arguments are summaries of the evidence presented to the jury by the plaintiff's attorney and the defendant's attorney.
17. Judge's instructions are given to the jury.
18. Jury's **verdict** occurs when the jury makes a decision.
19. Posttrial motions occur after the trial, for example, the motion for a new trial.
20. Judgment occurs when the judge declares which party prevailed at the trial (i.e., which party won the lawsuit) and the amount of recovery to be awarded.

Administrative Law

Administrative law is designed to ensure that as businesses seek profit, fairness and safety are maintained. Many federal and state agencies and executive departments are involved in administrative law and influence organizational loss prevention policies and programs. On the federal level, these include the Occupational Safety and Health Administration (OSHA), the National Labor Relations Board (NLRB), and the Equal Employment Opportunity Commission (EEOC), among others. (OSHA, NLRB, and EEOC are discussed in various sections of this book.) Likewise, on the state level, similar bodies exist to regulate various activities such as workplace safety and the security industry. Administrative agencies are formed because legislative and executive branches of government typically lack the expertise to regulate specialized areas. Independent agencies are formed that are less susceptible to direct political influence. **Administrative agencies** are government bodies that regulate various activities, make rules, conduct investigations, perform law-enforcement functions, issue penalties, and initiate criminal and civil litigation. Federal agencies document rules in the *Federal Register*, published by the General Services Administration. State agency manuals serve a similar function. Local governments follow generally accepted fire and building codes.

Inspections and Searches

Federal, state, and local administrative agency personnel visit business sites to conduct inspections and tests. An OSHA inspector may visit a workplace to ensure that safety regulations to protect employees are in force. An Environmental Protection Agency (EPA) inspector may test ground water near a facility to check on the quality of the water. Local health departments check restaurants for compliance with health codes. Inspections may cause issues related to invasion of privacy, even though most administrative inspections do not focus on crimes, but rather administrative violations.

Inspections and tests may fall under the 4th Amendment, which prohibits government from conducting unreasonable searches and seizures and requires warrants to be issued based on probable cause. However, courts recognize that **administrative inspections** differ from searches related to criminal cases because of the following rationale: administrative inspections are not usually conducted to gather evidence for prosecution, but if prosecution is sought, the 4th Amendment applies; because society is so complex, the 4th Amendment must be relaxed to facilitate effective monitoring of compliance with administrative law; and the expectation of privacy at a business is lower than at a home. Furthermore, administrative inspections differ from **administrative searches**. The latter require 4th Amendment protections. To illustrate, an inspector may inspect the public area of a restaurant (e.g., dining room), but a demand to visit the owner's office or closed kitchen falls under the category of a search because there is a greater expectation of privacy on those premises. One type of exception to the 4th Amendment requirement occurs when consent to search is provided by the owner. The legal standard for obtaining an administrative search warrant, also known as an *inspection warrant*, is lower than what is required to obtain a search warrant in a criminal case (*Camera v. Municipal Court* 387 US 523, 1967). Probable cause may not be required, as long as reasonable inspection standards were established by the legislature (Hall, 2006: 146–151).

Impact of Legislation on Private Security Operations

ASIS International–funded research showed that legislation directly impacts not only organizations in general, but also private security operations (Collins et al., 2005). The top three acts cited in this research as having the most impact on security policies and procedures were the Health Insurance Portability and Accountability Act of 1996 (HIPAA), the Sarbanes-Oxley (SOX) Act of 2002, and the USA Patriot Act of 2001. The Patriot Act and the USA Patriot Improvement and Reauthorization Act of 2005 are covered later in this book. HIPAA, SOX, and the Financial Modernization Act of 1999 are explained here.

Health Insurance Portability and Accountability Act of 1996 (HIPAA): This law is designed to improve healthcare services delivery, lower costs by reducing paper records and claims, enhance electronic transmission of documents, secure medical data and patient information, prevent errors in the healthcare system, and securely transfer funds. IT systems play a major role in fostering these activities. The act establishes compliance rules to reach the objectives of this legislation. The Department of Health and Human Services is the regulatory agency for this act, and it issues various standards, such as those for information security.

HIPAA has gone through changes from new legislation. For example, the Health Information Technology for Economic and Clinical Health Act of 2009 (HITECH) has made changes to the HIPAA Security Rule in which HIPAA-covered entities must provide notification to patients in the event of a breach of unsecured health information(e.g., from a hacker). HITECH provides an exception to the notification requirement if encryption technology is applied (i.e., data is scrambled and unscrambled to maintain privacy), even if a breach occurs with encryption (American Medical Association, 2010).

Sarbanes-Oxley (SOX) Act of 2002: SOX was enacted following several high-profile accounting scandals. The Securities and Exchange Commission, a federal government administrative agency, administers the act. The legislation seeks to prevent fraud and impacts the processes and accountability for financial reporting in publicly traded U.S. companies. SOX holds executives responsible for establishing, evaluating, and monitoring the effectiveness of internal controls over financial and operational processes. Senior executives must sign an attestation that they are responsible and that the internal controls meet the requirements of SOX. Some SOX compliance provisions are becoming standard operating procedure for not only publicly held businesses, but all businesses. Specifics of the law include using accounting controls, identifying how companies report financial results and disclose executive compensation, holding company executives and external auditors directly accountable for the accuracy of financial reports, and protecting employees who blow the whistle on suspected fraud. Corporate security and audit departments have become involved with SOX because these departments investigate internal fraud.

SOX is one of several laws that seek to govern the securities industry (i.e., stock, bonds, and other securities). Such laws date back to the Securities Act of 1933. This act prohibited deceit in the sale of securities (U.S. Securities and Exchange Commission, 2010). "Creative accounting" is used by offenders to make a company appear financial healthy when it is not, in order to attract investors. Fraud is a recurring problem that government tries to curb. Chapter 11 elaborates on SOX.

Financial Modernization Act of 1999: This act, also known as the *Gramm-Leach-Bliley Act,* or GLBA, includes provisions to ensure security and confidentiality of customer financial information held by financial institutions. The Federal Trade Commission is the primary administrator of GLBA. The act requires an assessment of internal and external risks and the sufficiency of existing security. The GLBA applies to "financial institutions," which include banks, securities firms, insurance companies, and many other businesses that provide financial products and services to consumers. These include lending, servicing consumer loans, transferring or safeguarding money, reporting credit, preparing individual tax returns, providing financial advice or credit counseling, providing residential real estate settlement services, collecting consumer debts, and many other activities. IT security practitioners in particular are involved in the compliance requirements of GLBA.

A broader perspective on the GLBA comes from critics who view this law as a major cause of the financial crisis of 2007–2009 because it removed legal barriers between commercial banking and investment banking in the United States. White (2010: 942) writes: "After the GLBA's enactment, commercial banks could transform their bank holding companies into financial holding companies, which could engage in investment banking activities through non-bank subsidiaries." Supporters and critics of the GLBA continue to debate its impact on the economy.

Labor Law

Historically, workers have sought improved working conditions, higher wages, and better benefits from their employers. At the same time, management seeks to control costs while increasing profits. The National Labor Relations Board (NLRB), an independent agency of the U.S. government, controls relations between management and labor. It is essential that management and loss prevention practitioners be familiar with labor laws to avert charges of unfair labor practices. Surveillance and investigation of union activities are violations of the **National Labor Relations Act** (NLRA), which makes it an unfair labor practice to interfere with, restrain, or coerce employees in the exercise of their rights to self-organize, to assist labor organizations, and to bargain collectively through representatives. One type of surveillance, photographing activities of striking workers, is unlawful unless there is a legitimate purpose, such as gathering evidence for the prosecution of criminal acts (e.g., assault or destruction of property). In one case, a security service company took 60,000 pictures and collected thousands of hours of video that were used in court against strikers who committed violations of law and ended up owing $64 million in fines (Sunoo, 1995: 58).

Care must be exercised when instituting loss prevention strategies during disputes between management and labor. Precautions to protect company property (e.g., additional officers, CCTV) may be construed as interference with union activities. For example, courts have declared illegal the observance by officers of who was going in and out of union meetings. Also, it has been held that even "creating the impression" of surveillance (e.g., management implying that surveillance is taking place) is illegal. The NLRB found that aiming CCTV on a company building in which a union meeting was held created the impression of surveillance. Undercover investigations that conduct labor surveillance are also illegal.

The NLRB periodically changes its position on recurring and important issues as illustrated by Bloom and Bryant (2005: 84–92) in reference to **Weingarten rights**. Weingarten rights originated in 1975, when the U.S. Supreme Court upheld a decision by the NLRB (*NLRB v. J. Weingarten Inc.*, No. 251, 1975). It stated that unionized employees have a legal right to insist on union representation during an investigatory interview conducted by their employer, if the employees reasonably believe that the interviews might result in disciplinary action. The U.S. Supreme Court explained that this right arises from federal labor laws that are a "guarantee of the right of employees to act in concert for mutual aid and protection." The right is limited to situations when employees specifically request representation. An employer is not required to state this right to employees. *Since this decision, the NLRB has frequently changed its position on whether Weingarten rights apply to employees who are not unionized.*

Criminal Justice Procedures

The following list describes procedural law for criminal cases. Because jurisdictional procedures vary, a generalization is presented.

1. The purpose of an **arrest** is to bring a person into the criminal justice system so that he or she may be held to answer criminal charges.
2. For less serious offenses a citation is often used instead of a formal arrest by public police (e.g., traffic violations). If the conditions set forth in the citation are not followed, a magistrate of the appropriate court will issue a misdemeanor arrest warrant.
3. All arrests must be based on probable cause, which is stated in arrest warrants. **Probable cause** is defined as reasonable grounds to justify legal action, rather than mere suspicion. Police officers, security officers, and able witnesses and victims typically provide the foundation of probable cause through their observations of offenders.
4. **Booking** takes place when an arrestee is taken to a police department or jail so that a record can be made of information such as the arrestee's name, the charge, and the arresting officer's name. Fingerprinting and photographing the arrestee are part of the booking process.

5. Because our system of justice has a high regard for civil liberties as expressed in the Bill of Rights, the accused is informed of his or her *Miranda rights*, by public police, prior to questioning and when the subject is in custody.
6. After booking, and without unnecessary delay, the accused is taken before a magistrate for an **initial appearance**. At this appearance, the magistrate has the responsibility of informing the accused of his or her constitutional rights, stating the charge, and setting bail (if necessary).
7. The arresting officer will then meet with the prosecutor, or a representative, to review evidence. A decision is made whether to continue legal action or to drop the case. A case may be dropped by the prosecutor for insufficient evidence or because the defendant is suffering from a problem better handled by a social service agency.
8. The prosecutor prepares an "information" when prosecution is initiated. It cites the defendant's name and the charge and is signed by the complainant. A judicial officer then prepares an arrest warrant. The defendant may already be in custody at this point.
9. At the initial appearance, the magistrate will inform the defendant about the right to have a **preliminary hearing**. The defendant and the defense attorney make the decision of whether or not to request one. This hearing is used to determine if probable cause exists for a trial. The courtroom participants in a preliminary hearing are judge, defendant, defense attorney, and prosecutor. The prosecutor has the "burden of proof." Witnesses may be called by the prosecutor to testify.
10. Federal law and the laws of more than half the states require that probable cause to hold a person for trial must result from grand jury action. The 5th Amendment of the Bill of Rights states such a requirement. When probable cause is established in an action ordered by the judge or prosecutor, the grand jury will return an "indictment" or "true bill" against the accused. A "presentment" results from an investigation initiated by the grand jury establishing probable cause. Based on the indictment or presentment, an arrest warrant is issued.
11. At an **arraignment** the accused enters a plea to the charges. The four plea options are guilty, not guilty, nolo contendere (no contest), and not guilty by reason of insanity.
12. Few defendants reach the trial stage. **Plea-bargaining** is an indispensable method to clear crowded court dockets. Essentially, it means that the prosecutor and defense attorney work out an agreement whereby the prosecutor reduces the charge in exchange for a guilty plea.
13. Pretrial motions can be entered by the defense attorney prior to entering a plea at arraignment. Some examples are a "motion to quash" an indictment or information because the grand jury was improperly selected; a "continuance" requested by the defense attorney because more time is needed to prepare the case; or a "change of venue" requested when pretrial publicity is harmful to the defendant's case and the defense hopes to locate the trial in another jurisdiction so that an impartial jury is more likely to be selected.
14. The accused is tried by the court (judge) or a jury. The system of justice is adversarial, involving opponents. This is apparent in a **trial**, where the prosecutor and defense

attorney make brief opening statements to the jury. The prosecutor then presents evidence. Witnesses are called to the stand to testify; they go through direct examination (i.e., questioning) by the prosecutor, followed by defense cross-examination. The prosecutor attempts to show the defendant's guilt "beyond a reasonable doubt." The defense attorney strives to discredit the evidence through cross-examination. Redirect examination rebuilds evidence discredited by cross-examination. Recross-examination may follow. After the prosecutor presents all the evidence, the defense attorney may move for acquittal. The judge commonly overrules this motion. The defense attorney then presents evidence. Defense evidence undergoes direct and redirect examination by the defense and cross- and recross-examination by the prosecutor.

15. The judge will then "charge the jury," which means that the judge briefs the jury on the charge and how a verdict is to be reached based on the evidence. In certain states, juries have a responsibility for recommending a sentence after a guilty verdict; the judge will brief the jury on this issue. Closing arguments are then presented by opposing attorneys.
16. The jury retires to the deliberation room; a verdict follows. A not guilty verdict signifies release for the defendant. A guilty verdict leads to sentencing. Motions and appeals may be initiated after the sentence.

How Much Power is in the Hands of Private Security Officers? Can They:

- make an arrest without probable cause?
- make an arrest for a felony?
- make an arrest for a misdemeanor?
- use force to complete an arrest?
- search an arrestee following an arrest?
- question a suspect or arrestee without public police being present?
- be barred from using seized evidence as the police are under the exclusionary rule?

These questions are answered in the following paragraphs.

Arrest Law

Because our justice system places a high value on the rights of the individual citizen, private security and public police personnel cannot simply arrest, search, question, and confine a person without cause. The Bill of Rights of the U.S. Constitution affords citizens numerous protections against government. The 4th and 5th Amendments of the Bill of Rights demonstrate how individual rights are safeguarded during criminal investigations.

> *Amendment IV—The right of the people to be secure in their persons, houses, papers, and effects, against unreasonable searches and seizures, shall not be violated, and no Warrants shall issue, but upon probable cause, supported by Oath or affirmation; and particularly describing the place to be searched, and the person or things to be seized.*

Amendment V—[N]or shall [any person] be compelled in any criminal case to be a witness against himself, nor be deprived of life, liberty, or property, without due process of law.

The 6th, 8th, and 14th Amendments are other important amendments frequently associated with our criminal justice process. Briefly, the 6th Amendment pertains to the right to trial by jury and assistance of counsel. The 8th Amendment states, "Excessive bail shall not be required, nor excessive fines imposed, nor cruel and unusual punishments inflicted." The 14th Amendment bars states from depriving any person of due process of law or equal protection of the laws.

The 4th Amendment stipulates guidelines for the issuance of warrants. Public police obtain arrest and search warrants from an impartial judicial officer. Sometimes immediate action (e.g., chasing a bank robber) does not permit time to obtain warrants before arrest and search. In such a case, an arrest warrant is obtained as soon as possible. Private security should contact public police for assistance in securing warrants and in apprehending suspects.

Knowledge of arrest powers is essential for those likely to exercise this authority. These powers differ from state to state and depend on the statutory authority of the type of individual involved. Generally, public police officers have the greatest arrest powers. They are protected from civil liability for false arrest, as long as they have probable cause that a crime was committed. Those in the private sector usually have arrest powers equal to citizen's arrest powers, which means that they are liable for false arrest if a crime was not, in fact, committed—regardless of the reasonableness of their belief. An exception is apparent if state statutes point out that security personnel have arrest powers equal to public police only on the protected property. If private sector personnel are deputized or given a special constabulary commission, their arrest powers are likely to equal those of public police.

Whoever makes an arrest must have the legal authority to do so. Furthermore, for those making arrests, the distinction between felonies and misdemeanors is of tremendous importance. Generally, public police can arrest an individual for a felony or a misdemeanor committed in his or her view; the viewing amounts to probable cause. Arrest for a felony not seen by public police is lawful with probable cause (e.g., an arrest warrant exists on an individual or a radio broadcast describes a robber). Arrest for a misdemeanor not seen by public police generally is unlawful; a warrant is needed based on probable cause. There are exceptions to this misdemeanor rule; for example, public police can arrest in domestic violence cases or cases of driving under the influence, when the offense is fresh although not observed by police. On the other hand, private security personnel have fewer powers of arrest (equal to citizen's arrest powers). Citizen arrest powers permit felony arrests based on probable cause but prohibit misdemeanor arrests.

A serious situation exists when, for example, a private security officer arrests and charges a person for a felony when in fact the offense was a misdemeanor and the jurisdiction does not grant security officers such misdemeanor arrest power. Many employers in the private sector are so fearful of an illegal arrest and subsequent legal action that they prohibit their officers from making arrests without supervisory approval. It is imperative that private sector personnel know state arrest law; proper training is a necessity.

Force

During the exercise of arrest powers, force may be necessary. The key criterion is **reasonable force**. This means force should be no more than what is reasonably necessary to carry out legitimate authority. If an arrestee struggles to escape but is subdued to the ground and stops resisting, it would be unreasonable for the arrestor to strike the arrestee. Once an arrestee stops resisting, force must stop. **Deadly force** is reserved for life-threatening situations (i.e., to protect a person's life), never to defend property. Unreasonable force can lead to difficulties in prosecuting a case, as well as civil and criminal litigation.

Searches

Typically, public police conduct a search of an arrestee right after an arrest. This has been consistently upheld by courts for the protection of the officer who may be harmed by a concealed weapon. However, *evidence obtained through an unreasonable search and seizure is not admissible in court;* this is known as the **exclusionary rule**.

The 4th Amendment prohibition against unreasonable searches and seizures applies only to government action. Searches by private citizens, including security officers, even if "unreasonable," are therefore not "unconstitutional" and the exclusionary rule does not apply, as ruled in *Burdeau v. McDowell*, 256 US 465, 1921 (Inbau et al., 1996: 54; Nemeth, 2005: 83–104). At the same time, the law of searches by private security officers is not clear and varies widely. Even though private security may not be restrained by the 4th Amendment, a lawsuit may result following a search. A search is valid when consent is given or where, in a retail environment, a state shoplifting statute permits the retrieval of merchandise. A search for weapons following an arrest may be justified through common law, which states that citizens have the right of self-defense. The recovery of stolen goods as the basis for a search is typically forbidden, except in some state shoplifting statutes. Whenever possible, private security personnel should let public police conduct searches in order to transfer potential liability.

Call It "Inspection," Not "Search and Seizure"

Norman M. Spain (1996: 4–7), an authority on legal issues in security, states that private security officers are generally not bound by constitutional constraints of search and seizure as are public police, unless they are "tainted by the color of law"—that is, jointly working with public police. Spain favors the term *inspection* instead of *search* for private security, because the 4th Amendment does not apply in most private settings. He cites various targets for inspections in private settings: a locker, a vehicle entering or leaving a facility, or an employee's belongings.

Spain recommends a formal inspection policy that is backed by common law—employers have the right to take reasonable measures to protect their property against theft. All parties (e.g., employees, contractors, and visitors) should be given notice through, for example, signs and publications. The policy should have four components:

1. A formal statement that the company reserves the right to inspect
2. Illustrations of types of inspections

3. A list of items that employees should not have in their possession (e.g., illegal drugs, weapons, company property removed without authorization)
4. A statement of penalties, including those for not cooperating

Spain cautions that a "pat-down" of a person's body or inspections of pockets may result in a civil action alleging invasion of privacy, unless the site requires intense security.

Questioning

An important clause of the 5th Amendment states that a person cannot be compelled in any criminal case to be a witness against him- or herself. Therefore, what constitutional protections does a suspect have on being approached by an investigator for questioning? Here again, the law differs with respect to public and private sector investigations.

A suspect who is in custody and is about to be questioned by public police about a crime must be advised of

1. The right to remain silent
2. The fact that statements can be used against the person in a court of law
3. The right to an attorney, even if the person has no money
4. The right to stop answering questions

These rights, known as the **Miranda warnings**, evolved out of the famous 1966 U.S. Supreme Court case known as *Miranda v. Arizona*. If these rights are not read to a person by public police before questioning, statements or a confession will not be admissible as evidence in court.

Are private security personnel required to read the Miranda warnings to a subject prior to questioning? The U.S. Supreme Court has not yet required the reading. Many private sector investigators read suspects the Miranda warnings anyway, to strengthen their cases. Any type of coercion or trick during questioning is prohibited by both private security and public police. A voluntary confession by a suspect is in the best interest of public and private investigators. If security personnel are working jointly with public police (as in a cooperative protection effort), or if a police officer is working part-time in security, then the warnings should be read.

Another topic concerning questioning of subjects pertains to disciplinary hearings. As written earlier, the NLRB extends to union employees the right of representation during an interview that might lead to disciplinary action.

Search the Internet

Here is a list of general-purpose websites pertaining to this chapter:

American Bar Association: www.americanbar.org
Cornell Law School: www.law.cornell.edu
Federal Register: www.gpoaccess.gov/fr/index.html
Lawyers and the Law: lawyers.com

Resource on law: www.lexisnexis.com/community/portal
Washburn University School of Law: www.washlaw.edu
U.S. Department of Justice: www.usdoj.gov/criminal/cybercrime/searching.html

Case Problems

4A. You are a loss prevention officer with three years of experience and a newly acquired college degree. The plant manager of loss prevention has selected you for a special assignment: design and conduct an eight-hour training program on law for company loss prevention (security) officers. Indications are that success on this project could lead to a promotion to training officer, a position vacant at this time. You are required to (1) formulate an outline of topics to be covered and the hours for each topic, (2) justify in one to three sentences why each topic is important, and (3) prepare an examination of 10 to 20 questions.

4B. You are a member of a jury on a civil case involving the question of negligent security. In this case, a young man was killed by a stray bullet at a restaurant parking lot during a weekend night as he approached a group of rowdy people about to fight. The restaurant has a history of three gun incidents prior to the murder: one month earlier, a shot was fired into the restaurant, without injury; three months earlier, police arrested a subject for assault in the parking lot and confiscated a pistol; and five months earlier, police found a revolver in the bushes in the parking lot following an arrest for public intoxication. Each incident occurred during a weekend night. Following these three incidents, the restaurant maintained the same level of security on the premises—good lighting and training employees to call police when a crime occurred. Does the business owe a duty to provide a safe environment to those who enter the premises? Did the restaurant breach that duty? If the duty was breached, was it the proximate cause of the criminal act and the victimization? Is the restaurant negligent? Why or why not?

4C. As a security manager at a chemical plant, you learn that a plant executive is receiving kickbacks from a shady toxic chemical disposal service company that contracts with the plant. Two months earlier, you heard the same information from another source, and you reported your investigative findings to your superior, who stated that he would handle the situation. You wonder why the same executive is in the same job and the same shady company is still a contractor for your employer. What do you do at this point and why?

4D. Research and prepare a report on the laws of arrest and search and seizure in your state as they apply to citizens, private security officers, private detectives, and public police.

4E. Create, in writing, situations in which each of the individuals in case problem 4D can make a legal arrest. Describe appropriate search and seizure guidelines for each situation.

The Decision for "You Be the Judge"

The appeals court said the trial court acted properly. Yates had failed to show that another security officer would have prevented the assault. [*Walsh v. Ramada Franchise Systems, Inc. et al.*, No. 98–5040, US Ct. of App. for the 6th Circ., 1999 US App. Lexis 11281.] Question: What methods do you use to arrive at staffing levels for your officers? How would you defend those methods in court?

The motel was lucky; Yates's expert failed to show that more security officers would have been more effective. If the expert had testified that similarly sized and located motels used more security officers, the motel would have been in the difficult position of defending its use of only one officer.

Benchmarking your practices against industry norms and against companies in your neighborhood is an excellent way to assess whether your company is doing enough to secure your grounds.

References

Administrative Office of the United States Courts. (2010). Judicial Business of the United States Courts. <www.uscourts.gov> retrieved April 20, 2011.

American Medical Association. (2010). HIPAA Security Rule: Frequently asked questions regarding encryption of personal health information. <www.ama-assn.org> retrieved April 26, 2011.

Bloom, H., & Bryant, M. (2005). Labor law's changing tides. *Security Management*, 49, August.

Collins, P., et al. (2005). *The ASIS foundation security report: Scope and emerging trends.* Alexandria, VA: ASIS, International.

Conference of State Court Administrators. (2010). Examining the Work of State Courts: An Analysis of 2008 State Court Caseloads. <www.courtstatistics.org> retrieved April 20, 2011.

Goode, E. (2011). Police lessons: Social network tools have two edges. *The New York Times* (April 6). <www.nytimes.com> retrieved April 7, 2011.

Gordon, C., & Brill, W. (1996). *The expanded role of crime prevention through environmental design in premises liability (April).* Washington, DC: National Institute of Justice.

Hall, D. (2006). *Administrative law: Bureaucracy in a democracy* (3rd ed.). Upper Saddle River, NJ: Pearson Education.

Inbau, F., Farber, B., & Arnold, D. (1996). *Protective security law* (2nd ed.). Boston: Butterworth-Heinemann.

Kaminsky, A. (1995). *A complete guide to premises security litigation.* Chicago, IL: American Bar Association.

McShane, L. (2005). GPS technology in workplaces raises privacy concerns. *Associated Press* (November 18).

Nemeth, C. (2005). *Private security and the law.* Burlington, MA: Elsevier Butterworth-Heinemann.

Purpura, P. (1997). *Criminal justice: An introduction.* Boston: Butterworth-Heinemann.

Purpura, P. (2003). *The security handbook* (2nd ed.). Boston: Butterworth-Heinemann.

Rassas, L. (2011). *Employment law: A guide to hiring, managing, and firing For employers and employees.* New York, NY: Aspen Pub.

Spain, N. (1996). Call It 'Inspection' (Not 'Search and Seizure'). *Security Management Bulletin* (May 10).

Sunoo, B. (1995). Managing strikes, minimizing loss. *Personnel Journal* (January).

Supreme Court of the United States. (2011). 2010 Year-End Report on the Federal Judiciary. <www.supreme court.gov> retrieved April 20, 2011.

U.S. Securities and Exchange Commission. (2010). The Laws That Govern the Securities Industry. <www.sec .gov> retrieved April 27, 2011.

White, L. (2010). The gramm-leach-bliley act of 1999: A bridge too far? Or not far enough? *Suffolk Law Review, 43.*

Wrennall, L. (2010). Confronting Judicial Misconduct. *Criminal Justice Matters: The Magazine of the Center for Crime and Justice Studies* (June). In, Roberson, C. and DiMarino, F. (2012). *American Criminal Courts.* Upper Saddle River, NJ: Prentice Hall.

Zachary, M. (2006). Labor law: New source of liability for employee internet use. *Supervision*, 67, June.

5 Internal and External Relations

OBJECTIVES

After studying this chapter, the reader will be able to:

1. Define internal and external relations
2. Explain methods of marketing security and loss prevention and the value of marketing
3. Discuss internal relations
4. Explain how an intranet, e-mail, and social media can assist security and loss prevention
5. Discuss external relations
6. Explain how to work with the media and the digital tools of today
7. Debate the issues of prosecution
8. Debate the issues of loss prevention attire

KEY TERMS

- internal relations
- external relations
- customer-driven
- marketing
- market segmentation
- target marketing
- return on investment
- intranet
- Facebook
- YouTube
- Twitter
- designated spokesperson
- prosecution threshold

Internal and External Relations

This chapter explains the importance and value of recruiting people and organizations to assist with security and loss prevention efforts. With practitioners being asked to handle increasingly complex problems, often with limited resources, it is vital that all possible sources of assistance be solicited. Strategies are delineated here for improving relations between the loss prevention department within an organization and those groups that loss prevention serves and works with in reducing losses. **Internal relations** refer to cooperative efforts with individuals and groups within an organization that a loss prevention department serves. **External relations** refer to cooperative efforts with external individuals and groups that assist in loss prevention objectives.

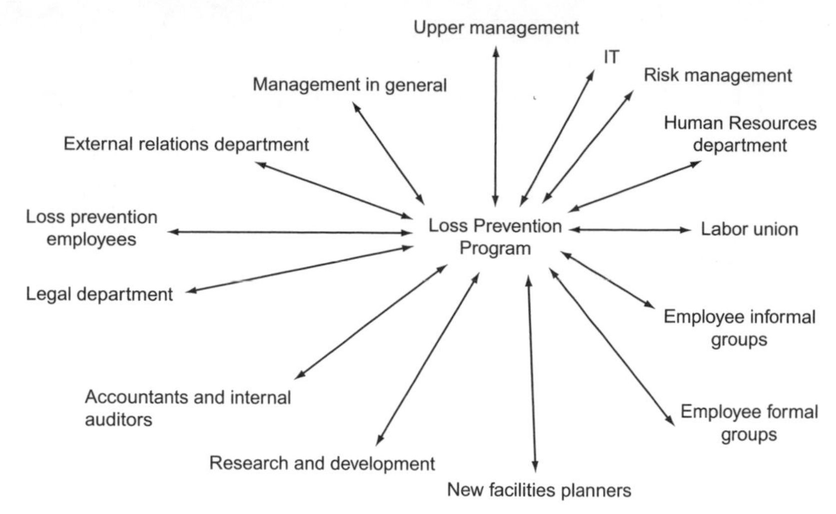

FIGURE 5-1 Internal relations.

In reference to examples in the internal relations chart (clockwise, Figure 5-1), upper management dictates loss prevention goals. Because of threats and hazards facing information technology (IT), IT and loss prevention departments are working closer together more frequently than before. A risk management department in an organization can provide the foundation for prioritizing risks and developing broad strategies to deal with them. The human resources and loss prevention departments should collaborate on several activities such as applicant screening, workplace violence prevention, and internal investigations. When labor problems are anticipated (e.g., unrest, strike), losses can be minimized through cooperative efforts. In addition, labor union contracts and labor laws stipulate limitations on loss prevention activities (e.g., the questioning of suspect employees). Loss prevention practitioners benefit from listening to feedback and criticism from formal and informal employee groups. By listening to and satisfying employee needs (e.g., clean restrooms, a well-run cafeteria, and recreational programs) , with other departments, an organization can reduce losses. When new facilities are being planned, architects, engineers, and loss prevention practitioners can jointly design prevention strategies into the plan and thereby save money by not having to install equipment after construction is completed. Trade secrets and other proprietary information must be kept under strict security, especially in research and development. Accountants and auditors can help plan cooperative strategies against losses. The loss prevention manager should be given an opportunity to review their findings. At various times, a loss prevention department may require legal assistance. Because loss prevention personnel are the backbone of the program, the manager should do everything possible to satisfy their needs (e.g., offer professional development, give praise and pay raises, and institute participatory management). Good rapport with the external relations department ensures that appropriate security and loss prevention information is released to outsiders and that this information is in the best interest of the organization. Communicating with management in general is vital for many reasons. For instance, the objectives of the loss prevention program can be transmitted to all

FIGURE 5-2 External relations.

employees via meetings with management. Feedback from management also assists in planning loss prevention strategies that meet business and organizational needs.

In reference to examples in the external relations chart (clockwise, Figure 5-2, beginning with law enforcement agencies), public safety agencies are essential ingredients of private sector loss prevention programs. Good communication and cooperation are important in case crimes occur or a disaster strikes. Business continuity planning should consider a variety of resources in the community (e.g., police, fire, medical). Educational institutions can provide expertise, research assistance, interns, and potential employees. The loss prevention manager may want to serve on a college advisory committee or speak to classes to spread the word about loss prevention. When possible, the news media should be recruited to aid in loss prevention objectives. Other businesses and loss prevention peers with similar problems can be a good source of ideas. Insurance companies can provide information to improve prevention strategies while reducing premiums. Independent auditors and certified public accountants can point out weaknesses in the loss prevention program to protect against fraud and other crimes. When people (e.g., customers) visit the premises, they should not be seriously inconvenienced by loss prevention systems (e.g., access controls), but at the same time, losses must be prevented. Friendly dialogue and cooperative efforts should be emphasized when government administrative agencies are involved in a loss prevention program. *To enhance all the preceding relations, the security and loss prevention practitioner is wise to emphasize two-way communication, understanding, and partnering for mutual benefits.*

Benefits of Good Relations

The following are reasons why good internal and external relations are important to a loss prevention program. Good relations:

1. Build respect for the loss prevention department, its objectives, and its personnel.
2. Reinforce compliance with policies and procedures to prevent losses.

3. Foster assistance with loss prevention activities, such as programs and investigations.
4. Provide a united front against risks, extend the impact of strategies, and save money.
5. Educate employees, community residents, and others.
6. Improve understanding of complex security problems.
7. Reduce rumors and false information.
8. Improve understanding of the loss prevention program.
9. Stimulate consciousness-raising relevant to loss prevention.
10. Make the loss prevention job easier.

Good internal and external relations enhance the effectiveness of security and loss prevention programs.

Internal Relations

Customer-Driven Security and Loss Prevention Programs

Like well-managed companies, security and loss prevention programs should be **customer-driven** through constant contact with customers who provide feedback and direction for improvements in performance. Employees should be held accountable for maximizing service to customers to retain business and generate additional business. Also, customers should be made aware that they are receiving value from the services provided. A customer-driven approach helps a business to distinguish itself from the competition. Marketing is a major avenue to facilitate a customer-driven business environment. It is also relevant to security and loss prevention programming.

Marketing

Marketing consists of researching a target market and the needs of customers and developing products and services to be sold at a profit. Marketing is also considered the study of consumer problems as opportunities.

The concepts of marketing are universally applicable. **Market segmentation** divides a market into distinct groups of consumers. When one or more of the market segments is chosen for a specific product or service, this is known as **target marketing**. To illustrate, for security programs, market segmentation within a corporation can yield the following groups: executives, women, salespeople, production workers, and truck drivers. Once the market is segmented, and if time and resources are available, a protection program can be designed for each target market. Research and risk analyses will produce a foundation from which to satisfy the protection needs of each type of customer; at the same time, statements or the impression that security programs or strategies will eliminate all victimizations must be avoided. For those security programs that are so general in nature that few customers find them appealing, marketing strategies may be the solution to generate interest. Loss prevention influence over a target group or groups is better than no influence at all (Purpura, 1989: 141–148).

The following seven strategies place a heavy emphasis on marketing and are designed to produce a high-quality security program (Wozniak, 1996: 25–28):

1. Identify all internal and external customers of security services and determine what the customers want and how security can tell whether they are satisfied with the quality of the services rendered. Surveys may be used to collect relevant information.
2. Focus on the customer and his or her perspective. Security should avoid an "us versus them" viewpoint.
3. Facilitate teamwork by meeting with representatives of all internal departments to discuss security issues and to seek solutions.
4. Listen and be receptive to customer concerns. In one hospital, employees complained about the difficulty of entering their departments outside of regular business hours. In response, the security department used its technology to customize access control for each department.
5. Develop definitions. Ensure that customers and security share each group's definitions of security and loss prevention.
6. Set priorities among customers. One hospital ranked its customers as follows: patients or outside customers, employees, and employees' families.
7. Take action and be prepared to adapt to new circumstances.

Do you think marketing concepts can assist a security manager, or does marketing consist of theories that have no practical value in the real world?

Surveys Improve Security Services and Add Value

McCoy (2006: 44–46), a director of security, writes of how he improved security with internal surveys. He notes that security departments are in a difficult position with customer relations because security must enforce policies and conduct investigations. However, he seeks positive views of security from customers by ascertaining customer needs and fulfilling them. McCoy recruited an outside firm to conduct surveys and collect data. The firm provided benchmarks for internal customer service because of its experience with hundreds of clients. As McCoy's department received feedback from the research, specific changes were put in place for continual improvement that was followed by more research. Survey questions focused on access, response time, expertise, courtesy, value, and other subjects. The response rate was about 30 percent.

Surveys often produce unexpected results, and in McCoy's case, he learned that background screening generated numerous complaints because of the length of time required for the check. The average time to complete a check was 14 days, and many managers resented the delay when bringing in a new employee. The security department processed about 4,000 screenings annually, with a denial rate of four percent. To improve performance, the security department began to use a management system model for continual improvement; background screening was reduced to five days by selecting a vendor that emphasized speed and accuracy. This accomplishment was highlighted in a security awareness letter prior to a survey. Survey data was used to justify requests for additional personnel and other resources, and to require security certification to enhance skills. In essence, through internal surveys, the security department improved performance and internal relations.

International Perspective: Market Research to Improve Crime Prevention

Crime prevention by residents in a community is not a unitary construct; it is composed of various types of preventive activities (Hope and Lab, 2001: 7–22). Research from a British crime survey revealed that citizens take part in five clear groupings of preventive activities. An understanding of the groupings, and the characteristics of citizens who choose the groupings, can assist those in the public and private sectors who market crime prevention programs. The groupings are as follows:

- *Evening Precautions:* Actions taken at night to protect against attack, such as special transportation arrangements
- *Self-Defense:* Taking self-defense classes and carrying weapons and personal alarms
- *Neighborhood Watch:* Membership in a watch group and marking property
- *Technological Security:* Burglar alarms and timed lights
- *Fortress Security:* Door and window locks and grills

This research showed that people's use of various preventive methods represents their different perceptions of crime, their routine activities, their aims and needs, and to a lesser extent, demographic variables. Only particular types of people will adopt certain crime prevention methods. More extensive research is needed on this topic.

Corporate security practitioners should consider surveying employees to ascertain what is on their minds (i.e., understanding customers) as input for selecting the best possible security methods. Survey questions can focus on perceptions of security and safety on and off the premises, their routine activities, and their needs. Such information provides a foundation for market segmentation and target marketing.

Human Relations on the Job

Getting along with others is a major part of almost everyone's job. Many say it is half of their job. The result of such effort is increased cooperation and a smoother working environment. These suggestions can help a person develop good human relations:

1. Set a conscious goal of getting along with others as well as possible. Cooperation increases productivity.
2. Say hello to as many employees as possible, even if you do not know them.
3. Smile.
4. Think before you speak.
5. Be aware that nonverbal communication such as body language, facial expression, and tone of voice may reveal messages not included in your oral statements.
6. Listen carefully.
7. Maintain a sense of humor.
8. Try to look at each person as an individual. Avoid stereotyping (i.e., applying an image of a group to an individual member of a group).
9. Personalities vary from one person to another.
10. People who are quiet may be shy, and these people should not be interpreted as being aloof.

11. Carefully consider rumors and those who gossip. Such information often is inaccurate.
12. Remember that when you speak about another person your comments are often repeated.
13. Avoid people with negative attitudes whenever possible. A positive attitude increases the quality of human relations and has an impact on many other activities (e.g., opportunities for advancement).
14. Do not flaunt your background.
15. Everybody makes mistakes. Maintain a positive attitude and learn from mistakes.

Management Support

Support from senior management is indispensable for an effective loss prevention program, and it can be enhanced through quality human relations. Frequent dialogue between loss prevention practitioners and senior management should be a high priority. One method of gaining the interest of management is to show that loss prevention has a **return on investment**. This means that money invested in security and loss prevention can generate not only savings, but also a profit. Knowledge of business principles and practices can aid the practitioner who must speak "business management language" for effective communications. A loss prevention practitioner who does not speak business management language would be wise to enroll in management and accounting courses at a local college.

In one large corporation, the strategy of the security department to gain management support is to quantify its worth and prove it can provide services more efficiently than an outside contractor can. This is the reality of proprietary security today as it struggles to survive and escape the outsourcing and downsizing trends. This corporation's security department tracks everything from the number of investigations conducted to the turnover rate of security officers. Such information (i.e., metrics) helps the security function evaluate and improve its services to its internal customers. Surveys are also used to measure customer satisfaction. Additionally, the department measures losses avoided because of security action. For example, an investigation revealed that a worker's compensation claim, which the corporation had been paying for several years, was fraudulent. The department took credit for the annual savings. Another avenue to gain management support was to show that a $50,000-a-year investigator recovers an average of $500,000 in lost inventory each year (Hollstein, 1995: 61–63).

Orientation and Training Programs

Beginning with a new employee's initial contact with a business, an emphasis should be placed on loss prevention. Through the employment interview, orientation, and training, an atmosphere of loss prevention impressed on the new people sets the stage for consciousness-raising about the importance and benefits of reduced losses through prevention strategies.

Specifically, orientation and training sessions should include a description of the loss prevention plan pertaining to crimes, fires, and accidents. Additional topics include how employees can help reduce losses, access controls and cards, information security, emergency procedures, and various loss prevention services, systems, policies, and procedures.

Involvement Programs

The essence of involvement programs is to motivate employees who are not directly associated with the loss prevention department to participate in loss prevention objectives. How can this be accomplished? Here are some suggestions: The employee responsible for the lowest shrinkage among several competing departments within a business wins $200 and a day off. The employee with the best loss prevention idea of the month wins $50. Those people who become involved should be mentioned in loss prevention reports to show others the benefits of participating in loss prevention activities. Another strategy is the use of anonymity and reward for reporting information helpful to loss prevention objectives.

Loss Prevention Meetings

Meetings with superiors, with employees on the same level as the loss prevention manager, and with subordinates all strengthen internal relations because communication is fostered. When the loss prevention manager meets with superiors, goals and objectives often are transmitted to the manager. Essentially, the manager does a lot of listening. However, after listening, the manager usually has an opportunity to explain the needs of the loss prevention program while conforming to senior management's expectations. This two-way communication facilitates mutual understanding.

Meetings with other managers vary, depending on the type of organization. For instance, a meeting between the human resources manager and the loss prevention manager helps to resolve problem areas. One typical conflict pertains to disciplinary decisions. Another problem area between both departments evolves from applicant screening. Both departments should meet, work together, and formulate a cooperative plan.

When a loss prevention manager meets with subordinates, internal relations are enhanced further. Morale and productivity often are heightened when subordinates are given an opportunity to express opinions and ideas. A manager who is willing to listen to subordinates fosters improved internal relations.

Intranet and E-Mail

The communications technology of today offers a security and loss prevention program superb opportunities to market its services (Huneycutt, 2000: 103–107; Richardson, 2000: 24–26). This involves informing customers of available services, providing helpful information, requesting assistance from customers, seeking input on the needs of customers to improve protection, and illustrating the value of security and loss prevention to the organization.

An **intranet** is an in-house, proprietary electronic network that is similar to the Internet. To prepare an internal web page for security, a committee can be formed composed of management, security personnel, a webmaster to take a leadership role, and an employee from the internal relations department. Next, ascertain through a survey what the customers of security would like to see on the web page. Graphics and other visual images should be used in moderation so that the site does not require excessive time to load to a customer's computer.

Visuals of police badges and handcuffs should be avoided to favor a partnering, loss prevention approach. To help market the site, it can be placed on the homepage of the organization's intranet.

An organization's security and loss prevention internal website can contain the following:

1. Goals, objectives, policies, and procedures
2. Services offered
3. Educational material on a host of topics such as personal protection, what to do if victimized by crime, property protection, IT security, and information security
4. Answers to frequently asked questions
5. Links to resources and information
6. Information on how to obtain parking privileges and access cards
7. Information on how to handle various emergencies
8. An incident form for reporting and avenues to report offline anonymously
9. A bomb threat form
10. A survey form for feedback to improve security
11. A quarterly report on incidents and accomplishments to show the value of security and security personnel
12. An e-mail feature for ease of interaction and to request information
13. A counter to record the number of "hits" at the website

E-mail is a convenient way to communicate. Management and other internal customers can be sent a quarterly report on vulnerabilities, incidents, direct and indirect losses, prevention strategies, and achievements. "Tooting one's own horn" should not be perceived as bragging, but a method to inform others that the past budget allocation was worthwhile. For example, achievements may include lowering shrinkage and increasing profits, foiling a criminal conspiracy, and quickly extinguishing a fire. Such a report submitted to superiors will better prepare them for meetings, budgets, and decisions associated with loss prevention.

Social Media

Although the improper use of social media can harm an individual or organization, and caution is advised, a practitioner can enlist it to benefit protection efforts. The information on social media that follows is a beginning point from which security and loss prevention practitioners can expand its applications (Reino, 2011; Davis, 2010).

Facebook is a popular social networking website owned and operated by Facebook Inc. Users build a personal profile, attach digital photos, add users as "friends," and interact via private or public messages. Users use privacy settings to choose who can access various parts of a profile. Facebook facilitates networking and learning as users join groups that have a special interest. Organizations also establish Facebook pages for marketing purposes. The capabilities of Facebook, as with many technologies, are forever expanding. Examples include live voice calls, integration with Twitter and other media, access from smartphones, and a blogging feature. A competitor of Facebook is MySpace, but it is not as popular.

Facebook is not without its detractors. Countries have blocked access to it and the privacy of users has been compromised. Employers have denied access because it interferes with job duties.

From the perspective of the security and loss prevention practitioner, Facebook can provide information, photos, videos, and news updates to customers (e.g., employees and outsiders). Topics can include new security programs, requests for information (e.g., as in an investigation), how to prepare for emergencies, employment announcements and screening requirements, and crime prevention initiatives with local police. Feedback from users can generate valuable information. Monitoring tools indicate which items are receiving the most attention.

YouTube, a subsidiary of Google Inc., is an Internet website enabling users to view, upload, and share videos. Much of the content on YouTube is made and posted by individuals. However, organizations also upload videos, and various entertainment is broadcast (e.g., films, TV programs, and sporting events). YouTube has been subject to substantial controversy over issues such as the uploading of copyrighted material and offensive content in certain videos. Several countries have blocked access to it. Organizations use YouTube to upload videos on recruiting and new initiatives that may be more difficult to communicate via traditional media. Furthermore, users control the content. Additionally, organizational websites and Facebook pages can include links to YouTube videos.

Twitter, an Internet social network website, owned and operated by Twitter Inc., has personal and business applications. On a personal level, it is used to contact friends and broadcast information. Businesses use it to broadcast company news, help customers, and generate communication among groups. Followers (or subscribers) view a user's text posts called "tweets" that are limited to 140 characters and, thus, users must be brief and "get to the point".

Reino (2011) offers suggestions for social media efforts by organizations:

- Develop a social media plan and strategies with the help of co-workers.
- Establish policies and procedures for employees. Note that social media is constantly evolving and policies and procedures must be updated continually.
- It is a 24/7 operation. Do you have the time to interact with users? Is the effort cost-effective? Does it show value?
- Understand that the "bad guys" also use it. Apply these tools to investigations.
- Once content is "out there" it is almost impossible to take it off the Internet. If you do remove it, someone may have copied it while it was still available.
- Embrace it because its use will grow and it is an opportunity to enhance security and loss prevention.

As a security manager, what types of information would you avoid placing on a company intranet, the Internet, and social media?

External Relations

Law Enforcement

Without the assistance of public law enforcement agencies, criminal charges initiated by the private sector would not be possible. Public police and prosecutors are the main components of public law enforcement. A prosecutor, who is often referred to as the chief law enforcement officer, frequently guides police. A prosecuting attorney has broad discretion to initiate criminal cases.

Unless private security officers adhere to the basic requirements within local jurisdictions, cases cannot be prosecuted successfully. An example can be seen in shoplifting cases. Depending on state law and local prosecutor requirements, a jurisdiction may require apprehension of a shoplifter right after the offender leaves the store with the item instead of apprehending within the store. Many prosecutors feel that this strengthens the case.

External relations with law enforcement agencies are vital for loss prevention programs. Sharing information is a major factor in enhancing this working relationship. Law enforcement agencies often provide information that aids the private sector. Intelligence information pertaining to the presence in an area of professional criminals, bad-check violators, counterfeiters, and con artists assists the private sector in preventing losses. Because of cybercrime, police agencies are increasingly working with the private sector to investigate and prosecute hackers and other cyber-offenders.

Public Safety Agencies

In addition to law enforcement, other public safety agencies such as fire departments, emergency medical services, rescue squads, and emergency management agencies are helpful to loss prevention programs. An analysis of these services is particularly important when a new facility is being planned. Factors relevant to emergency response time, equipment, and efficiency assist in planning the extent and expense of a loss prevention program.

Very little is accomplished by overtly criticizing local public safety agencies for their deficiencies. This action results in negative attitudes and strained relations. Moreover, in the future, the severity of losses may very well depend on action taken by these agencies.

Several strategies help create good relations:

1. Do not become aggressive when striving toward good external relations.
2. Speak on the same vocabulary level as the person with whom you are communicating.
3. Do not brag about your education and experience.
4. Do not criticize public safety practitioners, because these comments are often repeated.
5. Try to have a third party, such as a friend, introduce you to public safety personnel.
6. Join organizations to which public safety personnel belong.
7. Join volunteer public safety organizations (e.g., the local public police reserve or volunteer fire department), if possible.
8. Speak to civic groups.
9. Ask to sit in on special training programs and offer to assist in training.
10. Try to join local or regional criminal intelligence meetings, where information is shared and crimes are solved.
11. Create a softball league with private and public sector participation. Have a picnic at the end of the season. Try to obtain company funds for these activities.
12. Obtain a position on an advisory committee at a college that has a criminal justice, security and loss prevention, homeland security, or fire science program.

13. Form an external loss prevention advisory board and ask local public safety heads to volunteer a limited amount of time. Secure company funds to sponsor these as dinner meetings.
14. Accept official inspections and surveys by public safety practitioners.
15. Begin an organization, such as a local private sector–public sector cooperative association, that has common goals.

The Community

When a new industry moves to a community, external relations become especially important. Residents should be informed about safety plans. Safety is a moral obligation, as well as a necessity, to ensure limited losses and business survival. As shown repeatedly in the media, community resistance to certain industries is strong. Consequently, an industry with a history of adverse environmental impact must cultivate relations that assure residents that safety has improved to the point of causing minimal problems and an extremely low probability of accidents. Engineers, scientists, senior management, and other support personnel are needed to provide community residents, politicians, and resistance groups with a variety of information and answers. Another consideration involves reassuring the community that a proposed industry will not adversely drain public service resources while creating additional community problems.

To promote strong ties to the community, companies often become involved in community service projects. Employees volunteer time to worthy causes such as helping the needy and mentoring youth. ASIS International promotes a Security for Houses of Worship Project that offers security practitioners an opportunity to partner with college security and criminal justice academic programs to protect houses of worship (Purpura, 1999).

The Media

The media can help or hinder a loss prevention program. Efforts must be made to recruit the them. Difficulties often arise because members of the media are usually interested in information beyond that offered by interviewees. It is worthwhile to maintain positive relations; negative relations may create a vicious cycle leading to mutual harm.

In almost all large organizations, educated and experienced communications specialists maintain contact with the media. Loss prevention practitioners should take advantage of such a specialist. This, in effect, insulates the loss prevention department from the media. Policy statements should point out that comments to the media are released through a **designated spokesperson** who may be called a "Director of Corporate Communications." *When all loss prevention personnel clearly understand the media policy, mistakes and embarrassing situations are minimized.* An example of a mistake is the story about a young police officer who was confronted by an aggressive news reporter about a homicide investigation. Investigators had one lead: clear shoe prints under the window of entry to the crime scene. It was hoped that the suspect would be discovered with the shoes that matched the prints. Unfortunately, the news reporter obtained this information from the young officer. The local newspaper printed the story and mentioned the shoes. After reading the newspaper, the offender destroyed the shoes, and the case became more difficult to solve. There had never been a clear policy statement within the organization concerning how to deal with the media. A blunder may take place if the amount of valuables (e.g., cash) stolen in a burglary or robbery is revealed to the media, or the fact that the offenders overlooked high-priced valuables during the crime. Both types of information, broadcast by the media, have been known to cause future crimes at the same location.

Brundage (2010) stresses the need for a crisis communication plan. It includes a policy statement describing how an organization will respond to a crisis, such as notifying authorities during an emergency, conducting an investigation, and releasing information to the media. He recommends key employees meet to determine what potential crises could occur. However, several internal planning and research documents (e.g., risk management and risk analyses) can assist in this process, "rather than reinventing the wheel." Brundage writes of the need to anticipate questions that reporters may ask by focusing on organizational weaknesses and how the organization is working to correct them. He also recommends training and practice in preparation for interacting with the media.

If a security practitioner becomes the designated spokesperson, Sewell (2007: 4) recommends projecting sincerity and credibility, keeping it simple, understanding the reporter's questions before answering, and being ready to think spontaneously. An article in *Security Management*, "Stress from the Press—and How to Meet It" (1980: 8–11), provides many useful ideas concerning relations with the media. The article stresses the importance of preparation, knowing what you want to say, being prepared for tricky questions, and carefully phrasing answers. Numerous firms train executives in effective communication techniques. These courses vary, but most participants learn how to deal with hostile questioning and receive feedback and coaching with the assistance of audio-visual taping. The courses generally run for one to three days and can cost thousands of dollars per participant.

The article recommends several "don'ts" for interviews with the media:

1. Don't return any hostility from the interviewer.
2. Don't lose the audience. Speak clearly and use simple words.
3. Don't say "no comment." If you cannot answer a question, at least tell the audience why.
4. Don't make up answers. If you do not know an answer, say so.
5. Don't say "off the record." If you do not want it repeated, do not say it.
6. Don't offer personal opinions. You are on the air to represent your company, and everything you say will appear to be company policy.

Davis (2010) describes digital tools applied by both communications specialists in organizations and news media reporters. Reporters often subscribe to Really Simple Syndication (RSS) that automatically provides updated information from websites that are of interest to them. Such RSS feeds save time for reporters and can allow them to receive timely news on security topics. Security practitioners also use such tools, including Yahoo! and Google, for automatic alerts when information on a specific topic appears on the Internet. Topics may focus on people, threats, crimes, hate, and controversies, among others. Davis describes how

Twitter can be used by an organization as an alternative to a full news release; a single short message can be followed by regular updates to the media and others. Media outlets that broadcast news via the Internet often direct readers to YouTube if further information is available. Facebook provides an opportunity to release news and receive feedback.

For security and loss prevention practitioners, digital tools provide some degree of control over the accuracy of news stories as compared to the traditional reporter writing or typing news as they interpret it. The digital tools of today save time by eliminating the need to e-mail, fax, or telephone media outlets to promote news.

External Loss Prevention Peers

The practitioner who exists in a vacuum is like a student who does not pay attention. No knowledge is obtained. Through formal and informal associations with peers, the practitioner inevitably becomes involved in a learning experience. One result of relations with peers is gaining information that can improve loss prevention programming.

No individual is an expert about everything. If a group of experts comes together, a broad spectrum of ideas results. When a loss prevention practitioner does not know the answer to a particular question, contacting a peer can produce an answer.

Many formal organizations are open to practitioners. The website addresses at the end of chapters of this book provide direction.

Special Problems

Certain select areas of concern overshadow both internal and external relations. Two important areas are prosecution decisions and loss prevention attire.

Prosecution Decisions

Prosecution decisions concerning employee-offenders are often difficult for management. The difficulty arises because a wise manager considers numerous variables. No matter what decision is made, the company may suffer in some way. The situation is tantamount to "damned if we do and damned if we don't." The following are benefits of prosecuting:

1. Prosecuting deters future offenders by setting an example. This reduces the potential for future losses.
2. The company will rid itself of an employee-offender who may have committed previous offenses.
3. The company will appear strong, and this will generate greater respect from employees.
4. Morale is boosted when "rotten apple" employees are purged from the workforce.
5. If a company policy states that all offenders will be prosecuted, then after each prosecution, employees will know that the company lives up to its word and will not tolerate criminal offenses.
6. The company aids the criminal justice system and the community against crime.

7. Local law enforcement will feel that the company is part of the war against crime. This will reinforce cooperation.
8. A strong prosecution policy will become known inside and outside the company. Those seeking employment who are also thinking of committing crimes against the company will be deterred from applying for work.

The following are benefits of not prosecuting:

1. The company shows sympathy for the employee-offender. Therefore, employees say the company has a heart. Morale is boosted.
2. Not prosecuting saves the company time, money, and wages for those employees such as witnesses and the investigator, who must aid in the prosecution.
3. If the employee-offender agrees to pay back the losses to the company and he or she is not fired, the company will not lose an experienced and trained employee in whom it has invested.
4. The criminal justice system will not be burdened by another case.
5. Sometimes companies initiate prosecution, an arrest is made, and then management changes its mind. This creates friction with law enforcement agencies.
6. Possible labor trouble is avoided.
7. Possible litigation is avoided.
8. Prosecution sometimes creates friction among human resources, internal and external relations, and loss prevention departments.
9. Giving bad news to financial supporters, stockholders, customers, and the community is avoided.

Management must carefully word the policy statement concerning prosecution. There is no perfect statement that can apply to all incidents. A look at a strict policy statement illustrates this problem. Suppose a company policy states that all employees committing crimes against the business will be impartially and vigorously prosecuted. What if a 15-year employee, with an excellent work record and accumulated company-paid training, is caught stealing a box of envelopes? How would you handle this decision as a senior executive?

Employees Smoking Marijuana

An undercover investigation by public and private investigators at the Southern Manufacturing Plant revealed that seven experienced and well-trained employees smoke marijuana before work.

The company loss prevention and human resources managers met with senior management to decide what to do. One solution was to fire all the employees; however, these offending employees had good work records and considerable work experience, and the company had invested heavily in their training. The loss prevention and human resources managers suggested that the employees be confronted and threatened with firing unless they participate in an Employee Assistance Program through a local alcohol and drug abuse program. They also had to submit to drug testing. The managers stated that another local business had a similar problem and a contract had been formulated between the business and the government agency dealing with alcohol and drug abuse problems. Senior management at Southern favored this idea.

Later, the employees were confronted and surprised. They all admitted that they liked their jobs and were willing to participate in a drug abuse program on their own time to avoid being fired. By this time, the local alcohol and drug abuse commission had agreed to conduct a program for these employees. The local police chief and prosecutor favored the program as an innovative diversion from the criminal justice system.

Senior management was satisfied. Internal and external relations were improved. The company provided a second chance for the employees and did not lose experienced workers or the money invested in their training.

Many organizations would not choose this lenient path. Numerous organizations would terminate the substance-abusing employees who were under the influence at work, especially in industries that are regulated and/or involved in public safety and security.

Much economic crime is disposed of privately (Cunningham et al., 1991: 4). The *Report of the Task Force on Private Security* (U.S. Department of Justice, 1976: 128) adds interesting perspectives to the issue:

> *It would appear that a large percentage of criminal violators known to private security personnel are not referred to the criminal justice system. A logical conclusion would be that there is a "private" criminal justice system wherein employer reprimands, restrictions, suspensions, demotions, job transfers, or employment terminations take the place of censure by the public system. In many instances private action is more expedient, less expensive, and less embarrassing to the company. Fear of lawsuits or protecting the offender from a criminal record may be important. However, violations of due process, right to counsel, and other individual rights are more likely to occur under such a system. The criminal justice system is established for the purpose of resolving criminal offenses and can be a viable resource for the private security sector in this regard.*

Another factor affecting prosecution is the **prosecution threshold,** which is the monetary level of the alleged crime that must be met before prosecution. Trends indicate a rise in dollar amounts of individual instances of theft that are partly tied to employee perception of changes in company and prosecutor policies concerning prosecution thresholds. An article in *Security Management Bulletin* ["Finding Where the (Financial) Bodies Are Buried," 1993: 4–5] states: "In the 1970s, an employee who stole $10,000 would have been prosecuted. Today, in many places, that amount won't even produce a criminal referral."

Should company management seek to prosecute or not prosecute employee-offenders?

What if a company and management are subject to prosecution? What factors, outside of the weight of evidence, do prosecutors consider prior to deciding whether to prosecute? The U.S. Government Accountability Office (2009) reviewed U.S. Department of Justice prosecutions of corporate crimes and the use of deferred prosecution and non-prosecution agreements (DPAs and NPAs). Factors of consideration by federal prosecutors included a company's willingness to cooperate, harm to innocent parties, remedial measures taken by the company, and monetary gains to the company resulting from the crime. Most DPAs and NPAs require monetary payments ranging from $30,000 to $615 million to victims (for restitution) and the government (for penalties).

Loss Prevention Attire

The appearance of loss prevention personnel has an impact on how people perceive a loss prevention program. Attire is a vital part of appearance (i.e., uniforms). Wrinkled, messy uniforms send a message to observers that loss prevention objectives are not very important. A variety of people observe loss prevention personnel: employees, customers, visitors, salespeople, truck drivers, law enforcement personnel, and the community in general. What type of message to them is desired from an effective loss prevention program? Clearly, neat, good-looking attire is an asset and shows professionalism.

Attire can project two primary images: subtlety or visibility. Subtle attire consists of blazers or sports jackets (see Figure 5-3). It is generally believed that blazers project a warmer, less threatening, and less authoritarian image. Increased visibility and a stricter image are projected through traditional uniforms. However, it is vital that traditional uniforms do not look similar to those worn by public police; this can cause mistakes by citizens needing aid, public police resent private security officers wearing such uniforms, and state regulatory authorities over the private security industry may prohibit uniforms (and vehicles) that appear as belonging to public police.

All security and loss prevention personnel must realize that they have only one opportunity to create a first, favorable impression.

FIGURE 5-3 Security officers at a museum. *Courtesy of Wackenhut Corporation. Photo by Ed Burns.*

Those who plan uniforms for the private sector should consider research on the psychological influence of uniforms (Mendelson, 2008). It has been shown in psychological tests that individuals associate the color blue with feelings of security and comfort and the color black with power and strength. Other research shows black and brown being perceived as strong and passive, but also bad, and eliciting emotions of anger, hostility, and aggression. Darker police uniforms may send subconscious negative signals to citizens. Light colors such as white or yellow tend to be perceived as weak. One experiment showed that lighter-colored sheriff's uniforms were rated higher for warmth and friendliness than darker uniforms. Furthermore, a half-dark uniform (i.e., light shirt, dark pants) sends a more positive message than an all-dark uniform. Research on the traditional uniform versus the blazer showed mixed results. The Menlo Park, California, police tried the blazer for eight years but found that assaults on police increased and that it did not command respect. After 18 months of wearing blazers, the police of this department displayed fewer authoritarian characteristics when compared to police from other nearby agencies (Johnson, 2001: 27–32). For the private sector, research questions include: Do light-colored uniforms or blazers send more positive signals than dark-colored apparel? Do blazers create a less authoritarian atmosphere than do uniforms? What impact do blazers have on assaults and respect for private security officers?

Do you favor traditional uniforms or blazers for security officers? Why?

Search the Internet

Check out the following sites on the Internet:

American Marketing Association: www.marketingpower.com
Marketing virtual library: www.knowthis.com
Professional Associations and Institutes: ritim.cba.uri.edu/resources
Public Relations Society of America: www.prsa.org
Society for Industrial and Organizational Psychology: www.siop.org

Case Problems

5A. As a security supervisor, you have received reports that a security officer assigned to the employee parking lot is having repeated verbal confrontations with employees. One report described the officer's use of profane language and threats. Specifically, what do you say to this officer, and what actions do you take? How do you repair internal relations?

5B. You are a loss prevention supervisor, and two company employees report to you that a loss prevention officer is intoxicated while on duty. You approach the officer, engage in conversation, and smell alcohol on his breath. An added complication is that this particular officer has a brother in the local police department who has provided valuable aid during past loss prevention investigations. Furthermore, the officer has had a good work record while employed by the company for five years. After you carefully study the advantages and disadvantages of various internal and external ramifications to proposed actions, what do you do?

5C. As a security officer was routinely patrolling the inner storage rooms of a manufacturing facility, he accidentally stumbled on two employees engaged in sexual intercourse. All three were shocked and surprised. The man and woman should have been selecting orders for shipment. The woman cried and begged the officer to remain silent about this revelation. The man also pleaded for mercy, especially because both were married to other people. When the officer decided that he should report the matter, the man became violent. A fight developed between the officer and the man and woman. The officer was able to radio his location and within 10 minutes, other officers arrived. The pair has been brought to the security office. As the security manager, what do you do?

5D. At the end of Chapter 3 a case problem required the reader, as a security executive, to prepare at least 15 situational crime prevention strategies to curb the loss of computers in a company office building. As the security executive, how can you influence employees in following your strategies?

5E. You are a newly hired security manager for an electronics company of 2,000 employees. You were told by your supervisor, the vice president of human resources, "You are here to bring our security program into the 21st century." Apparently, the previous security manager did not meet management's expectations. As you speak with employees and public safety practitioners in the community, you get the impression that you need to "build many bridges." What are your ideas for improving internal and external relations?

5F. An unfortunate explosion and subsequent injuries took place at Smith Industries, and a company vice president designates you, a loss prevention supervisor, to speak to the media. It has been only one hour since the explosion, the investigation is far from complete, and you have no time to prepare for the media representatives who have arrived. As you walk to the front gate, several news people are anxiously waiting. The rapid-fire questions begin: "What is the extent of injuries and damage?" "Is it true that both high production quotas and poor safety caused the explosion?" "What dangers will the community face due to this explosion and future ones?" "What are your comments about reports that safety inspectors have been bribed by Smith Industries executives?" How do you, as a loss prevention supervisor with the authority and responsibility to respond to media questions, answer each question?

5G. Because of your experience and college education, you are appointed manager of loss prevention at a large manufacturing facility in a small city. It is your understanding that the local police chief is introverted and uncooperative with strangers. The chief has an 11th-grade education and is extremely sensitive about this deficiency. He refuses to hire college graduates. Despite these faults, he is respected and does a good job as a police administrator. The local fire chief is the police chief's brother. Both have similar backgrounds and character traits. Furthermore, the local prosecutor is a cousin of both chiefs. A "clannish" situation is apparent. As the loss prevention manager, you know that you will have to rely on as much cooperation as possible from these public officials. What do you do to develop good relations with them?

5H. Allen Dart has worked for the Music Manufacturing Company for 14 years. He has always done above-average work and was recently promoted to production supervisor. One afternoon, when Allen was leaving the facility, he dropped company tools from under his coat in front of a loss prevention officer. Allen was immediately approached by the officer, who asked Allen to step inside to the loss prevention office. Next, Allen broke down and began crying. Before anybody could say a word, Allen stated that he was very sorry and that he would not do it again. Later, the loss prevention and human resources managers met to discuss the incident. An argument developed because the loss prevention manager wanted to seek prosecution, whereas the human resources manager did not. How do you, as a vice president in this company, resolve the situation?

References

Brundage, R. (2010). Crafting crisis communications. *Security Management, 54* (June).

Cunningham, W., et al. (1991). *Private security: Patterns and trends.* Washington, D.C.: National Institute of Justice.

Davis, P. (2010). The public information officer and today's digital news environment. *FBI Law Enforcement Bulletin, 79* (July).

"Finding Where the (Financial) Bodies Are Buried." (1993). *Security Management Bulletin* (August 25).

Hollstein, B. (1995). Internal security and the corporate customer. *Security Management, June.*

Hope, T., & Lab, S. (2001). Variation in crime prevention participation: Evidence from the British Crime Survey. *Crime Prevention and Community Safety: An International Journal, 3.*

Huneycutt, J. (2000). Nothing but net. *Security Management, 44* (December).

Johnson, R. (2001). The psychological influence of the police uniform. *FBI Law Enforcement Bulletin, 70* (March).

McCoy, R. (2006). Better service through surveys. *Security Management, 50* (May).

Mendelson, D. (2008). Putting your best face forward: The psychology of uniforms for security firms proves you are what you wear. *Security Executive, 3* (June/July).

Purpura, P. (1989). *Modern security and loss prevention management.* Boston: Butterworth-Heinemann.

Purpura, P. (1999). *Security for houses of worship: A community service manual for ASIS chapters.* Alexandria, VA: ASIS International.

Reino, N. (2011). 10 Things your agency must know about social media. *Law Officer, January 11* <www.lawofficer.com> retrieved January 13, 2011.

Richardson, N. (2000). A blueprint for value. *Security Management, 44* (March).

Sewell, J. (2007). Working with the media in times of crisis: Key principles for law enforcement. *FBI Law Enforcement Bulletin, 76* (March).

"Stress from the Press—and How to Meet It", (1980). *Security Management, 24* (February).

U.S. Department of Justice, (1976). *Report of the task force on private security.* Washington, D.C.: U.S. Government Printing Office.

U.S. Government Accountability Office.(2009). "Corporate Crime: Preliminary Observations of DOJ's Use and Oversight of Deferred Prosecution and Non-Prosecution Agreements" (June 25). <www.gao.gov> retrieved May 15, 2011.

Wozniak, D. (1996). Seven steps to quality security. *Security Management,* March.

6

Applicant Screening and Employee Socialization

OBJECTIVES

After studying this chapter, the reader will be able to:

1. Define applicant screening and employee socialization
2. Summarize the legal guidelines to counter employment discrimination
3. Explain equal employment opportunity, affirmative action, quotas, diversity, and sexual harassment
4. List and explain at least four applicant screening methods and the legal guidelines for each
5. Explain the legal guidelines for "lie detector" tests
6. Describe how to enhance employee socialization

KEY TERMS

- applicant screening
- employee socialization
- infinity screening
- Equal Pay Act of 1963
- Civil Rights Act of 1964, Title VII
- Age Discrimination in Employment Act of 1967
- Equal Employment Opportunity Act of 1972
- Rehabilitation Act of 1973
- Pregnancy Discrimination Act of 1978
- Americans with Disabilities Act of 1990
- Civil Rights Act of 1991
- disparate impact
- disparate treatment
- Family and Medical Leave Act of 1993
- Uniformed Services Employment and Reemployment Rights Act of 1994
- Genetic Information Nondiscrimination Act of 2008
- Griggs v. Duke Power
- *Bakke v. University of California*
- equal employment opportunity
- affirmative action (AA)
- quotas
- diversity
- sexual harassment
- quid pro quo
- hostile working environment
- sexting
- negligent hiring
- diploma mills
- bona fide occupational qualification
- Sarbanes-Oxley (SOX) Act of 2002
- Fair Credit Reporting Act of 1971
- Fair and Accurate Credit Transaction Act of 2003
- job reference immunity statutes
- Employee Polygraph Protection Act of 1988
- agent of socialization

Introduction

The purpose of **applicant screening** is to find the most appropriate person for a particular job. Several methods are available for screening applicants (e.g., interviewing and testing). ASIS International (2009: 6) offers compelling reasons to conduct applicant screening: reducing risks, increasing productivity, gaining a competitive advantage, reducing turnover, and complying with law.

The culmination of the applicant screening process results in hiring an applicant. At this stage, an organization has already invested personnel, money, and time in making the best possible choice. The next step is to develop a productive employee. This can be accomplished through adequate **employee socialization**, which is a learning process that, it is hoped, produces an employee who will benefit the organization and show value. Two primary methods of socialization are employee training and examples set by superiors.

Both applicant screening and employee socialization are primary loss prevention strategies. If an organization can select honest, stable, and productive people, less has to be done to protect the organization, employees, and others (e.g., customers) from risky employees. In addition, if employees are socialized to act safely and protect company assets, loss prevention is enhanced further.

A broad perspective on applicant screening and employee socialization is essential for protection. Employers should be mindful of full-time, part-time and temporary workers, contractors, consultants, vendors, volunteers, and others on-site and/or connected to the organization remotely. Management must work to bar from the organization those who are at risk for violence, theft, illegal drug use, and other harmful behavior. Employee socialization should be an ongoing process of loss prevention education. Periodic rescreening, known as **infinity screening**, is wise to ensure that those in the organization continue to maintain ethical and legal behavior and have not become a liability and/or risk. There are many cases where either initial employment screening missed risky behavior or an individual began risky behavior after being hired.

The importance of the applicant screening process is illustrated through the following case. A large shopping mall, certified as a company police agency through its state attorney general's office, began to hire experienced police officers. Unfortunately, because of weak background checks by mall management and certain police agencies not reporting police officer transgressions to the state standards commission, the mall became a magnet for police officers with tainted backgrounds. Serious problems surfaced when a mall police lieutenant organized a youth club, sponsored by the mall, then committed statutory rape on one of the female members. He was arrested, prosecuted, and sentenced to prison. During civil litigation brought by the family of the victim, it was learned that the mall chief and the assistant chief had a tainted past involving on-the-job sexual improprieties while employed at police agencies. The lieutenant also had a blemished record. Mall management relied on the records of the state standards commission, rather than conducting a thorough background investigation at the numerous police agencies where these individuals had worked previously. Several of the police officers at the mall were replaced and applicant screening was strengthened.

Applicant screening and employee socialization are primary loss prevention strategies.

Employment Law

An understanding of employment law is essential for applicant screening. Rassas (2011: 7–9) writes that employment law is complex, much like other specialized areas of law. For instance, an employment law may only apply to employers who employ a minimum number of people, have a certain level of annual revenue, are in a specific industry, or are in a specific geographic location.

Government regulation affects the balance and working relationship between employers and employees in three major areas: (1) the prohibition of employment discrimination; (2) the promotion of a safe and healthy workplace; and (3) fair negotiation between management and labor concerning terms of employment (Mann and Roberts, 2001: 858). Here we begin with major federal laws prohibiting employment discrimination, followed by applicant screening methods. The laws on workplace safety and labor are covered in other chapters.

It is important to note that beyond the federal law emphasized here, and executive orders of presidents, are state and municipal laws that exceed federal legal requirements and state court decisions that interpret state laws. For instance, several states consider sexual orientation a "protected class." Also, many corporations have policies to protect employees on the basis of sexual orientation (DeCenzo and Robbins, 2005: 62).

Federal Legislation

- **Equal Pay Act of 1963:** This legislation requires that men and women be paid equally if they work at the same location at similar jobs. Exceptions include a seniority or merit system and earnings through quantity or quality of production. The Equal Employment Opportunity Commission (EEOC) enforces the act.
- **Civil Rights Act of 1964, Title VII:** This law, amended several times to expand its coverage of protected groups, prohibits employment discrimination based on race, color, religion, gender, or national origin. Title VII prohibits discrimination with regard to any employment condition, including recruiting, screening, hiring, training, compensating, evaluating, promoting, disciplining, and firing. It also prohibits retaliation against an individual who files a charge of discrimination. The law impacts both public and private sectors. Title VII requires that organizations go beyond discontinuing discriminatory practices and gives preferences to minority group members in employment decisions; this is referred to as affirmative action. Congress established the EEOC to enforce Title VII.
- **Age Discrimination in Employment Act of 1967:** The ADEA prohibits employment discrimination on the basis of age in areas such as hiring, firing, and compensating. It applies to private employers with 20 or more employees and all government units. This law protects employees between 40 and 65 years of age, but in 1978, the law was amended to afford protection to age 70. In 1986 the law was amended again to eliminate the upper age limit. Mandatory retirement is prohibited, absent a suitable defense. The EEOC enforces this act.
- **Equal Employment Opportunity Act of 1972:** The purpose of this federal law (EEO) is to strengthen Title VII by providing the EEOC with additional enforcement powers to file suits and issue cease-and-desist orders. Further, EEO expands coverage to employees of state

and local governments, educational institutions, and private employers of more than 15 persons. EEO programs are implemented by employers to prevent discrimination in the workplace and to offset past employment discrimination.

- **Rehabilitation Act of 1973:** This act requires government agencies and contractors with the federal government to take affirmative action to hire those with physical or mental handicaps. The Office of Federal Contract Compliance Procedures enforces this act.
- **Pregnancy Discrimination Act of 1978:** This law requires pregnancy to be treated as any other type of disability. In addition, EEO protection is afforded to pregnant employees.
- **Americans with Disabilities Act of 1990:** The ADA prohibits discrimination against individuals with disabilities and increases their access to services and jobs. The law requires employers to make reasonable accommodations for employees with a disability if doing so would not create an undue hardship for the employer. Reasonable accommodations include making existing facilities accessible and modifying a workstation. This law has had a significant impact on the security and safety designs of buildings. Access controls, doorways, elevators, and emergency alarm systems are among the many physical features of buildings that must accommodate disabled people. The EEOC enforces this act.
- **Civil Rights Act of 1991:** This legislation provides additional remedies to deter employment discrimination by codifying disparate impact concepts and allowing plaintiffs to demand a jury trial and seek damages. This act requires businesses to prove that the business practice that led to the charge of discrimination was not discriminatory but job related for the position and consistent with business necessity. The EEOC enforces this act.

  Ivancevich (2001: 74) writes: "**Disparate impact** or unintentional discrimination occurs when a facially neutral employment practice has the effect of disproportionately excluding a group based upon a protected category." (The U.S. Supreme Court expanded the definition of illegal discrimination to include disparate impact as illustrated in the *Griggs* case.)

  Disparate treatment is another type of discrimination whereby an applicant claims that he or she was not hired because of a discriminatory reason. Examples are asking only one applicant about age or asking only female applicants about childcare.
- **Family and Medical Leave Act of 1993**: This law requires employers to provide 12 weeks of unpaid leave for family and medical emergencies without employees suffering job loss. The U.S. Department of Labor enforces this law.
- **Uniformed Services Employment and Reemployment Rights Act of 1994**: Under this law, employees who serve in the military have certain rights to take leave from employment and return to employment following military obligations. The U.S. Department of Labor enforces this law.
- **Genetic Information Nondiscrimination Act of 2008 (GINA)**: This legislation prohibits discrimination resulting from genetic information. The EEOC enforces this law.

U.S. Supreme Court Decisions

When laws are passed, the courts play a role in helping to define what the legislation means. Such court cases evolve, for example, when the EEOC develops and enforces guidelines based on their interpretation of legislation. Confusion over how to interpret the legislation has led to many lawsuits and some conflicting court decisions. What follows here are two famous U.S. Supreme Court cases from an historical context to illustrate the evolution of issues and laws.

Griggs v. Duke Power (1971): In 1968, several employees of the Duke Power Company in North Carolina were given a pencil-and-paper aptitude test for manual labor. Willie Griggs and 12 other black workers sued their employer with the charge of job discrimination under the Civil Rights Act of 1964. Their contention was that the pencil-and-paper aptitude test had little to do with their ability to perform manual labor. The U.S. Supreme Court decided that a test is inherently discriminatory if it is not job related and differentiates on the basis of race, sex, or religion. Furthermore, employers are required to prove that their screening methods are job related.

Bakke v. University of California (1978): Reverse discrimination was the main issue of this case. Allan Bakke, a white man, sued the Davis Medical School under the "equal protection" clause of the 14th Amendment because it set aside 16 of 100 openings for minorities, who were evaluated according to different standards. The Court concluded that the racial quota system was unacceptable because it disregarded Bakke's right to equal protection of the law, and that affirmative action programs are permissible as long as applicants are considered on an individual basis and a rigid number of places has not been set aside. Race can be a key factor in the selection process; however, multiple factors must be considered.

What do all these laws and cases mean for those involved in applicant screening? Basically, all screening methods must be job related, valid, and nondiscriminatory. Included in this mandate are interviews, background investigations, and tests. Simply put, the EEOC regards all screening tools as capable of discriminating against applicants.

Equal Employment Opportunity Commission

The EEOC does not have the power to order employers to stop a discriminatory practice or to provide back pay to a victim. However, the EEOC has the power to sue an employer in federal court. The EEOC requires employers to report employment statistics annually. It investigates claims, collects facts from all parties, seeks an out-of-court settlement, and promotes mediation. According to the Equal Employment Opportunity Commission (2011), in FY 2010, 99,922 charges of discrimination were received and 104,999 were resolved (includes charges prior to FY 2010). The litigation statistics for the same year showed 271 enforcement suits filed in federal district courts and 315 resolved.

Denial of Equal Employment Opportunities Because of Sex

Women's access to employment opportunities continues to be obstructed by sex bias in some workplaces, particularly in jobs traditionally held by men. Two cases follow to illustrate this problem.

In a case brought against a nationwide manufacturer of household products, the EEOC alleged that Dial Corporation's use of a physical "work tolerance" test for production operator positions at a food processing plant in Iowa intentionally discriminated against female applicants and also had a disparate impact on women. In *EEOC v. Dial Corp* (S.D. Iowa Sept. 29, 2005), EEOC presented at trial the testimony of an expert witness that 97 percent of men pass the test while only 40 percent of women succeed, that the test is more difficult than the job, that the scoring is subjective, and that the test does not accomplish its stated objective of reducing injuries. EEOC also presented testimony from 10 of approximately 40 unsuccessful female applicants, focusing on their experience in performing jobs that require heavy lifting. The company presented two expert witnesses who testified that the production operator job is in the 99th percentile of all jobs in the economy with respect to the physical strength required, that the test is very like the job and therefore is content valid, and that the test had in fact reduced injuries.

The jury returned a verdict for EEOC, finding that the company's continued use of the work tolerance test since April 2001 (when the company became aware of the test's disparate impact on women) constituted intentional sex discrimination against women. The judgment provided approximately $3.38 million in back pay, benefits, prejudgment interest, and compensatory damages to 52 class members. It also prohibits the company from implementing any preemployment screening device for five years without first consulting EEOC, and provides job offers with rightful place wages to all class members.

In 2010, WalMart agreed to pay $11.7 million in back wages and other damages and offer jobs to women to settle a sex discrimination lawsuit filed by the EEOC. The case, *EEOC (Janice Smith) v. WalMart Stores, Inc.*, involved WalMart's London, KY, Distribution Center that denied jobs to female applicants. The EEOC argued that the company regularly hired male applicants for warehouse positions, excluded females who were equally or better qualified, told females that order-filling positions were not suitable for women, and that the gender exclusion practice violated Title VII of the Civil Rights Act of 1964 (Equal Employment Opportunity Commission, 2010).

EEO, AA, and Quotas

Equal employment opportunity, affirmative action, and quotas are important terms relevant to staffing organizations (Heneman et al., 1997: 62–64). **Equal employment opportunity (EEO)** refers to practices that are designed so that all applicants and employees are treated similarly without regard to protected characteristics such as race and sex. For example, suppose a vacant position requires applicants to undergo a written job knowledge test and an assessment interview. Anyone is free to apply for the position, and all who apply will be given both the test and the interview. How well each applicant performs on both screening methods determines who is hired. Thus, all applicants have an equal opportunity and the job will be offered following an unbiased assessment. **Affirmative action (AA)** focuses on procedures employers use to correct and abolish past discriminatory employment practices against minority group members, women, and those in other groups, while setting goals for hiring and promoting persons from under-represented groups. AA may be undertaken voluntarily by an employer or court ordered. **Quotas** are rigid hiring and promotion requirements. A hiring formula would be set that specifies the number or percentage of women and minorities to be hired. These concepts, as applied in the workplace, have raised considerable legal turmoil and controversy over whether in fact they have been successful in correcting discrimination. The issue of reverse discrimination has intensified the debate. Court decisions provide guidelines for employers.

Diversity

Diversity in the workforce encompasses many different dimensions, including sex, race, national origin, religion, age, and disability (Byars and Rue, 1997: 8). The workforce, historically dominated by white men, is being increasingly replaced with workers from diverse backgrounds. DeCenzo and Robbins (2005: 13) write that much of the workforce change is attributed to federal legislation prohibiting discrimination, and minority and female applicants are the fastest growing segments of the workforce. Projections for the workforce show that half of the new entrants into the workplace will be women, the average age of employees will climb, immigrant employees will have language and cultural differences, and as companies become more global, there will be an increasing demand to respond to the unique needs of individual employees, including their languages, values, and customs. Diversity facilitates tolerance of different behavioral styles and wider views, which can lead to greater responsiveness to diverse customers. The challenge of learning to manage a diverse workforce pays off as an investment in the future.

In August of 2000, ASIS International held a conference on "Women and Minorities in Security." The conference was noteworthy because white males have dominated this field since its inception. The speakers were straightforward with the challenges facing minorities and the security industry. With an increasingly diverse society, recruitment of women and minorities is essential; however, public and media perceptions of security—often in a negative light—make recruitment difficult. Women have played increasingly important roles in the industry, but more needs to be done to recruit additional women as well as African Americans and Hispanics (Hamit, 2000: 60–62).

ASIS International is in a key position to take the lead to meet the challenges of diversity in the security industry. Solutions include a recruitment campaign; improved collaborating among the ASIS, proprietary and contract security organizations, and colleges and universities; internships; and mentoring.

You Be the Judge*

Facts of the Case

When security officer Bronislav Zaleszny was passed over for promotion to supervisor, he decided that his Eastern European origin was at least one of the reasons (he spoke English with a heavy accent). Therefore, he went to the EEOC and filed a charge against his employer, Hi-Mark Home Products, alleging discrimination on the basis of national origin.

During the month after the EEOC notified the company of the charge, Zaleszny's troubles multiplied. Hi-Mark informed the police that products had been disappearing for months, that the disappearances evidently had occurred during Zaleszny's shift, and that Zaleszny was a reasonable suspect. Then the company terminated him on suspicion of theft. Police arrested Zaleszny, but a preliminary hearing resulted in dismissal of the charges against him.

Zaleszny fumed with anger at his former employer. He sued, alleging that Hi-Mark had prosecuted him maliciously and had fired him in retaliation for his EEOC complaint. In court, the company argued that Zaleszny's allegations really could not stand up to logical analysis. "We didn't prosecute him, maliciously or otherwise," Hi-Mark noted. "We truthfully told the police everything we knew about the disappearance of our products, and we said that on the basis of the facts, Mr. Zaleszny seemed to us to be a logical suspect. Then, completely on their own discretion, the police and the district attorney initiated charges against him. We disagree that the prosecution was malicious, but whether it was or not, we're not the ones who prosecuted."

In answer to Zaleszny's retaliatory-firing allegation, Hi-Mark's director of corporate security took the stand. "I'm the company official who recommended Mr. Zaleszny's termination," she testified, "and when I made the recommendation, I didn't know about his EEOC complaint. Yes, the Commission had notified our company, and yes, our HR department knew, but I didn't! And if I didn't know about the complaint when I fired him, then the firing obviously wasn't a retaliation."

Did Zaleszny win his suit against Hi-Mark Home Products?

Make your decision; then turn to the end of the chapter for the court decision.

Sexual Harassment

The EEOC defines **sexual harassment** as unwelcome sexual conduct that has the purpose or effect of unreasonably interfering with an individual's work performance or creating an intimidating, hostile, or offensive work environment. Although the Civil Rights Act was passed in 1964, only during the 1970s did courts begin to recognize sexual harassment as a form of gender discrimination under Title VII. Thereafter, the EEOC issued guidelines for determining what activities constitute sexual harassment These guidelines now influence courts.

The two theories upon which an action for sexual harassment may be brought are explained here. **Quid pro quo** involves an employee who is required to engage in sexual activity in exchange for a workplace benefit. For example, a male manager tells his female assistant that he will obtain a promotion for her if she engages in sex with him. A second theory of sexual harassment is **hostile working environment**, which occurs when sexually offensive behavior by one party is unwelcome by another and creates workplace difficulties. Examples include unwelcome suggestive remarks or touching and posted jokes or photos of a sexual nature.

Employers can be vicariously liable for sexual harassment and must take immediate and appropriate corrective action; otherwise, civil and criminal legal action can be devastating. Besides legal action for harassment, the following tort actions may be initiated: assault, battery, intentional infliction of emotional distress, and false imprisonment. In addition, criminal charges may be filed for assault, battery, and/or sexual assault.

Bryant (2006: 50–58) notes that harassment claims can go beyond sexual harassment and involve allegations of unlawful discrimination against members of any of the "protected classes." She writes that to file a discrimination lawsuit, most individuals obtain a right-to-sue letter from the EEOC and then contact a private attorney. However, in some cases the EEOC initiates action. Bryant categorizes actionable workplace harassment of a nonsexual nature into three groups:

- Harassment because of an individual's affiliation or association with a particular religious or ethnic group. Examples include harassing a person who is of the Islamic faith, paying an employee less because of being Hispanic, or intimidating an individual who associates with a particular religious or ethnic group.
- Harassment because of physical or cultural traits. Examples are harassment of a Muslim woman for wearing a headscarf or not hiring a man because he has an accent.
- Harassment because of perception pertains to bullying. The National Institute for Occupational Safety and Health defines bullying as "repeated intimidation, slandering, social isolation, or humiliation by one or more persons against another."

The following list offers guidance when taking action against the problem of sexual harassment and other types of workplace harassment (Hare and Kubis, 2012; Maatman, 2011: 68–76; Bryant, 2006: 50–58; Warfel, 2005: 14–20):

1. Refer to EEOC publications such as "Guidelines on Employer Liability for Sexual Harassment."
2. Ensure that top management takes the lead to establish a zero tolerance policy.
3. Provide relevant training and document it by recording the date of training and topics and requiring the signatures of attendees.
4. Communicate the policy and reporting procedures to all employees, including strong prohibitions against retaliation for reporting.
5. Ensure that reported incidents are taken seriously, are thoroughly and promptly investigated, and that corrective action is taken if the allegations are true.
6. Ensure that the human resources department is notified about each complaint.
7. Maintain confidentiality, providing information only on a "need to know" basis.

Cases of Sexual Harassment

Sexual harassment can occur in any work environment, from the fields to the factory floor to the boardroom. Teenagers are a group particularly vulnerable to sexual harassment. Many of EEOC's suits involving the harassment of young women occur in the settings of restaurants and retail establishments, typical part-time jobs for teens (Equal Employment Opportunity Commission, Office of General Counsel, 2005). In *EEOC v. Midamerica Hotels Corp. dba Burger King* (E.D. Mo. Dec. 7, 2004), at a Burger King franchise in Missouri, the EEOC found evidence that a restaurant manager subjected female employees, most of them teenagers, to repeated groping, sexual comments, and demands for sex. The women complained to their first line supervisors and to a district manager, but no action was taken until they learned how to contact the corporate office. Under a consent decree, the company is prohibited from future sex discrimination, agreed to pay a total of $400,000 to seven women, and will not rehire the restaurant manager. Also, managers will be required to attend sexual harassment training. In addition, the company, which operates 37 Burger King restaurants in four states, will distribute its sexual harassment policy, complaint procedure, and hotline information to all current employees and new hires at its restaurants.

In *EEOC v. Carmike Cinemas, Inc.* (E.D.N.C. Sept. 26, 2005), teenage boys were the victims in a case against a large movie theater chain. A 29-year-old male concessions manager in the chain's Raleigh, North Carolina, theater subjected 16- and 17-year-old boys he supervised to offensive verbal and physical sexual conduct over a nine-month period. The manager had previously served more than two years in prison after being convicted of two counts of taking indecent liberties with a minor. Several of the boys complained about the manager, but the theater failed to take corrective action. The manager was finally discharged only when he violated the company's "no call/no show" rule, occasioned by his arrest for failing to register as a sex offender. Under a consent decree, 14 victims shared $765,000 and the company is prohibited from future discrimination. In addition, the company will take the following actions at 13 theaters in North Carolina and Virginia: revise its sexual harassment policy, provide a copy to all new employees, display an 11- by 17-inch poster summarizing the policy, provide sexual harassment training to all new employees at the time of hire and annually to all managers and employees, and report semiannually to the EEOC on complaints of sexual harassment by employees, including the identities of the complainant and alleged harasser and the action taken by the company.

In *EEOC v. Taco Bell Corp.* (W.D. Tenn. August 27, 2009), the EEOC alleged that "a nationwide restaurant chain serving Mexican-style fast food" subjected two 16-year-old female employees to sexual harassment and discharged them. While in a walk-in cooler, one of the females was attacked by a general manager who grabbed her breasts and genital area. During the same day, police arrested him. The EEOC later learned that the same manager had sexually assaulted a different 16-year-old female a year earlier after forcing his way into her home by claiming he was delivering her paycheck. The EEOC suit resulted in monetary relief of $350,000 to the two females (Equal Employment Opportunity Commission, Office of General Counsel, 2009).

Social Media, Sexual Harassment, and Employee Venting

Another type of sexual harassment that employers must address is **sexting**, which is the sending of sexually oriented text messages to others. It includes sexually explicit photos, videos, and audio. The problem extends to e-mail, instant messaging (IM), tweeting, and other networking. Certain employees do not realize how such action affects others, besides their own career. As with other challenges facing employers, the importance of policies, procedures, and training cannot be overstated.

A further concern tied to social media is employees venting about their supervisors and work-related issues. In a groundbreaking case, the National Labor Relations Board (NLRB) accused a business of illegally firing a worker who criticized her supervisor on Facebook. Dawnmarie Souza, an emergency medical technician for American Medical Response of Connecticut, was angered because her supervisor did not let her seek the help of a union representative when responding to a complaint about her work. Ms. Souza vented her anger by criticizing her supervisor on Facebook, using vulgar language, and referring to the supervisor as a psychiatric patient. The Facebook posting resulted in additional negative postings by co-workers. The NLRB noted that the company's policy pertaining to Facebook was overly broad and constrained workers' rights to discuss working conditions. Federal law has consistently protected workers who discuss workplace issues, including critical statements about supervisors. The NLRB announced that the company agreed to change its broad policy and not discipline employees for work-related discussions. The NLRB emphasized that employers should review such broad policies whether employees are unionized or not. The issues in this case will continue to generate controversy. Comments that are defamatory, unrelated to the employer, or not factual can cause legal problems for an individual (Anderson, 2011: 74).

Screening Methods

Screening methods vary among organizations and depend on such factors as regulatory requirements of certain industries, budget, number of personnel available to investigate applicants, outsourcing to service firms, and the types of positions open. Employers are known to expend minimal efforts to properly screen, using the excuse that their hands are tied because of legal barriers. Others follow legal guidelines and screen thoroughly. The EEOC, the Office of Personnel Management, and the Departments of Justice and Labor have adopted and published the "Uniform Guidelines on Employee Selection Procedures," which is periodically updated and serves as a guide for determining the proper use of tests and other selection procedures for any employment decision such as hiring, promotion, demotion, retention, training, and transfers. These guidelines also contain technical standards and documentation requirements for the validation of selection procedures as described in the "Standards for Educational and Psychological Tests," prepared by the American Psychological Association and other groups. Courts rely on such guidelines in deciding cases.

Negligent hiring is a serious problem resulting from hiring an employee who was an unfit candidate for hiring and retention. The courts have established screening standards from negligence cases; awards have been made to victims who have sued, claiming the employer was negligent in not conducting a reasonable inquiry into the background of an employee who, for example, had a history of physical violence. The term *reasonable inquiry* has various definitions. The theory supporting negligent hiring involves foreseeability. It is defined as follows by Black (1991: 449): "The ability to see or know in advance; e.g. the reasonable anticipation that harm or injury is a likely result from certain acts or omissions. In tort law, the 'foreseeability' element of proximate cause is established by proof that an actor, as person of ordinary intelligence and prudence, should reasonably have anticipated danger to others created by his negligent act." An employer can take a number of steps to screen applicants and prevent negligent hiring.

First, careful planning is required. Input from a competent attorney can strengthen the legality of the screening process. *No single screening tool should be used to assess an applicant. Multiple measures are always best.*

It is important that the job duties and qualifications be clearly defined through a job analysis. Noe et al. (2006: 151–157), write that there is no "one best way" for analyzing jobs. They offer various methods of job analysis that include questionnaires focusing on topics such as work behaviors, work conditions, and job characteristics. An important point they make is that errors in the job analysis process result mostly from job descriptions (based on job analyses) being outdated because of our rapidly changing world.

Help-wanted advertisements should be worded carefully to attract only those who meet the requirements of the job. This also prevents expensive turnover and charges of discrimination.

To save money, the most expensive screening methods should be performed last. The time and labor spent reading application forms are less expensive than conducting background investigations.

An employer can be held liable for negligent hiring if an employee causes harm that could have been prevented if the employer had conducted a reasonable background check.

Resumes and Applications

Applications must be carefully studied. Job seekers are notorious for exaggerating and actually lying. The Port Authority of New York and New Jersey did a study by using a questionnaire to ask applicants if they had ever used certain equipment that did not really exist. More than one-third of the applicants said that they had experience with the nonexistent equipment ("Lying on Job Applications May Be Widespread," 1988: 13).

Diploma mills, which provide a "degree" for a fee, with little or no work, are another problem. Because employers may check whether a college is accredited, diploma mill con artists came up with bogus accreditation associations; watermarks, holographs, and encrypting on the diplomas and toll-free numbers so employers can "verify" the graduate. Today, the problem is compounded by online degree programs and the difficulty of distinguishing between quality online degree programs and bogus programs. Databases that list accredited schools may list only those that receive federal financial aid. There is no national accrediting body, only regional ones. Solutions include carefully studying transcripts, asking specific questions about course work, requesting samples of course work (e.g., research papers), and being cautious about credit for "life education" ("Fighting Diploma Mills by Degrees," 2005: 18).

Signs of deception on resumes and applications include inconsistencies in verbal and written statements and among background documents. Periods of "self-employment" may be used to hide institutionalization. Not signing an application may be another indicator of deception. Social Security numbers are issued by states and can assist in verifying past residence. A thorough background investigation is indispensable to support information presented by the applicant.

Employers are increasingly adding clauses and disclaimers to applications. Clauses include a statement on EEO and AA, employment at will (i.e., employer's decision to terminate employees) and the resolution of grievances through arbitration rather than litigation. Disclaimers warn an applicant of refusal to hire or discharge for misstatements or omissions on the application.

Today, many companies use the Internet to recruit applicants by including a recruitment section in their website. The use of the Internet to solicit applications has its advantages and disadvantages. Advantages, when compared to traditional recruitment methods, include the opportunity to attract more applicants globally, lower cost, convenience, and speed. Disadvantages include the workload of possibly screening numerous applications, ignoring other means of recruiting, and hiring too quickly without screening properly.

Applicants also establish their own websites containing their resume and other information. Because "Googling" a name is simple and millions use social media, applicants may place themselves at a competitive disadvantage if an employer finds online information, photographs, or video that is offensive.

Interview

When the applicant is asked general questions about work experience and education, open-ended questions should be formulated so the interviewee can talk at length. "What were your duties at that job?" elicits more information than short-answer questions requiring "yes" or "no" responses. Answers to questions should be compared to the application and resume.

Some employers ask the applicant to complete an application at home to be e-mailed to the employer before the interview. Before the interview, while the applicant is waiting in an office, he or she is asked to complete another application. Both applications are then compared for consistency before the interview.

The following information concerns questions prohibited during the entire screening process (Jensen, 2011: 81–82). Court rulings under EEO legislation have stressed repeatedly that questions (and tests) must be job related. This legal requirement is known as a **bona fide occupational qualification** (BFOQ). Complaints by applicants can result in an EEOC investigation.

Questions pertaining to arrest records generally are unlawful, but it depends on the position. An arrest does not signify guilt. The courts have stated that minority group members have suffered disproportionately more arrests than others have. A question that asks about a conviction, however, may be solicited. It is not an absolute bar to employment. Here again, minority group members have disproportionately more convictions. Certain offenses can cause an employer to exclude an applicant, depending on the particular job. Therefore, questions of arrest and conviction must be job related (e.g., related to loss prevention) and carefully considered (Giles and Devata, 2011: 65–70).

Unless a "business necessity" can be shown, questions concerning credit records, charge accounts, and owning one's own home are discriminatory because minority group applicants often are poorer than others are. Unless absolutely necessary for a particular job, height, weight, and other physical requirements are discriminatory against certain minority groups (e.g., Latino, Asian, and women applicants are often physically smaller than are other applicants).

Other unlawful questions, unless job related, include asking age, sex, color, or race; maiden name of applicant's wife or mother; and membership in organizations that reveal race, religion, or national origin.

Under the ADA, certain questions are prohibited. Do not ask: "Do you have any physical or mental disabilities?" Instead, ask: "Can you perform the functions of the job?"

Do not ask: "Do you plan to become pregnant?" or "What religious holidays do you observe?" Instead, ask: "Can you perform the functions of the job during the required schedule?"

Do not ask: "When are you retiring?" Instead, ask: "What are your short- and long-term career plans?"

The questions that can be asked of an applicant and on an application form, among others, are name, address, telephone number, Social Security number, past experience and salary, reasons for leaving past jobs, education, convictions, U.S. citizenship, military experience in U.S. forces, and hobbies.

Extensive research on the interview process shows that without proper care, it can be unreliable, low in validity, and biased against certain groups. In *Watson v. Fort Worth Bank and Trust*, 108 Supreme Court 2791 (1988), the Court ruled that subjective selection methods such as the interview must be validated by traditional criterion-related or content-validation procedures. Research has pointed to concrete steps that can be taken to increase the utility of the personnel selection interview. First, the interview should be structured, standardized, and focused on a small number of goals (e.g., interpersonal style or ability to express oneself). Second, ask questions dealing with specific situations (e.g., "As a security officer, what would you do if you saw a robbery in progress?"). Third, use multiple interviewers and ensure that women and minority group members are represented to include their perspectives on the applicants (Noe et al., 2006: 234–235).

Tests

The testing of applicants varies considerably. Here is a summary of various types of tests (Rassas, 2011: 214; Noe et al., 2006: 239–244).

- *Physical ability tests* may predict performance and occupational injuries and disabilities. These tests are likely to have an adverse impact on applicants with disabilities and women. However, key questions are as follows: Is the physical ability essential for the job and is it mentioned prominently in the job description? Is there a probability that the inability to perform the job would cause risk to the safety or health of the applicant or others?
- *Cognitive ability tests* measure a person's ability (e.g., verbal, quantitative, reasoning) to learn and perform a job. Highly reliable commercial tests that measure cognitive abilities are available. They are generally valid predictors of job performance.
- *Personality inventories* attempt to measure personality characteristics and categorize applicants by what they are like, such as agreeable and conscientious. When such tests ask job applicants to answer intimate questions, such as their sex practices, class action lawsuits can result.
- *Assessment center* is a method to test applicants on their ability to handle duties encountered on the job. Multiple raters evaluate applicant performance on exercises, such as how to respond to an e-mail from a customer who has a complaint. These tests are expensive to prepare, job content validity is high, and they are low in adverse impact.

- *Medical examinations* are given to determine whether applicants are physically capable of performing the job. The ADA requires employers to make medical inquiries directly related to the applicant's ability to perform job-related duties and requires employers to make reasonable accommodations to help handicapped individuals to perform the job. This act requires that the medical exam cannot be conducted until after the job offer has been provided to the applicant.
- *Honesty tests* are paper-and-pencil tests that measure trustworthiness and attitudes toward honesty. Thousands of companies have used this evaluation tool on millions of workers, and its use is increasing as employers deal with the legal restrictions of the polygraph.
- *Drug tests* have grown dramatically in our drug-oriented world. Employers expect workers to perform their jobs free from the influence of intoxicating substances, and accidents must be prevented. The opposing view favors protection from an invasion of an individual's right to privacy. Employers in regulated or safety-sensitive industries are required by law to test for alcohol or drugs. Numerous employers conduct such tests as a loss prevention measure. Drug tests vary in terms of cost, quality, and accuracy. A drug test can result in a "false positive," showing that a person tested has used drugs when they haven't. A "false negative" can show that the individual has not used drugs when, in fact, the opposite is true. Another problem with drug testing is cheating. Simply stated, if an observer is not present when a urine sample is requested, a variety of ploys may be used by an abuser to deceive an employer. For example, "clean" urine may be substituted. Such deception is a huge problem. Another strategy of drug testing is to measure drug usage from a sample of a person's hair. Experts view this method as more accurate than urine sampling.

Hire the Right Person, Not the Wrong One!

An error in hiring can bring crime to the workplace, loss of proprietary information, and litigation. Security managers have a duty to work with employers to avoid hiring an employee who:

- Has been convicted of embezzlement, but is handling accounts payable.
- Has a history of convictions for computer crimes, but is a corporate IT specialist.
- Has been convicted of securities violations and insider trading, but is working in the corporate communications department.
- Has a history of child molestation convictions, but is working in corporate daycare.
- Is a convicted rapist, but is working as a security officer escorting female employees to their vehicles at night.
- As a temporary employee is collecting trash throughout the premises and is really a news reporter seeking a story.
- While working in research and development, is really an industrial spy collecting information to sell to a competitor.
- Has been hired as a security officer, but is really a terrorist and the "inside person."

Background Investigations

Numerous laws pertain to background investigations. The **Sarbanes-Oxley (SOX) Act of 2002** requires publicly traded companies to conduct background investigations, especially for applicants for positions involving financial matters, trade secrets, IT systems, and other sensitive areas. Employers must also adhere to privacy laws pertaining to the acquisition and protection of sensitive background information (e.g., financial, health).

The **Fair Credit Reporting Act of 1971** (FCRA), enforced by the Federal Trade Commission, is a major law that seeks to protect consumers from abuses of credit reporting agencies while controlling many aspects of background and other types of investigations. Rassas (2011: 224) writes: "The information that falls within the scope of the FCRA is very broad, and despite the name of the law, the Act applies to far more information than what would be contained in an individual's credit report." If a company conducts investigations with in-house investigators, instead of contracting the work to a service firm, the impact of the FCRA may be less burdensome. However, most companies cannot afford in-house investigators.

State laws should also be considered when conducting background checks, especially because of the possibility of being more restrictive than the FCRA. Acohido (2011) writes that many state legislatures are studying proposed bills to ease pressure on job applicants faced with the challenges of unemployment who are being judged on their creditworthiness and honesty by employers using credit checks.

Under the FCRA, an employer is required to notify a job applicant that a background report will be obtained from an outside firm. The employer must receive written permission from the applicant prior to seeking a report. Some states require that a free copy be provided to the applicant. An employer who takes "adverse action" (e.g., not hiring) against the applicant, based on the report (credit, criminal, or otherwise), must do the following: notify the applicant about the development, show the applicant the report, provide information on the applicant's rights under the FCRA, and allow the applicant to dispute any inaccurate information in the report with the reporting agency. Following this process, if the employer still takes adverse action, the applicant must be notified of the action, with justification.

The EEOC has issued guidelines to protect applicants against discrimination from background investigations. For example, before an employer makes an adverse decision on hiring or promoting based on the candidate's personal financial data, the information should be job related, current, and severe. For instance, an employer may decide not to offer a financial position to a candidate who has serious, current debt. The FCRA prohibits the use of negative information that is older than seven years.

In an amendment to the FCRA in 2003, ASIS International and other groups were able to lobby for a provision in the law that removes workplace misconduct investigations (e.g., theft, violence, harassment) from the notice and disclosure requirements of the act. This occurred through the **Fair and Accurate Credit Transaction Act of 2003**, also called the FACT Act. Prior to this amendment, employers who used outside investigative firms for cases of employee misconduct were required to notify the suspect prior to the investigation, which could result in evidence or witness tampering.

Background Checks

Background checks should be conducted on applicants for not only full-time positions, but also those for part-time, temporary, and contract positions. Companies should ensure that when a service firm is contracted, contract language should include thorough applicant screening and liability coverage for offenses committed by contract employees.

In one case, a part-time worker was part of a crew cleaning executive offices for a service company. One day he followed a female executive home and raped and killed her. Later, it was learned that he had a record of convictions for sexual assault.

Quirke (2012: 60–66) writes about social media and its use as a source in background investigations and to make employment decisions. Employers may see information online that would be illegal to inquire about in an application or during an interview. Once an employer learns of such information (e.g., religion), it may be difficult to prove that the information was not applied in a discriminatory manner.

An applicant's criminal history, if any, is a prime concern of employers, especially when the applicant is applying for a security position. Asking about an applicant's arrest record is generally unlawful, but conviction records are legally obtainable in most jurisdictions; they are usually public records on file at court offices. If an applicant appears to have no convictions, it is possible that the background investigator did not search court records in other jurisdictions where the applicant has lived.

Employers spend almost $2 billion a year checking on applicants and the private company databases employers use contain errors. The case of Kathleen Casey offers an illustration of what can go wrong with errors in databases. She was offered a job in a pharmacy in Boston if she passed a background investigation. The investigation showed a criminal record of larceny and fraud against elderly victims. The truth of the matter was that the criminal record was that of a different Kathleen Casey who lived nearby and was much younger. Errors can result from such activities as input mistakes by police or other criminal justice system personnel or assigning records to the wrong individual. Companies that compile databases from public sources typically do not verify information; consequently, employers should verify information collected from databases (Associated Press, 2011).

The FBI's National Crime Information Center (NCIC) database holds an enormous amount of information on offenders and stolen items. However, its use is restricted to criminal justice agencies. For the screening of security officers, refer to Chapter 2 for the Private Security Officer Employment Authorization Act of 2004 that enables private security organizations to check on security applicants through the FBI.

Past employment is a crucial area of inquiry because it reveals past job performance. A customary response by employers is to provide dates of employment, positions, and salary. However, human resources offices may be reluctant to supply negative information because of the potential for a defamation suit. Many states have passed **job reference immunity statutes** that shield employers from lawsuits when sharing adverse information with other employers on employee or former employee work history. For protection, employer statements must be truthful, made in good faith, and made for a legitimate purpose. In those states with such laws, a copy of the law can be attached to the release (to be signed by the applicant) authorizing the background check. This may prompt the applicant and the former employer to release more information. In addition, the previous employer can be tactfully advised that withholding information could result in liability for negligent referral (Nixon, 2005).

The personal references supplied by the applicant are usually those of people who will make favorable comments about the applicant. If an investigator can obtain additional references from the references provided, more may be learned about the applicant.

Most colleges will verify an applicant's attendance and degree over the telephone. College transcripts can be checked out by mail as long as a copy of the applicant's authorization is enclosed. This conforms to privacy legislation. When educational records are received, the investigator should study characteristics and look for inconsistencies.

The private use of public records is on the increase for background investigations. As we know, conviction records are available in most jurisdictions. Records from state motor vehicle departments can reveal a history of careless driving behavior. A motor vehicle report (MVR) can serve as a crosscheck for name, date of birth, and physical description. Federal court records expose violations of federal laws, civil litigation, and bankruptcy. (Chapter 10 discusses online databases for acquiring information.)

Because individuals sometimes try to hide their true identity by using both a false name and background due to a criminal record or other problems, fingerprint-based checks are best, rather than name-based checks. Certain positions legally require a fingerprint-based check.

Nadell (2004: 108–116) offers seven steps to effective background checks to protect organizations from negligent hiring allegations while promoting a safe and secure environment:

1. Prepare and distribute to all employees a background screening policy that conforms to all state and federal laws. This lets employees know that promotions depend on background screening.
2. Communicate the policy by placing signs at select locations on the premises.
3. Place a notice about background screening and drug testing on the company website.
4. Disclose the screening methods to job applicants.
5. Use the job application process to ask all legally allowable questions.
6. Ensure that temporary employment agencies perform background checks and request a copy of the check.
7. Ensure that vendors and contractors perform background checks and request a copy of the check.

You Decide!

What is Wrong with This Employer's Applicant Screening Methods?

The Maxey Tool Company, a small manufacturer of specialized household tools and parts, is facing multiple issues related to its applicant screening process. These include theft, illegal drugs, excessive vehicle accidents involving delivery drivers, and high turnover. The Human Resources Department (HR) only checks the present state in which the applicant lives for conviction and driving records. Numerous employees and applicants have lived in multiple states. Management refuses to use credit checks because they view it as a "legal mine field." When conducting reference checks, HR personnel concentrate on closed-ended questions and state the reasons why the company favors the particular candidate. In addition, the HR Department has approved the hiring of candidates who have on their application five or more years of unexplained gaps in employment and education.

As a security specialist, you are assigned the task of working with a team to improve this company's applicant screening methods to reduce risks and losses. What is wrong with this employer's applicant screening methods and what are your suggestions?

History and Controversy: Polygraph and PSE

Background information on the polygraph and psychological stress evaluator (PSE) (also known as voice stress analysis) will assist the reader in understanding the controversy and subsequent legal restrictions on these devices. In 1895, Cesare Lombroso used the first scientific instrument to detect deception through changes in pulse and blood pressure. In 1921, Dr. John A. Larson developed the polygraph, which measured blood pressure, respiration, and pulse. By 1949, Leonard Keeler added galvanic skin response (i.e., electrical changes on the surface of the skin).

The PSE was developed for the U.S. Army in 1964 by Robert McQuiston, Allan Bell, and Wilson Ford. After the Army rejected it, McQuiston patented a civilian version and marketed it to the private sector.

When questions are asked during a polygraph exam, bodily changes are recorded on graph paper or a computer. The examiner interprets these readings with reference to questions asked. Persons have been known to try to "fool" the polygraph by biting their tongues or pressing a toe into a thumbtack previously hidden in their shoe. The PSE has a few variations, but it records voice stress as questions are asked. There is no hookup, so it can be used covertly.

A disadvantage of the PSE is that only one factor is being recorded, as opposed to the multiple factors of the polygraph. Training for administering and interpreting the PSE is shorter than for the polygraph.

The accuracy of either device is subject to considerable debate, especially concerning the PSE. University of Utah research concluded that the polygraph could be over 90 percent accurate (U.S. Department of Justice, 1978: 8). Gardner and Anderson (2007: 240) write that "in 2002 a panel of leading scientists confirmed a U.S congressional study done in 1983, with both studies reporting that lie detector tests do a poor job of identifying spies or other national security risks and are likely in security screening to produce false accusations about innocent people." Research by Damphousse (2008) found the PSE to be about 50 percent accurate or "no better than flipping a coin." He did note the possibility that having a device present during questioning may deter a person from answering falsely.

Much depends on the training and skill of the examiner behind the device. The polygraph has been responsible for eliminating undesirable job applicants, in addition to assisting with criminal and civil cases, but at the same time, abuses have occurred that resulted in passage of the Employee Polygraph Protection Act.

Employee Polygraph Protection Act of 1988

The **Employee Polygraph Protection Act of 1988** (EPPA) was passed by Congress and signed into law by then-president Ronald Reagan on June 27, 1988. It became effective on December 27. The act prohibits most private employers from using polygraph or "lie detector" tests to screen job applicants and greatly restricts the use of these instruments to test present employees. The EPPA defines the term *lie detector* to include any device that is used to render a diagnostic opinion regarding the honesty of an individual. The congressional Office of Technology Assessment estimated that two million polygraph exams had been conducted each year—90 percent by private employers.

The EPPA states that it is unlawful for an employer to directly or indirectly force an employee to submit to a polygraph test. Discrimination against those who refuse to be tested or who file a complaint under the EPPA is prohibited. Employers who violate the EPPA may be assessed a civil penalty up to $10,000 for each violation. In addition, the Secretary of Labor may seek a restraining order enjoining the employer from violating the act. The law provides individuals with the right to sue employers in federal and state courts for employment reinstatement, promotion, and payment of lost wages and benefits.

A few kinds of employees are exempt from the act and can be tested, including employees of the following types of organizations and companies:

- National security organizations or defense industries
- Federal, state, and local governments
- Businesses involved with controlled substances
- Certain security service firms, such as armored car or security alarm firms

In addition, a limited exemption exists for any employer who is conducting an ongoing investigation involving economic loss or injury; the suspect employee must have had access to the subject of the investigation, and reasonable suspicion must be present. Considerable justification and documentation is required. Chapter 10 contains proper testing procedures under the EPPA.

Should the EPPA be amended to permit more widespread use of polygraph testing in the workplace to screen job applicants for honesty?

Employee Socialization

Socialization, the learning process whereby an employee gains knowledge about the employer and how to become a productive worker, is broader in scope than orientation and training programs. Employers who understand the socialization process are likely to enhance the value of employees to the organization. Furthermore, losses can be reduced as employees adhere to loss prevention strategies. The following emphasizes orientation, training programs, examples set by superiors, and employee needs.

Loss Prevention Orientation for New Employees

When new employees begin to work for an organization, the orientation session plays a significant role in the socialization process. Employees learn about company policies, procedures, departments, products and services, customers, and the community (Noe et al., 2006: 314).

Examples set at the beginning can go a long way in preventing future problems and losses. The orientation program should be designed to acquaint new employees with the "big picture" of loss prevention. Such discussion can enhance employees' understanding of the objectives of the loss prevention program, how they can help, and the benefits of loss prevention to everyone.

Employee Training

In this discussion, the focus is on training protection personnel, but the principles that follow can apply to a variety of training programs for a broad spectrum of employees.

Despite the training problems of the security industry, as covered in Chapter 2, increasing numbers of practitioners in the field realize the importance of training. Consequently, training standards and programs are in a state of constant improvement. Although training costs money, the investment is well worth it. Training aims to provide up-to-date information and guidance to employees for daily tasks and critical incidents. It prevents disciplinary problems and litigation while enhancing morale and motivation. All these benefits are impossible unless there is management support for training.

Planning Training

Step 1: Training Needs

Several questions need to be answered when preparing a training program. Who are the recipients of the training (loss prevention personnel or regular employees; new or experienced employees)? What training programs are available? What deficiencies were noted in employee evaluations? What are the suggestions from supervisors and employees? *Of particular importance is to conduct a job analysis to pinpoint the skills required for the job; both skills and job duties change frequently.* Scaramella et al. (2011: 442), add that "the sophistication of today's culturally diverse, quick paced, ever-changing, technically advanced and globally influenced society demands a set of skills and competencies that are substantially different and more complex compared to the past."

When assessing training needs, research should be conducted on the training applied in the particular industry (e.g., transportation; retail). Sources include federal, state, and local government regulations and industry standards and guidelines. Examples of government sources include state regulators of the security industry and OSHA. ASIS International and the National Fire Protection Association, among other groups, publish industry standards and guidelines.

Step 2: Budget

Before the training program is prepared, an estimate of money available is necessary, since one cannot spend what one does not have.

Step 3: Behavioral Objectives

Each behavioral objective consists of a statement, usually one sentence, which describes the behavioral changes that the student should undergo because of the training. For example, loss prevention officers must explain how the 5th Amendment to the Bill of Rights relates to the private sector.

Step 4: Training Program Outline

With the use of the behavioral objectives, a training program outline is prepared. It can be considered a step-by-step sequence for training.

Step 5: Learning Medium

The method and mediums of presentation are described. Various strategies are available, such as lecture, discussion, demonstration, case method, role-playing, and e-learning (online learning).

Hipkiss (2006: 29) writes of learning through alternatives to the classroom. She notes that the classroom is just one way to learn and that people learn in many ways. Hipkiss argues that employers are realizing that experiential learning—by doing, reflection, and real-world application—has greater value than sitting in a classroom.

Many training programs apply a mixture of techniques. Scenario training or simulations, for example, create workplace situations that trainees will encounter on the job. For this to be successful, managers must determine what types of behavior or performance are desired. Then the training is designed around such objectives. Several trainees can participate in each scripted scenario, acting as security officers, employees, customers, visitors, and evaluators. As the scenarios change, so should the roles of the trainees. Examples of scenarios include assisting a visitor who is lost and upset about being late to a meeting, assisting a handicapped person who must deal with an inoperable elevator, barring access to an estranged spouse of an employee, and responding to a report of employee theft. A variation of scenario training is written scenario testing whereby the trainee reads a script and decides what to do and why. As with all training methods, feedback by the instructor is essential so performance can be improved (Dominguez, 1999: 29–30).

Technology-based training methods involving DVD, Internet, or intranet are popular and used by many organizations. Each is characterized by advantages and disadvantages. Major advantages are flexibility as to the location (i.e., globally) and time (i.e., 24/7) of training, cost savings on travel, and management's ability to track employee progress. Whereas the traditional classroom offers an instructor who can provide immediate guidance and feedback to students, technology-based training methods, such as computer-assisted, self-paced, and distance learning offer a mixture of approaches in providing guidance and feedback to students.

Step 6: Evaluation, Feedback, and Revision

After training is completed, students should provide valuable feedback to the instructor. An evaluation questionnaire, completed by students, can guide the instructor in revising the training program. Scaramella et al. (2011: 442), write that a pretest of student knowledge on topics to be presented followed by a posttest are increasingly being employed to evaluate training and to prevent the waste of training resources. Training is further validated through interviews of participants and supervisors a few months after the training to see if it helped participants to perform their tasks effectively and to identify topics requiring more or less attention. Audits can be used to further assess the success of training. This can entail observing an employee on the job, checking the quality of written reports, hiring external investigators to audit the courtesy of employees, or performing access penetration tests.

Learning Principles

1. Learning results in a behavioral change. Learning objectives are stated in terms of specific behaviors. When a student is able to perform a task that he or she was unable to perform prior to training, then behavior has changed.
2. Tests are used to measure the changed behavior.
3. If the proper conditions for learning are presented to students, learning will take place. The instructor should help the student to learn by facilitating learning through effective instructional methods.
4. An instructional program should begin with basic introductory information to develop a foundation for advanced information.
5. Feedback, an instructor informing the student whether a response was correct or incorrect, is vital to learning.
6. An instructional program must consider the learner's ability to absorb information.
7. A student will be more receptive to learning if information is job related.
8. Conditioning aids the learning process. Conditioning can be perceived as a method of molding or preparing a student for something through constant practice; for example, repetitive drills so that employees know exactly what to do in case of fire.
9. Increased learning will take place if the practice is spread out over time as opposed to a single, lengthy practice session.
10. Information that is learned and understood is remembered longer than that which is learned by rote.

Wasted Training

In an article in *Administrative Management*, "How Not to Waste Your Training Dollars," Donald J. Tosti (1980: 44) declares that American businesses and government agencies spend billions of dollars annually on training, and about half of that amount is wasted. Tosti describes "Seven Deadly Sins of Training," which are still applicable today:

1. Using training to solve motivational problems.
2. Making training more complicated than is necessary.
3. Training personnel at the wrong time, such as training all of an organization's employees for a program that will be active in two years.
4. Overtraining, such as instructing retail clerks on the theoretical aspects of their job before explaining important procedural aspects of retailing.
5. Failing to understand the true costs of training.
6. Failing to calculate training on a cost-effective basis. Evaluations of training programs help to predict benefits. Questions of concern are as follows: Did employees learn and apply the new information? Are losses reduced? Was the training worth the money?
7. Following fads in training.

Think about a training program or course you attended in the past. In what ways could it have been improved?

Examples Set by Superiors

Poor example setting is pervasive in many organizations. All organizations have informal rules that serve as guides to action. These rules are often transmitted from superiors to subordinates. The length of the 15-minute coffee break varies within organizations, as does the time when the 11:00 A.M. meeting begins; the amount of time allowed before an employee is considered late also varies; the number of minor safety violations permitted before strict disciplinary action is taken differs from one organization to another as well as from one superior to another. Clearly, superiors serve as teachers and role models. The actions of superiors greatly affect subordinate performance.

A supervisor can be perceived to be what sociologists call an **agent of socialization**, a person who plays a dominant role in the socialization of an individual. In society, parents, teachers, and clergy are agents of socialization. In a business organization, a supervisor becomes an agent of socialization by guiding subordinates in their job duties. Because first impressions are lasting, the initial part of the socialization process is important. A good example must be set in the beginning.

Based on your experience, can you think of any poor examples set by superiors in the workplace?

Poor Example Set by Ralph Marks, Loss Prevention Manager

The LOCOST retail company emphasized the importance of loss prevention policies and procedures as an aid to increased profits. The loss prevention manager at each store was expected to reinforce the loss prevention program. Each store's employees looked to the loss prevention manager for guidance.

At one store, Ralph Marks, the loss prevention manager, did not set a good example. All employees were permitted to make purchases and receive a 15 percent discount. Procedures dictated that items bought were to be recorded and then stored under a designated counter until the end of the day. When Ralph Marks bought a car stereo system, he did not follow the appropriate procedures. In addition, he installed the system during working hours, which compounded the poor example. By the end of the workday, all employees had seen or heard of the incident. This poor example caused many employees to lose respect for the manager and the loss prevention program.

Employee Needs

The way an organization responds to employee needs has an impact not only on the socialization process, but also on losses. When employee needs are met, workers learn about the employment environment. They learn that management and supervisors care; employees learn to respect and appreciate the employment environment while helping to reduce losses.

What are employee needs? Psychologist Abraham Maslow (1954) became famous for designing a "hierarchy of needs" in the early 1950s (Figure 6-1). Maslow's view is that people

Self-actualization Needs
Esteem and Status Needs
Societal Needs
Safety and Security Needs
Basic Physiological Needs

FIGURE 6-1 Maslow's hierarchy of needs.

are always in a state of want, but what they want depends on their level within the hierarchy of human needs.

Lower level needs must be satisfied before upper level needs. Maslow's hierarchy of needs is as follows:

- *Basic physiological needs.* Survival needs such as food, water, and the elimination of wastes can be satisfied with employer assistance. A well-run company cafeteria and clean lavatories are examples.
- *Safety and security needs.* These needs relate to order in one's life. A person needs to feel free from anxiety and fear. Adequate wages, medical insurance, and workplace safety and security help to satisfy these needs.
- *Societal needs.* The need to be loved and have friends and the need for esteem can be fulfilled by supervisors. A supervisor should praise a subordinate when appropriate. Employees should receive recognition or awards after completing a good job. Employee socials also are helpful.
- *Esteem and status needs.* A person needs to be competent, to achieve, and to gain approval and respect.
- *Self-actualization needs.* This need is at the top of the hierarchy of needs. It signifies that a person has reached his or her full potential, whether as a janitor, homemaker, doctor, or teacher. An organization and its superiors can do a lot in assisting an employee to fulfill this need (e.g., training, promotion).

Employees learn which needs are satisfied and which are not. Suppose a workplace has a terrible cafeteria, dirty lavatories, poor wages, an inadequate safety program, authoritarian supervisors, and poor training and promotional opportunities. What level of losses would be sustained at this workplace in comparison to another that adhered to Maslow's hierarchy of human needs?

How do you think corporate downsizing affects Maslow's hierarchy of human needs and loss prevention?

Search the Internet

Employers and job applicants are becoming increasingly knowledgeable about the validity and reliability of tests, especially because of the need to eliminate or to detect discrimination. The Buros Institute of Mental Measurement (www.unl.edu/buros) evaluates published tests and acts as a consumers' evaluation service.

Here are additional websites related to this chapter:
American Psychological Association: www.apa.org
ASIS International, "Preemployment Background Screening Guideline": www.asisonline.org
Federal Trade Commission: www.ftc.gov
National Association of Professional Background Screeners: www.napbs.com
Society for Human Resource Management: www.shrm.org
U.S. Department of Labor: www.dol.gov
Workplace Fairness: www.workplacefairness.org/federalagencies

Case Problems

6A. Plan and write a step-by-step screening process for applicants interested in uniformed loss prevention positions. Develop an application form. Pay particularly close attention to applicable laws.

6B. You are seeking a position as a security officer at a research and development company. Officers at this site wear blazers and focus on access controls and protecting people and information. The job pays well, with opportunities for advancement, so you strive to do your best at each stage of the applicant screening process. You must now complete an assessment center "in-basket" exercise while thinking as a security officer. You are to prioritize the following items, with justification for each, upon reaching the scene of an assault in the parking lot:

- A witness to the assault approaches you to offer information.
- Someone who is scaling the perimeter fence is screaming for help because of being stuck in the razor ribbon.
- An employee approaches you for help because he locked his keys in his car.
- You receive a radio transmission from your supervisor who wants to meet with you immediately.
- The victim is down and bleeding.
- You must complete an incident report for this case.
- A car alarm has been activated.

6C. You are a candidate for the position of security manager for a large shopping mall near a major city. The number of candidates has been narrowed to six, and the mall human resources manager has decided to use an "in-basket" exercise to further narrow the list of candidates. The "in-basket" exercise consists of a series of memoranda or e-mails, telephone calls, and radio transmissions that the mall security manager would encounter in the job. Your task is to read all items, set priorities among them, and write what action you would take and the reasoning for it for each item. The date is September 24 and you just returned to work from a vacation. The time to complete this assignment is 60 minutes. It is possible that all candidates will be handed additional memoranda or e-mails, telephone messages, or radio transmissions during the exercise. A review panel (police captain, firefighter, college educator, and mall security officer) will evaluate each candidate's work without knowing the identity of the writer. [*Source*: Philip P. Purpura, *Retail Security and Shrinkage Protection* (Boston: Butterworth-Heinemann, 1993), pp. 327–329.]

Item 1

TO: Mall Security Manager
FROM: Mall Manager
SUBJECT: Security Seminar
DATE: September 20

Several merchants would like a seminar on security before the busy holiday season. Please get back to me as soon as possible.

Item 2

TO: Mall Security Manager
FROM: Human Resources Manager, Bigmart Department Store
SUBJECT: Selection of Store Detective
DATE: September 24

Please walk over to review the applications for store detective. I have no idea who would be the best one.

Item 3

TELEPHONE MESSAGE: September 21

Mr. John Poston, a mall customer, called again. He is still irate about the damage to his car window when Security Officer Mallory broke into the vehicle after Mr. Poston left his keys in the ignition. Mr. Poston is threatening to sue.

Item 4

TELEPHONE MESSAGE: September 20

Mrs. Johnson, owner of the Befit Health Store, thinks someone is entering her store at night. She is very upset and worried and wants you to meet her at her store.

Item 5

TO: Mall Security Manager
FROM: Mall Manager
SUBJECT: Application Verification
DATE: September 22

The Westwood Mall office called to verify your application for their job opening in security. Are you planning to begin another job? Please let me know immediately. Let's talk.

Item 6

RADIO TRANSMISSION: September 24, 11:15 A.M.

"Four-year-old boy lost at south end of mall. We have not been able to locate for one hour."

Item 7

TELEPHONE MESSAGE: September 23

Attorney for the plaintiff who was assaulted in the parking lot last month wants you to call him right away.

Item 8

TO: Mall Security Manager

FROM: Mall Manager

SUBJECT: Emergency Plans

DATE: September 16

In speaking with other mall managers at a recent seminar, they mentioned updating their emergency plans. We probably need to do this, too. Please respond.

Item 9

RADIO TRANSMISSION: September 24, 11:20 A.M.

"Small fire in stock room of Smith's Department Store. We can put it out."

Item 10

TO: Mall Security Manager

FROM: Paula Reed, Security Officer

SUBJECT: Pay Raise

DATE: September 23

I am not pleased about my raise of only $0.50 per hour. I have been doing a good job and I really work hard when we get busy. The male security officers are making much more than my rate per hour. I believe that this difference is because I am a black woman. We have talked about this already, but you have not done anything about it. I want something done right away or I will contact the EEOC and take legal action.

Item 11

TELEPHONE MESSAGE: September 22

The manager of Hall Stuart Clothes wants to know why it took so long for security to respond to a shoplifting incident yesterday.

6D. You are a security manager who has just been given an assignment by the vice president of human resources to obtain a copy of an e-mail containing racial jokes that has been circulating in the company. She received complaints about the e-mail and wants you to bring it to a meeting and provide input for corrective action. A short time later, you obtain the e-mail, which contains the story of a young man named Boy. One of the sentences states: "I axed my mudder for some money. She had only too bucks so I axed my fiend, Kenya spare a quarter. He said no so Afro a chair at him." As the security manager, what do you suggest at the meeting?

6E. As a security officer at a large domestic manufacturing plant producing garden tools, you have been given a special assignment to plan and implement a training program for newly hired security officers. What are your plans? What resources will you apply? What topics and how many hours for each topic will be contained in the training program?

The Decision for "You Be the Judge"

Zaleszny did not win his suit against Hi-Mark Home Products. However, the company had to run a gauntlet before breaking out into the clear. There were conflicting judgments at two lower court levels, but finally a higher court ruled in favor of Hi-Mark. In the end, Zaleszny did not win anything. This case is based on *Griffiths v. CIGN*, 988 F 2nd 457 3rd Circuit Court (1993). The names in this case have been changed to protect the privacy of those involved.

Comment

The fact that Zaleszny's company had to go through appeals can give you pause. You might suppose that simple logic should have upheld Hi-Mark from the outset. If the company did not prosecute, how could it be guilty of malicious prosecution? If the manager who fired Zaleszny did not know about his EEOC complaint, how could the firing have been retaliation for the complaint? Logic does not always prevail in court. Often, other variables are in play, including the effectiveness with which a case is presented, a jury's understanding of a judge's instructions, the extent to which all parties understand the relevant law, and even the personalities in the courtroom.

References

Acohido, B. (2011). Limits sought to employers' use of credit reports. *USA TODAY* (April 8).

Associated Press. (2011). When your criminal past is not yours. *Morning News* (December 17).

ASIS International. (2009). Preemployment Background Screening Guideline. <www.asisonline.org> retrieved February 25, 2011.

Anderson, T. (2011). Legal report. *Security Management, 55* (May).

Black, H. (1991). *Black's Law Dictionary* (6th ed.). St. Paul, MN: West Pub.

Bryant, M. (2006). Harassment lawsuits and lessons. *Security Management, 50* (April).

Byars, L., & Rue, L. (1997). *Human resource management* (5th ed.). Chicago: Irwin Pub.

Damphousse, K. (2008). Voice stress analysis: only 15 percent of lies about drug use detected in field test. *NIJ Journal, no. 259* (March). <www.ojp.usdoj.giv/nij/journals/259/voice-stress-analysis.htm> retrieved May 31, 2011.

DeCenzo, D., & Robbins, S. (2005). *Fundamentals of human resource management* (8th ed.). Hoboken, NJ: John Wiley & Sons.

Dominguez, E. (1999). Training that triumphs. *Security Management, 43* (June).

Equal Employment Opportunity Commission. (2011). Enforcement and litigation, FY 1997–2010. <www.eeoc.gov> retrieved May 23, 2011.

Equal Employment Opportunity Commission, Office of General Counsel. (2005). FY 2005 annual report. <www.eeoc.gov/litigation/05annrpt/index.html> retrieved June 16, 2006.

Equal Employment Opportunity Commission, Office of General Counsel. (2009). FY 2009 annual report. <www.eeoc.gov/litigation/reports/09annrpt.cfm> retrieved May 25, 2011.

Equal Employment Opportunity Commission. (2010). Walmart to pay more than $11.7 million to settle EEOC sex discrimination suit. <www.eeoc.gov> retrieved May 24, 2011.

"Fighting Diploma Mills by Degrees" (2005). *Security Management*, 49 (May).

Gardner, T., & Anderson, T. (2007). *Criminal evidence: principles and cases* (6th ed.). Belmont, CA: Thomson Wadsworth.

Giles, F., & Devata, P. (2011). The matrix quandary. *Security Management, 55* (June).

Hamit, F. (2000). ASIS confronts a changing demographic. *Security Technology & Design, 10* (October).

Hare, G., & Kubis, K. (2012). Avoid harassment headaches. *Security Management, 56* (June).

Heneman, H., et al. (1997). *Staffing Organization* (2nd ed.). Middleton, WI: Irwin Pub.

Hipkiss, A. (2006). Learning moves out of the classroom. *Personnel Today* (May 16).

Ivancevich, J. (2001). *Human resources management* (8th ed.). New York: McGraw-Hill Pub.

Jensen, A. (2011). How to avoid discrimination charges. *Security Management, 55* (February).

"Lying on Job Applications May Be Widespread" (1988). *Security* (February).

Maatman, G. (2011). Delivering on diversity's promise. *Security Management, 55* (March).

Mann, R., & Roberts, B. (2001). *Essentials of business law* (7th ed.). Cincinnati, OH: West.

Maslow, A. (1954). *Motivation and personality*. New York: Harper & Row.

Nadell, B. (2004). The cut of his Jib doesn't Jibe. *Security Management, 48* (September).

Nixon, B. (2005). How to avoid hiring hazards. *Security Management, 49* (February). <www.securitymanagement.com/library/001706.html> retrieved February 4, 2005.

Noe, R., et al. (2006). *Human resource management: gaining a competitive advantage* (5th ed.). Boston, MA: McGraw-Hill Irwin.

Quirke, J. (2012). Social media and the workplace. *Security Management, 56* (February).

Rassas, L. (2011). *Employment law: a guide to hiring, managing, and firing for employers and employees*. New York, NY: Aspen Pub.

Scaramella, G., et al. (2011). *Introduction to policing*. Thousand Oaks, CA: Sage Pub.

Tosti, D. (1980). How not to waste your training dollars. *Administrative Management, 41* (February).

U.S. Department of Justice, (1978). *Validity and reliability of detection of deception*. Washington, D.C: U.S. Government Printing Office.

Warfel, W. (2005). SEX ED: insulating yourself from sexual harassment litigation. *Risk Management Magazine, 52* (February).

7 Internal Threats and Countermeasures

OBJECTIVES

After studying this chapter, the reader will be able to:

1. Explain the meaning of internal loss prevention and describe the broad spectrum of internal threats
2. Explain the internal theft problem
3. Outline at least five management countermeasures to prevent internal theft
4. List and explain the steps involved in confronting an employee suspected of internal theft
5. Explain integrated systems and convergence of IT and physical security
6. Outline access control methods and systems, including the types of cards used for access
7. List and describe at least three types of locks
8. List and describe at least five types of interior intrusion detection sensors
9. Describe CCTV technology, including IP-based network systems
10. Explain the characteristics of safes

KEY TERMS

- threat
- internal loss prevention
- theft of time
- telework
- universal threats
- employee theft
- pilferage
- embezzlement
- occupational fraud
- Edwin Sutherland
- differential association
- Donald R. Cressey
- employee theft formula
- accountability
- accounting
- auditing
- graph-based anomaly detection
- inventory system
- marking property
- metal detectors
- integrated system
- convergence of IT and physical security
- technology convergence
- organizational convergence

- identity management system
- Homeland Security Presidential Directive (HSPD) 12
- cloud computing
- access controls
- authentication
- authorization
- cryptography
- encryption
- common user provisioning
- interoperable
- digital certificate systems
- tailgating
- pass back
- near field communication
- biometric security systems
- mechanical locks
- electromechanical locks
- deadbolt
- latches
- cylinder
- lock picking
- lock bumping
- master key system
- intrusion detection system
- sensors
- control unit
- annunciator
- dual technologies
- operational zoning
- digital video recorders
- network video recorder
- Internet protocol (IP)-based network cameras
- analog technology
- digital technology
- compression
- charged coupled device (CCD) or "chip" camera
- megapixel
- high definition
- multiplex
- video motion detection
- video analytics
- fire-resistive (or record) safe
- burglary-resistive (or money) safe

Introduction

A **threat** is a serious, impending or recurring event that can result in loss and must be dealt with immediately. **Internal loss prevention** focuses on threats from inside an organization. Crimes, fires, and accidents are major internal loss problems. Catrantzos (2010) points to the insider, trusted employee who betrays their allegiance to their employer and commits workplace theft, violence, sabotage, espionage, and other harmful acts. The workplace can be subject to infiltration by spies, gangs, organized crime, and terrorists. Losses can result from full-time, part-time, and temporary employees; contractors; vendors; and other groups who have access to the worksite both physically and remotely.

Productivity losses also illustrate the range of internal losses. Such losses can result from poor plant layout or substance abuse by employees. Other productivity losses result from employees who loaf, arrive at work late, leave early, abuse coffee breaks, socialize excessively, use the Internet for nonwork-related activities, and prolong work to create overtime; these abuses are called **theft of time**.

Faulty measuring devices, which may or may not be known to employees, are another cause of losses. Scales or dispensing devices that measure things ranging from truck weight to copper wire length are examples.

We can see that the spectrum of internal threats is broad. Although this chapter concentrates on internal theft and associated countermeasures, the strategies covered also apply to many internal and external threats (e.g., burglary and robbery).

The Insider IT Threats

While the media often focuses on a few high-profile cyberattacks from outside of organizations, the greatest threat to corporate information technology (IT) systems is from within (i.e., from employees). Because news of many insider attacks is not released to the public, the frequency of the following scenario is impossible to gauge: A systems administrator in one hospital learned that she was about to be fired, so she arranged for a "severance package" for herself by encrypting a critical patient database. Her supervisor feared the worst, including loss of his job, so in exchange for the decryption key, the manager arranged for a termination "bonus" and an agreement that the hospital would not prosecute (Shaw et al., 2000: 62).

The dilemma facing the hospital, as to whether to meet the offender's demands or prosecute, can produce interesting debate. How long could the hospital function without the critical patient database? How much time would be required by the criminal justice system to resolve the case? As we know from previous chapters, there are several procedural steps to a criminal case, and the decision to prosecute has its advantages and disadvantages.

From a loss prevention perspective, the following methods would have placed the hospital in an improved position: maintain strict confidentiality about the impending firing of the employee, follow established policies and procedures pertaining to firing employees, exercise extreme caution, block the employee's access to the IT system and other vulnerable locations and systems, and always back up data. Technical solutions alone are not the answer because internal attacks are a "people problem" requiring personnel security solutions. The challenges include the expense of money and time for increased security. Criminal conviction checks may be ineffective with IT personnel because their misdeeds are likely to be unrecorded and, as in the case of the hospital systems administrator, unreported. At-risk behaviors, however, can lead to exposure by supervisors and coworkers. Examples include personnel who avoid procedures and hack into a system to fix problems, curious individuals who explore the system while violating security policies, and individuals who cause outages to facilitate their own travel or advancement.

A growing threat is the insider who steals proprietary or confidential information such as customer identifying information and financial information. These losses can also result from accidental losses of data or attacks by hackers.

Two additional concerns are the growing remote workforce and the devices used to work away from the traditional worksite. Laptop and handheld computers, high-speed Internet, wireless networks, and smart cell phones have facilitated **telework** (i.e., working away from the traditional worksite by transmitting information via communication technology). Many other devices also aid the mobile workforce, including personal digital assistants (PDAs), digital cameras, and USB memory sticks—all of which are high-capacity storage devices. Because of telework, traditional internal threats are also becoming external threats. For instance, an employee working off-site may have a company laptop computer stolen from home or while traveling. The employee may also be victimized by hacking while working off-site. Furthermore, because of technology, an employee can cause losses (e.g., embezzlement or theft of proprietary information) for an organization while off as well as when on the premises. Differentiating internal from external threats is becoming increasingly difficult and blurred, especially because we have entered the era of **universal threats**. In other words, employees and organizations face the same threats whether work is accomplished on or off the premises.

Jordan (2006: 16) writes that both public and private sectors are increasingly embracing telework because of efforts to ensure continuity of operations when a disaster strikes. He refers to the Federal Telework Survey that showed that 41 percent of responding federal employees indicated that they telework, up from 19 percent the previous year. The research also showed that federal IT professionals expanded their support for telework initiatives. Jordan's article adds that telework is not just a technological issue, it is also an organizational and cultural change issue, and agencies must share best practices. From a security perspective, it is also a socialization issue to prevent losses.

The U.S. Department of Homeland Security, Science and Technology Directorate and the Executive Office of the President, Office of Science and Technology Policy (2004: 42) warned that the greatest threat to critical infrastructure (e.g., food, water, electricity) is from the insider who performs actions that could destroy or degrade systems and services. Insider threats develop from individuals who have authorization to access information and infrastructure resources. These threats are difficult to guard against because the offenders are on the inside and are trusted. They exploit vulnerabilities and have advantages over outsiders in choosing the time, place, and method of attack.

To personalize the information presented in this chapter, three businesses are described: a retail lumber business (see Figure 7-1), a clothing manufacturing plant (see Figure 7-2), and a research facility (see Figure 7-3). Suppose you are a loss prevention specialist working for a corporation that has just purchased these three businesses. Your supervisor informs you that you are responsible for recommending modifications at these facilities to improve internal loss prevention. First, read this chapter and then proceed to the case problem pertaining to these businesses at the end of the chapter.

Internal Theft

How Serious is the Problem?

Internal theft also is referred to as *employee theft*, *pilferage*, *embezzlement*, *fraud*, *stealing*, *peculation*, and *defalcation*. **Employee theft** is stealing by employees from their employers.

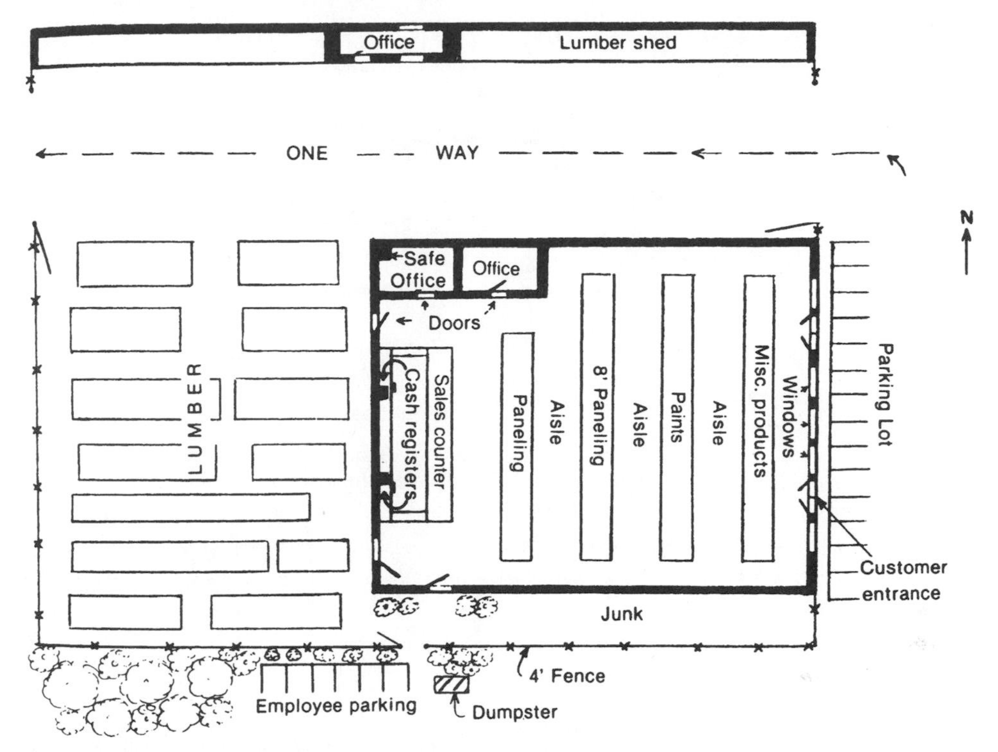

FIGURE 7-1 Woody's lumber company. Woody's Lumber Company has suffered declining profits in recent years. A recently hired manager quickly hired six people to replace the previous crew, which was fired for internal theft. Four additional people were quickly hired for part-time work. The process for conducting business is to have customers park their cars in the front of the store, walk to the sales counter to pay for the desired lumber, receive a pink receipt, drive to the rear of the store, pick up the lumber with the assistance of the yard crew, and then depart through the rear auto exit. At the lumber company, loss prevention is of minimal concern. An inoperable burglar alarm and two fire extinguishers are on the premises.

Pilferage is stealing in small quantities. **Embezzlement** occurs when a person takes money or property that has been entrusted to his or her care; a breach of trust occurs. *Peculation* and *defalcation* are synonyms for embezzlement. Whatever term is used, this problem is an insidious menace to the survival of businesses, institutions, and organizations. This threat is so severe in many workplaces that employees steal anything that is not "nailed down."

The total estimated cost of employee theft varies from one source to another, mainly because theft is defined and data are collected in so many different ways. An often-cited statistic, from The U.S. Chamber of Commerce, is that 30 percent of business failures result from employee theft. The Association of Certified Fraud Examiners (2010: 4–5) conducted research that found that the typical organization loses five percent of its annual revenue to **occupational fraud**, defined as follows: "The use of one's occupation for personal enrichment through the deliberate misuse or misapplication of the employing organization's resources or assets." They claim that if this figure were applied to the Gross World Product, global losses would translate to about $2.9 trillion annually. These figures may be higher when direct and

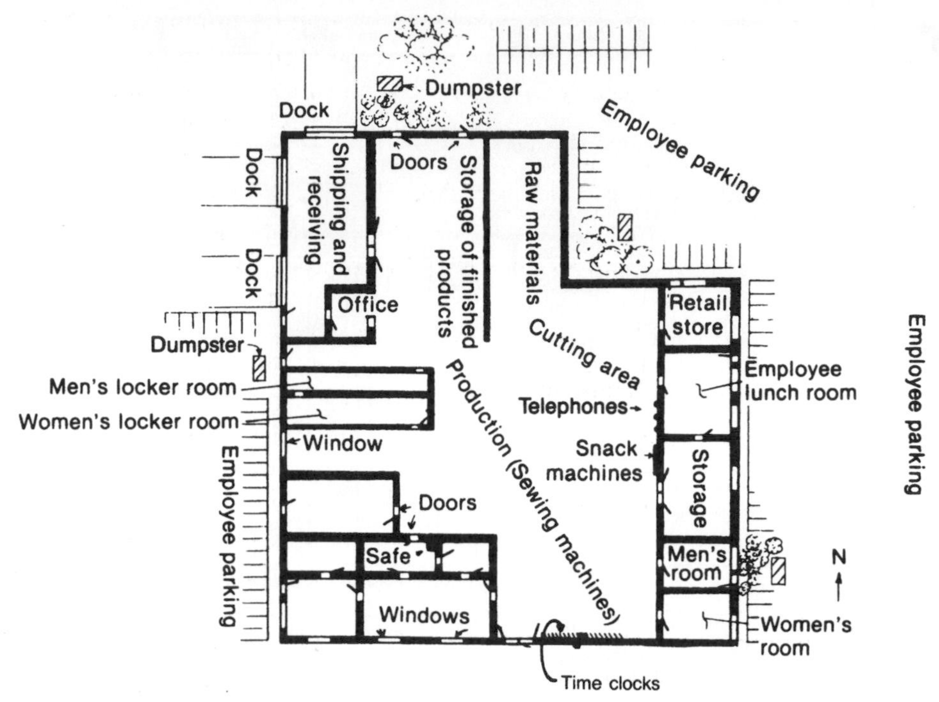

FIGURE 7-2 Smith shirt manufacturing plant. In the past two years, the Smith plant has shown declining profits. During this time, managers believed that employee theft might be the cause, but they were unsure of what to do and were worried about additional costs. Employees work one shift from 8 A.M. to 5 P.M. five days per week and are permitted to go to their cars to eat lunch from noon to 1 P.M. A total of 425 employees are divided as follows: 350 sewing machine operators, 15 maintenance personnel, 20 material handlers, 20 miscellaneous workers, two retail salespeople, five managers, and 13 clerical support staff members. A contract cleanup crew works from 6 to 8 A.M. and from 5 to 7 P.M. on Monday, Wednesday, and Friday; Sunday cleanup is from 1 to 4 P.M. The crewmembers have their own keys. Garbage dumpster pickup is 7 A.M. and 7 P.M. Monday, Wednesday, and Friday. The plant contains a fire alarm system and four fire extinguishers. One physical inventory is conducted each year.

indirect costs are combined. Indirect costs can include damage to brand, a slowing of production, lower employee morale, investigative expenses, and an insurance premium hike after a claim. The mean loss per case was about $160,000 and each crime lasted about 18 months before detection.

Why Do Employees Steal?

Two major causes of employee theft are employee personal problems and the environment. Employee personal problems often affect behavior on the job. Financial troubles, domestic discord, drug abuse, and excessive gambling can contribute to theft. It is inappropriate to state that every employee who has such problems will steal, but during trying times, the pressure to steal may be greater. Research by the Association of Certified Fraud Examiners (2010: 4–5) revealed that the most common behavioral "red flags" of offenders were living beyond their

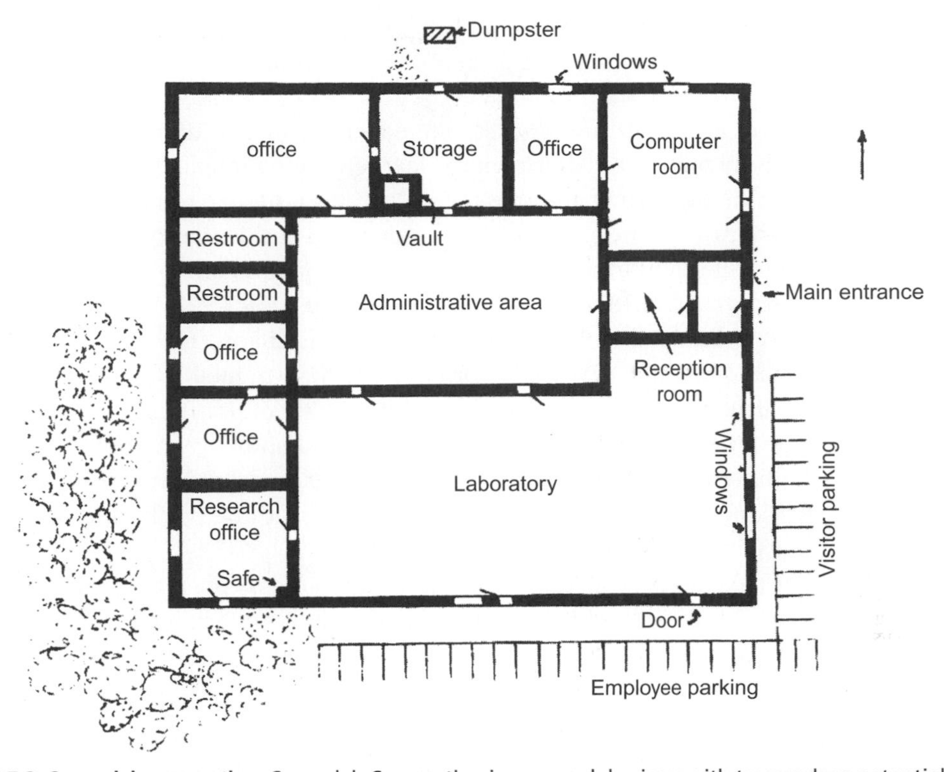

FIGURE 7-3 Compulab corporation. Compulab Corporation is a research business with tremendous potential. However, it seems that whenever it produces innovative research results, a competitor claims similar results soon afterward. Compulab employs 33 people, including a research director, two assistants, 10 scientist-researchers, eight computer specialists, and an assortment of office staff. The facility is open 24 hours a day, seven days per week, and employees work a mixture of shifts each month and remotely from their homes and other locations. Almost every employee has his or her own key for entrance into the building.

means (43%) and financial difficulties (36%). A wise employer should be alert to troubled employees and suggest referral to an Employee Assistance Program (see Chapter 18).

The environment also affects internal theft. Blades (2010: 35) writes that an individual's differences (e.g., ethnicity, accent, or hobbies) can result in bullying or tormenting by co-workers that can lead to alienation and thoughts of revenge, such as theft and violence. Management must ensure that the corporate culture facilitates respect for individual difference. A system of policies, procedures, awareness, and training are essential. Otherwise, litigation can result, besides other losses.

Politicians, corporate executives, and other "pillars of society" are constantly being found guilty of some form of crime, resulting in inadequate socialization. In other words, poor examples are set: employees may observe managerial illegalities and then act similarly. In many businesses, because so many people are stealing, those who do not steal are the deviants and outcasts; theft becomes normal and honesty becomes abnormal. Some managers believe that employee theft improves morale and makes boring jobs exciting. In certain workplaces, employees are actually instructed to be dishonest. This can be seen when receiving department workers are told by their supervisor to accept overages during truck deliveries without notifying the vendor.

Edwin Sutherland, a noted criminologist, offered his theory of **differential association** to explain crime. Simply put, criminal behavior is learned during interaction with others, and a person commits crime because of an excess of definitions favorable to violation of law over definitions unfavorable to violation of law. The implication of this theory for the workplace is that superiors and colleagues in a company are probably a more important determinant of crime than is the personality of the individual. Conklin (2001: 278–279) writes in his criminology textbook that a former head of the Securities and Exchange Commission's Division of Enforcement stated bluntly: "Our largest corporations have trained some of our brightest young people to be dishonest."

A study of college student knowledge of how to commit computer crimes found that threat of punishment had little influence on their misdeeds. In this study, the strongest predictor of computer crime was differential association with others who presented definitions favorable to violation of the law (Skinner and Fream, 1997: 495–518).

The implications for security from differential association theory point to the importance of ethical conduct by top management, who should set a good example in the socialization of all employees. In addition, since criminal laws can be impotent, preventive security strategies are essential.

"Let's Not Fire Him for Stealing—He's a Good Employee"

An undercover investigation at Smith's lumberyard #7 revealed that the yard boss, Joe Crate, was stealing. The undercover investigator, Jimmy Wilson, worked at yard #7 and found that Joe was stealing about $80 worth of building products per week. Each evening Joe would hide merchandise near the back gate, and when it was time to close up and lock the gate, he would quickly load his vehicle, which was conveniently parked nearby.

Before Jimmy was assigned to another yard, he met with a vice president and the manager of yard #7 at company headquarters. During the meeting, Jimmy asked, "Are you going to fire Joe Crate?" The VP stated, "Let's not fire him for stealing—he's a good employee." Then the VP explained: "Joe's salary is $10 per hour, which is equal to $400 per week. If Joe steals about $80 per week, then Joe's salary is about equal to $480 per week. If we hired a carpenter to build the lumber sheds that Joe is building at yard #7, it would cost us almost twice as much." Jimmy could not believe what he was hearing, especially from the VP. He did not say a word and listened to instructions for his next assignment.

What are your views of the way in which internal theft was handled at Smith's lumberyard #7 in the preceding box?

When employees steal, a hodgepodge of rationalizations (excuses) is mentally reviewed to relieve guilt feelings. Some of these rationalizations are "Everybody does it," "It's a fringe benefit," and "They aren't paying me enough." Research by Klenowski et al. (2011) found through interviews of white-collar offenders that they rely on gender themes of masculinity and femininity to justify their criminal behavior. The researchers show that men and women account for their crimes in different ways. Generally, both seek to minimize guilt and maintain a positive self-image; however, men used *self-reliance* (e.g., accomplish goals in the business world at the expense of all else) while women relied on *necessity* (e.g., self-defined distressed financial situation) to rationalize criminal behavior. The researchers concluded that it is easier for men to deny harm, condemn accusers, and argue that the behavior is normal than it is for women.

Donald R. Cressey (1971), in his classic study, analyzed thousands of offenders to ascertain common factors associated with inside thievery. He found three characteristics that must be present before theft would be committed. Cressey's **employee theft formula** is

Motivation + Opportunity+Rationalization=Theft

Motivation develops from a need for money to finance a debt or a drug problem or to win approval from others. Opportunity occurs at many unprotected locations, such as a loading dock. Cressey observed that embezzlers' financial problems are "nonshareable" because of embarrassment or shame, and they rationalize their illegal behavior. This formula illustrates the need for security and an honest environment.

The theoretical foundations explained in Chapter 3 are applicable to our discussion here as to why employees steal. *Deterrence* has its limitations because following a crime, the certainty of both swift action by authorities and punishment often do not occur. *Prevention* seeks proactive security methods to reduce the probability of harmful events and to mitigate losses if harmful events occur. *Rational choice theory*, related to deterrence theory, points out that a person studies the consequences of a crime against the benefits prior to committing the crime and chooses a criminal act if the rewards offset the consequences. *Routine activity theory* notes that crime occurs when three elements converge: (1) a motivated offender, (2) a suitable target, and (3) the absence of a capable guardian. Rational choice theory and routine activity theory are sometimes viewed as opportunity theories that seek to reduce opportunities for crime by changing physical features of the environment, implementing security strategies, and changing behavior. *Situational crime prevention* techniques offer practical strategies to reduce *opportunities* for crime. Examples include increasing the effort and risks confronting the offender. These theories have practical application to the employee theft problem.

Speed (2003: 31–48) offers insights into the complexity of employee dishonesty, what deters and motivates employee thieves, and management countermeasures. He focused his research on a major retailer in the United Kingdom to learn how loss prevention could be better targeted. Speed studied company records of employee offenders and surveyed attitudes of a sample of employees. He proposed a management strategy that divides employees into four groups, based on age and length of service, and then designed loss prevention strategies for each group. The four groups and the strategies for each are summarized next:

- *First group:* Employees 20 years of age or younger, new to the company
- *Second group:* Employees in their 20s, employed with the company for about two years
- *Third group:* Employees with greater length of service and experience than the first two groups
- *Fourth group:* Employees with considerably greater length of service or are much older

Speed's research shows that the first group presents great risk of theft because they are less likely to be deterred by disapproval by others or by losing their jobs. However, more of them fear being caught than the slightly more experienced employees. The first group commits the simplest types of offenses with the lowest values. Strategies for this group include restricted access to high-risk operations and ensuring they are complying with systems. The second group also presents great risk of theft because they are confident they will avoid detection. They commit high value offenses but are influenced more than the first group by the possibility of losing their jobs. The recommended strategy for this group is to portray the risks of criminality and the possibility of prosecution. Theft among the third group is less common, but more complex and less easy to detect. This group is more likely to be deterred by disapproval by others. Controls that remove opportunities are less likely to be successful with this group. A more successful strategy is to remind them of the status and benefits they maintain within the company and the financial impact of offending. The fourth group represents the lowest risk but the greatest confidence of not being caught. This group is similar to the third group on other characteristics.

How Do Employees Steal?

The methods used to steal from employers are limited only by employee imagination. Employees often pilfer items by hiding them under their clothing before leaving the workplace. Methods that are more sophisticated may involve the careful manipulation of computerized accounting records. Collusion among employees (and outsiders) may occur. Research by the Association of Certified Fraud Examiners (2010: 5) noted that asset misappropriation schemes were the most common forms of fraud by employees. Some employee theft methods follow:

1. Wearing stolen items while leaving the workplace. For example, wearing pilfered underwear or wearing scrap lead that has been molded to one's body contours
2. Smuggling out pilfered items by placing the items in a lunchbox, pocketbook, computer, bundle of work clothes, umbrella, newspaper, legitimate purchase, hat, or even one's hair
3. Hiding merchandise in garbage pails, dumpsters, or trash heaps to be retrieved later
4. Returning to the workplace after hours and helping oneself to goods
5. Truck drivers turning in fictitious bills to employers for fuel and repairs and then splitting the money with truck stops
6. Collusion between truck drivers and receiving personnel
7. Executives padding expense accounts

8. Purchasing agents receiving kickbacks from vendors for buying high-priced goods
9. Retail employees pocketing money from cash sales and not recording the transaction
10. Padding payrolls with hours and rates of pay
11. Maintaining nonexistent or fired employees on a payroll and then cashing the paychecks
12. Accounts payable employees paying fictitious bills to a bogus account and then cashing the checks for their own use

Possible Indicators of Theft

Certain factors *may* indicate that theft has occurred:

1. Inventory records and physical counts that differ
2. Inaccurate accounting records
3. Mistakes in the shipping and receiving of goods
4. Increasing amounts of raw materials needed to produce a specific quantity of goods
5. Merchandise missing from boxes (e.g., every pallet of 20 boxes of finished goods has at least two boxes short a few items)
6. Merchandise at inappropriate locations (e.g., finished goods hidden near exits)
7. Security devices found to be damaged or inoperable
8. Windows or doors unlocked when they should be locked
9. Workers (e.g., employees, truck drivers, repair personnel) in unauthorized areas
10. Employees who come in early and leave late
11. Employees who eat lunch at their desks and refuse to take vacations
12. Complaints by customers about not having their previous payments credited to their accounts
13. Customers who absolutely have to be served by a particular employee
14. An unsupervised, after-hours cleaning crew with their own keys
15. Employees who are sensitive about routine questions concerning their jobs
16. An employee who is living beyond his or her income level
17. Expense accounts that are outside the norm

Management Countermeasures

Management Support

Without management support, efforts to reduce losses are doomed. *A good management team sets both a foundation for strategies and an atmosphere in which theft is not tolerated.* Support for budget requests and appropriate policies and procedures are vital.

Effective Planning and Budgeting

Before measures are implemented against internal theft, a thorough analysis of the problem is essential. What are the losses and cost-effective countermeasures? What types of losses are occurring, where, how, by whom, when, and why?

Internal and External Relations

Good internal and external relations can play roles in preventing employee theft. Employees respect loss prevention practitioners who are professional and are often more willing to provide information and cooperate. With a heightened loss prevention atmosphere within a workplace, an external reputation is sure to follow.

Job Applicant Screening and Employee Socialization

The screening of job applicants is a major theft-prevention technique. Although screening is often touted as an effective strategy to prevent internal theft, research by the Association of Certified Fraud Examiners (2010: 5) found that 85 percent of offenders had not been previously charged or convicted of a fraud offense. Thus, infinity screening (see Chapter 6) is vital.

Accountability, Accounting, and Auditing

Accountability defines a responsibility for and a description of something. For example, John Smith is responsible (i.e., is held accountable) for all finished products in a plant, and he maintains accurate records of what is in stock. **Accounting** is concerned with recording, sorting, summarizing, reporting, and interpreting business data. **Auditing** is an examination or check of a system to uncover deviations. Personnel audit physical security by checking intrusion alarm systems, closed-circuit television (CCTV), and so on. An auditor audits the accounting records of a company to see if the records are reliable and to check for fraud.

Policy and Procedural Controls

Policy and procedural controls coincide with accountability, accounting, and auditing. In each of these three functions, policies and procedures are communicated to employees through manuals and memos. *Policies* are management tools that control employee decision making and reflect the goals and objectives of management. *Procedures* guide action to fulfill the requirements of policies.

As an example, a company policy states that, before trash is taken to outside dumpsters, a loss prevention officer must be present to check for stolen items. Procedures point out that, to conform to this policy, the head of the cleaning crew must call the loss prevention office and wait for an officer to arrive before transporting the trash outside.

Signs

Placing messages about loss prevention on the premises is another method to reduce internal losses. The message must be brief, to the point, and in languages for diverse readers. An example of a message is "Let's all work together to reduce losses and save jobs." Boba and Santos (2008: 257) reported that crime prevention signage at construction sites is cost-effective and shows management commitment to reduce theft.

Loss Reporting and Reward System

Numerous organizations have established loss reporting through such avenues as a toll-free number, suggestion box, website, or intranet. A reward system is a strategy to reinforce reporting; one method is to provide the anonymous informant with a secret number that is required to pick up reward money at a bank.

The Sarbanes-Oxley (SOX) Act of 2002 requires publicly traded companies to provide a system of reporting anonymously, with penalties for noncompliance. Research shows that the best avenue to encourage reporting is through a confidential, 24-hour hotline operated by a third party (Greene, 2004).

Research by Scicchitano et al. (2004: 7–19) found that, among the large retailers they surveyed, management encouraged employees to report dishonesty that they observed in the workplace. A hotline with rewards was effective in encouraging employees to report losses. The researchers emphasized that corporate climate plays an important role in facilitating peer reporting. Boba and Santos (2008: 255) found through their study that hotlines were cost-effective in controlling theft; employees are less likely to steal when they believe the probability of apprehension is high; and employee-offenders fear co-worker sanctions more than management sanctions.

Investigations

Employee thieves often are familiar with the ins and outs of an organization's operation and can easily conceal theft. In addition, a thorough knowledge of the loss prevention program is common to employee thieves. Consequently, an undercover investigation is an effective method to outwit and expose crafty employee thieves and their conspirators.

Businesses subject to theft should partner with police to investigate and disrupt stolen goods markets. Investigations should focus on pawnshops, flea markets, suspect wholesalers and retailers, online sites, fences, and organized crime groups.

Another investigative approach is **graph-based anomaly detection** that consists of mining of data sets, such as employee information, network activity, e-mail, and payroll that contain possible interconnected and related data for analysis and plotting on a graph to expose unusual activities that may indicate an insider threat. Eberle et al. (2011) explain that if we know what is normal behavior, deviations from that behavior could be an anomaly; however, they note that graph-based anomaly detection is challenging because an offender will try to act as close to legitimate actions as possible.

Property Losses and Theft Detection

To remedy property losses within an organization, several strategies are applicable. Closed-circuit television (CCTV), both overt and covert, and Radio Frequency Identification (RFID) are popular methods discussed in other parts of this book. In addition, for high-value assets, a global positioning satellite (GPS) locator chip can be imbedded in an asset to notify security or police and to track the movement of an asset via computer when it is moved without authorization. Here, an emphasis is placed on inventory system, marking property, and use of metal detectors.

An **inventory system** maintains accountability for property and merchandise. For example, when employees borrow or use equipment, a record is kept of the item, its serial number, the employee's name, and the date. On return of the item, both the clerk and the user make a notation, including the return date. Automated systems can include a microchip on the item that is read by a scanner for a digital record. Inventory also refers to merchandise for sale, raw materials, and unfinished goods.

Marking property (e.g., tools, computers, and furniture) serves several useful purposes. When property is marked with a serial number, a special substance, or a firm's name etched with an engraving tool, thieves are deterred because the property can be identified, it is more difficult to sell, and the marks serve as evidence. Marking also helps when locating the owner. Publicizing the marking of property reinforces the deterrent effect. Police departments have operated a program known as "Operation Identification" for many years in an effort to recruit citizens to mark their property in case of loss.

One popular covert surveillance method is the use of a pinhole lens camera; another popular investigative technique is to use fluorescent substances to mark property. An ultraviolet (black) light is necessary to view these invisible marks, which emerge as a surprise to the offender. To illustrate, suppose an organization's petty cash is not adequately secured and money is missing. To expose the offender, a fluorescent substance, in the form of powder, crayon, or liquid, is used to mark the money. A few employees who work after hours are the suspects. Before these after-hour employees arrive, the investigator handling the case places bills previously dusted with invisible fluorescent powder in envelopes at petty cash locations. The bills can also be written on with the invisible fluorescent crayon. Statements such as "marked money" can be used to identify the bills under ultraviolet light. Serial numbers from the bills are recorded and retained by the investigator. Before the employees are scheduled to leave, the "planted" bills are checked. If the bills are missing, then the employees are asked to show their hands, which are checked under an ultraviolet light. Glowing hands expose the probable thief, and identification of the marked money carried by the individual strengthens the case. The marked money should be placed in an envelope because the fluorescent substances may transfer to other objects and onto an honest person's hands. A wrongful arrest can lead to a false-arrest suit. A check of a suspect's bills, for the marked money, helps avoid this problem. Many cleaning fluids appear orange under an ultraviolet light. The investigator should analyze all cleaning fluids on the premises and select a fluorescent color for marking bills that is different from the cleaning substances. Other items that may fluoresce include lotions, plastics, body fluids, and some drugs.

Another method of marking property is by using microdots that contain a logo or ID number and are painted or sprayed onto property. A microscope is used to view the dots that identify the owner of the property. One utility company, for example, suffered millions of dollars of losses from the theft of copper wire and equipment, so it applied the dots to copper assets to help identify company property during investigations and recovery (Canada.com, 2007).

Walk-through **metal detectors**, similar to those at airports, are useful at employee access points to deter thefts of metal objects and to identify employee thieves. Such detectors also uncover weapons being brought into an area. Handheld metal detectors are also helpful. It is important to note that metal detectors may be overrated because certain firearms, knives, and other weapons are made primarily of plastic. Consequently, X-ray scanners are an expensive option with which to identify contraband. (The next chapter covers contraband detection.)

Insurance, Bonding

If insurance is the prime bulwark against losses, premiums are likely to skyrocket and become too expensive. For this reason, *insurance is best utilized as a supplement to other methods of loss prevention that may fail.* Fidelity bonding is a type of employee honesty insurance for employees who handle cash and perform other financial activities. Bonding deters job applicants and employees with evil motives. Some companies have employees complete bonding applications but do not actually obtain the bond.

Confronting the Employee Suspect

Care must be exercised when confronting an employee suspect. Maintain professionalism and confidentiality. The following recommendations, in conjunction with good legal assistance, can produce a strong case. The list of steps presents a *cautious approach*. Many locations require approval of management before an arrest.

1. Never accuse anyone unless absolutely certain of the theft.
2. Theft should be observed by a reliable person. Do not rely on hearsay.
3. Make sure intent can be shown: the item stolen is owned by the organization, and the person confronted removed it from the premises.

In steps 4 through 14, an arrest has not been made.

4. *Ask* the suspect to come to the office for an interview. Employees do not have a right to have an attorney present during one of these employment meetings. If the suspect is a union employee and requests a union representative, comply with the request.
5. Without accusing the employee, he or she can be told: "Some disturbing information has surfaced, and we want you to provide an explanation."
6. Maintain accurate records of everything. These records may become an essential part of criminal or civil action.
7. Never threaten a suspect.
8. Never detain the suspect if the person wants to leave. Interview for less than one hour.
9. Never touch the suspect or reach into the suspect's pockets.
10. Request permission to search the suspect's belongings. If left alone in a room under surveillance, the suspect may take the item concealed on his or her person and hide it in the room. This approach avoids a search.
11. Have a witness present at all times. If the suspect is female and the person confronting her is male, have another woman present.

12. If permissible under the Employee Polygraph Protection Act of 1988, ask the suspect to volunteer for a polygraph test and have the suspect sign a statement of voluntariness. Follow EPPA guidelines.
13. If a verbal admission or confession is made by the suspect, have him or her write it out, and have everyone present sign it. Do not correct the suspect's writing errors.
14. Ask the suspect to sign a statement stipulating that no force or threats were applied.
15. For the uncooperative suspect, or if prosecution is favored, call the public police, but first be sure there is sound evidence as in step 3.
16. Do not accept payment for stolen property because it can be construed as a bribe and it may interfere with a bond. Let the court determine restitution.
17. Handle juveniles differently from adults; consult public police.
18. When in doubt, consult an attorney.

Prosecution

Many feel strongly that prosecution is a deterrent, whereas others maintain that it hurts morale and public relations and is not cost effective. Whatever management decides, it is imperative that an incident of theft be given considerable attention so that employees realize that a serious act has taken place. Establish a written policy that is fair and applied uniformly.

As we transition here from management countermeasures to physical security countermeasures, keep in mind standards from such organizations as ASIS, NFPA, UL, and others (see Chapter 3) that enhance security. For example, the Physical Asset Protection Standard, ASIS/ANSI PAP.1-2012 provides a framework for planning, managing and improving physical security.

Physical Security Countermeasures

Integration and Convergence

The physical security strategies covered in subsequent pages are being increasingly combined into what is called integrated systems. Keener (1994: 6) offers this definition: "An **integrated system** is the control and operation by a single operator of multiple systems whose perception is that only a single system is performing all functions." These computer-based systems include access controls, alarm monitoring, CCTV, electronic article surveillance, fire protection and safety systems, HVAC, environmental monitoring, radio and video media, intercom, point-of-sale transactions, and inventory control. Traditionally, these functions existed separate from each other, but increasingly they are integrated and installed within facilities worldwide, controlled and monitored by operators and management at a centralized workstation or from remote locations.

The benefits of integrated systems include lower costs, a reduction in staff, improved efficiency, centralization, and reduced travel and time costs. For example, a manufacturing executive at corporate headquarters can monitor a branch plant's operations, production, inventory, sales, and loss prevention. Likewise, a retail executive at headquarters can watch a store and its sales floor, special displays, point-of-sale transactions, customer behavior, inventory, shrinkage, and loss prevention. *These "visits" to worldwide locations are conducted without leaving the office!*

Integration requires careful planning and clear answers to many questions, such as the following:

- Will the integrated system cost less and be easier to operate and maintain than separate systems?
- Does the supplier have expertise across all the applications?
- Is the integration software listed or approved by a third-party testing agency such as Underwriters Laboratories?
- Do authorities prohibit integration of certain systems? Some fire departments prohibit integrating fire alarm systems with other systems.

Robert Pearson (2000: 20) writes:

> *When attending a conference or trade show, it becomes obvious that every vendor and manufacturer claims to have the "total integrated solution." It would appear that one would only need to place an order at any number of display booths and all the security problems at a user's facility would simply vanish. The vendors and manufacturers freely use terms such as integrated systems, enterprise systems and digital solutions in an effort to convince end users to purchase systems and components.*

Pearson goes on to describe a typical security alarm system composed of sensors that connect to a data-gathering panel connected to a computer at a security control center. Integration would mean that sensors, card readers, and other functions would connect to the same data-gathering panel that reports to the same computer. Which multiple functions are integrated depends on the manufacturer. Some manufacturers began with energy management and added security alarm systems in later years; others began with security alarm systems and added access control. Pearson points out that integrating functions among different manufacturers via a single computer is often challenging and results in various approaches. However, there are integration firms that specialize in application-specific software that combines systems for a specific client.

Convergence of IT and physical security means that both specializations and related technologies unite for common objectives. Efforts to secure access to databases, e-mail, and organizational intranets are merging with access controls, fire and burglar alarm systems, and video surveillance. Physical security is increasingly relying on IT systems and related software. Both IT systems and physical security systems have sensors that generate managed data. As examples, an IT system will have an antivirus program and a physical security system will have motion detectors.

Bernard (2011: 34–37) notes that convergence continues to evolve; he distinguishes between technology convergence and organizational convergence. He writes about **technology convergence**, whereby voice, data, and video devices and systems interact with each other. This convergence requires a cable and wireless communications infrastructure with enough bandwidth to hold the enhanced level of data throughput. A second type of security convergence is **organizational convergence**, which aims to integrate IT and physical security.

Bernard illustrates this type by explaining that IT security protects information as does physical security; organizations should include both simultaneously when planning their information security.

Brenner (2010) offers scenarios of how the IT security side and the physical security side can work together: An offender steals a computer in the workplace. A security incident event management technology system detects a resource change (missing computer). A physical security information management system checks door access records and other physical security. All of the systems "talk to each other" (i.e., compare data) and notification technology triggers an alarm and response. Security practitioners have enhanced data to assist investigations.

In another scenario, data loss prevention (DLP) technology on a company computer detects a threat to an employee's spouse. A corporate physical security investigator checks the insider's background and finds that the insider is a domestic violence offender. IT security technology (e.g., telephony monitoring and DLP systems) is applied and security personnel monitor the case closely for action.

Bernard (2006: 28–32) refers to another aspect of convergence known as **identity management system** (IDMS). It is used to manage identities and privileges of computer systems and people. Bernard touts the benefits of IDMS by the following example: "Physical security can leverage the HR enrollment of employees by integrating the physical access control system with the IDMS, so the access control privileges are managed automatically along with IT privileges as HR enrolls, re-assigns and terminates employees." Bernard notes that the federal government is aware of the importance of IDMS in its personal identity verification (PIV) systems mandated by **Homeland Security Presidential Directive (HSPD) 12**. This mandate points to a single smart access card to be used for both physical and IT security among federal agencies.

Advantages of the convergence of IT and physical security include enhanced data, remote monitoring, less travel time, and fewer expenses. Disadvantages include a virus that may affect physical security when sharing a single server; downtime (from various causes, such as maintenance, a threat, or hazard), and an organization's bandwidth may reach its limit from the requirements of video surveillance. Sources for planners include best practice IT security standards such as those from the International Organization for Standardization (ISO) and the International Electrotechnical Commission (IEC).

IT specialists in organizations are playing a larger role in physical security decisions. They want to ensure that physical security technology is compatible with the network. An organization's physical security purchasing decisions often consist of a committee of personnel from security or loss prevention, IT, and operations. If IT managers can convince senior management that cybercrime is a greater threat than physical crime, then this will also influence the direction of the security budget.

The facility manager is another player in corporate management change. This individual, often an engineer, ensures that the company's infrastructure, which houses people and operations, functions at optimum efficiency to support business goals. The traditional security department is likely to feel a "pull" toward IT or the facility manager because its boundaries are dissolving due to information and communications technology. The process of management is increasingly dependent on information, who controls it, what is done with it, and its dissemination (Freeman, 2000: 10).

There are those who may claim the demise of the traditional security manager, who will be replaced by the IT manager or facility manager. The argument is that if an offender enters a facility and steals a computer, this crime is minor in comparison to, say, the potential harm from a hacker accessing a company's IT system. Such reasoning misses the broad, essential functions performed by the traditional security manager and staff. Examples are preventing crimes against people, responding to crimes, rendering first aid, conducting investigations, working with public police to arrest offenders, life safety, and fire protection. At the same time, traditional security practitioners must be put on notice to become involved in lifelong learning of IT systems, which touch all aspects of their traditional duties.

Cloud computing is a practice relevant to the convergence of IT and physical security. Simply put, cloud computing is the outsourcing of IT services to a contractor, with the customer accessing such services (e.g., stored data) through the Internet. The cybersecurity issues of cloud computing are growing. See Chapter 16 for additional information.

Access Controls

Access controls regulate people, vehicles, and items during movement into, out of, and within a building or facility. With regulation of these movements, assets are easier to protect. If a truck can enter a facility easily, back up to the shipping dock so the driver can load valuable cargo without authorization, and then drive away, that business will not survive. However, access controls such as the following prevent losses: the truck must stop at a gate, a security officer records identifying information on the truck and driver and runs a check through corporate IT for authorization to access, a pass is then issued, and documents are exchanged at the shipping dock under the watchful eyes of another officer. Radio Frequency Identification (RFID) technology can be applied whereby information in ID tags on the truck is uploaded (wirelessly) to a reader or access control system at the gate. If the truck is authorized to enter, the gate opens. As the truck enters, the reader automatically collects identifying information on the driver and the truck; when the truck departs, information is updated. Other options include CCTV and vehicle license plate recognition (the plate is scanned and then checked in a database for problems).

At one corporation, a security officer permitted two salespeople from another company to enter a restricted area involved in new product development. Management fired the officer.

Controlling Employee Traffic

Access control varies from simple to complex. A simple setup for employees includes locks and keys, officers checking identification badges, and written or digital logs of entries and exits. Systems that are more complex use a "smart card" containing computer memory that interacts with a reader for a host of functions and records; biometrics are used to deny or grant access. A person holding a card containing RFID can be monitored by readers throughout a facility, and if the person enters a sensitive area without authorization, security is notified and physical security features are activated (e.g., alarm sounded, doors locked, and camera zoomed in on person). Need is a prime factor influencing the kind of system employed. A research laboratory developing a new product requires strict access controls, whereas a retail business would require minimal controls.

The least number of entrances and exits are best for access control and lower costs. If possible, employees and others should be routed to the entrance closest to their destination and away from valuable assets.

Unauthorized exits locked from within create a hazard in case of fire or other emergency. To ensure safety and fewer losses, emergency exit alarms should be installed as required by codes. These devices enable quick exit, or a short delay, when pressure is placed against a horizontal bar that is secured across the door. An alarm is sounded when these doors are activated, which discourages unauthorized use.

Searching Employees

In the contract of employment, management can provide that reasonable detentions are permissible; that reasonable searches may be made to protect people and company assets; and that at any time searches may be made of desks, lockers, containers carried by employees, and vehicles (Inbau et al., 1996: 47 and 68; Nemeth, 2005: 84). Case law has permitted an employer to use a duplicate key, known to the employee, to enter a locker at will. On the other hand, an employee who uses a personal lock has a greater expectation of privacy, barring a written condition of employment to the contrary that includes forced entry. When a desk is assigned to a specific employee, an expectation of privacy exists, unless a contract states otherwise. If employees jointly have access to a desk to obtain items, no privacy exists.

Policies and procedures on searches should consider input from management, an attorney, employees, and a union if on the premises. Also, consider business necessity, what is subject to search, signed authorization from each employee, signs at the perimeter and in the workplace, and searches of visitors and others.

Should management and security have the right to search employees and others on the premises? Why or why not?

Visitors

Visitors include customers, salespeople, vendors, service people, contractors, and government employees. Any of these people can steal or cause other losses. Depending on need, a variety of techniques is applicable to visitor access control. An appointment system enables preparation for visitors. When visitors arrive without an appointment, the person at reception should lead him or her to a waiting room. Lending special equipment, such as a hardhat, may be necessary. A record or log of visits is wise. Relevant information would be name of the visitor, driver's license number and state, date of visit, times entering and leaving, purpose, specific location visited, name of employee escorting visitor, and temporary badge number. These records aid investigators. A kiosk with touch screen directory features (Figure 7-4) offers options such as visitor check in, printing a customized map, and creating a temporary access badge. Whenever possible, procedures should minimize employee–visitor contact. This is important, for instance, in the shipping and receiving department, where truck drivers may become friendly with employees and conspiracies may evolve. When restrooms and vending machines are scattered throughout a plant, truck drivers and other visitors who are permitted

FIGURE 7-4 Interactive kiosk that manages a variety of visitors. *Courtesy: Honeywell Security.*

easy access may cause losses. These services should be located at the shipping and receiving dock, and access to outsiders should be limited.

Controlling the Movement of Packages and Property

The movement of packages and property must be subject to access controls. Some locations require precautions against packaged bombs, letter bombs, and other hazards. Clear policies and procedures are important for incoming and outgoing items. To counter employee theft, outgoing items require both scrutiny and accountability. RFID tags on assets, in union with readers placed strategically at a facility, signal an alarm when an asset is moved to an unauthorized location. Uniformed officers can check outgoing items, while a property pass system serves the accountability function. At one distribution center, an employee was given permission and a property pass provided to take home cardboard. He tied the flat cardboard with string and while he exited, a security officer asked to search the cardboard. Although the employee strongly objected to the search because he had a property pass, a flat screen television was found.

Employee Identification System

The use of an employee identification (card or badge) system will depend on the number of employees that must be accounted for and recognized by other employees. An ID system not only prevents unauthorized people from entering a facility, but also deters unauthorized employees from entering restricted areas. For the system to operate efficiently, clear policies should state the use of ID cards, where and when the cards are to be displayed on the person, who should collect cards from employees who quit or are fired, and the penalties for non-compliance. A lost or stolen card should be reported so that the proper information reaches all interested personnel. Sometimes ID systems become a joke and employees refuse to wear the badges, or they decorate them or wear them in odd locations on their persons. To sustain an ID system, proper socialization is essential.

Simple ID cards contain employer and employee names. A more complex system includes an array of information, for example name, signature, address, employee number, physical characteristics (e.g., height, weight, hair and eye colors), validation date, authorized signature, location of work assignment, thumbprint, and color photo. ID cards often serve as access cards. They can be used for many other purposes, as well.

Contractors, visitors, and other nonemployees require an ID card that should be clearly distinguishable from employee ID cards. Temporary ID badges can be printed with a chemical that causes the word *void* to appear after a set period. If the ID card is an access card, an expiration date and time can be entered into the computer system.

A protective laminate coating increases the life of cards. It also discourages tampering; if an attempt is made to alter the card, it will be disfigured. Anticounterfeiting measures are always improving to counter offenders and include a variety of holographic (image) techniques, secret symbols or letters on the badge, and invisible alphanumeric type viewed by a laser.

The area where ID cards are prepared, and relevant equipment and supplies, must be secure. In addition, the equipment and software should be password protected.

Automatic Access Control

The Security Industry Association traces the development of automatic access control systems as described next (D'Agostino, 2005: 1–2). Traditionally, access control systems have been at the center of electronic security systems at buildings that include access control, ID badges, alarm systems, and CCTV. **Authentication** (i.e., verifying identity) and **authorization** (i.e., verifying that the identified individual is allowed to enter) have typically occurred as a single-step process in access control. Depending on security needs, access control has been designed for 1-factor authentication (e.g., card or personal identification number or biometric), 2-factor authentication (e.g., card-plus-PIN or card-plus-biometric), or 3-factor authentication (e.g., card-plus-PIN and biometric).

Cryptography (i.e., the study of coded or secret writings to provide security for information) became part of access control systems with the use of **encryption** (i.e., hardware or software that scrambles data, rendering it unintelligible to an unauthorized person intercepting it) to protect passwords and other information. Encryption is essential under the requirements of HSPD-12 (Hulusi, 2011: 39). It continues in importance as Ethernet networks (i.e., a trademark for a system of communications between computers on a local area network [LAN]) replace proprietary equipment connections and as security systems increasingly rely on Internet Protocol (IP) messages and shared networks with other businesses. Traditionally, because no security standards existed for these systems, manufacturers applied their own designs. However, according to the Security Industry Association, standards are now in place because of the following drivers:

- The convergence of physical and IT security
- **Common user provisioning** that permits a single point of employee registration and dismissal (usually in a human resources system) with assignment of physical and IT privileges
- Large customers (e.g., the federal government) require their facilities to be **interoperable** (i.e., products or systems working with other products or systems)
- Access controls enabling a single credential (i.e., smart card) to be used across an enterprise at buildings, facilities, and computer networks. This refers to HSPD-12, the federal government PIV program, and standards for cryptography that all serve as a model for the private sector (Kosaka, 2010, 58).
- **Digital certificate systems**, which are the electronic counterparts to driver licenses and other ID, are used to sign electronic information and serve as part of the foundation of secure e-commerce on the Internet, and are essential for physical access control system integration with IT

The traditional lock-and-key method of access control has its limitations. For instance, keys are difficult to control and easy to duplicate. Because of these problems, the need for improved access control, and technological innovations, a huge market has been created for electronic card access control systems. These systems contain wired and wireless components. The benefits of these systems include the difficulty of duplicating modern cards and cost savings because security officers are not required at each access point.

Modern access control systems are also Internet-based and offer numerous features for employees, visitors, and management. For example, employees can report a lost or stolen card, visitors can pre-register, management can see detailed reports of those who enter and leave, and CCTV and other physical security can be integrated into the system to aid investigations.

Before an automatic access control system is implemented, several considerations are necessary. *Safety must be a prime factor to ensure quick exit in case of emergency.* Another consideration deals with the adaptability of the system to the type of door presently in use. Can the system accommodate all traffic requirements? How many entrances and exits must be controlled? Will there be an annoying waiting period for those who want to gain access? Are additions to the system possible? What if the system breaks down? Is a backup source of power available (e.g., generators)?

Tailgating and pass back are other concerns. **Tailgating** means an authorized user is followed by an unauthorized user. To thwart this problem, a security officer can be assigned to each access point, but this approach is expensive when compared to applying CCTV, revolving doors, and turnstiles. Revolving doors can be expensive initially, and they are not an approved fire exit. Optical turnstiles contain invisible infrared beams to count people entering and leaving to control tailgating and pass back. These sensors can be installed in a doorframe and connected to an alarm system and CCTV. **Pass back** refers to one person passing an opening and then passing back the credential so another person can pass through the opening.

A summary of cards used in card access systems follows:

- *Smart cards* contain computer memory within the plastic that records and stores information and personal identification codes. Security is increased because information is in the card, rather than the reader. Zalud (2010: 86) writes that because of the memory in the card, it can require the user to supply a PIN or biometric to the reader before the card interacts with the reader; this feature prevents unauthorized use. In addition, cryptography features secure communications between the reader and the card. Smart cards permit a host of activities from access control to making purchases, while almost eliminating the need for keys or cash. This type of card is growing in popularity as its applications expand.
- *Proximity cards* (also referred to as RFID) need not be inserted into a reader but placed in its "proximity." A code is sent via radio frequency, magnetic field, or microchip-tuned circuit. This card is in wide use today.
- *Contact Memory Buttons* are stainless steel buttons that protect an enclosed computer chip used for access. The information in the button can be downloaded or updated with a reader like other systems. These buttons are known for their durability, serve to ensure accountability of security officers on patrol, and are applied as an asset tag. The buttons are used widely.
- *Magnetic stripe cards* are plastic, laminated cards (like credit cards) that have a magnetic stripe along one edge onto which a code is printed. When the card is inserted, the magnetically encoded data are compared to data stored in a computer and access is granted on verification. This card is widely used, but easy to clone.
- *Weigand cards* employ a coded pattern on a magnetized wire within the card to generate a code number. To gain access, the card is passed through a sensing reader. Other technologies have reduced the popularity of this type of card.
- *Bar-coded cards* contain an array of tiny vertical lines that can be visible and vulnerable to photocopying or invisible and read by an infrared reader. Other technologies have reduced the popularity of this vulnerable card.
- *Magnetic dot cards* contain magnetic material, often barium ferrite, laminated between plastic layers. The dots create a magnetic pattern that activates internal sensors in a card reader. This card is easy to clone and rarely used.

Access card systems vary in terms of advantages, disadvantages, and costs. Each type of card can be duplicated with a sufficient amount of knowledge, time, and equipment. For example, a magnetic stripe is easy to duplicate. A piece of cardboard with a properly encoded magnetic stripe functions with equal efficiency. The Weigand card has the disadvantage of wear and tear on the card that passes through a slot for access. Proximity cards have the advantage of the sensing element concealed in a wall, and the card typically can be read without removing it from a pocket. Smart cards are expensive, but they can be combined with other card systems; also, they are convenient because of the capability of loading and updating the card applications over the Internet (Morton, 2011: 28; Barry, 1993: 75; Garcia, 2006: 156–157; Gersh, 2000: 18; and Toye, 1996: 23).

Near field communication (NFC) is a technology with wide applications (Nosowitz, 2011; Jarvis, 2011: S-10). It is short-range (less than four inches), wireless (radio) communication between two devices by touching them together or placing the devices in close proximity. NFC evolved from RFID; the difference is that whereas RFID is one-way, NFC is two-way—an NFC-enabled device can send and receive information. A smart phone containing NFC capabilities can serve as a credit or debit card. A device with NFC can serve as a library card, transit pass, and access card to a secure door, computer or network, among other applications. A wallet and keys may someday become obsolete. NFC communication is subject to hacking. The RF signal can be captured with an antenna; information can be stolen or modified. Security includes use of a password, encryption, keypad lock, and anti-virus software.

Biometric Technologies

Biometric security systems have been praised as a major advance in access control because such systems link the event to a particular individual, whereas an unauthorized individual may use a key, card, PIN, or password. Di Nardo (2009: 195) defines biometrics as "the automated use of physiological or behavioral characteristics to determine or verify identity." He adds that it can also be defined as "the study of measurable biological characteristics."

These systems seek to verify an individual's identity through fingerprint scan, hand geometry (shape and size of hand) (see Figure 7-5), iris scan (the iris is the colored part around the pupil of the eye), facial scan, retinal scan (capillary pattern), voice patterns, signature recognition, vascular (vein) pattern recognition, palm print, ear shape, gait, and keystroke dynamics.

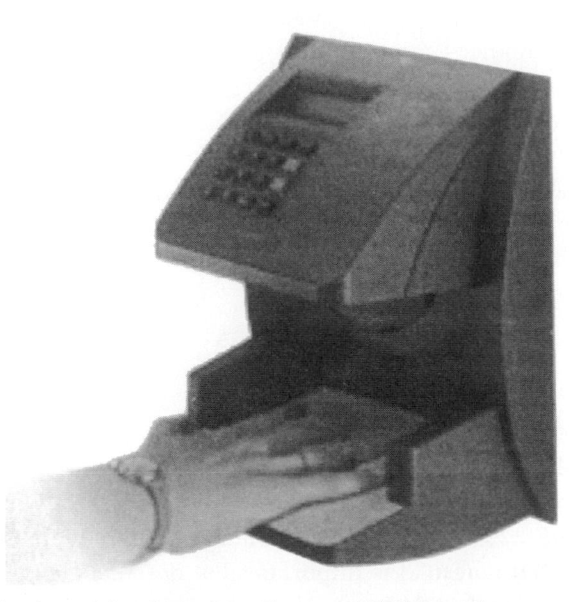

FIGURE 7-5 Verifying identity through hand geometry. *Courtesy: HID Corporation.*

The biometric leaders are fingerprint, hand geometry, iris, and face recognition (Di Nardo, 2009: 194–216; Piazza, 2005: 41–55).

Biometric security systems have disadvantages and can be circumvented. As examples, retinal can be affected by diseases, vascular is expensive and requires bulky equipment, drugs can affect gait, and keystroke has high error rates. Iris was defeated by using a glass eyeball, contact lenses, or high-resolution photos (Di Nardo, 2009: 194–216). Fingerprint can be circumvented by collecting an authorized person's latent fingerprint by lifting it with tape. Terrorists cut off the thumb of a bank manager to gain entry through a fingerprint-based access control system. Researchers constructed fake fingers by taking casts of real fingers and molding them into Play-Doh. The researchers developed a technique to check for moisture as a way to reduce this ploy (Aughton, 2005).

Spence (2011) writes in Locksmith Leger that certain locksmiths are hesitant to become involved with biometric fingerprint systems. One quipped: "They're one percent of my sales and 10 percent of my service calls." Failure rates were between three and 20 percent.

Research continues to improve biometrics. In the near term, we will not see facial scan pick a known terrorist out of a crowd, but the technology is evolving. Digitized photos shot at angles or in poor light can be flawed. The challenge with facial scan is being able to identify a person on the move (Philpott, 2005: 16–21).

Biometric systems operate by storing identifying information (e.g., fingerprints, photos) in a computer to be compared with information presented by a subject requesting access. Access controls often use multiple technologies, such as smart card and biometrics. One location may require a card and a PIN (see Figure 7-6), whereas another requires scanning a finger and a PIN. Many systems feature a distress code that can be entered if someone is being victimized. Another feature is an alarm that sounds during unauthorized attempted entry. Access systems can be programmed to allow select access according to time, day, and location. The logging capabilities are another feature to ascertain personnel location by time, date, and the resources expended (e.g., computer time, parking space, cafeteria). These features provide information during investigations and emergencies.

We are seeing an increasing merger of card access systems and biometric technology, and thus, missing or stolen cards are less of a concern. We will see more point-of-sale readers that accept biometric samples for check cashing, credit cards, and other transactions. As research continues to improve biometrics, these systems will become universal—banking, correctional facilities, welfare control programs, and so forth.

Locks and Keys

The basic purpose of a lock-and-key system is to hinder unauthorized entry. Attempts to enter a secure location usually are made at a window or door to a building or at a door somewhere within a building. Consequently, locks deter unauthorized access from outsiders and insiders. *Many see a lock only as a delaying device that is valued by the amount of time needed to defeat it.* Zunkel (2003: 32) notes: "It is important that designers know that a lock by itself is only part of a larger system that includes the door, the wall, the perimeter and a security plan."

FIGURE 7-6 Card reader and key pad. *Courtesy: Diebold, Inc.*

An offender may decide to avoid a high-security lock and break through a weak door, wall, ceiling, floor, roof, or window.

Standards related to locking systems include those from American National Standards Institute (ANSI), American Society for Testing and Materials (ASTM), Underwriters Laboratories (UL), and the Builders Hardware Manufacturers Association (BHMA). Local ordinances may specify requirements for locks.

Two general ways to classify locks are mechanical and electromechanical. **Mechanical locks** include the common keyed lock and the pushbutton lock that contains a keypad to enter an access code to release the lock. **Electromechanical locks** include an electronic keypad that is connected to an electric strike, lock, or magnetic lock. When the access code is entered, the strike or lock is released to open the door (Department of Defense, 2000: D-5).

There are many types of locks and locking systems ranging from those that use simple, ancient methods to those that apply modern technology, including electricity, wireless components, computers, and the Internet. Here, we begin with basic information as a foundation for understanding locks.

Locking devices are often operated by a key, numerical combination, card, or electricity. Many locks (except padlocks) use a deadbolt and latch. The **deadbolt** (or bolt) extends from a door lock into a bolt receptacle within the doorframe. Authorized entry is made by using an appropriate key to manually move the bolt into the door lock. **Latches** are spring loaded and less secure than a deadbolt. They are cut on an angle to permit them to slide right into the

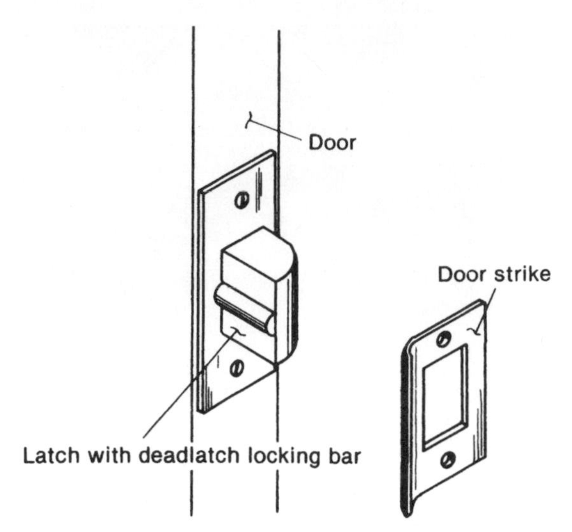

FIGURE 7-7 Latch and door strike.

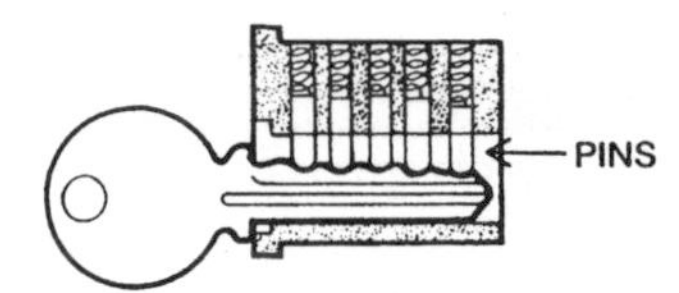

FIGURE 7-8 Cylinder.

strike when the door is closed (see Figure 7-7). Unless the latch is equipped with a locking bar (deadlatch), a knife can possibly be used to push the latch back to open the door.

The **cylinder** part of a lock contains the keyway, pins, and other mechanisms that permit the deadbolt or latch to be moved by a key for access (see Figure 7-8). Double-cylinder locks, in which a cylinder is located on each side of a door, are a popular form of added security as compared to single-cylinder locks. *Double-cylinder locks require a key for both sides; however, fire codes may prohibit such locks.* With a single-cylinder lock, a thief may be able to break glass or remove a wood panel and then reach inside to turn the knob to release the lock. For safety's sake, locations that use double-cylinder locks must prepare for emergency escape by having a key readily available.

Key-in-knob locks are used universally but are being replaced by key-in-the-lever locks (see Figure 7-9) to be ADA compliant. As the name implies, the keyway is in the knob or lever. Most contain a keyway on the outside and a button on the inside for locking from within.

Entrances for Handicapped

The Internal Revenue Service offers a tax credit to eligible businesses that comply with provisions of the ADA to remove barriers and promote access for individuals with disabilities (EEOC, 2011). The door hardware industry offers several products and solutions to aid the

FIGURE 7-9 Mechanical lock with lever requiring no wiring, electronics, or batteries. *Courtesy: Ilco Unican.*

disabled (see Figure 7-10). Electrified door hardware such as magnetic locks and electromechanical locks retracts the latch when energized.

Attacks and Hardware

The Internet, including YouTube, offers a wealth of information on attacking locks. In addition, lock picking has become a sport with club members and chapters in multiple countries (Loughlin, 2009). Google queries for "attacks on locks", "lock picking", and "lock bumping" result in millions of sources. These methods are explained next.

There are several ways to attack locks. One technique, as stated earlier, is to force a knife between the doorframe (jamb) and the door near the lock to release the latch. However, when a deadlatch or deadbolt is part of the locking mechanism, methods that are more forceful are needed. In one method, called "springing the door," a screwdriver or crowbar is placed between the door and the doorframe so that the bolt extending from the door lock into the bolt receptacle can be pried out, enabling the door to swing open (see Figure 7-11). A 1-inch bolt will hinder this attack.

In "jamb peeling," another method of attack, a crowbar is used to peel at the doorframe near the bolt receptacle so that the door is not stopped from swinging open. Strong hardware for the doorframe is helpful. In "sawing the bolt," a hacksaw is applied between the door and the doorframe, similar to the placement of the screwdriver in Figure 7-11. Here again, strong

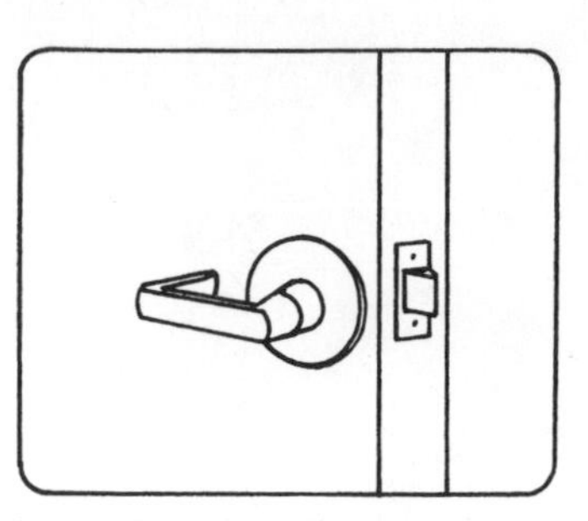

Lever trim reduces force required to unlatch a door.

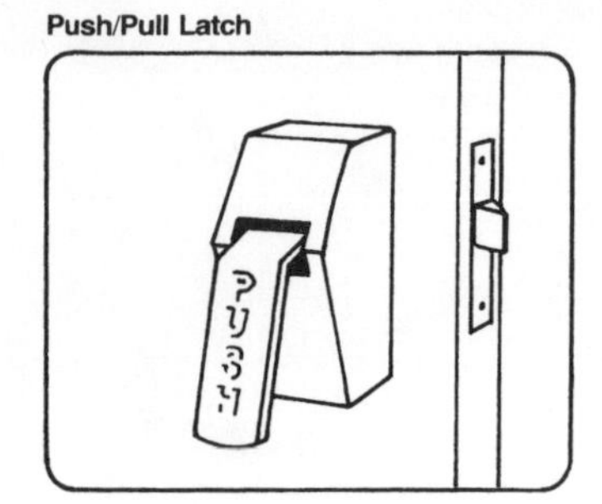

Push/Pull latches are popular on institutional doors because of ease of operation.

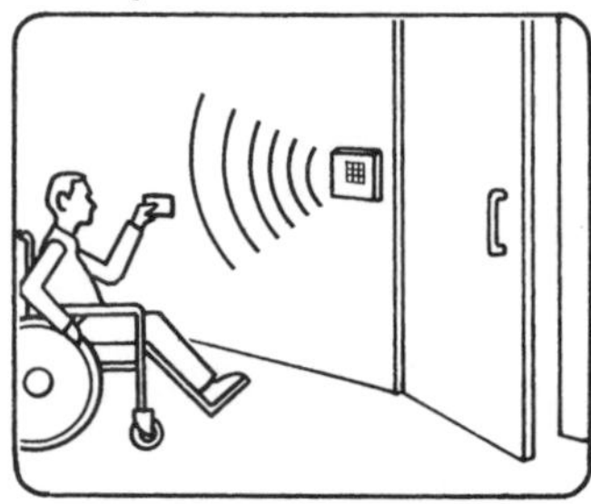

Proximity card reader requires only close presence of the user's card to activate door's automatic opener.

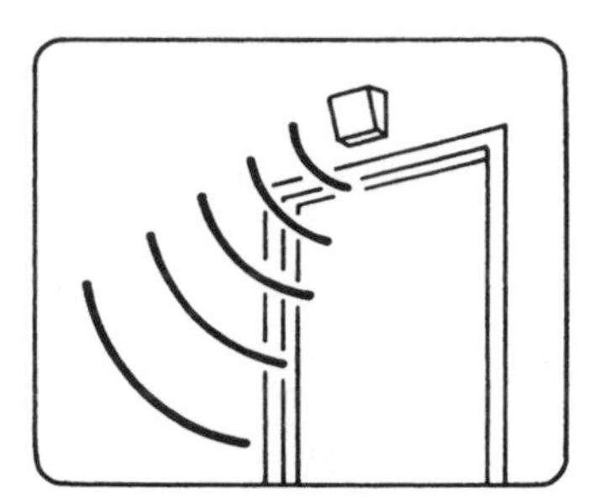

Presence detectors are popular with automatic exit doors and require no physical action.

FIGURE 7-10 Entrances for handicapped. *Courtesy: Von Duprin Division of Ingersoll-Rand Company.*

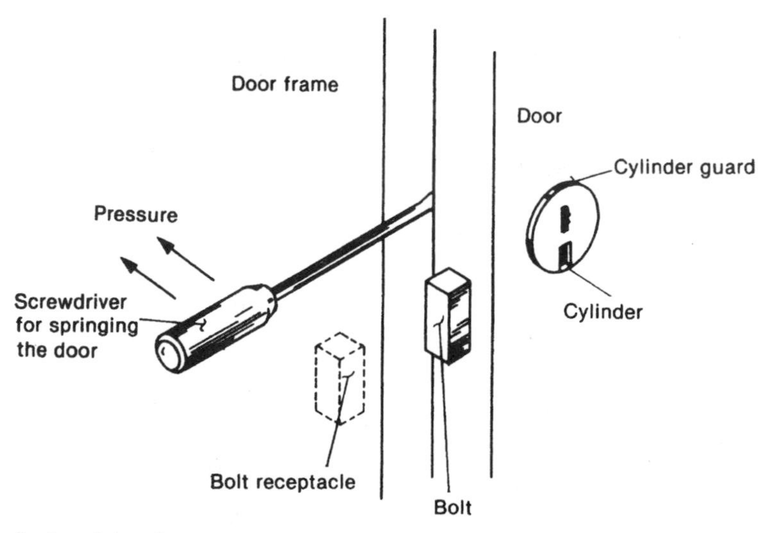

FIGURE 7-11 Deadbolt and door frame.

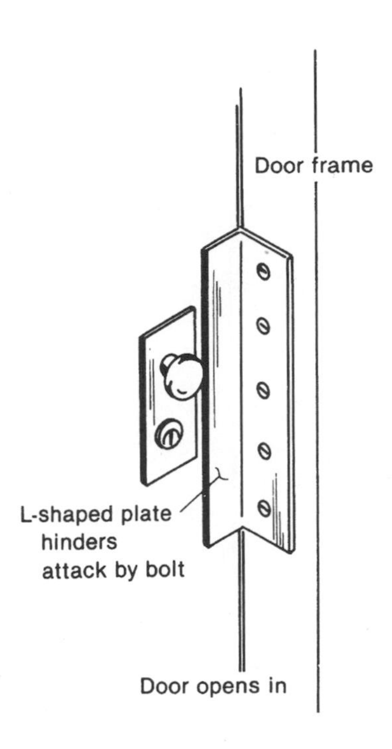

FIGURE 7-12 L-shaped plate.

hardware, such as a metal bolt composed of an alloy capable of withstanding a saw blade, will impede attacks. Some offenders use the cylinder-pulling technique: the cylinder on the door is actually ripped out with a set of durable pliers or tongs. A circular steel guard surrounding the cylinder (see Figure 7-11) will frustrate the attacker. Offenders also are known to use automobile jacks to pressure doorframes away from a door.

Both high-quality hardware and construction will impede attacks, but the door itself must not be forgotten. If a wood door is only 1/4-inch thick, even though a strong lock is attached, the offender may simply break through the door. A solid wood door 1 3/4 inches thick or a metal door is a worthwhile investment. Wood doorframes at least two inches thick provide durable protection. When a hollow steel frame is used, the hollow area can be filled with cement to resist crushing near the bolt receptacle. An L-shaped piece of iron secured with one-way screws will deter attacks near the bolt receptacle for doors swinging in (see Figure 7-12). When a padlock is used in conjunction with a safety hasp, the hasp must be installed correctly so that the screws are not exposed (see Figure 7-13).

Many attacks are by forced entry, which is easier to detect than when the use of force is minimal. Lock picking is one technique needing a minimum amount of force. It is used infrequently because of the expertise required, although picks are available on the Internet. **Lock picking** is accomplished by inserting a tension wrench (an L-shaped piece of metal) into the cylinder and applying tension while using metal picks to align the pins in the cylinder as a key would to release the lock (see Figure 7-8). The greater the number of pins, the more difficult

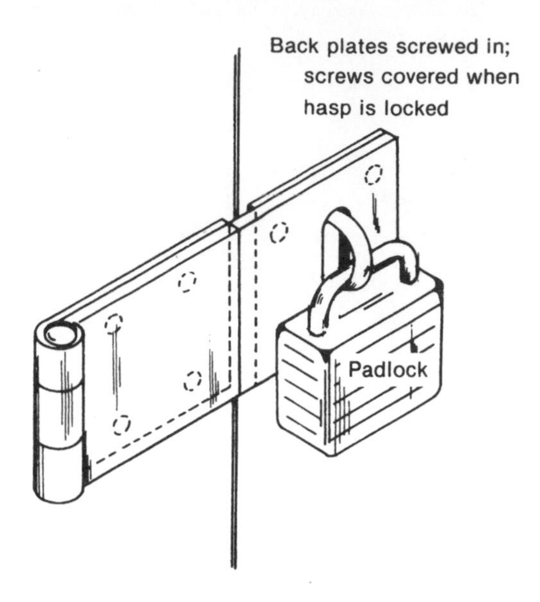

FIGURE 7-13 Safety hasp.

it is to align them. A cylinder should have at least six pins. **Lock bumping** is touted as an easy method of picking a pin tumbler lock by inserting a specially designed bump key into the keyway and applying a slight force while turning the key. Another type of attack (more difficult) utilizes a blank key, matches, and a file. The blank key is placed over a lighted match until carbon is produced on the key. Then the key is inserted into the cylinder. The locations where the pins have scraped away the carbon signify where to file. Needless to say, this method is time-consuming and calls for repeated trials. Offenders sometimes covertly borrow a key, quickly press it into a bar of soap or wax, return the key, and then file a copy on a blank key. This method illustrates the importance of key control.

After gaining access, an offender may employ some tricks to make sure nobody enters while he or she is busy. This is accomplished, for instance, by inserting a pin or obstacle in the keyway and locking the door from the inside.

Whatever hardware is applied, the longer it takes to attack a lock, the greater is the danger for the offender. One further point: most burglary insurance policies state that there must be visible signs of forced entry to support a claim.

Offenders may use other methods of entry. A thief may simply use a stolen key or a key (or access card) borrowed from another person. Unfortunately, intruders often enter restricted areas because somebody forgot to use a locking device. This mistake renders the most complex locks useless.

The methods of defeating lock-and-key systems do not stop here. Innovative thieves and various kinds of locks, keys, and access systems create a hodgepodge of methods that loss prevention practitioners should understand.

Types of Locks

Volumes have been written about locks. The following briefly summarizes simple and more complex locks:

- *Warded (or skeleton key tumbler) lock:* This older kind of lock is disengaged when a skeleton key makes direct contact with a bolt and slides it back into the door. It is an easy lock to pick. A strong piece of L-shaped metal can be inserted into the keyway to move the bolt. Warded locks are in use in older buildings and are recognized by a keyway that permits seeing through. Locks on handcuffs are of the warded kind and can be defeated by a knowledgeable offender.
- *Disc tumbler lock:* The use of this lock, originally designed for the automobile industry, has expanded to desks, file cabinets, and padlocks. Its operation entails flat metal discs, instead of pins, that align when the proper key is used. These locks are mass produced, inexpensive, and have a short life expectancy. More security is offered than warded locks can provide, but disc tumbler locks are subject to defeat by improper keys or being jimmied.
- *Pin tumbler lock:* Invented by Linus Yale in 1844, the pin tumbler lock is used widely in industry and residences (see Figure 7-8). Its security surpasses that of the warded and disc tumbler kinds.
- *Lever tumbler lock:* Lever locks vary widely. These locks disengage when the proper key aligns tumblers. Those found in luggage, cabinets, chests, and desks often provide minimal security, whereas those found in bank safe deposit boxes are more complex and provide greater security. The better quality lever lock offers more security than the best pin tumbler lock.
- *Combination lock:* This lock requires manipulating a numbered dial(s) to gain access. Combination locks usually have three or four dials that must be aligned in the correct order for entrance. These locks provide greater security than key locks because a limited number of people probably will know the lock combination, keys are unnecessary, and lock picking is obviated. They are used for safes, bank vaults, and high-security filing cabinets. With older combination locks, skillful burglars are able actually to listen to the locking mechanism to open the lock; more advanced mechanisms have reduced this weakness. A serious vulnerability results when an offender watches the opening of a combination lock with either binoculars or a telescope. Retailers sometimes place combination safes near the front door for viewing by patrolling police; however, unless the retailer uses his or her body to block the dial from viewing, losses may result. This same weakness exists where access is permitted by typing a PIN.
- *Combination padlock:* This lock is similar in operation to a combination lock. It is used on employee or student lockers and in conjunction with safety hasps or chains. Some of these locks have a keyway so they can be opened with a key.
- *Padlock:* Requiring a key, this lock is used on lockers or in conjunction with hasps or chains. Numerous types exist, each affording differing levels of protection. More secure ones have disc tumbler, pin tumbler, or lever characteristics. Serial numbers on padlocks are a security hazard similar to combination padlocks.

Other kinds of locks include devices that have a bolt that locks vertically instead of horizontally. Emergency exit locks with alarms or “panic alarms” enable quick exit in emergencies while deterring unauthorized door use. Sequence locking devices require locking the doors in a predetermined order; this ensures that all doors are locked because the outer doors will not lock until the inner doors are locked.

The use of interchangeable core locks is a method to deal with the theft, duplication, or loss of keys. Using a special control key, one core (that part containing the keyway) is simply replaced by another. A different key then is needed to operate the lock. This system, although more expensive initially, minimizes the need for a locksmith or the complete changing of locks.

Automatic locking and unlocking devices also are a part of the broad spectrum of methods to control access. Digital locking systems open doors when a particular numbered combination is typed. If the wrong number is typed, an alarm is sounded. Combinations can be changed when necessary. Electromagnetic locks use magnetism, electricity, and a metal plate around doors to hold doors closed. When the electricity is turned off, the door can be opened. Remote locks enable opening a door electronically from a remote location.

Trends taking place with locks and keys include increasing use of electronics and microchip technology. For example, hybrids have been developed whereby a key can serve as a standard hardware key in one door and an electronic key in another door. “Smart locks” have grown in popularity. These locks combine traditional locks with electronic access control; read various types of access cards for access; use a tiny computer to perform multiple functions, including holding data (e.g., access events); and can be connected to an access control system for uploading and downloading data. Aubele (2011) refers to “intelligent key systems” whereby keys are programmable like access control cards to limit access according to specific times and doors. In addition, these keys store data on use.

Wireless locking systems and RF online locking systems make use of modern technology, although care must be exercised in the evaluation and purchasing process. A pilot project helps to ensure reliability. Signals are hindered by metallic materials (e.g., steel buildings).

Although card access systems are used universally, locks and keys are still used to protect a variety of assets.

Master Key Systems

In most instances, a lock accepts only one key that has been cut to fit it. A lock that has been altered to permit access by two or more keys has been *master keyed*. The **master key system** allows a number of locks to be opened by the master key. This system should be confined to high-quality hardware utilizing pin tumbler locks. A disadvantage of the master key system is that if the master key is lost or stolen, security is compromised. A *change key* fits one lock. A *submaster key* will open all locks in, for instance, a wing of a building. The master key opens locks covered by two or more submaster systems.

Key Control

Without adequate key control, locks are useless and losses are likely to climb. Accountability and proper records are necessary, as with access cards. Computerized, online record-keeping programs are available for key control, similar to software used in electronic card access control systems (Truncer, 2011: 79–85). Keys should be marked with a code to identify the corresponding lock; the code is interpreted via a record stored in a safe place. A key should never be marked, "Key for room XYZ." When not in use, keys should be positioned on hooks in a locked key cabinet or vault. The name of the employee, date, and key code are vital records to maintain when a key is issued. These records require continuous updating. Employee turnover is one reason why precise records are vital. Departing employees will return keys (and other valuables) if their final paycheck is withheld. Policies should state that reporting a lost key would not result in punitive action; an investigation and a report will strengthen key control. If key audits check periodically who has what key, control is further reinforced. To hinder duplication of keys, "do not duplicate" may be stamped on keys, and company policy can clearly state that key duplication will result in dismissal. It is wise to change locks every eight months and sometimes at shorter intervals on an irregular basis. Key control is also important for vehicles such as autos, trucks, and forklifts. These challenges and vulnerabilities of traditional lock and key systems have influenced organizations to switch to modern access control and biometric systems.

Intrusion Detection Systems

An **intrusion detection system** detects and reports an event or stimulus within its detection area. A response to resolve the reported problem is essential. The emphasis here is on interior sensors. Sensors appropriate for perimeter protection are stressed in Chapter 8. We must remember that intrusion detection systems are often integrated with other physical security systems and rely on IT systems with Internet capabilities.

What are the basic components of an intrusion detection system? Three fundamental components are sensor, control unit, and annunciator. **Sensors** detect intrusion by, for example, heat or movement of a human. The **control unit** receives the alarm notification from the sensor and then activates a silent alarm or **annunciator** (e.g., a light or siren), which usually produces a human response. There are a variety of intrusion detection systems; they can be wired or wireless. Several standards exist for intrusion detection systems from UL, ISO, the Institute of Electrical and Electronics Engineers, and other groups. Types of interior sensors are explained next (Garcia, 2006: 104–122; Honey, 2003: 48–94).

Interior Sensors

In the electronics field, a "switch" is a component that can interrupt an electrical circuit (e.g., as with a light switch). A *balanced magnetic switch* consists of a switch mounted to a door (or window) frame and a magnet mounted to a moveable door or window. When the door is closed, the magnet holds the switch closed to complete an electrical circuit. An alarm is triggered when the door is opened and the circuit is interrupted. An ordinary magnetic switch is similar to the balanced type, except that it is simpler, less expensive, and provides a lower level of security. Switches can be visible or hidden and afford good protection against opening a door; however, no security method is ever foolproof.

Mechanical contact switches contain a pushbutton-actuated switch that is recessed into a surface. An item is placed on it that depresses the switch, completing the alarm circuit. Lifting the item interrupts the circuit and signals an alarm.

Pressure-sensitive mats contain two layers of metal strips or screen wire separated by sections of foam rubber or other flexible material. When pressure is applied, as by a person walking on the mat, both layers meet and complete an electrical contact to signal an alarm. These mats are applied as internal traps at doors, windows, and main traffic points, as well as near valuable assets. The cost is low, and these mats are difficult to detect. If an offender detects the mat, he or she can walk around it.

Grid wire sensors are made of fine insulated wire attached to protected surfaces in a grid pattern consisting of two circuits, one running vertical, the other horizontal, and each overlapping the other. An interruption in either circuit signals an alarm. This type of sensor is applied to grill work, screens, walls, floors, ceilings, doors, and other locations. Although these sensors are difficult for an offender to spot, they are expensive to install, and an offender can jump the circuit.

Trip wire sensors use a spring-loaded switch attached to a wire stretched across a protected area. An intruder "trips" the alarm (i.e., opens the circuit) when the wire is pulled loose from the switch. If an offender spots the sensor, he or she may be able to circumvent it.

Vibration sensors detect low-frequency energy resulting from the force applied in an attack on a structure (see Figure 7-14). These sensors are applied to walls, floors, and ceilings. Various sensor models require proper selection.

Capacitance sensors create an electrical field around metallic objects that, when disturbed, signals an alarm (see Figure 7-15). These sensors are applied to safes, file cabinets, grills at openings (e.g., windows), fences, and other metal objects. One sensor can protect many objects; however, it is subject to defeat by using insulation (e.g., heavy gloves).

Infrared photoelectric beam sensors activate an alarm when an invisible infrared beam of light is interrupted (see Figure 7-16). If the system is detected, an offender may jump over or

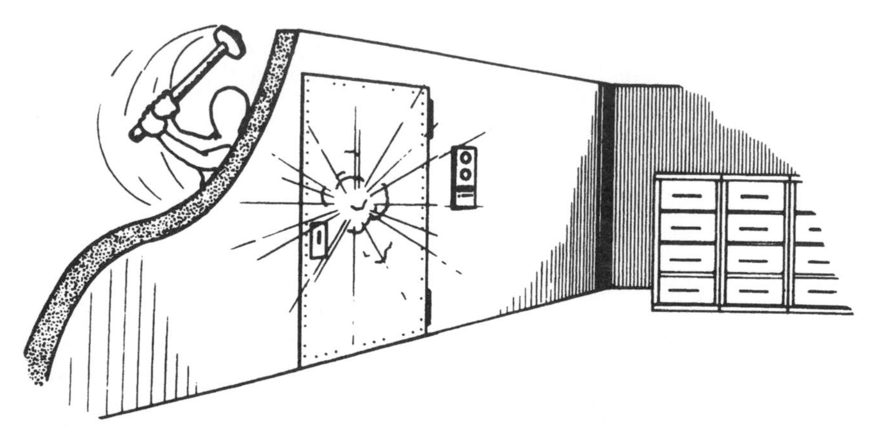

FIGURE 7-14 Vibration sensor.

crawl under the beam to defeat it. To reduce this vulnerability, tower enclosures can be used to stack sensors.

Ultrasonic motion (UM) detectors focus on sound waves to detect motion. Active UM detectors create a pattern of inaudible sound waves that are transmitted into an area and monitored by a receiver. This detector operates on the *Doppler Effect,* which is the change in frequency that results from the motion of an intruder. Passive UM detectors react to sounds (e.g., breaking glass). These detectors are installed on walls or ceilings or used covertly (i.e., disguised

FIGURE 7-15 Capacitance sensor.

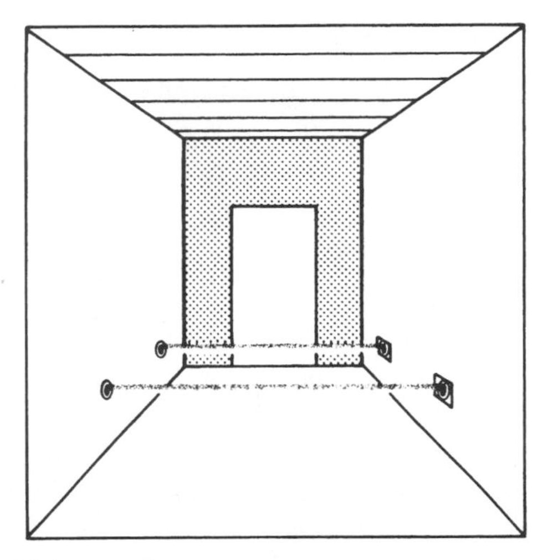

FIGURE 7-16 Infrared photoelectric beam system.

within another object). The sensitive abilities of these detectors result in many false alarms, which limit use.

Microwave motion detectors also operate on the Doppler frequency-shift principle. An energy field is transmitted into an area and monitored for a change in its pattern and frequency, which results in an alarm. Because microwave energy penetrates a variety of construction materials, care is required for placement and aiming. However, this can be an advantage in protecting multiple rooms and large areas with one sensor. These sensors can be defeated (like UM) by objects blocking the sensor or by fast or slow movement.

Passive infrared (PIR) intrusion sensors are passive in that they do not transmit a signal for an intruder to disturb. Rather, moving infrared radiation (from a person) is detected against the radiation environment of a room. When an intruder enters the room, the level of infrared energy changes and an alarm is activated. Although the PIR is not subject to as many nuisance alarms as ultrasonic and microwave detectors, it should not be aimed at sources of heat or surfaces that can reflect energy. The PIR can be defeated by blocking the sensor so it cannot pick up heat.

Passive audio detectors listen for noise created by intruders. Various models filter out naturally occurring noises not indicating forced entry. These detectors can use public address system speakers in buildings, which can act as microphones to listen to intruders. The actual conversation of intruders can be picked up and recorded by these systems. To enhance this system, CCTV can provide visual verification of an alarm condition, video in real time, still images digitally to security or police, and evidence. The audio also can be two-way, enabling security to warn the intruders. *Such audiovisual systems must be applied with extreme care to protect privacy, confidentiality, and sensitive information, and to avoid violating state and federal wiretapping and electronic surveillance laws.*

Fiber optics is used for intrusion detection and for transmission of alarm signals. It involves the transportation of information via guided light waves in an optical fiber. This sensor can be attached to or inserted in many things requiring protection. When stress is applied to the fiber optic cable, an infrared light pulsing through the cable reacts to the stress and signals an alarm.

Intrusion detection systems only detect and report an alarm condition. These systems do not stop or apprehend an intruder.

Trends

Two types of sensor technologies often are applied to a location to reduce false alarms, prevent defeat techniques, or fill unique needs. The combination of microwave and passive infrared sensors is a popular example of applying **dual technologies** (see Figure 7-17). Reporting can be designed so an alarm is signaled when both sensors detect an intrusion (to reduce false alarms) or when either sensor detects an intrusion. Sensors are also becoming "smarter" by sending sensor data to a control panel or computer, distinguishing between humans and animals, and activating a trouble output if the sensor lens is blocked. *Supervised wireless sensors* have become a major advancement because sensors can be placed at the best location without the expense of running a wire; these sensors are constantly monitored for integrity of the radio

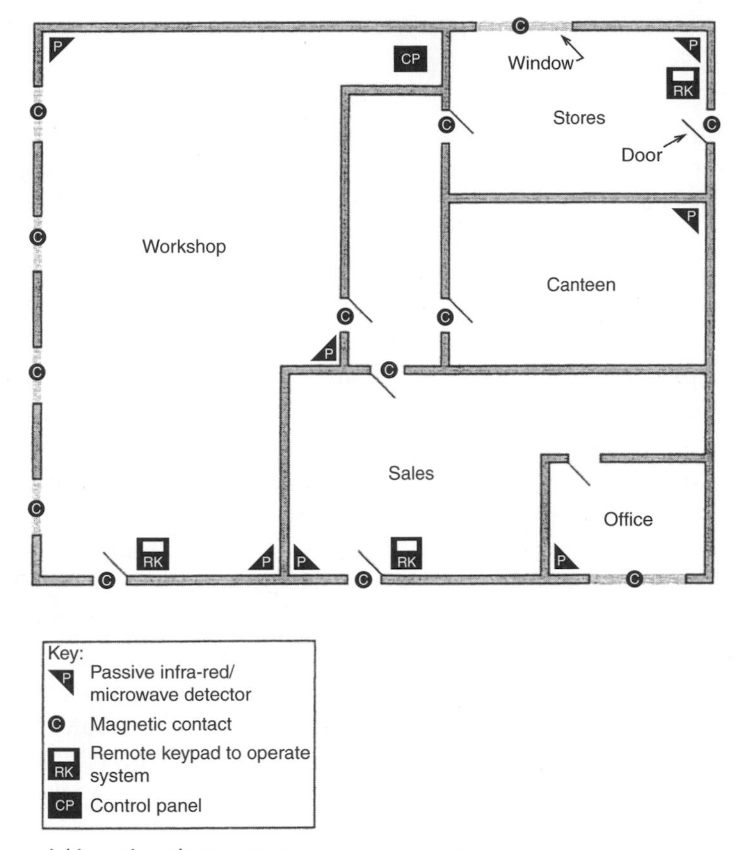

FIGURE 7-17 Commercial intrusion alarm system.

frequency link between the sensor and panel, status of the battery, and whether the sensor is functioning normally (Garcia, 2006: 104; O'Leary, 1999: 36–48).

Operational Zoning

Operational zoning means that the building being protected has a segmented alarm system, whereby the alarm can be turned on and off within particular zones depending on usage. For example, if an early morning cleaning crew is in the north end of a plant, then that alarm is turned off while other zones still have the alarm on. Furthermore, zoning helps to pinpoint where an intrusion has occurred.

Alarm Monitoring

Today, many entities have an alarm system that is monitored by an in-house station (e.g., a console at a secure location) or from a central station (contract service) located off the premises. These services can easily supply reports of unusual openings and closings, as well as those of the regular routine. Chapter 8 covers alarm signaling systems.

Closed-Circuit Television

Closed-circuit television, or CCTV (see Figure 7-18), assists in deterrence, surveillance, apprehension, and prosecution. This technology is also helpful in civil cases to protect an organization's interests. The applications go beyond security and justice. For instance, CCTV can yield a greater ROI by serving as a tool to understand production problems or customer behavior. Although it may be costly initially, CCTV reduces personnel costs because it allows the viewing of multiple locations by one person. For instance, throughout a manufacturing plant, multiple cameras are installed, and one security officer in front of a console monitors the cameras. Accessories include pan (i.e., side-to-side movement), tilt (i.e., up-and-down movement), and zoom lenses, referred to as "PTZ" in the industry, which are mechanisms that permit viewing mobility and opportunities to obtain a close look at suspicious activity. Additional system capabilities include recording incidents and viewing when limited light is present. Modern technology has greatly altered CCTV capabilities, as described in subsequent paragraphs.

Standards for CCTV systems are from several sources. These include ANSI, SIA, National Electrical Manufacturers Association, American Public Transportation Association, government agencies, ISO/International Electrotechnical Commission, and the International Code Council. England and Australia are especially active preparing CCTV standards.

CCTV began its development during the 1950s. The traditional CCTV system that came into greater use in the 1970s consisted of analog recording systems, solid-state cameras, and coaxial cable (Siemon Company, 2003; Suttell, 2006: 114). This older technology applied multiple cameras connected through cabling to a camera control unit and a multiplexer that fed several videocassette recorders (VCR) in a central control room. The images were viewed real time via several monitors. The disadvantages of this technology include the following: the control room is a single point of failure within the security infrastructure; if a camera is moved, cable is required for the connection; the use of VCRs results in numerous cassette tapes requiring storage space; and humans are necessary to change and store tapes.

Older technology, such as the VCR that could record for a limited number of hours, was followed by time-lapse recorders (i.e., single frames of video are stored at intervals over an

FIGURE 7-18 Closed-circuit television (CCTV) sign. Camera at top.

extended period of time) with recording capabilities up to several hundred hours, plus an alarm mode in which the recorder reverts to real time when an alarm condition exists. Real-time setting records 30 frames a second; time-lapse video may record between one frame a second and one frame every eight seconds. Time-lapse recorder features included a quick search for alarm conditions during playback, the playing of recorded video frames according to the input of time by the user, and the interface with other security systems such as access controls to ensure a video record of all people entering and departing.

A new generation CCTV system developed with unshielded twisted-pair (UTP) cabling (i.e., a cable with multiple pairs of twisted insulated copper conductors in a single sheath) that enabled cameras to run on the existing infrastructure. **Digital video recorders** (DVR) were introduced in the mid-1990s, and with them, several advantages over analog, including recording on hard disk drives, as a file is stored on a personal computer. Other advantages are avoiding tape storage, remote viewing, easy playback and searches, improved quality of images, and longer life of recordings. Another advance is digital recording in networking, which is referred to as **network video recorder** (NVR). Rather than many DVRs networked together, an NVR is the camera system. An NVR comprises digital cameras managed by specially designed computer operating software designed to manage video surveillance (Alten, 2005: 8).

Internet protocol (IP)-based network cameras permit IP networking of video to be shared where the network reaches, including offsite storage. IP video can be controlled and viewed from a PDA, phone, laptop computer, or other mobile device. It is also encrypted. IP-based CCTV systems, including IP cameras, IP video servers, and IP keyboards, can be located almost anywhere. In addition, the IP keyboard can control the PTZ and other management functions such as recording and searching. When the existing infrastructure in a building is used, a building can become automated on one cable system and include not only CCTV, but also access control, fire/safety systems, voice, network traffic, and other systems.

Not everyone is happy with IP-based network cameras. Pfeifle, L. (2010) reports of one retail security practitioner who argued that with IP a lot of coordination is required, especially with IT employees, installation is difficult, enough bandwidth must be available, and with several cameras, a platform to manage them is necessary.

It is important to distinguish between the older **analog technology** and the newer **digital technology**. Analog signals are used in their original form and placed, for example, on a tape. Most earlier electronic devices use the analog format (e.g., televisions, record players, cassette tape recorders, and telephones). Analog technology is still applied today. With digital technology, the analog signals are sampled numerous times, turned into numbers, and stored in a digital system. Today, many devices contain digital technology (e.g., high-definition TV, CDs, fiber-optic telephone lines, and digital telephones).

Even with the shift to IP-based network systems for CCTV, video is still transmitted over coaxial cable, twisted pair wire, fiber-optic cable, microwave, radio frequency, and telephone lines. What we have is the opportunity (as with other electronic security systems) for, say, an executive in New York to monitor inside a business in Hong Kong.

The choice of wireless video transmission (e.g., microwave or radio frequency) is an option under certain circumstances. Examples include flexible deployment whereby cameras must be moved periodically (e.g., changing exhibition hall), covert surveillance requiring quick and easy installation, at emergency sites, and historical buildings where a cable route is not possible. Careful planning is required prior to the installation of transmitters and receivers to prevent the radio signal from being blocked. *Line of sight* is an important issue. Interference can result from environmental conditions such as metallic buildings, aluminum siding, solar flares, lightning, heavy rain, snow, and high wind (Chan, 2005: 46–48).

When IT personnel are approached about including CCTV on a network, they are often concerned about how much bandwidth the video will use. To allay fears, one option at a multibuilding facility is to maintain a DVR at every building for storage of video so all video is not transmitted to the central computer.

For those end users using traditional analog technology while moving toward an IP-based retrofit, options include using "hybrid" products that accommodate both analog and IP-based signals. Lasky (2006: 38) advises against a full IP retrofit unless there is a clear understanding of the amount of bandwidth required on a network with numerous IP cameras.

Organizations that employ CCTV systems may consider streaming video surveillance from remote sites to regional centers. Although this approach can be challenging, it can also reduce plant and personnel costs. A key factor in this decision is compression because bandwidth limitations affect the amount of video that can be exchanged between transmitting and receiving sites. Similar to a roadway tunnel, only a certain number of vehicles can enter the tunnel at any one time. However, if the vehicles are made smaller, more can fit. **Compression** is the amount of redundant video that can be stripped out of an image before storage and transmission, and there are various compression techniques (Mellos, 2005: 34).

As physical security personnel increasingly rely on a network another concern is protecting access to the network for a variety of security-related information. In this case, the IT personnel have the option of placing such security information on a subnet to prevent access to the whole network.

Changing technology has brought about the **charged coupled device (CCD) or "chip" camera**, a small, photosensitive unit designed to replace the tube in the closed-circuit camera. CCD technology is found in camcorders. CCD cameras have certain advantages over tube cameras: CCD cameras are more adaptable to a variety of circumstances, they have a longer life expectancy, "ghosting" (i.e., people appearing transparent) is less of a problem, there is less intolerance to light, less power is required, and less heat is produced, thereby requiring less ventilation and permitting installation in more locations.

Another technology for capturing images digitally is the complementary metal oxide semiconductor (CMOS). Teledyne DALSA (2011), a global leader in digital imaging, offering both CCD and CMOS, notes that each has a bright future and there are advantages and disadvantages depending on application. Both types convert light into electric charge and process it into electronic signals. CCD costs less and is not as complex as CMOS.

Digital cameras are replacing analog cameras. Although analog signals can be converted into digital signals for recording to a PC, quality may suffer. Digital cameras use digital signals that are saved directly to hard drive, but space on a hard drive is limited for video. Network cameras are analog or digital video cameras connected to the Internet with an IP address.

Megapixel and high definition (HD) security cameras are part of the more recent evolution of video surveillance (Nilsson, 2011: 40–43). Both provide an enhanced picture image over analog. **Megapixel** refers to the number of million pixels in an image. We often identify this term with still picture photography. A greater number of pixels results in increased image detail. This is especially helpful during investigations if, for example, a person needs to be positively identified, or when a playing card requires identification in a casino. Keys (2010: 27) extols the benefits of megapixel cameras. He notes that with megapixel technology, by zooming in on a recorded image that may initially appear useless, it can show important details. Keys concedes that there are lighting issues with these cameras. He also writes that they are "memory hogs" and planners must study the memory needed to archive data. **High definition** could be considered a subset or type of megapixel camera. It complies with industry standards to ensure excellent color and it produces a wider image than megapixel. If campus police were seeking disorderly students in a large crowd, HD would be appropriate over megapixel. Which camera is better depends on the application. Combining megapixel, HD, and other types of cameras can result in an effective, multipurpose IP-based network video system.

Increasing "intelligence" is being built into CCTV-computer-based systems. **Multiplex** means sending many signals over one communications channel. Video multiplex systems minimize the number of monitors security personnel must watch by allowing numerous cameras to be viewed at the same time on one video screen. The pictures are compressed, but a full view is seen of each picture. If an alarm occurs, a full screen can be brought up. The digital multiplex recorder enables users to record events directly to a hard drive, reducing storage space.

The prolonged watching of CCTV monitors (i.e., screens) by personnel, without falling asleep, has been a challenge since the origin of these systems. Personnel that are not rotated periodically become fatigued from watching too much TV. This serious problem is often overlooked. People may "test" the monitoring of the system by placing a bag or rag over a camera or even spraying the lens with paint. If people see that there is no response, CCTV becomes a hoax. The use of dummy cameras is not recommended because, when people discover the dummy, CCTV can be perceived as a deceitful farce.

Users of CCTV systems are especially interested in the recording capabilities of their systems, knowing their personnel are often occupied with multiple tasks (e.g., answering questions for customers, providing information over the telephone) and unable to watch monitors continuously. When an event does occur, these systems permit a search of recordings by date, time, location, and other variables.

CCTV capabilities can be enhanced by using **video motion detection** (VMD). A video motion detector operates by sending a picture from a camera to a memory evaluator. The memory evaluator analyzes the image for pixel changes. Any change in the picture, such as movement, activates an alarm. These systems assist security officers in reacting to threats and reduce the problem of fatigue from watching monitors. Tse (2006: 42) refers to a study by an Australian firm that found that after 12 minutes of continuous watching of monitors, an operator would often miss up to 45 percent of scene activity, and after 22 minutes, up to 95 percent is overlooked.

The integration of VMD and **video analytics**, also referred to as intelligent video systems (IVS), is a technology that offers a variety of functions that aim to precisely define alarm conditions, enhance the capabilities of CCTV systems, and reduce the problem of humans missing important events on monitors. These systems enable the user to preselect actions that are programmed into the digital video system, and this software signals an alarm when such an event takes place. Examples of events triggering an alarm include stopped or moving vehicles, objects that are abandoned or removed, and loitering of people (Aubele, 2011: 48–53 and Duda, 2006: 48–50).

Cameras are commonly placed at public streets, access points, passageways, shipping and receiving docks, merchandise storage areas, cashier locations, parts departments, and overlooking files, safes, vaults, and production lines. In the workplace, the location of cameras requires careful planning to avoid harming employee morale. A key restriction on the placement of cameras is that they must not be applied to an area where someone has a reasonable expectation of privacy (e.g., restrooms, locations where individuals change clothes).

The extent of the use of hidden surveillance cameras is difficult to measure, especially because many individuals are unaware of the existence of these cameras in workplaces. Pinhole lenses are a popular component of hidden surveillance cameras. They get their name from the outer opening of the lens, which is 1/8 to 1/4 inch in diameter and difficult to spot. Cameras are hidden in almost any location, such as in clocks, file cabinets, computers, sprinkler heads, and mannequins.

It is important to note that when the network is down, IP cameras, NVR, and other technology tied to the network are down; therefore, emergency plans are essential to maintain business continuity.

You Decide!

What is Wrong with This Facility?

The Tedson Manufacturing Corporation operates a thriving complex of office buildings and manufacturing facilities. Several protection issues have surfaced and action is required to correct vulnerabilities. To begin with, in two separate cases, employees appeared to take company property off the premises without authorization. In both cases, the company investigator was unable to prove that the property belonged to the company, the prosecutor dropped each case, and lawsuits are pending. In reference to the facility's access control system, Tedson employs 4,000 people, but almost 6,000 access cards are active. In the past three months, twelve people were able to enter the facility with an access card and PIN that did not belong to them. Last summer, on two separate occasions, teenagers were able to access the premises at night and stole two company vehicles that had keys in the ignition. When the CCTV system was used to try to identify the teenagers, megapixel cameras made identification impossible. The false alarm problem is increasing, possibly because ultrasonic motion detectors were installed near HVAC vents in buildings. As a newly hired security specialist, you have been assigned the task of identifying what is wrong with the security in place and correcting the problems.

Security Officers

Security officers play an important role in countering internal losses. They must be integrated with technology, and this entails quality training and supervision. When uniformed officers patrol on foot inside a facility—through production, storage, shipping, receiving, office, and sales floor areas—an enhanced loss prevention atmosphere prevails. Unpredictable and irregular patrols deter employee theft (among other losses). A properly trained officer looks for deviations, such as merchandise stored or hidden in unusual places and tampered security devices. Thoroughly searching trash containers deters employees from hiding items in that popular spot. Losses also are hindered when officers identify and check people, items, and vehicles at access points.

Safes, Vaults, and File Cabinets

Safes

Protective containers (see Figure 7-19) secure valuable items (e.g., cash, confidential information). These devices are generally designed to withstand losses from fire or burglary. Specifications vary, and an assessment of need should be carefully planned. Management is frequently shocked when a fire-resistive safe in which valuable items are "secured" enables a

FIGURE 7-19 Safe with electronic lock. *Courtesy: Sargent & Greenleaf, Inc.*

burglar to gain entry because the safe was designed only for fire. The classic **fire-resistive (or record) safe** often has a square (or rectangular) door and thin steel walls that contain insulation. During assembly, wet insulation is poured between the steel walls; when the mixture dries, moisture remains. During a fire, the insulation creates steam that cools the safe below 350°F (the flash point of paper) for a specified time. The FBI maintains safe insulation files to assist investigators. Record safes for computer media require better protection because damage can occur at 125°F, and these records are more vulnerable to humidity. Fire safes are able to withstand one fire; thereafter, the insulation is useless.

The classic **burglary-resistive (or money) safe** often has a thick, round door and thick walls. Round doors were thought to enhance resistance, but today many newer burglary-resistive safes have square or rectangular doors. The burglary-resistive safe is more costly than the fire-resistive safe.

Better quality safes have the Underwriters Laboratories (UL, a nonprofit testing organization) rating (see Table 7–1). This means that manufacturers have submitted safes for testing by UL. These tests determine the fire- or burglary-resistive properties of safes. For example, a fire-resistive container with a UL rating of 350–4 can withstand an external temperature to 2000°F for four hours while the internal temperature will not exceed 350°F. The UL test actually involves placing a safe in an increasingly hot furnace to simulate a fire. In reference to burglary-resistive containers, a UL rating of TL15, for example, signifies weight of at least 750 pounds and resistance to an attack on its door by common tools for a minimum of 15 minutes. UL-rated burglary-resistive safes also contain UL-listed combination locks and other UL-listed components. When selecting a safe, consider recommendations from insurance companies and peers, how long the company has been in business, and whether or not safe company employees are bonded.

Attacks

Before a skilled burglar attacks a safe, he or she studies the methods used to protect it. Inside information (e.g., a safe's combination) is valuable, and scores of employees and former employees of attacked firms have been implicated in burglaries. Listed next are major attack techniques of two types: with force and without force. Attack methods using force include the following:

- *Rip or peel:* Most common, this method is used on fire-resistive safes that have lightweight metal. Like opening a can of sardines, the offender rips the metal from a corner. The peel technique requires an offender to pry along the edge of the door to reach the lock.
- *Punch:* The combination dial is broken off with a hammer. A punch is placed on the exposed spindle, which is hammered back to enable breakage of the lock box. The handle then is used to open the door. This method is effective against older safes.
- *Chop:* This is an attack of a fire-resistive safe from underneath. The safe is tipped over and hit with an ax or hammer to create a hole.
- *Drill:* A skillful burglar drills into the door to expose the lock mechanism; the lock tumblers are aligned manually to open the door.
- *Torch:* This method is used against burglar-resistive safes. An oxygen-acetylene cutting torch melts the steel. This equipment is brought to the safe, or the offender uses equipment from the scene.

Table 7-1 UL Testing of Safes

Class	Resistance to attack	Attack time	Description
Fire			
350-*	Not tested	N/A	For paper and document storage
150-*	Not tested	N/A	For storage of magnetic computer tapes and photographic film
125-*	Not tested	N/A	See hour rating below.
Burglary			
TL-15	Door or front face	15 min	Resists against entry by common mechanical and electrical tools or combination of these means, Group 2 M, 1, or 1R combination lock, or Type 1 High Security Electronic Lock.**
TL-15×6	6 sides	15 min	Same as above
TRTL-15×6	6 sides	15 min	Resists against entry by common mechanical, electrical tools and cutting torches or combination of each, Group 1 or 1R combination lock or, Type 1 High-Security Electronic Lock.**
TL-30	Door or front face	30 min	Same tools as TL-15, plus abrasive cutting wheels and power saws. Group 2 M, 1 or 1R combination lock, or Type 1 High Security Electronic Lock.**
TL-30×6	6 sides	30 min	Same as TL-30.
TRTL-30	Door or front face	30 min	Same tools as TRTL 15×6, Group 1 or 1R combination lock or Type 1 High Security Electronic Lock, body construction requirement of 1 in. thick steel min. 50,000 psi (or equivalent) encased in minimum 3 in. concrete, 4000 psi minimum.
TRTL-30×6	6 sides	30 min	Same tools as TRTL 15×6, Group 1 or 1R combination lock or Type 1 High Security Electronic Lock.**
TRTL-60×6	6 sides	60 min	Same tools as TRTL-15, Group 1 or 1R combination lock or Type 1 High Security Electronic Lock.
TXTL-60×6 *(UL no longer provides testing to or supports the TXTL-60×6 safe rating.)*	6 sides	60 min	Same tools as TRTL-15 + up to 8 oz. of nitroglycerine with a maximum of 4 oz., per test, Group 1 or 1R combination lock or Type 1 High-Security Electronic Lock, minimum weight 1000 lbs., wall thickness not specified.

*Hour rating 4, 2 or 1. Before inside temperature reaches 350, 150, or 125°F as shown by class designation.

**Minimum weight 750 lbs., body 1″ steel, minimum tensile strength of 50,000 PSI. UL 2058: new standard for Type 1 High-Security Electronic Lock for safes.

Combination locks.

(These products are tested in accordance with UL 768)

Group 1. Highly resistant to expert or professional manipulation. Used in safes designated as TRTL-15×6, TRTL-30, TRTL-30×6, and TRTL-60×6.

Group 1R. These locks meet all of the requirements of Group 1 and are resistant against radiological methods of manipulation.

Group 2M. Moderately resistant to skilled manipulation, these are found in TL-15, TL-15×6, TL-30, and TL30×6 safes, ATM safes, gun safes, and fire-rated record containers.

Group 2. Resistant to semiskilled manipulation, these locks are found in non-Listed safes, insulated record containers, and residential security containers.

Source: Correspondence (June 20, 2011) with UL, 1285 Walt Whitman Rd., Melville, NY 11747.

- *Carry away:* The offender removes the safe from the premises and attacks it in a convenient place.

Attack methods using no force include the following:

- *Office search:* Simply, the offender finds the safe combination in a hiding place (e.g., taped under a desk drawer).
- *Manipulation:* The offender opens a safe without knowing the combination by using sight, sound, and touch—a rare skill. Sometimes the thief is lucky and opens a safe by using numbers similar to an owner's birth date, home address, or telephone number.
- *Observation:* An offender views the opening of a safe from across the street with the assistance of binoculars or a telescope. To thwart this, one should place the numbers on the top edge of the dial, rather than on the face of the dial.
- *Day combination:* For convenience, during the day, the dial is not completely turned each time an employee finishes using the safe. This facilitates an opportunity for quick access. An offender often manipulates the dial in case the day combination is still in effect.
- *X-ray equipment:* Metallurgical x-ray equipment is used to photograph the combination of the safe. White spots appear on the picture that helps to identify the numerical combination. The equipment is cumbersome, and the technique is rare.

The following measures are recommended to fortify the security of safes and other containers:

1. Utilize alarms (e.g., capacitance and vibration), CCTV, and adequate lighting.
2. Secure the safe to the building so it is not stolen. (This also applies to cash registers that may be stolen in broad daylight.) Bolt the safe to the foundation or secure it in a cement floor. Remove any wheels or casters.
3. Do not give a burglar an opportunity to use any tools on the premises; hide or secure all potential tools (e.g., torch).
4. A time lock permits a safe to be opened only at select times. This hinders access even if the combination is known. A delayed-action lock provides an automatic waiting period (e.g., 15 minutes) from combination use to the time the lock mechanism activates. A silent signal lock triggers an alarm when a special combination is used to open a safe.
5. At the end of the day, turn the dial several times in the same direction.
6. A written combination is risky. Change the factory combination as soon as possible. When an employee who knows the combination is no longer employed, change it.
7. Maintain limited valuables in the safe through frequent banking.
8. Select a safe with its UL rating marked on the inside. If a burglar identifies the rating on the outside, an attack is made easier.
9. Consider modern features of safes: remote access management, reports of cash flow, and traceable deposits.

Vaults

A walk-in vault is actually a large safe; it is subject to similar vulnerabilities from fire and attack. Because a walk-in vault is so large and expensive, typically, only the door is made of steel, and the rest of the vault is composed of reinforced concrete. Vaults are heavy enough to require special support within a building. They commonly are constructed at ground level to avoid stress on a building.

File Cabinets

Businesses that sustain loss of their records from theft, fire, flood, or other threats or hazards face serious consequences, such as the possibility of business failure and litigation. Certain types of records require protection according to law. Some vital records are customer-identifying information, accounts receivable, inventory lists, legal documents, contracts, research and development, and human resources data. Records help to support losses during insurance claims.

File cabinets that are insulated and lockable can provide fair protection against fire and burglary. The cost is substantially lower than that of a safe or vault, but valuable records demanding increased safety should be placed in a safe or vault and copies stored off-site. Special computer safes are designed to protect against forced entry, fire, and moisture that destroys computer media.

Search the Internet

Use search engines to see what vendors have to offer and prices for the following products: access control systems, locks, interior intrusion detection systems, CCTV, and safes.

Also, check out the following sites:
American National Standards Institute: www.ansi.org
American Society for Testing and Materials: www.astm.org
Association of Certified Fraud Examiners: www.acfe.com
Builders Hardware Manufacturers Association: www.buildershardware.com
International Organization for Standardization: www.iso.org
National Fire Protection Association: www.nfpa.org
National White Collar Crime Center: www.nw3c.org
Security Industry Association: www.siaonline.org
Underwriters Laboratories (UL): www.ul.com/global/eng/pages

Case Problems

7A. Consult the floor plans for Woody's Lumber Company, the Smith Shirt manufacturing plant, and Compulab Corporation (Figures 7-1, 7-2, and 7-3). Draw up a priority list of 10 loss prevention strategies for each company that you think will reduce risks from internal losses. Why did you select your first three strategies in each list as top priorities?

7B. As a corporate security manager, you learn that an IT specialist at the same company is extremely upset because he did not receive a promotion and raise he was expecting. This very intelligent young man told his supervisor that he would get back at the company for the injustice before he quits. What do you do?

7C. You are a security officer at a manufacturing plant where an employee informs you about observing another employee hiding company property near a back door. You check the area near the door and find company property under boxes. What actions do you take?

7D. As a security officer, you learn that officers on your shift and your immediate supervisor have secretly installed, without authorization, a pinhole lens camera in the women's restroom. You refuse to be involved in peeping. The officers have been your friends since high school, and you socialize with them when off duty. One day the chief security officer summons you to her office and questions you concerning the whereabouts of the pinhole lens camera. What do you say?

7E. As a security manager for a manufacturing company, you are scheduled for a meeting with the Director of IT and an outside vendor on the subject of transitioning to an IP-based network camera system. List ten questions that you will ask during this meeting to ensure the planning and transition are successful and security is improved.

References

Alten, J. (2005). Shhh...Don't tell anyone that DVRs are becoming obsolete. *Security Director News, 2* (March).

Association of Certified Fraud Examiners. (2010). Report to the Nations on Occupational Fraud and Abuse, 2010. <www.acfe.com> retrieved June 11, 2011.

Aubele, K. (2011). Checking out security solutions. *Security Management, 55* (December).

Aughton, S. (2005). Researchers Crack Biometric Security with Play-Doh. PC PRO. <www.pcpro.co.uk/news/81257> retrieved December 14, 2005.

Barry, J. (1993). Don't always play the cards you are dealt. *Security Technology & Design*, July–August.

Bernard, R. (2011). The state of converged security operations. *Security Technology Executive, 21* (April).

Bernard, R. (2006). Web services and identity management. *Security Technology & Design, 16* (January).

Blades, M. (2010). The insider threat. *Security Technology Executive, 20* (November/December).

Boba, R., & Santos, R. (2008). A review of the research, practice, and evaluation of construction site theft occurrence and prevention: Directions for future research. *Security Journal, 21* (October).

Brenner, B. (2010). How Physical, IT Security Sides Can Work Together. Computerworld (September). <www.computerworld.com> retrieved January 20, 2011.

Canada.com. (2007). Hydro Lost Millions from Theft, Damage Last Year. Vancouver Sun (February 7). <www.canada.com> retrieved February 9, 2007.

Catrantzos, N. (2010). No dark corners: A different answer to insider threats. *Homeland Security Affairs, 6* (May). <www.hsaj.org> retrieved May 12, 2010.

Chan, H. (2005). Overcoming the challenges of wireless transmission. *Security Technology & Design, 15* (October).

Conklin, J. (2001). *Criminology* (7th ed.). Boston: Allyn & Bacon Pub.

Cressey, D. (1971). *Other people's money: A study in the social psychology of embezzlement.* Belmont, CA: Wadsworth.

D'Agostino, S. et al. (2005). The Roles of Authentication, Authorization and Cryptography in Expanding Security Industry Technology. <www.siaonline.org> retrieved May 30, 2006.

Department of Defense, (2000). *User's guide on controlling locks, keys and access cards.* Port Hueneme, CA: Naval Facilities Engineering Service Center.

Di Nardo, J. (2009). Biometric technologies: Functionality, emerging trends, and vulnerabilities. *Journal of Applied Security Research, 4*.

Duda, D. (2006). The ultimate integration—video motion detection. *Security Technology & Design, 16* (June).

Eberle, W., et al. (2011). Insider threat detection using a graph-based approach. *Journal of Applied Security Research, 6* (1).

EEOC. (2011). "How to Comply with the American with Disabilities Act: A Guide for Restaurants and other Food Service Employers" (January 19). <www.eeoc.gov> retrieved June 21, 2011.

Freeman, J. (2000). Security director as politician. *Security Technology & Design, August*.

Garcia, M. (2006). *Vulnerability assessment of physical protection systems*. Burlington, MA: Butterworth-Heinemann.

Gersh, D. (2000). Untouchable Value. *iSecurity, November*.

Greene, C. (2004). *Hang up on fraud with confidential hotlines*. Chicago, IL: McGovern & Greene. Fraud Alert.

Honey, G. (2003). *Intruder alarms* (2nd ed.). Oxford, UK: Newnes.

Hulusi, T. (2011). Creating a trusted identity. *Security Technology Executive, 21* (May).

Inbau, F., et al. (1996). *Protective security law* (2nd ed.). Boston: Butterworth-Heinemann.

Jarvis, B. (2011). The next generation of access control: Virtual credentials. *Access Control Trends and Technology*, June.

Jordan, B. (2006). Telework's growing popularity. *Homeland Defense Journal, 4* (June).

Keener, J. (1994). Integrated systems: What they are and where they are heading. *Security Technology & Design*, May.

Keys, R. (2010). What is intelligent video? *Law Officer Manazine, 6* (March).

Klenowski, P., et al. (2011). Gender, identity, and accounts: How white collar offenders do gender when making sense of their crimes. *Justice Quarterly, 28* (February).

Kosaka, M. (2010). Public goes private. *Security Products, 14* (March).

Lasky, S. (2006). Video from the top. *Security Technology & Design, 16* (June).

Loughlin, J. (2009). Security through transparency: An open source approach to physical security. *Journal of Physical Security, 3*(1).

Mellos, K. (2005). A choice you can count on. *Security Products, 9* (October).

Morton, J. (2011). Top smart card blunders. *Buildings, 105* (April).

Nemeth, C. (2005). *Private security and the law*. Burlington, MA: Elsevier Butterworth-Heinemann.

Nilsson, F. (2011). The resolution to your confusion. *Security Technology Executive, 21* (March).

Nosowitz, D. (2011). Everything You Need to Know About Near Field Communication. Popular Science (March). <www.popsci.com> retrieved June 18, 2012.

O'Leary, T. (1999). New innovations in motion detectors. *Security Technology & Design, 9* (November).

Pearson, R. (2000). Integration vs. Interconnection: It's a matter of semantics. *Security Technology & Design, 11* (November).

Pfeifle, L. (2010). McDonald's, saks begin first install of HDCCTV. *Security Director News (January 25)* <www.securitydirectornews.com> retrieved January 12, 2011.

Philpott, D. (2005). Physical security—biometrics. *Homeland Defense Journal, 3* (May).

Piazza, P. (2005). The smart cards are coming…really. *Security Management, 49* (January).

Scicchitano, M., et al. (2004). Peer reporting to control employee theft. *Security Journal, 17* (April).

Shaw, E., et al. (2000). Managing the threat from within. *Information Security, 3* (July).

Siemon Company.(2003). Video over 10G ip™. <www.siemon.com> retrieved July 24, 2006.

Skinner, W., & Fream, A. (1997). A social learning theory analysis of computer crime among college students. *Journal of Research in Crime and Delinquency, 34* (November).

Speed, M. (2003). Reducing employee dishonesty: In search of the right strategy. *Security Journal, 16* (April).

Spence, B. (2011). Advances in fingerprint biometric technology. *Locksmith Ledger, June* <www.locksmithledger.com> retrieved June 14, 2011.

Suttell, R. (2006). Security monitoring. *Buildings, 100* (May).

Teledyne D.A.L.S.A. (2011). CCD vs. CMOS. <www.teledynedalsa.com> retrieved June 23, 2011.

Toye, B. (1996). Bar-Coded security ID cards efficient and easy. *Access Control*, March.

Truncer, E. (2011). Controlling access system performance. *Security Management, 55* (March).

Tsė, A. (2006). The real world of critical infrastructure. *Security Products, 10* (May).

U.S. Department of Homeland Security, Science and Technology Directorate and the Executive Office of the President, Office of Science and Technology Policy. (2004). The National Plan for Research and Development in Support of Critical Infrastructure Protection. <www.dhs.gov> retrieved June 13, 2005.

Zalud, B. (2010). Higher level credentials leave footprint on card printers. *Security, 47* (October).

Zunkel, D. (2003). A short course in high-security locks. *Security Technology & Design, 13* (February).

8 External Threats and Countermeasures

OBJECTIVES

After studying this chapter, the reader will be able to:

1. Explain the meaning of external loss prevention and describe methods of unauthorized entry
2. List and define the five "Ds" of security
3. Explain how environmental security design can enhance security
4. Define sustainability and green security
5. Discuss perimeter security and list and define five types of barriers
6. List and explain methods to protect buildings against terrorism
7. Explain window and door protection
8. Describe the application of intrusion detection systems to perimeter protection
9. Explain lighting illumination and at least five types of lamps
10. Describe parking lot and vehicle controls
11. Explain the deployment and monitoring of security officers
12. Discuss the use of protective dogs
13. Explain the importance of communications and the control center

KEY TERMS

- external loss prevention
- forced entry
- smash and grab attacks
- surreptitious entry
- aura of security
- redundant security
- layered security
- environmental security design
- "Broken Windows" theory
- sustainability
- green security
- perimeter
- clear zones
- natural barriers
- structural barriers
- human barriers
- animal barrier
- energy barriers
- passive vehicle barriers
- active vehicle barriers

- common wall
- land use controls
- target-rich environment
- keep out zones
- stand-off distance
- blast and antiramming walls
- fiber optics
- bypass
- spoofing
- point protection
- spot or object protection
- area protection
- perimeter protection
- local alarm
- central station
- remote programming
- lumens
- illuminance
- foot-candle (FC)
- lux
- color rendition
- traffic calming strategies
- stationary post
- foot or vehicle patrols
- contraband

Introduction

External loss prevention focuses on threats from outside an organization. This chapter concentrates on countermeasures to impede unauthorized access from outsiders. If unauthorized access is successful, numerous losses are possible from such crimes as assault, burglary, robbery, vandalism, arson, and espionage. Naturally, employees as well as outsiders or a conspiracy of both may commit these offenses. Furthermore, outsiders can gain legitimate access if they are customers, repair technicians, and so on.

Internal and external countermeasures play an interdependent role in minimizing losses; a clear-cut division between internal and external countermeasures is not possible because of this intertwined relationship. In addition, as explained in the preceding chapter, we are in an era of universal threats. This means that because of telework, employees and organizations face the same threats whether work is accomplished on or off the premises.

The IT perspective is important to produce comprehensive security. IT specialists use terms such as *denial of access* and *intrusion detection*, as do physical security specialists; however, IT specialists apply these terms to the protection of information systems. As IT and physical security specialists learn from each other, a host of protection methods will improve. Examples include integration of systems, investigations, and business continuity planning.

Many organizations have developed formidable perimeter security to prevent unauthorized entry, while not realizing that the greatest threat is from within.

Methods of Unauthorized Entry

One avenue to begin thinking about how to prevent unauthorized entry is to study the methods used by offenders. Both management (to hinder penetration) and offenders (to succeed in gaining access) study the characteristics of patrols, fences, sensors, locks, windows, doors, and the like. By placing yourself in the position of an offender (i.e., *think like a thief*) and then that of a loss prevention manager, you can see, while studying Woody's Lumber Company, the Smith Shirt manufacturing plant, and Compulab Corporation (discussed in Chapter 7), that a combination of both perspectives aids in the designing of defenses. (Such planning is requested in a case problem at the end of this chapter.) Furthermore, keep in mind the theories and practical applications of the theories from Chapter 3; these include rational choice and routine activity theories and situational crime prevention techniques to reduce opportunities for crime.

Forced entry is a common method used to gain unauthorized access, especially at windows and doors. Offenders repeatedly break or cut glass (with a glasscutter) on a window or door and then reach inside to release a lock or latch. To stop the glass from falling and making noise, an offender may use a suction cup or tape to remove or hold the broken glass together. A complex lock may be rendered useless if the offender is able to go through a thin door by using a hammer, chisel, and saw. Forced entry also may be attempted through walls, floors, ceilings, roofs, skylights, utility tunnels, sewer or storm drains, and ventilation vents or ducts. Retail stores may be subject to **smash and grab attacks**: a store display window is smashed, merchandise is quickly grabbed, and the thief immediately flees.

Atlas (2010: 52) describes burglars who targeted a retail chain by rattling storefronts to create false alarms; this resulted in slow police response and retail management turning off the burglar alarm systems to avoid fines. Police and retail management "bit the bait" and the burglars, disguised with masks and gloves, shattered drive-through windows with a tire iron, climbed in, and stole unsecured cash boxes prepared for morning business.

Berube (2010: 326-381) writes of conflicting research results on the deterrent effect of burglar alarm systems. Certain burglars do not care if they activate an alarm because they only spend one or two minutes at the crime scene and police are unlikely to respond quickly enough to make an arrest. Because of the false alarm problem, police may be hesitant to respond quickly or not respond at all. When we consider rational choice and routine activity theories (see Chapter 3), slow police response to alarm activations reduce risk for burglars and increase opportunities. Berube also cites research that supports the deterrent effect of alarm systems.

Unauthorized access also can be accomplished *without force*. Wherever a lock is supposed to be used, if it is not locked properly, access is possible. Windows or doors left unlocked are a surprisingly common occurrence. Lock picking or possession of a stolen key or access card renders force unnecessary. Dishonest employees are known to assist offenders by unlocking locks, windows, or doors and by providing keys and technical information. Offenders sometimes hide inside a building until closing and then break out following an assault or theft. Tailgating and pass back are other methods of gaining access without force, as covered in the preceding chapter. Sly methods of gaining entry are referred to as **surreptitious entry**.

Countermeasures

Countermeasures for external (and internal) threats can be conceptualized around the five "Ds":

- *Deter:* The mere presence of physical security can dissuade offenders from committing criminal acts. The impact of physical security can be enhanced through an **aura of security**. An aura is a distinctive atmosphere surrounding something. Supportive management and security personnel should work to produce a professional security image. They should remain mum on such topics as the number and types of intrusion detection sensors on the premises and security system weaknesses. Security patrols should be unpredictable and never routine. Signs help to project an aura of security by stating, for example, PREMISES PROTECTED BY HIGH-TECH REDUNDANT SECURITY. Such signs can be placed along a perimeter and near openings to buildings. The aura of security strives to produce a strong psychological deterrent so offenders will consider the success of a crime to be unlikely. It is important to note that no guarantees come with deterrence. (Criminal justice policies are in serious trouble because deterrence is faulty; criminals continue to commit crimes even while facing long sentences.) In the security realm, deterrence must be backed up with the following four "Ds."

- *Detect:* Offenders should be detected and their location pinpointed as soon as they step onto the premises or commit a violation on the premises. This can be accomplished through observation, CCTV, intrusion sensors, duress alarms, weapons screenings, protective dogs, and hotlines.
- *Delay:* Security is often measured by the time it takes to get through it. **Redundant security** refers to two or more similar security methods (e.g., two fences; two types of intrusion sensors). **Layered security** refers to multiple security methods that follow one another and are dissimilar (e.g., perimeter fence, strong doors, a safe). Both redundant and layered security creates a time delay. Thus, the offender may become frustrated and decide to depart, or the delay may provide time for a response force to arrive to make an apprehension.

  Nunes-Vaz, et al. (2011: 372) favor distinguishing between "layered security" and "security-in-depth" to enhance research and to guide investments in security. They write: "Risk minimization is best achieved by strengthening the layer that may already be the most effective, and by focusing on the weakest function within that layer." In addition, they argue that "security-in-depth" aims to produce not only effective layers, but also coherent integration of layers.
- *Deny:* Strong physical security, often called *target hardening,* can deny access. A steel door and a safe are examples. Frequent bank deposits of cash and other valuables extend the opportunity to deny the offender success.
- *Destroy:* When someone believes his or her life or another's will be taken, he or she is legally permitted to use deadly force. An asset (e.g., proprietary information) may require destruction before it falls into the wrong hands.

Which "D" do you view as most important? Which "D" do you view as least important? Explain your answers.

Environmental Security Design

When a new facility is planned, the need for a coordinated effort by architects, fire protection and safety engineers, loss prevention practitioners, local police and fire officials, and other specialists cannot be overstated. Furthermore, money is saved when security and safety are planned before actual construction rather than accomplished by modifying the building later.

Years ago, when buildings were designed, loss prevention features were an even smaller part of the planning process than today. Before air conditioning came into widespread use, numerous windows and wide doors were required for proper ventilation, providing thieves with many entry points. Today's buildings also present problems. For example, ceilings are constructed of suspended ceiling tiles with spaces above the tiles that enable access by simply pushing up the tiles. Once above the tiles, a person can crawl to other rooms on the same floor. Roof access from neighboring buildings is a common problem for both old and new buildings. Many of these weak points are corrected by adequate hardware such as locks on roof doors and by intrusion sensors.

Architects are playing an increasing role in designing crime prevention into building plans. **Environmental security design** includes natural and electronic surveillance of walkways and parking lots, windows and landscaping that enhance visibility, improved lighting, and other architectural designs that promote crime prevention. Additionally, dense shrubbery can be cut to reduce hiding places, and grid streets can be turned into cul-de-sacs by using barricades to reduce ease of escape. Chapter 3 covered Oscar Newman's work with defensible space and C. Ray Jeffery's work with Crime Prevention Through Environmental Design (CPTED).

An illustration of how CPTED is applied can be seen with the design of Marriott hotels (Murphy, 2000: 84–88). To make offenders as visible as possible, traffic is directed toward the front of hotels. Lobbies are designed so that people walking to guest rooms or elevators must pass the front desk. On the outside, hedges are emphasized to produce a psychological barrier that is more appealing than a fence. Pathways are well lit and guide guests away from isolated areas. Parking lots are characterized by lighting, clear lines of sight, and access control. Walls of the garage are painted white to enhance lighting. On the inside of hotels, the swimming pool, exercise room, and vending and laundry areas have glass doors and walls to permit maximum witness potential. One application of CCTV is to aim cameras at persons standing at the lobby desk and install the monitor in plain view. Since people can see themselves, robberies have declined. CPTED enhances traditional security methods such as patrolling officers and emergency call boxes.

Research from the United Kingdom has extended the reach of CPTED. The UK Design Against Crime (DAC) Program seeks a wide group of design professionals to develop creative and often subtle design solutions to combat crime and fear of crime. The DAC is a holistic, human-centered approach that facilitates crime prevention without inconveniencing people or creating a fortress environment. Examples include the following: a fence with a top rail that is angled to discourage young people from sitting on the fence and "hanging out"; the playing of classical music to prevent youth from congregating in certain areas; and the "antitheft handbag" that has a short strap, a carefully located zipper, thick leather, and an alarm (Davey et al., 2005: 39–51).

CPTED is enhanced through the **"Broken Windows" theory** of James Q. Wilson and George Kelling (1982). This theory suggests that deteriorated buildings that remain in disrepair and disorderly behavior attract offenders and crime while increasing fear among residents. If someone breaks a window and it is not repaired, more windows may be broken, and a continuation of dilapidated conditions may signal that residents do not care. Minor problems, such as vandalism, graffiti, and public intoxication, may grow into larger problems that attract offenders and destroy neighborhoods. However, residents can increase safety and security when they take pride in the conditions of their neighborhood.

Sustainability and Green Security

Sustainability impacts security and security professionals should develop an understanding of what this term implies. The U.S. Environmental Protection Agency (2011) describes **sustainability** as "policies and strategies that meet society's present needs without compromising the ability of future generations to meet their own needs." Wroblaski et al. (2011, 8) write that many businesses and organizations are involved in sustainability initiatives such as energy, lighting and water conservation, waste reduction, and recycling. However, they note that cost-effective initiatives are a top priority because funding is often limited.

The term "green building" is broad in scope and refers to an environmentally responsible building, including planning, construction, operation, and use. The "green building industry" maintains standards and a rating system. The American Society of Heating, Refrigerating, and Air Conditioning Engineers, the U.S. Green Building Council (USGBC), and the Illuminating Engineering Society jointly prepare standards with such topic areas as site selection, light pollution, energy efficiency, and water use. The Leadership in Energy and Environmental Design (LEED), developed by USGBC, and Green Globes, both maintain a rating system for green buildings (Burton, 2010: 14).

Green security is defined as security planning, designs, operations, and use that conform to sustainability. Suppose that as a security planner you selected a particular geographic site for a new corporate business location because of low crime in the area. You also chose a lighting system that would enhance security at the site, a CCTV system that is state-of-the-art, and a beautiful water fountain for the front of the building that also serves as a security barrier. To your surprise, your plans are rejected because they are not "green security." The site you selected confronts the "heat island effect". This refers to an island of higher temperatures, often in urban areas, as a building replaces vegetation and permeable and moist surfaces. The lighting system you chose will create light pollution, the CCTV system will consume too much energy, and the water fountain will not use water efficiently.

Perimeter Security

Perimeter means outer boundary, and it is often the property line and the first line of defense against unauthorized access (see Figure 8-1). Building access points such as doors and windows also are considered part of perimeter defenses at many locations. Typical perimeter security begins with a fence and gate and may include multiple security methods (e.g., card access, locks, sensors, lighting, CCTV, and patrols) to increase protection (see Figure 8-2). Technology can extend security surveillance beyond the perimeter, as illustrated with radar that is applied at a facility near a waterway (see Figure 8-3).

The following variables assist in the design of perimeter security:

1. Whatever perimeter security methods are planned, they should interrelate with the total loss prevention program and business objectives. In addition, green security should be considered.
2. Perimeter security needs to be cost effective. When plans are presented, management is sure to ask: "What type of return will we have on our investment?"

FIGURE 8-1 Perimeter security. *Courtesy: Wackenhut Corporation. Photo by Ed Burns.*

FIGURE 8-2 Multiple security methods increase protection.

FIGURE 8-3 Radar extends security surveillance beyond the perimeter at a facility near a waterway. *Courtesy: Honeywell Security.*

3. Although the least number of entrances strengthens perimeter security, the plan must not interfere with normal business and emergency events.
4. Perimeter security has a psychological impact on potential intruders. It signals a warning to outsiders that steps have been taken to block intrusions. Offenders actually "shop" for vulnerable locations (i.e., opportunities).
5. Even though a property line may be well protected, the possibility of unauthorized entry cannot be totally eliminated. For example, a fence can be breached by going over, under, or through it.
6. Penetration of a perimeter is possible from within. Merchandise may be thrown over a fence or out of a window. Various things are subject to smuggling by persons walking or using a vehicle while exiting through a perimeter.
7. The perimeter of a building, especially in urban areas, often is the building's walls. An offender may enter through a wall (or roof) from an adjoining building.
8. To permit an unobstructed view, both sides of a perimeter should be kept clear of vehicles, equipment, and vegetation. This allows for what is known as **clear zones**.
9. Consider integrating perimeter intrusion sensors with landscape sprinkler systems. Trespassers, protesters, and other intruders will be discouraged, and, when wet, they are easier to find and identify.
10. Perimeter security methods are exposed to a hostile outdoor environment not found indoors. Adequate clothing and shelter are necessary for security personnel. The selection

of proper security systems prevents false alarms from animals, vehicle vibrations, and adverse weather.

11. Perimeter security should be inspected and tested periodically.

International Perspective: Physical Security Proves its Value

Forty hooded demonstrators seemed to have appeared out of nowhere at the front gate of a breeding farm in the English countryside, where a pharmaceutical giant breeds animals for government-mandated testing of new medicines. A video recording of the incident showed protesters rocking the perimeter fence and harassing employees. What follows here is a description of how this business responded to its protection needs (Gips, 1999: 42–50).

For simplicity's sake, we will refer to this actual company as "PC" for pharmaceutical company. One threat facing the PC was the 50 or so incidents from animal activists in one year. Consequently, protection against sabotage, terrorism, and infiltration by animal rights activists became top priorities. Measures included physical security and access control, internal theft countermeasures, information safeguards, and bomb threat response. Protection was afforded not only to 2,000 scientists, support personnel, intellectual property, and physical assets, but also to the company image.

The PC favors a layered approach to physical security, which begins with strong perimeter protection. At the breeding farm, a seven-foot-high fence bounds the site and security officers monitor the farm from a gatehouse that doubles as a control room for intrusion and fire detection and CCTV. Because no police are nearby, a PC facility 12 miles away provides backup. The PC's response to protestors is low-key in part because in England simple trespass is a civil, not a criminal, matter. Protestors, even if verbally abusive, can be arrested only if they are violent; then police will make the arrests. Protestors generally want media attention, so they usually surrender to security when found on the premises. They know they will not be arrested, and no civil action will be initiated.

At another PC facility, security integration is shown through CCTV cameras, mounted every 75 yards along the perimeter, which work with video motion detection and infrared sensors. Although continuous recording occurs, when motion is detected, the action appears on a monitor for evaluation in the control room. This facility requires vehicles to pass through a raising-arm barrier. Pedestrians must register at a gatehouse, and employees use an access card as they pass through a full-height antipassback turnstile. Doors are alarmed, and windows are treated with anti-bandit glazing to delay an offender.

To reduce internal theft from employees and contractors, personnel are reminded of their responsibility to secure valuables, vulnerable areas have restricted access, doors are kept locked, and a crime prevention day is held. Information is protected through an awareness course, security bulletins, secure fax and videoconferencing facilities, a high priority on IT security, technical surveillance sweeps, and tours under close controls.

The animal activist threat is handled through counterintelligence (i.e., a database of information), vetting (i.e., examination of all personnel to prevent infiltration or the planting of devices to collect information), and public relations (i.e., outreach to explain the importance of research with animals). To deal with bomb threats, PC facilities are too large for a dedicated team to conduct a search, so each employee is responsible for checking for anything unusual in his or her work area. In addition, all incoming mail passes through an X-ray scanner. One lesson from all this protection is that losses can be much more expensive than security.

Barriers

Post and Kingsbury (1977: 502–503) state, “The physical security process utilizes a number of barrier systems, all of which serve specific needs. These systems include natural, structural, human, animals, and energy barriers.” **Natural barriers** are rivers, hills, cliffs, mountains, foliage, and other features difficult to overcome. Fences, walls, doors, and the architectural arrangement of buildings are **structural barriers. Human barriers** include security officers who scrutinize people, vehicles, and things entering and leaving a facility. The typical **animal barrier** is a dog. **Energy barriers** include protective lighting and intrusion detection systems.

The most common type of barrier is a *chain-link fence* topped with barbed wire (Figure 8-1). A search of the Internet shows many industry standards for fences from ASTM, UL, ISO, and other groups from the United States and overseas. For example, ASTM F567 focuses on materials specifications, design requirements, and installation of chain-link fencing.

One advantage of chain-link fencing is that it allows observation from both sides: a private security officer looking out and a public police officer looking in. Foliage and decorative plastic woven through the fence can reduce visibility and aid offenders. Opposition to chain-link fencing sometimes develops because management wants to avoid an institutional-looking environment. Hedges are an alternative.

It is advisable that the chain-link fence be made of at least 9-gauge or heavier wire with 2″ × 2″ diamond-shaped mesh. It should be at least seven feet high. Its posts should be set in concrete and spaced no more than 10 feet apart. The bottom should be within two inches of hard ground; if the ground is soft, the fence can become more secure if extended a few inches below the ground. Recommended at the top is a *top guard*—supporting arms about one or two feet long containing three or four strands of taut barbed wire six inches apart and facing outward at 45 degrees.

Anticlimb fences (Figure 8-4) are an alternative to the chain-link fence. Applied in Europe, with growing interest and application in the United States, these fences are more attractive and more difficult to climb than the chain-link fence. The mesh openings are small which prevents fingers and shoes from being inserted into the fence to climb. As with all other security measures, anticlimb fences have vulnerabilities.

Barbed wire fences are used infrequently. Each strand of barbed wire is constructed of two 12-gauge wires twisted and barbed every four inches. For adequate protection, vertical support posts are placed six feet apart, and the parallel strands of barbed wire are from two to six inches apart. A good height is eight feet.

Concertina fences consist of coils of steel razor wire clipped together to form cylinders weighing about 55 pounds. Each cylinder is stretched to form a coil-type barrier three feet high and 50 feet long. The ends of each 50-foot coil need to be clipped to the next coil to obviate movement. Stakes also stabilize these fences. This fence was developed by the military to act as a quickly constructed barrier. When one coil is placed on another, they create a six-foot-high barrier. One coil placed on two as a base provides a pyramid-like barrier that is difficult to penetrate. Concertina fences are especially helpful for quick, temporary repairs to damaged fences.

Razor ribbon and *coiled barbed tape* are increasing in popularity. They are similar to concertina fencing in many ways. Every few inches along the coil are sharp spikes, looking something like a small, sharpened bow tie.

FIGURE 8-4 Anticlimb fence with vulnerability.

Gates are necessary for traffic through fences. The fewer gates the better because, like windows and doors, they are weak points along a perimeter. Gates usually are secured with a chain and padlock. Uniformed officers stationed at each gate and fence opening increase security while enabling the observation of people and vehicles.

Vehicle barriers control traffic and stop vehicles from penetrating a perimeter. The problems of vehicle bombs and drive-by shootings have resulted in greater use of vehicle barriers. These barriers are assigned government-certified ratings based on the level of protection; however, rating systems vary among government agencies. One agency, for example, tests barriers against 15,000-pound trucks traveling up to 50 miles per hour, while another agency tests 10,000-pound trucks traveling the same speed. **Passive vehicle barriers** are fixed and include decorative bollards, large concrete planters, granite fountains, specially engineered and anchored park benches, hardened fencing, fence cabling, and trees. An alternative to bollards is a *plinth wall*—a continuous low wall of reinforced concrete with a buried foundation (U.S. Department of Homeland Security, 2003: 2–33). Moore (2006) notes alternatives to bollards, including *tiger traps* (i.e., a path of paving stones over a trench of low-density concrete that will collapse under a heavy weight) and *NOGOs* (i.e., large, heavy bronze blocks). **Active vehicle barriers** are used at entrances and include gates, barrier arms, and pop-up type systems that are set underground and, when activated, spring up to block a vehicle (True, 1996: 49–53). Factors to consider when planning vehicle barriers include frequency of traffic, type of road (e.g., a curved road slows vehicles), aesthetics, and how the barrier is integrated with other physical security and personnel (Morton, 2011: 30). As we know, no security method is foolproof, and careful security planning is vital, including ADA requirements. In 1997, to protest government policy, the environmental group Greenpeace penetrated government security in Washington, D.C., and dumped four tons of coal outside the Capitol building. The driver of the truck drove the wrong way up a one-way drive leading to the building!

Walls are costly and are a substitute for fences when management is against the use of a wire fence. Attractive walls can be designed to produce security equal to fences while blending into surrounding architecture. Walls are made from a variety of materials: bricks, concrete blocks, stones, or cement. Depending on design, the tops of walls six or seven feet high may contain barbed wire, spikes, or broken glass set in cement. Offenders often avoid injury by throwing a blanket or jacket over the top of the wall (or fence) before scaling it. An advantage of a wall is that outsiders are hindered from observing inside. However, observation by public police during patrols also is hindered; this can benefit an intruder.

Hedges or shrubbery are useful as barriers. Thorny shrubs have a deterrent value. These include holly, barberry, and multiflora rose bushes, all of which require a lot of watering. The privet hedge grows almost anywhere and requires minimal care. A combination of hedge and fence is useful. Hedges should be less than three feet high and placed on the inside to avoid injury to those passing by and to create an added obstacle for someone attempting to scale the fence. Any plants that are large and placed too close to buildings and other locations provide a climbing tool, cover for thieves, and a hiding place for contraband.

Municipal codes restrict the heights of fences, walls, and hedges to maintain an attractive environment devoid of threatening-looking barriers. Certain kinds of barriers may be prohibited (e.g., barbed wire) to ensure conformity. Planning should encompass research of local standards.

The following list can help a security manager eliminate weak points along a perimeter or barrier.

1. Utility poles, trees, boxes, pallets, forklifts, tools, and other objects outside a building can be used to scale a barrier.
2. Ladders left outside are an offender's delight. Stationary ladders are made less accessible via a steel cage with a locked door.
3. A **common wall** is shared by two separate entities. Thieves may lease and occupy or just enter the adjoining building or room and then hammer through the common wall.
4. A roof is easy to penetrate. A few tools, such as a drill and saw, enable offenders to cut through the roof. Because lighting, fences, sensors, and patrols rarely involve the roof, this weakness is attractive to thieves. A rope ladder often is employed to descend from the roof, or a forklift might be used to lift items to the roof. Vehicle keys should be hidden and other precautions taken.
5. Roof hatches, skylights, basement windows, air-conditioning and other vent and duct systems, crawl spaces between floors and under buildings, fire escapes, and utility covers may need a combination of locks, sensors, steel bars, heavy mesh, fences, and inspections. A widely favored standard is that any opening greater than 96 square inches requires increased protection.

Protecting Buildings against Terrorism

To help justify security and loss prevention expenditures, executives should refer to the *Reference Manual to Mitigate Potential Terrorist Attacks against Buildings* (U.S. Department of Homeland

Security, 2003: iii), here referred to as FEMA 426. This publication notes that building designs can serve to mitigate multiple hazards. For example, hurricane window design, especially against flying debris, and seismic standards for nonstructural building components apply also to bomb explosions. Next, Purpura (2007) describes protection methods from FEMA 426.

FEMA 426 refers to site-level considerations for security that include land use controls, landscape architecture, site planning, and other strategies to mitigate risks of terrorism and other hazards. **Land use controls**, including zoning and land development regulations, can affect security because they define urban configurations that can decrease or increase risks from crime and terrorism. For instance, managing storm water on-site can add security through water retention facilities that serve as a vehicle barrier and blast setback. This reduces the need for off-site pipes and manholes that can be used for access or to conceal weapons. FEMA 426 offers several building design suggestions to increase security (see Figure 8-5).

A **target-rich environment** is created when people, property, and operations are concentrated in a dense area. There are advantages and disadvantages to a dense cluster. An advantage is the possibility to maximize standoff (i.e., protection when a blast occurs) from the perimeter. Additional security benefits are a reduction in the number of access and surveillance points and a shorter perimeter to protect. A dense cluster of buildings can possibly save energy costs through, for instance, heat transfer from heat-producing areas to heat-consuming areas. In addition, external lighting would not be dispersed over a large area, requiring more lights and energy. In contrast, dispersed buildings, people, and operations spread the risk. However, dispersal can increase the complexity of security (e.g., more access points), and it may require more resources (e.g., security officers, CCTV, lighting, perimeter protection).

Sustainability and green security issues should be considered when planning protection against terrorism.

FEMA 426 recommends that designers consolidate buildings that are functionally compatible and have similar threat levels. For instance, mailrooms, shipping and receiving docks, and visitor screening areas, where people and materials are often closely monitored prior to access, should be isolated and separated from concentrations of people, operations, and key assets.

Keep out zones help to maintain a specific distance between vehicles or people and a building. This is accomplished through perimeter security. If terrorists plan to attack a specific building, they will likely use surveillance to study security features, look for vulnerabilities, and try to penetrate access controls and defenses through creative means; security planning should include surveillance and other methods to identify individuals who may be gathering such information from off or on the premises.

Here are other suggestions for buildings from FEMA 426:

- Provide redundant utility systems to continue life safety, security, and rescue functions in case of an emergency.

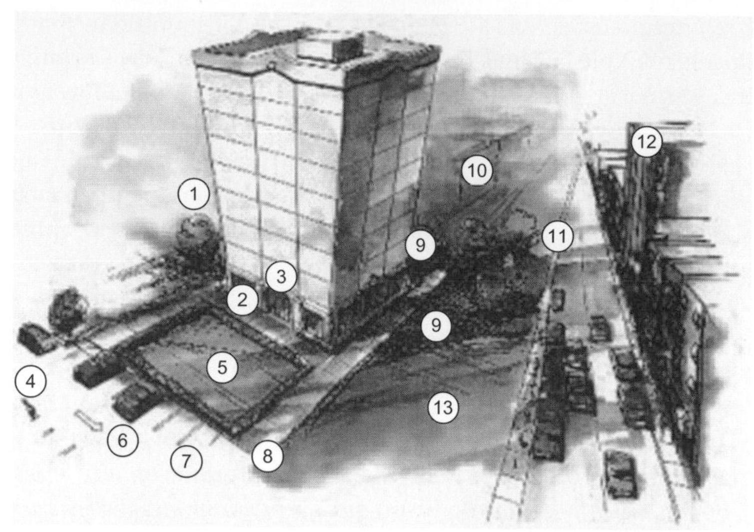

1 Locate assets stored onsite but outside of the facility within view of occupied rooms in the facility

2 Eliminate parking beneath facilities

3 Minimize exterior signage or other indications of asset locations

4 Locate trash receptacles as far from the facility as possible

5 Eliminate lines of approach perpendicular to the building

6 Locate parking to obtain stand-off distance from facility

7 Illuminate building exteriors or sites where exposed assets are located

8 Minimize vehicle access points

9 Eliminate potential hiding places near facility; provide an unobstructed view around facility

10 Site facility within view of other occupied facilities on the installation

11 Maximize distance from facility to installation boundary

12 Locate facility away from natural or man-made vantage points

13 Secure access to power/heat plants, gas mains, water supplies, and electrical service

FIGURE 8-5 Summary of site mitigation measures. *Source: U.S. Department of Homeland Security (2003). Reference Manual to Mitigate Potential Terrorist Attacks against Buildings, FEMA 426 (December). Washington, D.C.: FEMA.*

- Since hardened glazing may cause windows not to blow out in a blast, a system for smoke removal is essential.
- When possible, elevate fresh-air intakes to reduce the potential of hazardous materials entering a building from ground level. The intakes should be sloped down and have screens in case a device is thrown toward the opening.
- Manipulation of the HVAC system could minimize the spread of a hazardous agent. Filtration systems are another option, although expensive.

Mitigation for Explosive Blasts

Standoff distance is the distance between an asset and a threat. FEMA 426 views *distance* as the most effective and desirable strategy against a blast because other methods may vary in effectiveness, be more costly, and result in unintended consequences. A blast wall can become a part of the fragmentation if a bomb is detonated close to it. Urban environments create challenges when designing standoff distance because land is often expensive and it may be unavailable. There is no ideal standoff distance; numerous variables take part in planning, such as the type of threat or explosive, construction characteristics and target hardening, and desired level of protection.

Blast and antiramming walls provide an expensive option for protecting buildings, especially in urban areas. Revel (2003: 40) writes that a test of a blast wall conducted by the U.S. Government's Technical Support Working Group (TSWG) showed the effectiveness of this security method. The blast wall sustained an explosion more powerful than the one that destroyed the Murrah Federal Building (Oklahoma City bombing) and the effects on the test building behind the blast wall were reduced by about 90 percent. The blast wall was constructed by first inserting in the ground 18-foot blast posts, with nine feet extending above the ground. Then steel-jacketed concrete and rebar-filled panels were lowered between the posts in an interlocking pattern. When the explosion occurred, the posts twisted and deflected the blast above and back from the panels, directing the force up and beyond the lower structural steel of the building and around the ends of the wall. The blast wall is also capable of absorbing large vehicle impact at high speeds.

Although several building design features can mitigate explosive blasts, many factors enter into the design of buildings, including cost, purpose, occupancy, and location. A high-risk building should incorporate more mitigation features than a low-risk building. Significant changes to existing buildings may be too expensive; therefore, lower cost changes must be sought. Bollards and strong gates are less expensive than making major structural changes to a building. In addition, trees, vegetative groupings, and earth berms offer some degree of blast shielding. Examples of mitigation features from FEMA 426 are as follows:

- Avoid "U" or "L" shaped building designs that trap the shock waves of a blast. Circular buildings reduce a shock wave better than a rectangular building because of the angle of incidence of the shock wave.
- Avoid exposed structural elements (e.g., columns) on the exterior of a facility.
- Install as much glazing (i.e., windows) as possible away from the street side.
- Avoid locating doors across from one another in interior hallways to limit the force of a blast through the building.
- High-security rooms should be blast- and fragment-resistant.
- Provide pitched roofs to permit deflection of launched explosives.

Glazing

Annealed glass, also called *plate glass*, is commonly used in buildings. It has low strength, and upon failure, it fractures into razor sharp pieces. *Fully thermally tempered glass* (TTG) is four to five times stronger than annealed glass, and upon failure, it will fracture into small cube-shaped fragments. Building codes generally require TTG anywhere the public can touch (e.g., entrance doors). *Wire-reinforced glass* is made of annealed glass with an embedded layer of wire mesh. It is applied as a fire-resistant and forced entry barrier. All three types of glass present a dangerous hazard from a blast (U.S. Department of Homeland Security, 2003).

Traditionally, window protection focused on hindering forced entry. Today, we are seeing increasing designs that mitigate the hazardous effects of flying glass from a variety of risks, besides explosion. Experts report that 75 percent of all damage and injury from bomb blasts results from flying and falling glass. Vendors sell *shatter-resistant film*, also called *fragment retention film* (FRF), which is applied to the glass surface to reduce this problem. Conversely, a report on the 1993 World Trade Center attack claimed that the destroyed windows permitted deadly gases to escape from the building, enabling occupants to survive. A balanced design (i.e., type of glass, glass frame, and frame to building) means that all the window components have compatible capacities and fail at the same pressure levels. The U.S. General Services Administration publishes glazing protection levels based on how far glass fragments would enter a space and cause injuries. It is important to note that the highest level of protection for glazing may not mitigate the effects from a large explosion (U.S. Department of Homeland Security, 2003).

Blast curtains are window draperies made of special fabrics designed to stop glass window shards that are caused by explosions and other hazards. Various designs serve to catch broken glass and let the gas and air pressure dissipate through the fabric mesh. The fibers of these curtains can be several times as strong as steel wire. The U.S. General Services Administration establishes criteria for these products (Owen, 2003: 143–144).

Glass can be designed to block penetration of bullets, defeat attempted forced entry, remain intact following an explosion, and protect against electronic eavesdropping. The Internet shows many standards for glazing from the American Architectural Manufacturers Association (AAMA), ANSI, UL, ASTM, Consumer Product Safety Commission, ISO, and overseas groups. Security glazing should be evaluated on comparative testing to an established national consensus standard such as ASTM F1233, Standard Test Method for Security Glazing Materials and Systems. Important issues for glazing include product life cycle, durability, installation, maintenance, and framing (Saflex, Inc., 2007).

Underwriters Laboratories classifies *bullet-resistant windows* into eight protection levels, with levels 1 to 3 rated against handguns and 4 to 8 rated against rifles. Level 4 or higher windows usually are applied by government agencies and the military. Protective windows are made of either glass or plastic or mixtures of each.

Laminated glass absorbs a bullet as it passes through various glass layers. The advantage of glass is in its maintenance: it is easy to clean and less likely to scratch than plastic. It is less expensive per square foot than plastic but heavier, which requires more workers and stronger frames. Glass has a tendency to spall (i.e., chip) when hit by a bullet. UL752-listed glass holds up to three shots, and then it begins to shatter from subsequent shots.

Two types of plastic used in windows are acrylic and polycarbonate. Both vary in thickness and are lighter and more easily scratched than glass. *Acrylic windows* are clear and monolithic, whereas glass and polycarbonate windows are laminates consisting of layers of material bonded one on top of another. Acrylic will deflect bullets and hold together under sustained hits. Some spalling may occur. *Polycarbonate windows* are stronger than acrylics against high-powered weapons. Local codes may require glazing to pop out in an emergency.

In addition to protective windows, wall armor is important because employees often duck below a window during a shooting. These steel or fiberglass plates also are rated.

Burglar-resistant windows are rated (UL 972, Burglary Resisting Glazing Material); available in acrylic and polycarbonate materials; and protect against hammers, flame, "smash and grab," and other attacks. Combined bullet- and burglar-resistant windows are available. Although window protection is an expense that may be difficult to justify, insurers offer discounts on insurance premiums for such installations.

Electronic security glazing, containing metalized fabrics, can prevent electromagnetic signals inside a location from being intercepted from outside, while also protecting a facility from external electromagnetic radiation interference from outside sources. Standards for this type of glazing are from the National Security Agency, NSA 65–8.

Window Protection

Covering windows with grating or security screens is an additional step to impede entrance by an intruder or items being thrown out by a dishonest employee. *Window grating* consists of metal bars constructed across windows. These bars run horizontally and vertically to produce an effective form of protection. Although these bars are not aesthetically pleasing, they can be purchased with attractive ornamental designs. *Security screens* are composed of steel or stainless steel wire (mesh) welded to a frame. Screens have some distinct advantages over window grating. Employees can pass pilfered items through window bars more easily than through a screen. Security screens look like ordinary screens, but they are much heavier in construction and can stop rocks and other objects.

When planning window protection, one must consider the need for emergency escape and ventilation. To ensure safety, certain windows can be targeted for the dismantling of window protection during business hours.

Window Locks

Businesses and institutions often contain windows that do not open. For windows that do open, a latch or lock on the inside provides some protection. The *double-hung window,* often applied at residences, is explained here as a foundation for window protection. It consists of top and bottom windows that are raised and lowered for user convenience. When the top window is pushed up and the bottom window pushed down, a crescent sash lock containing a curved turn knob locks both parts of the whole window in place (Figure 8-6). By inserting a knife under the crescent sash lock where both window sections meet, an offender can jimmy the latch out of its catch. If an offender breaks the glass, the crescent sash lock can be unlocked by reaching inside. With such simple techniques known to offenders, defenses that are more complicated are necessary. Nails can be used to facilitate a quick escape while maintaining good window security: one drills a downward-sloping hole into the right and left sides of the window frame where the top and bottom window halves overlap and inserts nails that are thinner and longer than the holes. This enables the nails to be quickly removed during an emergency escape. If a burglar attacks the window, he or she cannot find or remove the nails (Figure 8-6). Another method is to attach a window lock requiring a key (Figure 8-6). These locks are capable of securing a window in a closed or slightly opened position. This can be done with the nail (and several holes) as well. The key should be hidden near the window in case of emergency.

Electronic Protection for Windows

Four categories of electronic protection for windows are foil, vibration, glass-breakage, and contact-switch sensors. *Window foil,* which has lost much of its popularity, consists of lead foil tape less than 1-inch wide and paper thin that is applied directly on the glass near the edges of a window. In the nonalarm state, electricity passes through the foil to form a closed circuit. When the foil is broken, an alarm is sounded. Window foil is inexpensive and easy to maintain. One disadvantage is that a burglar may cut the glass without disturbing the foil. *Vibration sensors* respond to vibration or shock. They are attached directly on the glass or window frame. These sensors are noted for their low false alarm rate and are applicable to fences, walls, and valuable artwork, among other things. *Glass-breakage sensors* react to glass breaking. A sensor the size of a large coin is placed directly on the glass and can detect glass breakage several feet away. Some types operate via a tuning fork, which is tuned to the frequency produced by glass breaking. Others employ a microphone and electric amplifier. *Contact switches* activate an alarm when opening the window interrupts the contact. In Figure 8-7, this sensor protects a door and roof opening.

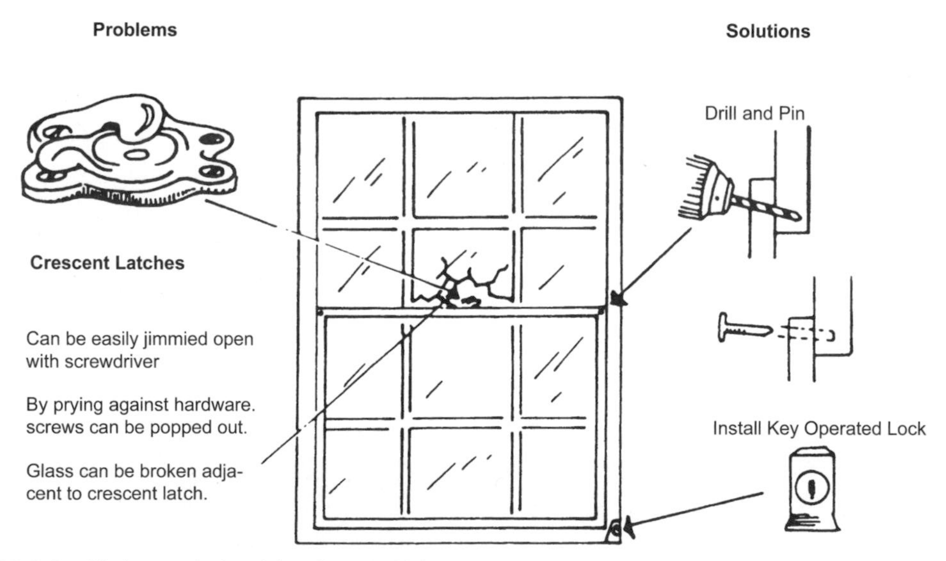

FIGURE 8-6 Double-hung window (view from inside).

Additional ideas for window protection follow:

1. A strong window frame fastened to a building prevents prying and removal of the entire window.
2. First floor windows are especially vulnerable to penetration and require increased protection.
3. Consider tinting windows to hinder observation by offenders.
4. Windows (and other openings) that are no longer used can be bricked.
5. Expensive items left near windows invite trouble.
6. Cleaning windows and windowsills periodically increases the chances of obtaining clear fingerprints in the event of a crime.

Doors

Many standards apply to doors, from the AAMA, ANSI, ASTM, BHMA, National Association of Architectural Metal Manufacturers (NAAMM), NFPA, Steel Door Institute (SDI), UL, and ISO. In addition, other countries have standards. Aggleton (2010: 24) writes that because of life

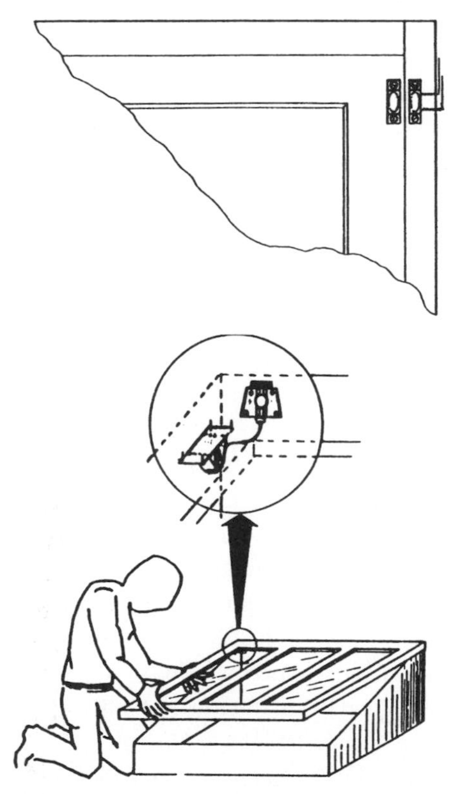

FIGURE 8-7 Switch sensors have electrical contacts that make or break an electrical circuit in response to a physical movement.

safety issues (e.g., quick egress in case of emergency), door hardware and locks are subject to more stringent codes and standards than other physical security, such as intrusion sensors.

Doors having fire ratings must meet certain frame and hardware requirements. Decisions on the type of lock and whether electronic access will be applied also affect hardware. Decisions on doors are especially crucial because of their daily use and the potential for satisfying or enraging users and management (Schumacher, 2000: 40).

Businesses and institutions generally use aluminum doors. Composed of an aluminum frame, most of the door is covered by glass. Without adequate protection, the glass is vulnerable, and prying the weak aluminum is not difficult. The all-metal door improves protection at the expense of attractiveness.

Hollow-core doors render complex locks useless because an offender can punch right through the door. Thin wood panels or glass on the door are additional weak points. More expensive, *solid-core doors* are stronger; they are made of solid wood (over an inch thick) without the use of weak fillers. To reinforce hollow-core or solid-core doors, one can attach 16-gauge steel sheets, via one-way screws.

Whenever possible, door hinges should be placed on the inside. Door hinges that face outside enable easy entry. By using a screwdriver and hammer, one can raise the pins out of the hinges to enable the door to be lifted away. To protect the hinge pins, it is a good idea to weld them so they cannot be removed in this manner. Another form of protection is to remove two screws on opposite sides of the hinge, insert a pin or screw on the jamb side of the hinge so that it protrudes about half an inch, and then drill a hole in the opposite hole to fit the pin when the door is closed. With this method on both top and bottom hinges, even if the hinge pins are removed, the door will not fall off the hinges (Figure 8-8).

Contact switches applied to doors offer electronic protection. Greater protection is provided when contact switches are recessed in the edges of the door and frame. Other kinds of electronic sensors applied at doors include vibration sensors, pressure mats, and various types of motion detectors aimed in the area of the door.

More hints for door security follow:

1. A wide-angle door viewer within a solid door permits a look at the exterior prior to opening a door.
2. Doors (and windows) are afforded extra protection at night by chain closures. These frequently are seen covering storefronts in malls and in high-crime neighborhoods.
3. To block "hide-in" burglars (those who hide in a building until after closing) from easy exit, require that openings such as doors and windows have a key-operated lock on the inside as well as on the outside.
4. Almost all fire departments are equipped with power saws that cut through door locks and bolts in case of fire. Many firefighters can gain easy access to local buildings because building owners have provided keys that are located in fire trucks. Although this creates a security hazard, losses can be reduced in case of fire.
5. All doors need protection, including garage, sliding, overhead, chain-operated, and electric doors.

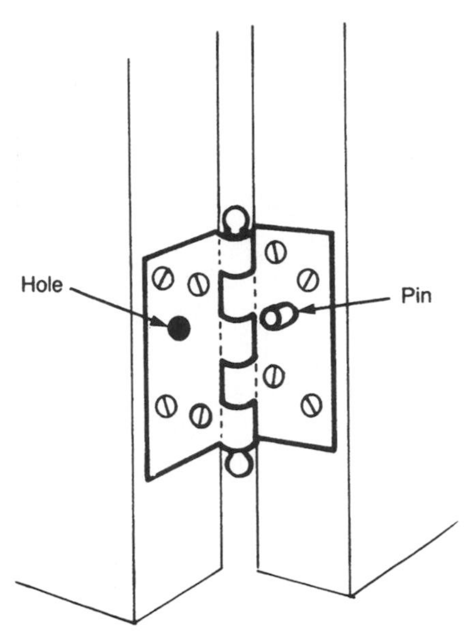

FIGURE 8-8 Pin to prevent removal of door.

Intruding Neighbors

The Finch Brothers Supermarket Company maintained a busy warehouse stocked with hundreds of different items for local Finch supermarkets. The company leased the large warehouse to accommodate the increasing number of supermarkets. After 18 months at this location, managers were stumped as to why shrinkage was over four percent. Several precautions were taken to avert losses: perimeter security consisted of intrusion sensors, lighting, and a security officer. A perpetual inventory was maintained.

Eventually, Finch's loss prevention manager's job was on the line, so he began a secret, painstaking, and continuous surveillance of the warehouse at night. After an agonizing week went by, he made an astonishing discovery. A printing company building next door was only seven feet away from the warehouse, and printing company employees on the late shift were able to slide a 12'×16"×2" board from a third-story window to a window of the same height at the warehouse. Within 30 minutes, the group of thieves hauled and threw many burlap sacks of items from one building to another. With camera equipment, the manager recorded the crime. Police were later notified and arrests made.

The thieves confessed that, when they worked the 11 P.M. to 8 A.M. shift, they stole merchandise from the warehouse. They stated that a maintenance man, who visited the warehouse each day, left the window open so the board could be slipped in. They added that dim lighting and the fact that intrusion sensors existed only on the first floor were factors that aided their crimes.

Intrusion Detection Systems

Standards for intrusion detection systems are from UL, the Institute of Electrical and Electronics Engineers (IEEE), and ISO, plus other groups in the United States and overseas. UL, for example, "lists" installation companies that are authorized to issue UL Certificates on each installation. This means that the installer conforms to maintenance and testing as required by UL, which conducts unannounced inspections.

Table 8-1 describes intrusion detection systems; these systems have gone through several generations, leading to improved performance. Not in the table is *magnetic field*, which consists of a series of buried wire loops or coils. Metal objects moving over the sensor induce a current and signal an alarm. Research shows that the vulnerability to defeat (VD) for magnetic field and infrared photo beam is high. Microwave, electric field, fence disturbance, seismic sensor cable, taut wire, and video motion systems all have a medium VD. The VD for ported coaxial cable systems is low. Visible sensors are relatively easy to defeat but cost effective for low-security applications. Multiple sensors, and especially covert sensors, provide a higher level of protection (Clifton and Vitch, 1997: 57–61; Reddick, 2005: 36–42; Shelton, 2006: 80–82).

Fiber optics is a growing choice for intrusion detection and transmission. **Fiber optics** refers to the transportation of data by way of guided light waves in an optical fiber. This differs from the conventional transmission of electrical energy in copper wires. Fiber optic applications include video, voice, and data communications. Fiber optic data transmission is more secure and less subject to interference than older methods.

Fiber optic perimeter protection can take the form of a fiber optic cable installed on a fence. When an intruder applies stress on the cable, an infrared light source pulsing through the system notes the stress or break and activates an alarm. Optical fibers can be attached to or inserted within numerous items to signal an alarm, including razor ribbon, security grills, windows, and doors, and it can protect valuable assets such as computers.

Dibazar et al. (2011) writes of their research on the development and deployment of "smart fence" systems consisting of multiple sensor technologies. Their research illustrates the direction and capabilities of "smart fence" systems. Included in their design are "(a) acoustic based long range sensor with which vehicles' engine sound and type can be identified, (b) vibration based seismic analyzer which discriminates between human footsteps and other seismic events such as those caused by animals, and (c) fence breaching vibration sensor which can detect intentional disturbances on the fence and discriminate among climb, kick, rattle, and lean."

Garcia (2006: 83–84) views intrusion sensor performance based on three characteristics: probability of detection of the threat, nuisance alarm rate, and vulnerability to defeat. The *probability of detection* depends on several factors including the desired threat to be detected (e.g., walking, tunneling), sensor design, installation, sensitivity adjustment, weather, and maintenance/testing. According to Garcia, a *nuisance alarm rate* results from a sensor interacting with the environment, and a sensor cannot distinguish between a threat and another event (e.g., vibration from a train). A *false alarm rate* results from the equipment itself, and

Table 8-1 Types of Intrusion Alarm Systems*

System	Graphic idea	Concept	Advantages	Disadvantages
Fence-mounted sensor		Detection depends on movement of fence	Ease of installation; early detection on interior fence; relatively inexpensive; requires little space; follows terrain easily	Frequent false alarms (weather and birds); conduit breakage; dependent on quality, rigidity of fence, and type of installation
Seismic sensor cable (buried)		Detection depends on ground movement (intruder walking over buried movement sensors, or other seismic disturbances)	Good for any site shape, uneven terrain; early warning; good in warm climate with little rain	False alarms from ground vibrations (vehicles, thunderstorms, heavy snow); not recommended for heavy snow regions; difficult installation and maintenance
Balanced capacitance		Detection depends on touching of cable, interfering with balance of cable	Few false alarms; good for selected areas of fence, rooftops, curves, corners, any terrain	Not to be used independently; for selected areas only
Taut wire		Detection depends on deflecting, stretching, or releasing the tension of wire that triggers alarming mechanism	Good for any terrain or shape; can be used as interior fence; extremely low false alarm rates	Relatively expensive; possible false alarms from snow, ice, birds, etc.; temperature changes require adjustments
Microwave sensor		Based on line of sight; detection depends on intrusion into volumetric area above ground between transmitter and receiver	Does not require a great deal of maintenance	Not good on hilly or heavily contoured terrain; costly installation; potential false alarms caused by weather (snow, ice, wind, and rain); vegetation must be removed
Infrared photo beam sensor		Based on line of sight; detection depends on intrusion into beam(s) stacked vertically above ground	Good for short distances, building walls, and sally ports	Distances between transmitter and receiver must be short, requiring more intervals; potential false alarms by animals and weather conditions (fog, dust, snow); voltage surges
Ported coaxial cable		Detection depends on interruption of field in terms of mass, velocity, and length of time	Adaptable to most terrains	False alarms caused by heavy rain (pooling of water), high winds, tree roots; relatively expensive installation and maintenance
Video motion detection		Detection depends on change in video-monitor signal	Good for enhancing another system; good for covering weak spots	Lighting is a problem
Electric field sensor		Detection depends on penetration of volumetric field created by field wires and sensor wires	Good on hilly or heavily contoured terrain; can be freestanding or fence-mounted	Requires more maintenance; sensor wires must be replaced every three years; vegetation must be controlled

*_Sources_: Information from New York State Department of Corrections, Pennsylvania Department of Corrections, South Carolina Departmentof Corrections, and Federal Bureau of Prisons. Reproduced from U.S. Dept. of Justice, National Institute of Justice, _Stopping Escapes: Perimeter Security_ (U.S. Government Printing Office, August 1987), p. 6.

FIGURE 8-9 Point protection.

it is caused by inadequate design, failure, or poor maintenance. *Vulnerability to defeat* varies among systems. **Bypass** means the adversary circumvented the intrusion detection system. **Spoofing** means the adversary traveled through the detection zone without triggering an alarm; depending on the sensor, one strategy is by moving very slowly. Garcia emphasizes the importance of proper installation and testing of intrusion detection systems.

No one technology is perfect; many protection programs rely on dual technology to strengthen intrusion detection. In the process of selecting a system, it is wise to remember that manufacturers' claims often are based on perfect weather. Security decision makers must clearly understand the advantages and disadvantages of each type of system under a variety of conditions.

Applications

Intrusion detection systems can be classified according to the kind of protection provided. There are three basic kinds of protection: point, area, and perimeter. **Point protection** (Figure 8-9) signals an alarm when an intrusion is made at a special location. It is also referred to as **spot or object protection**. Files, safes, vaults, jewelry counters, and artwork are targets for point protection. Capacitance and vibration systems provide point protection and are installed directly on the object. These systems often are used as a backup after an offender has succeeded in gaining access. **Area protection** (Figure 8-10) detects an intruder in a selected area such as a main aisle in a building or at a strategic passageway. Microwave and infrared systems are applicable to area protection. **Perimeter protection** (Figure 8-11) focuses on the

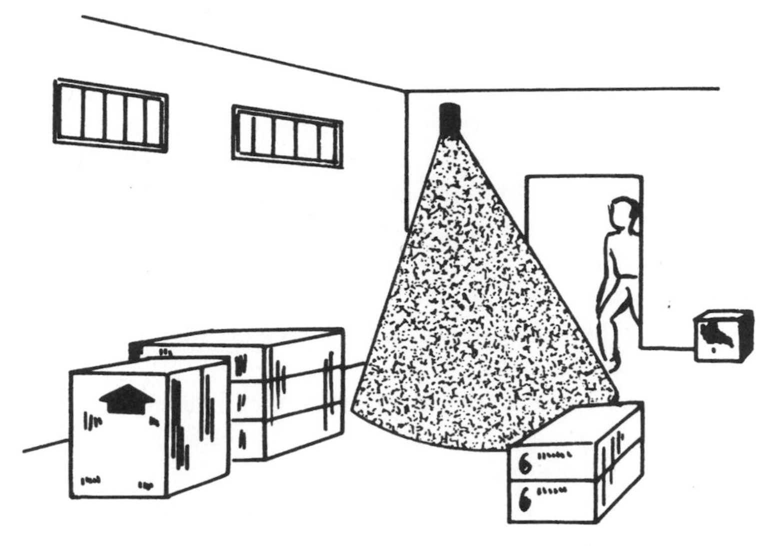

FIGURE 8-10 Area protection.

FIGURE 8-11 Perimeter protection.

outer boundary of the premises. If doors and windows are part of the perimeter, then contact switches, vibration detectors, and other devices are applicable.

Alarm Signaling Systems

Alarm signaling systems transmit data from a protected area to an annunciation system. Local ordinances and codes provide guidelines and restrictions on these systems.

Local alarm systems notify, by sound or lights, people in the hearing or seeing range of the signal. This includes the intruder, who may flee. Typically, a siren or bell is activated outside a building. Often, local alarms produce no response—in urban areas responsible action may not be taken, and in rural areas, nobody may hear the alarm. These alarms are less expensive than other signaling systems but are easily defeated. If a local alarm is used during a robbery, people may be harmed. Research from the UK (Coupe and Kaur, 2005: 53–72) points to the benefits of delayed-audible alarms during a burglary that are triggered as the offender enters the premises but sound a few minutes later so cameras have an opportunity to record the offender and police still have time to respond prior to the offender's escaping if they are notified promptly. Combining these strategies with an immediate silent alarm to a central station increases the opportunity for an arrest.

A **central station** alarm system receives intrusion, fire, medical, and environmental signals at a computer console located and monitored a distance away from the protected location. When an alarm signal is received, central station personnel contact police, firefighters, or other responders. Central station services employ sales, installation, service, monitoring, and response personnel. Proprietary monitoring systems are similar to central station systems, except that the former does the monitoring and the system is operated by the proprietary organization. Resources for central station design are available from UL, NFPA, and the Security Industry Association (Patterson, 2000: 80).

Technology drives advances in central station capabilities. Remote video monitoring enables a central station operator to view what triggered an alarm to verify the need for a human response. Global positioning system (GPS) permits real-time tracking (e.g., location, direction, and speed) and archiving of moving assets and people. Off-site video storage, especially at a UL-listed central station, affords increased protection and backup for video recordings. It also helps to prevent the problem of offenders taking recording equipment with them as they leave the crime scene and, thus, destroying evidence (Evans, 2005: 44–46).

WeGuardYou (2011), a security vendor, describes how its technology is applied to shopping mall security as explained next. Each security vehicle functions as a central station that offers real-time local and remote video, alarm, and data monitoring while transferring information over a secure wireless system. In one scenario, a woman leaves a mall one evening, packages and pocketbook in hand. Suspected muggers are in the parking lot. Cameras follow the woman in real time as the images are sent to mobile units (i.e., security vehicles), besides showing on TV monitors inside the mall central station control room. Mobile units containing emergency lights, loudspeaker, and enhanced lighting converge on the "hot spot" to prevent victimization. If an emergency occurs, security officers take action, all security personnel are notified via radio, police and EMS are notified, and fixed cameras and those on mobile units record the incident with images remotely accessible.

Various data transmission systems are utilized to signal an alarm. Here, the older technology is covered first before the modern technology. As with fire alarm systems, security alarm systems are using less traditional phone lines to transmit an alarm as digital systems involving cellular networks, fiber optics, and Voice over IP are advancing (Morton, 2012: 32).

Automatic telephone dialer systems include the tape dialer and digital dialer. Tape dialer systems are seldom used today. They deliver a prerecorded or coded message to an interested party (e.g., central station, police department) after that party answers the telephone. Digital dialers use coded electronic pulses that are transmitted and an electronic terminal decodes the message. Digital dialers, often called digital communicators, are still applied today, although the technology is more advanced than in earlier years. Local codes typically prohibit tape dialers or similar automatic devices connected to authorities (e.g., police and fire) because of false alarms, wasted resources, and the need for authorities to ask questions about the emergency. The central station evolved to serve as a buffer between the site of the emergency and authorities so information can be gathered and verified prior to contacting authorities.

Today, there is a variety of automatic voice/pager dialer systems on the market that contact a central station or individual when a sensor is activated. The technology is also applied in sales, such as the use of software enabling calls through a computer.

Radio frequency (RF) and *microwave* data transmission systems often are applied where telephone lines are not available or where hardwire lines are not practical. The components include transmitter, receiver, repeaters to extend range, battery backup, and solar power.

Fiber optic data transmission systems, as discussed earlier, transport data by way of light waves within a thin glass fiber. These cables are either underground or above ground. The components include transmitter, receiver, repeaters, battery backup, and solar power. Fiber optic systems are more secure than direct wire.

Signals should be backed up by multiple technologies. Options for off-site transmission of activity include satellite, local area network (LAN), wide area network (WAN), cellular, and the Internet. Cellular is especially useful for backup, since it is more likely to remain in operation in certain disasters. It can also be used as a primary transmission method (Zwirn, 2003: 74–83).

Among the advances in alarm monitoring is **remote programming**. Using this method, a central station can perform a variety of functions without ever visiting the site. Capabilities include arming and disarming systems, unlocking doors, performing diagnostics and corrections, and, with access systems, adding or deleting cards.

Alarm systems may also be multiplexed or integrated. *Multiplexing* is a method of transmitting multiple information signals over a single communications channel. This single communications channel reduces line requirements by allowing signal transmission from many protected facilities. Two other advantages are that information that is more detailed can be transmitted, such as telling which detector is in an alarm state, and transmission line security is enhanced with encoding. *Integrated systems*, as covered in Chapter 7, combine multiple systems (e.g., alarm monitoring, access controls, and CCTV).

CCTV

CCTV allows one person to view several locations (Figure 8-12). This is a distinct advantage when protecting the boundaries of a facility, because it reduces personnel costs.

Television programs and movies sometimes portray an intruder penetrating a perimeter barrier by breaking through when a CCTV camera had momentarily rotated to another

FIGURE 8-12 CCTV.

location. Usually, the camera just misses the intruder by returning to the entry point right after the intruder gains access. Such a possibility can be averted via overlapping camera coverage. If cameras are capable of viewing other cameras, personnel can check on viewing obstructions, sabotage, vandalism, or other problems. Smoked domes prevent an offender from identifying the direction of the camera. In addition, covert CCTV surveillance should be considered for outdoor applications in conjunction with overt CCTV surveillance.

Tamper-proof housings will impede those interested in disabling cameras. Different models are resistant to vandalism, bullets, explosion, dust, and severe weather. Housings are manufactured with heaters, defrosters, windshield wipers, washers, and sun shields.

Low-light-level cameras provide the means to view outside when very little light is available. When no visible light is available, an infrared illuminator creates light, invisible to the naked eye, but visible to infrared-sensitive cameras. Another option is *thermal imaging cameras,* which sense heat from an intruder and are especially helpful to spot them in darkness, fog, smoke, foliage, and up to several miles away (Spadanuta, 2011: 56–66; Pierce, 2006: 24–28).

An essential aspect of CCTV usage is proper monitoring. Although video motion detection and video analytics (Chapter 7) apply technology to identify anomalies that human observers may miss, security management should take action to reduce fatigue and ensure good-quality viewing. Suggestions include rotating personnel every two hours, limit TV monitors to fewer than 10, arrange monitors in a curved configuration in front of the viewer, control the lighting over the console to avoid glare on the monitor screens or tilt the monitors if necessary, place the monitors in an order that permits easy recognition of camera locations, provide a swivel chair that hampers the opportunity for sleeping, and assign tasks to the viewer (e.g., communications and logging).

Hacking into Physical Security Systems and Planting "Evidence"

Bernard (2010: 15) writes that a company that conducts network penetration tests for clients was able to hack into a brand-name system and "feed back copied video into its video display and recording stream. They picked up an object off a table, but the video system showed the object as still being there." Another version of this attack is injecting captured video of a theft, violent incident, or other event later (or possibly earlier) than the time of the actual occurrence. The system's time-stamped video and video management software watermark would provide erroneous evidence, possibly helping to convict the wrong person while supporting an alibi for the devious offender.

Bernard also wrote of a researcher's successful penetration of a brand-name networked access control system. He writes: "From now on *it will be the rule rather than the exception* that hacker conferences will include sessions on how to hack physical security systems—just like they contain sessions about hacking telephones, web servers, information systems, and so on."

Lighting

From a business perspective, lighting can be justified because it improves sales by making a business and merchandise more attractive, promotes safety and prevents lawsuits, improves employee morale and productivity, and enhances the value of real estate. From a security perspective, three major purposes of lighting are *to create a psychological deterrent to intrusion, to enable detection, and to enhance the capabilities of CCTV systems.* Good lighting is considered such an effective crime control method that the law, in many locales, requires buildings to maintain adequate lighting.

Painter and Farrington (1999) conducted a major study on the effect of lighting on the incidence of crime in England. Three residential areas were selected. One was the experimental area that contained improved lighting. The second was labeled the adjacent area. In addition, the third served as the control area. Lighting in the adjacent and control areas remained unchanged. The research included the question of whether improved lighting might result in a reduction of crime in the adjacent area. The research results showed a marked reduction in a variety of crimes in the experimental area, whereas crime in the adjacent and control areas remained the same.

One way to study lighting deficiencies is to go to the premises at night and study the possible methods of entry and areas where inadequate lighting will aid an offender. Before the visit, one should contact local police as a precaution against mistaken identity and to recruit their assistance in spotting weak points in lighting.

Three sources for information on lighting are the Illuminating Engineering Society of North America (IESNA), the National Lighting Bureau, and the International Association of Lighting Management Companies. The IESNA provides information on recommended lighting levels for a variety of locations.

What lighting level will aid an intruder? Most people believe that under conditions of darkness a criminal can safely commit a crime. However, this view may be faulty, in that one generally cannot work in the dark. Three possible levels of light are *bright light, darkness,* and *dim light.* Bright light affords an offender plenty of light to work but enables easy observation by others; it will deter crime. Without light—in darkness—a burglar finds that he or she cannot see to jimmy a door lock, release a latch, or perform whatever work is necessary to gain access; a flashlight is necessary, which someone may observe. However, dim light provides just enough light to break and enter while hindering observation by authorities. Support for this view was shown in a study of crimes during full-moon phases, when dim light was produced. This study examined the records of 972 police shifts at three police agencies, for a two-year period, to compare nine different crimes during full moon and non-full-moon phases. Only one crime, breaking and entering, was greater during full-moon phases (Purpura, 1979: 350–353). Although much case law supports lighting as an indicator of efforts to provide a safe environment, security specialists are questioning conventional wisdom about lighting (Berube, 1994: 29–33). Because so much nighttime lighting goes unused, should it be reduced or turned off? Should greater use be made of motion-activated lighting? How would these approaches affect safety and cost-effectiveness? These questions are ripe for research.

What are your views on nighttime lighting? Should certain locations turn it off?

Illumination

Lumens (of light output) per watt (of power input) are a measure of lamp efficiency. Initial lumens-per-watt data are based on the light output of lamps when new; however, light output declines with use. **Illuminance** is the intensity of light falling on a surface, measured in foot-candles (English units) or lux (metric units). The **foot-candle (FC)** is a measure of how bright the light is when it reaches one foot from the source. One **lux** equals 0.0929 FC. For measures of illuminance, values not labeled as vertical are generally assumed to be horizontal FC (or lux). The light provided by direct sunlight on a clear day is about 10,000 FC; an overcast day would yield about 100 FC; and a full moon, about 0.01 FC. A sample of outdoor lighting illuminances recommended by the Illuminating Engineering Society of North America (2003) are as follows: guarded facilities, including entrances and gatehouse inspection, 10 FC (100 lux); parking facilities, garages, and covered parking spaces, 6 FC (60 lux) on pavement and 5 FC (50 lux) for stairs, elevators, and ramps; and for fast-food restaurant parking, general parking at schools, hotels, motels, and common areas of multifamily residences and dormitories, 3 FC (30 lux).

Care should be exercised when studying illuminance. Horizontal illuminance may not aid in the visibility of vertical objects such as signs and keyholes. FC vary depending on the distance from the lamp and the angle. If a light meter is held horizontally, it often gives a different reading than if it is held vertically. Are the FC initial or maintained? Maintenance and bulb replacement ensure high-quality lighting (National Lighting Bureau, n.d.: 1–36; Smith, 1996: 1–4).

Lamps

The following lamps are applied outdoors (National Fire Protection Association, 2005: 12–13; National Lighting Bureau, n.d.: 1–36; Smith, 1996: 1–4):

- *Incandescent* lamps are at residences. Electrical current passes through a tungsten wire enclosed in a glass tube. The wire becomes white-hot and produces light. These lamps produce 17 to 22 lumens per watt, are the least efficient and most expensive to operate, and have a short lifetime of from 500 to 4,000 hours. *Compact fluorescent light bulbs* are replacing incandescent bulbs because they are "earth friendly," use less energy, last longer (10,000 hours), and generate less heat.
- *Halogen* and *quartz halogen* lamps are incandescent bulbs filled with halogen gas (like sealed-beam auto headlights) and provide about 25 percent better efficiency and life than ordinary incandescent bulbs.
- *Fluorescent* lamps pass electricity through a gas enclosed in a glass tube to produce light, producing 67 to 100 lumens per watt. They create twice the light and less than half the heat of an incandescent bulb of equal wattage and cost five to ten times as much. Fluorescent lamps do not provide high levels of light output. The lifetime is 9,000 to 17,000 hours. They are not used extensively outdoors, except for signage.
- *Mercury vapor* lamps also pass electricity through a gas. The yield is 31 to 63 lumens per watt, and the life is over 24,000 hours with good efficiency compared to incandescent lamps. Because of their long life, these lamps are often used in street lighting.
- *Metal halide* lamps are also of the gaseous type. The yield is 80 to 115 lumens per watt, and the efficiency is about 50 percent higher than mercury vapor lamps, but the lamp life is about 6,000 hours. They often are used at sports stadiums because they imitate daylight conditions, and colors appear natural. Consequently, these lamps complement CCTV systems, but they are the most expensive light to install and maintain.
- *High-pressure sodium* lamps are gaseous, yield about 80 to 140 lumens per watt, have a life of about 24,000 hours, and are energy efficient. These lamps are often applied on streets, parking lots, and building exteriors. They cut through fog and are designed to allow the eyes to see more detail at greater distances.
- *Low-pressure sodium* lamps are gaseous, produce 150 lumens per watt, have a life of about 15,000 hours, and are even more efficient than high-pressure sodium. These lamps are expensive to maintain.
- A light source growing in popularity is LED (light-emitting diode). Each LED contains a semiconductor diode (i.e., an electronic device that permits current to flow through it) that emits light upon receiving voltage. LEDs, used as indicator lights for many electronic devices, have progressed to street lighting, interior lighting, and security lighting. Although LED products vary, manufacturers claim superior life, low maintenance, and green characteristics (e.g., no mercury as in many types of bulbs).

Each type of lamp has a different **color rendition**, which is the way a lamp's output affects human perceptions of color. Incandescent, fluorescent, and certain types of metal halide lamps provide excellent color rendition. Mercury vapor lamps provide good color rendition but are heavy on the blue. High-pressure sodium lamps, which are used extensively outdoors, provide poor color rendition, making things look yellow. Low-pressure sodium lamps make color unrecognizable and produce a yellow-gray color on objects. People find sodium vapor lamps, sometimes called *anticrime lights*, to be harsh because they produce a strange yellow haze. Claims are made that this lighting conflicts with aesthetic values and that it affects sleeping habits. In many instances, when people park their vehicles in a parking lot during the day and return to find their vehicle at night, they are often unable to locate it because of poor color rendition from sodium lamps; some report their vehicles as being stolen. Another problem is the inability of witnesses to describe offenders accurately.

Mercury vapor, metal halide, and high-pressure sodium take several minutes to produce full light output. If they are turned off, even more time is required to reach full output because they first have to cool down. This may not be acceptable for certain security applications. Incandescent, halogen, and quartz halogen have the advantage of instant light once electricity is turned on. Manufacturers can provide information on a host of lamp characteristics including the "strike" and "restrike" time.

Lighting Equipment

Fresnel lights have a wide flat beam that is directed outward to protect a perimeter, glaring in the faces of those approaching. A floodlight "floods" an area with a beam of light, resulting in considerable glare. Floodlights are stationary, although the light beams can be aimed to select positions. The following strategies reinforce good lighting:

1. Locate perimeter lighting to allow illumination of both sides of the barrier.
2. Direct lights down and away from a facility to create glare for an intruder. Make sure the directed lighting does not hinder observation by patrolling officers.
3. Do not leave dark spaces between lighted areas for offenders to move within. Design lighting to permit overlapping illumination.
4. Protect the lighting system: locate lighting inside the barrier, install protective covers over lamps, mount lamps on high poles, bury power lines, and protect switch boxes.
5. Photoelectric cells will enable lights to go on and off automatically in response to natural light. Manual operation is helpful as a backup.
6. Consider motion-activated lighting for external and internal areas.
7. If lighting is required near navigable waters, contact the U.S. Coast Guard.
8. Work to reduce light pollution such as wasting energy and disturbing neighbors with light trespass.
9. Maintain a supply of portable, emergency lights and auxiliary power in the event of a power failure.
10. Good interior lighting also deters offenders.
11. If necessary, join other business owners to petition local government to install improved street lighting.

Parking Lot and Vehicle Controls

Employee access control at a building is easier when the parking lot is on one side of a building rather than surrounding the building. Vehicles should be parked away from shipping and receiving docks, garbage dumpsters, and other crime-prone locations.

Employees should have permanent parking stickers, whereas visitors, delivery people, and service groups should be given a temporary pass to be displayed on the windshield. Stickers and passes allow uniformed officers to locate unauthorized vehicles.

Parking lots are more secure when these specific strategies are applied: CPTED, access controls, signs, security patrols, lighting, CCTV, panic buttons, and emergency phones. Crimes often occur in parking lots, and these events can harm employee morale and result in lawsuits, unless people are protected. Hospitals, for example, supply escorts for nurses who walk to their vehicles after late shifts. Employee education about personal safety, locking vehicles, and additional precautions prevent losses.

Certain types of equipment can aid a parking lot security/safety program. Cushman patrol vehicles, capable of traveling through narrow passageways, increase patrol mobility. Bicycles are another option. A guardhouse or security booth is useful as a command post in parking lots (Figure 8-13).

FIGURE 8-13 Access controls at a parking lot.

Various technologies can be applied to controlling vehicles at access points. One example is *automatic license plate recognition systems* that apply image-processing technology that reads a vehicle's license plate and uses infrared light to illuminate a plate in the dark. A high-speed camera is used to photograph a plate, and then the recorded information is compared to a database. Besides access controls, the applications include fleet management, locating stolen vehicles, and border security (National Law Enforcement and Corrections Technology Center, 2006: 1).

The threat of terrorism has influenced the design of parking lots and vehicle controls. Different types of parking lots present various security issues. Surface lots keep vehicles away from buildings, consume large amounts of land, and may add to storm water runoff volume. On-street parking provides no setback. A garage may require blast resistance. If the garage is under a building, a serious vulnerability exists, since an underground bomb blast can be devastating.

A designer can propose minimizing vehicle velocity because, for example, a bollard that can stop a 15,000-pound truck moving at 35 mph may not be able to stop the same truck moving at 55 mph (FEMA 426). *The road itself can become a security measure by avoiding a straight path to a building.* A straight road enables a vehicle to gather speed to ram a barrier, penetrate a building, and then detonate a bomb. Approaches should be parallel to the building and contain high curbs, trees, or berms to prevent vehicles from leaving the road. Curving roads with tight corners offer another strategy.

Traffic calming strategies are subtler and communicate appropriate speed. Examples are speed humps and raised crosswalks. A speed hump is not as rough as a speed bump. The latter is often used in parking lots. All these strategies reduce speed and liability while increasing safety. Drawbacks are that the response time of first responders increases and snow removal may become difficult.

Security Officers

Officers normally are assigned to stationary (fixed) posts or to patrol. A **stationary post** is at a door or gate where people, vehicles, and objects are observed and inspected. Stationary posts also involve directing traffic or duty at a control center where communications, CCTV, and alarms are monitored. **Foot or vehicle patrols** conducted throughout the premises, in parking lots, and along perimeters identify irregularities while deterring offenders. Examples of unusual or harmful conditions that should be reported are damaged security devices, holes in perimeter fences or other evidence of intrusion, hidden merchandise, unattended vehicles parked inappropriately, keys left in vehicles, employees sleeping in vehicles or using drugs, blocked fire exits, cigarette butts in no-smoking areas, accumulations of trash, and odors from fuels or other combustibles. In contrast to public police officers, private security officers act in primarily a preventive role and *observe and report.*

Before security officers are employed, far-sighted planning ensures optimum effectiveness of this service. What are the unique needs and characteristics of the site? How many people and what assets require protection? What are the vulnerabilities? How many hours per day is the facility open? How many employees? How many visitors and vehicles enter and exit daily? What are the specific tasks of security officers, how much time will be expended on each task, and how many officers are required?

Security officers are expensive. Costs include wages, insurance, uniforms, equipment, and training. If each officer costs $40,000 per year for a proprietary force and five officers are required for the premises at all times, to maintain all shifts seven days per week requires approximately 20 officers. The cost would be about $800,000 per year. To reduce costs, many companies switch to contract security services and/or consider technological solutions.

Several specific steps can be taken to improve the effectiveness of officers. Three of the most critical are *careful applicant screening, sound training,* and *proper supervision.* Management should ensure that officers know what is expected of them. Policies, procedures, and day-to-day duties are communicated via verbal orders, memos, and training programs. Courtesy and a sharp appearance command respect that enhances security.

Policies should ensure that supervisors check on officers every hour. Rotating officers among duty assignments reduces fatigue while familiarizing them with a variety of tasks. Providing inspection lists for adverse conditions will keep them mentally alert. The formal list should be returned with a daily report. Miller (2010), in an ASIS CRISP Report entitled, "Fatigue Effects and Countermeasures in 24/7 Security Operations," offers shift work strategies to counter fatigue, such as "smart" scheduling and ensuring adequate time between schedule changes for proper sleep.

Armed versus Unarmed Officers

The question of whether to arm officers is controversial. Probably the best way to answer this question is to study the nature of the particular officer's assignment. If violence is likely, then officers should be armed. Officers assigned to locations where violent crimes are unlikely do not need firearms, which, if worn by officers, could be offensive. The trend is toward unarmed officers because of liability issues and costs for training and equipment. If weapons are issued to officers, proper selection of officers and training are of the utmost importance. Training should include use of force and firearms safety, as well as practice on the firing range every four months.

Monitoring Officers

Lower burglary and fire insurance premiums result from monitored patrols and insurance personnel subject the records to inspection. Early technology used *watch clocks* to monitor officer patrols along preplanned routes. The officer on patrol carried this old technology, consisting of a timepiece that contained a paper tape or disc divided into time segments. A watch clock was operated by an officer via keys mounted in walls at specific locations along a patrol route. These keys were often within metal boxes and chained to walls. When inserted into the watch clock, the key made an impression in the form of a number on the tape or disc. Supervisors examined the impressions to see whether the officer visited each key location and completed the scheduled route. Keys were located at vulnerable locations (e.g., entry points, flammable storage areas). Good supervision prevented officers from disconnecting all the keys at the beginning of the shift, bringing them to one location for use in the watch clock (and, thus, avoiding an hourly tour), and returning the keys at the end of the shift.

Automatic monitoring systems are another way to monitor patrols and keep records. Key stations are visited according to a preplanned time schedule and route. If an officer does not visit a key station within a specific time, a central monitoring station receives a transmitted signal, and if contact cannot be made, personnel are dispatched.

Bar code or *touch button* technology provides other avenues for monitoring patrols. A security officer carries a wand that makes contact with a bar code or touch button to record data that is later downloaded into a computer. Bar codes or buttons are affixed at vulnerable locations for a swipe by the wand to record the visit by the officer, who can also swipe bar codes or buttons that represent various conditions (e.g., fire extinguisher needs recharging). Supervision of these systems ensures that officers are patrolling properly and conditions are being reported (Arnheim, 1999: 48–58). To improve the efficiency of a security officer, the officer can use a wireless tablet PC (Figure 8-14), which enables the officer to leave a monitoring post and take the workstation with him or her. If, for example, an officer must leave a control center to investigate an incident, the officer can bring the tablet PC and continue to watch CCTV, monitor alarms, and open doors for employee access. Levine (2010: 35) describes a customized PDA system that grew into a digital incident-reporting tool, then into a mobile phone with GPS tracking, as well as containing a camera, e-mail and text messaging functions, a panic alarm, time and attendance recording, and the means to read bar codes for monitoring patrols. He explained that in one case a client at a meeting complained about never seeing a security officer on patrol, so the security executive took out his computer, accessed the website, pulled up the previous day, and showed the client the tracking of the officer on patrol.

You Decide!

What Is Wrong with This Facility?

You are a newly hired security specialist for a large corporation that is involved in a restoration project for its headquarters. You replaced a former security specialist who was fired for incompetence. Your first assignment is to help evaluate the physical security presently in place and the physical security plans your predecessor prepared.

At the present time, infrared photo beams are in place around the perimeter and microwave/PIR sensors are in buildings. Fluorescent lamps serve the exterior and the illuminance at the gatehouse at the entrance is 2 FC.

Because of the slow response by police to the facility intrusion detection system, the predecessor planned a data transmission system with a direct connection to the local 911 emergency dispatch system. He also planned to arm security officers and provide them with watch clocks so their patrols can be monitored. To save money, he planned to suggest extensive use of annealed glass. It was noted in his plans that senior management requires an alternative to both chain link fence and anything resembling barbed wire.

Your present tasks are to identify what is wrong with the security in place, plan improvements, and modify and improve the plans of your predecessor.

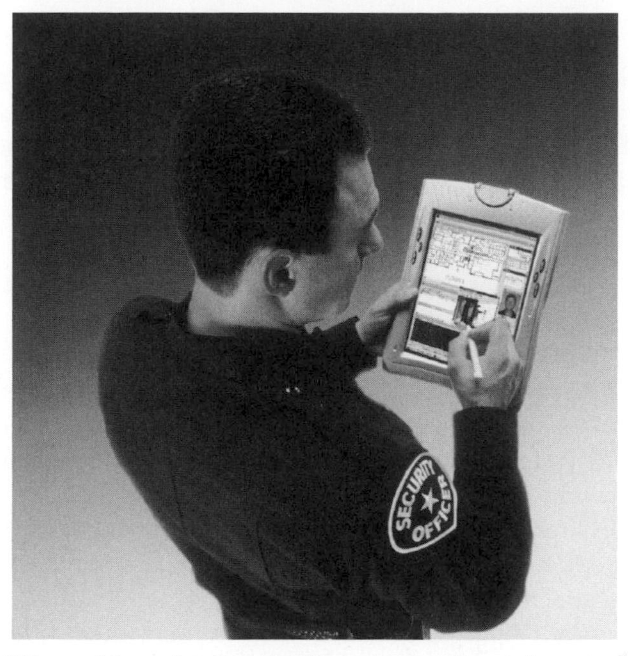

FIGURE 8-14 The tablet PC is a mobile workstation enabling a security officer to leave a post and do many things while mobile that are done from a desktop PC such as view CCTV, monitor alarms, and open doors. *Courtesy: Hirsch Electronics, Santa Ana, CA.*

Contraband Detection

Contraband is an item that is illegal to possess or is prohibited from being brought into a specific area. Examples are weapons, illegal drugs, and explosives. Security officers and government personnel play a crucial role in spotting contraband at many locations. They use special devices and canine services to locate contraband, and these devices and services are as good as the personnel behind them.

Various types of devices detect contraband. Metal detectors transmit a magnetic field that is disturbed by a metallic object, which sets off a light or audio signal. Two types of metal detectors are handheld and walkthrough. X-ray scanners use pulsed energy to penetrate objects that are shown on a color monitor. These devices are mobile and stationary and can inspect such things as mail, packages, loaded trucks, and shipping containers. Since the 9/11 attacks, ports, border checkpoints, airports, and other locations have intensified efforts to detect contraband; research and development and businesses selling detection devices have increased. Although vendors are prone to overpromise and under deliver, contraband detection technology is improving. Reputable research groups applying scientific research methodologies improve the likelihood that devices operate as touted. For example, the National Institute of Standards and Technology (2010) focuses on customer needs (e.g., police and military) and conducts research on various technologies. The focus of this group's research includes real-time imaging systems to detect large concealed objects for identifying suicide bombers and microwave electromagnetic signatures to identify dangerous liquids.

Protective Dogs

Besides serving to detect contraband and protect people, canine (K-9) is classified as an animal barrier that can strengthen security at a protected site. An *alarm dog* patrols inside a fenced area or building and barks at the approach of a stranger, but does not attempt to attack. These dogs retreat when threatened but continue to bark. Such barking may become so alarming to an intruder that he or she will flee. A *guard or attack dog* is similar to an alarm dog, with an added feature of attacking an intruder. To minimize the possibility of a lawsuit, a business should selectively apply and adequately fence in these dogs, and warning signs should be posted. An experienced person on call at all times is needed to respond to emergencies. Another type of attack dog is the *sentry dog*. This dog is trained, kept on a leash, and responds to commands while patrolling with a uniformed officer. The advantages are numerous. These animals protect officers. Their keen senses of hearing and smell are tremendous assets when trying to locate a hidden offender (or explosives or drugs). Dogs can discern the slightest perspiration from people under stress, enabling the dogs to sense individuals who are afraid of them. An ingredient in stress perspiration irritates dogs, which makes frightened persons more susceptible to attack. When an "attack" command is given, a German shepherd has enough strength in its jaws to break a person's arm.

In addition to the possibility of a lawsuit if a dog attacks someone, there are other disadvantages to the use of dogs. If proprietary dogs are part of the protection team, personnel and kennel facilities are needed to care for the dogs. These costs and others include the purchase of dogs and their training, medical care, and food. Using a contract service would probably be more feasible. Another disadvantage is the possibility that dogs may be poisoned, anesthetized, or killed. An offender also may befriend a dog. Dogs should be taught to accept food only from the handler. Neighbors near the protected premises often find dogs noisy or may perceive them as offensive for other reasons.

Since the 9/11 attacks, interest in canines has increased. At the same time, there is a need for consistent standards for training, quality assurance, kenneling, selection of handlers, and presentation of evidence. Definitions also present a problem. For example, there is no consistent definition as to what constitutes an explosive detection canine. The Bureau of Alcohol, Tobacco, Firearms and Explosives has developed the National Odor Recognition Testing initiative, which could be a standard to which dogs could be certified. Research is being conducted on the use of chemical warfare agent detector dogs and GPS technology in conjunction with remote commands for searches, surveillance, and tracking of persons (Harowitz, 2006: 36–38).

Communications and the Control Center

As emergency personnel know, the ability to communicate over distance is indispensable. Every officer should be equipped with a portable two-way radio; this communication aid permits officers to summon assistance and notify superiors about hazards and impending disasters. Usually, officers on assignment communicate with a control center that is the hub of the loss prevention program. FEMA 426 (U.S. Department of Homeland Security 2003: 3–45) recommends redundant communications. The control center is the appropriate site for a console containing alarm indicators, CCTV monitors, door controls, the public address system, and an assortment of other components for communication and loss prevention (Figure 8-15).

Because of the convergence of IT and physical security, the traditional security control center may be within a network operations center. Some organizations may choose to outsource a portion of operations. Since these operations are critical, they must be secure both electronically and physically (Milne, 2005).

Because personnel will seek guidance from a control center in the event of an emergency, that center must be secure and operational at all times. A trend today is automated response systems programmed into the control center because so many decisions and actions are required for each type of emergency (Patterson, 2000: 76–81). The control center is under increased protection against forced entry, tampering, or disasters when it contains a locked door, is located in a basement or underground, and is constructed of fire-resistant materials. An automatic, remotely operated lock, released by the console operator after identifying the caller, also enhances security. Bullet-resistant glass is wise for high-crime locations. FEMA 426 (U.S. Department of Homeland Security, 2003: 3–47) recommends a backup control center, possibly at an off-site location. Whoever designs the control center should be well versed in ergonomics, which deals with the efficient and safe partnership between people and machines.

FIGURE 8-15 Security officer at console managing access control, CCTV, alarm monitoring, and video imaging. *Courtesy: Diebold, Inc.*

Search the Internet

Use search engines to see what vendors have to offer and prices for the following: fences, window protection or glazing, door protection, exterior intrusion detection systems, security lighting, and protective dogs.

Check out the following sites on the Internet:

American Institute of Architects, Security Resource Center: www.aia.org
American National Standards Institute: www.ansi.org
ASIS International: www.asisonline.org
American Society for Testing and Materials: www.astm.org
American Society of Heating, Refrigerating, and Air Conditioning Engineers: www.ashrae.org
American Society of Landscape Architects: www.asla.org
Builders Hardware Manufacturers Association: www.buildershardware.com
Green Globes: www.greenglobes.com
Illuminating Engineering Society of North America: www.iesna.org
International Association of Lighting Management Companies: www.nalmco.org
International CPTED Association: www.cpted.net
International Organization for Standardization: www.iso.org
National Crime Prevention Council: www.ncpc.org
National Fire Protection Association: www.nfpa.org
National Institute of Standards and Technology: www.nist.gov
National Lighting Bureau: www.nlb.org
Security Industry Association: www.siaonline.org
Technical Support Working Group: www.tswg.gov
Underwriters Laboratory (UL): www.ul.com
United Nations Crime and Justice Information Network: www.uncjin.org
U.S. General Services Administration: www.gsa.gov
U.S. Green Building Council: www.usgbc.org

Case Problems

8A. Study the characteristics of Woody's Lumber Company, the Smith Shirt manufacturing plant, and Compulab Corporation (refer to Figures 7-1, 7-2, and 7-3). Establish a priority list of what you think are the 10 most important countermeasures for each location to prevent unauthorized entry. Why did you select your first three strategies in each list as top priorities?

8B. As a newly hired security manager for an office building, a site for research and development, you are faced with three immediate challenges: (1) some employees are not wearing required ID badges from the time they first enter the building to when they depart; (2) during off-hours there are too many security system false alarms; and (3) public police are responding to about half of these alarms. What do you do?

8C. You are a physical security specialist for a corporation with locations worldwide. Your next big assignment from your supervisor, the VP of Loss Prevention, is to work with a corporate IT specialist to apply Internet technology to the integration of access controls, intrusion detection, and CCTV systems. Basically, the VP wants to "visit" all corporate locations from her office. The IT specialist that you must work with constantly complains, feels that the corporate IT infrastructure has reached its maximum capacity, and seems to be distrustful of loss prevention personnel. How do you gain his cooperation, promote harmony, and get the job done?

8D. You are a physical security specialist for a major global corporation based in the United States. Upon conducting a security survey of a large corporate building in a medium-size city in the United States, you listed the following vulnerabilities. Prepare solutions for each vulnerability and prioritize the list so the most serious items are corrected as soon as possible (Purpura, 2007).

Item A

The building contains an underground garage with minimal controls (i.e., an access gate opened by employee access card).

Item B

The front of the building is on Main Street, close to the street, and any vehicles can park at the front on the street. The three other sides of the building contain parking lots close to the building and accessible through an access gate opened by employee access card.

Item C

The rear lobby of the building is at ground level and at the ending point of a straight road 1/8 mile long.

Item D

In the last 12 months, two employees were robbed at night outside the building in the parking lots.

Item E

Corporate offices, functions, assets, and utilities are clearly marked by signs inside and outside the building.

Item F

Air intakes for the building are at ground level at the rear of the building.

Item G

Although employees use card keys to access the building, tailgating is a problem.

Item H

A minimum number of security officers on the premises results in gaps in security and at access points as they are called off-post to obtain mail and conduct other errands.

Item I

Executive staff have their names on designated parking spaces.

Item J

Two garbage dumpsters are located up against the rear of the building.

Item K

The rear of the building faces nearby hills containing a variety of buildings.

References

Aggleton, D. (2010). The latest innovations in door hardware. *Security Technology Executive, 20* (May).

Arnheim, L. (1999). A tour of guard patrol systems. *Security Management* (November).

Atlas, R. (2010). Fast food, easy money. *Security Management, 54* (December).

Bernard, R. (2010). The security industry world has changed. *Security Technology Executive, 20* (May).

Berube, H. (2010). An examination of alarm system deterrence and rational choice theory: The need to increase risk. *Journal of Applied Security Research, 5*(3).

Berube, H. (1994). New notions of night light. *Security Management* (December).

Burton, R. (2010). A new standard for high-performance green buildings. *Buildings, 104* (March).

Clifton, R., & Vitch, M. (1997). Getting a sense for danger. *Security Management* (February).

Coupe, T., & Kaur, S. (2005). The role of alarms and CCTV in detecting non-residential burglary. *Security Journal*, 18.

Davey, C., et al. (2005). Design against crime: Extending the reach of crime prevention through environmental design. *Security Journal*, 18.

Dibazar, A., et al. (2011). Intelligent recognition of acoustic and vibration threats for security breach detection, close proximity danger identification, and perimeter protection. *Homeland Security Affairs* (March) <www.hsaj.org/?special:article=supplement.3.4>.

Evans, R. (2005). "Remote monitoring. *Security Products, 9* (March).

Garcia, M. (2006). *Vulnerability assessment of physical protection systems.* Burlington, MA: Butterworth-Heinemann.

Gips, M. (1999). A pharmacopoeia of protection. *Security Management, 43* (March).

Harowitz, S. (2006). Dog use dogged by questions. *Security Management, 50* (January).

Illuminating Engineering Society of North America, (2003). *Guideline for security lighting for people, property, and public spaces.* New York, NY: IESNA.

Levine, D. (2010). Armed and ready. *Security Technology Executive, 20* (October).

Miller, J. (2010). Fatigue Effects and Countermeasures in 24/7 Security Operations (CRISP Report). <www.asisonline.org> retrieved July 28, 2011.

Milne, J. (2005). Build your own security operations center. *Secure Enterprise* (August 1) <www.secureenterprisemag.com> retrieved September 26, 2005.

Moore, M. (2006). Defensive devices designed to blend in with New York. *USA Today* (July 31) <www.usatoday.com/news/nation/2006-07-31-ny-security_x.htm> retrieved August 1, 2006.

Morton, J. (2011). Access denied. *Buildings, 105* (May).

Morton, J. (2012). Upgrade your fire alarms with IP reliability. *Buildings, 106* (March).

Murphy, P. (2000). Grounds for protection. *Security Management, 44* (October).

National Fire Protection Association. (2005). *NFPA 730, guide for premises security* (2006 ed.). Quincy, MA: NFPA.

National Institute of Standards and Technology. (2010). Concealed Weapon and Contraband Detecting, Locating, and Imaging. <www.nist.gov/oles/diet-conceal.cfm> retrieved July 29, 2011.

National Law Enforcement and Corrections Technology Center, (2006). No license to steal. *TECHbeat, Spring*

National Lighting Bureau. (n.d.). *Lighting for Safety and Security.* Washington, D.C.: National Lighting Bureau.

Nunes-Vaz, R., et al. (2011). A more rigorous framework for security-in-depth. *Journal of Applied Security Research, 6*(3).

Owen, D. (2003). *Building security: Strategies & cost.* Kingston, MA: Reed.

Painter, K., & Farrington, D. (1999). Street lighting and crime: Diffusion of benefits in the stoke-on-trent project. In K. Painter & N. Tilly (Eds.), *Crime prevention studies*. Monsey, NY: Criminal Justice Press.

Patterson, D. (2000). How smart is your setup? *Security Management, 44* (March).

Pierce, C. (2006). Thermal video for the mainstream? *Security Technology & Design, 16* (May).

Post, R., & Kingsbury, A. (1977). *Security administration: An introduction* (3rd ed.). Springfield, IL: Charles C. Thomas.

Purpura, P. (1979). Police activity and the full moon. *Journal of Police Science and Administration, 7* (September).

Purpura, P. (2007). *Terrorism and Homeland Security: An Introduction with Applications*. Burlington, MA: Elsevier Butterworth-Heinemann.

Reddick, R. (2005). What you should know about protecting a perimeter. *Security Products, 9* (April).

Revel, O. (2003). Protective blast and anti-ramming wall development. *Security Technology & Design* (November).

Saflex, Inc. (2007). Architectural Glazing. <www.saflex.com> retrieved July 17, 2011.

Schumacher, J. (2000). How to resolve conflict with proper systems integration. *Security Technology & Design, 10* (October).

Shelton, D. (2006). The new and improved moat. *Security Technology & Design, 16* (March).

Smith, M. (1996). *Crime prevention through environmental design in parking facilities*. Washington, D.C: National Institute of Justice. (April).

Spadanuta, L. (2011). How to improve your image. *Security Management* (March), 55.

True, T. (1996). Raising the ramparts. *Security Management* (October).

U.S. Department of Homeland Security, (2003). *Reference manual to mitigate potential terrorist attacks against buildings, FEMA*. Washington, D.C: FEMA. 426 (December).

U.S. Environmental Protection Agency. (2011). Sustainability. <www.epa.gov> retrieved July 14, 2011.

WeGuardYou. (2011). Mall Security Scenario. <weguardyou.com/applications-mallsecurity.html> retrieved July 21, 2011.

Wilson, J., & Kelling, G. (1982). Broken windows: The police and neighborhood safety. *Atlantic Monthly* (March).

Wroblaski, K., et al. (2011). The great debate: 2011's key sustainability issues. *Buildings* (January), 105.

Zwirn, J. (2003). Alarm design that rings true. *Security Management* (April).

Services and Systems

Methods for Making Wise Purchasing Decisions

OBJECTIVES

After studying this chapter, the reader will be able to:

1. Discuss pitfalls when purchasing security services and systems
2. List seven specific purchasing rules and nine sources of information
3. List guidelines and inquisitive questions that improve purchasing decisions when seeking security services
4. Discuss contract undercover investigations, consultants, and certifications
5. List guidelines and inquisitive questions that improve purchasing decisions when seeking security systems
6. Name and explain three types of bids in the purchasing process
7. Explain outsourcing

KEY TERMS

- security services
- security systems
- vendors
- key performance indicators
- "observe and report"
- competitive bidding
- request for information
- request for quotation
- request for proposal
- work breakdown structure
- outsourcing
- contract lifecycle management
- due diligence

Introduction

One topic often neglected in the security literature is how to make wise purchasing decisions when obtaining security services and systems. The best security plans are useless when poor purchasing decisions are made to implement those plans. **Security services** include duties performed by personnel to further the goals of security and loss prevention. Security officers represent a large part of available security services. **Security systems** include hardware and software that protect people and assets. An example is an intrusion detection system. This chapter emphasizes security services and systems, keeping in mind that fire protection and safety are integral components of an effective loss prevention program.

Most business executives and institutional administrators do not know how to select security services or systems or even what questions to ask vendors. Frequently, money is wasted and the results after the purchases are disappointing. A specialist in the field who is not a salesperson can improve decision-making.

During their careers, security and loss prevention practitioners are most likely to purchase all kinds of services and systems. Purchasing decisions have a definite impact on careers and on the success of protection programs. Care is required during decision-making to obtain the best services and systems for the money at hand.

Pitfalls when Purchasing Security Services and Systems

Suppliers of services and systems are not immune to the temptation of unethical and illegal activities for profit. **Vendors** (i.e., sellers or salespeople) are known to misrepresent information, exaggerate, and fail to deliver on promises. The rotten-apple syndrome is prevalent in this industry just as it is in other facets of life—there are unscrupulous vendors as well as honest ones.

Because the contract security officer service industry, often called "guard services," is so competitive, some companies bid very low, knowing that they will have to bill the client for phantom services (services not rendered) to make a profit. Other companies lie to clients about training and experience. Promised supervision may not take place. Liability insurance coverage may be exaggerated or nonexistent. Investigative companies that conduct overt and undercover investigations are known to deceive clients about excessive losses through scary weekly reports in order to lengthen investigations and thereby reap greater profits.

Practices employed by people selling security systems include selling outdated technology and unneeded equipment. Intrusion detection systems, which are overstocked in inventory, may be pushed on customers who do not realize that offenders can easily defeat these systems. Salespeople conveniently delete information concerning extra personnel needed, additional hardware required, software problems, and expensive maintenance. Sales information may portray systems operating under ideal conditions of perfect weather and lighting (Strauchs, 2001: 98). Some security system installers set sensors to detect minor intrusions during a test and then, following the test, lower the sensitivity of the sensor to reduce false alarms. *One specific tactic involves reinforcing the purchaser's fear.* Crime, fire, and accident dangers are intertwined within high-pressure sales pitches.

McCumber (2006: 55–58) notes that hype is also a problem in the IT security industry. He offers the example of an IT salesperson who exaggerated a problem that a customer would pay a high price to avoid. The salesperson demonstrated the vulnerability on his laptop and showed how his company's software solution eradicated the problem. When the salesperson offered to load the software onto the customer's IT system to show its effectiveness, and emphasized how he could easily delete it, the salesperson was escorted to the exit.

Many vendors adhere to ethical conduct. The point of the industry criticism here is that the buyer should be aware of these practices when confronted with a purchasing decision.

The *Report of the Task Force on Private Security* (U.S. Department of Justice, 1976: 146–147) and the first Hallcrest report (U.S. Department of Justice, National Institute of Justice, 1985: 71) stated several recommendations for improving the industry with the consumer in mind. For example, both reports favor certified training for alarm service personnel.

The following seven cardinal rules, designed with the consumer in mind, can put the buyer on the road toward making wise purchasing decisions:

1. Buyer beware.
2. Use a team approach for decision-making. Bring together various specialists.
3. Properly evaluate the *needs of the organization*, not the needs of the vendor.
4. Acquire information and know the state of the art.
5. Study the advantages and disadvantages of each service or system. Apply critical thinking skills.
6. Ask vendors to demonstrate the value and ROI on what they are selling.
7. Avoid panic buying.

A risk analysis will assist the buyer in pinpointing weaknesses and evaluating needs. If the buyer has a list of specific weaknesses and needs, the salesperson will be hindered from influencing the buyer into purchasing something unnecessary. To acquire information about services and systems, nine beneficial sources are the Internet, standards and related organizations, peers, customers with experience with the service or system, independent consultants not affiliated with any vendors, seminars, courses, trade publications, and salespeople.

Is the security industry the only industry where the buyer should beware? Support your answer.

Purchasing Security Services

Questions when Considering Contract Security Officers

1. Does the company conform to state and local regulatory law, such as registration, licensing, training, and bonding?
2. What are the contract company's liability and other insurance coverage (e.g., worker's compensation)? Request copies of policies. Are lawsuits pending against the company?
3. Has the vendor been vetted? What is the company's Dun and Bradstreet financial rating?
4. Is the company willing to customize service for client company needs?
5. Does the company have the ability to provide extra officers?
6. Can the company perform expanded services, such as investigations?
7. Where are the company's offices?
8. What type of background screening is conducted on applicants? Can you set up an agreement whereby your company interviews officers before assignment? Do you require personnel folders to help you select the best candidates?
9. How often during a shift does a supervisor visit? What type of monitoring system is used for patrols?

10. How does the company ensure the honesty of its officers?
11. How are disciplinary problems handled?
12. In general, how is morale? What is the turnover rate?
13. What is the wage-to-rate ratio? This shows the portion of the total rate received by each officer.

Many security firms draft their contracts so that much of the risk is on the customer. If, for example, an assault takes place on the 10th floor of a building, the security company may go back to the contract and read that an officer was assigned only to the lobby and that more services were available, but the client refused them (Finnerty, 1996: 4–7). An attorney should review contracts and negotiate changes. The contract should specify number of officers and hours, rates of pay (including overtime pay), minimum required background and physical condition of officers, post orders, location of officers, supervision, uniforms and equipment (e.g., vehicles, weapons, and radios), and training.

Heil (2006: 57–64) recommends that customers of contract security services carefully consider the low bidder before final selection. A low bidder may result in poor quality officers and increased liability. He also favors ongoing expectations for performance during the contract through metrics (also called **key performance indicators**) that can be tied to financial rewards or penalties for the vendor. These metrics include the following: no-show rates, missed tours, missed supervisory visits, complaints against officers, violations of policies and procedures, and inappropriate uniform and appearance. Campbell (2011: 17) emphasizes the importance of metrics showing *results achieved*. For example, rather than just reporting the number of supervisory inspections, the percent of officers performing required tasks should be reported.

Levine (2012: 20–21) adds that a client should select a vendor that embraces technology so security officers are more efficient and valuable to the client. An example is software that aids reporting of conditions and incidents for management analysis. Another factor of consideration is the resilience capabilities of the vendor in case of emergencies. Professionalism is also important and it includes opportunities for officers to be certified and to attend college.

Those in the contract security business know that the above requirements from clients have a cost. Contractors face a constant challenge of satisfying clients while striving for successful bids and cost containment.

When one is dealing with contract companies, it is good to know the views of these business people. A portion of the managers in contract guard companies refer to their vocation as a "nickel-and-dime business" with a "never-ending turnover of bodies." The former comment refers to the awarding of contracts by clients based on slim differences in bids. The latter comment refers to the high turnover of officers because of low wages.

You Decide!

What is Wrong with this Facility?

You are a security consultant working jointly with two clients: a contract security company and a shopping mall. Your assignment is to improve security and prevent litigation. Both the contract company and the mall were defendants in two serious lawsuits in the last few years.

In one case, an employee of a large department store was shot and killed by her estranged husband in the parking lot. Five months prior to the shooting, the offender had been banned from the mall for two years for disorderly conduct and resisting police arrest. When he was banned, the offender's photo was taken and placed in a ban book that required review by security officers prior to each shift, as stated in policy. The offender was also issued a ban notice stating that if he entered mall property, public police would be contacted to make an arrest for trespassing. On the day of the shooting, two security officers on patrol passed the victim and her estranged husband arguing at the mall and did not know the estranged husband was banned. In addition, they ignored the domestic dispute.

In another case, a female customer was robbed of her purse in the parking lot, dragged by the robbers as they drove away, and is now paralyzed. When the robbery occurred, all five unarmed security officers on duty were eating their evening meal at the food court.

Prior to the two serious incidents, there was tension between mall management and contract security officers because mall management required security officers to bus tables at the food court, return shopping carts in the parking lot to stores, take food orders to mall management, and operate the mall garbage compactor. Since the contract security company did not want to jeopardize the security contract, they strived to satisfy mall management requests. At the same time, such duties lowered morale among security officers and the multi-tasking resulted in less time to perform patrol duties. Experienced security officers resigned, turnover increased, and the quality of security and safety decreased. The training of the security officers consisted of on-the-job training where a supervisor would patrol with a new officer for one shift. In addition, new officers were required to read a policy and procedure manual; no testing was done. Training, required by state law, was ignored. All these conditions still characterize mall security. At this mall, five officers are assigned to the first shift (8:00 AM–4:00 PM) and five are assigned to the second shift (4:00 PM–midnight). During these shifts, three officers patrol inside the mall and two patrol outside in vehicles. Two officers patrol inside the mall on third shift (midnight–8:00 AM). All of the above issues surfaced during the lawsuits.

Soon you will meet with mall management and contract security managers. As a security consultant, what is wrong at this facility and what suggestions do you offer?

Critical Management Decision

Whether to Go beyond "Observe and Report"

Security officers are often directed by their employer to **"observe and report."** This means that when they observe an incident, they are to collect information and then report the incident to their supervisor, who may or may not contact 911 for assistance from public safety agencies. This approach has advantages and disadvantages. The advantages are lower costs for salaries, training, equipment, and insurance, as well as less risk of liability in case an officer makes an error when taking action. On the other hand, "observe and report" does not offer immediate protection for people and assets. For instance, what if one employee is beating another employee in the workplace? If a security officer does not intervene, the employee's life may be in jeopardy and the company may suffer from litigation and reputational harm.

For employers to go beyond "observe and report" and call for officers to take action, costs increase and a whole gamut of requirements must be implemented. These include recruiting and paying officers who are capable of taking action; additional training, supervision and equipment (e.g., weapons); and adequate insurance to cover risks. Once management decides to go beyond "observe and report," there is no middle ground. If, say, a security officer takes physical action to defend a victim of assault, the offender may attempt to kill the officer, and if the officer is only trained and equipped for minimum use of force, the officer and the victim are placed in a dangerous position. An option is to hire off-duty police officers.

What are your views on "observe and report"?

Contract Undercover Investigations

A common scenario in businesses is the panic atmosphere after the discovery of a high inventory-shrinkage statistic following an inventory. When this happens, management wants immediate action even when it is predisposed to avoid panic buying. Management often recruits an outside firm specializing in undercover investigations. The undercover investigator secretly infiltrates employee informal groups, as a regular employee, to gather information about theft, illegal drug usage, production problems, employee behavior toward customers, and other issues.

With an understanding of how contract undercover investigation services operate, the client will obtain better results. When speaking to a service representative, the client needs to find out the cost and probable length of the investigation. These investigations last from six to eight weeks but may require months to yield success. The cost may be thousands of dollars per week, depending on the background of the investigator. The client should ask the representative about the backgrounds of investigators, selection methods (for employment and assignment), training, and supervision. Are investigators bonded? Proof should be provided. Are they prepared to testify in court, if necessary? How many reports will the client receive each week? Undercover investigators send reports to their immediate supervisor, who edits them before sending a report to the client. The reporting phase of the investigation is the time when unscrupulous activity by the service company may take place. Supervisors are known to withhold good information from clients to submit later during "dry weeks," when no substantial information is uncovered. Frightening reports about losses can scare clients into paying for unnecessarily lengthy investigations. When clients become impatient and ask why the investigation is taking so long, sometimes the service company's response is that "a break in the case is right around the corner and we just need a few more weeks." If put under excessive pressure by supervisors, investigators may succumb to exaggerated reports and may even invent information. Although these practices produce a negative image of undercover investigative services, not all of these companies are unethical. Undercover investigations are a widely used and effective method to combat losses.

Consultants

Why would an executive require the services of a loss prevention consultant? Three major reasons are (1) the executive lacks knowledge about loss prevention and the company doesn't retain a proprietary loss prevention professional; (2) the executive is a loss prevention professional but lacks expertise in a specialized field; or (3) the executive does not have the time to deal with a task or problem.

Consultants work on assignments such as a baffling shrinkage problem, the loss of trade secrets, and executive protection. They can be a tremendous asset when an organization is contemplating a loss prevention program for the first time or when an established program suffers from morale or training problems. New ideas can stimulate greater value and ROI. A consultant can also act as a company's representative in negotiating, purchasing, and implementing security services and systems.

A client interested in a successful consulting experience will be involved in three specific phases: (1) selection, (2) direction, and (3) evaluation. The objective of the selection phase is to hire the most appropriate person for the job. The client must first *clearly define the problem* and then search for the individual with the required background. Select a consultant who is independent, that is, not affiliated with any particular service or system. Also, ask how much money and time are required to complete the work. Consider peer review of the consultant's work for improved results and discuss this topic with the primary consultant.

In the second phase, the client assists the consultant in becoming familiar with the business and the problem. The consultant is introduced to select personnel. A tour of the premises is another part of what is known as the startup time, which can easily consume a day. The consultant will be preoccupied with collecting information via interviews, observation, and records. He or she will ask many questions. A previously prepared survey or checklist form is brought by the consultant as a reminder of what specific questions to ask or areas to check. Clients may request a one-day survey followed by verbal advice, whereas other assignments may last for weeks, months, or years. A schedule of periodic reports is essential for the client.

When sufficient information is collected, and the consultant has a good grasp of the problem and possible solutions, he or she presents a final report of findings and recommendations. Naturally, the consultant is in an advisory capacity, and the executives in charge have the authority and must accept the responsibility of instituting the recommendations.

The third and final phase for the client is evaluating the consultant. A standard personnel evaluation form provides several relevant questions (e.g., works well with others, is flexible, and communicates clearly). However, the primary question is whether the problem was uncovered and satisfactorily remedied.

Certifications in Security

Most states have no regulation for security and loss prevention consultants. When registration or licensing is required, it often is accomplished via the laws regulating security officers and private investigators. With such minimal controls, almost anybody can call him- or herself a consultant. Hence, charlatans appear who tarnish the reputation of the field and create a bad image that reflects on competent professionals.

In light of the scarcity of regulations or standards for security and loss prevention consultants and managers, and to reinforce professionalism in this field, ASIS International created the Certified Protection Professional (CPP) program in 1972. To qualify for CPP certification, the applicant must meet certain education and experience requirements, affirm adherence to the CPP Code of Professional Responsibility, receive endorsement by a person certified as a CPP, and achieve a passing grade on a written examination. ASIS International also offers two other certifications: Physical Security Professional (PSP) and Professional Certified Investigator (PCI).

Another source for a security consultant, although small, is the International Association of Professional Security Consultants (IAPSC), founded in 1984. Members of this professional association are required to have education and experience, and the CPP is accepted as a component of the qualifications for membership. Like ASIS, the IAPSC requires its members to adhere to a code of ethics. The IAPSC offers the Certified Security Consultant program.

As a practitioner, what are the benefits of obtaining certifications in security?

Purchasing Security Systems

Questions When Considering a Security System

1. Does the system meet the customer's needs? What can the system do?
2. What is the *total cost* of the system, including hardware, software, installation, additional personnel, training, maintenance, finance charges, and so on? When the purchaser buys in large volume, the greater profit for the vendor may result in lower prices.
3. Is the total price competitive with other vendors?
4. Can the system adapt to new technology?
5. How long will it take the system to be installed and become operational? If the manufacturer contracts the installation to a subcontractor, how are standards maintained and inspections conducted, and what are the related stipulations in the contract?
6. What is the system life?
7. What does the warranty cover?
8. What maintenance is required and by whom? Since a system is useless when it fails and is not repaired, are preventive maintenance and emergency maintenance part of the contract?
9. Has the vendor been vetted? What is the company's Dun and Bradstreet rating?
10. Does the vendor have appropriate business licenses? Does the vendor conform to applicable state laws requiring training or registration?
11. Is the vendor a member of trade associations that help members stay current on new technology and industry standards?
12. What is the background of personnel involved with the product? Are professional engineers employed? Do licensed contractors do electrical work? Is pre-employment screening conducted? Are employees trained and certified?
13. Have customers using the same system been contacted?
14. Does the manufacturer freely release information about its systems to outsiders? (Offenders are known to pose as writers, reporters, or customers to acquire system information.)

15. Have the system and its components been evaluated by an independent testing organization? If applicable, does it meet NFPA and other standards?

16. How will the system be evaluated? Can security personnel, in a controlled test, "trip" a sensor to ensure reliability?

Rather than purchase a security system, a company may choose the option of leasing. Security vendors, manufacturers, and integrators are not only offering products, technical support, and installation, but also equipment leasing, rentals, and specialized services. For security executives on a tight budget, and for those searching for value-added services and an improved return on investment, creative avenues should be researched to acquire security systems and services. Among the benefits of leasing is the opportunity to keep up with newer technology, more quickly, rather than having to retain older, purchased equipment to recoup the investment (Whittemore, 2007: 20). A financial analysis of purchasing versus leasing is best.

System Acquisition

Many organizations (e.g., government agencies and large corporations) require a formal competitive bidding process to acquire a new security system (or service). Competitive bidding offers such benefits as obtaining alternate prices from vendors, learning new ideas from experts, and gaining improved value for the budget. Large organizations seeking to improve efficiency with numerous projects are increasingly using online procurement and placing contract, design requirements, and questionnaires online for vendors (Aggleton, 2012: 40–44).

Khairallah (2006: 159–272) explains the **competitive bidding** process. Once a security need is established, a preliminary design is prepared, and when management approves the plan and funding, the next step is to prepare a solicitation for the market to obtain pricing and availability of the product or service. Three types of bids are explained in subsequent paragraphs: request for information (RFI), request for quotation (RFQ), and request for proposal (RFP). The type of bid to employ depends on such factors as the complexity of the project and the purchasing policies of the company. In addition, elements of multiple types of bids may be used.

A **request for information** serves to gather information for an RFQ or RFP. The RFI is especially helpful when security needs are unusual and the security practitioner is searching for guidance. The RFI is not specific as to what will be purchased; it broadly describes mission and functions. Essentially, the market helps with problem solving. One drawback to the nonspecific nature of the RFI is that vendors typically recommend products that they sell instead of meeting the customer's need.

A **request for quotation** seeks the costs of the required components of the security system. It also contains other elements including project management, scheduling and cost controls, system demonstration, freight and insurance, installation, system testing, user training, warranty, and post-installation support. The RFQ provides an opportunity to compare bidders.

A **request for proposal** contains the elements of an RFQ plus equipment performance criteria. There are several variations of the RFP. The National Contract Management Association offers a listing of contract types and conditions. Khairallah notes that even for a trained professional it is nearly impossible to know all the features of all security products from the numerous manufacturers. He recommends that the RFP be well organized and indexed. Khairallah suggests using a technique from federal government contracting called **work breakdown structure** (WBS). This process separates parts of a project into clearly definable tasks, and it includes features and capabilities from the preliminary design. The security practitioner can use the WBS to check each bid to ensure that all components of the project are included and vendors can respond to each section. For the security practitioner, this process facilitates ease of comparison and helps with the ranking of bids.

Jensen (2010: 44) notes that when a customer has a very specific challenge and need, the RFP should be detailed so vendors can offer precise solutions. He also recommends including existing standards applied by the customer and consideration of hiring a trusted consultant to work on the project.

Following the evaluation of the written proposals, the next step is to request a product demonstration from the most likely vendors that can meet the requirements of the RFP. As the evaluation of vendor proposals ends, the security practitioner prepares a written report for management to pinpoint the most favorable bidder. Once the contract is awarded, there are specific stages of installation and implementation. They include inspections, testing, and training.

You Be the Judge[1]

Carl Simpson, the security director at Southeast Tool Company, was fighting mad. The plant had been burglarized again, but that was nothing new. What had him angry was that the alarm system the company had recently leased had failed to detect the intrusion.

"Get Security Systems International on the phone," snapped Simpson at his secretary.

Once SSI came on the line, Simpson began his attack: "We had $135,000 worth of precision machine tools stolen last night, and your people are responsible. According to our contract, you're supposed to make sure this alarm system works. It doesn't, so your company had better come up with $135,000."

The manager of SSI just laughed. "Calm down and reread your contract," he said. "It has what's called an exculpatory clause, which says that SSI is not responsible for any loss caused by burglary."

"But this was your fault!" Simpson cried.

"It doesn't matter," replied the manager. "The clause covers us even if we're negligent." He chuckled, "I can tell you've never dealt with a burglar alarm company before. Almost all alarm contracts have an exculpatory clause in them."

But Simpson refused to give up without a fight. He had Southeast sue SSI for breach of contract, breach of warranty, and negligence, despite the exculpatory clause in the contract. "It's not fair," Simpson argued. "When we contracted with SSI, we put the safety of our company in their hands. They took on the responsibility, and they shouldn't be able to use a catch-all clause to escape liability for their negligence."

Did the court agree?

Make your decision; then turn to the end of the chapter for the court's decision.

[1]Reprinted with permission from *Security Management—Plant and Property Protection*, a publication of Bureau of Business Practice, Inc., 24 Rope Ferry Road, Waterford, CT 06386.

Outsourcing

Outsourcing is purchasing, from outside companies, services that were previously performed in-house. In *Business at the Speed of Thought,* Bill Gates writes: "An important reengineering principle is that companies should focus on their core business and outsource everything else."

Internet connectivity has produced a global economy with intense competition requiring management to concentrate on new challenges and opportunities or be left behind. For success and profit, management staff is learning that they must focus on what the company does best and outsource support functions. Outsourcing improves a company's focus, frees internal resources for other purposes, reduces costs, and shares risks. Consequently, activities that do not contribute directly to the bottom line and are part of "the cost of doing business" are ripe targets for outsourcing. Examples are human resources, risk management, environmental management, safety, and security (Caldwell, 2001: 18–20).

Here, we consider some outsourcing decisions. A frequent outsourcing decision concerns choosing between in-house and contract security officer services. Consider the following generalized advantages and disadvantages. With in-house, the advantages are lower turnover and increased control over hiring, training, and quality. In addition, officers often have greater loyalty and are familiar with unique needs. The disadvantages are higher costs and total responsibility for security. With contract services, the advantages are lower costs, fewer human resources duties, and shared responsibilities. The disadvantages are less direct control, less impact on hiring and loyalty, and higher turnover. It is not uncommon for a contract security officer to work for months without the client knowing that the officer has a felony record. One solution is for the contractor to maintain a personnel folder on-site for each officer and for it to contain a copy of the application, background check, training, and regulatory papers (Maurer, 2000: 14–18).

Another outsourcing decision involves choosing between in-house and off-site access control. For many years, central stations have serviced clients by monitoring intrusion and fire detection systems. In the late 1970s and early 1980s, central stations began to offer access control services. This avenue works well in multitenant facilities with both limited entry points and visitors, and a preference to avoid the presence of a security officer. Off-site access controls typically include audio and video systems. Conversely, facilities characterized by many visitors and exception-based entry requests can result in access delays if access is controlled off-site. Delay becomes more problematic if people must wait in adverse weather. The screening of packages presents another problem for off-site access controls; personnel on-site may be necessary for this task (Friedenfelds, 2006: 42–46).

In the IT sphere, "managed security" is a growing trend involving outsourcing of security technologies, infrastructure, and services. Many companies with in-house staff simply cannot keep up with all the pressing IT security issues (DeJesus, 2001: 34–49).

Adler et al. (2005: 61–65) write about the popularity of outsourcing and the importance of the process being well managed to meet corporate objectives. They note that the key to success is to have a plan for every step of the process of outsourcing, beyond selecting the vendor and integrating the service into company operations. They suggest a model known as **contract lifecycle management** (CLM) that contains four primary steps: (1) contract governance and oversight, (2) RFP, (3) due diligence, and (4) contract negotiation and execution. Governance consists of an internal council of specialists from such areas as procurement, legal, finance, operations, and IT who participate in all phases of the CLM. The second step is the RFP. Step three, **due diligence**, is the attention and care expected in checking the accuracy of information and omissions. This step can begin with an RFI from vendors and may include vendor financial information, licensure requirements, insurance, and a list of clients. The contract negotiation and execution is the last step. It focuses on the final terms between the client and the vendor and the signing of the contract.

Is outsourcing a good or bad idea? Justify your answer.

Career: Security Sales, Equipment, and Services

This security specialty can be stimulating, challenging, and financially rewarding. New security-related products and services have resulted from emerging threats and evolving technology, and the number of companies offering various security services has grown as a result. Sales positions can range from products such as barriers, alarm systems, biometrics, CCTV, and risk management software, to uniformed security services. Sales and service personnel may be employed by a product manufacturer or by an independent dealer who represents a variety of products. Entry-level positions may entail making sales calls, handling advertising queries, organizing sales booths, demonstrating products, and providing input on proposal requests. In addition to typical management functions, mid-level management responsibilities may include directing and motivating sales personnel, organizing sales and marketing campaigns, preparing and presenting proposals, and conducting briefings.

Entry-level management positions may call for a college degree, depending on the size and nature of the employer. It is recommended that broad-based education and experience be achieved in areas such as accounting, industrial engineering, management, marketing, human resources, communications, and statistics.

Mid-level management positions require the experience to perform a variety of functions as indicated previously. The ability to effectively deal with a range of people and the capability to present information verbally and in writing are particularly important.

Certifications that can enhance professionalism and advancement opportunities in the above vocations include the Physical Security Professional (PSP) from ASIS International (2012) and the Certified Security Project Manager (CSPM) from the Security Industry Association (2012). In addition, a variety of relevant training courses are offered from the Electronic Security Association (2012).

Modified from source: Courtesy of ASIS International (2005). "Career Opportunities in Security." www.asisonline.org

Search the Internet

Use search engines to view what is online for "security services" and "security systems." Also, check out the following sites:

ASIS International: www.asisonline.org
International Association of Professional Security Consultants: www.iapsc.org.

Case Problem

9A. Select a specific security service or system. Study the state of the art. Then list, in order of priority, 10 questions you would ask vendors. Explain why the first three questions are placed at the top of the list.

The Decision for "You Be the Judge"

The court did not agree with Simpson's reasoning and dismissed Southeast's lawsuit against SSI. The court held that the clause was valid and clear in totally absolving SSI from liability for the burglary. If your company has leased a burglar alarm system, check the contract to see if it has an exculpatory clause. If it does, take heart. Not all courts have upheld the validity of such clauses in all circumstances. Check with your company's lawyer to find out where your state courts stand on this issue.

This case is based on *L. Luria & Son v. Alarmtech International*, 384 So2d 947. The names in this case have been changed to protect the privacy of those involved.

References

Adler, S., et al. (2005). The inside story on outsource planning. *Security Management, 49* (August)

Aggleton, D. (2012). Security system procurement. *Security Technology Executive, 22* (January/February)

ASIS International. (2012). Board Certifications in Security. <www.asisonline.org/certification/index.xml> retrieved April 12, 2012.

Caldwell, G. (2001). Do it yourself or outsource it? *Security Technology & Design, 12* (March)

Campbell, G. (2011). Measuring guard force operations. *Security Technology Executive, 21* (May)

DeJesus, E. (2001). Managing managed security. *Information Security, 4* (January)

Electronic Security Association. (2012). Courses. esa.digitalnts.org/course_list retrieved April 12, 2012.

Finnerty, J. (1996). Who's liable, the security firm or you? *Risk Management Advisor* (May)

Friedenfelds, L. (2006). Is outsourcing right for you? *Security Technology & Design, 16* (February)

Heil, R. (2006). Guarding against poor performance. *Security Management, 50* (June)

Jensen, J. (2010). It's time for a proposal. *Security, 47* (February)

Khairallah, M. (2006). *Physical security systems handbook: The design and implementation of electronic security systems*. Boston, MA: Butterworth-Heinemann.

Levine, D. (2012). Checklist: Getting the most from guard services. *Security Technology Executive, 22* (April)

Maurer, R. (2000). Outsourcing: An option or a threat? *Security Technology & Design, 10* (August)

McCumber, J. (2006). Truth in advertising. *Security Technology & Design, 16* (April)

Security Industry Association. (2012). Certified Security Project Manager Course. <www.siaonline.org> retrieved April 12, 2012.

Strauchs, J. (2001). Which way to better controls? *Security Management, 45* (January)

U.S. Department of Justice, (1976). *Report of the task force on private security*. Washington, D.C.: U.S. Government Printing Office.

U.S. Department of Justice, National Institute of Justice, (1985). *Crime and protection in america (Executive Summary of the Hallcrest Report)*. Washington, D.C.: U.S. Government Printing Office.

Whittemore, S. (2007). Leasing offers attractive benefits to security directors. *Security Director News, 4* (January)

10 Investigations

OBJECTIVES

After studying this chapter, the reader will be able to:

1. List and discuss the six basic investigative questions
2. Describe at least five types of investigations in the private sector
3. Differentiate among proprietary and contract investigations, private and public investigations, and overt and undercover private investigations
4. List and explain at least five subject areas important to the investigative process
5. Explain both digital evidence and digital forensics
6. Explain how to prepare quality reports and testimony

KEY TERMS

- proprietary investigation
- contract investigation
- private investigation
- public investigation
- overt investigation
- undercover investigation
- counter social media techniques
- qualified privilege
- attorney/client privilege
- direct evidence
- circumstantial evidence
- hearsay evidence
- chain of custody of evidence
- interview
- interrogation
- open-ended questions
- close-ended questions
- pretexting
- Freedom of Information Law
- modus operandi
- artificial intelligence
- data mining
- link analysis
- Identi-Kit
- digital evidence
- digital forensics
- Global Positioning System (GPS)
- surveillance
- deposition

Introduction

An investigation is a search for information. Information is obtained from many sources, including the following: people, such as victims, witnesses, suspects, and informants; physical evidence, such as fingerprints, DNA, shoe prints, and tool marks; and information technology, such as the Internet and databases.

There are six basic questions to ask in an investigation:

1. *Who?* Who are the individuals involved in the particular incident being investigated? Names, addresses, and telephone numbers are important.
2. *What happened?* What is the story of the incident? For instance, what happened before, during, and after a theft incident?
3. *Where?* The location of the incident and the movement of people and objects are important. For example, where exactly were witnesses and the suspect when the theft occurred?
4. *When?* A notation of the times of particular activities during an incident is necessary for a thorough investigation. If a theft occurred between 7 and 8 P.M. on April 9, and Joe Doe is a suspect, he can later be exonerated because he really was at another location at that time.
5. *How?* The focus of this question is how the incident was able to take place in the face of (or absence of) loss prevention measures. After a theft, investigators often attempt to find out how the thief was able to circumvent security. In the case of an industrial accident, investigators study how the accident occurred while safety equipment was supposedly in use.
6. *Why?* This question can be difficult to answer. However, the answer can lead to the discovery of a pressing problem that may not be obvious. An example is seen with numerous losses in a manufacturing plant brought about by low employee morale. In this case, theft and destruction of company property can be reduced by, for example, increasing management's concern for employees through praise, a sports program, contests, and high-quality meals in the cafeteria.

The answer to the "why" question helps to establish the motive for the loss activity. Once the motive is established, suspects can be eliminated. A recently fired employee would have a motive of retribution for setting fire to his or her former place of employment. Investigations are not always criminal in nature, and not all investigations require answering all six investigative questions.

Once an investigator gathers sufficient information, a report is often written and submitted to a supervisor. After that, the information in the report results in action or inaction by supervisors and management. Investigative reports may lead to either punitive or nonpunitive action. Punitive action can include discipline, or prosecution, or both. Nonpunitive action includes exonerating a suspect, hiring an applicant, promoting an employee, or corrective action such as an improved loss prevention program.

Investigations are unique to each type of business, institution, or organization served. The varieties and depths of investigations vary; they reflect management's needs, objectives, and budget. The personnel involved also vary and may include company investigators, auditors, IT specialists, contract investigators, public police, and attorneys.

The following steps guide the process of an investigation:

1. The decision to begin an investigation
2. Selection of a supervisor and investigator
3. Planning, objectives, and methods of investigation

4. Gathering accurate facts
5. Accurate, factual report to management
6. Management decision making

Types of Investigations

There are several types of investigations in the private sector. The following paragraph illustrates some of the more common ones, categorized according to the target of the investigation.

The laws or "ground rules" for each type of investigation vary. In addition, the policies of organizations guide investigations, although are subservient to law. The term *workplace investigation* is a generic term focusing on a wide variety of allegations. Examples are misconduct, substance abuse, threats, and violence. An *applicant background* investigation requires adherence to the laws discussed in Chapter 6. A private investigation of an alleged *criminal offense* requires knowledge of criminal law, evidence, and local police and prosecutor requirements. The notification of public police depends on such factors as whether company investigators arrested the suspect and whether management seeks to prosecute. A *financial* investigation checks on alleged fraud or embezzlement. *Computer crime* investigations focus on a variety of acts involving IT systems. The Internet has expanded the scope and complexity of these investigations. In an *accident* investigation, the investigator usually is knowledgeable about safety, OSHA regulations, and workers' compensation insurance. A major objective of a *fire or arson* investigation is to determine the probable cause of the fire. Sophisticated equipment may be applied to locate and identify any substance employed to accelerate the fire. A *civil or negligence and liability* investigation involves, among other things, gathering evidence to determine whether a failure to exercise reasonable care in a situation caused harm to someone or something. For example, a customer (plaintiff) may seek to show that a retail store (defendant) with poor housekeeping caused the customer's injury. Both the plaintiff initiating the suit and the defendant are likely to conduct investigations and present evidence in court. *Insurance* investigations, which at times result in litigation, are conducted to determine losses and their causes, and to assist in deciding on indemnification. Both the insurer and the insured may conduct separate investigations before an insurance claim is settled. Investigations of *labor* matters (e.g., workers' activities during a strike) are often sensitive. Legal counsel is necessary to guide the investigator because of associated federal and state laws. *Due diligence* is the attention and care legally expected in checking the accuracy of information and omissions. It can range from determining a customer's financial status to assessing the desirability of acquisition targets.

Proprietary and Contract Investigations

A **proprietary investigation** is undertaken by an in-house company employee who performs investigative work. A **contract investigation** requires the contracting of an outside company (e.g., private investigation agency) to supply investigative services for a fee.

To protect the public and clients, there are varied laws regulating contract investigative firms and employees. Regulations include requirements of an application, a fee, fingerprints, and a photo for license approval, training, experience, no felony convictions, examination, and insurance.

For the most part, proprietary investigators are not subject to government regulation. Many businesses, institutions, and organizations maintain their own investigators. Large corporations, utilities, insurance companies, and banks are some of the many concerns that rely on a staff of proprietary investigators.

Numerous companies utilize both contract and proprietary investigators. If a large corporation has a particular season when the investigative workload is heavy, some of the extra workload can be assigned to a contract investigation company. This approach frees proprietary investigators for more pressing and specialized problems.

Citizens frequently obtain a distorted picture of private investigations via television, which produces misconceptions that falsify the various kinds of investigative work—work that is interesting, exciting, requires a lot of "paper work," and is boring at times. Another misconception is that many investigators in the private sector are armed. Few are armed, and those who are rarely use their weapons.

Private and Public Investigations

Private investigations serve the private sector (e.g., businesses and private citizens). **Public investigations** feature public police agencies, for the most part, serving the public. Both investigative efforts often are entwined. This can be seen, for example, when an office building is burglarized and company investigators call local police in a joint effort to solve the crime. However, how much time and effort can the public police devote to the burglary in comparison to company investigators? Public police can devote only limited resources to such a crime. Typically, a uniformed public police officer arrives for a preliminary investigation and an incident report is completed. Next, the incident report is transferred to a detective unit. Within a day or two, a detective arrives at the scene and conducts a follow-up investigation that involves gathering additional information and perhaps some placation (i.e., a public relations effort that assures the victim that the police are doing everything possible to solve the crime). Generally, public police devote more time and effort to crimes against people than crimes against property.

The inability of the criminal justice system to assist adequately in private loss prevention efforts is also illustrated by clearance rates, which are the proportion of cases solved by an arrest or exceptional means (e.g., victim's refusal to cooperate; death of offender). According to the Federal Bureau of Investigation (2010), in 2009, clearance rates were higher (47.1 %) for crimes against people (e.g., murder, rape, robbery, and aggravated assault) and lower (18.6 %) for crimes against property (e.g., burglary and larceny-theft). The success of private-sector investigations is difficult to gauge, since many firms do not prosecute and the outcome of investigations is usually confidential and unpublished.

Overt and Undercover Private Investigations

An **overt investigation** is an obvious investigation. People encountering the overt investigator know that an investigation is taking place. A common scenario would be a company investigator, dressed in a conservative suit, arriving at the scene of a loss to interview employees and collect evidence.

An **undercover investigation** (UI), on the other hand, is a secret investigation. In a typical approach, an undercover investigator is hired as a regular employee, a truck driver, for example, and collects information by associating with employees who are not knowledgeable about the undercover investigation.

Since undercover investigators are often assigned a fictitious background, care must be exercised because social media leaves a digital footprint available to anybody through the Internet and an undercover investigator may be exposed and subject to harm. Consequently, **counter social media techniques** (CSMT) are necessary. This is an emerging concept that can include an investigative supervisor conducting a thorough Internet search on the undercover investigator, studying what is available online, merging what is available to the new identity, and adding new identity items, among other techniques. An option is selecting a different UI who has less exposure online. If the undercover investigator assumes a new name, CSMT will change; however, online photos and video are other challenges. Consideration should also be given to the devices used by the UI. For instance, what if the UI's smart phone is lost, stolen, or hacked?

Each type of investigation serves many useful purposes. An overt investigation that begins immediately after a loss shows that the loss prevention staff is on the job. This in itself acts as a deterrent. An overt investigation does not have to be in response to a loss; for example, pre-employment investigations prevent losses.

An example can illustrate the usefulness of a UI. The XYZ warehouse is losing thousands of dollars of merchandise every week. Management believes that employees are stealing the merchandise. In an effort to reduce losses, management decides to hire a private investigator. The investigator interviews numerous employees over a two-week period, but the case remains unsolved. An executive of the company argues to management that the private investigator idea is a waste. The executive points out that the private investigator is unfamiliar with warehouse operations, cannot penetrate the employee informal organizations, has no informants, is wasting time during surveillance from another building, and has no substantial leads. After three weeks, the private investigator is terminated. The vocal executive argues for a UI. A loss prevention service company is contacted, and an undercover investigator is placed in the warehouse. After three weeks, the new investigator penetrates the informal employee organization. Four key employees are implicated and fired. Losses are reduced. The company decides to conduct an annual UI as a loss prevention strategy.

Accident at Hardy Furniture Plant

The Hardy Furniture Plant provided more than 700 jobs to Clarkston residents as sales boomed. Most workers at the plant were satisfied with their jobs. Unfortunately, a group of young forklift drivers was becoming increasingly bored while transporting furniture throughout the plant. One day, two forklifts collided, and both drivers were hospitalized with broken arms and legs. The forklifts were damaged slightly; furniture on the forks was a total loss. Immediately, the company loss prevention staff began an investigation. The forklift drivers and witnesses were interviewed. All interviewees reported that neither forklift operator saw the other because excessive furniture on the forks obstructed their views. Investigators and a forklift mechanic inspected the forklifts and found no irregularities.

The investigators became suspicious when all of the interviewees produced identical stories during questioning. Later, another round of questioning included nonwitnesses. Finally, after two weeks of persistent interviewing, the case broke when an older employee, loyal to the company, informed investigators that several bored employees bet on forklift drivers who were "playing chicken" while racing toward each other. It was learned that five previous contests had taken place involving hundreds of dollars in cash. In addition, one minor accident had occurred resulting in damaged furniture. Two supervisors were involved in covering up the loss of the damaged furniture.

When the investigation was complete, loss prevention investigators filled out appropriate reports and presented their findings to management. The two forklift drivers and the two supervisors were fired.

Undercover Investigation of Missing Shirts

The Chester Garment Company, headquartered in New York City, experienced a loss of hundreds of men's shirts at its plant in North Carolina. Company executives were concerned and worried because they had no experience with such losses and no loss prevention program.

The plant manager in North Carolina had notified the New York office that 900 shirts were missing. Because of their limited knowledge of loss prevention strategies, company executives decided to contact Harmon Lorman Associates, a company specializing in loss prevention and investigations. A meeting between managers of the investigative company and the garment company decided that an undercover investigator, assigned to the North Carolina plant, would be a wise strategy. The UI would cost $1,500 per week for an unspecified time period.

One week later, the UI, Gary Stewart, arrived at the Chester Garment Plant in North Carolina to seek employment. The plant manager, who was the only plant employee knowledgeable about the UI, hired Gary and assigned him to shipping and receiving.

Gary was from New York and educated at a college in North Carolina. The investigative company felt that his experience, a college degree in criminal justice, and his living experience in the South would add up to an appropriate background for this assignment. Anyway, he was the only company investigator who had lived in the South.

Harmon Lorman executives told Chester executives that Gary had been working with them for a year and a half. In addition, they told them that Gary had extensive experience and loss prevention training. The truth was that Gary had been recently hired with six months of previous investigative experience. Gary had no previous loss prevention training; he had a criminal justice degree with no loss prevention or business courses.

After two weeks at the plant, Gary had established numerous contacts. His fictitious background ("cover") pointed out that he grew up in Maryland and that he arrived at the Chester Garment plant because a friend said that he could get a job there while taking a semester off from college. Gary obtained North Carolina license plates as soon as he entered North Carolina from New York.

Gary sent three to five reports per week to his supervisor at Harmon Lorman Associates. The first few reports contained background information on the plant, such as the plant layout, and the names, addresses, telephone numbers, description of autos, and plate numbers of select employees.

Thereafter, the reports contained information pertaining to loss prevention features and loss vulnerabilities. Janitorial service, employee overtime, Saturday activities, and any unusual events were also reported.

By the third month, Gary had made close contacts with employees and had worked in numerous assignments throughout the plant. His findings showed numerous instances of pilferage by many employees. Women sewing-machine operators were hiding several manufactured shirts under their outer clothes immediately before the workday ended.

Gary's reports were "edited" by his superiors and then sent to the home of one of Chester Garment's executives. After three months, Chester Garment executives became impatient. They wanted better quality results and threatened to terminate the investigation.

Harmon Lorman executives assured Chester Garment executives that "a break in the case was imminent." Increased pressure was put on investigator Gary Stewart, who realized that the investigation was being prolonged for profit. He responded by withholding information from reports for dry spells when good information was unavailable. As the investigation went on, the report quality went down.

Finally, Chester Garment executives ordered an inventory at the plant. Surprisingly, half of the missing shirts were accounted for and a previous inventory was criticized as inaccurate. The undercover investigation was terminated. Gary returned to New York.

What type of investigative work do you think you would prefer as a career?

Important Considerations

The following considerations relate to investigations in general:

1. The supervision of investigators must be adequate to produce tangible results. Rarely will a supervisor/investigator ratio of 1 to 20 prove adequate. Investigators may require close supervision by attorneys and other specialists.
2. Sensitive and confidential information must be safeguarded.
3. Information resulting from investigations can be used to improve loss prevention efforts and reduce vulnerabilities. For example, if machine shop workers are constantly stealing company tools, then an option is to require workers to provide their own tools.
4. Investigations can support the loss prevention budget by documenting vulnerabilities and losses, admissions of losses by offenders, and recoveries of company assets.
5. Investigations should be cost effective. If an investigation costs more than the loss, the expense of the investigation may not be worthwhile. For example, the loss of a box of pencils is not worth an investigator's time when losses that are more serious are occurring.
6. Although computer crimes can be committed remotely, modern technology also permits investigations to be conducted remotely. For example, instead of an investigator visiting stores to check CCTV video of cashier–customer transactions and other store activities, web-based systems enable investigators to view store activities online.
7. An investigation may be required of senior management because of criminal, civil, or regulatory misconduct. Authorization and direction for such a sensitive investigation may come from the corporate board of directors.

Law

Knowledge of law is indispensable to the investigative process. As covered in Chapter 4, ASIS International-funded research showed that legislation affects organizations in general and private security operations (Collins et al. 2005). The top three acts cited in this research as having the most impact on security policies and procedures were the Health Insurance Portability and Accountability Act of 1996 (HIPAA), the Sarbanes-Oxley (SOX) Act of 2002, and the USA Patriot Act of 2001. SOX, for example, has forced businesses to develop internal investigation teams to comply with regulations and improve the protection of assets and stockholder investments. This includes increased attention to employee theft and fraud, workplace violence, sexual harassment, due diligence, and other issues (Daniels, 2006: 22–23). Chapter 6 referred to the Fair Credit Reporting Act of 1971 and the Fair and Accurate Credit Transaction Act of 2003 for investigations of employment applicants and incidents involving employees in the workplace.

Another important area of law pertains to electronic surveillance (electronic devices used to listen to conversations) and wiretapping (listening to telephone communications). The U.S. Supreme Court has called these techniques a "dirty business." Generally, both are prohibited unless under court authority. However, because of the difficulty of detection and the advantages in information gathering, some private (and public) sector investigators violate the law. Although private conversations of two parties cannot be recorded unless under court order, if one of the parties approves, recordings are legal. Examples include one party cooperating with an investigation or serving as an UI. An employee in the workplace, using a company telephone (or computer), does not generally have an expectation of privacy, especially if the employer informs employees of monitoring.

Conflict of interest is another concern. An example is when a full-time public police officer works part-time for a private investigative firm. Generally, states prohibit one person from holding dual commissions (i.e., public police commission and private investigator license). The *Report of the Task Force on Private Security* (U.S. Department of Justice, 1976: 238) states that "a citizen might file a defamation-of-character suit against a city, law enforcement agency, or officer by claiming that surveillance conducted by an off-duty law enforcement officer working as a private investigator gave others the impression he was the target of a law enforcement criminal investigation." Other issues can evolve, such as a suspect's rights during questioning by a public police officer working part-time in private security. The task force recommended that public police officers should be "strictly forbidden" from performing private-sector investigative work.

Poorly executed investigations can result in liability. In one case, a bank investigator believed that a loan manager had mob connections and an investigation resulted in termination of the manager. Police brought charges, but the judge dismissed the case on groundless charges, and the manager sued and was awarded damages. In another case, an oil company terminated an agent who was allegedly manipulating prices, but the prosecutor declined to bring charges because of insufficient evidence. When a truck driver stated to someone that the agent was a "thief," the agent sued for slander. Investigators can protect themselves from slander and libel by the use of privilege when issuing reports. **Qualified privilege** permits defamatory statements if made in the discharge of duty and without malice. At one company, an employee was fired for a security violation and sued by claiming defamation from a security report, but the court held that the investigator was responsible for reporting security breaches, that she acted within her authority, and that her report was privileged. **Attorney/client privilege** (also referred to as confidential communications) is another form of protection. It involves statements between persons who have a necessary relation of trust to help communication. The statements cannot be disclosed. To facilitate this privilege, investigators should ensure that reports are issued directly to counsel. However, the best defense is accurate reports (Ray, 2000: 94).

Roman (2005), an attorney, offers suggestions of what to avoid during workplace investigations. Failure to investigate cases of discrimination or harassment can result in liability under antidiscrimination laws. Courts have held that no investigation was evidence that the employer acted with "malice or reckless indifference." Another point is that using a biased investigator can harm a case. Examples include assigning an investigator from the accused's chain of command and following up on leads from the accused, but not from the accuser. Roman emphasizes the importance of conducting thorough, objective, and unbiased investigations. To avoid claims for defamation and invasion of privacy, the employer and investigator should disclose information only on a "need to know" basis.

Reibold (2005: 18–29), an attorney, writes of the legal liabilities of hiring private investigators (PIs). Those who hire PIs can be held vicariously liable for torts committed by PIs, directly liable for actions of PIs, and negligent in hiring and supervising them. A lawsuit may be directed at an employer, a law firm, and a PI agency. Lawsuits against PIs and those who hire them may involve trespass, invasion of privacy, infliction of emotional distress, assault and battery, and conversion after gathering garbage from a subject's trash container. The defense argument that the PI was an independent contractor has been rejected by courts.

Reibold offers the following suggestions when screening a PI firm: ensure the PI is properly licensed; request references; determine if the PI is a member of a professional association that promotes ethics; use a Freedom of Information Act request to the state regulatory authority to check on the PI; require the PI to contractually indemnify the entity hiring the PI; and upon verifying the PI's liability insurance, request that the PI add the entity hiring the PI to the policy.

Evidence

Laws govern the introduction of evidence into judicial proceedings to ensure fairness and due process. Evidence law is contained in federal and state constitutions, statutes, and case law.

There are several classifications of evidence. **Direct evidence** directly proves or disproves a fact without drawing an inference (i.e., conclusion). Examples are a confession, an eyewitness identification of a suspect, CCTV video of a suspect committing a crime, and physical evidence (e.g., contraband in possession of the accused). **Circumstantial evidence** indirectly proves or disproves a fact and an inference must be drawn. Examples are a statement by the accused that he or she was with the victim right before the victim was murdered, and physical evidence, such as fingerprints and DNA found at a crime scene (Gardner and Anderson, 2007: 62–63).

Another type of evidence is **hearsay evidence**. It is second-hand information or what someone heard. Courts favor first-hand information through personal observation or the use of other senses. Hearsay evidence is generally inadmissible in court, although there are exceptions, such as a dying declaration or a spontaneous declaration by an offender following a crime.

The accountability of physical evidence, especially before it reaches a court of law, can have a definite impact on a case. The **chain of custody of evidence** must be maintained. This refers to the written accountability of persons having possession of evidence from its initial discovery, through judicial proceedings, and to its final location. A loss scene should be protected, photographed, videotaped, and sketched before evidence is touched. The proper labeling, packaging, and storage of evidence is equally important. The accountability of evidence often is brought forth in court. Attorneys are sure to scrutinize the paperwork and procedures associated with the evidence. The following questions are among the many frequently asked in court. Who saw the evidence first? Who touched it first? Where was it taken? By whom? How was it stored? Was the storage area locked? Who had the keys or access code?

International Perspective: Overseas Investigations

Global business has resulted in increasing demand for overseas investigations. Ast (2011: 110) describes three types of international investigations: background screening, criminal, and due diligence. He describes a due diligence investigation of a potential business partner for his employer which showed the potential partner's address in an Asian city as a dilapidated building surrounded by laundry and livestock.

The following information provides tips and guidelines for overseas investigations (Ast, 2011: 109–110; Van Nostrand and Luizzo, 1995: 33–35).

- *U.S. Department of State* ensures that passports are current and valid for all countries not off limits to U.S. citizens. Some countries require a special visa. Fact sheets online are provided on many countries and include information on such topics as political stability and crime. A State Department regional security officer, or legal attaché (often an FBI agent/lawyer) with a U.S. embassy or consulate, may offer assistance in locating a reputable foreign investigative firm.
- *Central Intelligence Agency* maintains fact books on countries.
- *U.S. Department of Health and Human Services* determines whether travel to particular countries requires immunization.
- *U.S. Department of Commerce* publishes information to alert U.S. citizens to countries that may be dangerous to U.S. travelers.
- Because legal systems vary among countries, a major rule in conducting a foreign investigation is to study the legal and policing system of the respective country. In addition, consult with a competent attorney.
- Investigations are fundamentally about personal interaction, so another major rule is to study and understand the host culture and try to speak some of the language.
- The international investigator who travels to many countries and freely investigates is actually Hollywood fantasy. Foreign countries do not permit such activities. Without careful research, an investigator can find himself or herself in jail in a foreign country.

- An option is to work with an official of the foreign country, such as a police official or attorney, or contract the investigation to an investigator in that country. The key is to select someone who has experience in the country, has reliable contacts, and speaks the language.
- Avoid bringing a firearm to a foreign country. Illegal possession of a firearm aboard any U.S. airline is a felony.

Another source is the Overseas Security Advisory Council (n.d.). OSAC "is a Federal Advisory Committee with a U.S. Government Charter to promote security cooperation between American business and private sector interests worldwide and the U.S. Department of State. OSAC currently encompasses the 34-member core Council, an Executive Office, over 100 Country Councils, and more than 3,500 constituent member organizations and 372 associates." The objectives of OSAC are to facilitate cooperation between State Department security functions and the private sector, exchange information on the overseas security environment and suggestions for security planning, and recommend methods to protect the competitiveness of American businesses operating worldwide.

Interviewing and Interrogation

Interviewing and interrogation are methods of gathering information from people. During an **interview**, the suspect supplies information willingly; but during an **interrogation**, the suspect is often unwilling. Experienced investigators know the techniques associated with each type of situation. Although a few recognized training programs focus on conducting interviews and interrogations, no single method holds all the answers and is applicable in all situations. Investigators often use a variety of methods. (See "Confronting the Employee Suspect" in Chapter 7 for additional guidelines.)

Why are interviews or interrogations conducted? A primary reason is to learn the truth. Other reasons are to obtain evidence or a confession to aid in prosecution, eliminate suspects, recover property, and obtain information that results in corrective action. This chapter emphasizes investigations in the private sector, although many of the ideas presented here are applied in public-sector investigations.

The preliminaries include:

1. Maintaining records.
2. Planning the questioning.
3. Making an appointment, if necessary.
4. Consulting with a superior or an attorney if a procedure or legal question arises.
5. Questioning in privacy, if possible, and ensuring that the interviewee can leave when he or she chooses.
6. Making sure someone of the same sex as the interviewee is present.
7. Ensuring the interview room is safe from potentially dangerous items (e.g., paperweights and scissors).
8. Openly recording the questioning, if possible.
9. Identify yourself to the interviewee. Remember that it is illegal to pretend to be someone in authority such as a police officer.

Regarding the interviewee:

1. Consider the interviewee's background, intelligence, education, biases, and emotional state.
2. Communicate on the same level.
3. Watch for nervousness and perspiration.
4. Reluctance to talk may indicate that the interviewee feels the need to protect himself/herself or others.
5. Responding freely may indicate that the interviewee may need to relieve guilt or he/she may want to cause problems for an enemy not involved in the loss.

The objectives of the investigator include:

1. Establishing good rapport (e.g., asking, "How are you?").
2. Maintaining good public relations.
3. Maintaining eye contact.
4. Not jumping to conclusions.
5. Maintaining an open mind.
6. Listening attentively.
7. Being perceptive to every comment and any slips of the tongue.
8. Maintaining perseverance.
9. Controlling the interview.
10. Carefully analyzing hearsay.
11. Not discussing the case with anyone unless they have a need to know.
12. Helping the suspect tell the truth and confess to wrongdoing.

Strategies of the investigator include:

1. Asking **open-ended questions**, those questions that require lengthy answers—for example, "What happened at the plant before the accident?" **Close-ended questions** require short "yes" or "no" answers that limit responses—for example, "Were you close to the accident?"
2. Maintaining *silence* makes many interviewees feel uncomfortable. Silence by an investigator, after an interviewee answers an open-ended question, may cause the interviewee to begin talking again.
3. Building up interviewee memory by having the interviewee relate the story of an incident from its beginning.
4. To test honesty, asking questions to which you know the answers.
5. Using trickery. For example, falsely stating that the suspect's fingerprints were found at the crime scene. In *Frazier v. Cupp* 394 US 731 (1969), the U.S. Supreme Court upheld the use of false information to obtain a confession if the technique does not shock the conscience.

The reader is probably familiar with movies and television programs that portray the interrogation process as a "third degree," in which one bright light hangs over the seated suspect in a dark room and investigators stand around constantly asking questions and using violence as they try to "break" the suspect. Court action against police has curbed this abuse. However, because of the unpleasant connotations associated with interrogations, for the private sector, to prevent litigation, a less threatening term such as *intensive interview* is more appropriate.

During interrogation or *intensive interview* (an extension of the interview):

1. Discuss the seriousness of the incident.
2. Request the story several times. Some investigators request the story backward to catch inconsistencies.
3. Appeal to emotions; for example, "Everybody makes mistakes. You are not the first person who has been in trouble. Don't you want to clear your conscience?"
4. Point out inconsistencies in statements.
5. Confront the interviewee with some of the evidence.

Deslauriers-Varin et al. (2011: 113–145) researched factors that influence suspects' decisions to confess to police. Their research provides direction for similar studies in the security profession. They conclude from their research that interviewer abilities in convincing suspects to confess are integral to cases and "such abilities might be most effective with suspects who have no prior record, who are single at the time of the interrogation, who feel guilty about their crime, and who do not use legal advice." In addition, the researchers noted that when suspects perceive that police evidence is strong, they are most likely to confess. They caution that false confessions might be more likely to occur with vulnerable individuals who lack social support and experience with the justice system and who do not seek assistance of counsel.

Kieffer and Sloan (2009: 317–330) examined techniques of neutralization used by white-collar suspects to minimize guilt and justify their crimes. Examples of techniques of neutralization include denial of responsibility (e.g., ignorance or accident), denial of injury (e.g., no one was harmed), and denial of victim (e.g., the employer deserved it). Kieffer and Sloan suggest that investigators apply neutralization theory to guide interviews of white-collar suspects to obtain confessions. Rather than the investigator criticizing the suspect, a more productive approach is to appear sympathetic, minimize a moral tone, reduce the suspect's apprehension about punishment, and introduce or reinforce neutralizations stated by the suspect; these factors create comfortable conditions for a confession.

LaBruno (2010: 28), an experienced in-house loss prevention counsel, writes that she would "break out in a cold sweat at the notion of obtaining coerced or false admissions from employees under investigation." She illustrates the cost of third degree tactics by noting that a jury in an AutoZone case awarded a former employee $7.5 million because he was coerced into falsely confessing he stole $800. Virgillo (2010: 52–58) explains the problem of management imposing admissions quotas on investigators to measure performance; this may result in unethical and illegal conduct to "successfully" close cases. Solutions to this dilemma include quality screening of applicants for investigative positions, training, and supervision; holding investigators accountable for applying "best practices;" and using multiple performance indicators (e.g., shrinkage and recoveries), in addition to the number of admissions.

Interestingly, scientists have disproven certain indicators of lying that are applied by investigators today. Matsumoto et al. (2011: 1–8) note that no scientific evidence supports eye behavior (e.g., shifty eyes) or gaze aversion as indicators of lying. They also write that research has only weakly associated fidgety feet and hands, voice stress, and body posture as associated with deception. Caution is advised with interviewing techniques because there is no one indicator of lying and investigators should consider multiple sources of evidence.

Why would a private security investigator choose to "interview" rather than "interrogate"?

Polygraph: Proper Testing Procedures under the Employee Polygraph Protection Act

In the course of a workplace investigation, an employer cannot suggest to employees the possible use of a polygraph instrument until these 10 conditions are satisfied:

1. *Economic loss or injury.* The employer must administer the test as part of an investigation of a *specific incident* involving economic loss or injury to the business, such as theft or sabotage.
2. *Access.* The employee who is to be tested must have had access to the property that is the subject of the investigation.
3. *Reasonable suspicion.* The employer must have a reasonable suspicion of the worker's involvement in the incident under investigation.
4. *Before the test,* an employer's failure to adhere to guidelines can void a test and subject the employer to fines and liability. The employee who is to be tested must be notified in writing at least 48 hours prior to the test (not counting weekends and holidays):
   - Where and when the examination will take place
   - The specific matter under investigation
   - The basis for concluding that the employee had access to the property being investigated
   - The reason the employer suspects the employee of involvement
   - The employee's right to consult with legal counsel or an employee representative before each phase of the test

   Also before the test, the employee must be provided with the following:
   - Oral and written notice explaining the nature of the polygraph, its physical operation, and the test procedure
   - Copies of all questions that will be asked during the test
   - Oral and written notice, in language understood by the employee and bearing the employee's signature, advising the worker of his or her rights under the EPPA
5. Procedural requirements for polygraph examinations include the following:
   - The test must last at least 90 minutes unless the examinee terminates the test.
   - Either party, employer or employee, can record the test with the other's knowledge.
   - Questions cannot pertain to religious, political, or racial matters; sexual behavior; or beliefs, affiliations, or lawful activities related to unions or labor organizations; and they cannot be asked in a degrading or needlessly intrusive manner.
   - A worker can be excused from a test with a physician's written advisement that the subject suffers from a medical or psychological condition or is undergoing treatment that might cause abnormal responses during the examination.
   - An employee has the right to consult with counsel before, during, and after the examination but not to have counsel present during the actual examination.
   - An employee must be advised that his or her confessions may be grounds for firing or demotion and that the employer may share admissions of criminal conduct with law enforcement officials.

- A worker can terminate or refuse to take a test and cannot be demoted or fired for doing so. However, the employer can demote or fire the worker if he or she has enough separate supporting evidence to justify taking that action.

6. *After the test*, the employee has a right to a written copy of the tester's opinion, copies of the questions and corresponding replies, and an opportunity to discuss the results with the employer before the employer takes action against the worker based on the test results. An employee may be disciplined, fired, or demoted based on the test results if the employer has supporting evidence, which can include the evidence gathered to support the decision to administer the test, to justify such action. Test results cannot be released to the public, only to the employee or his or her designate; the employer; a court, government agency, arbitrator, or mediator (by court order); or appropriate government agency if disclosure is admission of criminal conduct (without court order). The examiner may show test results, without identifying information, to other examiners in order to obtain second opinions.
7. *Qualifications of examiners*. An employer can be liable for an examiner's failure to meet requirements, which cover licensing, bonding, or professional liability coverage; testing guidelines; and formation of opinions.
8. *Waiving employee rights*. A worker cannot be tested—even at his or her insistence—if the employer cannot meet procedural requirements and prove reasonable suspicion and access. Employees may not waive their rights under the EPPA except in connection with written settlement of a lawsuit or pending legal action.
9. *State law and collective bargaining agreements*. The EPPA does not preempt any state or local law or collective bargaining agreement that is more restrictive than the act.
10. *Record-keeping requirement*. Records of polygraph exams should be kept for at least three years by the employer and the examiner, who must make them available—within 72 hours upon request—to the Department of Labor.

Source: U.S. Chamber of Commerce. See Chapter 6 for the Employee Polygraph Protection Act of 1988.

Information Sources

Traditional information sources include interviewing people in person; traveling to a government agency, library, or other location to comb through records and information; obtaining information over the telephone; and conducting surveillance. Today, the Internet has made it easier than ever to investigate personal and business information. Search engines such as Google, and alternatives that use different algorithms than Google, including Mamma, AltaVista, Gigablast, and Exalead offer vast information (Reino, 2011). Google maps enable the user to "visit" sites many miles away. Social media sites such as Facebook offer additional information (especially when users do not apply privacy settings) as people broadcast their lives and include photos and videos (Hetherington, 2010: 44–52). Other rich sources of information are YouTube, Flickr, LinkedIn, Twitter, Loopt Star, and blogs. Quirke (2012: 60–66) recommends caution with sites that have privacy settings (e.g., Facebook) in comparison to sites that permit public access (e.g., YouTube). Privacy rights must be respected. Action or decisions based upon information acquired by violating privacy can result in litigation.

Social Network Site Helps to Close $365,000 Case

In one investigation, a case was closed in three days with the aid of a social network site. Two women entered a retail store and escaped with over $1,000 is merchandise. Retail employees were able to obtain a vehicle description and a plate number. When police checked with the DMV, a DMV photo of the registered owner did not match either of the shoplifters, nor other suspects in similar crimes at other stores. Police closed the investigation. Loss prevention investigators checked online auction sites for sales of the items that were shoplifted, but no leads developed. However, a local resale store employee described two women selling items for twenty cents on the dollar. Shoplifting cases with similar suspects were studied, and in one case, a vehicle was identified whose owner had a daughter who fit the description of one of the suspects. A study of MySpace provided photos of the suspects that were identified by store employees. Using whitepages.com and LexisNexis, the suspects were located, interviewed, and they signed written confessions to shoplifting about $365,000 in items from several stores. The case was shared with police and the offenders were convicted and sentenced (Accardi, 2010: 21).

For a nominal fee, a variety of information can be obtained with a computer from the comfort of one's office or home, as illustrated here:

- **PIMall** offers a wealth of information and links to information brokers, private investigators (PIs), a PI store, PI associations, training, and publications: www.pimall.com
- **USSearch** can be used to locate someone, conduct a background check, trace phone numbers and addresses, and seek the following records: criminal, Department of Corrections, court, real estate, bankruptcies, marriage, divorce, and death. Business searches include due diligence and verification of professional licenses: www.ussearch.com
- **LexisNexis** offers an extensive search of legal, news, and business sources: global.lexisnexis.com
- **DNB** offers a variety of business information: www.dnb.com

One problem with these databases is selecting the most appropriate information broker. Another problem is that data may not be verified. In addition, the scope of the search must be considered. What geographic area and what months or years were searched? Information brokers must adhere to legal restrictions and should provide such guidelines to clients.

Despite the problems with databases, security practitioners use such information sources in a variety of ways. One airport security director uses such databases to locate owners of vehicles abandoned in the airport's parking lots. Another security practitioner conducts asset searches of employees suspected of fraud; a database may show that an employee earning $45,000 per year has purchased a $600,000 house.

Legal Restrictions when Collecting Information

Investigator ability to obtain usable information in today's privacy-protected environment has been reduced. This was illustrated in Chapter 6 in the coverage of background investigations.

In earlier years, the "old boy" network was in greater use. It consists of employees in both the public and private sectors who informally assist each other with information. For example, a retired police officer is hired by an investigative firm and contacts friends from the police agency where he or she was employed to obtain criminal history information on individuals subject to a private-sector investigation. There have been indictments against people involved in acquiring nonpublic information.

The reality of information acquisition is that no information is totally secure from unauthorized acquisition. Although difficult to measure, private information is periodically obtained in an unethical or illegal manner. **Pretexting** is a method of seeking information under false pretenses. An example is a private investigator stating that he or she is a police officer to convince an interviewee to release information. A variety of Federal and state laws prohibit pretexting to obtain personal information and it should be avoided (Anderson, 2010: 65–73).

Another problem for the investigator results when he or she mistakenly collects information that is not usable in court. In litigation, information improperly obtained and not authenticated or certified by the respective agency could subject a litigation team to civil or criminal action, unless the records are subpoenaed or are part of a court action (civil or criminal). Another challenge is waiting for a Freedom of Information Law request to obtain documents, which can take as long as a year. The **Freedom of Information Law** (FOIL) grants citizens access to public documents because an informed electorate is essential to safeguard democracy and because publicity is a protection against official misconduct. This law requires all federal agency documents to be publicly disclosed, unless exempted. This law also recognizes the need to restrict intrusions into a private individual's affairs. States also have Freedom of Information laws.

Investigative Leads

Investigative work requires patience and perseverance, and difficult cases often tax the abilities of investigators as they search for answers. Investigative leads are aids to the investigator.

Scene of the Loss

A search of the scene of a loss can provide answers to investigative questions. Offenders at a crime scene often leave something (e.g., fingerprints, DNA) or take something with them (e.g., stolen item), either of which ties them to the crime scene.

Because bar codes and RFID tags are attached to consumer goods, evidence (e.g., knife) at a crime scene may contain such a code or tag that can possibly be used to trace merchandise through the supply chain to a retailer. If the merchandise was purchased with a bankcard or other digital ID, the customer can possibly be identified. Bar codes and RFID tags also help police and security personnel maintain accountability of evidence while assisting with chain of custody of evidence requirements.

The loss scene requires protection from unauthorized persons. Photographs, video, and sketches should be made without disturbing the characteristics of the scene and before evidence is removed.

Offenders at a crime scene often leave something or take something with them, either of which ties them to the crime scene.

Victims

Care and empathy for victims are essential in the investigative process. However, sometimes the victim is the offender. Good leads can be obtained by checking the background of the victim. The victim may be a person, business, or organization. A person owning a failing business may have perpetrated an accident, arson, or other crime to collect on an insurance policy. A male employee may falsely claim that his wife's lover attacked him at work. Sometimes employees are hurt off the job, but are able to go to work, claim injury on the job, and seek compensation. In most instances, the victim is not an offender; however, the investigator must maintain an open mind.

Motive

The motive behind a loss is an important consideration. Questions of concern include: Who will gain from the loss? Are there any ulterior motives? What types of persons would create such a loss? Why? The investigator also must recognize that the human factor may not be involved in the loss. Equipment malfunction or weather may be the cause.

Witnesses

Investigative leads frequently are acquired from witnesses. Good interviewing is important and can turn up valuable leads.

Informants

Why do informants divulge information? Sometimes, they do so because they are seeking favors or money, because they see it as their duty, or because they want to get someone in trouble (e.g., competitor, unfaithful lover). Informants often supply misinformation to investigators. An investigator can test an informant by asking questions to which he or she knows the answers. An investigator must never become too involved with informants or perform any unethical or criminal activity to acquire information. Obviously, an informant's identity must be protected, unless a court requires otherwise. Many investigators (private and public) have money in their budgets specifically designed to pay informants for information. (Chapter 7 explains loss reporting and reward system.) A common practice of investigators is to catch an individual in violation of a rule or law but not seek punishment (e.g., prosecute) if that individual supplies the investigator with useful information.

Modus Operandi

MO stands for **modus operandi**, or method of operation. An investigator may ask, "What method was used by the burglar?" Because people differ, they commit crimes in different ways. Police and security organizations maintain MO files on offenders. When a crime takes place, investigators may check an MO file for suspects who fit the crime at hand. A particular offender may use a specific tool during a burglary. A robber may wear a unique style of clothing during robberies. A saboteur at a manufacturing plant may be using a particular type of wire cutter. Sometimes, a rare MO is discovered—for example, a burglar who defecates at the crime scene.

Computers and Software

Computers and various types of software assist investigators. **Artificial intelligence** uses software and databases containing a variety of stored investigative information to analyze data to link crime scene evidence to a suspect. **Data mining** seeks useful information from a large amount of information or databases to develop patterns and anomalies. With enormous amounts of data (e.g., travel records, expense accounts, and telephone and e-mail records) available to investigators, data mining saves time. **Link analysis** is especially helpful with complex investigations. It also involves large amounts of information and seeks commonalties and relationships among, for example, people, organizations, and geographic locations.

Harold (2006: 66–72) emphasizes the value of data mining. He writes that an investigator, who is able to ask the right questions, and a computer programmer, who is able to translate the questions into effective code, can produce answers for investigations. He explains the case of an employee thief who stole valuables in an office building of 5,000 employees. The 10 victims had worked all hours of the day and night, weekends, and one holiday. The building could be entered only with an access card, and there were 65 access points. All entries were recorded in the access control computer. One and a half million access control records were used to form a database. Then a table was created containing the distinct dates, times, victims, and building locations of the thefts. With millions of records to review, there was only one query for the database: Who was in the building when all the thefts occurred? One employee was revealed who was targeted for further investigation and eventually prosecuted.

The **Identi-Kit**, which police have used for years, consists of hundreds of overlays of facial features (e.g., eyes, noses, chins) that a victim or witness selects so a drawing of the suspect can be created. Computer-aided identification software uses the same principle and stores more than 100,000 facial features.

Digital Investigations and Evidence

Digital evidence is electronically stored records or information located on a computer, server, storage network, or other media. Investigators are finding that the locations of digital evidence are varied. Besides computers, laptops, and personal digital assistants (PDAs), digital evidence may also be found in such devices as USB drives and other memory devices, digital cameras, cell phones, fixed telephony, answering machines, DVD players that contain hard drives, GPS units, and even a mouse that has features similar to a flash storage drive. Ordinary items such as watches and pens can also be suspect and hold memory cards. In addition, because of wireless technology, a wireless hard drive can be hidden almost anywhere (Smith, 2013: 181–184; Bartolomie, 2005: 5).

Maras (2012: 169–197) explains that hard drives in computers contain files such as e-mails, spreadsheets, calendars, password-protected and encrypted files, log files, and deleted files.

Another source is metadata, which is "data about data." Examples are the author of a document, the date and time it was created, and the last time it was saved, modified, or printed. Maras notes that metadata can yield considerable evidence. She writes that peripheral devices such as printers, copiers, scanners, and fax machines may also contain valuable information (e.g., logs of usage).

Donnelly (2005: 59–61) writes that iPods—portable music players—provide offenders with huge storage capacity as portable hard drives. These devices enable offenders to copy sensitive data. Donnelly describes one case in which police served a search warrant at the home of a suspect bookkeeper. As police collected computers, hard drives, and other related equipment, they noticed an iPod and it was also seized. Forensic examiners at a lab examined the suspect's equipment, except the iPod, and found no evidence of embezzlement. Careful to treat the iPod as any other external hard drive and to preserve its integrity, an examiner made a complete copy of the iPod's hard drive, to be used for examination. It contained two data partitions: one held music and the other held data files of illegal financial transactions (i.e., the evidence used for prosecution).

Interestingly, an iPod can be used as a mobile forensic device once it is configured and loaded with forensic software. Such a device avoids bringing bulky equipment into the field. Add-ons include voice-recording capability and a camera connection to download photos to the iPod for storage (Donnelly, 2005: 61–62).

Digital forensics is the application of investigative and scientific skills and specialized tools and software to examine digital media to collect evidence. Investigators preserve a digital crime scene by making a copy of the data and then search for evidence from such sources as e-mails and files, which may be hidden or deleted. Digital forensics is a growing field supporting investigators who seek evidence in cases involving Internet abuse, fraud, extortion, cyberattacks, theft of confidential information, intelligence, sexual harassment, child pornography, online defamation, breach of contract, negligence, insurance claims, and other areas. Acohido (2010) writes of the growth of the global cyberforensics industry. He notes that organizations need specialists to focus on data security and integrity and preserving and extracting digital records for litigation and regulatory audits.

The following list offers guidelines for digital evidence (Lang, 2005: 55–74; Mallery, 2005: 44–50):

- For the workplace, policies and procedures are important to ensure that employees know that the company owns workstations, company information, IT systems, data, and hard-copy files, and that monitoring and searches may occur.
- Avoid assigning the IT department to investigate because the investigation should be objective and the IT staff should not examine its own department.
- Begin cybercrime investigations internally instead of contacting public police immediately.
- Large entities often maintain a digital forensic department, separate from IT. Smaller concerns outsource.
- The International Association of Computer Investigative Specialists and the International Society of Forensic Computer Examiners both offer certifications that enhance professionalism and competence.

- Management must decide whether to contact police during the investigation. Cases involving embezzlement at a public company or child pornography are examples of cases requiring police notification.
- If an employee refuses to give up personal property (e.g., iPod) that contains evidence, seek assistance from public police. A search warrant may be the next step.
- Once items are seized, maintain a chain of custody. Company policies and procedures should ensure proper handling and preservation of digital evidence so it is not altered or destroyed.
- Opposing counsel in a case is sure to attack the evidence and how it was handled and examined. To comply with Federal Rules of Evidence (e.g., Rule 702), a digital forensic examiner must build case testimony from a foundation of documented science, accepted procedures, and reliability. The National Institute for Standards and Technology tests commercial forensic products, and the results can assist experts.

Rules involving cell phones, camera phones, and other personal electronic devices should be included in workplace policies and procedures and communicated to employees. Generally, security personnel cannot seize a personal device such as a cell phone that may contain evidence. Public police can be called for assistance, and they would follow such guidelines as listed next (Dunnagan and Schroader, 2006: 46–53):

- Public police must have a search warrant prior to seizing and searching a device, or obtain a signed consent form.
- Do not change the condition of evidence. If the cell phone is on or off, do not change it.
- Bring the device, power and other cables, and all accessory devices to a lab for processing. Use forensically sound software and tools and validate evidence. No one software package can examine all cell phones.
- Text messages that pass through a messaging center may remain on a server for only a few days, so process the device as soon as possible.
- Place a seized cell phone in a Faraday bag to block all wireless signals. This prevents data from being remotely erased.

Nagosky (2005: 1–9), of the FBI, writes about digital photos. He notes that film-based photos can be manipulated by, for example, selection of exposure times, or by crop and splice, whereby two different negatives are combined. To counter such manipulation, a negative can be requested. Digital photos are also subject to manipulation by, for example, adding and deleting items. Detection of manipulation is through, for example, density of the image based on light exposure and splice lines. Nagosky offers these recommendations for digital photos as evidence: preserve the original; concentrate on integrity of the image (i.e., chain of custody); store it on a storage device that can be written to only once and then is only readable; and limit access to files.

Another topic related to digital evidence is the eDiscovery amendments to the Federal Rules of Civil Procedure that became effective December 1, 2006. These rules address corporate electronically stored information (ESI) that may be subpoenaed in a civil case. Such information includes e-mail, voice mail, and data on computer drives. To avoid court sanctions and an adverse settlement, a corporate team (e.g., attorney, IT specialist, CSO) must adhere to the rules by identifying, preserving, and producing ESI. The rules require parties to meet as soon as possible on discovery of ESI and allow a request for limiting ESI discovery if sufficient detail and costs are provided as to why compliance is burdensome (Plante, 2007: 20).

Global Positioning System

Global Positioning System (GPS) is a navigation system consisting of satellites placed in orbit by the U.S. Department of Defense that is used by the military and available to the public. The satellites circle and transmit information to earth. A GPS device receives the information to compute the device's location. LeMere (2011) writes that investigators typically use GPS evidence in the form of "trackpoints" (i.e., a record in the GPS that pinpoints where it has been). The applications are varied, such as placement in a vehicle or asset, or on a person (e.g., parolee), and tracking can be accomplished on the Internet. LeMere emphasizes that GPS can assist with many types of investigations including accident reconstruction and search and rescue. He notes that these devices contain much more information than "trackpoints," as described next. A "track log" is a list of "trackpoints" that can be applied to retrace and navigate to previous positions. A "waypoint" is a site that a user enters into the device. A "route" consists of "waypoints" that the user visits in a particular order. "Smart" GPS devices are USB mass storage devices and may contain maps, photos, videos, and other data that a computer can store.

Surveillance

Surveillance, watching or observing, is an investigative aid used widely to acquire information. Among the kinds of cases in which surveillance is helpful are these examples: assembly line workers are suspected of stealing merchandise; an employee is suspected of passing trade secrets to another company; truck drivers, while on their routes, experience unexplainable losses between company facilities; and an employee, claiming to be unable to work because of an on-the-job accident, is observed building an extension on his home.

Two major kinds of surveillance are stationary and moving. *Stationary surveillance* requires the investigator to remain in one spot while observing—for example, an investigator sitting in an auto watching a suspect's house. This type can be tedious and frustrating. In *moving surveillance*, investigators follow a suspect—for example, tailing a truck driver whose cargo was "lost." GPS is an aid to moving surveillance.

During surveillance, an investigator must be careful not to attract attention. The person being watched usually has the advantage and can attempt to lose the investigator through a variety of quick moves (e.g., going out a back door, driving through a red light). Therefore, the investigator must blend into the environment to prevent detection.

Another type of surveillance is audio surveillance. This includes wiretapping and eavesdropping, which are restricted by law.

Equipment used during surveillance includes binoculars, telescopes, communication equipment, cameras, listening devices, video and audio recorders, and global positioning system (GPS) tracking devices. Investigators must keep informed about these devices and related legal restrictions on usage.

Concealed pinhole lens cameras are a popular method of surveillance. What is your opinion of this technique?

Information Accuracy

An investigation is essentially an information-gathering process. The accuracy of the information not only reflects on the investigator but also has a direct bearing on the consequences of the investigation. The following guidelines are helpful in obtaining accurate information:

1. Double-check information whenever possible.
2. Ask the same questions of several people. Compare the results.
3. To check on the reliability of a source, ask questions to which you know the answers.
4. Cross-check information; for example, if you have a copy of a person's employment application and college transcript, cross-check name, date of birth, and so on.
5. Read information back to a source to check for accuracy.
6. If possible, check the background of a person providing information or check the accuracy of a records system.
7. Maintain accurate notes, records, photographs, and sketches; do not depend on memory for details.
8. If you are unable to write notes or a report because, for instance, you are driving a vehicle, record the information on a tape recorder.
9. Provide adequate security for information and records to prevent tampering or loss.

Report Writing

Report writing usually begins after the investigator has invested time and energy in collecting sufficient information on the basic investigative questions. How well these reports are prepared will have a definite impact on the investigator's career. Many supervisors get to know their subordinates more through reports than from any other means of communication. Furthermore, many supervisors consider report writing a major skill when promoting investigators.

Reports have a variety of uses aside from punitive results. They are used by management to analyze critical problems (e.g., thefts or accidents). Summations of many reports can assist planning and budgetary efforts. Reports also may be used in litigation.

Investigators usually record information in a small notebook before completing a report. An investigator has many thoughts in mind during an investigation that prevents the report from being written as the investigation proceeds. Eventually, the investigator has an overall view of the incident; this assists in the development of an outline that improves the structure of the report.

Many investigators use standard reports. These reports are formulated by management, often computer-based, and guide investigators in answering important questions. A typical standard report begins with a heading that includes the type of incident, date, time, and location. Next is a list of persons involved in the incident along with their addresses, telephone numbers, ages, and occupations. Another section can include a list of evidence. The narrative, sometimes called "the story," follows, written in chronological order. The end of the report contains a variety of information such as the investigator's name and the status of the investigation. Diagrams and photographs may be attached. Report characteristics vary depending on need.

During report writing, the investigator should get to the point in easily understood language. An impressive vocabulary is not an asset to a report. Neatness and good grammar are important. Supervisors often complain about poor narratives written by subordinates. Let us look at some blunders that have reflected on the investigator:

- When the employee was approached by loss prevention staff he had a switchblade he had bought in his lunch box.
- A telephone pole of manufacturing plants within our corporation showed that 15 percent of employees were ignoring loss prevention rules.
- The woman caused the loss because her newborn son was branded as illiterate.
- The sick employee was honestly in bed with the doctor for two weeks even though he did not give her any relief.

Career: Investigations

The process of investigation is an important function in both the public and private sectors. It is a very broad field and includes many subspecialties. Investigators use a variety of tools and techniques such as interviewing, evidence collection and processing, physical and technical surveillance, computer forensics, database searches, and crime analysis algorithms. Like most security measures, an effective investigations program serves both as a deterrent to crime and a response once a crime has been committed.

Entry-level management positions generally prefer a four-year degree. Criminal justice and criminology are popular, but a business degree is acceptable. Two-year degrees can help the entry-level aspirant. Depending on the position, no experience may be required, but specialty areas usually require at least one to two years in the respective area. Experience from police or security investigations is taken into consideration.

Mid-level management positions, requiring expertise in multiple investigative and business disciplines, generally require a degree in an appropriate discipline, as well as five or more years of demonstrated success in the field.

Professional certifications such as the Professional Certified Investigator (ASIS International, 2012), the Certified Fraud Examiner (Association of Certified Fraud Examiners, 2012), and the Certified Computer Examiner (International Society of Forensic Computer Examiners, 2012) are often desired as an indicator of professionalism and qualifications.

Modified from source: Courtesy of ASIS International (2005). "Career Opportunities in Security." www.asisonline.org.

Refer to Chapter 2 under "Occupational Outlook" for additional career information.

Testimony

Security practitioners periodically testify in depositions or in court. A **deposition** is a pretrial discovery method whereby the opposing party in a case asks questions of the other party (e.g., victim, witness, expert) under oath, usually in an attorney's office, and while a word-for-word transcript is recorded. Depositions help to present the evidence of each side of a case and assist the justice system in settling cases before the expensive trial stage. Most civil and criminal cases never make it to trial.

Well-prepared testimony in both criminal and civil cases can be assured most readily by the following suggestions. (See Table 10-1 for additional advice.)

1. Prepare and review notes and reports. Recheck evidence that has been properly labeled and identified. Confer with attorney.
2. Dress in a conservative manner, if not in uniform. Appear well groomed.
3. Maintain good demeanor (conduct, behavior). Do not slouch, fidget, or put hands on your face. Do not argue with anyone. Remain calm (take some deep breaths without being obvious).
4. Pause and think before speaking. Do not volunteer information beyond what is requested. Never guess. If you do not know an answer, say so.
5. If you bring notes, remember that the opposing attorney can request that the notes become part of the evidence. Recheck notes to prevent any unwanted information from entering the case.
6. Request feedback from associates to improve future performance.

Search the Internet

Refer to the websites in this chapter to check the variety of personal and business information available.

Check "(CompanyName)sucks.com" sites to view sites that are in contention with corporate America.

Here are additional sites relevant to this chapter:

ASIS International, Professional Certified Investigator: www.asisonline.org/certification/pci/pciabout.xml
Association of Certified Fraud Examiners: www.acfe.com
International Association of Computer Investigative Specialists: www.iacis.com
International Society of Forensic Computer Examiners: www.isfce.com
Overseas Security Advisory Council (OSAC): www.osac.gov

Table 10-1 Brief Review of Common Tactics of Cross-Examination

Counsel's tactic	Example	Purpose	Officer's response
Rapid-fire questions	One question after another with little time to answer.	To confuse you; attempt to force inconsistent answers.	Take time to consider the question; be deliberate in answering, ask to have the question repeated, remain calm.
Condescending counsel	Benevolent in approach, over-sympathetic in questions to the point of ridicule.	To give the impression that you are inept, lack confidence, or may not be a reliable witness.	Firm, decisive answers, asking for the questions to be repeated if improperly phrased.
Friendly counsel	Very courteous, polite; questions tend to take you into his confidence.	To lull you into a false sense of security, where you will give answers in favor of the defense.	Stay alert; bear in mind that the purpose of defense is to discredit or diminish the effect of your testimony.
Badgering, belligerent	Counsel staring you right in the face, shouts "That is so, isn't it, officer?"	To make you angry so that you lose the sense of logic and calmness. Generally, rapid questions will also be included in this approach.	Stay calm; speak in a deliberate voice, giving prosecutor time to make appropriate objections.
Mispronouncing officer's name; using wrong rank	Your name is Jansen, counsel calls you Johnson.	To draw your attention to the error in pronunciation rather than enabling you to concentrate on the question asked, so that you will make inadvertent errors in testimony.	Ignore the mispronunciation and concentrate on the question counsel is asking.
Suggestive question (tends to be a leading question allowable on cross-examination)	"Was the color of the car blue?"	To suggest an answer to his or her question in an attempt to confuse or to lead you.	Concentrate carefully on the facts, disregard the suggestion. Answer the question.
Demanding a yes or no answer to a question that needs explanation	"Did you strike the defendant with your club?"	To prevent all pertinent and mitigating details from being considered by the jury.	Explain the answer to the question; if stopped by counsel demanding a yes or no answer, pause until the court instructs you to answer in your own words.
Reversing witness's words	You answer, "The accident occurred 27 feet from the intersection." Counsel says, "You say the accident occurred 72 feet from the intersection?"	To confuse you and demonstrate a lack of confidence in you.	Listen intently whenever counsel repeats back something you have said. If counsel makes an error, correct him or her.
Repetitious questions	The same question asked several times slightly rephrased.	To obtain inconsistent or conflicting answers from you.	Listen carefully to the question and state, "I have just answered that question."

Table 10-1 Brief Review of Common Tactics of Cross-Examination

Counsel's tactic	Example	Purpose	Officer's response
Conflicting answers	"But Officer Smith, Detective Brown just said ..."	To show inconsistency in the investigation. This tactic is normally used on measurements, times, and so forth.	Remain calm. Conflicting statements have a tendency to make a witness extremely nervous. Be guarded in your answers on measurements, times, and so forth. Unless you have exact knowledge, use the term "approximately."
Staring	After you have answered, counsel stares as though there were more to come.	To have a long pause that one normally feels must be filled, thus saying more than necessary. To provoke you into offering more than the question called for.	Wait for the next question.

Source: Reproduced from *The Training Keys* with permission of the International Association of Chiefs of Police.

Case Problems

10A. The Safeparts Company, a California-based distributor of automobile parts, recently conducted an inventory at its Los Angeles facility that showed inventory losses of $850,000. Concerned about the losses, management at Safeparts sought immediate action. You are a partner at Klein and Smith Loss Prevention Associates, a consulting firm specializing in loss problems. Safeparts executives contact you for assistance. A meeting is arranged. After competition with two other security and loss prevention firms, Safeparts executives decide on a two-month contract for your firm's services. You are in charge. What are your specific plans and actions?

10B. You are senior investigator for the Bolt Corporation, which is a top 100 corporation with large holdings in electrical supplies. Because you have an excellent record and 11 years of varied investigative experience with Bolt, you are selected by the director of loss prevention to train five newly hired college-educated investigators. The director stresses that you will design a 105-hour training program to span three weeks. After three weeks, the investigators will be assigned to various divisions within Bolt, where they will receive specialized training while working with experienced investigators. The director states that your typed curriculum design is due tomorrow for a 4 P.M. loss prevention meeting. She requires that you list the topics, hours for each topic, and why the particular topics and hours were chosen.

10C. As a corporate investigator based in the United States, you are assigned to investigate a serious internal theft problem at a company industrial plant near Reynosa, Mexico. How do you proceed with this investigation? What resources do you rely on? What are your plans to solve this case?

10D. As a corporate investigator you and your supervisor have been contacted by the human resources department about a female employee who claims that a male employee is sending her sexually suggestive e-mails and pornography via the corporate IT system. How do you investigate this case?

References

Accardi, N. (2010). Social networking: A double-edged sword. *Loss Prevention, 9* (July–August)

Acohido, B. (2010). *Security needs drive cyberforensics industry. USA TODAY.* (November 23).

Anderson, T. (2010). Pretexting: What you need to know. *Security Management, 54* (June).

ASIS International. (2012). Board Certifications in Security. <www.asisonline.org/certification/index.xml> retrieved April 12, 2012.

Association of Certified Fraud Examiners. (2012). Membership & Certification. <www.acfe.com> retrieved April 12, 2012.

Ast, S. (2011). Investigating abroad. *Security Management, 55* (June).

Bartolomie, J. (2005). Hide and seek. *TECHbeat* (Summer).

Collins, P., et al. (2005). *The ASIS foundation security report: Scope and emerging trends.* Alexandria, VA: ASIS International.

Daniels, R. (2006). Compliance forcing firms to take investigative stance. *Security Director News, 3* (March).

Deslauriers-Varin, N., et al. (2011). Confessing their crime: Factors influencing the offender's decision to confess to police. *Justice Quarterly, 28* (February).

Donnelly, D. (2005). iPods sing for investigators. *Security Management, 49* (March).

Dunnagan, K., & Schroader, A. (2006). Dialing for evidence: Finding and protecting forensic treasures in mobile phones. *Law Officer Magazine* (January/February).

Federal Bureau of Investigation. (2010). Crime in the United States 2009. <www2.fbi.gov/ucr/cius2009/documents/clearancetopic.pdf> retrieved August 28, 2011.

Gardner, T., & Anderson, T. (2007). *Criminal evidence: Principles and cases* (6th ed.). Belmont, CA: Thomas Higher Education.

Harold, C. (2006). The detective and the database. *Security Management, 50* (March).

Hetherington, C. (2010). Investigating in facebook and other parlor tricks. *Loss Prevention, 2* (March–April).

International Society of Forensic Computer Examiners. (2012). Forensics certs: Certified computer examiner. <www.isfce.com> retrieved April 12, 2012.

Kieffer, S., & Sloan, J. (2009). Overcoming moral hurdles: Using techniques of neutralization by white-collar suspects as an interrogation tool. *Security Journal, 22*

LaBruno, L. (2010). The risks of a bad admission. *Loss Prevention, 9* (July–August).

Lang, D. (2005). Dos and don'ts for digital evidence. *Security Management, 49* (June).

LeMere, B. (2011). Enhancing investigations with GPS evidence. *Digital Forensic Investigator News* (April 13)

Mallery, J. (2005). Cyberforensics: The ultimate investigative tool. *Security Technology & Design, 15* (December).

Maras, M. (2012). *Computer forensics: Cybercriminals, law, and evidence.* Sudbury, MA: Jones & Bartlett Pub.

Matsumoto, D., et al. (2011). Evaluating truthfulness and detecting deception. *FBI Law Enforcement Bulletin, 80* (June).

Nagosky, D. (2005). The admissibility of digital photographs in criminal cases. *FBI Law Enforcement Bulletin, 74* (December).

Overseas Security Advisory Council. (n.d.). About OSAC. <www.osac.gov/About/index.cfm> retrieved February 10. 2007.

Plante, W. (2007). New rules for your electronically stored information: FRCP's eDiscovery rules. *Security Technology & Design, 49* (April).

Quirke, J. (2012). Social media and the workplace. *Security Management, 56* (February).

Ray, D. (2000). When bad things happen to good businesses. *Security Management, 44* (October).

Reibold, R. (2005). The hidden dangers of using private investigators. *South Carolina Lawyer, 17* (July).

Reino, N. (2011). 10 Things your agency must know about social media. *Law Officer* (January 11) <www.lawofficer.com> retrieved January 13, 2011.

Roman, G. (2005). Ten things to avoid during workplace investigations. *RJL Newsletter* <www.rothgerber.com/newslettersarticles/le0054.asp> retrieved March 7, 2005.

Smith, R. (2013). *Elementary information security*. Burlington, MA: Jones & Bartlett Pub.

U.S. Department of Justice, (1976). *Report of the task force on private security*. Washington, D.C.: U.S. Government Printing Office.

Van Nostrand, G., & Luizzo, A. (1995). Investigating in a new environment. *Security Management* (June).

Virgillo, A. (2010). Are we creating liars? Part 2: Performance measures and controls to keep investigators honest. *Loss Prevention, 9* (July–August).

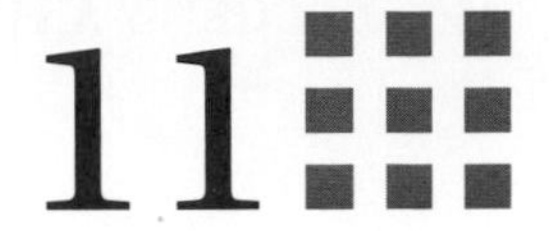

11 Accounting, Accountability, and Auditing

OBJECTIVES

After studying this chapter, the reader will be able to:

1. Define and explain accounting, accountability, and auditing
2. Describe how accountability is applied to the areas of purchasing and inventory
3. Describe the functions of auditors
4. Explain the problem of fraud and the Sarbanes-Oxley Act of 2002
5. Define and explain the importance of governance, risk management, and compliance

KEY TERMS

- accounting
- accountability
- auditing
- purchase requisition
- purchase order
- invoice
- receiving report
- kickback
- inventory
- shrinkage
- periodic inventory system
- perpetual inventory system
- attest function
- Association of Certified Fraud Examiners
- forensic accounting
- internal control questionnaire
- American Institute of Certified Public Accountants
- fraud
- Enron Corporation
- Ponzi scheme
- Sarbanes-Oxley (SOX) Act of 2002
- Public Company Accounting Oversight Board
- governance
- risk management
- compliance

Introduction

Accounting, often referred to as the language of business, is concerned with recording, sorting, summarizing, reporting, and interpreting data related to business transactions. Accounting information assists executives, auditors, investors, regulators, and others in decision making. Virtually every type of concern requires accounting records. Bookkeepers (also referred to as accounting technicians or accounting clerks) perform the day-to-day recording of financial data such as sales, income, purchases, and accounts payable. Accountants design accounting systems, prepare and interpret reports based on the financial transactions recorded by bookkeepers, and prepare and file reports with government agencies.

Accounting is the language of business.

Accountability defines a responsibility for and a description of something. For example, John Smith is responsible (i.e., is held accountable) for all finished products in a plant, and he maintains accurate records (i.e., an inventory) of what is in stock. Another example would be a loss prevention officer keeping a log of people entering and leaving a restricted area.

Auditing is an examination or check of something; the major purpose of an audit is to uncover deviations. An audit can be simple or complex. For example, a loss prevention officer audits a CCTV system to ensure that it is working properly. Alternatively, an auditor examines the financial records of a company and reports that they are fair, reliable, and conform to company policies and procedures.

Accounting

Within a business, for example, the accounting department has control over financial matters that are vital to business operations. Common components of an accounting department are cashiering operations, accounts receivable, accounts payable, payroll, and company bank accounts. Each component of an accounting department has the responsibility for maintaining records that are scrutinized by management to ascertain the financial position of the business. Without adequate loss prevention strategies or controls in financial transactions and records, organizations could not survive.

Potential losses are possible throughout the accounting department. A cashiering operation must be protected, not only from burglary and robbery, but also from employee theft. Accounts receivable must be protected from opportunities that allow employees to destroy bills and pocket cash. Accounts payable also needs protection; employees in collusion with supply company employees have been known to alter invoices to embezzle money. A scheme by some payroll clerks is to maintain fictitious employees on the payroll and cash their paychecks.

A record of an individual transaction by a bookkeeper does not have as much impact as the summation of transactions in a financial statement or business report prepared by an accountant (see Table 11-1).

Accounting statements assist management in answering questions and making decisions.

What is the financial condition of the organization?
What is the financial value?
Was there a profit or loss?
Which part of a firm is doing well (or poorly)?
How serious are losses from hazards?

Because security and loss prevention practitioners often investigate financial matters and manage a budget, they are well advised to study accounting at the college level to prepare for their careers.

Table 11-1 Financial Statements of Two Separate Companies

Trico Corporation Balance Sheet June 30, 20_

Assets		**Liabilities**		
Cash	4,000	Accounts payable	44,000	
Accounts receivable	100,300	Notes payable	100,000	
Inventory	100,000			144,000
Equipment	34,000			
Land	80,000	**Capital**		
Buildings	300,400			
	618,700	Preferred stock	74,700	474,700
		Common stock	400,000	
Total assets	618,700	Total liabilities and capital		618,700

Simple examples of an income statement and a capital statement follow.
Note that "expenses" and "net income" are two additional major categories of accounting besides assets, liabilities, and capital.

Quality Loss Prevention Service Income Statement for Month Ended October 31, 20_

Sales and service		11,800
Operating expenses:		
Salary expenses	6,000	
Supplies expense	1,100	
Rent expense	1,400	
Miscellaneous expense	1,300	–9,800
Net income		2,000

Quality Loss Prevention Service Capital Statement for Month Ended October 31, 20_

Capital, October 2, 20_		10,000
Net income for the month	2,000	
Less withdrawals	−1,000	
Increase in capital		1,000
Capital, October 31, 20_		11,000

Accountability

The definition of formal accountability points to the documentation or description of something. Informal accountability usually is verbal and results in no documentation; for example, a loss prevention manager asks a subordinate if a fire extinguisher was checked (audited). The subordinate states that it was audited. Thus, a basic audit of a loss prevention device is accomplished. What if two weeks pass, a fire takes place near the particular fire extinguisher, and it is found to be inoperable? An employee who tries to extinguish the fire with the inoperable extinguisher complains to management. Superiors ask the loss prevention manager if the extinguisher was checked. The manager states that it was audited. The superiors ask for documentation to support the statement. Because of the verbal accountability, no record exists.

From that point on, the loss prevention manager realizes the value of formal accountability and develops an excellent system of records. Practitioners are familiar with the initials "CYA" ("cover your ass"), which refers to taking steps, such as formal accountability and documenting and communicating vulnerabilities to senior management, so when incidents occur, the practitioner has some foundation of support.

The importance of accountability must not be underestimated. It is a key survival strategy. Documentation can result from many types of loss prevention activities. Examples are a variety of investigative reports (e.g., crimes, accidents); security surveys; security system maintenance; alarm activations; visitor logs; crime prevention, fire protection and life safety plans; meetings; policies and procedures; and training. As well as assisting a loss prevention practitioner when supporting a contention, documentation can assist in planning, budgeting, preparing major reports, and providing general reference.

Accountability is a key survival strategy.

Purchasing

Because procedures vary and various types of computer software are available to enhance purchasing systems, a generalized approach to purchasing is presented here. Four forms are discussed in the subsequent purchasing system: purchase requisition, purchase order, invoice, and receiving report.

When a company orders merchandise, equipment, or supplies, for example, the order should be documented to avoid any misunderstandings. Suppose a maintenance department head at a plant seeks to order an item. Generally, the documentation process begins when the order is written on a standard form known as a **purchase requisition**. This form may also be completed in the company computerized accounting information system. The purchase requisition lists, among other things, the originator (who placed the order), the date, the item, a description, justification for need, and cost. Once the originator completes the purchase requisition and makes a copy for filing, superiors approve the purchase and sign the requisition. Then, the purchasing department reviews the purchase requisition and selects the best vendor. The purchasing staff completes a prenumbered purchase order. Copies of the purchase order are sent to the originator, the receiving department, and the accounts payable department; the purchasing department retains the original purchase order. The **purchase order** includes information on the originator, the item, quantity, possibly an item code number from a vendor catalog, and the cost. The purchase order is sent to the vendor. Upon fulfillment of the order, the vendor sends an **invoice** to the buyer's accounts payable department. An invoice contains the names and addresses of the buyer and the vendor, cost, item, quantity, date, and method of shipment.

When the accounts payable department receives the invoice, it checks the invoice for accuracy by comparing it with a copy of the purchase order. Cost, type of item, proper quantity, and address of buyer are checked.

The buyer's receiving and purchasing departments receive copies of the invoice to check them for accuracy. To decrease the possibility of mistakes (or collusion), the purchase order and invoice sent to the receiving department may have the number of items deleted. When the merchandise arrives, the receiving person records the number of items and type, and checks for irregularities (e.g., damage). This form often becomes a **receiving report**. Copies are sent to the purchasing and accounts payable departments. The purchasing department compares the receiving report with the invoice. The accounts payable department makes payment after examining the purchase order, invoice, and the receiving report. These three documents and a copy of the check constitute the inactive file for this purchase (see Figure 11-1).

This purchasing system may appear complicated; however, without such accountability, losses can increase. For example, in one company, accounting employees in collusion with outside supply company employees altered records so items paid for were never delivered, but sold on the black market for illegal gain (Mann and Roberts, 2001: 470).

Another widespread vulnerability in purchasing results from a **kickback**. This means that the purchaser receives something of value from the seller for buying the seller's product or service. Losses occur if the product or service is inferior and overpriced in comparison to the competition. For example, in a secret deal, John Doe Forklift Company agrees to pay Richard Ring, purchaser for Fence Manufacturing, $1000 cash for each forklift purchased at an inflated cost. After the forklifts are delivered, it is discovered that the forklift tires are too smooth for the outside gravel and dirt grounds of the manufacturing company. With limited traction, the forklifts frequently are stuck, and employees are unable to work until delivery trucks return and pull the forklifts free. The losses include both cash and lost time.

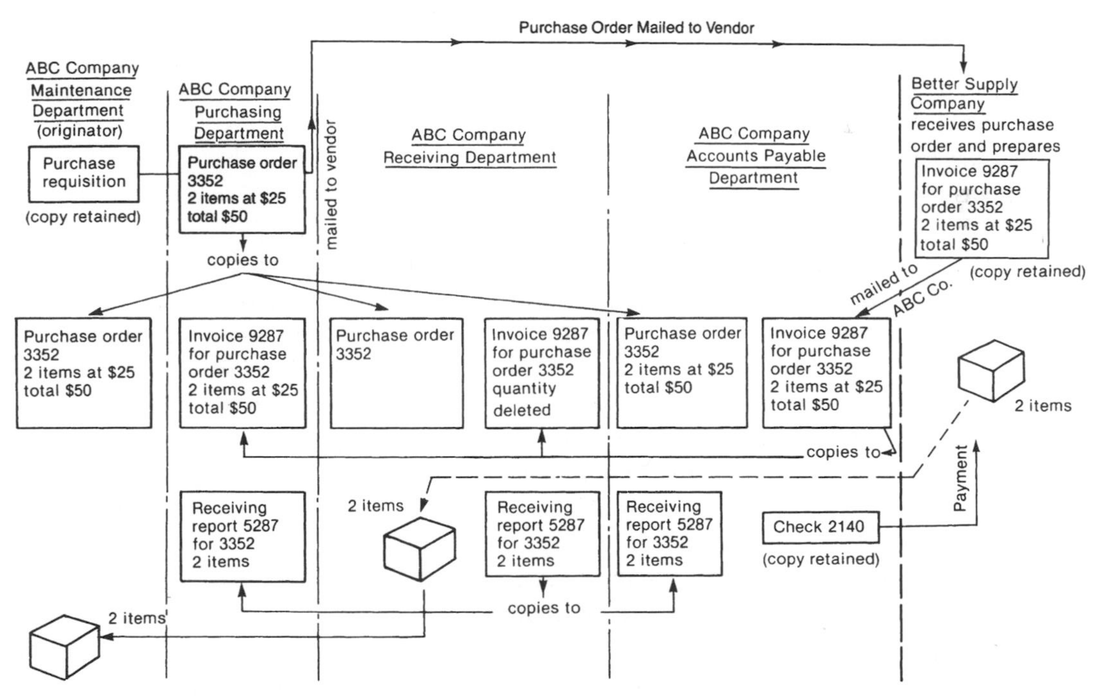

FIGURE 11-1 Accountability and paper trail for purchase of two items by ABC company.

Accounting controls and loss prevention strategies in purchasing include the following suggestions:

1. Centralize all purchasing through a purchasing department.
2. Maintain accountability through documents (standard forms), signatures, and carefully designed computer software.
3. Separate duties and responsibilities so that each person and department can check on the others' work.
4. Test by deliberate error.
5. Use unalterable paper to prevent alterations or erasures.
6. Prenumber purchase order forms (and other forms when needed).
7. Conduct loss prevention checks without notice.
8. Conduct periodic audits.
9. Scrutinize the purchasing department to prevent favoritism and kickbacks. Use competitive bidding.
10. Prohibit gifts or favors from vendors (sellers).
11. Screen applicants for employment.
12. Develop clear policies and procedures.

Inventory

In a wholesale or retail business, merchandise is continually purchased and sold. This sale of merchandise is the primary source of revenue. A substantial amount of a business's resources is invested in saleable merchandise, and this merchandise is the largest asset. Therefore, this asset must be protected. The term **inventory** includes merchandise for sale, raw materials, and unfinished goods. Inventory is reported on the balance sheet as an asset.

Shrinkage is the amount of merchandise that has disappeared through theft, has become useless because of breakage or spoilage, or is unaccounted for because of sloppy recording. It is often expressed as a percentage. Levy and Weitz (2001: 548) define shrinkage as "the difference between the recorded value of inventory (at retail prices) based on merchandise bought and received and the value of the *actual* inventory (at retail prices) in stores and distribution centers divided by retail sales during the period." For example, if accounting records indicate the inventory should be $1,500,000, the *actual* count of the inventory reveals $1,236,000, and sales were $4,225,000, the shrinkage is 6.2 percent, or ($1,500,000 – $1,236,000)/$4,225,000.

In many businesses, shrinkage of three percent or more is a serious loss problem. Loss prevention managers frequently express the objective of their job as lowering shrinkage. An accurate measurement of shrinkage depends on the quality of the inventory system; both have a definite impact on the loss prevention program and its manager.

Two primary inventory systems are the periodic and the perpetual systems. The **periodic inventory system** results in a physical count of merchandise only at specific intervals, usually once per year. When this system is used, daily revenue from sales is recorded in accounting records, but no transaction is recorded to adjust the inventory account to reflect the fact that a sale was made. The periodic system makes it difficult to measure shrinkage accurately. To make matters worse, when a monthly or quarterly financial statement is necessary for a particular business using the periodic inventory system, managers sometimes estimate the inventory without taking a physical count.

The **perpetual inventory system** uses accounting records that maintain an up-to-date inventory count. These systems typically are computerized. Handheld microcomputer technology and point-of-sale (POS) computers capture data through bar code scanning. RFID is increasingly replacing bar codes. In addition to recording daily revenue from sales, an individual inventory record is maintained for each type of merchandise sold, which enables a continuous count. Thus, the accounting records reflect cost of goods sold and the inventory quantity. This information provides a better opportunity to measure shrinkage than that available with the periodic system.

To increase the accuracy of an inventory and the shrinkage statistic, these strategies are recommended:

1. Maintain a careful inventory system.
2. Establish accountability.
3. Standardize forms and procedures for the count.
4. Make sure employees can count accurately.
5. If possible, do not subject employees to extensive inventory counts at any one time.
6. Automate the process by using handheld microcomputer technology that captures data.
7. Conduct surprise counts of a sample of the merchandise at erratic time intervals. Compare manual counts with computer data.
8. If possible, require prenumbered requisition forms for merchandise taken out of inventory.
9. Prohibit unnecessary people (e.g., truck drivers, service people, and other employees) from entering merchandise storage areas.
10. Use an undercover investigator to participate in the inventory count.
11. The loss prevention manager should have an opportunity to examine the methods used to formulate the shrinkage statistic, especially because it will reflect on him or her and on the loss prevention program.

Radio Frequency Identification: Great Potential, But Vulnerable

RFID was applied as early as World War II when it was used to identify aircraft. Today, it has wide applications beyond protecting library books and merchandise at retail stores. The system has three basic components: tags, readers, and a host computer.

RFID tags contain tiny semiconductor chips and antennas. Many tags appear as paper labels, while others are embedded into such items as containers or wristbands. Each tag is programmed with a unique identifier that permits wireless tracking of the tag and the object holding the tag. Tags can hold a large amount of data, including serial numbers, time stamps, travel history, and technical data. Similar to television and radio, RFID uses various frequency bands. A tiny battery powers active tags, and passive tags are powered by a reader that "wakes up" the tag when it is within range of the reader.

Readers contain an antenna that communicates with tags and an electronics module networked to a computer. The reader performs security functions including encryption/decryption and user authentication. Software connects the RFID system to the IT system where supply chain management or asset management databases are located.

Supply chains benefit from RFID technology through increased efficiency and decreased labor costs and losses. To illustrate, readers installed at loading dock doors detect tags on pallets of merchandise passing by. The reader signals the tags to transmit their identities and other data to the reader that forwards the data to a computer. The computer then credits or debits the inventory depending on whether the merchandise is entering or leaving.

The applications of RFID are broad and include tracking of property, evidence, passports, visas, inmates in prison, and visitors at facilities. RFID may prove more efficient than bar codes for recording, locating, and tracking. RFID technology is also applicable to access controls (National Law Enforcement and Corrections Technology Center, 2005: 6–8).

RFID is not without vulnerabilities. Johnston and Warner (2005: 116), who conducted tests on RFID at Los Alamos National Laboratory, warn that low-end tags can be counterfeited and that readers can be tampered with, replaced, or controlled remotely. Tags can also be removed from one object and placed on another. RFID signals can be blocked or jammed. Organizations should test these systems at unpredictable times. High-end systems that use cryptography, challenge-response protocols, rotating passwords, or tamper detection technology afford increased protection but are used infrequently. Another vulnerability is that a tag could contain a software virus that, when scanned by a reader, could spread through an RFID system.

Auditing

Auditors

An auditor examines business accounting records to check for irregularities. These irregularities may include (1) deviations from generally accepted accounting methods, (2) errors, and (3) criminal activity. Organizations typically employ internal auditors. External auditors are also important. During an audit of financial records by an independent (external) auditor, known as a certified public accountant (CPA), guidance is provided by state and federal statutes, court decisions, a contract with the client, and professional standards as established by Generally Accepted Auditing Standards and Generally Accepted Accounting Principles. Because it is impossible to check every financial record and transaction, a CPA narrows an audit to certain records, such as financial reports and areas where problems are common to the particular concern. How accounting data are recorded and summarized is frequently studied.

At times, a CPA may encounter misleading financial information that attempts to make a business look better than its true financial position. The misleading information often is an attempt by management to attract investors. To counter this problem, cautious investors are more likely to favor a business that has had an audit by an outside independent CPA, as opposed to no audit or one performed by an internal auditor.

When an independent CPA completes an audit, a report is prepared. If a business's financial records are dependable and credible, then the CPA expresses this favorable opinion in the audit report. This is known as the **attest function**.

CPAs, like other skilled professionals, are liable for damages proximately caused by their negligence. A CPA is liable to a client when he or she negligently fails to detect or fraudulently conceals signs that an employee of the client is embezzling. In addition, the CPA is liable for not detecting and reporting to the client that internal audit controls are lax (Twomey et al., 2001: 926).

Public (e.g., federal, state, and local) and private investigation practitioners have expanded their competency in accounting. This is in response to increased investigations into the white-collar crime arena.

Cross-training can be used to reduce the knowledge gap between auditor and criminal investigator. Cross-training involves the auditor being trained in criminal investigation and the criminal investigator being trained in auditing. An auditor's training could include criminal law, evidence, interviewing, and interrogation. A criminal investigator's training could include accounting principles and procedures and auditing. Both should have training in IT systems and related investigative methods.

The amount of digital financial data that must be analyzed during fraud investigations can be overwhelming and the number of financial transactions can amount to thousands for individuals and millions for businesses. However, well-trained fraud investigators apply numerous tools and methods to enhance efficiency. Computers contain standard office software used by businesses. This software contains metadata such as logs of changes to financial transactions. In addition, various commercial products are available to investigators to analyze financial data. Financial data can be loaded into a spreadsheet program of thousands of rows of data. Databases offer another tool to analyze data sets. Hybrid applications combine spreadsheet and database programs to perform tasks such as linking multiple data sets for analysis, creating subsets of tables from searches and queries, and applying formulas to aid analysis. Examples of clues investigators search for include looking for transactions occurring on non-business days (e.g., weekend), transactions with rounded amounts (since many businesses price products under the next dollar amount, say, $249.99), number of payments to vendors beyond the typical twelve times per year (i.e., monthly), duplicate invoice numbers and check numbers, and invalid social security numbers in payroll data (Kardell, 2011: 1–7).

The **Association of Certified Fraud Examiners** promotes professionalism, training, and certification (CFE). Bodnar and Hopwood (2004: 105) write: "**Forensic accounting** is one of several terms that is used to describe the activities of persons who are concerned with preventing and detecting fraud. The terms 'fraud examiner,' 'fraud auditor,' and 'loss prevention professional' are also descriptive of this type of activity."

Internal Control Questionnaire

A popular and convenient way to conduct an audit is through an **internal control questionnaire**. These questionnaires are used by public accounting firms, internal audit departments, and other organizations that are involved in reviews of internal controls (Bodnar and Hopwood, 2004: 131–132). A typical questionnaire contains a list of questions to remind the person conducting the audit to focus attention on specific areas of concern. Questionnaire results provide feedback that help to pinpoint and correct deviations and deficiencies. Here are sample questions from the **American Institute of Certified Public Accountants** (1978: 54) for a small business. Many of the questions are relevant to other organizations.

- Are accounting records kept up to date and balanced monthly?
- Are monthly or quarterly financial reports available to the owner?
- Are the personal funds of the owner and his or her personal income and expenses completely segregated from the business?
- Does the company practice "separation of functions" (e.g., separate check preparation from check authorization) so accountability is assigned and each employee can check on the other?
- Are employees rotated periodically among financial jobs and are they required to take vacations?
- Are employees who handle funds bonded?
- Do two different people reconcile the bank records and make out the deposit slip?
- Are prenumbered checks used?
- Is the owner's signature required on checks?
- Does the owner review the bank reconciliation?
- Does the owner never sign blank checks?
- Do different people reconcile the bank records and write the checks?
- Is credit granted only by the owner?
- Is the person responsible for inventory someone other than the bookkeeper?
- Are periodic physical inventories taken?
- Are perpetual inventory records maintained?
- Are there detailed records available of property assets and allowances for depreciation?
- Does someone other than the bookkeeper always do the purchasing?
- Does the owner approve payroll checks?

Endless Techniques of Fraud by Employees

Hubbs (2011: 83–84) describes various techniques applied by employees to commit fraud. In reference to expense report fraud, he explains that two employees on company business may share a motel room and both submit the room receipt for reimbursement. Alternatively, an employee may register for five nights at a hotel, stay one night, register at a less expensive hotel for the other four nights, and submit the original registration document for reimbursement. Websites are available that enable employees to create bogus receipts for hotels, meals, and so forth. Another scheme occurs when an employee fills an online shopping cart with products or services, prints the screen showing the shopping cart, does not place the order, and seeks reimbursement. In reference to benefits fraud, an employee may create bogus transcripts and diplomas for training and education reimbursements. The techniques of fraud are endless. Hubbs recommends policies, increased reviews of expense reports, and analytical tools

IT Staffers Influenced to Go Bad

In what may be a trend, the U.S. Securities and Exchange Commission (SEC), which enforces laws to reduce fraudulent financial reporting, is increasingly charging IT directors with securities fraud. Sensormatic Electronics Corp., for example, was in trouble because its home security systems firm had its IT personnel roll back computer clocks so sales could be booked sooner to inflate revenue figures. In another company, Bio Clinic Corp., more than 400 invoices with a value of $6 million, which had already been paid, were added into the ledger a second time (to falsify revenue) by reprogramming the accounting software. The SEC notes that IT directors are responsible for the accuracy and integrity of the documents and data generated by a company computer system. In addition, they should know if unauthorized changes have been made in general ledger, accounts receivable, and other accounting software (Nash, 1999: 20).

Fraud

Fraud is a broad term that includes a variety of offenses that share the elements of deceit or intentional misrepresentation of fact, with the intent of unlawfully depriving a person or organization of property or legal rights. It is a multi-billion-dollar problem. Although subsequent paragraphs emphasize fraud by top executives, fraud can occur at any level of an organization. Chapter 7 describes internal theft and fraud.

As the lead federal agency investigating corporate fraud, the FBI partners with many other government agencies (e.g., Securities and Exchange Commission; Internal Revenue Service) and private sector organizations (e.g., AICPA). At the conclusion of 2009, 592 corporate fraud cases were being investigated by the FBI throughout the United States, several of which involved losses to public investors that individually exceeded $1 billion. Through FY 2009, FBI fraud cases resulted in 153 indictments/informations and 156 convictions. During FY 2009, the FBI obtained $6.1 billion in restitution orders and $5.4 million in fines from corporate offenders (Federal Bureau of Investigation, 2009a). It is important to note that state and local law enforcement agencies expend enormous resources investigating fraud and work with federal agencies.

In 2001, one of the most infamous corporate scandals began to unfold from the Houston-based energy company known as **Enron Corporation**. It filed for bankruptcy protection with $63 billion in assets, while its stock closed under $1 a share. A year earlier one share sold for $75. Unfortunately, investors lost billions of dollars. Many company employees had received compensation in the form of company stock, and they lost billions of dollars in their retirement and life savings, besides losing their jobs.

The collapse of Enron resulted from years of "creative accounting," whereby top executives hide billions of dollars in debt and made failing ventures appear successful and profitable. In 2006, former Enron founder, Kenneth Lay, age 64, was convicted of fraud, conspiracy, and lying to banks. A month later, he died of a heart attack. Interestingly, a judge followed legal precedent and vacated Lay's conviction. Other former Enron executives were convicted and sentenced to prison. Andrew Fastow, former chief financial officer, cooperated with prosecutors and was sentenced to six years in prison. He helped to secure the prosecution's case against former chief executive officer, Jeffrey Skilling, who was sentenced in 2006 to 24 years in prison following his conviction for fraud, conspiracy, insider trading, and lying to auditors. The judge also approved a settlement requiring Skilling to surrender $43 million in assets for a restitution fund for Enron retirees and shareholders. Skilling's sentence was close to the 25-year sentence that former WorldCom chief executive officer Bernard Ebbers received for his involvement in an $11 billion accounting fraud. In 2002, the accounting firm that audited Enron, Arthur Andersen, was forced to surrender its CPA licenses pending prosecution by the U.S. Department of Justice.

In 2009, in what has been referred to as the largest Ponzi scheme in history, Bernard L. Madoff was charged with 11 felony charges including securities fraud, investment advisor fraud, money laundering, and perjury (Federal Bureau of Investigation, 2009b). A **Ponzi scheme** is a fraudulent investment plan that pays investors with either its own money or money from subsequent investors. Early investors receive "high rates of return on their investments," and "when the word gets out," other investors practically beg the con to take their money for a similar high rate of return. Eventually, the scheme collapses. In the Madoff case, he was the founder of Madoff Investment Securities LLC, in New York City, and served as a broker-dealer. Since the 1980s and until his arrest in 2008, he solicited, under false pretenses, billions of dollars from customers, promising high rates of return with limited risk while failing to invest the money in stocks and other securities as promised while converting the money to his and others' own benefit. Madoff made tens of thousands of fraudulent certified financial statements, such as statements of income and statements of cash flow, plus bogus reports of internal controls—all of which were sent to clients. In addition, he made false filings of documents to the SEC. Thousands of investors lost billions of dollars. In 2009, Madoff was sentenced to 150 years in federal prison. His accountant pleaded guilty to nine charges and two former computer programmers who worked for Madoff were indicted for falsifying records.

Unfortunately, as the memory of Enron, WorldCom, and the Madoff Ponzi scheme are still fresh in the minds of executives, some continue to falter and face indictments. Consequently, as written in Chapter 2, *the impact of deterrence is questionable and prevention is a key strategy to reduce losses.*

Why do you think fraud is such a huge problem?

The Enron case sparked legislation seeking to control the problem of fraudulent accounting practices. As explained in Chapter 4, the **Sarbanes-Oxley (SOX) Act of 2002** seeks to prevent fraud and affects the processes and accountability for financial reporting in publicly traded U.S. companies. It redesigned the federal regulations and reporting requirements of public companies. The act makes executives responsible for establishing, evaluating, and monitoring the effectiveness of internal controls over financial and operational processes. SOX emphasizes the importance of an audit committee as an essential component of a public company's board of directors. Some SOX provisions are becoming standard operating procedure for all businesses, besides publicly held ones. The law includes accounting controls and how companies report financial results and executive compensation. It holds company executives and external auditors directly accountable for the accuracy of financial reports and protects employees who blow the whistle on suspected fraud. *Security and audit departments in corporations have become involved with SOX because these departments investigate internal fraud.*

SOX established the **Public Company Accounting Oversight Board** (PCAOB) which investigates and disciplines public accounting firms in conjunction with the Securities and Exchange Commission (SEC) for noncompliance with SOX. The SEC is a federal administrative law enforcement agency that seeks to protect investors. [Its effectiveness has been criticized, as in the Madoff Ponzi scheme.] SOX also increased criminal sentences for fraud. It broadened the scope of violations pertaining to obstruction of justice and includes acts of knowingly destroying, altering, or falsifying documents to interfere with any federal investigation. In addition, SOX restricts nonaudit services of outside auditors, such as bookkeeping services related to financial statements, IT financial system guidance, and investment banking services (Public Company Accounting Oversight Board, 2011).

Wilson (2004: 7) provides a view of how a security director in a corporation participated in SOX, as described next. The security director lobbied for and was assigned to the corporate SOX task force. It focused on SOX, section 404, on internal controls, which pertains to management's assessment of internal controls, while requiring executives and auditors to confirm the effectiveness of those controls. The task force prepared fraud policies, internal controls, and best practices. In addition, data analysis techniques were implemented to search for anomalies in a variety of departments and reporting documents. A financial integrity department was established with the following objectives: determining risk potential, improving internal controls, identifying technology solutions, communicating the code of conduct and fraud policies, and implementing prevention programs.

Although privately held businesses and nonprofit organizations are not required by law to conform to SOX, increased concern over accounting irregularities is causing management in a variety of entities to consider some provisions of SOX. This concern impacts decisions of public companies planning to merge with private companies and of private companies planning to become public companies.

Another point about SOX is that compliance is expensive and time-consuming. Companies are spending millions of dollars to comply with the law.

Do you think there is too much pressure on executives to reach business goals? Explain your answer.

Governance, Risk Management, and Compliance

Zoellick and Frank (2005: 1) note that corporate boards of directors and senior management are generally aware of the importance of setting objectives and managing programs that involve governance, risk management, and compliance (GRC). They write of seven operational concerns, drawn from the Federal Sentencing Guidelines, to promote effective compliance and ethics programs to strengthen GRC operations. The next chapter includes risk management. Here, GRC is defined from the perspectives of Zoellick and Frank (2005: 3–6), followed by the seven operational concerns.

Governance is "the process by which the board sets the objectives for an organization and oversees progress toward those objectives." "**Risk management** means different things in different contexts." "The old view of risk [is that it was] managed by buying insurance." "What has happened [is that] board members and senior management are using the language and techniques of risk management to address a much broader range of organizational concerns." "Risk management is part of the process of making decisions." "Risk management supports risk taking and the organization's ability to compete." (**Compliance** refers to an organization adhering to numerous laws, regulations, and initiatives.) "Monitoring and supporting compliance is not just a matter of keeping the regulators happy; it is the way that the organization monitors and maintains its health."

Here is a list of core GRC operations (Zoellick and Frank, 2005: 13–16):

1. *Establish and support policies, procedures, and controls.* This creates the foundation to ensure that the GRC program works.
2. *Maintain centralized oversight.* The key is high-level oversight of the program from top management.
3. *Maintain decentralized administration and accountability.* Ensure accountability for every unit of the organization.
4. *Establish communication channels across all organization levels.* Communications should provide up-to-date policies, procedures, and controls.
5. *Audit, monitor, and report.* Methods must be implemented to ensure that policies and procedures are being employed.
6. *Provide uniform support, remediation, and enforcement.* Ensure consistent incentives to comply and discipline for violators.
7. *Implement continuous process improvement.* Management must respond to feedback to improve GRC.

Search the Internet

To learn more about preventing and detecting fraud, embezzlement, and other white-collar crimes, check these websites:

American Institute of Certified Public Accountants, Anti-Fraud and Corporate Responsibility Center: www.aicpa.org
Association of Certified Fraud Examiners: www.acfe.com
Federal Bureau of Investigation: www.fbi.gov/about-us/investigate/white_collar
Institute of Management & Administration: www.ioma.com
National White Collar Crime Center: www.nw3c.org
Securities and Exchange Commission: www.sec.gov

Case Problems

11A. With reference to the purchasing accountability section of this chapter and Figure 11-1, design an accountability system to strengthen control and prevent losses when merchandise travels from the receiving department to the originator. Look for any other weaknesses and suggest controls.

11B. You are a CSO for a medium-sized corporation that manufactures computer components. Your boss asks you to prepare a plan of antifraud strategies. Prepare a list of what you think are the top five strategies that will be the heart of your plan.

11C. As a corporate IT director, you have been asked by a top executive to program software so revenue will appear greater than it actually is because of a slowing economy. What are your choices in this matter, and how do you respond to the executive?

11D. As director of loss prevention for a corporation, you learn that the company president and the head of IT have conspired to alter financial records to show revenue higher than what is expected. What do you do?

References

American Institute of Certified Public Accountants. (1978). A small business internal control questionnaire. *The Journal of Accountancy, July*

Bodnar, G., & Hopwood, W. (2004). *Accounting information systems* (9th ed.). Upper Saddle River, NJ: Pearson Prentice Hall.

Federal Bureau of Investigation, Financial Crimes Section. (2009a). 2009 Financial crimes report. <www.fbi.gov/stats-services/publications/financial-crimes-report-2009> retrieved September 28, 2011.

Federal Bureau of Investigation. (2009b). Bernard L. Madoff charged in 11-count criminal information. <www.fbi.gov/newyork/press-releases/2009/nyfo031009.htm> retrieved September 28, 2011.

Hubbs, R. (2011). Preventing employee fraud. *Security Management, 55* (January).

Johnston, R., & Warner, J. (2005). The Dr. Who conundrum. *Security Management, 49* (September).

Kardell, R. (2011). Analysis of digital financial data. *FBI Law Enforcement Bulletin, 80* (August).

Levy, M., & Weitz, B. (2001). *Retailing management* (4th ed.). New York: McGraw-Hill Irwin.

Mann, R., & Roberts, B. (2001). *Essentials of business law* (7th ed.). Cincinnati, OH: West.

Nash, K. (1999). IT staffers charged in accounting frauds. *Computerworld* (December).

National Law Enforcement and Corrections Technology Center, (2005). Technology primer: Radio frequency identification. *TECHBeat, Summer*

Public Company Accounting Oversight Board. (2011). PCAOB oversees the auditors of companies to protect investors. <pcaobus.org> retrieved September 27, 2011.

Twomey, D. (2001). *Anderson's business law & the regulatory environment* (14th ed.). Cincinnati, OH: West.

Wilson, R. (2004). Sarbanes-oxley: Lessons from security directors. *Security Director News, 1* (August).

Zoellick, B., & Frank, T. (2005). Governance, risk management, and compliance: an operational approach. (May 16). <gilbane.com/publications/GRCOperationalApproachPD10050512.pdf> retrieved October 28, 2006.

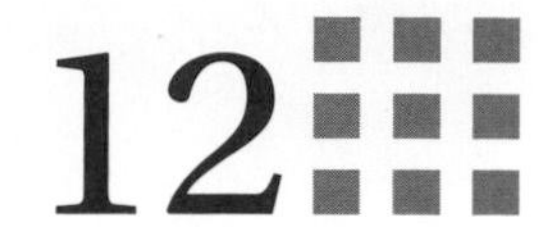

12 Resilience, Risk Management, Business Continuity, and Emergency Management

OBJECTIVES

After studying this chapter, the reader will be able to:

1. Define resilience and think critically about it
2. Explain the purpose of risk management, define it, and summarize two theories relevant to it
3. Describe the role of the risk manager
4. Explain the risk management process, risk modeling and simulation, and risk management tools
5. Explain enterprise risk management and enterprise security risk management
6. List and explain at least eight types of insurance
7. Elaborate on the insurance claims process
8. Define business continuity planning and explain PS-Prep
9. Define emergency management and explain relevant theories
10. Explain risk management in government, all-hazards preparedness, and emergency management
11. Explain the National Response Framework, the National Incident Management System, and the National Preparedness Goal
12. Describe the role of the military when disasters strike

KEY TERMS

- resilience
- risk
- risk management
- risk perception theory
- risk communication theory
- insurance brokers
- risk management process
- first party risks
- third party risks
- risk financing
- risk control
- Delphi approach
- game theory
- enterprise risk management
- event risk management
- financial risk management
- enterprise security risk management
- insurance
- shared risk
- commercial package policy

- bond
- fidelity bond
- surety bond
- Terrorism Risk Insurance Act of 2002
- cyber insurance
- claims management
- business continuity
- PS-Prep
- supply chain
- emergency management
- systems theory
- chaos theory
- decision theory
- management theory
- organizational behavior theory
- risk perception and communication theory
- social constructionist theory
- Weberian theory
- Marxist theory
- all-hazards preparedness concept
- generic emergency management
- specialized emergency management
- Civil Defense Programs
- community-based mitigation
- Federal Emergency Management Agency (FEMA)
- mitigation
- response
- Stafford Act
- recovery
- preparedness
- interoperability
- National Response Framework (NRF)
- National Incident Management System (NIMS)
- National Preparedness Goal
- logistics
- U.S. Northern Command (NORTH-COM)
- Posse Comitatus Act

Introduction

This chapter begins with an explanation and critical thinking perspective of the increasingly popular term "resilience." The resilience movement is covered first because it has gained much attention in the public and private sectors and confusion has developed over its meaning and how it relates to risk management, business continuity planning, emergency management, critical infrastructure protection, and other concepts and efforts applied to the challenges of threats and hazards. Once a critical thinking approach is applied to the budding concept of resilience, the more entrenched concepts (e.g., risk management, business continuity planning, etc.) that contain a wealth of history and information are explained.

Resilience: A Critical Thinking Perspective

"Resilience" is a buzzword that is gaining momentum globally. The question is: does "resilience" have value? Does it hold answers to protect people, assets, organizations, and nations?

Resilience is not a panacea, but it contains seeds for advancing our efforts in the face of threats and hazards. As with many other terms, there is disagreement over the definition and concept of resilience. Additionally, several issues relevant to resilience will see years of development and refinement. These include the following: What are the goals, objectives, and standards of resilience? How will resilience be integrated with protection within and among government, industries, and other organizations? What groups will take the lead to facilitate resilience? What incentives will be offered to facilitate resilience? How will it be funded? How will it be managed and audited? How will it be evaluated? The Homeland Security Policy Institute (2011: 1) notes: "There is general agreement that resilience is a good thing, but we lack a shared vision of how to achieve it." The Institute argues for "definitions and frameworks" to motivate behaviors and actions for a resilient nation.

Presidential Policy Directive-8 (U.S. Department of Homeland Security, 2011a) defines **resilience** as "the ability to adapt to changing conditions and withstand and rapidly recover from disruption due to emergencies." This is a commendable goal with great potential and all sectors must work diligently to reach it. At the same time, devastating disasters require enormous funding and years to recover. The Katrina disaster and the Eastern Japan Great Earthquake Disaster of 2011 serve as examples (see the end of this chapter).

The U.S. Government Accountability Office (2010: 29) notes that Congress and executive branch agencies of the federal government view resilience as vital to the recovery of our nation's critical infrastructure following a disaster, and "most of the current focus is on assets, systems, and networks rather than agencies or organizations." A question posed here by the author is: What about people?

Here are illustrations of how resiliency is achieved (U.S. Government Accountability Office, 2010: 29–31):

> *Among these are an organization's robustness (based on protection, for example better security or hardening of facilities); the redundancy of primary systems (backups and overlap offering alternatives if one system is damaged or destroyed); and the degree to which flexibility can be built into the organization's culture (to include continuous communications to assure awareness during a disruption, distributed decision-making power so multiple employees can take decisive action when needed, and being conditioned for disruptions to improve response when necessary).*

An important question is how risk management, business continuity planning, emergency management, critical infrastructure protection, and other related efforts are linked to resilience. Bendixen (2011: 2) writes that these efforts have "designed in redundancy, developed inventories of spare parts, trained back-ups, and set up alternative work sites. We do not want to throw out all these efforts; rather we want to build on and leverage them for greater efforts." She offered suggestions such as sharing best practices and lessons learned, and partnering and sharing resources and technology among several companies and/or government entities. These suggestions from Bendixen are trite; however, she should be commended for explaining the unique aspects of the resilience movement that set it apart from other initiatives. These include thinking more broadly, optimizing at a higher level, and governments at all levels, small and large businesses, entire industries, and individuals pursuing a mix of separate and joint efforts at building relations and partnerships aimed at embracing resilience as a way of life and making it a part of our culture.

The U.S. Department of Homeland Security (DHS), in cooperation with various federal agencies (named Sector-Specific Agencies), partner with critical infrastructure (CI) sectors (e.g., agriculture, energy, transportation, etc.) to enhance protection from threats and hazards. The U.S. Government Accountability Office (2010) reported that the DHS, in its 2009 National Infrastructure Protection Plan (NIPP), increased its encouragement of resilience, besides emphasizing the importance of protection. DHS stated that the purpose of including more discussion of resilience in the 2009 NIPP was to raise awareness of resilience. At the same time, a major policy shift away from protection and toward resilience has not occurred. It is important to note that the National Security Strategy (White House, 2010: 18), which faces global threats to United States interests, points to the need to "strengthen security and resilience at home."

Interestingly, as the resilience movement progresses, we are seeing two camps develop: those in government and industries who view protection as more important than resilience, and those in government and industries who view resilience as more important than protection. Since both concepts are vital, we will see a mixture of both in government and industry, with each specific industry applying the best components of protection and resilience to meet individual risks and needs.

Since protection and resilience are complementary concepts and should be integrated within each organization and between organizations, a new employment title may emerge (e.g., Chief Protection and Resilience Officer or Vice President of Protection and Resilience), requiring skill sets from security, risk management, business continuity planning, and other specialties. Then, relevant academic and training programs may develop to refine the skill sets required for this new role. Associations and certifications (e.g., Certified Protection and Resilience Professional) will likely fill the needs of organizations and professionals. In addition, producers of software will possibly market their products as "the best" for assessing protection and resilience.

Another perspective on understanding the resilience movement is that government and private sector initiatives and programs often have a life cycle of birth, growth, maturation, and then possible decline. Consequently, a new term and concept may be required in order to renew the enthusiasm to facilitate a vital goal of our nation and systems (e.g., surviving and recovering as fully as possible following a disaster). Bendixen (2011: 16) writes, "There is no need to force them to call something resilience if they already think of it in terms of business continuity. On the other hand, relabeling business continuity efforts as part of a new resilience initiative could breathe life into a plan or effort that has not been widely accepted within an organization." As critical thinkers, we should seek to understand the meaning of a new term, its justification, the concept attached to the new term, and the life cycle and success or failure of the new concept.

What are your views on resilience and its future?

Risk Management

Risk management is at the foundation of resilience, business continuity planning, emergency management, critical infrastructure protection, and other efforts applied to the challenges of threats and hazards. Risk management helps to estimate future risks and offers strategies and tools as a beginning point from which comprehensive action can be taken, such as resilience, business continuity planning, etc.

Brown (2011: 11) writes of the link between resilience and risk management. He points out that "resilience is the result of effective risk management" and that "being resilient simply means being able to assess (identify, analyze, and evaluate) and treat potential risk events." (The reader can refer back to Chapter 3 for a multi-step risk analysis process.) Brown looks to ISO 31000:2009 Risk Management Principles and Guidelines for a risk management framework. However, he argues that although risk management is conceptually simple, it is difficult to apply in "the real world" because "assessing potential risk events is part science, part art, and determining how to treat a potential risk event involves multiple factors that must be balanced against organizational resources and objectives."

Risk management is at the foundation of resilience, business continuity planning, emergency management, critical infrastructure protection, and other efforts applied to the challenges of threats and hazards.

Here we begin with a basic foundation of risk management. As defined in Chapter 2, **risk** is the measurement of the frequency, probability, and severity of losses from exposure to threats or hazards (e.g., crime, fire, accident, and natural disaster). Management in organizations should be knowledgeable about all exposures. In the private sector, business interruption, for example, results from crime, fire, accident, flood, tornado, and the like. Another exposure is liability. A customer might become injured on the premises after falling or be harmed in some way when using a product manufactured by a business. Government is especially concerned about the exposure to natural disasters and terrorism.

The most productive way of handling unavoidable risks is to manage them as well as possible. Hence, the term *risk management* has evolved. **Risk management** makes the most efficient before-the-loss arrangement for an after-the-loss continuation of a business. (Notice that this definition is somewhat similar to the definition of resilience; however, resilience seeks adaptation to change and enhanced survivability and recovery.)

Insurance is a major risk management tool. Leimberg et al. (2002: 6), writing from an insurance perspective, defines risk management as follows: "A preloss exercise that reflects an organization's postloss goals; a process to recognize and manage faulty and potentially dangerous operations, trends, and policies that could lead to loss and to minimize losses that do occur."

Risk management and loss prevention are naturally intertwined. Loss prevention is another tool risk managers can use to make their jobs easier. Insurance is made more affordable through loss prevention methods. Speight (2011: 536) writes: "Many insurance companies are no longer prepared to expose themselves without first asking what their clients have done, or what they could have reasonably done, to prevent losses."

Both loss prevention and risk management originated in the insurance industry. Soon after the Civil War, fire insurance companies formed the National Board of Fire Underwriters, which was instrumental in reducing loss of life and property through prevention measures. Today, loss prevention has spread throughout the insurance industry and into the business community. Risk management is also an old practice. The modern history of risk management is said by many insurance experts to have begun in 1931, with the establishment of the insurance section of the American Management Association. The insurance section holds conferences and workshops for those in the insurance and risk management field.

Risk management theory draws on probability and statistics, mathematics, engineering, economics, business, and the social sciences, among other disciplines. The study of risk has expanded to include the understanding of the psychological, cultural, and social context of risk. The expanding nature of the study of risk is illustrated by the two theories that follow (Borodzicz, 2005: 14–47).

- **Risk perception theory** focuses on how humans learn from their environment and react to it. Psychologists apply a cognitive research approach to understand how humans gain knowledge through perception and reasoning. For instance, risk can be researched by isolating a variable and simulating it in a laboratory with a group of subjects in an experiment involving risk decision making (e.g., gambling). The psychometric approach is another method of researching risk; it involves a survey to measure individual views of risks. Research on risk perception shows that people find unusual risks to be more terrifying than familiar ones and, interestingly, the familiar risks claim the most lives; voluntary risk (e.g., smoking) is preferred over imposed risk (e.g., a hazardous industry moving near one's home); and people have limited trust in official data.
- **Risk communication theory** concerns itself with communication perceptions of experts and lay citizens. Although experts work to simplify information for lay citizens, communication of simplified information and behavior change may not be successful. Research on this topic focuses on lay citizen perceptions of risk within the context of psychological, social, cultural, and political factors. Risk communication theory is important because it holds answers for educating and preparing citizens for emergencies.

The Role of the Risk Manager

Traditionally, businesses purchased insurance through outside **insurance brokers**. Generally, a broker brings together a buyer and a seller. Insurance brokers are especially helpful when a company seeking insurance has no proprietary risk manager to analyze risks and plan insurance coverage. Not all businesses can afford the services of a broker or a proprietary risk manager; however, risk management tools are applicable to all entities.

The risk manager's job varies with the company served. He or she may be responsible for insurance only; or for security, safety, and insurance; or for loss prevention, insurance, investments, and business continuity. *One important consideration in the implementation of a risk management (or loss prevention) program is that the program must be explained to top executives in financial terms.* Is the program cost effective? What is the return on investment?

Financial benefits and financial protection are primary expectations of top executives that the risk manager must consider during decision-making. Leimberg et al. (2002: 4) write:

> *It is extremely difficult to measure tangible benefits against a nonevent (i.e., the catastrophic loss did not occur due to our highly effective risk management program). Yet this is the challenge facing all risk managers. Ultimately, the metrics that senior management uses to gauge success—for example, earnings growth and return on invested capital—must also serve as the yardsticks used to measure the effectiveness of risk management.*

In England, research of the activities of risk managers in 30 different organizations showed five major factors influenced risk managers' roles. *Top management* had a major influence on the risk manager in the form of direct instruction on primary tasks. *External influences* included recommendations from outside groups to increase attention to risk management within businesses and requirements for risk reporting. The *nature of the business, corporate developments* (e.g., expansion and exposures), and *characteristics of the risk management department* (e.g., resources available) also influenced the role of the risk manager (Ward, 2001: 7–25).

Among the many activities of the traditional risk manager are to develop specifications for insurance coverage wanted following a study of risks, meet with insurance company representatives, study various policies, and decide on the most appropriate coverage at the best possible price. Coverage may be required by law or contract, such as workers' compensation insurance and vehicle liability insurance. Plant and equipment should be reappraised periodically to maintain adequate insurance coverage. In addition, the changing value of buildings and other assets, as well as replacement costs, must be considered in the face of depreciation and inflation (Glasser, 2011: 62–63; Bieber, 1987: 23–30).

It is of tremendous importance that the expectations of insurance coverage be clearly understood. The risk manager's job could be in jeopardy if false impressions are communicated to top executives, who believe a loss is covered when it is not. Certain things may be excluded from a specific policy that might require special policies or endorsements. Insurance policies state what incidents are covered and to what degree. Incidents not covered are also stated. An understanding of stipulations concerning insurance claims, when to report a loss, to whom, and supporting documentation are essential in order not to invalidate a claim.

During this planning process, loss prevention measures are appraised in an effort to reduce insurance costs. Because premium reductions through loss prevention are a strong motivating force, risk managers may view strategies, such as security officers, as a necessary annoyance.

Deductibles are another risk management tool to cut insurance expenses. There are several forms of deductibles, but generally the policyholder pays for small losses up to a specified amount (e.g., $1,000, $10,000), while the insurance carrier pays for losses above the specified amount, less the deductible.

A major concern for the risk manager in the planning process is *what amount of risk is to be assumed by the business beyond that covered by insurance and loss prevention strategies.* A delicate balance should be maintained between excessive protection and excessive exposure.

Today, the risk manager's job has become more complicated, and executives throughout corporations—from finance to human resources to corporate boards of directors—are increasingly concerned about risks. Financial failures and corporate scandals (e.g., Enron and WorldCom) illustrate the variety of risks businesses face. Regulatory laws, such as the Sarbanes-Oxley (SOX) Act of 2002, reinforce the legal obligations of businesses to properly identify, assess, and manage risks. In addition, country-specific risks are important as international business operations and the global economy grow (Moody, 2006).

The Risk Management Process

As the risk management discipline developed, a need arose for a systematic approach for evaluating risk. Leimberg et al. (2002: 2–3) describe what is known as the **risk management process**.

Step 1, Risk Identification:

Risks must be identified prior to being managed. Risks are divided into various categories. For example, **first party risks** involve owned assets. Damage to a company truck is a first party risk. **Third party risks** pertain to liability resulting from business operations. If a company truck is involved in a traffic accident, the company may be liable for property damage and injury to others not connected to the company. Risk identification is challenging because many threats and hazards face organizations, as listed in Chapter 2.

Step 2, Quantitative Analysis:

Risk quantification applies probability, statistics, tools, and software to anticipate the maximum and expected financial loss from each identified risk. If a fire destroys a building costing $10 million to replace, the maximum loss is $10 million. However, loss prevention methods (e.g., fire-resistive construction and a sprinkler system) can reduce the loss significantly. Risk quantification for physical assets is usually easier than for public liability and worker safety.

Step 3, Evaluate Treatment Options:

This step uses identified risks and quantitative analysis as a foundation to prepare measures to reduce exposure. Two methods included in this step are risk financing and risk control. **Risk financing** is very broad and can be categorized as "on-balance sheet" and "off-balance sheet." The former includes insurance policy deductibles and self-insured loss exposures; essentially, a business absorbs losses. The latter includes insurance policies and contractual transfers of risk. Transferring risks "off-balance sheet" is not free, and costs can increase. For instance, if insured losses increase in frequency, insurance premiums and deductibles are likely to rise. **Risk control**, also known as loss prevention, involves precautions (e.g., security and safety methods) to reduce risk.

Step 4, Implementation:

Once the treatment options are studied and planned, the next step is to put the selected options into practice.

Step 5, Monitoring and Adjusting:

The risk management process concludes with program oversight, analysis, and modifications.

Risk Modeling and Simulation

The process of selecting which exposures deserve increased attention is difficult. Although no person, software, or technology can predict the future, methodologies are available to help estimate risk, while serving as a foundation for prioritizing and planning. Risk modeling and simulation offer various methodologies to estimate risk (Pinto and Kirchner, 2011: 56–61). However, these methodologies are not a crystal ball. The Rand Corporation, for example, developed the **Delphi approach** during World War II. It consists of sending a structured questionnaire to a group of experts and then conducting a statistical analysis to generate probabilistic forecasts. To enhance the process, the experts may play the roles of an adversary making decisions. Another type of risk modeling is **game theory**. In reference to the problem of terrorism, Starner (2003: 32) states: "Game theory suggests that the likelihood—and targets—of a future terrorist attack can be modeled by understanding the operational and behavioral characteristics of the terrorist organization." It involves the concept that adversaries are rational and make choices based on their information and rules. Game theory is a method to get inside the minds of terrorists. The key is to know adversaries and their rules to anticipate their actions.

Pinto and Kirchner (2011: 60) emphasize that the gateway in the modeling and simulation process is to "gather, map, and document everything known about a risk event and how the enterprise may be affected." The information gathered is then portrayed visually in a chart and various "what if" scenarios are plotted to illustrate potential impact. This process serves to aid risk management decisions.

Modeling and simulation help insurance companies to understand risks and set premiums. Since the 9/11 attacks, for example, insurers have been asking specific questions about the number of employees at individual buildings at specific geographic locations so they can compute the risk they are accepting. For instance, if 3,000 employees are at one location, it represents a $3 billion workers' compensation exposure in a state that puts a $1 million price tag on a life (Stoneman, 2003: 18).

Risk Management Tools

Within the risk management process, and before a final decision is made on risk management measures, the practitioner should consider the following tools (also referred to as "risk treatment") for dealing with risk:

- *Risk avoidance:* This approach asks if the risk should be avoided. For example, the production of a proposed product is canceled because the danger inherent in the manufacturing process creates a risk that outweighs potential profits. Or, a bank avoids opening a branch in a country subject to political instability or terrorism.

FIGURE 12-1 Florida hotel faces risk of beach erosion from Hurricane Irene. *Courtesy: Ty Harrington/FEMA.*

- *Risk transfer:* Risk can be transferred to insurance. The risk manager works with an insurance company to tailor a coverage program for the risk. This approach should not be used in lieu of loss prevention measures but rather to support them. *Insurance should be last in a series of defenses.* Another method of transferring risk is to lease equipment rather than own it. This transfers the risk of obsolescence.
- *Risk abatement:* In abatement, a risk is decreased through a loss prevention measure. Risks are not eliminated, but the severity of loss is reduced. Sprinklers, for example, reduce losses from fire. Sand bags assist in decreasing erosion (Figure 12-1).
- *Risk spreading:* Potential losses are reduced by spreading the risk among multiple locations. For example, a copy of vital records is stored at a remote, secure location. In another example, following the 9/11 attacks, companies have spread operations among multiple locations to facilitate business continuity.
- *Risk assumption:* In the assumption approach, a company makes itself liable for losses. Not obtaining insurance is an example. This tool may be applied because the chance of loss is minute. Another path, self-insurance, provides for periodic payments to a reserve fund in case of loss. Risk assumption may be the only choice for a company if insurance cannot be obtained. With risk assumption, prevention strategies become essential.

Enterprise Risk Management and Enterprise Security Risk Management

A trend today in the risk management field is **enterprise risk management** (ERM). Leimberg et al. (2002: 6) define it as "a management process that identifies, defines, quantifies, compares, prioritizes, and treats all of the material risks facing an organization, whether or not it is insurable." ERM takes risk management to the next level. It refers to a comprehensive risk management program that addresses a variety of business risks. Examples are risk of profit or loss; uncertainty regarding the organization's goals as it faces its strengths, weaknesses, opportunities, and threats; and risk of accident, fire, crime, and disasters. When all of these risks are packaged into one program, planning is improved and overall risk can be reduced. Because risks frequently are uncorrelated (i.e., all of them causing loss in the same year), insurance costs are lower. For instance, a company is unlikely to face the following losses in the same year: fire, adverse movement in a foreign currency, and homicide in the workplace (Rejda, 2001: 64–66).

Leimberg et al. (2002: 6) describe the trend of two separate and distinct forms of risk management. **Event risk management** focuses on traditional risks (e.g., fire) that insurance covers. **Financial risk management** protects the financial assets of a business from risks that insurers generally avoid. Examples are foreign currency exchange risk, credit risk, and interest rate movements. Various capital risk transfer tools are available to protect financial assets. ERM seeks to combine event and financial risk for a comprehensive approach to business risks.

Mehta (2010) differs from Leimberg by arguing for a more holistic approach to risks by including intangible assets (e.g., brand and customer relationships) that are typically not protected by traditional risk management. He notes that ERM is not always about reducing risks; it can address over-managing risk or not taking enough risk and exploiting business opportunities. Mehta writes that although much has been written about ERM, not all organizations have embraced the concept and some prefer the term "risk management" because adding "enterprise" creates a distraction about its meaning while managing risk is the important goal.

Another term with the word "enterprise" attached is **enterprise security risk management** (ESRM). Straw (2010: 58) writes that ERM includes ESRM, and similar to ERM, ESRM is holistic in its approach. He espouses the importance of interdependencies. For example, the risks resulting from a labor dispute disrupting supply chains and how all the units of a company work together to address all risks.

ASIS International (2010a: 4) research showed that top security leaders from major organizations are "deeply involved with evaluating and mitigating nonsecurity risks in their organizations." Top nonsecurity risks included the economy, competition, regulatory pressure, and failure of IT systems. Skill sets required to succeed at ESRM focused on business management, leadership, and communication skills.

As explained in Chapter 18, ESRM also includes human resources protection (HRP). This is a broad concept that protects all employees and those linked to them (e.g., family and customers). Depending on organizational requirements, HRP can include workplace violence prevention, executive protection, safety, health, use of technology and social media, and personal and family protection. HRP is vital because people are the most valued asset to an organization and, depending on the type of harm to them, the consequences can be devastating.

Should a security and loss prevention executive or a CSO in a company be part of a company enterprise risk management committee? Why or why not?

International Perspective: Risk Management in a Multinational Business

Morris (2001: 22–30) writes about overseas business operations, risks, and the need for answers to specific questions about each country in which business will be conducted. She begins with the following questions: How is business conducted in comparison to the United States? How strong is the currency? How vulnerable is the area to natural disasters, fire, and crime? What are the potential employment practices liability issues? What is the record of accomplishment of shipments to and from the area?

Political risks are especially challenging in overseas operations. Are terrorist groups or the government hostile to foreign companies and their employees? Does the host government have a record of instability and war, seizing foreign assets, capping increases in the price of products or adding taxes to undermine foreign investments, and imposing barriers to control the movement of capital out of the country?

Eighty percent of the terrorist acts committed against U.S. interests abroad target U.S. businesses, rather than governmental or military posts. These threats include kidnapping, extortion, product contamination, workplace violence, and IT sabotage.

The concept of enterprise risk management can be especially helpful with multinational businesses because of a multitude of threats and hazards. A key challenge for the risk manager is to bring together a full range of resources and network in the United States and overseas prior to potential losses so, if a loss occurs, a speedy and aggressive response helps the business to rebound.

Options for insurance include buying it in the home country and arranging coverage for overseas operations; however, this may be illegal in some countries that require admitted insurance. Another approach is to let the firm's management in each country make the insurance decision, but this means that the corporate headquarters has less control of risk management. A third avenue is to work with a global insurer who has subsidiaries or partner insurers in each country; this approach offers uniform coverage globally. A key question in these approaches is: *Is the insurer financially solvent to pay the insured following a covered loss?*

Insurance

Insurance is the transfer of risk from one party (the insured) to another party (the insurer), in which the insurer is obligated to indemnify (compensate) the insured for economic loss caused from an unexpected event during a period of time for which the insured makes a premium payment to the insurer. This permits the insured to avoid holding a large amount of liquid capital (cash) in reserve to pay for huge losses; the liquid capital can be invested. The essence of insurance is the sharing of risks; insurance permits the insured to substitute a small cost (the premium) for a large loss under an arrangement whereby the fortunate many who escape loss will indirectly assist in the compensation of the unfortunate few who experience loss. For an insurance company to function properly, a large number of policyholders are required. This creates a **shared risk**.

The technical aspects of the insurance industry involve the skills of statisticians, economists, financial analysts, engineers, attorneys, physicians, and, of course, risk managers and loss prevention specialists, among others. Insurance companies must carefully set rates, meticulously draft contracts, establish underwriting guidelines (i.e., accepting or rejecting risks for an insurance company), and invest funds prudently.

Insurance rates are dependent on two primary variables: the frequency of claims and the cost of each claim. When insurance companies periodically review rates, the "loss experience" of the immediate past is studied.

The insurance industry is subject to two forms of control: competition among insurance companies and government regulation. Competition enables the consumer to compare rates and coverage for the best possible buy. Government regulatory authorities in each state or jurisdiction have a responsibility to the public to assure the solvency of each insurance company so policyholders will be indemnified when appropriate. Furthermore, rates should be neither excessive nor unfairly discriminatory. Problems with the state system of regulation came to light following the case of Martin Frankel, who looted $200 million from insurance companies he owned during the 1990s (Gurwitt, 2001: 18–24). Since the case affected insurers in five states, Congress asked the U.S. General Accounting Office (GAO) to investigate. A fall 2000 report by the GAO blamed the states for inadequate regulatory policies, procedures, practices, investigations, and information sharing. There are calls for a federal regulatory system, but the states will fight for the revenue—$10.2 billion in insurer premium taxes and fees while spending $839 million to regulate the industry. Insurance industry executives argue, "Fifty monkeys are better than one gorilla."

To check on the financial health of an insurer, contact an insurance company rating service and look for a rating of A+ or better. Examples of these firms are A.M. Best, Standard & Poor's, and Moody's Investor Service.

Types of Insurance

Insurance can be divided into two broad categories: government and private. Government insurance programs include Social Security, Medicare, unemployment insurance, workers' compensation, retirement, insurance on checking and savings accounts in banks, flood insurance, and numerous other programs on the federal and state levels.

In the United States, the private insurance industry is divided into property and liability (casualty) insurance and life and health insurance. This industry employs millions in the United States, among thousands of insurance companies, while administering trillions of dollars in assets. Outside the United States, the private insurance industry is divided into life and nonlife (or general) insurance.

Property and liability insurance covers fire, ocean marine, inland marine (i.e., goods shipped on land), and liability insurance, which is a broad field including general liability (e.g., from sales of products, professional services), automobile, crime, workers' compensation, boiler and machinery, glass, and nuclear, among other types. The property and liability field also includes multiple-line insurance (i.e., two or more perils covered under one policy) and fidelity and surety bonds.

Advisory organizations have been established in the property and liability field to offer a variety of research services to insurance companies, develop policy forms, and pool loss statistics among insurance companies to increase the accuracy of rates. Two major advisory organizations are the Insurance Services Office (ISO) and the American Association of Insurance Services (AAIS).

Because types of insurance are numerous, varied and confusing, groups such as the ISO develop insurance contracts and forms and seek standardization. This group's effort is illustrated through the **commercial package policy** (CPP) that contains multiple coverage in a single policy, fewer gaps in coverage, and lower premiums because individual policies are not purchased. Retail stores, office buildings, manufacturers, motels, hotels, apartments, schools, churches, and many other organizations use the CPP. CPP coverage commonly contains two or more coverage parts. A business may select, for example, coverage focusing on commercial property, general liability, auto, and crime (Insurance Services Office, 2011a; Rejda, 2001: 266–268).

Crime Insurance and Bonds

Two basic kinds of protection against crime losses are (1) fidelity and surety bonds and (2) burglary, robbery, and theft insurance. The first covers losses caused by dishonesty or incapacitation from persons entrusted with money or other property that violate this trust. The second type of protection covers theft by persons who are not in a position of trust.

What are the differences between insurance and a bond? A **bond** is a legal instrument whereby one party (the surety) agrees to indemnify another party (the obligee) if the obligee incurs a loss from the person bonded (the principal or obligor). Although a bond may seem like insurance, there are differences between them. Generally, a bonding contract involves three parties, whereas an insurance contract involves two. With a bond, the surety has the legal right to attempt collection from the principal after indemnifying the obligee; collection would be absurd by an insurer against an insured party, unless fraud was evident. Another difference is that insurance is easier to cancel than a bond. The insured can cancel insurance by simply notifying the insurer or by nonpayment of premium. Breach of the insurance contract by the insured or nonpayment of premium are the insurer's frequent reasons for cancellation and a legal defense by the insurer to avoid liability. On the other hand, with a bond, the surety is liable to the beneficiary even though breach of contract or fraud occurred by the principal.

Fidelity Bond

Generally, a **fidelity bond** requires that an employee(s) be investigated to limit the risk of dishonesty for the insured. If the bonded employee violates the trust, the insurer (bonding company) indemnifies the employer (insured) for the amount of the policy.

Fidelity bonds may be of two kinds: (1) those in which an individual is specifically bonded, by name or by position, and (2) "blanket bonds," which cover a whole category of employees.

Surety Bond

A **surety bond** essentially is an agreement providing for compensation if there is a failure to perform specified acts within a certain period. One of the more common surety bonds is called a *contract construction bond*. It guarantees that the contractor(s) involved in construction will complete the work that is stipulated in the construction contract, free from debts or encumbrances.

Several types of surety bonds are used in the judiciary system. A *fiduciary bond* ensures that persons appointed by the court to supervise the property of others will be trustworthy. A *litigation bond* ensures specific conduct by defendants and plaintiffs. A *bail bond* ensures that a person will appear in court; otherwise, the entire bond is forfeited.

Burglary, Robbery, and Theft Insurance

Understanding the definitions for burglary, robbery, and theft is important when studying insurance contracts. In reference to businesses, a valid *burglary* insurance claim requires the unlawful taking of property from a closed business that was entered by force. In the absence of visible marks showing forced entry, a burglary policy is inapplicable. *Robbery* is the unlawful taking of property from another by force or threat of force. Without force or threat of force, robbery has not occurred. *Theft* is a broad, catchall term that includes all crimes of stealing, plus burglary and robbery.

Despite the availability of insurance, crime against property is one of the most underinsured perils. Risk assumption remains the often-used tool to handle the crime peril.

Federal Crime Insurance

The *Federal Crime Insurance Program*, established by Congress, began operation in 1971 to counter the difficulty of obtaining adequate burglary and robbery insurance, particularly in urban areas. The program was discontinued in 1995. Private insurers and their agents administered the coverage, and the federal government, through the *Federal Insurance Administration*, was the bearer of the risk.

Kidnapping and Extortion Insurance

Another form of crime insurance covers losses from a ransom paid in a kidnapping or through extortion. During the 1970s, an upsurge in domestic and international kidnappings and terrorism created a need for this form of insurance. U.S. corporations with overseas executives are especially interested in this coverage. These policies cover executives, their families, expenses for a crisis management team, ransom money during delivery to extortionists, and corporate negligence during negotiations, among other areas of coverage.

Although accurate statistics are difficult to compile, Zuccarello (2011) reports that the number of reported kidnappings worldwide is growing, with estimates of 12,500 to 25,000 annually. Many of these crimes are unreported. What was once a problem concentrated in Latin America has become a global criminal enterprise. In the United States, Phoenix, Arizona, is the kidnapping capital. Mexico is a hotbed of kidnappings. Unfortunately, in certain regions, a portion of ransom payments goes to police or security officials. Government no-ransom pronouncements are often followed by denials that a ransom was paid to free a hostage. The U.S. government has a policy of not paying ransom. However, American companies and individuals produce ransom and negotiate through intermediaries hired by insurers.

Terrorism Insurance

The **Terrorism Risk Insurance Act of 2002** (TRIA) was passed by Congress to calm the insurance industry that faced claims resulting from the 9/11 attacks and concern over subsequent attacks. The act requires insurance companies to provide terrorism coverage to businesses willing to purchase it. Participating insurance companies pay out a claim (a deductible) before TRIA pays for the loss. TRIA losses are capped at $100 billion. The act is viewed as important to the U.S. economy to support recovery in the event of attacks. Congress and the President extended this Act multiple times. At the time of this writing TRIA will expire on December 31, 2014. This latest extension includes coverage for terrorism perpetrated by domestic persons or interests, whereas earlier the Act covered only terrorism perpetrated by foreign persons or interests (Ball, 2008).

The number of businesses purchasing terrorism insurance has increased steadily since 2003. Large organizations with assets in metropolitan areas while doing business globally view terrorism as another catastrophic exposure. These firms seek broad coverage to manage the risk (Tuckey, 2010).

Cyber Insurance

Traditional insurance products often do not cover cyber risks. Thus, insurers offer businesses **cyber insurance**. Basic policies include coverage for unauthorized website access, data privacy loss, and repairs to company databases following victimization. Broader policies cover such losses as lost income from system failure and the cost of notifying customers of business interruption. Insurers are less likely to refuse coverage if a business employs IT security specialists and maintains security standards (Risen, 2010). Because this insurance field is developing as technology advances, the insured must be careful in selecting an insurer and understanding the definitions, terms, and limitations of these policies.

You Be the Judge #1*

Cliff Hawkins, the newest member of Conway Excavation's repair crew, pulled his rolling tool chest to a stop and extended his hand to his new supervisor.

"Well," said Dave Greco, smiling and shaking Hawkins's hand, "it looks like you brought everything but the kitchen sink."

"A good mechanic can't do much without a good set of tools," replied Hawkins, patting the chest gently. "It took me five years and almost $3,000 to build up this set. Which reminds me"—he glanced around the garage—"if you expect me to leave these tools here, you'd better have some kind of security."

"You've got nothing to worry about," replied Greco. "We lock up at night, and nothing has ever been stolen yet."

However, there is a first time for everything. A short time after Hawkins started working for Conway Excavation the garage was broken into. Hawkins's tools were stolen.

"I thought you said my tools would be safe here," Hawkins fumed when he faced Greco.

"I never said that," Greco corrected him. "I said this garage had never been broken into. And it hadn't."

"Yeah, well, I hope this company is prepared to reimburse me," Hawkins said.

Greco sat up in his chair, surprised. "Reimburse you?" he echoed. "No way! You knew our security wasn't very extensive, but you chose to leave your tools here anyway."

"I had to leave my tools here," Hawkins said angrily.

Greco shrugged. "Still, they were your tools and their loss isn't this company's responsibility."

"We'll see about that," Hawkins said as he stormed out of the office.

Hawkins went to court to try to force Conway Excavation to reimburse him for his stolen tools. Did Hawkins get his money?

Make your decision; then turn to the end of the chapter for the court's decision.

---

*Reprinted with permission from *Security Management—Plant and Property Protection*, a publication of Bureau of Business Practice, Inc., 24 Rope Ferry Road, Waterford, CT 06386.

Fire Insurance

Historically, the fire policy was one of the first kinds of insurance developed. For many years, it has played a significant role in assisting society against the fire peril. Prior to 1873, fire insurance contracts were not standardized. Each insurer developed its own contract. Omissions in coverage, misinterpretations, and conflicts between insurer and insured resulted in considerable problems. These individualized contracts and resultant ambiguities caused the state of Massachusetts, in 1873, to establish a standard contract. Seven years later, the standard contract became mandatory for all insurance companies in the state. Today, except for minor variations in certain states, the wording of fire insurance contracts is very similar. However, in recent years, these standard fire policies have diminished in importance as broad coverage policies have increased in number.

An understanding of insurance rating procedures provides risk managers and loss prevention managers with the knowledge to propose investments in fire protection that can show a return on investment. Factors that influence fire insurance rates include the ability of the community's fire alarm, fire department, and water system to minimize property damage once a fire begins. Class 1 communities have the greatest suppression ability, whereas Class 10 has the least. Strategies such as convincing the community to take steps to improve its grade and installing sprinkler systems in buildings can produce a return on investment (Insurance Services Office, 2011b; Williams et al., 1995: 341–345).

Property and Liability Insurance

Business Property Insurance

The building and personal property coverage form is one of several property forms developed by the ISO program. It covers physical damage loss to commercial buildings, business personal property (e.g., furniture, machinery, and inventory), and personal property of others in the control of the insured. Additional coverage includes debris removal, pollutant removal, and fire department service charge. A cause-of-loss form is added to the policy to have a complete contract; it lists specific causes of loss to be covered. A basic cause-of-loss form covers fire, lightning, explosion, windstorm or hail, smoke, aircraft or vehicles, riot, vandalism, sprinkler leakage, and other perils. A broader form can be selected to expand coverage to, for example, glass breakage and earthquake (International Risk Management Institute, 2011; Rejda, 2001: 268–274).

Another important kind of insurance is business income insurance (formerly called *business interruption insurance*). It indemnifies the insured for profits and expenses lost because of damage to property from an insured peril.

Liability Insurance

Legal liability for harm caused to others is one of the most serious risks. Negligence can result in a substantial court judgment against the responsible party. There are several kinds of exposures for businesses in the liability area. Relevant factors are the functions performed, relationships involved, and care for others required, such as the employee–employer relationship, a contract situation, consumers of manufactured products, and professional acts. Examples of liability exposures are bodily injury or death of customers, product liability, completed operations (i.e., faulty work away from the premises), environmental pollution, personal injury (e.g., false arrest, violation of right of privacy), sexual harassment, and employment discrimination.

In many jurisdictions, the law views the failure to obtain liability insurance against the consequences of negligence as irresponsible financial behavior. Mandatory liability insurance for automobile operators in all states is a familiar example.

Several kinds of business liability insurance are available. The commercial general liability (CGL) policy, developed by ISO, is widely used and can be written alone or as part of a CPP.

Workers' Compensation Insurance

Workers' compensation coverage includes loss of income and medical and rehabilitation expenses that result from work-related accidents and occupational diseases. An employer can obtain the coverage required by law through three possible avenues: (1) commercial insurance companies, (2) a state fund or a federal agency, or (3) self-insurance (i.e., risk assumption).

Risk and Insurance Management Society, Benchmark Survey

The Risk and Insurance Management Society (RIMS) and Advisen (an insurance information service) work together to publish an annual *Benchmark Survey* for risk managers and other specialists in this profession who rely on this data to help justify their insurance coverage recommendations to senior executives. The data also assist risk managers in measuring their organization's risk management performance, structuring insurance programs, and making cost comparisons to others in the same industry. Survey respondents are from nearly one third of the Fortune 500 companies and 70 industry categories, such as industrials, banks, energy, healthcare, utilities, telecommunications, education, IT, and government.

A major gauge in the survey is the total cost of risk (TCOR) per $1,000 of revenue. As an illustration, we refer to industrials from the 2011 *Benchmark Survey*. Industrials manufacture finished goods in a $10 trillion global industry. Industrials showed an average TCOR of $15.59 per $1,000 of revenue in 2010, significantly higher than the average of $10.02 for all companies (Risk and Insurance Management Society, 2011: 66).

Following the 9/11 attacks, insurance costs increased through 2003. The 2005 *Benchmark Survey* showed that the TCOR for 2004 was down for the first time since 2001; almost every industry showed a decrease in the median premium per $1,000 of revenue. In 2005, the median TCOR fell about 11 percent. Despite Hurricanes Rita, Wilma, and Katrina, the property and casualty insurance industry earned about $55 billion in 2005 (Insurance Newsnet, 2006).

You Be the Judge #2*

A vice president had been embezzling money from the Michigan Mining Corporation for several years, but Security Director Steve Douglas finally caught him. It was something of a Pyrrhic victory, however—the culprit was nabbed, but the company was out $135,000. Luckily, MMC had comprehensive business insurance that Douglas was sure would cover most of the loss.

The security director looked over the two policies, but they were poorly written and very confusing, so he called Lester Blank, the agent who handled the policies.

"I limped through the policies," Douglas explained, "and I think I get the gist of them. MMC's covered for $100,000, right?"

"Wrong," Blank said. "The second policy replaced the first. You're only covered for $50,000."

Douglas was stunned, but he recovered quickly. "Now, wait a minute," he said. "I may not have caught every mixed-up word in these policies, but the second one says we can collect on the first one for up to a year after its expiration date, provided the loss occurred during the time the first policy was in effect."

"But the total limit is still $50,000," insisted the insurance agent. "You'll find a clause to that effect in the second policy, if you read carefully."

"If I read carefully!" Douglas cried. "This second policy is so full of spelling and clerical errors that it's anybody's guess what it means. One look at this piece of slipshod writing and any court will side with us."

Therefore, MMC went to court, claiming that because the policy was so complicated and poorly written, it should be interpreted in the company's favor.

Did the court agree with MMC?

Make your decision; then turn to the end of the chapter for the court's decision.

*Reprinted with permission from *Security Management—Plant and Property Protection*, a publication of Bureau of Business Practice, Inc., 24 Rope Ferry Road, Waterford, CT 06386.

Claims

When an insured party incurs a loss, a claim is made to the insurer to cover the loss as stipulated in the insurance contract. For an insurance company, the settling of losses and adjusting differences between itself and the policyholder is known as **claims management**. Care is necessary by the insurer because underpayments can lead to lost customers, yet overpayments can lead to bankruptcy.

Tuohy (2010) writes about third-party administrators (TPAs), rather than companies (the insured) or their insurance carrier (the insurer), handling insurance claims. He views TPAs as a check-and-balance or second opinion of the carrier and an avenue to avoid a strained relationship between the insured and the insurer. Companies that operate in many jurisdictions find that having a TPA is less expensive and more efficient than processing claims internally. Drawbacks include having another layer in the insured-insurer relationship and additional work (e.g., transition to a TPA and communications) for the company risk manager. Tuohy notes the successful use of a TPA by PetSmart Inc., with over 1,000 pet stores and over 100 animal boarding facilities in the United States and Canada. Each year the company has about 15,000 claims (e.g., medical, dog bites, and workers' compensation) and $35 million in loss payments on self-insured losses. Once a claim is reported to the TPA, PetSmart is assured the claim is properly recorded and filed.

Generally, an investigation of a claim includes (1) a determination that there has been a loss, (2) a determination that the insured has not invalidated the insurance contract, (3) an evaluation of the proof of loss, and (4) an estimate of the amount of loss. An example of item 2 occurs when the insured has not fulfilled obligations under the insurance contract, such as not protecting property from further damage after a fire or not adequately maintaining loss prevention measures. Most insurance contracts specify that the insured party must give immediate notice of loss. The purpose is to give the insurer an opportunity to study the loss before evidence to support the claim has been damaged. The insured is expected to provide accounting records, bills, and so on that might help in establishing the loss.

Before a settlement is reached, the insurer checks the coverage, the claim is investigated, and loss reports and claim papers are prepared. Then the insurance company claims department studies the loss, the policy is interpreted and applied to the loss, and a payment is approved or disapproved.

Insurance companies employ different classifications of adjusters to settle claims. An *insurance agent* (i.e. salesperson) may serve as an adjuster for small claims up to a certain amount. A *company adjuster* is more experienced about claims and handles larger losses. *Independent adjusters* offer services to insurance companies for a fee. *Public adjusters* represent the insured party for a fee.

From the insurance industry's perspective, the work of an adjuster is demanding. Some claimants make honest mistakes in estimating losses. They may place a value on destroyed property that is above the market value. Exaggerations are common. Insurance fraud is pervasive and costly. Confusion may arise when claimants have not carefully read their insurance policy. Consequently, a process of education and negotiation often takes place between the adjuster and the claimant. Once the claimant signs the proof of loss papers or cashes the settlement check, this signifies that the claimant is satisfied and that further rights to pursue the claim are waived. In a certain number of claims, an agreement is not reached initially. The policy states the terms for settling claims. Typically, arbitration results. Each party appoints a disinterested party to act as arbitrator. The two arbitrators then select a third disinterested party. Agreements between two of the three arbitrators are binding. In liability cases, the court takes the place of arbitrators.

Arson is unfortunately a popular way of defrauding insurance companies. Generally, a property owner sets a fire to collect on an insurance policy.

Claims for Crime Losses

When a person or business takes out an insurance policy to cover valuables, the insurance agent may not require proof that the valuables exist. However, when a claim is filed, the insurer becomes very interested in not only evidence to prove that the valuables were stolen (e.g., police report), but also evidence that the valuables in fact existed. Without proof, indemnification may become difficult. To avoid this problem, several steps are useful. First, the insured should prepare an inventory of all valuables. Accounting records and receipts are good sources for the inventory list. The list should include the item name, serial number, date it was purchased, price, and a receipt. Photographs and video of valuables are also useful. Copies should be located in two separate safe places.

Bonding Claims

Businesspeople have attempted to use fidelity bond claims to cover losses from mysterious disappearance and general inventory shortages, rather than for their intended purpose—coverage for internal theft. Care must be exercised throughout the interaction with the insurer so as not to in any way invalidate the contract. Furthermore, the burden of proof of losses rests entirely on the insured.

Before 1970, the insurer investigated applicants and notified the insured of any criminal history of the applicant that would bar coverage. Because of economy measures and the difficulty of checking into a person's background, many insurers have made a shift to the insured for verifying the applicant's past. Bonds stipulate that past dishonesty by the employee justifies an exclusion from bonding from the day the information is discovered; if a loss occurs and the insurer can prove that this information was known to the insured company but not reported to the insurer, the bond is likely to be invalid.

Another way to invalidate a bond is through restitution by the employee to the employer without notifying the insurer. In many cases, the employee is eager to pay back what was stolen but makes only a few payments before absconding. Thereafter, if a claim is made, the bond is useless.

In reference to the burden of proof, the loss prevention practitioner should have considerable expertise when dealing with the insurer on behalf of his or her employer. Confusion often arises from the "exclusionary clause" of the fidelity bond policy. This clause essentially states that the bond does not cover losses that are dependent on proof from inventory records or a profit and loss computation. Prior to 1970, these records were not even allowed to establish the extent of losses even though employees had confessed. However, in the early 1970s, courts began to be more flexible in limiting the exclusionary clause and thus allowing inventory records and associated computations to establish the amount of loss when independent proof also was introduced to establish that there was loss due to employee theft. A confession is of prime importance to bonding claims. A guilty verdict in a criminal court or a favorable labor arbitration ruling are additional assets for the claimant.

What do you think are the most difficult challenges of a risk manager's job?

Business Continuity

Business continuity is a term applied in the private sector (business) for planning and action against risks. *Emergency management* is the term applied in the public sector (government) for planning and action against risks. Here, we begin with a discussion of business continuity, and then we turn to emergency management to understand the role of government in emergencies.

Business continuity is defined by ASIS International (2010b: 52) as follows: "Strategic and tactical capability of the organization to plan for and respond to incidents and business disruptions in order to continue business operations at an acceptable predefined level."

The essence of business continuity is an up-to-date, comprehensive plan to increase the survivability of a business when it faces an emergency. Examples of topics in the plan are employee safety, IT backup, customer support, and limited, if any, recovery time.

There is a lack of consistency with business continuity planning in the private sector. Top executives view and support it differently among businesses. Pandolfi (2011) writes of confusion created by disagreement over terminology such as business continuity planning (BCP), disaster recovery (DR), and resiliency planning. He argues that this confusion results in obstacles when practitioners seek to justify funding and explain the value of BCP. Pandolfi also notes that when many corporations re-brand BCP, DR, or resiliency, they do not restructure the organization, vision, mission, processes, or technologies into a holistic enterprise BCP program to support the new branding.

Garris (2005: 28–32) writes that in our litigious society, organizations that fail to plan for emergencies could be held liable for injuries or deaths. He recommends learning about codes, regulations, and laws that are essential foundations of plans. Sources include the Department of Homeland Security, the Occupational Safety and Health Administration, the National Fire Protection Association, the Environmental Protection Agency, the Americans with Disabilities Act, as well as city and state authorities. Garris writes that 44 percent of all businesses that suffer a disaster never reopen.

Guidance for Business Continuity Planning and Private Sector Preparedness

Following the 9/11 terrorist attacks, among the recommendations of the 9/11 Commission was to implement a program to improve private sector preparedness for disasters and emergencies. In 2010, the U.S. Department of Homeland Security (DHS) (2010) announced the adoption of standards for the Voluntary Private Sector Preparedness Accreditation and Certification Program. Also known as **PS-Prep**, these standards represent a partnership between the DHS and the private sector to enable private sector organizations to seek emergency preparedness certification from a DHS accreditation system coordinated with the private sector. Included in the private sector are critical infrastructure and key resources, a variety of businesses, and **supply chain** (i.e., the people, assets, and systems that move a product or service from its source to customers). The standards are from the National Fire Protection Association (NFPA), the British Standards Institution, and ASIS International.

NFPA 1600: 2007 and 2010 Standard on Disaster/Emergency Management and Business Continuity Programs. This standard has been acknowledged by the U.S. Congress, the American National Standards Institute (ANSI), which approves and accredits standards, and the 9/11 Commission. The Federal Emergency Management Agency, the National Emergency Management Association, and others have endorsed it. This standard has been referred to as the "National Preparedness Standard" for all organizations, including government and business. NFPA 1600 serves as a benchmark of basic criteria for a comprehensive program, and it contains elements from emergency management and business continuity. The NFPA 1600 includes standards on program management, risk assessment, mitigation, training, and logistics, among other standards.

British Standard Institution BS 25999-2:2007 Business Continuity Management. This standard seeks to minimize risks from disruptions. It promotes resilience—protecting employees, preserving reputation, and continuing business operations. It contains comprehensive requirements and best practices.

ASIS International Standard: ASIS SPC 1-2009, Organizational Resilience: Security, Preparedness, and Continuity Management Systems-Requirements with Guidance for Use. This standard, approved by ANSI, focuses on enterprise risk management and is compatible with ISO management system standards. It offers general auditable criteria—rather than specifics—so an entity can implement its own criteria for resilience. Topics in the standard point to an organization's own risks and recovery requirements. The following questions provide a paraphrased and abbreviated sample from this standard (ASIS International, 2009: 12–13):

- What are the plans for prevention, preparedness, and response to emergencies and which personnel are responsible and have the authority to act in emergencies?
- What types of risks, threats, and hazards does the entity face?
- What is the nature of onsite hazards (e.g., toxic materials) and/or nearby hazards?
- What actions are planned to minimize human casualties, mitigate harm to assets, and protect the environment?
- Are evacuation routes and assembly points prepared?
- What training is provided for emergencies?
- How will IT be protected?
- Of the various possible emergencies, how will critical infrastructure (e.g., electricity and transportation) affect the entity?

- What are the plans for command and control at an emergency operations center and at alternative work sites?
- What are the procedures to declare an emergency?
- What are the internal and external communication plans for an emergency?
- What are the plans for mutual assistance from other organizations?
- How will post-event evaluation be conducted for corrective and preventive action?

Critical Thinking: PS-Prep

One perspective on PS-Prep is that businesses and government agencies may refrain from conducting business with non-certified organizations. This can result in a competitive advantage for those businesses that are certified.

D'Addario (2010: 55) offers other perspectives. He writes that although it has value for those working on an all-hazards risk mitigation plan, he urges caution for those who view it as law because PS-Prep is voluntary. In addition, he explains that for those who seek to advance a PS-Prep agenda for their employer, resistance may result from those who are skeptical about it, budget limitations, and the difficulty of showing a ROI. D'Addario also notes that the voluntary nature of PS-Prep is in competition with federal and state requirements.

Research by Dunaway and Shaw (2010) explored the willingness of the private sector to commit resources for business continuity programs. They cite study after study showing many businesses facing serious risks and exposures without adequate protection. Dunaway and Shaw make an excellent point by noting that the fundamental challenge for business continuity programs is not necessarily establishing standards, but to motivate individual businesses to take action.

DiMaria (2010) shows his concern about organizations claiming to have the credentials to certify companies to standards. He refers to a long history of organizations that seek to sell services contrary to accepted practices of the standards accreditation business. History repeats itself and DiMaria recommends, "Buyer beware." Companies can waste time and money through training and certification programs from a provider who does not possess appropriate credentials. DiMaria suggests demanding from a vendor proof of accreditation and registration to the specific standard desired.

Methodology for Business Continuity

Persson (2005), writing from an IT perspective, sees the need for increased effort to provide quantitative, accurate, empirical reporting on how business continuity planning is working. He favors the use of metrics and lists the following benefits:

- Metrics present activities in objective terms.
- Metrics help focus on issues. For example, if recovery time objective (RTO) is six hours on paper, but testing shows it is 24 hours, this issue needs to be addressed.
- Metrics can be included in personnel objectives (i.e., responsibilities) and measured.

Here is a sample of metrics applicable to continuity planning from Persson:

- Disaster recovery planning (DRP) as a percentage of total IT budget. This may be between two percent and eight percent.
- Anticipated recovery time.
- Total number of outages per year by category (e.g., power, network, virus).

Drawing on Persson's work, metrics can be applied more broadly. Examples include the business continuity budget as a percentage of assets or revenue; the number of full-time employees (may be less than 1.0) who work on business continuity per $10 million of assets or revenue; and metrics can be compared within and among industries.

What do you think influences senior executive interest in business continuity planning?

Emergency Management

Bullock et al. (2006: 1–2) write that there is no single definition of emergency management because the discipline "has expanded and contracted in response to events, the desires of Congress, and leadership styles." They note that emergency management "has clearly become an essential role of government" and the "Constitution entrusted the states with responsibility for public health and safety—hence, responsibility for public risks—and assigns the federal government to a secondary, supportive role. The federal role was originally conceived such that it intervenes when the state, local, or individual entities are overwhelmed."

Bullock et al. (2006: 2) define emergency management as "the discipline dealing with risk and risk avoidance." Here, we define **emergency management** as preparation for potential emergencies and disasters and the coordination of response and resources during such events.

McEntire (2004: 14–18) explains the ongoing transformation of emergency management. He refers to issues such as definitions of terms, what hazards to focus on, what variables to explore (e.g., location of buildings, building construction, politics, critical incident stress), and what disciplines should contribute to emergency management. McEntire notes that there is no single overarching theory of emergency management because it would be impossible to develop a theory that would contain every single variable and issue involving disasters. He sees systems or chaos theory as gaining recognition because they incorporate many causative variables. McEntire also offers other theories and concepts (below), which he views as relevant to emergency management. Notice the similarities in some of the following theories to those explained under risk management (i.e., risk perception theory; risk communication theory):

- **Systems theory** involves diverse systems that interact in complicated ways and impact vulnerability. These systems include natural, built, technological, social, political, economic, and cultural environments.
- **Chaos theory** has similarities to systems theory in that many variables interact and affect vulnerability. This theory points to the difficulty of detecting simple linear cause-and-effect relationships and it seeks to address multiple variables simultaneously to mitigate vulnerability.

- **Decision theory** views disasters as characterized by uncertainty and limited information that results in increased vulnerability to causalities, disruption, and other negative consequences. This theory focuses on perceptions, communications, bureaucracy, politics, and other variables that impact the aftermath of disasters.
- **Management theory** explains disasters as political and organizational problems. Vulnerability to disasters can be reduced through effective leadership and improved planning. Leaders have the responsibility to partner with a wide variety of players to reach objectives that reduce vulnerability.
- **Organizational behavior theory** sees agencies concerned about their own interests and turf, without understanding how their action or inaction affects others. Cultural barriers, a lack of communications, and other variables limit partnering to increase efficiency and effectiveness. Improved communications during the 9/11 attacks in New York City would have saved some of the lives of firefighters, police, and others.
- **Risk perception and communication theory** focuses on the apathy of citizens prior to disasters. Vulnerability can increase if citizens do not understand the consequences of disasters and do not take action for self-protection (e.g., evacuation). Citizens may be more likely to reduce their vulnerability if authorities communicate risk accurately and convincingly.
- **Social constructionist theory** shifts from hazards that we cannot control to the role of humans in disasters, who determine through decisions the degree of vulnerability; for example, citizen decisions to reside in areas subject to mudslides.
- **Weberian theory** looks to culture—including values, attitudes, practices, and socialization—as contributing to increased vulnerability. Other factors under this theory resulting in greater vulnerability are weak emergency management institutions and a lack of professionalization among emergency managers.
- **Marxist theory** explains economic conditions and political powerlessness as playing roles in disaster vulnerability. The poor and minorities are more likely to reside in vulnerable areas and are often unable to act to protect themselves. The Hurricane Katrina disaster serves as an example; thousands of citizens were trapped in flooded New Orleans, which was built below sea level.

Risk Management in Government

Risk management is as important in the public sector as it is in the private sector. Risk management helps government executives study and manage risks so they can prioritize risks and then plan and take action under limited budgets. For government, exposures and risks are enormous and broad—affecting local, state, national, and global levels. Examples include liability, crime, workers' compensation, fire, war, terrorism, proliferation of weapons of mass destruction (WMD), disease and pandemics, natural disasters, conflicts over natural resources, and issues of migration.

Although the private sector, especially the insurance industry, has played a lead role in advancing the discipline of risk management, local and state governments are increasingly employing risk management methods. The Public Risk Management Association (PRIMA), the largest network of public risk management practitioners, states: "Risk management in the public sector finds itself at a challenging moment in time." "There is a very high level of interest in the subject, but a lack of clarity and consistency as to its meaning, form and purpose." The group sees pockets of advanced risk management practice in the United States; however, most small government entities do not have a formal risk management unit. PRIMA sees various segments of the discipline spread among departments (e.g., insurance, purchasing, and finance) in government (Public Risk Management Association, 2003).

On the federal government level, the U.S. Government Accountability Office (GAO), a federal "watchdog," is a strong proponent of risk management. It has argued for risk management to be applied at all levels of government, including the Departments of Homeland Security and Defense and the FBI. The U.S. Government Accountability Office (2005: 124–125) stated: "A vacuum exists in which benefits of homeland security investments are often not quantified and are almost never valued in monetary terms." The GAO argues for criteria for quantifiable and nonquantifiable benefits of homeland security spending so analysts can develop information to inform management.

The U.S. Department of Homeland Security (2004b: 54) stated: "We will guide our actions with sound risk management principles that take a global perspective and are forward-looking. Risks must be well understood, and risk management approaches developed, before solutions can be implemented."

The SAS Institute (2011) explains that enterprise risk management (ERM) is gaining attention in government to enhance efficiency and effectiveness. However, government risk managers face more than showing the value and ROI of ERM. They must satisfy the needs of politicians and their constituents as never-ending news events alter ERM priorities.

All-Hazards Preparedness Concept

"All-hazards" include natural disasters (e.g., hurricanes, earthquakes) and human-made events (e.g., inadvertent accidents, such as an aircraft crash, and deliberate events, such as the terrorist bombing of an aircraft). Another category is technological events (e.g., an electric service blackout resulting from a variety of possible causes). FEMA (2010: 1–2) states: "Planning considers all-hazards and threats." "Planners can address common operational functions in their basic plans instead of having unique plans for every type of hazard or threat." FEMA (2004) defines this approach as follows: "The **all-hazards preparedness concept** is simple in that how you prepare for one disaster or emergency situation is the same for any other disaster." This strategy is important because it plays a role in preventing our nation from over-preparing for one type of disaster at the expense of other disasters (Figure 12-2).

It is argued here that the all-hazards approach seeks to maximize the efficiency and cost effectiveness of emergency management efforts through a realization that different disasters contain similarities that can benefit, to a certain degree, from generic approaches to emergency management. This includes generally similar emergency services, equipment, and products. Conversely, there is a point at which **generic emergency management** must divide

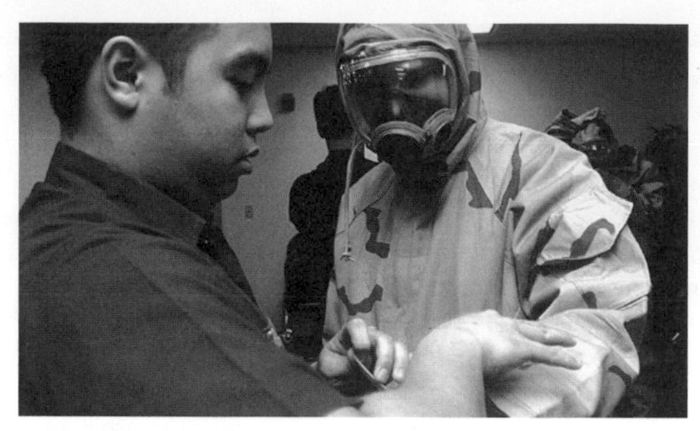

FIGURE 12-2 The U.S. government is working to provide an all-hazards response capability. *Courtesy: U.S. Department of Homeland Security.*

into **specialized emergency management**. For example, the detonation of a "dirty-bomb" or an approaching hurricane necessitates evacuation; however, the victims and the scene of these disasters require markedly different expertise, treatment, and equipment.

The History of Emergency Management

Haddow and Bullock (2003: 1–13) traced the development of emergency management as described next. The first example of the federal government becoming involved in a disaster was when, in 1803, a Congressional Act was passed to provide financial assistance to a New Hampshire town that was destroyed by fire. During the Great Depression (1930s), President Franklin D. Roosevelt spent enormous federal funds on projects to put people to work and to stimulate the economy. Such projects had relevance to emergency management. The Tennessee Valley Authority was created to not only produce hydroelectric power, but to reduce flooding.

During the 1950s, as the Cold War brought with it the threat of nuclear war, **Civil Defense Programs** grew. Communities and families building bomb shelters as a defense against attack from the Soviet Union characterized these programs. Air raid drills became common at schools; children practiced going under their desks or kneeling down in the hall while covering their heads. Civil defense directors, often-retired military personnel, were appointed at the local and state levels, and these individuals represent the beginning of emergency management in the United States. The Federal Civil Defense Administration (FCDA) provided technical assistance to local and state governments. Another agency, the Office of Defense Mobilization (ODM), was established in the Department of Defense (DOD) to quickly amass materials and production in the event of war. In 1958, both the FCDA and the ODM were merged into the Office of Civil and Defense Mobilization.

A series of destructive hurricanes in the 1950s resulted in ad hoc Congressional legislation to fund assistance to the affected states. The 1960s also saw its share of natural disasters that resulted in President John F. Kennedy's administration, in 1961, creating the Office of Emergency Preparedness. The absence of flood insurance on the standard homeowner policy prompted Congress to enact the National Flood Insurance Act of 1968, which created the National Flood Insurance Program (NFIP). This noteworthy act included the concept of **community-based mitigation**, whereby action was taken against the risk prior to the disaster. When a community joined the NFIP, the program offered federally subsidized, low-cost flood insurance to citizens in exchange for the community enacting an ordinance banning future development in its floodplains. As a voluntary program, the NFIP was not successful. The Flood Insurance Act of 1972 created an incentive for communities to join the NFIP. This act required mandatory purchase of flood insurance for homeowner loans backed by federal mortgages, and a significant number of mortgages were federally backed.

The 1970s brought to light the fragmentation of federal, state, and local agencies responsible for risk and disasters. Over 100 federal agencies added to the confusion and turf wars. Unfortunately, the problems were compounded during disasters. The administration of President Jimmy Carter sought reform and consolidation of federal emergency management. The **Federal Emergency Management Agency (FEMA)** was established by Executive Order 12127 on March 31, 1979. The goal of this effort was to consolidate emergency preparedness, mitigation, and response into one agency, with a director who would report directly to the president. John Macy, who was in the Carter cabinet, became the director of FEMA. He emphasized similarities between natural hazards preparedness and civil defense and developed an "all-hazards" approach. Through the 1980s, FEMA experienced numerous troubles and criticism. The top priority of FEMA became preparation for nuclear attack, at the expense of other risks. Environmental cleanup (e.g., Love Canal) became another priority. States saw their funding for emergency management decline. By the end of the 1980s, FEMA was in need of serious reform. FEMA's problems included tension with its partners at the state and local level over priorities and funding, and difficulty in responding to natural disasters. In 1989, Hurricane Hugo became the worst hurricane in a decade and caused 85 deaths and damage of more than $15 billion as it slammed into South Carolina, North Carolina, and other locations. The FEMA response was so poor that Senator Ernest Hollings (D-SC) called the agency the "sorriest bunch of bureaucratic jackasses." The FEMA problems with Hurricane Andrew, in 1992, in Florida, further eroded confidence in the agency.

Reform in FEMA finally occurred in 1993 when President William Clinton nominated James L. Witt, of Arkansas, to be director. Witt was the first director with emergency management experience, and he was skilled at building partnerships and serving customers. Inside FEMA, Witt fostered ties to employees, reorganized the agency, and supported new technologies to deliver disaster services. Externally, he improved relationships with state and local governments, Congress, and the media. Subsequent disasters, such as the Midwest floods in 1993 that resulted in nine states being declared disaster areas and the Northridge, California, earthquake in 1994, tested the reforms FEMA had made. Witt was elevated to be a member of President Clinton's cabinet, illustrating the importance of emergency management. Witt then sought to influence the nation's governors in elevating state emergency management directors to cabinet posts.

The World Trade Center bombing in 1993 and the Oklahoma City bombing in 1995 brought the issue of terrorism to the forefront of emergency management. A major question surfaced:

FIGURE 12-3 In 2005, Hurricane Katrina flooded sections of New Orleans, left hundreds of thousands of people homeless, and overwhelmed government response. *Courtesy: Gary Nichols, DOD.*

Which agency would be in charge following a terrorist attack? Prior to the 9/11 attacks, several federal agencies sought leadership against terrorism. Major agencies included FEMA, the Department of Justice (DOJ) (it contains the FBI), the DOD, the National Guard, and the Department of Health and Human Services (HHS). Coordination was poor as agencies pursued their own agendas. The DOD and the DOJ received the most funds. State and local governments felt unprepared and complained about their vulnerabilities and needs. The 9/11 attacks resulted in a massive reorganization of the federal government and the creation of the Department of Homeland Security (DHS) to better coordinate both actions against terrorism and homeland security.

In September of 2005, FEMA experienced another setback when its director, Michael Brown, was relieved as commander of Hurricane Katrina (Figure 12–3) relief efforts along the Gulf Coast, and he was sent back to Washington, D.C. Three days later, he resigned. Critics argued that he was the former head of an Arabian horse association and that he had no background in disaster relief when his friend and then-FEMA Director, Joe Allbaugh, hired him in 2001 to serve as FEMA's general counsel. Although finger-pointing occurred by politicians at all levels of government as the relief effort became overwhelmed, Michael Brown became the classic scapegoat. Following the Katrina disaster, there were calls for reform of FEMA.

Emergency Management Disciplines

Haddow and Bullock (2003) divide emergency management into the disciplines of mitigation, response, recovery, preparedness, and communications. Although they emphasize natural hazards, their writing provides a foundation for multiple risks.

Mitigation

Haddow and Bullock (2003: 37) define **mitigation** as "a sustained action to reduce or eliminate risk to people and property from hazards and their effects." It looks at long-term solutions to reduce risk, as opposed to preparedness for emergencies and disasters.

Mitigation programs require partners outside traditional emergency management. Examples are land-use planners, construction and building officials, and community leaders. The tools of mitigation include hazard identification and mapping and building codes.

Response

Response includes activities that address the short-term, direct effects of an incident, and it seeks to save lives, protect property, and meet basic human needs. Depending on the emergency, response can entail applying intelligence to lessen the effects or consequences of an incident; increased security; investigations; public health and agricultural surveillance and testing processes; immunizations, isolation, or quarantine; operations to disrupt illegal activity; and arresting offenders. First responders at emergencies and disasters are usually local police, fire, emergency medical specialists, and not state or federal personnel. Besides local, state, and federal responses, volunteer organizations and the business community respond to emergencies and disasters.

All 50 states and six territories of the United States operate an office of emergency management. The names of these offices vary, as does the placement in the organizational structure of each jurisdiction. National Guard Adjutant Generals lead these offices in most states, followed by leadership by civilian employees. Governors rely primarily on their state National Guard for responding to disasters. National Guard resources include personnel, communications systems, air and road transport, heavy construction equipment, mass care and feeding equipment, and assorted supplies (e.g., tents, beds, and blankets). When a state is overwhelmed by a disaster, the governor may request federal assistance under a presidential disaster or emergency declaration.

Although there is tension between federal and state authorities during disasters (e.g., the Katrina disaster), Gaines (2011) sees the need to provide more authority to the federal government for disaster coordination purposes, even prior to a governor's request. Present law includes the Robert T. Stafford Disaster Relief and Emergency Assistance Act of 1988 (**Stafford Act**) that provides a central statutory framework for the federal government's response to disasters. Through this law, the President can issue a Disaster Declaration and assistance is released through FEMA which coordinates responses from various agencies. The John Warner National Defense Authorization Act of 2006 (Warner Act) provides the President with power to deploy federal troops to a disaster without state approval. The Homeland

Security Act of 2002 established the DHS, which contains FEMA, to reduce our nation's vulnerability to terrorism, natural disasters, and other emergencies, and to assist in recovery. The Secretary of Homeland Security is responsible for coordinating federal operations in response to emergencies and disasters.

Recovery

Recovery involves actions to help individuals and communities return to normal. It serves to restore infrastructure, basic services, government, housing, businesses, and other functions in a community. Besides action by all levels of government, recovery includes the role of the insurance industry, which provides financial support. However, the federal government plays the largest role in technical and financial support for recovery. The range of grant and assistance programs that the federal government has been involved in over the years, especially through FEMA, has been enormous and cost tens of billions of dollars.

Preparedness

Before a disaster strikes, preparation is vital. **Preparedness** includes risk management to identify threats, hazards, and vulnerabilities. Then all levels of government and the private sector plan and prepare for disasters through planning, training, exercises, and identifying resources to prevent, protect against, respond to, and recover from disasters. Preparedness is the job of everyone and all sectors of our society.

Communications

DHS emphasizes the importance of communications and information sharing. Also important are a common operating picture and systems interoperability to disseminate warnings, communicate operational decisions, prepare for requests, and maintain overall awareness across jurisdictions (U.S. Department of Homeland Security, 2004a: 49).

During the 9/11 attacks at the World Trade Center, an unforeseen enemy of first responders was their own communications systems. During the months following the attacks, a serious flaw was found in the police and fire radio systems: they were incompatible. When the first WTC tower collapsed, police were notified to evacuate the second tower, but firefighters (on a different system) did not hear the warning and hundreds of first responders died. This serious problem can be solved by what has been called **interoperability**—the ability of multiple agencies to communicate using technology.

National Response Framework, National Incident Management System, and National Preparedness Goal

The United States has faced numerous emergencies and disasters and the risks and exposures will not end. Government agencies are in a constant state of improving planning and response efforts. We see new initiatives with new names and strategies as government continually seeks to refine and invigorate efforts at protection.

The **National Response Framework (NRF)** builds upon and replaces the National Response Plan. It provides guiding principles to unify all partners (e.g., government at all levels and the private sector) in their efforts to develop an all-hazards national response to emergencies and disasters. Rather than being a federal-centric plan, the NRF is a national effort that includes local and state jurisdictions. The NRF defines roles and structures that organize response.It also describes specific authorities and best practices. The NRF ties together emergency management activities to include prevention, mitigation, preparedness, response, and recovery. It groups capabilities and resources that are most likely to be needed during an incident. Examples include firefighting, search and rescue, and mass care. It outlines core responsibilities and expertise for specific contingencies (e.g., radiological, biological, cyber) (U.S. Department of Homeland Security, 2011b; U.S. Department of Homeland Security, 2008).

Homeland Security Presidential Directive (HSPD)-5, Management of Domestic Incidents, February 28, 2003, focused on establishing a single, comprehensive response plan (now the NRF) and the **National Incident Management System (NIMS)**. This directive designates the Secretary of Homeland Security to administer these programs. The NIMS was developed with input from many groups in the public and private sectors. *NIMS emphasizes standardization*. Examples include standardization in incident command structure, planning, training, exercises, personnel qualifications and certification, equipment, resource management, forms, and information management. It represents best practices and contains a core set of doctrines, concepts, principles, terminology, and organizational processes. *HSPD-5 requires all federal agencies and departments to adopt the NIMS and make it a requirement for state and local organizations to receive federal preparedness assistance.* The NIMS Integration Center publishes standards, guidelines, and compliance protocols. The DHS offers no-cost, on-line training for the NIMS; thousands have completed the training.

The **National Preparedness Goal (NPG)**, under Presidential Policy Directive (PPD) 8, defines the core capabilities required to prepare for the greatest risks to the security of the United States. It aims to achieve an integrated, "all-of-Nation" preparedness approach (i.e., a broad spectrum of partners). The NPG reflects the NIMS and policies from the National Security Strategy, Homeland Security Presidential Directives, and other National Strategies and Directives. The Secretary of DHS is responsible for coordinating the Goal, which includes a system of guidance for planning, organization, equipment, training, and exercises (U.S. Department of Homeland Security, 2011c).

The Military

The military has the capability to provide significant resources to civilian authorities in the event of a disaster (Figures 12-4 and 12-5). Federal troops can provide communications, transportation, food, water, shelter, and other assets under what is referred to as **logistics**. Specifically, **U.S. Northern Command (NORTH-COM)**, in the DOD, is charged with protecting the United States from foreign attacks and is responsible for supplying military resources to emergency responders. The DOD emphasizes that it understands its role in assisting civilian authorities—serve to support, not take the lead.

A state's governor is responsible for disaster response, with the option of requesting assistance from FEMA. Then, if necessary, FEMA can contact thc DOD for assistance. Once the

FIGURE 12-4 U.S. Navy Search and Rescue specialist prepares to bring a Hurricane Katrina victim onboard a helicopter. *Courtesy: Jay C. Pugh, USN.*

FIGURE 12-5 U.S. Army soldier directs Hurricane Katrina victims away from helicopter following rescue. *Courtesy: Robert McRill, USN.*

request arrives, DOD officials must determine whether the request is legal under the **Posse Comitatus Act** (PCA), a post-Civil War-era act that generally prohibits the military from engaging in civilian law enforcement. The PCA does not apply to the U.S. Coast Guard in peacetime or to the National Guard in Title 32 (United States Code) or State Active Duty status.

Prohibiting direct military involvement in civilian law enforcement is in keeping with long-standing U.S. law and policy limiting the military's role in domestic affairs. However, this law has been labeled as archaic because it limits the military from responding to disasters. Consequently, a modification of the law is an option.

Congress enacted a number of exceptions to the PCA that allow the military, in certain situations, to assist civilian law enforcement agencies in enforcing the laws of the United States. The most common example is counterdrug assistance (Title 10 USC, Sections 371–381).

An example of the DOD assisting in past disasters includes its response during and immediately after Hurricane Katrina, in August of 2005, when New Orleans was flooded and city residents were forced to evacuate. Because Gulf-state and FEMA resources were overwhelmed, the DOD provided massive aid to assist victims, conduct search-and-rescue missions, maintain order, and deal with the flooding.

The Katrina Disaster

Despite the best intentions and efforts of the specialists who formulated the "all-hazards" National Response Plan, the National Incident Management System, and the National Preparedness Goal, Hurricane Katrina showed the difficulty of planning for disasters. The hurricane caused about 1,300 deaths, the flooding of New Orleans, and enormous destruction along Gulf-coast states. The media, reporting from New Orleans, offered vivid coverage of looters, snipers, fires, bloated corpses, victims sitting on rooftops or swimming through toxic water, and the awful conditions at the Superdome where people sought shelter and help from the government. Today, victims of Hurricane Katrina are still living in cities across the United States and are unable or unwilling to return to their communities while rebuilding goes on.

One major lesson from the Hurricane Katrina disaster came from the collapse of intergovernmental relations. The NRP called for an immediate activation of a joint operations center to bring together all levels of government. Why did this NRP requirement occur so slowly? Why did each level of government appear inept with its authority? Why were the resources inadequate? Did the federal government place too much emphasis on the threat of terrorism at the expense of emergency management? Local, state, and federal officials blamed each other for the response failures, and the issues were politicized. Interestingly, officials received warning of the approaching hurricane. A WMD attack is unlikely to be preceded by a warning. If a warning preceded Hurricane Katrina and government reacted so inadequately, what type of government reaction can citizens expect from a surprise WMD attack?

Another major lesson learned was that chaos can result from a serious disaster, whether from a WMD attack, pandemic, or other cause. Leaders should learn from the Katrina disaster and enhance planning at all levels of government.

In 2006, following congressional hearings that exposed the shortcomings of government response to Hurricane Katrina, a revised NRP was formulated to react more quickly with improved resources. U.S. Government Accountability Office (2006) staff visited the affected areas, interviewed officials, and analyzed a variety of information. It found widespread dissatisfaction with the preparedness and response, and many of the lessons that emerged from Hurricane Katrina were similar to those the GAO identified earlier with Hurricane Andrew. For example, the president should designate a senior official in the White House to oversee federal preparedness and response to major disasters. Again, in 2006, the GAO recommended that, prior to disasters, the leadership roles, responsibilities, and lines of authority for all levels of government must be clearly defined prior to disasters. The GAO also emphasized the need for strong advance planning among responding organizations, as well as robust training and exercises to test plans in advance of disasters. Changes are inevitable for all plans, because disasters are often unpredictable and public and private sector plans typically fall short of need during major disasters.

Following Hurricanes Katrina and Rita, as of mid-December 2005, FEMA had distributed nearly $5.4 billion in assistance to more than 1.4 million registrants under the Individuals and Households Program (IHP) to meet necessary expenses and serious needs, such as temporary housing and property repair. Because of the devastation, FEMA also activated expedited assistance to provide fast-track money to victims. GAO investigators found rampant fraud with expedited assistance. FEMA made millions of dollars in payments to thousands of registrants who used Social Security numbers that were never issued or belonged to deceased or other individuals, and several hundred registrants used bogus damaged property addresses. FEMA made duplicate expedited assistance payments of $2,000 to about 5,000 of the 11,000 debit card recipients—once through the distribution of debit cards and again through a check or electronic funds transfer. Although most of the payments were applied to necessities, some funds were used for adult entertainment, to purchase bail bond services, to pay traffic tickets, and to purchase weapons. The GAO recommended improvements by FEMA to automate the validation of identities and damaged properties (U.S. Government Accountability Office, 2006).

What are your suggestions for improving government response to disasters?

International Perspective

Eastern Japan Great Earthquake Disaster of 2011

On March 11, 2011, an undersea 9.0 magnitude earthquake about 45 miles off the coast of Japan became the most powerful earthquake ever to strike that country. It caused tsunami waves up to 133 feet high that traveled as far as six miles inland in the Sendai area of Japan. Tsunami walls built for protection were not high enough, residents did not expect the tsunami waves to be so high, and entire towns were destroyed. There were over 16,000 deaths, mostly by drowning. Because of damaged crematoriums, morgues, and limited capacity, governments and the military buried many bodies in mass graves. Thousands were injured or missing. Over 125,000 buildings were damaged or destroyed and damage to infrastructure was extensive. About 230,000 vehicles were damaged or destroyed. Approximately 4.4 million households in northeastern Japan were without electricity and 1.5 million were without water.

What made the disaster especially horrific was that three nuclear reactors exploded and radioactive leakage occurred, which added to the massive evacuation. Airborne radiation was detected by the U.S. Navy at bases in the Tokyo area, 200 miles to the south of the Sendai area. The Japanese food supply and other necessities were being tested for radiation. The disaster resulted in a huge humanitarian crisis. Hundreds of thousands of people were displaced and shortages of food, water, shelter, and other necessities occurred. Estimates of the cost of the disaster exceed $300 billion.

Since this chapter includes business continuity planning, it is emphasized next in conjunction with this disaster. To begin with, many manufacturers around the world depend on Japanese industries for essential parts in the supply chain. Without essential parts, products cannot be manufactured and businesses and economies suffer. Brennan (2011: 22–26) explained that the shock of the disaster caused risk managers, supply chain managers, and others to work on mitigating the impact of the disaster on production and revenue. In addition, businesses questioned their crisis management and business continuity plans for improved effectiveness. Brennan writes: "When the lack of availability of a one dollar part prevents a company from making a $30,000 product, something needs to change." He explained some of the work and challenges faced by those in manufacturing worldwide who were seeking to fill voids in the supply chain caused by the disaster. Teams worked late into the night, then communicated tasks to colleagues in other time zones. The teams alternated sleeping so work could continue without interruption. Some companies were unable to locate manufacturers because all they had was an invoice address; this poor planning hindered supply chain repair. Brennan noted that both government and the business sector need to do more to reduce the impact of major disasters.

Search the Internet

Following is a list of websites relevant to this chapter:

Insurance and Risk Management:

A.M. Best: www.ambest.com
American Association of Insurance Services: www.aaisonline.com
American Insurance Association: www.aiadc.org
American Risk and Insurance Association: www.aria.org
Coalition Against Insurance Fraud: www.insurancefraud.org
General advice: FREEADVICE.com
Insurance Committee for Arson Control: www.arsoncontrol.org
Insurance Services Office: www.iso.com
International Organization for Standardization: www.iso.org
International Risk Management Institute: www.irmi.com
National Alliance for Insurance Education & Research: www.TheNationalAlliance.com
Public Risk Management Association: www.primacentral.org
Risk and Insurance Management Society: www.rims.org
Risk World: www.riskworld.com
Society for Risk Analysis: www.sra.org
Standard & Poor's: www.standardandpoors.com
State Risk and Insurance Management Association: www.strima.org
Surety & Fidelity Association of America: www.surety.org

Business Continuity and Emergency Management:

ASIS International: www.asisonline.org
Business Continuity Institute: www.thebci.org

Centers for Disease Control and Prevention: www.bt.cdc.gov
DRI International: www.drii.org
Emergency Management Accreditation Program: www.emaponline.org/index.cfm
Federal Emergency Management Agency: www.fema.org
International Association of Emergency Managers: www.iaem.com
National Emergency Management Association: www.nemaweb.org
National Fire Protection Association: www.nfpa.org
The International Emergency Management Society: www.tiems.org
U.S. Department of Homeland Security: www.dhs.gov
U.S. Department of Homeland Security: www.ready.gov
U.S. Northern Command: www.northcom.mil

Case Problems

12A. As a loss prevention manager, your superior mentions that she is hearing a lot about resilience and wants to introduce the concept into corporate planning. How do explain this concept and what do you suggest?

12B. As a loss prevention manager, you will soon explain to top management why they should provide support and funds to initiate a risk management program by hiring a risk manager. Research and answer the following questions to prepare for your meeting.

(a) How will a risk manager help to perpetuate the business?
(b) How can a risk manager produce a return on investment?
(c) How are five risk management tools applied?
(d) How will the risk manager and loss prevention manager work together, and what will each do for the company?

12C. You are a corporate risk management executive responsible for insurance, business continuity, and the corporate safety and security departments for all corporate locations. You are challenged by the following list of tasks upon returning to your office. Prioritize these tasks and explain why you placed them in your particular order of importance.

Item A

E-mail message. At your request, an insurance company representative replies that she can meet with you at your convenience to explain why the corporate liability insurance premium will rise by 10 percent next year.

Item B

Telephone voice message. An accident on the premises between a truck and a forklift has resulted in two injured employees. The director of human resources wants to meet with you immediately.

Item C

E-mail message. As treasurer of a regional risk management association, you are assigned the task of arranging the next meeting, including location, meal, and speaker.

Item D

Your "to do list" states that you must re-evaluate the risks facing the corporation and implement an improved risk management plan that maximizes risk management tools and is more financially sound.

Item E

Telephone call. A security officer on the premises calls and states that he cannot locate a security supervisor; there is a fire in the warehouse; he can extinguish it; and you do not have to call the fire department.

Item F

Telephone voice message. Your boss wants to meet with you immediately because insurance covered only 50 percent of losses from an accident at another corporate plant.

Item G

Telephone voice message. An attorney representing a plaintiff/employee in a sexual harassment suit against the corporation wants to speak with you.

Item H

Telephone voice message. The emergency management director for the local county wants to meet with you about cooperation on joint planning and training.

Item I

E-mail message. The information technology (IT) director needs to meet with you as soon as possible. One of the insurers of the corporation rejected the corporate IT business continuity plan.

The Decision for "You Be the Judge #1"

Hawkins was reimbursed for the stolen tools. The court held that when Hawkins left his tools in the work area overnight with the knowledge and consent of his employer, his employer accepted temporary custody of the property. This situation is known as a *bailment,* and in such a situation, the party accepting custody of the property is usually responsible for its care and safekeeping. Under the laws of the state in which this case was tried, Conway Excavation might have escaped liability for the theft of Hawkins's tools if it had taken more extensive steps to make the garage secure. Instead, it had to pay him some $3,000.

This case is based on *Harper v. Brown & Root* 398 Sa2d 94. The names in this case have been changed to protect the privacy of those involved.

The Decision for "You Be the Judge #2"

The court disagreed with MMC, concluding that although the policy was complicated and full of errors, it was not ambiguous in spelling out the limits of its liability—$50,000. MMC would have to absorb the rest of the loss itself. Security Director Douglas *could have saved his company a lot of money if he had taken the time to read the insurance policy when it was first issued.* That's the time to ask questions and demand clarification. If you cannot understand the policy, find someone who can; the insurance agent or your company's attorney are two of the best people to ask. After the company puts in a claim, it may be too late to clear up the ambiguities.

This case is based on *Davenport Peters v. Royal Globe Ins.*, 490 FSupp 286. The names in this case have been changed to protect the privacy of those involved.

References

ASIS International. (2009). *ASIS International Standard: ASIS SPC 1-2009, Organizational Resilience: Security, Preparedness, and Continuity Management Systems-Requirements with Guidance for Use.* <www.asisonline.org> retrieved June 4, 2009.

ASIS International. (2010a). "Enterprise security risk management: how great risks lead to great deeds." <www.asisonline.org> retrieved March 24, 2011.

ASIS International. (2010b). *Business Continuity Management Systems: Requirements with Guidance for Use, ASIS/BSI BCM.01-2010, American National Standard.* <www.asisonline.org> retrieved October 10, 2011.

Ball, R. (2008). Terrorism risk insurance act (TRIA) extended to expire December 31, 2014. <www.integrogroup.com> retrieved October 26, 2011.

Bendixen, L. (2011). What is resilience? *The CIP Report, 9* (February) <cip.gmu.edu> retrieved October 12, 2011.

Bieber, R. (1987). The making of a risk manager—part one. *Risk Management* (September)

Borodzicz, E. (2005). *Risk, crisis & security management.* West Sussex, England: John Wiley & Sons.

Brennan, P. (2011). Lessons learned from the Japan earthquake. *Disaster Recovery Journal, 24* (Summer)

Brown, J. (2011). Risk management and resilience. *The CIP Report, 9* (February) <cip.gmu.edu> retrieved March 20, 2011.

Bullock, J., et al. (2006). *Introduction to homeland security* (2nd ed.). Burlington, MA: Elsevier Butterworth-Heinemann.

D'Addario, F. (2010). Compliance, care and the long view. *Security Technology Executive, 20* (September)

DiMaria, J. (2010). Buyer beware: PS-Prep auditor training and organization certification. *Continuity Central* (November 2) <www.continuitycentral.com> retrieved October 31, 2011.

Dunaway, W., & Shaw, G. (2010). The influence of collaborative partnerships on private sector preparedness and continuity planning. *Journal of Homeland Security and Emergency Management, 7*(1)

FEMA. (2004). "All-Hazards preparedness." <www.fema.gov/preparedness/hazards_prepare.shtm> retrieved March 1, 2005.

FEMA. (2010). "Developing and maintaining emergency operations plans" (November). <www.fema.gov> retrieved November 3, 2011.

Gaines, B. (2011). Law in crisis: a look at governmental powers in the face of a public health disaster. *Journal of Biosecurity, Biosafety and Biodefense Law, 1* <www.bepress.com/jbbbl/vol1/iss1/3> retrieved January 25, 2012.

Garris, L. (2005). Make-or-break steps for disaster preparation. *Buildings, 99* (February)

Glasser, B. (2011). Exploring business interruption insurance in the wake of disaster. *Disaster Recovery Journal, 24* (Summer)

Gurwitt, R. (2001). The riskiest business. *Governing, 14* (March)

Haddow, G., & Bullock, J. (2003). *Introduction to emergency management.* Boston: Butterworth-Heinemann.

Homeland Security Policy Institute. (2011). *Preparedness, Response, and Resilience Task Force: Interim Task Force Report on Resilience.* <www.homelandsecurity.gwu.edu> retrieved June 1, 2011.

Insurance Newsnet. (2006). RIMS benchmark survey book records total cost of risk down in 2005 (May 25). <www.insurancenewsnet.com> retrieved November 14, 2006.

Insurance Services Office. (2011a). About ISO. <www.iso.com> retrieved October 24, 2011.

Insurance Services Office. (2011b). ISO's public protection classification program. <www.isomitigation.com/ppc/0000/ppc0001.html> retrieved October 26, 2011.

International Risk Management Institute. (2011). Causes of loss forms (ISO). <www.irmi.com> retrieved October 26, 2011.

Leimberg, S., et al. (2002). *The tools & techniques of risk management & insurance*. Cincinnati, OH: The National Underwriter Co.

McEntire, D. (2004). The status of emergency management theory: issues, barriers, and recommendations for improved scholarship. Paper presented at the FEMA Higher Education Conference, Emmitsburg, MD, June 8. <highpapers.asp> retrieved February 14, 2006.

Mehta, S. (2010). It's time for ERM. *Financial Executive, 26* (November) <find.galegroup.com> retrieved January 28, 2011.

Moody, M. (2006). Risk management: a board issue yet? *Rough Notes* (April) <www.roughnotes.com/rnmagazine/2006/april06/04p034.htm>, retrieved May 8, 2006.

Morris, B. (2001). Risk takes on the world. *Risk & Insurance, 12* (April 16)

Pandolfi, G. (2011). Demonstrating the value of resiliency in an enterprise business continuity program. *Disaster Recovery Journal, 24* (Summer)

Persson, J. (2005). The time has come for DRP metrics. *Disaster Recovery Journal, 18* (Winter) <www.drj.com> retrieved January 28, 2005.

Pinto, C., & Kirchner, T. (2011). Innovative enterprise risk management tools. *Disaster Recovery Journal, 24* (Winter)

Public Risk Management Association. (2003). Core competency statement: a framework for public risk management. <www.primacentral.org> retrieved February 18, 2005.

Rejda, G. (2001). *Principles of risk management and insurance* (7th ed.). Boston, MA: Addison Wesley.

Risen, T. (2010). Can insurers protect the U.S. from cyber-attack? *National Journal* (November 7) <www.nationaljournal.com> retrieved October 26, 2011.

Risk and Insurance Management Society. (2011). *2011 RIMS Benchmark Survey*. <www.RIMS.org/benchmark> retrieved October 27, 2011.

SAS Institute. (2011). "Risk management's role in government1." <www.sas.com>, retrieved October 30, 2011.

Speight, P. (2011). Business continuity. *Journal of Applied Security Research, 6* (December)

Starner, T. (2003). Modeling for terrorism. *Risk & Insurance* (April 1)

Stoneman, B. (2003). An Aversion to dispersion. *Risk & Insurance* (September 15)

Straw, J. (2010). Bouncing back after a disruption. *Security Management, 54* (July)

Tuckey, S. (2010). Terrorism risk stays at bay. *Risk and Insurance Online* (September 1) <www.riskandinsurance.com> retrieved January 28, 2011.

Tuohy, C. (2010). Claims management in-depth series: the lure of the third party. *Risk and Insurance Online* (February 1) <www.riskandinsurance.com> retrieved January 28, 2011.

U.S. Department of Homeland Security. (2004a). *National Incident Management System*. <www.dhs.gov> retrieved February 10, 2005.

U.S. Department of Homeland Security. (2004b). *Securing Our Homeland: U.S. Department of Homeland Security Strategic Plan*. <www.dhs.gov>, retrieved October 1, 2004.

U.S. Department of Homeland Security. (2011a). Presidential Policy Directive/PPD-8: National Preparedness (March 30). <www.dhs.gov> retrieved October 31, 2011.

U.S. Department of Homeland Security. (2011b). National Response Framework (February 17). <www.dhs.gov> retrieved October 12, 2011.

U.S. Department of Homeland Security. (2011c). National Preparedness Goal (September). <www.dhs.gov> retrieved October 31, 2011.

U.S. Department of Homeland Security. (2008). National Response Framework (January). <www.dhs.gov> retrieved October 12, 2011.

U.S. Department of Homeland Security. (2010). Voluntary private sector preparedness accreditation and certification program (PS-Prep) resource center. <www.fema.gov> retrieved October 31, 2011.

U.S. Government Accountability Office. (2005). "Agency Plans, Implementation, and Challenges Regarding the National Strategy for Homeland Security (January). <www.gao.gov/cgi-bin/getrpt?GAO-05-33> retrieved February 15, 2005.

U.S. Government Accountability Office. (2006). Statement by comptroller general david m. walker on gao's preliminary observations regarding preparedness and response to hurricanes Katrina and Rita (February 1). <www.gao.gov/new.items/d06365r.pdf> retrieved February 3, 2006.

U.S. Government Accountability Office. (2010). Critical Infrastructure Protection: Update to National Infrastructure Protection Plan Includes Increased Emphasis on Risk Management and Resilience (March). <www.gao.gov/new.items/d10296.pdf> GAO-10-296, retrieved October 15, 2011.

Ward, S. (2001). Exploring the role of the corporate risk manager. *Risk Management: An International Journal*, 3.

White House. (2010). *National Security Strategy* (May). <www.wwhitehouse.gov> retrieved October 15, 2011.

Williams, C., et al. (1995). *Risk management and insurance* (7th ed.). New York: McGraw-Hill.

Zuccarello, F. (2011). Kidnapping for ransom: a fateful international growth industry. *Rough Notes* (May) <www.roughnotes.com> retrieved October 25, 2011.

13 Life Safety, Fire Protection, and Emergencies

OBJECTIVES

After studying this chapter, the reader will be able to:

1. Explain the significance of life safety and name two sources for life safety planning
2. Discuss the problems posed by fire
3. Describe the roles of private organizations and public fire departments in fire protection
4. List and explain five fire prevention strategies
5. List and explain five fire suppression strategies
6. Describe the roles of public police and emergency medical services in public safety
7. List and describe three types of human-made emergencies and three types of natural disasters

KEY TERMS

- life safety
- Occupational Safety and Health Administration (OSHA)
- NFPA 101 Life Safety Code
- authority having jurisdiction
- building codes
- fire codes
- fire prevention
- fire suppression
- hazmat
- floor wardens
- areas of refuge
- sheltering-in-place
- integrated fire suppression systems
- Internet Protocol Digital Alarm Communicators/Transmitters (IPDACT)
- standpipes
- fire brigade
- performance-based fire protection
- public safety
- true first responders
- personal protective equipment
- triage
- apps
- weapon of mass destruction (WMD)
- mass notification systems
- pandemic
- zoonotic diseases
- quarantine enforcement

Life Safety

Life safety pertains to building construction design that increases safety, what organizations and employees can do in preparation for emergencies, and what they can do once an emergency occurs. *Citizens cannot depend on government during the initial stages of an emergency.* Citizens facing emergencies—in their home, workplace, and in public—must take steps to protect themselves for a duration that may last from a few minutes to possibly days until government, first responders, and assistance arrive. Although this chapter focuses on life safety in the workplace, protection methods for a variety of emergencies, at home or in public, are available from numerous sources (see "Search the Internet" at the end of the chapter).

Standards, Regulations, and Codes

Standards, regulations, and codes provide a foundation for professionals involved in life safety programs. Refer to Chapter 3 for an explanation of these terms.

OSHA Regulations

The **Occupational Safety and Health Administration (OSHA)** is a federal agency established to administer the law on safety and health in the workplace. Its regulations are contained in the Code of Federal Regulations (CFR), as are other federal regulations. A widely known example of federal regulations is found in Title 29 CFR, OSHA. The following regulations were in effect when researched on November 14, 2011.

An example of an OSHA regulation that promotes life safety is 1910.38 Emergency Action Plans. It includes these minimum elements:

- An emergency action plan must be in writing, kept in the workplace, and available to employees for review.
- Procedures for reporting a fire or other emergency.
- Procedures for emergency evacuation, including type of evacuation and exit route assignments.
- Procedures to account for all employees after evacuation.
- Procedures to be followed by employees performing rescue or medical duties.
- An employer must have and maintain an employee alarm system that uses a distinctive signal for each purpose.
- An employer must designate and train employees to assist in a safe and orderly evacuation of other employees.
- An employer must review the emergency action plan with each employee.

Another OSHA regulation oriented toward life safety is 1910.39 Fire Prevention Plans. It includes these minimum elements:

- A fire prevention plan must be in writing, kept in the workplace, and available to employees for review.
- A list of all major fire hazards, proper handling and storage procedures for hazardous materials, potential ignition sources and their control, and the type of fire protection equipment necessary to control each major hazard.
- Procedures to control accumulations of flammable and combustible waste materials.

- Procedures for regular maintenance of safeguards installed on heat-producing equipment to prevent the accidental ignition of combustible materials.
- The name or job title of employees responsible for maintaining equipment to prevent or control sources of ignition or fires.
- An employer must inform employees upon initial assignment to a job of the fire hazards to which they are exposed and review with each employee those parts of the fire protection plan necessary for self-protection.

NFPA 101 Life Safety Code

Tragedy is often followed by action to enhance safety. National attention focused on the importance of life safety (e.g., adequate emergency exits) following the 1911 Triangle Shirtwaist Factory Fire in Manhattan. This fire spread from the eight to the tenth floor as young female garment workers, unable to escape because of a locked door, jumped to their deaths, some while their clothes were on fire, others while holding hands. Although firefighters rushed to the scene, their ladders reached only to the sixth floor. The fire resulted in 146 deaths and prompted New York to draft numerous laws to promote safety (Matthews, 2011). Another notable fire resulting in calls for safety was the Coconut Grove Night Club Fire in Boston in 1942, resulting in 492 deaths.

The history of NFPA 101 began with a presentation by R. H. Newbern at the 1911 Annual Meeting of the NFPA. The following year, his presentation resulted in the Committee on Safety to Life publishing "Exit Drills in Factories, Schools, Department Stores and Theaters." This committee studied notable fires and causes of loss of life, and prepared standards for the construction of stairways, fire escapes, and egress routes. The publication of additional pamphlets, which were widely circulated and put into general use, provided the foundation for the NFPA 101 Life Safety Code.

The **NFPA 101 Life Safety Code** is used in every state. It is also used by numerous federal agencies. The state fire marshal's office serves as a resource if one seeks to find out if the code has been adopted in a particular locale and which edition is being used.

The state fire marshal often serves as the **authority having jurisdiction** (AHJ), meaning the person or office charged with enforcing the code. In some jurisdictions, the AHJ is the fire department or building department. For some occupancy, there is more than one AHJ. A hospital, for example, may require approval for life safety from multiple authorities having jurisdictions.

Here are samples from the NFPA 101 Life Safety Code:

Chapter 1, Administration

1.1.2 Danger to Life from Fire. The *Code* addresses those construction, protection, and occupancy features necessary to minimize danger to life from fire, including smoke, fumes, or panic.

1.1.3 Egress Facilities. The *Code* establishes minimum criteria for the design of egress facilities to allow prompt escape of occupants from buildings or, where desirable, into safe areas within buildings.

Chapter 7, Means of Egress

7.1.10.1 Means of egress shall be continuously maintained free of all obstructions or impediments to full instant use in the case of fire or other emergency.

7.2.1.5.1 Doors shall be arranged to be opened readily from the egress side whenever the building is occupied.

7.2.1.5.2 Locks, if provided, shall not require the use of a key, a tool, or special knowledge or effort for operation from the egress side.

7.2.1.6.1 Delayed-Egress Locks. Approved, listed, delayed egress locks shall be permitted to be installed on doors serving low and ordinary hazard contents in buildings protected throughout by an approved, supervised automatic fire detection system.

7.2.1.7.1 Where a door is required to be equipped with panic or fire exit hardware, such hardware shall meet the following criteria:

- **(1)** It shall consist of a cross bar or a push pad, the actuating portion of which extends across not less than one-half of the width of the door leaf.
- **(2)** It shall be mounted as follows:
  - **(a)** New installations shall be not less than 865 mm (34in.), nor more than 1220 mm (48in.), above the floor.
  - **(b)** Existing installations shall be not less than 760 mm (30in.), nor more than 1220 mm (48in.), above the floor.
- **(3)** It shall be constructed so that a horizontal force not to exceed 66N (15 lbf) actuates the cross bar or push pad and latches.

7.9.1.1 Emergency lighting facilities for means of egress shall be provided.

Chapter 9 Building Service and Fire Protection Equipment

9.7.1 Automatic Sprinklers.

Chapter 11 Special Structures and High-Rise Buildings

11.8.2 Extinguishing Requirements.

11.8.2.1 High-rise buildings shall be protected throughout by an approved, supervised automatic sprinkler system in accordance with Section 9.7. A sprinkler control valve and a waterflow device shall be provided for each floor.

11.8.4 Emergency Lighting and Standby Power.

9/11 Transcripts Released

The following is from a *USA TODAY* news story (Cauchon et al., 2003) on the radio transmissions and telephone calls during the 9/11 attacks:

> *A man on the 92nd floor called the police with what was—though he did not know it—the question of his life. "We need to know if we need to get out of here, because we know there's an explosion," said the caller, who was in the south tower of the World Trade Center. A jet had just crashed into the Trade Center's north tower. "Should we stay or should we not?" The officer on the line asked whether there was smoke on the floor. Told no, he replied: "I would wait 'til further notice." "All right," the caller said. "Don't evacuate." He then hung up. Almost all the roughly 600 people in the top floors of the south tower died after a second hijacked airliner crashed into the 80th floor shortly after 9A.M. The failure to evacuate the building was one of the day's great tragedies. The exchange between the office worker and the policeman was one of many revealed when the Port Authority of New York and New Jersey, which owned and patrolled the office complex, released transcripts of 260 hours of radio transmissions and telephone calls on 9/11.*

Morgan Stanley

The following is from the U.S. Department of Homeland Security, READYBusiness (n.d.):

> *In 1993, when terrorists attacked the World Trade Center for the first time, financial services company Morgan Stanley learned a life-saving lesson. It took the company four hours that day to evacuate its employees, some of whom had to walk down 60 or more flights of stairs to safety. While none of Morgan Stanley's employees was killed in the attack, the company's management decided its disaster plan just was not good enough. Morgan Stanley took a close look at its operation, analyzed the potential disaster risk and developed a multi-faceted disaster plan. Perhaps just as importantly, it practiced the plan frequently to provide for employee safety in the event of another disaster. On September 11, 2001, the planning and practice paid off. Immediately after the first hijacked plane struck One World Trade Center, Morgan Stanley security executives ordered the company's 3800 employees to evacuate from World Trade Center buildings Two and Five. This time, it took them just 45minutes to get out to safety!*
>
> *The crisis management did not stop at that point, however. Morgan Stanley offered grief counseling to workers and increased its security presence. It also used effective communications strategies to provide timely, appropriate information to management and employees, investors and clients, and regulators and the media. Morgan Stanley still lost 13 people on September 11, but many more could have died if the company had not had a solid disaster plan that was practiced repeatedly. In making a commitment to prepare its most valuable asset, its people, Morgan Stanley ensured the firm's future.*

Does the U.S. Department of Homeland Security Place Enough Emphasis on Protecting Civilian Facilities?

Sternberg and Lee (2006: 1–19) write that when disasters strike, people's safety often depends on the buildings they occupy. They argue that the U.S. Department of Homeland Security is misdirected by placing too much emphasis on protecting infrastructures (e.g., transportation, utilities, and government buildings) at the expense of improving protection at civilian facilities (e.g., apartment and office buildings). Sternberg and Lee offer specific suggestions to improve the protection of civilian facilities:

- The protection of civilian facilities necessitates a category of homeland security that deserves discrete federal, state, and local attention.
- The federal government should do more to facilitate incentives, support, information, training, technology, and research to protect civilian facilities.
- The challenge is not for government to find a central solution for protecting civilian facilities, but rather to find policy solutions that motivate more managers in civilian facilities to assess risks and search for solutions.
- Governments at all levels must work to improve and clarify regulations so that investments in protection at civilian facilities increase.

Building Design and Building Codes

Making changes in building design and building codes is a slow process, and new codes often do not apply to existing buildings. Experts (e.g., structural engineers), industry groups (e.g., real estate), and government leaders frequently disagree over what changes, if any, should be made.

Guidance for local governments for building codes is found with the International Code Council (ICC) or the NFPA. The ICC's International Building Code (IBC) has been widely adopted in the United States. The NFPA writes the Life Safety Code (LSC) and the Building Construction and Safety Code (BCSC). The LSC has been adopted by far more jurisdictions than the BCSC.

The IBC establishes minimum standards for the design and installation of building systems, and it addresses issues of occupancy, safety, and technology. It plays a major role in transforming lessons from the 9/11 attacks into updated building codes, and it relies on research from the U.S. Department of Commerce, National Institute of Standards and Technology (NIST). The NIST detailed reports on the collapse of the World Trade Center (WTC) towers, and its reports also include the impact of natural hazards (e.g., hurricanes) on tall buildings.

NIST research found the following factors in the collapse sequence of the WTC towers: each aircraft severed perimeter columns, damaged interior core columns, and knocked off fireproofing from steel; jet fuel initially fed the fires, followed by building contents and air from destroyed walls and windows; and floor sagging and exposure to high temperatures caused the perimeter columns to bow inward and buckle. NIST recommendations that could have improved the WTC structural performance on 9/11 are as follows: fireproofing less susceptible to being dislodged; perimeter columns and floor framing with greater mass to enhance thermal and buckling performance; compartmentation to retard spread of fire; windows with improved thermal performance; fire-protected and hardened elevators; and redundant water supply for standpipes (National Institute of Standards and Technology, 2005).

Fire Protection

The Problems of Fire

As written in Chapter 2, the U.S. Fire Administration publishes statistics on the enormous losses from fire, including thousands of deaths and injuries and billions of dollars in property losses. Most human casualties in fires result from smoke and the toxic fumes or gases within it.

High-rise structural fires are particularly challenging because smoke and flame movement is very different from other structures. High-rises often contain multiple types of occupancies such as residential, commercial, restaurant, and underground parking. Each type presents challenges that must be approached differently. Exits from high-rises are limited, and emergency evacuation is difficult. High-rise fires require significantly more personnel and equipment to extinguish than do other types of fires.

Besides high-rise fires, many other types are challenging. Examples are wild fires, fires at industrial plants, and at vehicles containing fuels or chemicals.

Private Organizations Involved in Fire Protection

As explained in Chapter 3, a number of private organizations assist public and private sector efforts at security, safety, and fire protection. This assistance is through research, standards, and publications. Examples of these private organizations include NFPA and UL. Another related private organization is Factory Mutual Global. This organization works to improve the effectiveness of fire protection systems and new fire suppression chemicals, as well as cost evaluation of fire protection systems. It tests materials and equipment submitted by manufacturers. An approval guide is published, and like UL, it issues labels to indicate that specific products have passed its tests.

Fire Departments

Fire departments are local institutions. There are about 31,000 local fire departments in the United States and over one million firefighters, of whom about 750,000 are volunteers. Over half of firefighters protect small, rural communities of fewer than 5,000 residents, and these locales rely on volunteer departments with scarce resources. One of the best strategies for these locales, outside major metropolitan areas, is to develop mutual aid agreements to share resources. This approach improves preparation and response to fires and emergencies (U.S. Department of Homeland Security, n.d.).

Historically, fire departments focused on fire prevention and fire suppression. However, in the past 30 years, fire departments expanded services to emergency medical needs, hazardous material incidents, terrorism response, natural disasters, specialized rescue, and other community needs. Because of these additional community requirements and the call for research on the changing nature of the fire service, the U.S. Fire Administration (USFA) partners with many federal agencies, the International Association of Fire Chiefs, and the International Association of Fire Fighters (IAFF). Research initiatives include study of virtual reality training of fire behavior to avoid harm to firefighters, protective clothing and equipment, emergency vehicle safety, and cancer among firefighters (U.S. Fire Administration, 2011).

The "Emergency Support Function Annex," within the *National Response Framework* (U.S. Department of Homeland Security, 2008: ESF #4-1), states that the primary agency for coordination during a significant fire is the U.S. Department of Agriculture/Forest Service, with support from the Departments of Commerce, Defense, Homeland Security, Interior, State, and the Environmental Protection Agency. The function is to detect and suppress wildland, rural, and urban fires resulting from an incident requiring a coordinated federal response and provide personnel and equipment to supplement all levels of government in firefighting.

Fire Department Protection Efforts

Facility Planning

Public fire protection tasks often involve the private sector. In many locales, it is legally required that public fire personnel review construction plans for new facilities. This may entail consultation with architects, engineers, and loss prevention practitioners on a number of subjects ranging from fire codes to water supplies for sprinkler systems. On-site inspections by fire personnel ensure compliance with plans. Municipal fire and water department officials often prepare recommendations to local government bodies for water supply systems for new industrial plants to ensure an adequate water supply in the event of a fire. Cooperation and planning among stakeholders enhances fire protection.

Prefire Planning

Plans assist fire personnel in case of fire. An on-site survey is made of a particular building with the aid of a checklist. Then the actual prefire plans are formulated for that structure. Drawings and computer-aided designs are applied to identify the location of exits, stairs, firefighting equipment, hazards, and anything else of importance. Additional information is helpful: construction characteristics and that of adjacent buildings, types of roofs, number of employees, and the best response route to the building. Prefire plans also serve as an aid to training. Naturally, firefighting personnel do not have the time to prepare prefire plans for all structures in their jurisdiction. One- and two-family residential structures are omitted in favor of structures that are more complex where greater losses can occur, such as schools, hospitals, theaters, hotels, and manufacturing plants.

Public Education

This strategy involves educating the public about the fire problem. The public can become a great aid in reducing fires if people are properly recruited through education campaigns. Public education programs utilize mass media, social media, contests, lectures, and tours of firehouses. Building inspections also educate the public by pointing out fire hazards.

Zimmerman (2012: 373) writes that firefighters often top the polls in surveys as the most trusted profession and this popularity can be used to promote safety. He also notes that firefighters should ensure the success of public education through NFPA 1035, Standard for Professional Qualifications for the Fire and Life Safety Educator, Public Information Officer, and Juvenile Firesetter Intervention Specialist.

Codes

Years ago, as the United States was evolving into an industrial giant, buildings were constructed without proper concern for fire prevention. Building codes in urban areas either did not exist or were inadequate to ensure construction designed to prevent fire-related losses. In fact, a year before the Great Chicago Fire in 1871, Lloyd's Insurance Company of London halted the writing of policies in that city because of fire-prone construction practices.

Prompted by the difficulty of selling insurance because of higher rates for hazardous buildings and the losses incurred by some spectacular fires, insurance companies became increasingly interested in fire prevention strategies. Improvements in both building construction and fire departments slowly followed.

Although insurance associations played an important role in establishing fire standards, government support was necessary to enforce fire codes. Today, local governments enforce state regulations and local ordinances that support fire codes. Fire department personnel inspect structures to ensure conformance to codes. To strengthen compliance by owners of buildings, penalties (e.g., fines) are meted out for violations so that fire hazards are reduced.

Codes can be in the form of fire codes and building codes. Frequently, there is disagreement about what should be contained in each and what responsibility and authority should be given to fire inspectors as opposed to building inspectors.

Generally, construction requirements go into **building codes**, and these codes are enforced by building inspectors. Most of the United States has adopted the International Building Code (IBC) from the International Code Council (ICC). This code includes many requirements pertaining to fire prevention, such as the number of exits and related construction characteristics. A building code in competition with the IBC is the Comprehensive Consensus Code from the NFPA and other groups.

Fire codes, enforced by firefighting personnel, deal with the maintenance and condition of various fire prevention and suppression features of buildings (e.g., sprinkler systems). In addition, fire codes cover hazardous substances, hazardous occupancies, and general precautions against fire.

In some localities, codes are of poor quality. A prime factor is construction cost. Interest group pressure (e.g., the construction industry) on government officials who stipulate codes has been known to weaken codes. It is a sad case when a high-rise building catches fire and people perish because no sprinklers were installed on upper floors, and the fire department was not equipped to suppress a fire so high up. Later, the media broadcasts the tragedy, and government officials meet to satisfy the public outcry. Stronger codes often emerge. A similar scenario occurs with weak codes following natural disasters (e.g., a hurricane).

Inspections

The primary purpose of building inspections by firefighting personnel is to uncover deviations from the fire code. The frequency and intensity of these inspections vary. Because of budget constraints and a shortage of personnel, many fire departments are not able to conduct enough inspections to equal national standards of several inspections per year for hazardous buildings. The *NFPA Inspection Manual* outlines methods for conducting inspections.

A press release from the City of New York, Department of Investigation (2011), stated that 110 individuals were arrested on outstanding warrants for failing to appear in criminal court for fire and building code violations. This effort was coordinated with the Fire Department and Department of Buildings to track down violators who refused to remedy code violations issued by inspectors who seek safety for occupants of buildings and responding firefighters.

Legal Implications

Fire marshals are provided with broad powers to ensure public safety. This is especially evident in fire inspections and investigations, in rights to subpoena records, and in fire marshal's hearings. Most courts have upheld these powers.

In almost all local jurisdictions, the state has delegated police powers so that local officials regulate safety conditions through ordinances. Fire ordinances stipulate inspection procedures, number of inspections, violations, and penalties. When differences of opinion develop over individual rights (e.g., of a building owner) versus fire department police powers, the issue is often resolved by the courts. Chapter 4 differentiates administrative inspection and administrative search warrant. It also notes that the legal standard for obtaining an administrative search warrant, also known as an inspection warrant, is lower than what is required to obtain a search warrant in a criminal case (Hall, 2006: 146–151).

Fire Prevention and Fire Suppression Strategies

Businesses and institutions can do a lot to protect against fire. **Fire prevention** focuses on strategies that help to avoid the inception of fires. **Fire suppression** applies personnel, equipment, and other resources to suppress fires.

The following strategies are emphasized in subsequent pages:

Fire Prevention	Fire Suppression
Inspections	Integrated systems
Planning	Detection of smoke and fire
Safety	Contact the fire department
Good housekeeping	Extinguishers
Storage and transportation of hazardous substances and materials	Sprinklers
	Standpipe and hose systems
Evacuation	Fire-resistive buildings
Training	Training and fire brigades

The fire triangle (see Figure 13-1) symbolizes the elements necessary for a fire. Fire requires heat, fuel, oxygen, and then a chemical chain reaction. When all three characteristics, plus a chemical chain reaction, are present, there will be fire. If any one is missing, either through prevention strategies (e.g., good housekeeping, safety) or suppression (i.e., extinguishment), fire will not occur. Heat is often considered the ignition source. A smoldering cigarette, sparks from a welder's torch, or friction from a machine can produce enough heat to begin a fire.

Almost every working environment has fuels, heat, and oxygen. Loss prevention practitioners and all employees in general must take steps to reduce the chances for fire by isolating fuels and controlling heat. Not much can be done about pervasive oxygen, but fuels such as gasoline and kerosene should be stored properly away from sources of heat.

Fire Prevention Strategies

Inspections

Inspections seek to uncover deficiencies and deviations from codes. Then corrective action becomes the heart of the inspection prevention strategy. Checklist questions include the following: Are new facility designs and manufacturing processes being submitted to appropriate

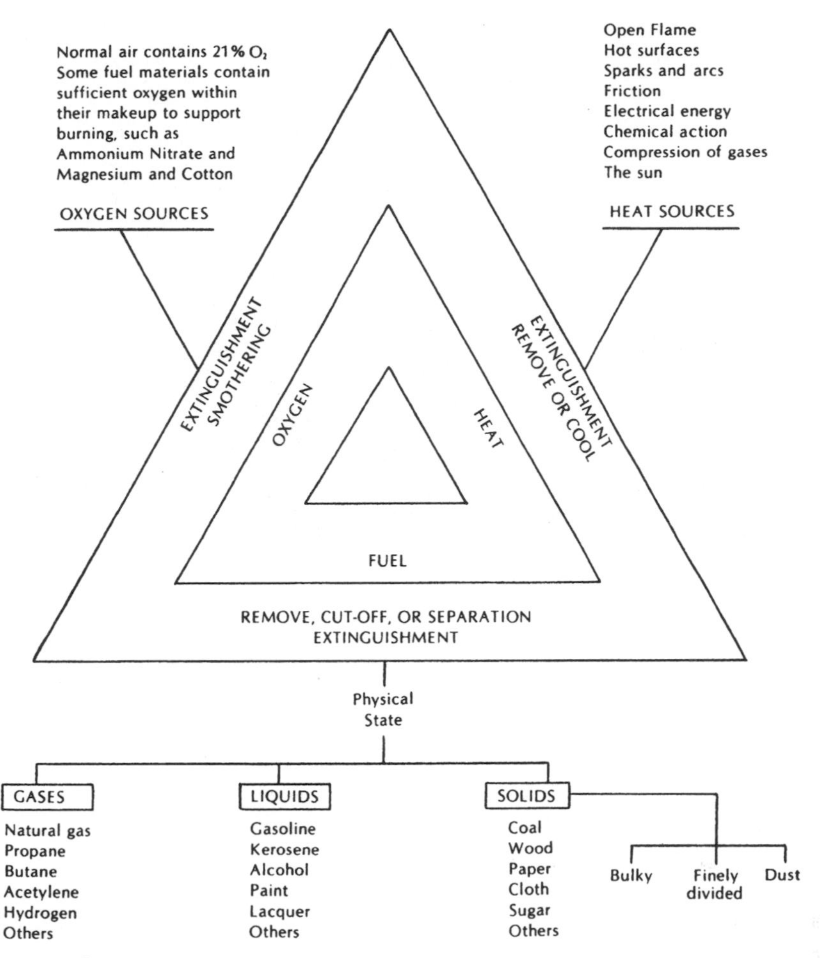

FIGURE 13-1 Fire triangle.

personnel for fire protection review? Do employees receive periodic training on fire prevention and suppression systems, policies, and procedures (see Figure 13-2)? What is the condition of fire suppression equipment?

Planning

Feedback from inspections helps to plan strategies against fire losses. An interdisciplinary planning group is often an excellent source for plans. Fire department personnel, architects, engineers, insurance specialists, loss prevention practitioners, and others can provide a multitude of ideas. Management support is an important ingredient in the planning process. By supplying adequate personnel and funding and supporting policies and procedures, management can strengthen the fire protection program. Earlier, OSHA regulations pertaining to fire protection plans were explained.

FIGURE 13-2 Training and inspections improve fire protection.

Safety

Some safety strategies for a fire prevention program follow:

1. Set up smoking and no smoking areas that are supervised, safe, and clearly marked with signs.
2. In smoking areas, provide cigarette butt and match receptacles or sand urns.
3. When fire protection equipment or systems are planned, select those that have been approved by a reputable testing organization (e.g., UL).
4. In the use of heating systems, such as boilers, maintain safety when lighting up, during usage, and when shutting down.
5. Examine motors frequently to ensure safe operation and to prevent overheating.
6. Never overload electrical circuits.
7. Maintain lightning protective devices (e.g., lightning rods).
8. Employ an electrician who is safety conscious.
9. Prohibit the use of welding equipment near flammable substances or hazardous materials.
10. Watch sparks during and after welding.
11. Train employees to create an atmosphere of safety.
12. Conduct inspections and correct deficiencies.
13. Ensure fire protection standby for hazardous operations.

Good Housekeeping

Good housekeeping is another fire prevention strategy. It consists of building care, maintenance, cleanliness, proper placement of materials, careful waste and garbage disposal, and other general housekeeping activities.

Hazardous Materials (Hazmat) Incidents

Hazmat refers to substances that are toxic, poisonous, radioactive, flammable, or explosive and can cause injury or death from exposure (Schottke, 2007: 450). The prevention of disasters from hazardous substances and materials is extremely important. *Such materials are used in every community and transported by trucks, railcars, ships, barges, and planes.* Examples of hazardous substances and materials are plastics, fuels, and corrosive chemicals (e.g., acids). A tremendous amount of information exists concerning their physical and chemical properties, methods for storage and transportation, and the most appropriate strategies in the event of fire or accident. Hazmat safety is promoted through numerous federal agencies (e.g., OSHA and Department of Transportation), state agencies, laws, and regulations.

Hazmat responses require *caution* because the effects can be deadly and first responders do not want to become victims themselves. Approaches, with proper protective clothing, should be made from upwind, uphill, and upstream, and vehicles should be parked facing out for quick escape. Top priorities during a hazmat incident are to protect yourself and others from exposure and contamination, isolate the area, identify the substance for subsequent action, and deny entry to reduce exposure.

Evacuation and Medical Services

A key factor in fire prevention planning is the prevention of injuries and deaths. Two vital considerations are *evacuation* and *medical services.* Evacuation plans and drills help people prepare for the risk of fire and other emergencies (Figure 13-3). Smoke and fire alarms often provide warning for escape. Emergency exit maps and properly identified emergency doors also prevent injuries and deaths. Employees should turn off all equipment, utilize designated escape routes, avoid elevators, and report to a predetermined point on the outside to be counted. While employees are evacuating, firefighters may be entering the premises. At this point, a coordinated traffic flow is crucial. Firefighting equipment and personnel need to be directed to the fire location. Personnel should also be assigned to crowd control.

If injuries do take place, the quickness and quality of emergency medical services can save lives and unnecessary suffering. Preplanning will improve services. Specific employees should be trained to administer first aid while waiting for public emergency medical services.

Here is a list of suggestions to promote safety and fire protection in buildings (American National Standards Institute Homeland Security Standards Panel, 2010; Spadanuta, 2010; Carter et al., 2004; U.S. Fire Administration, 2004; Azano and Gilbertie, 2003; and Quinley and Schmidt, 2002).

- A committee should be formed to plan life safety.
- Life safety plans should be coordinated with public safety agencies and a variety of groups on the premises (e.g., management, employees in general, safety and security officers, and others).
- Training, education, drills, and exercises are an investment in life safety. Records should be maintained of such activities for regulators, insurers, and others. The records should be secured in multiple locations.

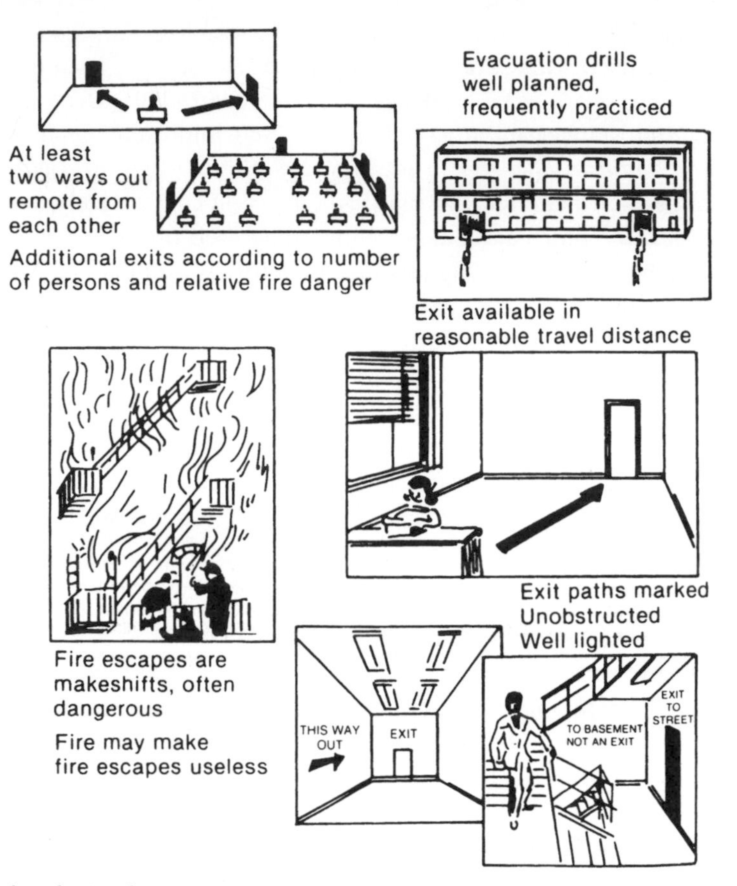

Evacuation drills
well planned,
frequently practiced
At least
two ways out
remote from
each other
Additional exits according to number
of persons and relative fire danger
Exit available in
reasonable travel distance
Exit paths marked
Unobstructed
Well lighted
Fire escapes are
makeshifts, often
dangerous
Fire may make
fire escapes useless
THIS WAY
OUT
EXIT
TO BASEMENT
NOT AN EXIT
EXIT
TO
STREET

FIGURE 13-3 Principles of exit safety.

- A chain of command and hierarchy of life safety volunteers should be maintained. For example, a building executive can choose **floor wardens** who can choose assistant floor wardens and searchers. These trained individuals can train others and lead drills. Searchers ensure no one is left behind.
- Training for floor wardens and assistants should include these topics: the variety of risks (e.g., fire, workplace violence), the evacuation plan, escape routes, safety, means of communication, what to do if someone is injured, where to assemble outside the building, and counts.
- Wardens can be supplied with evacuation kits containing flashlight, reflective vest, clipboard for recording employees outside the building, and a flag to mark the rendezvous point.
- Planning should consider assignments for security officers (e.g., stationary posts at access points and traffic control).
- Evacuation plans should include occupants with special needs, and employees should be assigned to assist them.

- Evacuation of high-rise buildings is complicated because of the number of occupants and the time needed to evacuate, especially those in upper floors. A minor fire may require evacuation of the floor containing the fire and two floors both above and below the fire. A serious fire would require a total evacuation.
- Evacuation plans should include **areas of refuge** (e.g., oversized landing at a stairwell or sealed smoke or fire compartments on a floor) and how they will be used. A method of communication should be installed at the area of refuge.
- Before opening a door during a fire emergency, feel the door with the back of your hand to gauge temperature. If it is warm, opening it may cause smoke and fire to enter. This means you may be forced to stay in your office or apartment. Stuff the cracks around the door with towels. Telephone the fire department to inform them about your location and go to the window with a flashlight or wave a sheet.
- If the door is not warm, brace your body against it, stay low, and slowly open it. If there is no smoke or fire, evacuate.
- If smoke is around you, stay low because smoke rises. Place a cloth over your mouth and nose.
- Employee information is crucial during an emergency, and it should be up-to-date. Also important is a contact list of public safety agencies, medical services, regulators, government officials, critical suppliers, and others.
- Building and site maps are important to indicate such features as utility shutoffs, water lines, electrical and other utilities, floor plans, exits, stairways, hazards, and high-value assets.
- An airborne hazard (e.g., WMD) outside a building may necessitate **sheltering-in-place**. For example, a plume of hazardous material may be approaching a high-rise building with no time to evacuate, and traffic would block escape. Ideal rooms to shelter-in-place would be on the opposite side of the approaching plume, above the first floor, windowless, and with a minimum number of vents and doors. Duct tape or other materials can be used to seal the room, and the HVAC system should be turned off. Because many hazardous agents are heavier than air, occupants should shelter-in-place in upper floors.
- Practitioners responsible for safety and fire protection programs should seek feedback to improve performance. Feedback can be gathered from training, drills, and exercises. In addition, audits can expose deficiencies in the expertise of personnel, equipment, systems, communications, and coordination with others.

Training

Through training, fire prevention becomes everybody's responsibility. Employees must first understand the disastrous effects of serious fire losses. This includes not only harm to humans, but also lost productivity and jobs. Topics within training can include good housekeeping, hazardous substances and materials, evacuation, and first aid. Knowledge is transmitted via online courses, lectures, videos, demonstrations, drills, visits by public fire prevention personnel, pamphlets, and posted fire prevention signs. Incentive programs, whereby employees compete for prizes, also can increase fire prevention awareness.

Haphazard Fire Protection at Best Buy Service Company

The Best Buy Service Company was a unique and rapidly growing business that sold numerous consumer items similar to those in department stores. Best Buy's success was due to a no-frills store design and customer self-service. Each store was essentially a warehouse located away from main roads. Loss prevention was of minimal concern to management. Strategies against crime, fire, and accidents were haphazard.

One store, which also served as a distribution center, had an unfortunate experience. Late one afternoon, before closing, a salesperson threw a lighted cigarette butt into a trash container. The trash and then some boxes nearby caught fire. When the spreading fire surprised store personnel, they panicked. The first thing they did was to run out of the warehouse with the customers. While the employees watched the burning warehouse in amazement, the manager asked if anybody had called the fire department. Nobody responded, so he used his cell phone to call the fire department. The manager continued to watch the fire and remembered that the automatic sprinklers were turned off because of freezing temperatures. Also in his thoughts were the thousands of dollars' worth of merchandise burning up.

For the Best Buy Company, the store and its contents were a tremendous loss. Insurance covered only a small part of the losses, especially because many insurance company recommendations went unheeded: the local fire department was never contacted for prefire plans and employees were never trained for fire protection. Senior management was clearly at fault.

What do you think can improve fire prevention in the United States?

Fire Suppression Strategies

The success of fire suppression strategies depends primarily on planning, preparation, equipment and system quality, and the readiness of personnel.

Integrated Systems

Integrated fire suppression systems perform a variety of functions (Spadanuta, 2010: 58–60; Suttell, 2006: 78–80). Detectors can measure smoke and the rate of temperature rise. If danger is evident, an alarm is sounded at the earliest stages of a fire. The fire can be extinguished automatically by water from sprinklers. Other functions include displaying CCTV video and floor plans on a computer screen to pinpoint the fire, notifying the fire department of the fire and its exact location, activating a public address (PA) system to provide life safety messages to occupants, starting up emergency generators for emergency lights and other equipment, unlocking doors for escape, detecting changes in sprinkler system water pressure, turning off certain electrical devices and equipment (e.g., shutting down fans that spread fire and smoke), venting specific areas, creating safe zones for occupants, and returning elevators to ground level to encourage the use of emergency stairways. If a human being were to analyze the fire threat and make these decisions, the time would obviously be greater than the split second needed by a computer.

Because buildings often contain integrated fire and security systems, it is important that engineers with verified skills design such systems. In addition, the local fire department can be contacted for their input. It is often recommended that the fire alarm system be planned as the first priority—as opposed to, for example, a security system—because of the need to deal with the fire authority's jurisdiction and codes and standards.

Pearson (2010: 30–32) writes of **Internet Protocol Digital Alarm Communicators/ Transmitters (IPDACT)** that allows fire alarm control panel communications to a central station over the internet. (IP systems were explained in Chapter 7.) He advises caution because of code requirements, AHJ approval of the IP fire system, and down times of IT systems.

Detection of Smoke and Fire

Many businesses utilize a combination of the following detectors for increased protection:

- *Smoke detectors* are widely used, especially since most human casualties in fires result from smoke and the toxic fumes or gases within smoke. These detectors operate with photoelectric light beams and react when smoke either blocks the beam of light or enters a refraction chamber where the smoke reflects the light into the photo cell.
- *Ionization detectors* are sensitive to invisible products of combustion created during the early stages of a fire. These detectors are noted for their early warning capabilities.
- *Thermal detectors* respond to heat either when the temperature reaches a certain degree or when the temperature rises too quickly. The latter is known as a *rate-of-rise detector.* Thermal detectors are made with either feature or a combination of both.
- *Flame detectors* detect flame and glowing embers. These detectors are sensitive to flames not visible to the human eye. The infrared kind is responsive to radiant energy that humans cannot see.
- *Sprinkler water flow detectors* contain a seal that melts when heat rises to a specific temperature. Then water flows from the sprinkler system. An alarm is activated when the water flow closes pressure switches.
- *Carbon monoxide detectors* protect against what is often called the "silent killer," because carbon monoxide is difficult to detect. In fact, victims, in their drowsy state, may be wrongly diagnosed as being substance abusers.
- *Gas detectors* monitor flammable gases or vapors. These devices are especially valuable in petroleum, chemical, and other industries where dangerous gases or vapors are generated.
- *Combination detectors* respond to more than one fire-producing cause or employ multiple operating principles. Examples include smoke/heat detector or rate-of-rise/fixed temperature heat detector.

Contact the fire department. Sometimes simple steps are overlooked. When a serious fire begins, the local fire department must be contacted as soon as possible to reduce losses. The best strategy to prevent a situation in which everybody thought somebody else had contacted the fire department is to ask: "Who called the fire department?" Another problem develops when people think that they can extinguish a fire without outside assistance. It is not until precious time has elapsed and serious danger exists that the fire department is contacted.

Alarm signaling systems are automatic or manual. With *automatic systems,* an attachment of a siren or a bell to a smoke or fire detection device or sprinkler system will notify people in the immediate area of a smoke or fire problem. This type of alarm is called a local alarm. Unless incorporated into this system, a local alarm will not notify a fire department. Automatic systems also consist of a local alarm and an alarm that notifies a central station or a fire department. Many large industries have a central, proprietary monitoring station that checks smoke, fire, burglar, and other sensors. Manual fire alarm signaling systems use a pull station fixed to a wall. This is a local alarm unless an alarm signal is transmitted to a central station or fire department.

Portable Extinguishers

The following classes of fires provide a foundation for firefighting and use of portable extinguishers:

- *Class A* fires consist of ordinary combustible materials such as trash, paper, fibers, wood, drapes, and furniture.
- *Class B* fires are fueled by a flammable liquid, such as gasoline, oil, alcohol, or cleaning solvents.
- *Class C* fires occur in live electrical circuits or equipment such as generators, motors, fuse boxes, computers, or copying machines.
- *Class D* fires, the rarest of the five types of fires, are fueled by combustible metals such as sodium, magnesium, and potassium.
- *Class K* fires involve combustible cooking fuels such as vegetable or animal oils or fats (Carter, 2004: 186).

Portable fire extinguishers are used to douse a small fire by directing onto it a substance that cools the burning material, deprives the flame of oxygen, or interferes with the chemical reactions occurring in the flame. Ratings and effectiveness of these extinguishers are in NFPA 10, Standard for Portable Fire Extinguishers (National Fire Protection Association, 2010).

Employees and loss prevention practitioners must be knowledgeable about the proper use of extinguishers. If the wrong extinguisher is used, a fire may become more serious. Water must not be used on a flammable liquid such as gasoline (Class B fire) because the gasoline may float on the water and spread the fire. Neither should one spray water on electrical fires (Class C fires) because water conducts electricity, and electrocution may result. Many locations use multipurpose dry chemical extinguishers that can be applied to A, B, or C fires. This approach reduces confusion during a fire. Class D fires are extinguished with dry powder extinguishers. Class B extinguishers have been used on Class K fires with limited effectiveness. Class K extinguishing agents are usually wet chemicals—water-based solutions of potassium carbonate, potassium acetate, or potassium citrate (Carter, 2004: 187).

Fire extinguishers should be checked at least every week during an officer's patrol. An alternative is digital monitoring connected to a security or fire alarm system; a device is attached to each extinguisher to check pressure, tampering, or if it is missing. A bar code can also be attached to each extinguisher. These technologies save labor costs, help to manage inventory, and can be applied to other equipment, such as defibrillators (Morton, 2012: 26).

Service companies recharge extinguishers when necessary. A seal is attached to the extinguisher that certifies its readiness.

An Angry Ex-Employee's Revenge

Albert Drucker had been warned numerous times about pilfering small tools from the maintenance department at Bearing Industries. When he was caught for the third time, via a strict inventory system, management decided to fire him. When Drucker was informed, he went into a rage and stormed out of the plant. While leaving, he vowed, "I'm gonna get you back for this." Management maintained that it made the right decision.

Two weeks later, Drucker was ready with his vindictive plan. At 2:00 one morning, he entered the Bearing plant by using a previously stolen master key. No loss prevention systems or services hindered his entrance. Within 15 minutes, he collected three strategically located fire extinguishers. When Drucker arrived home, he quickly emptied the contents of the extinguishers and then filled each one with gasoline. By 5:30 A.M., Drucker had replaced the three extinguishers and he was home sleeping.

At 2:00P.M., two days later, when the Bearing plant was in full production, Drucker sneaked into the plant unnoticed and placed, on a pile of rags, a book of matches with a lighted cigarette underneath the match heads. By the time Drucker was a few miles away, the old rags and some cardboard boxes were on fire. When employees discovered the fire, they were confident that they could extinguish it. They reached for the nearest extinguishers and approached the fire. To their surprise, the fire grew as they supplied it with gasoline. Their first reaction was to drop the extinguishers and run; one extinguisher exploded while the fire intensified. The fire caused extensive damage but no injuries or deaths. The police and management suspected arson. When police investigators asked management if there was anybody who held a grudge against the company, Albert Drucker's name was mentioned. He was arrested a week later and charged with arson.

Sprinklers

A sprinkler system consists of pipes along a ceiling that contain water under pressure, with an additional source of water for a constant flow. Attached to the pipes, automatic sprinklers are placed at select locations. When a fire occurs, a seal in the sprinkler head ruptures at a preestablished temperature, and a steady stream of water flows.

Research compiled by Hall (2011) shows that sprinklers are an effective and reliable fire suppression strategy for buildings. However, he writes that sprinklers are still rare in most places where people are most exposed to fire, including educational buildings, stores and offices, public assembly properties, and especially homes, where most fire deaths occur. Hall notes that sprinklers are 87 percent effective. When a sprinkler system fails, the most frequent reason (65%) is that the system was turned off prior to the fire. Other reasons include manual intervention that defeated the system (16%), maintenance issues (7%), and inappropriate system for the type of fire (5%).

A sprinkler system is a worthwhile investment for reducing fire losses. Lower insurance premiums actually can pay for the system over time. NFPA 13, Standard for the Installation of Sprinkler Systems, provides best practices for system design, installation, water supplies, and equipment. Also helpful as a "standard of care" and for protection in case of litigation is adhering to NFPA 25, Standard for the Inspection, Testing, and Maintenance of Water-Based

Fire Protection Systems. Failing to update inspection documentation with an insurer can affect coverage (Morton, 2011: 30). The AHJ may also be interested in such documentation.

There are several kinds of automatic sprinkler systems. Two popular ones are the wet-pipe and the dry-pipe systems. With the *wet-pipe system* (Figure 13-4), water is in the pipes at all times and is released when heat ruptures the seal in the sprinkler head. This is the most common system and is applicable where freezing is no threat to its operation. Where freezing temperatures and broken pipes are a problem, the *dry-pipe system* is useful. Air pressure is maintained in the pipes until a sprinkler head ruptures. Then the air escapes, and water enters the pipes and exits through the opened sprinklers; because of this delay, dry-pipe systems are not as effective as wet-pipe systems during the early stages of a fire (Naffa, 2009: 28).

Older buildings may have pipes that apply fire-suppressant chemicals such as carbon dioxide or Halon. Fire codes now prohibit these chemicals. The former absorbs oxygen, creating a danger to humans, whereas the latter depletes the earth's ozone layer.

Standpipe and Hose Systems

Standpipe and hose systems enable people to apply water to fires in buildings. **Standpipes** are vertical pipes that allow a water supply to reach an outlet on each floor of a building. In multiple-story structures, standpipes often are constructed within fire-resistant fire stairs as an added defense for the standpipes, hoses, and firefighting personnel. The typical setup is a folded or rolled 2½" hose enclosed in a wall cabinet and identified with fire emergency information (Figure 13-5). A control valve, which looks like a small spoked wheel, enables water to flow. Automatic extinguishing systems (e.g., sprinklers) often are the preferred system; however, the standpipe and hose systems are advantageous when the automatic system fails or is not present, when sections of a building are not accessible to outside hose lines and hydrants, and when properly trained employees are capable of fire suppression. (See NFPA and OSHA.)

Training is essential for employees if they are to have the responsibility of fighting a fire. With hoses, two people are required: one to stretch the hose to its full length and another to turn the water flow valve. Without training, employees may be injured if they do not understand the danger of turning the valve before the hose is stretched. This could cause the coiled section of the hose to react to the water pressure by acting like a whip and possibly striking someone.

Fire Walls and Doors

Fire walls are constructed in buildings to prevent the spread of fire. These walls are made of materials that resist fire and are designed to withstand fire for several hours. Fire walls are weakened by openings such as doorways. Therefore, fire doors at openings help to strengthen the fire wall when resisting fire. Fire doors often are designed to close automatically in case of fire. Nationally recognized testing laboratories study the reliability of these doors.

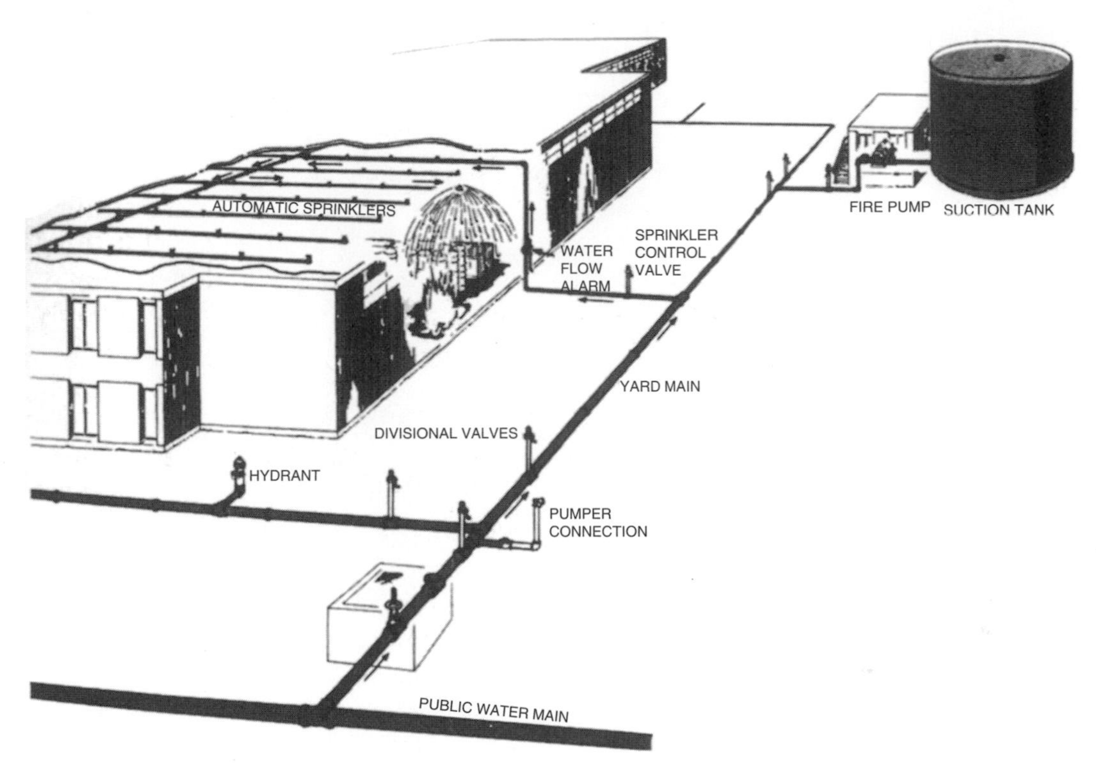

FIGURE 13-4 Total concept of the wet-pipe automatic sprinkler system.

FIGURE 13-5 Standpipe system including wall cabinet, hose with nozzle, and valves. A portable fire extinguisher is also contained in the cabinet.

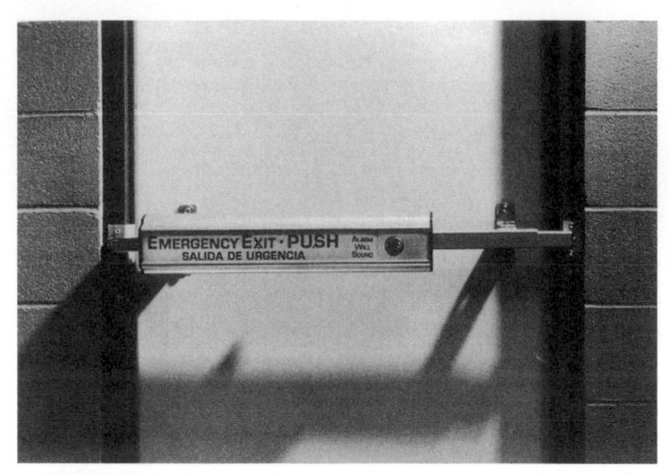

FIGURE 13-6 Emergency exit. *Courtesy: Sargent & Greenleaf, Inc.*

Stairwells

During evacuation, stairwells (made of masonry construction) provide a fire-resistant path for escape. Fire codes often require stairwells to withstand a fire for at least two hours. Stairwells may be equipped with fans to reduce smoke during evacuation.

Access Control

During an emergency, electronically controlled doors should be connected to the life safety system to permit escape. From a security perspective, this presents a problem because an alarm condition may provide an opportunity for an offender to enter or exit with ease. To deal with such vulnerability, CCTV and security officers can be applied to access points.

A locked door along the perimeter of a building may be unlocked from the inside to allow theft or unauthorized passage. Therefore, the door needs to be secured from both sides while permitting quick escape in case of an emergency as required by codes. The solution is a controlled exit device (Figure 13-6). One type of controlled exit device stays locked for a fixed time, usually 15 seconds, after being pushed, while sounding an alarm. Signage and Braille are required to alert people of the delay as it provides time for security to respond. In a true emergency, such a device should be unlocked immediately through a tie-in with the building's fire protection system as specified in NFPA 101.

The Las Vegas MGM Grand Hotel fire in 1980 provides a graphic example of security (locked exit doors) being one of the major reasons for a large loss of life (85 deaths). All the exit doors to the stairwell had a controlled exit device (panic hardware). However, once the occupants were in the stairwell, they encountered smoke. Unfortunately, the doors were locked on the stairwell side to ensure security for each floor. A person had to exit the building on the first floor to regain access, or go to the roof (where about 1,000 people were rescued by helicopters). With heavy smoke rising in the stairwell, and no access to floors, occupants were trapped and died, mostly from smoke inhalation (Moore, 1997: 61–62).

Fire-Resistive Buildings

If a person stops to think about the tons of combustible materials transported into a building during construction and as it becomes operational, he or she may be hesitant to enter. Wood in construction and furnishings, cloth and fibers in curtains and carpets, paper, cleaning fluids, and other combustibles are hazardous. However, reinforced concrete or protected steel construction and fire-resistive roofs, floors, ceilings, walls, doors, windows, carpets, furniture, and so on all help to produce greater fire protection.

Training and Fire Brigades

With the threat of fire, employee training is of tremendous importance. Of top priority during a fire is to safeguard lives and then to secure valuable assets. If specific employees are responsible for fire suppression, thorough training is necessary. This is especially true when local public firefighting capabilities are incompatible with the type of fire that may develop at a site. In this case, a private fire department at the site is appropriate and referred to as a **fire brigade**. OSHA 1910.156 contains requirements for organizing, operating, training, and equipping industrial fire brigades when established by an employer. Fire brigades are expensive, and all brigade members must be trained and equipped as professional firefighters.

Hazards Encountered by Firefighters

The following incidents are from Carter et al. (2004): In one incident, as a fire spread at a multi-story senior citizen center, a self-closing fire door remained ajar and a sprinkler system did not operate on the 10th-floor hallway. When strong winds pushed the flames like a blowtorch into the 10th-floor hallway, a search and rescue team was engulfed in the fireball and three firefighters died. This case illustrates the importance of properly functioning fire protection systems.

In another incident, firefighters responded to the smell of smoke at a restaurant and encountered a slight haze and a pungent odor upon entering, but no smoke or fire. Thinking a food preparation process caused the odor, the firefighters continued walking into the building and experienced irritation in their eyes and throat, tightness in the chest, headache, and nausea. Upon opening doors and windows to vent the building and remove the haze, numerous dead insects and rodents were seen. The firefighters heard on their radio a call for EMS to assist a firefighter who had entered the building earlier. Eventually, the firefighters realized that the building contained an aerosolized insecticide/pesticide. This case illustrates the importance of training (e.g., hazmat), proper equipment, and investigating the call (e.g., contact the building owner).

Performance-Based Design

A **performance-based fire protection** approach, found in a variety of codes and standards, is an alternative to following a rigid set of guidelines by evaluating hazards and planning the most appropriate protection in innovative ways to meet performance goals. This approach is driven by computer software that runs fire models, technical analysis that is less subjective than traditional planning, and performance-based design that offers comparisons of products. Chapter 5 of NFPA 101 explains the performance-based option: "Performance-based design is an engineering approach to fire protection that is based on a specific design goal as opposed to the normal prescriptive code requirements based on occupancies." For example, a fire code might require exit travel distance at no more than 250 feet. This requirement may not be appropriate in all cases. Performance-based design would seek fire safety through a goal, such as evacuation of the building within four minutes (Hurley, 2009; Jelenewicz, 2006: 34).

What do you think can improve fire suppression in the United States?

Public Safety Agencies

Public safety involves primarily government employees who plan, train, and equip themselves for emergencies. They respond as quickly as possible to the scene of an emergency to save lives, care for the injured, protect people and property, and restore order. During an emergency at a corporation or institution, when "**true first responders**" (e.g., security officers, floor wardens, and employees trained in fire fighting and first aid) perform their duties, their work will "merge" with the duties of arriving public safety specialists (e.g., police, firefighters). Because of this interrelationship, joint planning and training prior to an emergency can improve coordination, life safety, and public safety.

All responders face serious safety issues. Examples are traffic accidents, disease, emotionally upset family members, violent offenders, and hazardous materials. **Personal protective equipment** (PPE) is essential; it depends on the type of incident. At a minimum, PPE must include universal precautions (e.g., gloves, mask, protective eyewear) to prevent disease transmission, but it can range to a bullet-resistant vest for high-risk incidents or locations.

Thus far, in this chapter, emphasis has been placed on public fire departments within public safety. However, other agencies of government play an important role in public safety and responding to emergencies. They include emergency management (see the preceding chapter), police, and emergency medical services.

Police

The U.S. Bureau of Justice Statistics (2011a) reported that there were 18,000 state and local law enforcement agencies in the United States in 2008. The total included 12,501 local police departments, 3,063 sheriffs' offices, 50 state primary police agencies (includes Hawaii which was previously a special jurisdiction agency), and other agencies. Over 1.1 million people were employed in these agencies, including 765,000 sworn officers (i.e., those with arrest powers). The U.S. Bureau of Justice Statistics (2011b) surveyed 65 federal law enforcement agencies in 2004, including the FBI, CIA, Secret Service, ATF, military branches, and 27 offices of inspector general. Federal agencies employed 105,000 full-time personnel authorized to make arrests and carry firearms.

Public police play a crucial role in responding to "all-hazard" emergencies. At the scene of an emergency, police direct traffic, maintain order, keep crowds back, assist victims, protect potential crime scenes, and conduct investigations.

As with public fire departments, public police are also involved in the integration of life safety and public safety prior to emergencies. They may meet with private security management to coordinate efforts at protecting people and assets. Police often maintain floor plans of banks and other high-risk locations to assist them when responding to crimes, such as a robberies and hostage incidents. Violence in the workplace has caused public police to plan, train, and drill with institutions (e.g., schools) and businesses.

Emergency Medical Services (EMS)

The National Association of Emergency Medical Technicians (2011) reported the following:

- 17,000 transporting ambulance services (includes fire departments)
- 26,000 fire departments (most of which provide EMS and about half offering ambulance transport)
- 52,000 ambulances
- 600,000 EMTs
- 142,000 paramedics
- 1,009,000 firefighters (many cross-trained in EMS)

Carter et al. (2004: 691) write: "Emergency medical responses constitute more than 50 percent of total emergency responses for many fire departments across the country. In some jurisdictions, emergency medical calls make up 70 to 80 percent of the fire department's total emergency responses per year."

Firefighters are increasingly improving their skills at emergency medical services, and they can deliver lifesaving techniques to stabilize patients while waiting for EMTs and paramedics. Some firefighters are also EMT qualified. In comparison to EMTs, paramedics receive more training and perform more procedures during a medical emergency. States vary on training and qualification for these positions.

During a medical emergency, numerous critical, time-sensitive decisions must be made (Angle, 2005: 118). Examples include the safest and quickest route to the scene, medical treatment techniques, drug dosages, and the most appropriate method of transporting the patient. **Triage** (Figure 13-7) is a term relevant to sites of medical emergencies, and it is defined as follows: "A quick and systematic method of identifying which patients are in serious condition and which patients are not, so that the more seriously injured patients can be treated first" (Carter, 2004: 954).

Homeland Security

Emergency Services Sector

Homeland Security Presidential Directive 7 identifies the Emergency Services Sector (ESS) as one of our nation's critical infrastructure and key resources (CIKR) essential for our survival. The ESS serves as our nation's first line of defense during a disaster to save lives, protect property, and

FIGURE 13-7 U.S. Air Force firefighter triages victims during mass casualty exercise at Jacksonville International Airport, FL. *Courtesy: U.S. Air Force photo by Staff Sgt. Shelley Gill.*

assist in recovery. It also works to protect and ensure resilience of other CIKR sectors (e.g., food, water, energy, transportation, and others). ESS disciplines are law enforcement, fire and emergency services, emergency management, emergency medical services, and public works. The U.S. Department of Homeland Security (DHS) serves as the Sector-Specific Agency (SSA) to lead the protection of the ESS through a Sector-Specific Plan (SSP) that coordinates goals and seeks to enhance performance under the National Infrastructure Protection Plan. (More on these topics in subsequent chapters.)

Apps

The term "**app**" refers to application software designed to assist the operator in achieving a specific task. There are many types of apps that range from accounting to star gazing. Wagley (2012) refers to mobile device programs (i.e., apps) that assist first responders and security professionals. One app lets a user inform it that he or she is CPR-trained. The app will notify the user if a person in the vicinity is going into cardiac arrest and the 911 operator has been contacted. The app provides the location of the patient and the nearest public defibrillator. Other apps include legally approved versions of Miranda warnings (including Spanish translations) and field sobriety instructions. Other apps assist disaster responders, such as to assess damage following a disaster (Wagley, 2012).

Although apps are helpful for many tasks, the software presents security vulnerabilities. For example, certain apps can access data and information in mobile devices. Questions and liability issues will evolve as the use of apps increases.

Emergencies

Planning and training are two key strategies to mitigate losses when emergencies occur. Employees must know what to do to protect lives and assets. Business continuity plans are essential for business survival. Businesses must apply risk management and risk analysis tools to identify vulnerabilities and dependencies. "What if" scenarios help businesses plan. What if a major supplier is destroyed by a natural disaster? What if an electric service blackout occurs? What if a pandemic necessitates restricted travel? How would these events affect business and what can be done to strengthen business continuity?

Pre-deployment training is a growing issue to help those who respond to disasters. Responders are at risk of developing physical and mental health problems. Pearson and Weinstock (2011) recommend an emphasis on prevention (e.g., PPE) and training on survival skills, hazardous conditions, and stress and fatigue management.

Human-Made Emergencies

Accidents

Although many accidents are caused by human error, other factors, such as weather conditions or poor equipment design, may cause accidents. Thousands of lives, hundreds of thousands of injuries, and billions of dollars of losses are sustained each year because of accidents.

Bomb Threats and Explosions

Because of past bombings and terrorism, bomb threats and the possibility of bombings are taken very seriously today. Even the commonly circulated statistic that 98 percent of bomb threats are hoaxes makes decision makers more concerned than ever about the other two percent. The Bureau of Alcohol, Tobacco and Firearms (ATF) reports hundreds of bombing incidents in the U. S. each year. Pipe bombs and Molotov cocktails are often encountered in these incidents. Accurate statistics are difficult to gather on bomb threats, attempts, and actual bombings. Organizations may not report threats. Police agencies may "play down incidents" to prevent copycat threats and bombings (Estenson, 1995: 120). During FY 2010, in the United States, the Bureau of Alcohol, Tobacco and Firearms (2011) initiated criminal investigations in 564 arson and explosives cases, including bombings and attempted bombings.

Here are basic strategies for protection against bomb threats and explosions:

1. Seek management support and prepare a plan and procedures.
2. Ensure that employees know what to ask if a bomb threat is made (Figure 13-8).
3. Establish criteria and procedures for evacuation. Post routes. The evacuation decision can be especially difficult for management because of safety concerns versus the thousands of dollars in productivity lost due to evacuation of large numbers of employees.

Smith Corporation Bomb Threat Form

Date of Threat: ______________________________

Time: ______________ Number of minutes on telephone: ______________

Exact words of caller: ______________________________

____

____

____

____

____

____

Ask the caller these questions:

When will the bomb explode? ______________________________

Where is bomb? ______________________________

What type of bomb? ______________________________

What does it look like? ______________________________

Why did you place bomb? ______________________________

Description of caller's voice: Age: ______________ Accent: ______________

Sex: __________ Background noise: ______________

Tone of voice: __________

Additional comments: ______________________________

____

Employee receiving call: ______________ Telephone number: ______________

FIGURE 13-8 Smith Corporation bomb threat form.

4. Recruit and train *all* employees to observe and report suspicious behavior, items, or vehicles. In addition, never approach or touch something suspicious.
5. Control access and parking.
6. Control and verify outsiders (e.g., service personnel) prior to access.
7. Screen mail and deliveries. Route mail and deliveries to a specific location or building for screening instead of permitting direct access to employees (see Figure 13-9).
8. Maintain unpredictable patrols to avoid patterns that can be studied by offenders.

SUSPICIOUS MAIL ALERT

If you receive a suspicious letter or package:

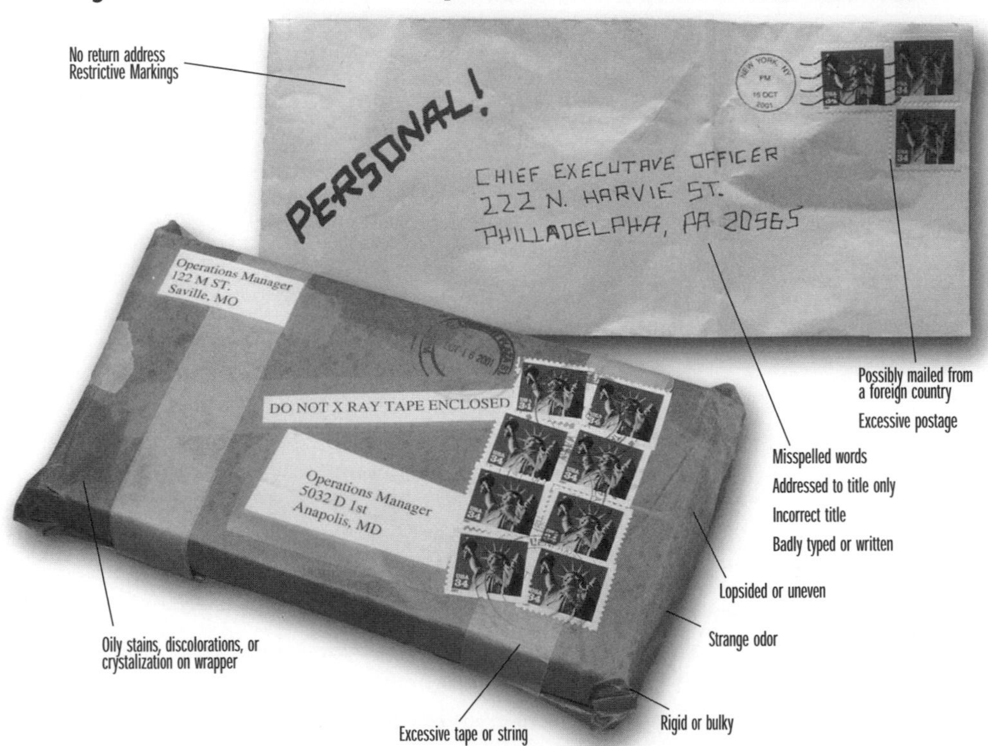

1 Handle with care. Don't shake or bump.

2 Isolate it immediately

3 Don't open, smell, touch or taste.

4 Treat it as suspect. Call local law enforcement authorities

If a parcel is open and/or a threat is identified . . .

For a Bomb:
Evacuate Immediately
Call Police
Contact Postal Inspectors
Call Local Fire Department/HAZMAT Unit

For Radiological:
Limit Exposure - Don't Handle
Evacuate Area
Shield Yourself From Object
Call Police
Contact Postal Inspectors
Call Local Fire Department/HAZMAT Unit

For Biological or Chemical:
Isolate - Don't Handle
Evacuate Immediate Area
Wash Your Hands With Soap and Warm Water
Call Police
Contact Postal Inspectors
Call Local Fire Department/HAZMAT Unit

FIGURE 13-9 Suspicious mail alert. *Source: www.idph.state.il.us/pdf/uspoalert.pdf*

9. Inspect the exterior and interior of buildings.
10. Ensure that emergency plans, response teams (e.g., fire brigades), and public safety agencies are coordinated for action. What are the qualifications, experience, and response time of the nearest bomb squad?

The important topic of "active shooter" is discussed in Chapter 18 with workplace violence.

Weapon of Mass Destruction (WMD)

A **weapon of mass destruction (WMD)** is something capable of inflicting mass casualties and/or destroying or rendering high-value assets as useless. Although chemical, biological, nuclear, and radiological weapons often serve as examples of WMD, many things can be used as a WMD. This became painfully true from the September 11, 2001, attacks when airliners, loaded with fuel, became missiles, killed thousands of people, leveled the 110-story Twin Towers, and hit the Pentagon. The range of potential WMD attacks depends on the creativity of the enemy. Examples include blowing up a train filled with toxic chemicals as it travels near a town while the wind is blowing toward the town, or placing a radiological device, disguised as HVAC equipment, near the intake vents of a crowded building. The number of "soft" targets capable of being targeted is enormous. The threat of WMD and response are explained in Chapter 15.

Strikes

The direct and indirect costs of a strike can be astronomical. Major losses include productivity, profits, employees and customers who never return, vandalism, additional loss prevention services, and legal fees.

The best defense for a strike is early preparation. When a labor contract is close to expiring, a strike is possible. A company that anticipates a strike should build up inventory and oversupply customers. Management should contact local law enforcement agencies to ensure that peace is maintained and property protected. Security, in cooperation with police, is likely to focus on perimeter protection for the facility, the protection of key executives and their families, and evidence gathering. Chapter 4 covers labor law and issues relevant to security.

Civil Disturbances

Demonstrations and precipitating incidents have evolved into destructive riots. Demonstrations are common as groups display their views. A typical precipitating incident is a public police arrest of a minority group member for a minor charge. Sometimes, all it takes for a crowd to go into a contagious frenzy is a rock thrown into the crowd. Deaths, injuries, and extensive property damage can result from a civil disturbance. Public police may not be able to contain rioters, and the private sector must be prepared for the worst. Summer months and high temperatures, when people gather out of doors, may precede a civil disturbance. Rumors also are dangerous. The social and political climate can be analyzed for possible predictive indicators.

Technological Emergencies

A technological emergency is the interruption of a vital service such as electricity. These events have varied causes. A prolonged electric service blackout results in extreme hardship in communities. For businesses, it affects not only production, but also security and loss prevention measures such as security systems. Secondary power sources (e.g., batteries and generators) should be in place for all vital services.

Mass Notification Systems

Mass notification systems, also referred to as "emergency communications systems," refer to "one or more modes of communication to as many people as possible in a specific geographic area to warn them to take precautions to protect themselves from threats or hazards, or to assist authorities in investigative or public safety duties" (Purpura, 2011: 195). Traditionally, these systems focused on warnings about fire in buildings so occupants could escape. However, today mass notification systems are designed to provide warnings for a multitude of threats and hazards, including bomb threats, persons with a weapon, tornadoes, tsunamis, hazmat incidents, serious traffic accidents, and abductions. Codes have been written for these systems by the NFPA (NFPA-72), the National Electrical Manufacturer's Association, and the Department of Defense. Mass notification systems provide information on the nature of the emergency and instructions for safety. Means (2011: 54–56) writes of the expensive challenges of upgrading and retrofitting an old fire alarm system to meet new codes. He views the costs about equal to installing a new mass notification system that would be ADA compliant and contain voice and strobe notification. In addition, software serves to integrate, from a central location, multiple modes of communication such as sirens, texting, cell calling, e-mail, IP phones, paging, and electronic signage. Research by Sattler et al. (2011) showed that text and e-mail messages are effective methods of notifying people and providing instructions during emergencies.

Natural Disasters

Windstorms

Hurricanes, typhoons, tornadoes, and cyclones are regionally specific names characterized by extremely violent and destructive winds. In the United States, the Atlantic and Gulf coasts are more susceptible to hurricanes, whereas tornadoes occur in many parts of the country.

Smith (2011: 28–30) notes that the 2011 tornado season in the United States was the worst since 1953. In 2011, over 500 deaths (about the same number in the previous 10 years combined) and over 1,500 confirmed tornadoes occurred. Smith writes of lessons learned from the 2011 tornado season. Communications are essential so people can receive warnings. On April 27, 2011, in Mississippi and Alabama, storms knocked out electricity to over a million people who were without television, Internet, and in some cases, cell service, local radio, and weather radio. Over three hundred people died. To avoid losing service, Smith recommends satellite phone. Another lesson is the importance of underground shelters. A third lesson is that too many siren activations may "train" people to ignore sirens.

Loss prevention efforts should concentrate on studying local climate conditions and then preparing contingency plans. The following measures are useful:

1. Design buildings for maximum wind velocity expected.
2. Closely follow weather reports.
3. Close down business if necessary.
4. Establish a safe area for people.
5. Instruct people to stay away from windows.
6. Open doors and windows on the side of the building away from the storm. This will help to equalize pressure and prevent building collapse.
7. Acquire emergency power sources.
8. Set up a communications system.
9. Anchor and protect company property from being damaged or blown away.
10. Cooperate with local officials.
11. Take steps to hinder looting.

Floods

Lowlands along bodies of water are subject to flooding. Dams and dikes have reduced this problem, but flooding is still a threat to many locales. Some practical remedies follow:

1. Work with the U.S. Army Corps of Engineers, which provides assistance to those areas subject to frequent flooding.
2. Work with local officials to enhance safety and to hinder looting.
3. Provide a safe place for valuable assets.
4. Exit from the flood area in time to avoid being marooned. Secure adequate gasoline for vehicles.
5. If a team remains at a business site, ensure adequate supplies, such as food, blankets, a boat, life preservers, rope, communications equipment, first-aid kit, sandbags, and pumps.
6. Store valuable records, equipment, tools, and chemicals above the expected flood level. Shut off utilities.

Blizzards

Blizzards are snowstorms accompanied by high winds and very cold temperatures. Sufficient warning is usually obtained from weather forecasts. Injuries and deaths occur because people do not reach proper shelter. If a blizzard is forecast, employees should be sent home. Planning is essential if employees have a chance of being stranded at work. Food, bedding, a heat supply, radios, communications equipment, and televisions will aid those stranded.

Earthquakes

Stringent building codes and adequate building design in suspect areas (e.g., California) are vital for reduced losses. Building collapse, damage to bridges, and falling debris are major causes of injuries and deaths. If indoors, one should take cover in a basement or under reinforced floors or doorways. If outdoors, one should watch out for falling objects and electrical wires.

The previous chapter described the Eastern Japan Great Earthquake Disaster of 2011. Seven years earlier, the December 2004 Sumatra-Andaman earthquake caused a series of tsunamis (i.e., large destructive waves) that killed almost 300,000 people and affected 18 countries around the Indian Ocean. The human and financial strain on nations to prepare and respond to such natural disasters is overwhelming. The challenges require global cooperation.

You Decide!

What is Wrong with This Facility?

You are a loss prevention specialist for a large corporation that has recently purchased a printing plant in the northwest United States The huge plant has a flat roof and is located in a rural area 15 miles from the nearest fire department. Fire protection at the plant consists primarily of Class B and C fire extinguishers and a wet-pipe sprinkler system. During the winter, weather averages are as follows: temperature of 30 degrees Fahrenheit, snowfall of 55 inches, and wind gusts occasionally to 50 mph.

As winter approaches, your present tasks are to identify what is wrong with this facility from a loss prevention perspective and recommend corrective action and plans.

International Perspective: The U.S. View of Hazards: Thoughts from Europe

Here is a critical thinking perspective of the U.S.-driven paradigm that social and technological approaches to mitigation can act, overall, to reduce the incidence or severity of disastrous events. Rockett (2001: 71–74) counters the U.S. paradigm by arguing that disasters are not the result of external forces, but rather the way we live our lives. Social, political, and economic forces are as much the cause of disasters as the natural environment. We already know where earthquakes, hurricanes, and flooding are likely to occur. Furthermore, we also know that many technologies (e.g., automobiles, which cause pollution) carry with them the power of mass disruption. Consequently, disaster reduction will result only from fundamental changes in political and social mores.

Rockett's viewpoint is illustrated when people, businesses, and organizations insist on congregating on a well-known tectonic fault when predictable earthquakes cause multibillion-dollar losses. Yes, we are speaking of Los Angeles as one example. However, this is a global problem. Other types of predictable disasters occur when, for example, a Yangtze or Bangladesh flood kills tens of thousands of people. Such hazards of the environment are impossible to eliminate, and continuing to have children and building in these areas load the costs of disasters.

What Rockett suggests is to change social structure, not social reaction. "Progress in disaster reduction will occur when we accept that we are the creators of disaster rather than its victims." Rockett argues that as we build stronger structures and invent more efficient mitigation systems, we run the risk of engendering a false sense of safety and security that will, when the bigger event arrives, lead to an even greater catastrophe.

What is your opinion of Rockett's contentions?

Pandemics

A **pandemic** is an outbreak of an infectious disease that affects humans over a large geographic area. Pandemics have occurred throughout history, with many from zoonotic diseases resulting from the domestication of animals. **Zoonotic diseases** are diseases transmitted from animals to humans. Examples are the West Nile virus (that can cause encephalitis), avian influenza, and human monkeypox.

Examples of pandemics occurring throughout history include the bubonic plague, also called the "Black Death," that killed about 20 million Europeans during the 1300s, seven cholera pandemics through the 1800s and 1900s, and numerous influenza pandemics. The 1918–1919 "Spanish flu" caused the highest number of known influenza deaths. Over 500,000 people died in the United States, and up to 50 million may have died worldwide. Many people died within the first few days of infection, and nearly half of those who died were young, healthy adults (Centers for Disease Control and Prevention, Department of Health and Human Services, 2006).

The CDC and the World Health Organization (WHO) have extensive surveillance programs to monitor and detect influenza activity worldwide. The CDC is involved in preparedness activities through vaccine development and production, stockpiling of medications, research, risk communications, and unified initiatives with other organizations.

A vaccine is unlikely to be available in the early stages of a pandemic. Scientists must select the virus strain that will provide the best protection. Then manufacturers use the strain to produce a vaccine. Antiviral medications are available, but they may not work because virus strains can become resistant to the medications.

According to the CDC, the severity of the next influenza pandemic cannot be predicted. Modeling studies suggest that the impact could be substantial. Between 15 percent and 35 percent of the U.S. population could be affected, and the economic impact could range between $71 and $166 billion (Centers for Disease Control and Prevention, Department of Health and Human Services, 2006).

Reissman et al. (2006: 2) write that pandemic influenza preparedness plans from the United States and WHO focus on (1) surveillance and early detection, (2) community containment (e.g., movement restrictions and facility closure), and (3) vaccines and antiviral medication. However, they note that "relatively little attention has been paid to identifying and managing psychological and social factors likely to influence human behavior during a pandemic." Examples are numerous and include how citizens will respond to containment, the need for healthcare and mental health services, and the impact on religious and cultural rituals surrounding burial and grieving.

Pandemics present numerous issues for businesses, besides homeland security and emergency management. *How will businesses function when large numbers of customers stay home and employees call in sick?*

A pandemic is likely to last longer than most emergencies and may include "waves" separated by months. The number of healthcare workers and first responders available to work may be reduced because of exposure and the need to care for ill family members. In addition, resources, vaccines, and medications in many locations could be limited (Centers for Disease Control and Prevention, Department of Health and Human Services, 2006).

During a pandemic, people who have had contact with symptomatic persons may be asked to quarantine themselves. Those with and without symptoms may wear facemasks to control the spread of disease. Government may decide to close schools and other locations of mass gatherings. If a pandemic is particularly lethal, healthcare workers would implement corpse-management procedures to safely destroy infected human remains, and government would apply emergency powers to enforce measures such as rationing and curfews (Tucker, 2006). The Centers for Disease Control and Prevention, Department of Health and Human Services (2011), offers on its website a variety of software tools for professionals planning for pandemic flu.

A serious pandemic can result in **quarantine enforcement**. It is the use of government authorities, such as police, healthcare specialists, or the military, to restrict the movement of people, by force if necessary, for purposes of public health and safety. Quarantine enforcement can create controversy over protection of public health versus civil liberties.

Employers should prepare plans for the risk of a pandemic. Here are ideas to begin a checklist.

- Prepare a business impact study.
- Support vaccination.
- Use infection control measures (e.g., cough etiquette, hand washing, facemasks, and sick leave).
- Cross-train so healthy employees can perform the duties of sick employees.
- Plan for teleworking.
- Maintain contact with customers through various technologies.

What do you think is the most serious potential emergency facing your community or organization? What protection methods do you recommend?

Search the Internet

Here are websites relevant to this chapter:

American National Standards Institute: www.ansi.org
American Society for Testing and Materials: www.astm.org
Centers for Disease Control and Prevention: www.cdc.gov
Factory Mutual Global: www.fmglobal.com
Federal Emergency Management Agency: www.fema.gov
Federal Register: www.fdsys.gov
International Association of Chiefs of Police: www.theiacp.org
International Association of Fire Chiefs: www.iafc.org
International Association of Fire Fighters: www.iaff.org
International Code Council: www.iccsafe.org
International Organization for Standardization: www.iso.org
National Association of Emergency Medical Technicians: www.naemt.org
National Fire Protection Association: www.nfpa.org
National Oceanic & Atmospheric Administration: www.noaa.gov/index.html
OSHA: www.osha.gov
Standards: standards.gov
Underwriters Laboratories (UL): www.ul.com/global/eng/pages
U.S. Department of Homeland Security: www.dhs.gov
U.S. Department of Transportation, Office of Hazardous Materials Safety: hazmat.dot.gov
U.S. Fire Administration: www.usfa.dhs.gov

Case Problems

13A. You are a loss prevention manager at company headquarters. Your superior tells you to help design a fire protection program for a new window and door manufacturing plant. A lot of woodcutting with electric circular saws will take place at this plant. Workers then will assemble the windows and doors using electric drills. A large stock of wood products will be stored in the plant. What will you recommend to the architects who will design the building? What are your fire prevention plans? What are your fire suppression plans?

13B. You are a loss prevention manager at an office building containing 800 employees. Your superior requests that you prepare criteria for evacuation of the building in case of a bomb threat. List criteria for management to consider and estimate the cost of a two-hour evacuation if the average employee earns $50,000 annually.

13C. As a corporate loss prevention manager, you have been assigned the task of designing a protection plan against windstorms for manufacturing plants located along the coast of Florida and those located in Kansas. What general plans do you have in mind?

References

American National Standards Institute Homeland Security Standards Panel. (2010). Emergency Preparedness for Persons with Disabilities and Special Needs (October). <publicaa.ansi.org> retrieved November 19, 2011.

Angle, J. (2005). *Occupational safety and health in the emergency services* (2nd ed.). Clifton Park, NY: Thomas Delmar.

Azano, H., & Gilbertie, M. (2003). Making planning a priority. *Security Management, May*.

Bureau of Alcohol, Tobacco and Firearms. (2011). Fact Sheet (March). <www.atf.gov/publications/factsheets/2010-factsheet-facts-and-figures.html> retrieved November 26, 2011.

Carter, W., et al. (2004). *Firefighter's handbook: Essentials of firefighting and emergency response* (2nd ed.). Clifton Park, NY: Thomas Delmar.

Cauchon, D., et al. (2003). Just-Released transcripts give voice to the horror. *USA TODAY* (August 28) <www.usatoday.com> retrieved August 29, 2003.

Centers for Disease Control and Prevention, Department of Health and Human Services. (2011). CDC Resources for Pandemic Flu. <www.cdc.gov/flu/pandemic-resources> retrieved November 27, 2011.

Centers for Disease Control and Prevention, Department of Health and Human Services. (2006). Key Facts about Pandemic Influenza. <www.cdc.gov/flu/pandemic/keyfacts.htm> retrieved January 24, 2006.

City of New York, Department of Investigation. (2011). DOI Arrests 110 on Warrants for Fire and Building Code Violations (Press Release, June 29). <www.nyc.gov/html/doi/downloads/pdf/june11/pr45warrants_62911.pdf> retrieved November 19, 2011.

Estenson, D. (1995). Should bomb blasts be kept quiet? *Security Management, November*.

Hall, D. (2006). *Administrative law: Bureaucracy in a democracy* (3rd ed.). Upper Saddle River, NJ: Pearson Education.

Hall, J. (2011). *U.S. experience with sprinklers*. Quincy, MA: National Fire Protection Association. <www.nfpa.org/assets/files/PDF/OS.sprinklers.pdf> retrieved November 21, 2011.

Hurley, M. (2009). Fire protection engineering. *Whole Building Design Guide* (June 2). <www.wbdg.org/design/dd_fireprotecteng.php> retrieved November 22, 2011.

Jelenewicz, C. (2006). Performance-Based design. *Buildings, 100* (May).

Matthews, K. (2011). 1911 Triangle fire remembered as spur to unions, safety laws. *USA Today* (March 22). <www.usatoday.com> retrieved March 23, 2011.

Means, T. (2011). Getting the word out. *Security Technology Executive, 21* (October).

Moore, W. (1997). Balancing life safety and security needs. *Security Technology and Design* (January–February).

Morton, J. (2012). Digitally monitor fire extinguishers. *Buildings, 106* (April).

Morton, J. (2011). Do your sprinklers pass the test? *Buildings, 105* (September).

Naffa, I. (2009). A guide to automatic sprinkler systems. *Buildings, 103* (October).

National Association of Emergency Medical Technicians. (2011). About EMS. <www.naemt.org> retrieved November 25, 2011.

National Fire Protection Association. (2010). NFPA 10: Standard for Portable Fire Extinguishers." <www.nfpa.org> retrieved November 20, 2011.

National Institute of Standards and Technology. (2005). Latest Findings from NIST World Trade Center Investigation. <www.nist.gov> retrieved April 8, 2005.

Pearson, B. (2010). IP fire systems. *Security Technology Executive, 20* (May).

Pearson, J., & Weinstock, D. (2011). Minimizing safety and health impacts at disaster sites: The need for comprehensive worker safety and health training based on an analysis of national disasters in the U.S. *Journal of Homeland Security and Emergency Management, 8* <www.bepress.com/jhsem/vol8/iss1/46> retrieved October 12, 2011.

Purpura, P. (2011). *Security: An introduction.* Boca Raton, FL: CRC Press. Taylor & Francis Group.

Quinley, K., & Schmidt, D. (2002). *Businesses at risk: How to assess, mitigate, and respond to terrorist threats.* Cincinnati, OH: The National Underwriter Co.

Reissman, D., et al. (2006). Pandemic influenza preparedness: Adaptive responses to an evolving challenge. *Journal of Homeland Security and Emergency Management, 3.*

Rockett, J. (2001). The U.S. view of hazards and sustainable development: A few thoughts from europe. *Risk Management: An International Journal, 3.*

Sattler, D., et al. (2011). Active shooter on campus: Evaluating text and e-mail warning message effectiveness. *Journal of Homeland Security and Emergency Management, 8.*

Schottke, D. (2007). *First responder: Your first response in emergency care.* Sudbury, MA: Jones and Bartlett.

Smith, M. (2011). Deadliest tornado season in 50 years. *Disaster Recovery Journal, 24* (Summer).

Spadanuta, L. (2010). Fast track to fire safety. *Security Management, 54* (May).

Sternberg, E., & Lee, G. (2006). Meeting the challenge of facility protection for homeland security. *Journal of Homeland Security and Emergency Management, 3*

Suttell, R. (2006). Fire protection system design. *Buildings, 100* (June).

Tucker, P. (2006). Preparing for pandemic. *The Futurist* (January–February).

U.S. Bureau of Justice Statistics. (2011b). *Census of Federal Law Enforcement,* 2004. bjs.ojp.usdoj.gov/index.cfm?ty=tp=tp&tid=74, retrieved November 25, 2011.

U.S. Bureau of Justice Statistics. (2011a). *Census of State and Local Law Enforcement Agencies,* 2008. <bjs.ojp.usdoj.gov/content/pub/pdf/csllea08.pdf> retrieved November 25, 2011.

U.S. Department of Homeland Security. (n.d.). About First Responders. <www.dhs.gov/dhspublic> retrieved February 11, 2005.

U.S. Department of Homeland Security. (2008). *National Response Framework* (January). <www.fema.gov/pdf/emergency/nrf/nrf-annexes-all.pdf> retrieved November 17, 2011.

U.S. Department of Homeland Security, READYBusiness. (n.d.). Testimonials. <www.ready.gov/business/testimonials.html> retrieved January 3, 2005.

U.S. Fire Administration. (2004). Danger Above: A Factsheet on High-Rise Safety. (November 23). <usfa.fema.gov/safety/atrisk/high-rise/high-rise.shtm> retrieved April 4, 2005.

U.S. Fire Administration. (2011). The U.S. Fire Administration's Research Program—Science Saving Lives. <www.usfa.fema.gov> retrieved November 17, 2011.

Wagley, J. (2012). Life safety apps. *Security Management, May*

Zimmerman, D. (2012). *Firefighter safety and survival.* Clifton Park, NY: Delmar Cengage Pub.

14 Safety in the Workplace

OBJECTIVES

After studying this chapter, the reader will be able to:

1. Explain the importance of safety, accident statistics, and direct and indirect losses from accidents
2. Discuss the history of safety legislation and workers' compensation
3. Explain OSHA's development, strategic plan, jurisdiction, standards, and inspections
4. Describe at least four strategies for improving safety in the workplace

KEY TERMS

- safety in the workplace
- safety
- accident
- injury
- workers' compensation laws
- Occupational Safety and Health Administration (OSHA)
- William Steiger Occupational Safety and Health Act of 1970
- National Institute for Occupational Safety and Health (NIOSH)
- hazard communication standard
- Globally harmonized system of classification and labeling of chemicals
- Hazardous waste operations and emergency response standard
- bloodborne pathogens standard
- lockout/tagout
- OSHA Whistleblower Protection Program
- accident proneness theory
- safety performance assessment

Introduction

Safety in the workplace is a broad loss prevention concept that includes a host of methods that aim to reduce risks from a variety of threats and hazards such as accidents, violence, and natural disasters. This chapter focuses on workplace accidents and related safety methods.

Webster's *New Collegiate Dictionary* defines **safety** as "the condition of being safe from undergoing or causing hurt, injury, or loss" and "to protect against failure, breakage, or accident." **Accident** is "an unfortunate event resulting from carelessness, unawareness, ignorance, or a combination of causes." **Injury** is "hurt, damage, or loss sustained."

Examples of hazards in the workplace are as follows:

- Inadequate training (Figure 14-1)
- Noncompliance with safety policies and procedures
- Failure to use safety devices or equipment

U.S. Department of Labor
Occupational Safety & Health Administration
www.osha.gov MyOSHA Search GO Advanced Search

ACCIDENT SUMMARY No. 15

Accident Type:	Crushed by Dump Truck Body
Weather Conditions:	Clear, Warm
Type of Operation:	General Contractor
Size of Work Crew :	N/ A
Collective Bargaining	Yes
Competent Safety Monitor on Site:	Yes
Safety and Health Program in Effect:	Yes
Was the Worksite Inspected Regularly:	Yes
Training and Education Provided:	No
Employee Job Title:	Truck Driver
Age & Sex:	25-Male
Experience at this Type of Work:	2 Months
Time on Project:	2 Weeks at Site

BRIEF DESCRIPTION OF ACCIDENT

A truck driver was crushed and killed between the frame and dump box of a dump truck. Apparently a safety "overtravel" cable attached between the truck frame and the dump box malfunctioned by catching on a protruding nut of an air brake cylinder. This prevented the dump box from being fully raised, halting its progress at a point where about 20 inches of space remained between it and the truck frame. The employee, apparently assuming that releasing the cable would allow the dump box to continue up-ward, reached between the rear dual wheels and over the frame, and disengaged the cable with his right hand. The dump box then dropped suddenly, crushing his head. The employee had not received training or instruction in proper operating procedures and was not made aware of all potential hazards in his work.

INSPECTION RESULTS

Following its inspection, OSHA issued one citation for one alleged serious violation of its construction standards. Had the required training been provided to the employee, this fatality might have been prevented.

ACCIDENT PREVENTION RECOMMENDATIONS

Employees must be instructed to recognize and avoid unsafe conditions associated with their work (29 CFT 1926.21 (b)(2)).

SOURCES OF HELP

- Construction Safety and Health Standards (OSHA 2207) which contains all OSHA job safety and health rules and regulations (1926 and 1910) covering construction. OSHA-funded free consultation services.
- Consult your telephone directory for the number of your local OSHA area or regional office for further assistance and advice (listed under U.S. Labor Department or under the state government section where states administer their own OSHA programs).

NOTE: The case here described was selected as being representative of fatalities caused by improper work practices. No special emphasis or priority is implied nor is the cases necessarily a recent occurrence. The legal aspects of the incident have been resolved, and the case is now closed.

FIGURE 14-1 Accident report, crushed by dump truck body. *Courtesy: OSHA.*

- Dangerous storage of toxic or flammable substances
- Poorly supported ladders and scaffold
- Electrical malfunctions
- Blocked aisles, exits, and stairways
- Poor ventilation
- Insufficient lighting
- Excessive noise
- Horseplay; running
- Violence

Accident Statistics and Costs

Since the early 1900s, great strides have been made to increase safety in the workplace. Safer machines, improved supervision, and training all have helped to prevent accidents. If one were to apply the industrial fatality rate that existed in 1910 to the present workforce in the United States, more than 1.3 million workers would lose their lives each year from industrial accidents. In the past, a manual worker's welfare was of minimal concern to management; the loss of life or limb was "part of the job" and "a normal business risk." In the construction of tall buildings, it was expected that one life would be lost for each floor. A 20-story building would yield 20 lost lives. During tunnel construction, two worker deaths per mile was the norm. Coal mining experienced exceedingly high death rates (Anderson, 1975: 5–6).

The U.S. Department of Labor (2011a) reported 4,547 fatal occupational injuries in the United States during 2010. These fatalities resulted from transportation incidents (1,766), violent acts (808), contact with objects and equipment (732), falls (635), exposure to harmful substances and environments (409), and fires and explosions (187). The number of nonfatal workplace injuries and illnesses reported from private industry during the same year was nearly 3.1 million.

The Centers for Disease Control and Prevention (2011) reported that society-wide there are over 180,000 deaths each year from injuries; millions more are injured and survive. In addition, over $406 billion is expended annually in medical costs and lost productivity.

Direct and indirect losses from accidents are costly. For example, the death of a worker at a manufacturing plant immediately creates a tremendous direct loss to family and friends. Direct losses also include an immediate loss of productivity, medical costs, and insurance administration. Indirect costs include continued grief by family and friends, continued loss of productivity, profit loss, selection and training of a new employee, overtime for lost production, and possible litigation. In addition, internal and external relations may suffer. The lowering of employee morale can result from the belief that management is incompetent or does not care. Rumors frequently follow. In the eyes of the community, the company may appear to have failed.

History of Safety Legislation

In 18th century England, as the Industrial Revolution progressed, a number of statutes governing working conditions were passed. One of the first statutes for safety resulted from a serious outbreak of fever at cotton mills near Manchester in 1784. Because child labor was involved, widespread attention added to public concern and government pressure to improve the dangerous and unsanitary conditions in factories. A few years later, additional legislation dealt with hours, conditions of labor, prevention of injury, and government inspectors. In 1842, the *Mines Act* provided for punitive compensation for preventable injuries caused by unguarded mining machinery. During this time, "strong evidence" pointed to incompetent management and the neglect of safety rules. More and more trades were brought under the scope of expanding laws on safety.

In the United States, textile factories increased in number between 1820 and 1840. Massachusetts, first of the United States to follow England's example, passed laws in 1876 and 1877 that related to working children and inspections of factories; important features pertained to dangerous machinery and necessary safety guards. As years passed, some industries increasingly realized that hazards were potentially harmful to workers, production, and profits. Consequently, more and more industrialists became safety conscious. However, serious hazards still existed in the workplace through the 20th century (Grimaldi and Simmonds, 1975: 33–43).

Workers' Compensation

Increasing concern for workers led to **workers' compensation laws**. In essence, these laws require employers to compensate injured employees. England passed such laws in 1897, and the United States followed in 1902, when Maryland passed this country's first workers' compensation law—although, essentially, the Maryland law was so restrictive that it was almost useless. In 1911, Wisconsin passed the first effective workers' compensation law. Seven other states passed similar legislation during that year. Amid controversy between businesspeople and groups interested in the welfare of laborers, the Supreme Court upheld the constitutionality of these laws in 1916. Businesspeople argued that they could never bear such compensation costs nor could they control accidents. They predicted that the cost of goods would rise considerably. Those who favored workers' compensation laws believed that these laws would provide the impetus for greater safety, because business owners and managers would want to control losses. Wade (2011) writes that many railroad, mining, and manufacturing businesses began requiring workers to use protective equipment. This movement resulted in the sharing of safety information and data among companies and the creation of a group that became known as the National Safety Council that has advocated worker safety since 1913.

Workers' compensation laws provided insurance companies with a new opportunity. As states enacted these laws, business owners became concerned about their ability to pay for workers' compensation. Therefore, insurance companies sold casualty insurance policies to businesses that needed the security from a possible workers' compensation burden. *A concurrent benefit of such insurance was that insurance companies were willing to reduce premiums if a company instituted accident prevention measures.* To remain competitive, insurance companies provided safety specialists who would survey a business and recommend prevention strategies (e.g., safeguards on machines). Businesspeople became increasingly interested in safety; insurance companies developed safety expertise.

Today, all states have workers' compensation laws. These laws vary from state to state, but each requires the reporting of injuries that are compensable. Private insurance companies insure most employers for workers' compensation judgments. Other employers are self-insured (insurance provided by the employer and not purchased through a private insurance company; it is regulated by a state insurance commissioner) or place this insurance with state insurance funds. When an employee is injured, medical benefits usually are granted, and benefits for wages lost are granted when an employee is incapacitated and cannot work.

The workers' compensation system of today has its problems. Because dealings with the insurance industry are frequently adversarial, injured employees often believe that the system helps them only when they obtain the services of an attorney. Insurers, on the other hand, claim that they must protect their interests and that fraud (i.e., worker malingering) is a very serious problem.

Are workers' compensation laws necessary today? Support your viewpoint.

The Development of OSHA

With the advent of improved safety conditions, accidents and injuries declined until the 1950s. In the late 1950s, rates leveled off until the late 1960s, when accidents and injuries began to increase. This upward trend caused the federal government to become increasingly concerned about safety. Several safety-related laws were passed during the 1960s, but none was as monumental as the federal law creating OSHA. OSHA stands for the **Occupational Safety and Health Administration**, a federal agency, under the U.S. Department of Labor, established to administer the law on safety and health resulting from the **William Steiger Occupational Safety and Health Act of 1970**. This federal legislation was signed into law by then-president Richard Nixon and became effective on April 28, 1971. The basic purpose of OSHA is to provide a safe working environment for employees engaged in a variety of occupations.

The OSHA act was significant because it was the first national safety legislation applying to every business connected with interstate commerce. Mason (1976: 21) notes: "The need for such legislation was clear. Between 1969 and 1973 [in the United States] more persons were killed at work than in the Vietnam war."

The OSHA act of 1970, as amended, states the following (Occupational Safety and Health Administration, 2011a): "To assure safe and healthful working conditions for working men and women; by authorizing enforcement of the standards developed under the Act; by assisting and encouraging the States in their efforts to assure safe and healthful working conditions; by providing for research, information, education, and training in the field of occupational safety and health; and for other purposes."

The Secretary of Labor via ten regional offices administers OSHA. The secretary has the authority and responsibility to establish occupational safety and health standards. Workplace inspections can result in citations issued to employers who violate standards (Occupational Safety and Health Administration, 2011b).

The **National Institute for Occupational Safety and Health** (NIOSH) performs numerous functions that aid OSHA and those striving for worker safety. These functions relate to research, the development of criteria and standards for occupational safety and health, training OSHA personnel and others (e.g., employers and employees), and providing publications dealing with both toxic substances and strategies on how to prevent occupational injuries and illnesses. NIOSH is under the Centers for Disease Control and Prevention (CDC), U.S. Department of Health and Human Services.

Occupational Safety and Health Administration

OSHA's Strategic Plan

OSHA's (Draft) Strategic Plan, FY 2010–2016 (U.S. Department of Labor, 2011b), includes the following:

- Assure fair and high quality work-life environments.
- Measure success by reducing fatalities and increasing the number of targeted hazards abated.
- Increase worker and employer awareness of OSHA rights and responsibilities.
- Strengthen regulatory and enforcement capabilities.

The Act's Jurisdiction

OSH Act of 1970, Section 4 - Applicability of the ACT, applies this law to all employers and their employees in the 50 states, the District of Columbia, Puerto Rico, and other U.S. possessions. An employer is anyone who maintains employees and engages in a business affecting commerce. This broad coverage involves a multitude of fields: manufacturing, construction, agriculture, warehousing, retailing, longshoring, education, and so on. The act does not cover self-employed persons, family-owned and operated farms, and workplaces protected by other federal agencies. The Act permits states to establish their own state plans subject to OSHA evaluation. Although federal agencies are not covered by OSHA, agencies are required to maintain a safe working environment equal to those groups under OSHA's jurisdiction. OSHA affects state and local government employees.

Workplace Violence

Employers and security practitioners should be attuned to OSHA's increased efforts in reducing workplace violence. Under the OSH Act "general duty clause," employers are required to protect employees against recognized hazards (e.g., workplace violence) that are likely to cause serious injury or death. OSHA's efforts to curb workplace violence include publications on awareness and prevention, a directive to OSHA staff establishing uniform procedures for enforcement and investigating incidents, targeting specific businesses, and issuing citations for failure to protect employees. OSHA supports security strategies such as training, reporting and investigating incidents, risk analyses, and physical security. Workplace violence is covered in Chapter 18.

OSHA Standards

OSHA promulgates legally enforceable standards to protect employees in the workplace. Since OSHA standards are constantly being updated and reviewed, it is the employer's responsibility to keep up-to-date. Two sources for standards and changes are the websites of OSHA and the Federal Register.

There are actually thousands of OSHA standards. Some pertain to specific industries and workers, whereas others are general and practiced by most industries. Examples include safety requirements for machines, equipment, and employees, such as requiring face shields or safety glasses during the use of certain machines; prevention of electrical hazards (Figure 14-2); adequate fire protection; adequate lunchrooms, lavatories, and drinking water; precautions against infectious diseases; and monitoring of employee exposure to chemical or toxic hazards.

The following standards illustrate OSHA's concern for worker safety and health (Occupational Safety and Health Administration, 2011c).

Subpart 1—Personal Protective Equipment

1910.132 General requirements.

(a) Application. Protective equipment, including personal protective equipment for eyes, face, head, and extremities, protective clothing, respiratory devices, and protective shields and barriers, shall be provided, used, and maintained in a sanitary and reliable condition wherever it is necessary by reason of hazards of processes or environment, chemical hazards, radiological hazards, or mechanical irritants encountered in a manner capable of causing injury or impairment in the function of any part of the body through absorption, inhalation or physical contact.

Subpart K—Medical and First Aid

1910.151 Medical services and first aid.

(a) The employer shall ensure the ready availability of medical personnel for advice and consultation of matters of plant health.

(b) In the absence of an infirmary, clinic, or hospital in near proximity to the workplace which is used for the treatment of all injured employees, a person or persons shall be adequately trained to render first aid. Adequate first aid supplies shall be readily available.

(c) Where the eyes or body of any person may be exposed to injurious corrosive materials, suitable facilities for quick drenching or flushing of the eyes and body shall be provided within the work area for immediate emergency use.

OSHA Hazard Communication Standard

The **Hazard Communication Standard** (HCS) (1910.1200) covers millions of employees and workplaces and thousands of hazardous chemical products, with hundreds of new ones being introduced annually. Also known as a "right to know law," the HCS requires all employers who have employees who may be exposed to hazardous substances on the job to inform them about such substances and how to deal with them. Employers are required to write and

ACCIDENT SUMMARY No. 57

Accident Type:	Electrocution
Weather Conditions:	Clear/Hot/Humid
Type of Operation:	Window Shutter Installers
Size of Work Crew:	2
Collective Bargaining	N/A
Competent Safety Monitor on Site:	No
Safety and Health Program in Effect:	Partial
Was the Worksite Inspected Regularly:	No
Training and Education Provided:	Some
Employee Job Title:	Helper
Age & Sex:	17-Male
Experience at this Type of Work:	One Month
Time on Project:	One Month

BRIEF DESCRIPTION OF ACCIDENT

One employee was climbing a metal ladder to hand an electric drill to the journeyman installer on a scaffold about five feet above him. When the victim reached the third rung from the bottom of the ladder he received an electric shock that killed him.

The investigation revealed that the extension cord had a missing grounding prong and that a conductor on the green grounding wire was making intermittent contact with the energizing black wire thereby energizing the entire length of the grounding wire and the drill's frame. The drill was not double insulated.

INSPECTION RESULTS

As a result of its investigation, OSHA issued citations for violations of construction standards.

ACCIDENT PREVENTION RECOMMENDATIONS

1. Use approved ground fault circuit interrupters or an assured equipment grounding conductor program to protect employees on construction sites [29 CFR 1926.404(b)(1)].
2. Use equipment that provides a permanent and continuous path from circuits, equipment, structures, conduit or enclosures to ground [29 CFR 1926.404(d)(6)].
3. Inspect electrical tools and equipment daily and remove damaged or defective equipment from use until it is repaired [29 CFR 1926.404(b)(iii)(c)].

SOURCES OF HELP

- OSHA General Industry Standards [CFR parts 1900-1910] and OSHA Construction Standards [CFR Part 1926] which together include all OSHA job safety and health rules and regulations covering construction.
- OSHA-funded free consultation services listed in telephone directories under U.S. Labor Department or under the state government section where states administer their own OSHA programs.
- OSHA Safety and Health Training Guidelines for Construction (Available from the National Technical Information Service, 5285 Port Royal Road, Springfield, VA 22161; 703/487-4650; Order No. PB-239-312/AS): a set of 15 guidelines to help construction employers establish a training program in the safe use of equipment, tools, and machinery on the job.
- Courses in construction safety are offered by the OSHA Training Institute, 1555 Times Drive, Des Plaines, IL 60018, 708/297-4810.

NOTE: The case here described was selected as being representative of fatalities caused by improper work practices. No special emphasis or priority is implied nor is the case necessarily a recent occurrence. The legal aspects of the incident have been resolved, and the case is now closed.

FIGURE 14-2 Accident report, electrocution. *Courtesy: OSHA.*

implement a hazard communication program, conduct a chemical inventory, ensure that a *Material Safety Data Sheet* (MSDS) is available for each chemical, label chemical containers, and train employees on the safe use of chemicals (e.g., protective equipment, procedures). The information shared under the HCS requirements provides the foundation for a chemical safety and health program in the workplace.

OSHA is involved in the development of the **Globally Harmonized System of Classification and Labeling of Chemicals** (GHS). It includes uniform provisions for classification of chemicals, labels, and safety data sheets. The GHS has been adopted by the United Nations and an international goal is to recruit as many countries as possible to participate in the GHS (Occupational Safety & Health Administration, 2011d).

OSHA HAZWOPER Standard

The **Hazardous Waste Operations and Emergency Response Standard (HAZWOPER)** (1910.120 and 1926.65) applies to a variety of public and private employers and employees engaged in operations that include clean-up of hazardous substances required by a government body; voluntary clean-up operations at hazardous waste sites; operations involving hazardous wastes that are conducted at treatment, storage, and disposal facilities; and emergency response operations related to hazardous substances. Training is essential to comply with this standard. It must focus on the duties employees perform without endangering themselves or others. The planning of training must also include consideration of hazards to the community and what capabilities personnel need to respond to the hazards, including a worst-case scenario. Training and equipment may include first responder capabilities and use of PPE. These issues are serious, as illustrated by the multi-billion dollar lawsuits filed by New York City firefighters who claimed lung damage due to inadequate PPE during their response to the collapse of the Twin Towers. One further note is that hazmat events are regulated by not only OSHA, but also the Environmental Protection Agency (EPA).

OSHA Bloodborne Pathogens Standard

The **bloodborne pathogens standard** (1910.1030) limits exposure to blood and other potentially infectious materials, which could lead to disease or death. The standard covers all employees facing potential exposure. Employers are required to establish an exposure control plan covering safety procedures, protective equipment, and the control of waste. The hepatitis B vaccination is to be made available to all employees who have occupational exposure to blood. Postexposure evaluation and follow-up is to be made available to all employees who have had an exposure incident, including laboratory tests at no cost to the employee. Exposure records must be confidential and kept for the duration of employment plus 30 years. Training is required on all aspects of this standard, and the training records must be maintained for three years.

Lockout/Tagout

OSHA standard 1910.147 is designed to control hazardous energy. Better known as **lockout/ tagout**, the aim is to prevent the accidental startup of machines or other equipment during maintenance and servicing. The rule requires that hazardous energy sources must be isolated and rendered inoperative before work can begin. Elements of a lockout/tagout program are written procedures, training, and audits.

An example of OSHA citations involving this standard include Anchor Hocking, a glass manufacturer in Lancaster, Ohio, where a worker's right index finger became caught in a machine during maintenance and was amputated. Workers had not been trained in "lockout/tagout" procedures and the company faced $113,800 in fines (Occupational Safety & Health Administration, 2011e).

OSHA Recordkeeping and Reporting

Before the development of OSHA's centralized recordkeeping system, workplace statistics on injuries and illnesses were kept by some states and private organizations. No uniform, standardized system existed. Today, with the help of OSHA's comprehensive statistics, it is easier for employers to pinpoint serious hazards and work toward improvements. In addition, employee awareness and safety precautions are enhanced through knowledge about injuries, illnesses, and hazards.

The OSHA website provides recordkeeping and reporting guidance for employers. For instance, employers in a Federal OSHA jurisdiction must report to OSHA the death of any employee from a work-related incident or the in-patient hospitalization of three or more employees because of a work-related incident within eight hours.

Confined Area Entry

Hazards that are not easily seen, smelled, or felt can be deadly risks to people who work in confined areas. For instance, storage tanks may reduce oxygen or leak combustible or toxic gases. The cardinal rule for entry into a confined area is, "Never trust your senses!" A harmless-looking situation may indeed be a potential threat. Some of the deadliest gases and vapors have no odor. Before entry, the following safety strategies are recommended: proper training, equipment to identify hazards, and an entry permit issued by a safety specialist. Schroll (2006: 44) emphasizes the importance of equipment. He writes that personnel should be able to select appropriate monitoring equipment and know how to use it. He also refers to personal protective equipment (PPE), especially respiratory protection. In addition, personnel should know the operation of harnesses and retrieval devices. OSHA's standard for confined spaces, Title 29 Code of Federal Regulations (CFR), Part 1910.146, contains the requirements for practices and procedures to protect employees from the hazards of entry into permit-required confined spaces.

In the petroleum industry, for example, a storage tank had been rinsed and vented for several days. When it was checked with gas detection equipment, no flammable gases were measured. However, after workers removed loose rust, scale, and sediment, the percentage of flammable gas rose, and the gas ignited.

As another example, two employees of a fertilizer company descended into an old 35-foot well to repair a pump. The well was covered with a concrete slab and entry was made through a covered manhole. About six feet below the opening was a plank platform. When the first worker dropped to the platform, he was immediately overcome and fell unconscious into the water below. His partner sought help quickly. When two helpers entered the well, they, too, fell unconscious into the water below. A passerby, in an attempt to save the drowning men, jumped into the water and drowned also. By this time, the fire department had arrived. The fire chief, wearing a self-contained breathing apparatus, went to rescue the victims. On the platform, he removed his facemask to give instructions to those above and was overcome. Subsequent tests revealed that the well atmosphere contained a lethal concentration of hydrogen sulfide. Five men died from pulmonary paralysis (Chacanaca, 1996: 61–65).

What would you say and do if your supervisor ordered you to enter a confined area without first taking safety precautions?

Additional Employer Responsibilities

An employer is required to post specific OSHA-related material for employee review. For example, the OSHA poster "Job Safety and Health" informs employers and employees about rights and responsibilities (Figure 14-3). If an employer receives a Citation and Notification of Penalty, the employer must post the citation, or a copy, near the place where each violation occurred to inform employees of hazards.

The **OSHA Whistleblower Protection Program** helps to ensure that employees are free to report safety and health violations. The OSH Act prohibits any person from discharging or retaliating against any employee because the employee exercised rights under the Act.

OSHA Inspections

Originally, an important priority of OSHA compliance inspectors was to view the workplace as it functions on a typical day. To attain this goal and to prevent an employer from altering typical workplace characteristics by concealing unsafe conditions, inspections were frequently made unannounced. However, only a few years after this practice began, it was challenged in the courts. In 1975, the president of a utility installation company, who posted a copy of the Bill of Rights on his office wall, sued while claiming that the Fourth Amendment restricts warrantless searches. The federal appellate court upheld the employer's contention. The case was appealed, and in May 1978 the Supreme Court, in *Marshall v. Barlow*, ruled that the Fourth Amendment protection against unreasonable searches protects commercial establishments as well as private homes. Therefore, OSHA inspectors must obtain a warrant before making an inspection, unless employers consent. Such a warrant is to be based on administrative probable cause: the inspector is required to show a judicial officer that the inspection is part of OSHA's general administrative plan to enforce safety and health laws, or upon evidence of a violation (Hall, 2006: 146–151; Dreux, 1995: 53).

When an inspection takes place, the employer and even employee representatives may join the inspector. The inspector is obligated to show credentials that contain a photograph and serial number. The number can be verified via the nearest OSHA office. Typically, machinery, equipment, and other workplace characteristics are examined. The inspector can interview the

Job Safety and Health
It's the law!

OSHA®
Occupational Safety
and Health Administration
U.S. Department of Labor

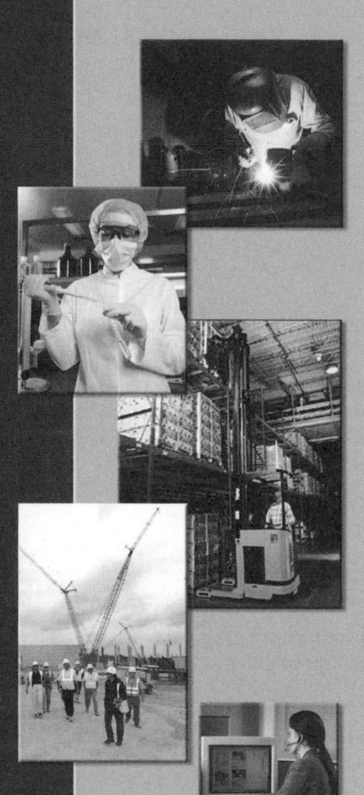

EMPLOYEES:

- You have the right to notify your employer or OSHA about workplace hazards. You may ask OSHA to keep your name confidential.
- You have the right to request an OSHA inspection if you believe that there are unsafe and unhealthful conditions in your workplace. You or your representative may participate in that inspection.
- You can file a complaint with OSHA within 30 days of retaliation or discrimination by your employer for making safety and health complaints or for exercising your rights under the *OSH Act*.
- You have the right to see OSHA citations issued to your employer. Your employer must post the citations at or near the place of the alleged violations.
- Your employer must correct workplace hazards by the date indicated on the citation and must certify that these hazards have been reduced or eliminated.
- You have the right to copies of your medical records and records of your exposures to toxic and harmful substances or conditions.
- Your employer must post this notice in your workplace.
- You must comply with all occupational safety and health standards issued under the *OSH Act* that apply to your own actions and conduct on the job.

EMPLOYERS:

- You must furnish your employees a place of employment free from recognized hazards.
- You must comply with the occupational safety and health standards issued under the *OSH Act*.

This free poster available from OSHA –
The Best Resource for Safety and Health

Free assistance in identifying and correcting hazards or complying with standards is available to employers, without citation or penalty, through OSHA-supported consultation programs in each state.

1-800-321-OSHA (6742)
www.osha.gov

OSHA 3165-12-06R

FIGURE 14-3 OSHA poster.

employer and employees in public or in private. Any interference with the inspector's duties can result in stiff penalties. The employer is wise to document the inspection and comments by the inspector through note taking. This information may become useful if a disagreement or a dispute of a citation or penalty evolves.

Because millions of workplaces are subject to inspections, OSHA has established priorities. Obviously, workplaces with serious accidents will be subject to inspections.

During FY 2010, OSHA conducted 40,993 inspections in U.S. work sites and identified 96,742 violations of OSHA's standards and regulations, a 15.3 percent increase since FY 2006. Under the Whistleblower Protection Program, during FY 2010, 1,177 investigations were conducted, and 27 State OSHA programs conducted 1,039 such investigations (Occupational Safety and Health Administration, 2011f). The FY 2012 budget request for OSHA was $583.4 million with 2,387 full-time employees, an increase of $24.8 million and 52 employees over the FY 2010 level (Occupational Safety & Health Administration, 2011g).

The National Federation of Independent Business recommends the following strategies during an OSHA inspection (Gaudio, 2005):

- Manage the inspection process to minimize operational disruptions
- Control the flow of information to OSHA so that when the inspection has ended, the employer understands the significance of the evidence OSHA has gathered
- Present the worksite in the best possible light
- Be proactive by addressing OSHA's compliance concerns during the inspection so OSHA does not issue citations.

By understanding OSHA's concerns early in the inspection process, the employer can provide information that refutes OSHA's factual findings or interpretation of the OSHA standard or regulations. This information may ultimately convince OSHA not to issue citations or to at least minimize penalties.

Each inspection is different, so the need for counsel will depend on individual circumstances. On one hand, a routine inspection that does not stem from a significant injury or fatality may be handled without an attorney. On the other hand, an employer is strongly advised to have counsel directly involved during an OSHA inspection when a major accident or event such as a catastrophic accident, serious injury or employee fatality has occurred.

OSHA: Criticism and Controversy

Through the years since its inception, OSHA has been the target of considerable criticism and controversy. Most of the OSHA battles have taken place on Capitol Hill, when different interest groups pressure legislators either to maintain and expand OSHA or to reduce or eliminate it. The forces in favor of OSHA are primarily OSHA itself, the AFL-CIO labor organization, and select legislators. Those opposed to OSHA consist mainly of businesspeople, business organizations, and select legislators. Some say the controversy essentially is between "big labor" and "big business."

Another factor in the controversy is the value of a life. As government agencies increase the price of a life, this cost is used to justify more costly government regulations. Businesspeople argue that increased regulations harm economic growth. For instance, the EPA set the price of a life at $9.1 million when proposing tighter regulations on air pollution. Politicians study costs (of regulations) versus benefits (how many lives can be saved) as input when making sensitive decisions on regulations (Appelbaum, 2011).

Arguments against OSHA are as follows:

1. "Regulatory overkill" is a major theme of those against OSHA. Many business organizations believe that OSHA has gone beyond what is necessary to foster a safe and healthy workplace.
2. The costs of OSHA requirements—resulting in reduced productivity, lost jobs and higher prices for goods—add to inflation and are a threat to companies' competitive position.
3. OSHA and other government bureaucrats—who are appointed rather than elected and thus are not accountable to the people—are making decisions that businesspeople should be making for themselves.
4. OSHA has had an impact on the labor-management process that has compounded labor troubles. Many unions have become involved in worker safety and employees' rights under OSHA to the point where productivity is hindered.

Arguments in favor of OSHA are as follows:

1. OSHA is essential to reinforce a safe and healthy workplace for employees.
2. Many deaths and injuries have been prevented because of OSHA.
3. In today's technologically complex business world, employees need protection that only government regulation can provide.
4. Employers who oppose OSHA are too interested in the costs of safety and health, and in productivity and profits, and they are not concerned enough about employees.
5. There are employers who have a favorable attitude toward OSHA and benefit from its existence. These employers actively work with OSHA in a joint effort to prevent and reduce safety and health problems. To these businesspeople, OSHA compliance is cost effective.

Assistance with Problems

There are many sources of assistance for employers concerned about workplace safety or health problems:

1. Many insurance companies provide specialists who visit, inspect, and recommend stategies for preventing and eliminating hazards at client workplaces.
2. Trade associations and employer groups have become more conscious about safety and health.
3. Trade unions and employee groups are often interested in coordinated activities for preventing and eliminating hazards.
4. The National Safety Council has an extensive information service.

5. Local doctors may be willing to provide information on a consulting basis about workplace medical matters. The Red Cross is a source of first-aid training.
6. The Internet and libraries contain a wealth of information on safety and health matters.
7. Local colleges and universities may have educational programs in the field of occupational safety and health. If an employer contacts the relevant departments, educators may provide useful information.
8. Free on-site consultation is offered in many states through agreements between OSHA and either a state or private contractor. These consultants do not write citations but expect cooperation or OSHA will be contacted. Enforcement action is rare, especially because the employer requests the consultant and shows a concern for safety and health.

Although this chapter emphasizes accidents in the workplace, other safety issues should be considered for a comprehensive loss prevention program. Examples include natural disasters and workplace violence. Furthermore, OSHA guidelines go beyond preventing accidents. For instance, in cooperation with the U.S. Department of Homeland Security, OSHA has published guidelines for emergencies, including terrorist events.

Safety Strategies

The OSHA *Small Business Handbook* states four basic elements to all good safety and health programs (Occupational Safety and Health Administration, 2005: 8–10):

1. Management commitment and employee involvement.
2. Worksite analysis to identify all hazards.
3. Hazard prevention and control methods are put in place and maintained.
4. Managers, supervisors, and employees are trained to understand and deal with worksite hazards.

Safety and Health Committee

A safety and health committee can be an important part of an effective loss prevention program. When employees jointly communicate about and work toward increased safety and health, they develop a greater awareness of associated problems and solutions. Topics to discuss at meetings can include past accidents and illnesses, OSHA standards and inspections, and cases of accidents, illnesses, and remedies that occurred at similar facilities. Kilbourne (2010) recommends that managers and supervisors set a positive example, correct hazards promptly, and reward safe performance.

Socialization and Incentive Programs

Training is a prime strategy for accident prevention. The objective of training is to change the behavior of the employee. He or she should think and act in a safe manner. Training is useless unless employees are motivated to continually act in a safe manner. Even if familiar with safety procedures and other relevant information, an employee will not necessarily "practice what is preached." A method for stimulating the employee to act safely and to use safety knowledge is necessary. This objective is accomplished through incentive programs. Businesses may develop such programs on their own or outsource the work to a service firm. Examples of incentives are as follows: Every time a manufacturing plant reaches a million hours without a lost-time accident, every employee receives a gift. Another approach is safety bingo: Numbers are drawn every workday for a week and posted on a sign in the employee parking lot. If an employee completes a row, $25 is awarded. If the whole card is filled, $1,000 is won. When lost-time accidents occur, the game is halted until a month later. Companies often set safety goals for individual departments, which facilitate a competitive spirit to win prizes. Safety, rewarded by merit pay increases for employees, will also reduce accidents.

What are the results of such incentive programs? Safety incentives provide a powerful, cost-effective management tool to prevent accidents. Enthusiasm and safety awareness increase. Employees are more vigilant about other workers' safety. Incentive programs provide fun in the workplace, which results in higher morale. As accidents decline, so do insurance premiums for workers' compensation. Money is saved. Fewer accidents mean fewer production interruptions and greater profits. It is feasible to use incentives for a comprehensive loss prevention effort involving not only safety but also crime and fire prevention.

The Dangers of Safety Incentive Programs

There are three basic types of safety incentive/reward programs as described here, along with concerns (Atkinson, 2000: 32–38).

Traditional programs offer rewards to employees for going a certain period without a recordable injury. The focus is on results. OSHA, labor unions, and some employers argue that these programs may be used by employers to take the place of formal programs. In addition, employees may feel pressured not to report injuries because coworkers would be upset about not winning incentives, and so a serious workplace hazard might go undetected by management.

Behavior-based programs offer rewards to employees for behaviors (e.g., wearing personal protective equipment) that promote safety. Labor groups say that this approach places safety on the shoulders of employees; management then blames them when accidents occur. Furthermore, employers watch actions and may ignore hazards that can be corrected.

Safety activities offer rewards to employees for suggestions to improve safety, identifying and correcting hazards, participating in inspections, and serving on a committee, among other activities.

Whereas the first two programs often involve a greater level of participation, safety activities are frequently voluntary; however, everyone should be involved in safety. Since each type of program can create concern, the employer and employees should seek to understand each one and capitalize on the beneficial aspects of each.

Investigations

After an accident, an investigation is vital to prevent future accidents; the cause of an accident can be pinpointed so that corrections can follow. Established procedures important for well-planned accident investigations include the following:

1. Respond quickly to reinforce that loss prevention personnel are "on the job." This will also show employees that management cares. Ensure that the injured receive proper medical attention.
2. Protect the scene of the accident to prevent unintentional or intentional harm to evidence.
3. If required, contact appropriate regulatory agencies and insurers.
4. Find out the following: who was involved, where did it occur, when did it happen, who was injured, what occurred, what was damaged, and how did the accident take place?
5. Try to pinpoint the cause. Investigate possible direct and indirect causes. Study equipment, work procedures, the environment, and the employees involved. Is a drug test required?
6. Estimate injuries and direct and indirect costs.
7. Use a standard accident report that fits management's requirements and aids the investigative inquiry.
8. Maintain an open mind, remaining aware that some employees attempt fraudulent workers' compensation claims by staging an accident or by providing false information.
9. Complete appropriate forms (e.g., company forms, workers' compensation, OSHA).
10. Prepare a presentation for the safety and health committee concerning the accident. Solicit feedback from the committee to solve problems.
11. Follow up on corrective action to ensure safety.

One of the most difficult questions to answer during an accident investigation is the *cause*; considerable controversy is often generated. Opinions vary, but facts are necessary. *There are two primary causes of accidents: unsafe conditions and unsafe acts by people.* Frequently, unsafe conditions (e.g., unguarded moving parts, poor lighting) are known. However, corrections may not be made because of inaction or costs. Unsafe acts by people can result from ignorance, poor training, negligence, drugs, fatigue, emotional upset, poor attitude, and high production demands. Other circumstances also can cause accidents. One of the oldest and most controversial theories of accident causation is the **accident proneness theory**. This theory suggests that people who repeatedly have accidents are accident prone. Many experts agree that about 20 percent of the people have most of the accidents, whereas the remaining 80 percent have virtually no accidents.

Haaland (2005: 51–57) writes that the literature in the discipline of safety shows that 90 percent of accidents are attributed to *human error*. He notes specific human traits that have been linked through empirical research to accident proneness. One of these traits is *conscientiousness*; Haaland defines it as "an individual's degree of organization, persistence, and motivation in goal-directed behavior." Those who are low in this trait tend to ignore safety rules and get into more accidents. *Extreme extroversion* and *extreme introversion* are other traits linked to accident proneness. The former is characterized by being sociable and outgoing, but also easily distracted and having a shorter attention span. The latter is characterized by limited interpersonal interactions and possibly hesitation before seeking assistance. Haaland adds that unsafe employees are also characterized by other behaviors harmful to the bottom line, such as arriving late to work, more absences, and poorer levels of productivity and quality. Consequently, employers can systematically hire applicants who are less likely to become involved in accidents and injuries. A **safety performance assessment** measures personality traits that influence safety in the workplace. It consists of written statements and an applicant can complete the assessment at a kiosk or through a website.

Harry Nash, Machinist, Is Injured Again

As Harry Nash was cutting a piece of metal on a band saw, he accidentally cut off his thumb. Workers in the surrounding area rushed to his aid. The foreman and the loss prevention manager coordinated efforts to get him and his thumb to the hospital as quickly as possible. After Harry was in the care of doctors, the foreman and the loss prevention manager began an investigation. A look at Harry's record showed that he had been working for the company for eight years. During the first year, he slipped on some oil on the workplace floor and was hospitalized for three weeks with a sprained back. The third year showed that he accidentally drilled into his finger with a drill press. The report stated that Harry did not receive any training from a now-retired foreman, who should have instructed Harry on drill press safety techniques. By the fifth year, Harry had had another accident. While carrying some metal rods, he fell and broke his ankle because he forgot about climbing one step to enter a newly constructed adjoining building.

Even though Harry had had no training in operating the band saw, the loss prevention manager and the foreman believed that Harry was accident prone. The foreman, angry with Harry spoiling the department's safety record and incentive gifts, wanted Harry fired. The loss prevention manager was undecided about the matter.

Later, management, the foreman, and the loss prevention manager decided to assign Harry to a clerical position in the shipping department. Harry eventually collected workers' compensation for his lost thumb.

Do you think certain people are accident prone? Justify your answer.

Additional Safety Measures

Other safety measures are to display safety posters, create a safety-by-objectives program, and recognize employees with excellent safety records. Those employees with multiple accidents in their records should be reassigned or retrained.

Safety posters or signs are effective if certain guidelines are used. Research indicates that if the safety message is in negative terms (e.g., "Don't let this happen to you," followed by a picture of a person with a physical injury), it causes fear, resentment, and sometimes anger. Posters with positive messages (e.g., "Let's all pitch in for safety") produce better results.

FIGURE 14-4 Safety signs.

Posters and signs (see Figure 14-4) are more potent when they reflect the diversity of employees (e.g., multiple languages), are located in appropriate places, are not too numerous, and have attractive colors.

A safety-by-objectives program is a derivative of management by objectives. In a manufacturing plant, department heads formulate safety objectives for their departments. Management makes sure that objectives are neither too high nor too low. Incentives are used to motivate employees. After a year, the objectives are studied to see if they were reached.

Search the Internet

Here are websites relevant to this chapter:

American Red Cross: www.redcross.org
Environmental Protection Agency: www.epa.gov
National Institute for Occupational Safety and Health (NIOSH): www.cdc.gov/niosh
National Safety Council: www.nsc.org
Occupational Safety and Health Administration (OSHA): www.osha.gov
U.S. Government Printing Office, Federal Digital System (FDsys): www.fdsys.gov

Case Problems

14A. As a loss prevention practitioner for a large corporation, you are asked by a local college professor to lecture on the pros and cons of OSHA. What will your comments be to the class?

14B. If you were a company safety and loss prevention manager, how would you react to an OSHA inspection? What conditions in your company do you think would influence your reaction?

14C. In reference to the case describing Harry Nash, do you feel that the loss prevention manager and the foreman were justified in labeling Harry as accident prone? Support your answer.

References

Anderson, C. (1975). *OSHA and accident control through training.* New York, NY: Industrial Press.

Appelbaum, B. (2011). As U.S. Agencies put more value on a life, businesses fret. *The New York Times* (February 16).

Atkinson, W. (2000). The dangers of safety incentive programs. *Risk Management, 47* (August).

Centers for Disease Control and Prevention. (2011). Injury and violence prevention and control. <www.cdc.gov/injury> retrieved December 13, 2011.

Chacanaca, M. (1996). Specialty-confined-space rescues. *Emergency*, February.

Dreux, M. (1995). When OSHA knocks, should an employer demand a warrant? *Occupational Hazards*, April.

Gaudio, B. (2005). OSHA inspection ... Can it happen to You? *NFIB Business Toolbox* (January 21). <www.nfib.com/object/IO_19802.html>, retrieved December 17, 2006.

Grimaldi, J., & Simmonds, R. (1975). *Safety Management.* Homewood, IL: Richard D. Irwin.

Haaland, D. (2005). Who's the safest bet for the job? *Security Management, 49* (February).

Hall, D. (2006). *Administrative law: Bureaucracy in a democracy* (3rd Ed.). Upper Saddle River, NJ: Pearson Education.

Kilbourne, C. (2010). Six strategies to improve worker safety attitudes. *Safety Daily Advisor* (October 7) <safetydailyadvisor.blr.com> retrieved February 10, 2011.

Mason, J. (1976). OSHA: Problems and prospects. *California Management Review, 19* (Fall).

Occupational Safety and Health Administration. (2011b). About OSHA. <www.osha.gov/about.html> retrieved December 14, 2011.

Occupational Safety and Health Administration. (2011a). OSH Act of 1970. <www.osha.gov> retrieved December 14, 2011.

Occupational Safety and Health Administration. (2011f). OSHA Enforcement. <www.osha.gov/dep/2010_enforcement_summary.html> retrieved December 17, 2011.

Occupational Safety & Health Administration. (2011g). OSHA statement to U.S. House of representatives (April 14). <www.osha.gov/pls/oshaweb/owadisp.show_documentdsg/hazcom/global.html> retrieved December 17, 2011.

Occupational Safety and Health Administration. (2011c). Regulations. <www.osha.gov> retrieved December 17, 2011.

Occupational Safety and Health Administration. (2005). *Small Business Handbook*, <www.osha.gov/Publications/smallbusiness/small-business.html> retrieved December 17, 2011.

Occupational Safety & Health Administration. (2011d). The globally harmonized system for hazard communication. <www.osha.gov/dsg/hazcom/global.html> retrieved December 17, 2011.

Occupational Safety & Health Administration. (2011e). U.S. labor department's OSHA cites anchor hocking in lancaster, Ohio, for 12 safety violations after maintenance employee suffers amputation injury. *OSHA Regional News Release* (September 1).

<www.osha.gov/pls/oshaweb/owadisp.show_document?p_table=NEWS_RELEASES&p_id=20604> retrieved December 17, 2011.

U.S. Department of Labor. (2011a). Census of fatal occupational injuries, 2010. <www.bls.gov> retrieved December 13, 2011.

U.S. Department of Labor. (2011b). Department of labor (Draft) strategic plan, FY 2010–2016. <www.dol.gov/_sec/stratplan/2010/osha/index.htm> retrieved December 15, 2011.

Schroll, C. (2006). Confined space training: Follow these 17 key steps. *Industrial Safety & Hygiene News, 40* (April).

Wade, J. (2011). Time line: The lives of workers. *Risk Management, June* <www.rmmag.com> retrieved September 12, 2011.

Special Problems and Countermeasures

15

Terrorism and Homeland Security

OBJECTIVES

After studying this chapter, the reader will be able to:

1. Explain the definition of terrorism
2. Discuss the history of terrorism
3. Describe the causes of terrorism
4. Discuss the measurement of terrorism
5. Describe international terrorism and domestic terrorism
6. Discuss terrorist methods and weapons
7. Define and describe homeland security
8. Differentiate national security and homeland security
9. Describe government action against terrorism
10. Define intelligence and explain its role in efforts to protect against terrorism
11. Explain methods to protect against weapons of mass destruction
12. Discuss the role of the private sector in homeland security

KEY TERMS

- terrorism
- Sicarii
- Zealots
- Assassins
- French Revolution
- "Reign of Terror"
- Ku Klux Klan
- state terrorism
- guerilla warfare
- Palestinian Liberation Organization
- Arab Spring
- Crusades
- Islam
- Prophet Muhammad
- Muslims
- Sunni
- Shiite
- fundamentalism
- psychological causes of terrorism
- rational choice causes of terrorism
- structural causes of terrorism
- international terrorism
- globalization
- asymmetrical warfare
- 9/11 attacks
- al-Qaida
- al-Qaida in the Arabian Peninsula
- Osama bin Laden
- domestic terrorism
- anarchism

- ecoterrorism
- Animal Liberation Front (ALF)
- Earth Liberation Front (ELF)
- sovereign citizen extremist movement
- extremism
- Ruby Ridge incident
- Christian Identity
- Waco incident
- Oklahoma City bombing
- Patriot movement
- homegrown terrorists
- lone wolf terrorists
- cellular model of organization
- leaderless resistance
- "new terrorism"
- secondary explosion
- improvised explosive device (IED)
- weapon of mass destruction (WMD)
- weapon of mass effect (WME)
- Aum Shinrikyo cult
- Anthrax
- radiological dispersion device
- homeland security
- National Security Strategy of the United States
- National Strategy for Homeland Security
- U.S. Northern Command (NORTH-COM)
- Defense Against Weapons of Mass Destruction Act of 1996
- Antiterrorism and Effective Death Penalty Act of 1996
- USA Patriot Act of 2001
- Homeland Security Act of 2002
- Department of Homeland Security (DHS)
- stovepiping
- connecting the dots
- intelligence
- intelligence process
- intelligence community
- The Intelligence Reform and Terrorism Prevention Act of 2004
- first responders
- National Terrorism Advisory System
- Information Sharing and Analysis Centers (ISACs)
- Fusion Centers
- Overseas Security Advisory Council (OSAC)
- American Red Cross
- Salvation Army
- National Volunteer Organizations Against Disasters
- Citizen Corps
- USA Freedom Corps

Terrorism

Terrorism Defined

Poland (2005: 2–5) writes of the complexity of defining terrorism. Many definitions exist and there is no international agreement on how to define the term. This results in methodological problems for both research and data collection and conflicts over efforts at international counterterrorism.

Hoffman (2012a: 5) explains that terrorism is so difficult to define "because the meaning of the term has changed so frequently over the past two hundred years." For example, during the French Revolution, terrorism was viewed in a positive light (see the subsequent description). Hoffman offers a contemporary perspective of terrorism and writes that it is "fundamentally and inherently political," and about the pursuit, acquisition, and use of power to achieve political change, while using threats and violence.

The United Nations has struggled with the term for many years. It has been argued that one state's "terrorism" is another state's "freedom fighter." Or, simply put, terrorism is a "dirty word" drenched in emotion, and it describes what the "other guy" has done.

The National Institute of Justice (2011) refers to Title 22 of the U.S. Code, Section 2656f(d) which defines **terrorism** as "premeditated, politically motivated violence perpetrated against noncombatant targets [e.g., civilians; military personnel who are unarmed and/or not on duty] by subnational groups or clandestine agents, usually intended to influence an audience." The National Institute of Justice writes of the Federal Bureau of Investigation definition of terrorism as "the unlawful use of force or violence against persons or property to intimidate or coerce a government, the civilian population, or any segment thereof, in furtherance of political or social objectives."

History

Dyson (2012: 5–6) writes of the difficulty of determining the exact role of terrorism in shaping history. He adds: "It appears that throughout most of history, terrorism has been more of a bothersome irritant to the governments in power than an actual threat to them." Dyson notes exceptions to this point by writing that terrorists, in some instances, have facilitated drastic change. What follows here are several accounts of terrorism that are subject to debate as to the impact on government and history. Purpura (2007: xi) writes:

> *Unfortunately, the 9/11 attacks were immensely successful and cost-effective for the terrorists. With a loss of nineteen terrorists and expenses between $400,000 and $500,000, the attackers were able to kill about 3,000 people, cause hundreds of billions of dollars in economic damage and spending on counterterrorism, and significantly impact global history. With such a huge kill ratio and investment payoff, governments and the private sector must succeed in controlling terrorism.*

Terrorism has existed for centuries. Ancient terrorists were "holy warriors" as we see with certain terrorists today. During the first century in the Middle East, the **Sicarii** and the **Zealots**, Jewish groups in ancient Palestine, fomented revolution against the occupying forces of Rome. The Sicarii instilled fear by using a dagger to stab Romans and Roman sympathizers during the day at crowded holiday festivities. The throngs provided cover for the killers and heightened terror because people never knew when an attacker would strike. The Zealots-Sicarii believed that by confronting the Romans, the Messiah would intervene and save the Jewish people. Between 66 and 70 A.D., revolution became a reality for the Zealots-Sicarii. However, it ended in disaster. With thousands of Jews killed and the Jewish state in shambles, the survivors fled to the top of Masada where their abhorrence to being subjected to control by the Romans, who surrounded them, resulted in mass suicide. Today, we have the term "zealot," meaning fanatical partisan.

In analyzing the methods of the Zealot-Sicarii, and drawing parallels to modern-day terrorism, Poland (2005: 26–27) writes that "the primary purpose of the Sicarii terrorist strategy, like so many terrorist groups today, seems to be the provocation of indiscriminate countermeasures by the established political system and to deliberately provoke repression, reprisals, and counterterrorism." Poland refers to Northern Ireland where, for hundreds of years, Catholics battled Protestants. Simonsen and Spindlove (2004: 70–75) explain that during the 16th century, James I, King of England, offered land in Ireland to Scottish settlers for the purpose of establishing the Protestant church in Ireland. Conflict ensued with Catholics, but the Protestant landowners prevailed. Catholics were regulated to a life of poverty. Poland cites incidents during the 20th century when Catholics, protesting peacefully against Protestants and British rule, were killed and wounded when police and British Security Forces overreacted.

Poland draws another parallel from the Zealot-Sicarii movement. The Irgun Zvai Leumi-al-Israel, led by Menachem Begin, terrorized the British military government of Palestine between 1942 and 1948 in an effort to establish a Jewish state. The Irgun perpetrated many bombings and assassinations, leading to a cycle of terror and counter-terror. Manifestos of the Irgun argued "No Masada." Irgun fighters had actually studied terrorist methods of the Sicarii and the Irish Republican Army. Eventually, the British turned the problem over to the United Nations, and in 1948, the country of Israel was born.

The **Assassins** were another religious sect that used terrorism to pursue their goals (Weinzierl, 2004: 31–32). This group gave us the term "assassin," which literally means "hashish-eater," a reference to the drug-taking that allegedly occurred (perhaps rumor) prior to murdering someone. During the 11th and 12th centuries, this group evolved from the Shiites of the present-day Mideast. They believed that the Muslim community needed the purified version of Islam to prepare for the arrival of the Inman, the Chosen of God and leader of humanity. The Assassins waged a war against the majority Sunni Muslim population. Their terror strategies consisted of using unconventional means, establishing mountain fortifications from which terror attacks were launched, and using daggers like the Sicarii. *Although the Assassins were not successful in reforming the Islamic faith or in recruiting many converts, they are remembered for their innovations in terrorist strategies, namely the suicide mission and using disguise and deception.* Acting under God, they were promised a place in paradise for their ultimate sacrifice as they sought out and killed Sunni religious and political leaders. Even today, we see conflict between Shiites and Sunni, as in Iraq.

Through history and today, we see the deep roots of religion within terrorism.

Secular motivations for terrorism were within the **French Revolution**, a major historic milestone in the history of terrorism. The French Revolution ended absolute rule by French kings and strengthened the middle class. The **"Reign of Terror"** (1793–1794), from which the word "terrorism" evolved, saw leadership from Maximilien Robespierre, who led the Jacobin government in executing, by guillotine, 12,000 people declared enemies of the Revolution. During the "Reign of Terror," terrorism was viewed in a positive light and Robespierre declared that terrorism is prompt justice and a consequence of democracy applied during a time of urgent need. In 1794, when Robespierre prepared a new list of traitors of the revolution, those fearing their names were on the list executed him at the guillotine, and thus, the "Reign of Terror" ended. By this time, terrorism was beginning to be characterized in a negative light. Edmund Burke, a British political leader and writer, criticized the French Revolution and popularized the term "terrorism" in the English language as a repulsive action. This state terrorism of the French government was significant in the history of terrorism because the terror was not justified by religion or God, but by the masses to promote a political ideology (Weinberg and Davis, 1989: 24–25). Greenberg (2001:1) writes: "Over the next century the Jacobin spirit infected Russia, Europe and the United States." Anarchist terrorist groups worked to foment revolution. In Russia, the Narodnaya Volya (meaning "people's will") was born in 1878 to destroy the Tsarist regime. They assassinated Alexander II in 1881, but the success sealed their fate because they were crushed by the Tsarist regime. This group did inspire other anarchists. During an anarchist-led labor protest at Chicago's Haymarket Square in 1886, a bomb was thrown into the crowd as police intervened. A riot ensued and several police and demonstrators were killed or injured. This incident hurt the labor movement. The anarchists were tried and sentenced to be hanged or imprisoned. Because there had been insufficient evidence, a few years later the Governor of Illinois pardoned the surviving anarchists.

As the 19th century gave way to the 20th century, anarchists and others dissatisfied with established order adopted terrorism as a strategy to reach their goals (Laqueur, 1999). Terrorist attacks and assassinations occurred in France, Italy, Spain, Germany, Ireland, India, Japan, the Balkans, and the Ottoman Empire, among other areas. Bombs were detonated in cafes, theaters, and during parades. In the United States, two presidents were assassinated—Garfield (1881) and McKinley (1901). This era also saw an increase in nationalism and struggles for statehood as native groups revolted and used terrorism to free themselves from imperial control and European colonialism. The assassination of Austrian heir Archduke Ferdinand, in 1914, is a famous illustration of how terrorism was used for nationalistic goals. Gavrilo Princip, a member of the Pan-Serbian secret group called the Black Hand, planned to kill Ferdinand to help southern Slavs. The assassination actually became the precipitating incident that triggered World War I and doomed the Austrian Empire. We can see that terrorism can sometimes have a profound impact on world affairs.

Although global terrorism was primarily left-wing oriented through history, during the late 18th century and early 20th century, conservative, right-wing groups formed to maintain the status quo and prevent change. In the United States, the **Ku Klux Klan** (KKK), a right-wing group, was formed in 1865 by Confederate Army veterans who believed in the superiority of whites (Chalmers, 1981). The major goal of the KKK is to hinder the advancement of blacks, Catholics, Jews, and other minority groups. Klan methods include wearing white robes and hoods, burning crosses (at meetings and to frighten nonmembers), intimidation, tar-and-feathering, lynching, and murder. The KKK evolved through periods of strength and weakness. During the mid-20th century, it was revived to oppose racial integration and it applied terrorist strategies, such as murder and bombing. Because of infighting, splits, court cases, FBI infiltration, and imprisonment, membership has declined.

Because of the spread of communism, especially the Bolshevik Revolution of 1917 in Russia that overthrew the Czar, many political leaders worldwide became fearful and took brutal steps to eliminate communists, socialists, and other left-wing activists within their own countries. Repressive regimes included the Nazis led by Adolph Hitler in Germany and the fascists led by Benito Mussolini in Italy, both becoming allies during World War II. Violence by a government to repress dissidence among its own population, or supporting such violence in another country, is called **state terrorism**.

Following the end of World War II in 1945, a resurgence of nationalism among European colonies in the Middle East, Asia, and Africa resulted in violent uprisings. A combination of **guerilla warfare** and terrorism occurred. The former was characterized by larger units resembling the military and operating from a geographic area over which they controlled, and the latter involving smaller groups blending into the population and planning spectacular attacks for maximum media attention.

According to Greenberg (2001), it was Algeria's Front de Liberation Nationale (FLN), fighting for independence from France in the 1950s, that set the tone for terrorism to come. The FLN raised the cost to France for its colonialism by bombing coastal tourist resorts, thereby randomly killing French families on vacation. By the 1960s, this precedent was copied worldwide by Palestinian and Irish nationalists, Marxists in Africa and Latin America, the Weather Underground in the United States, the Marxist Baader-Meinhoff Gang in the former West Germany, and the Red Brigades in Italy. Terrorist methods were being applied by not only nationalists seeking independence, but also by those with ethnic and ideological agendas.

It was during the 1970s, that Yasser Arafat's **Palestinian Liberation Organization** (PLO), and its splinter groups, while locked in a cycle of shocking violence with Israel, pioneered the hijacking of jet airlines to publicize their goal of Palestinian statehood. Black September, as one splinter group was known, staged an outrageous attack at the 1972 Munich Olympics in Germany and murdered 11 Israeli athletes. This attack became a major, global media event and inspired other terrorists.

During the mid-1980s, state-sponsored terrorism grew again and supported terrorist attacks against Western targets in the Middle East. Sponsors of terrorism included Syria, Libya, Iran, Iraq, and North Korea.

The **"Arab Spring"** of 2011 consisted of political uprisings, protests, and revolution in the north of Africa (e.g., Tunisia, Egypt, Libya, and other countries) that resulted in the ousting of autocratic rulers. The citizenry of each country demanded democracy, freedom, civil liberties, and government accountability (Sibalukhulu, 2011). In Syria and Egypt, the protests and violence continued into 2013. Hamid (2011) writes that if truly democratic governments evolve following these uprisings a variety of Islamist groups will assert their influence, which may be contrary to U.S. interests. Consequently, Hamid argues that the United States should enter into dialogue with various groups and parties early on prior to their seizure of power and government.

Were citizens who participated in the "Arab Spring" terrorists? Were George Washington and his supporters, who had somewhat similar demands as those who participated in the "Arab Spring," terrorists during the American Revolution against the British?

Religion and Politics

Confusion may arise over whether a terrorist group has a religious agenda, a political agenda, or both. Religious conflict has a long history and it characterizes several groups today. At the same time, religious conflict may be used as a front for a hidden political agenda.

The **Crusades** helps us to understand the long history of religion and violence. These eight major military expeditions originated in Western Europe between 1096 and 1270 (during the Middle Ages) with the purpose of recapturing Palestine from the Muslims because it was the area where Jesus had lived. In fact, Jerusalem is considered Holy Land to Muslims, Christians, and Jews. This was an era when Western Europe was expanding its economy and Christianity. Kings, nobles, knights, and peasants joined the Crusades and fought, not only for Christianity, but also for territory and wealth. Battles were won and lost, Jerusalem changed hands multiple times, and there was no significant impact, except for expanded trade. Subsequent attempts at organizing crusades failed as Europe turned its attention westward to the Atlantic Ocean toward the New World. During this time, the Holy Land was left to the Muslims (Queller, 1989).

The violence between the Israelis and the Palestinians today is an example of religious and political conflict. Several wars have been fought between Israel and its Arab neighbors. Israeli military superiority resulted in the capture of land from its Arab neighbors. Because the PLO saw that Arab allies were unable to drive Israeli armed forces from occupied lands, the PLO began a campaign of terrorism, as did other groups. The goals are to destroy Israel and form an Arab state in Palestine. Today, a political solution is slowly developing as Israeli forces continue to withdraw from occupied lands, although the withdrawals are marred by violence from both sides.

Wilkinson (2003: 124) writes that the al-Qaida network portray themselves as fighting a holy war and use religious language to legitimize their terrorist attacks. However, their political agenda is to force the U.S. military to withdraw from the Middle East, overthrow regimes that side with the West and fail to follow "true" Islam, and seek to unite all Muslims. Al-Qaida uses the Muslim communities and mosques in the West to recruit, seek aid, and conduct covert operations, even though the majority of Muslims who live in the West reject terrorism.

The conflict among some worshippers of Islam helps us to understand some of the religious violence of today. **Islam** is one of the world's largest religions. It refers to the religious doctrine preached by the **Prophet Muhammad** during the A.D. 600s. Those who believe in this doctrine are called **Muslims**. Over one-fifth of the global population is Muslim and they reside in countries throughout the world (Figure 15-1). Conflicts among Muslims over the rightful succession of Muslim rulers led to disputes within Islam. Civil war resulted in two Muslim sects that remain today, along with a history of conflict. The **Sunni** branch is the larger of the two sects and members dominate the Middle East. They are followers of the teachings of Muhammad. Al-Qaida is primarily Sunni. **Shiite** is the minority sect. They believe that Ali, cousin and son-in-law of Muhammad, is the Prophet of Islam. Shiite activism promoted the Iranian revolution of 1979 and emphasized that Islam should be lived as a tool of the oppressed, besides being a religious doctrine. Although Iraq is mostly Shiite, as is Iran, Iraq was ruled by the Sunni minority under Saddam Hussein, until his regime was toppled by U.S.-led forces in 2003. The conflict that followed pitted the formerly oppressed Shiites (supported by coalition forces to establish a new government) against Sunnis.

FIGURE 15-1 A mosque in the United States. Most Muslims are not violent, and stereotyping them is wrong.

Ford (2001) writes about opinions and perceptions of the United States by Arabs and others. Many Arabs see the carnage of 9/11 (Figure 15-2) as retribution for unjust policies by America. The Israeli-Palestinian conflict is a primary topic of contention. The perception is that Israel can get away with murder and the Unites States will turn a blind eye. Arab media shows countless photos of Israeli soldiers killing and wounding Palestinians and Israeli tanks plowing through Palestinian neighborhoods. Ford refers to the dominance of state-run media in the Middle East and how it often fans the flames of anti-American and anti-Israeli feelings because it helps to divert citizens away from the shortcomings of their own government.

Ford notes that the United States has provided billions of dollars of military and economic aid to Israel and other regional allies to strengthen U.S. interests in the oil-rich region. He writes: "It is this double standard that creates hatred." According to Ford, Middle Easterners argue that if the United States does not rethink the policies that cause anti-American sentiment, trying to root out terrorism will fail.

Greenberg (2001) writes about **fundamentalism** that refers to a strict adherence to the basic tenants of a belief system to reach purity in that belief system as defined by its leaders; it fosters intolerance of others' beliefs and it can generate terrorism. He adds that Islamic fundamentalists fear the spread of "satanic Western values and influence" and they believe that

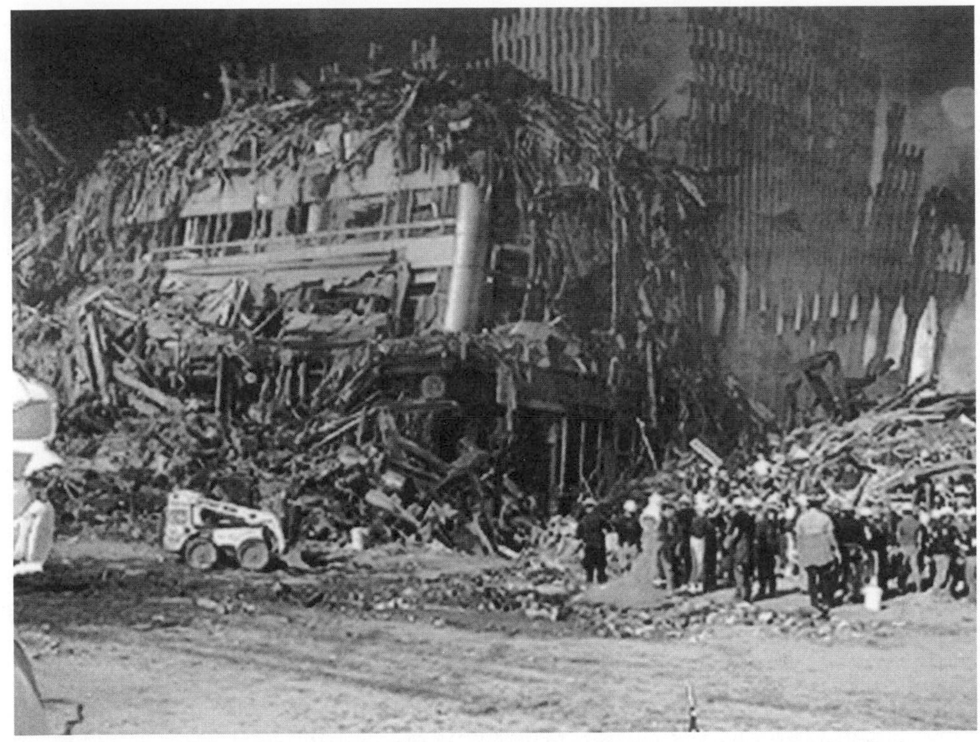

FIGURE 15-2 Destruction of the World Trade Center in New York City following the September 11, 2001 terrorist attack.

"to destroy America is to do God's work." Of great concern is how dangerous this mixture of religion and terrorism has become, as evidenced by the al-Qaida attacks of September 11, 2001. Future attacks might be more devastating and include weapons of mass destruction.

Causes of Terrorism

An understanding of the causes of terrorism is an important foundation for planning successful solutions. However, before we examine causes, we must define terrorism (see the beginning of the chapter for the difficulty of defining terrorism) and decide on which type of terrorism we want to study. As examples, are we going to study individual acts of terrorism, terrorism by a repressed group, or state terrorism? Here, an emphasis is placed on causes from the perspective of a repressed group.

Roach et al. (2005: 7–25) have developed a rough framework that diagnoses terrorism through an analysis of immediate causes and prior causes from individual, group, societal, or national history. Then, each family of causes can be tackled by an equivalent family of intervention principles, using specific targeted methods. Examples include target hardening, disrupting financing, attempting to resolve ethnic conflict, and socialization (e.g., working with institutions to promote diversity and tolerance).

Kegley (2003: 10–11) writes that while one perspective of terrorism sees it like a disease requiring a remedy, an opposing view sees it as a legitimate response to unjustified repression. The causes include perceived political, social, and economic inequities, as well as the intrusion of one culture on another.

The explanations and causes of terrorism are offered from many perspectives. Three major categories are psychological, rational choice, and structural. What follows here is a summary of research, including the work of Clayton et al., (2003).

Psychological Causes of Terrorism

Psychological causes of terrorism look to factors within an individual that influence acts of terrorism. Researchers often note that terrorists are not mentally disturbed. In fact, the recruiting process to obtain new recruits is selective since psychopathology can result in friction within a terrorist group and when completing missions (Hudson, 1999). In addition, it is not valid to apply one set of generalized psychological characteristics to all terrorists. The research on the personal backgrounds of terrorists shows conflicting results. Research shows that terrorists were unsuccessful in their personal lives, jobs, and educations. At the same time, research shows terrorists from middle-class or upper-class backgrounds with university educations (Clayton et al., 2003). Palestinian suicide bombers who terrorized Israel were often poor, with limited education and hope for the future. Their salvation was paradise after killing Jews. Conversely, the 9/11 terrorists were well-educated professionals living middle-class lives.

Silke (1998: 51–69) writes that most terrorists feel that they are doing nothing wrong when they kill, injure, and destroy property. He claims that terrorists are not mentally ill, but they share a psychological condition known as psychopathic personality disorder (also called antisocial personality disorder), which is the absence of empathy for others. He cites the case of Nezar Hindawi who, in 1986, sent his pregnant Irish girlfriend on an El Al flight to Israel, claiming that he would marry her soon. She apparently was not aware that Hindawi had hidden a bomb from the Abu Nidal Organization in her belongings.

Terrorists typically rationalize and minimize their actions by referring to historical events and injustices. They plan carefully and are patient until everything is in place. Violent behavior develops through a process of gradual moral disengagement.

Social learning models offer explanations for terrorism. Bandura (1977) emphasizes learning through observing and modeling the behaviors of others. Many terrorists grow up in locations where violence is both the norm and a daily event.

Rational Choice Causes of Terrorism

Rational choice causes of terrorism view terrorism as a logical political choice among alternative actions (Crenshaw, 1990). Rational choice explanations help us to understand the circumstances surrounding terrorist decisions to choose violence. For instance, global media exposure of terrorist violence and its aims can be beneficial to the terrorist group.

Another aspect of rational choice is the timing of terrorist attacks. Terrorists may attack on an anniversary of a significant event or on an enemy's holy day or national holiday. Timothy McVeigh blew up the Alfred P. Murrah Federal Building in Oklahoma City on the second anniversary of the deaths of the Branch Davidians in Waco, Texas, a clash he blamed on the FBI.

Structural Causes of Terrorism

Structural causes of terrorism explain terrorism as a response to social-structural conditions that groups are powerless to deal with through conventional political or military methods. These conditions include citizen access to services, rights, and protection. The social structures include the laws and policies of government, the role of police and military forces, and the geographic location of the group.

Relative deprivation theory helps us to understand terrorism (Barkan and Snowden, 2001: 17). Basically, a group's rising expectations and disappointments lead to feelings of deprivation and frustration. When group members compare their plight with more fortunate groups, the ill feelings are compounded. Violence is seen as a viable option. The living conditions of Palestinians in and near Israel serve as an example.

Terrorism may also be shaped by efforts to maintain cultural identity and autonomy. An example is the past violence between Catholics and Protestants in Northern Ireland.

The international environment can facilitate terrorism through, for example, open borders that enable terrorists to seek havens and travel to targets. States around the world—either through ignorance or intent—offer safe houses, training, communications systems, and financial opportunities that support terrorist operations.

International Terrorism

The FBI (n.d.) defines **international terrorism** as "the unlawful use of force or violence committed by a group or individual, who has some connection to a foreign power or whose activities transcend national boundaries, against persons or property to intimidate or coerce a government, the civilian population, or any segment thereof, in furtherance of political or social objectives." According to federal law, Title 22 USC 2656f(d), "the term international terrorism means terrorism involving citizens or the territory of more than one country."

Martin (2003: 228) seeks objectivity by comparing perceptions of international terrorism by those from the West and those from developing nations. Western governments are critical of international terrorism because (1) democratic justice is the norm and terrorism is seen as criminal behavior, (2) the West is often a target of terrorism, and (3) the West finds specific methods of warfare to be acceptable and terrorism is not one of them. Developing nations may find terrorism to be acceptable because (1) anticolonial revolutionaries (also called terrorists) have become national leaders, (2) terrorism was the best choice to freedom from colonialism, and (3) many revolutionaries crafted an effective fusion of ideology and terrorism that became justified and legitimate.

The concept of globalization helps us to understand international terrorism. **Globalization** is a process of worldwide changes that are increasingly integrating and remolding the lives of the people of the world (Bailey, 2003: 75). Economic changes are the driving forces behind globalization. Citizens of countries throughout the world are seeing their products and services sold globally, and they are buying from other countries. Modern technology and communications have enabled American movies, television programs, and music to reach almost anywhere in the world. McDonald's hamburgers and Kentucky Fried Chicken are sold in cities throughout the world. Likewise, the United States is a huge market for other countries. The ideology of capitalism and its goal of wealth are often seen as the driving forces behind the global economy. This ideology originated from Western (European) cultural beliefs and values. The process of globalization involves not only economic changes, but also political, social, and cultural changes. The United States is at the forefront of globalization through its military might, huge economy, power, and influence. Citizens of less developed nations perceive threats to their native culture and society from Western ideas, democratic values, and differences in religion. People in conservative cultures are angered by inequities, injustice, and the age-old problem of the "haves and the have-nots." Western industrialized nations are ripe targets for discontent. Samuel Huntington (1996), who wrote *The Clash of Civilizations and the Remaking of the World Order*, has often been quoted following the 9/11 attacks. He predicted that in the 21st century, wars would be fought, not between nations, but between different cultures and religions.

The Measurement of Terrorism

Data collected on terrorism influence not only government action against it—such as laws, plans, policies, budgets, and strategies—but also evaluations of government action. Measurements of terrorism also impact the business of counterterrorism, decisions on the geographic locations of business investment, and tourism.

Disagreement exists over the methodology applied to the measurement of terrorism, and methodological differences influence both conclusions on terrorism and policy. Ideally, measurements of terrorism should be value free, providing an objective foundation for government policy. Official government data has been a major source of information on terrorism. *Do measurements of terrorism really reflect the administration of counterterrorism (e.g., how terrorism is defined; number of intelligence analysts and enforcement officers; number of investigations and arrests; and data collection and software), rather than actual terrorism?*

According to van Dongen (2011: 357–371), there are difficulties in measuring the effectiveness of counterterrorism strategies and policies by studying the number of attacks and victims. He argues that indirect factors influence effectiveness. His examples include skillful diplomacy and a government's generous welfare policy. What is needed, van Dongen contends, is counterterrorism being divided into its discrete components and each evaluated separately.

Besides government measurements of terrorism, other sources of research on terrorism are universities and the private sector. The topics of research on terrorism that influence policy are varied. They include surveys of citizens' perceptions of terrorism and their fear of terrorism, and the role of the media in influencing perceptions of terrorism and terrorism itself.

The methodological problems of measuring terrorism include inconsistency in terminology, varied criteria of what acts to include and exclude from data, researcher bias, difficulty in making comparisons among different sources of data, political interference, and excessive delay in publishing data.

Grosskopf (2006) writes that "although terrorist attacks are often measured in loss of life and destruction, the immeasurable toll and intended consequence of terrorism is fear. The power of terror lies not only in attacks themselves, but also in the expectation and unpredictability of an attack." Grosskopf calls for more attention to fear and its consequences.

The U.S. Department of State (DOS), Office of Inspector General (OIG) (2005), reported that in April of 2005, less than one month before the DOS was to issue its 2004 mandated annual report, *Patterns of Global Terrorism*, the DOS ceased publishing it and instead published the first of its *Country Reports on Terrorism* (CRT). This new report *excluded* the statistical data on significant international terrorist incidents that were found in the *Patterns* reports. Simultaneously, the National Counterterrorism Center (NCTC) published a separate report titled *A Chronology of Significant International Terrorism 2004* that included statistical data that previously had been in the *Patterns* reports. The NCTC's 2004 report showed an increase in the number of reported significant international terrorist incidents between 2003 and 2004. The NCTC stated that the increase was "primarily the result of a modified reporting methodology as well as an increase in staff devoted to identifying terrorist incidents."

The NCTC provides the DOS with statistical information on international terrorism as described next. The U.S. Department of State (2011: 248–252), in *Country Reports on Terrorism 2010*, showed 11,604 terrorist incidents worldwide in 2010. There were 10,969 such incidents in 2009 and 11,662 in 2008. In 2010, there were 50, 000 victims, including almost 13,200 deaths. More than 75 percent of the 2010 incidents and deaths occurred in South Asia and the Near East (e.g., Syria, Turkey, Jordan, and Northeast Africa). Sunni extremists committed almost 60 percent of all global terrorist attacks. Five private U.S. citizens (i.e., not acting in a government position) worldwide were killed in 2010 as a result of terrorism.

National Counterterrorism Center

The NCTC was established by The Intelligence Reform and Terrorism Prevention Act of 2004. The Director of the Center is appointed by the president with advice and consent of the Senate. Staffed by personnel from various agencies, the purposes of the Center are detection, prevention, disruption, preemption, and mitigation of transnational terrorism against the people and interests of the United States The Center analyzes and integrates global intelligence, excluding purely domestic information, and it facilitates the exchange of information among government agencies. It conducts strategic operational planning for counterterrorism activities, including diplomatic, financial, military, homeland security, and law enforcement. The Center does not direct the execution of operations. The Center works with the CIA to prepare an annual report to the president (National Counterterrorism Center, 2011).

International Terrorists

When terrorists stage an attack domestically that has domestic implications, the attention they receive for their cause may not be as intense as a cross-border attack in another country. Terrorists may also select domestic targets that have international connections to enhance global attention to their cause. Examples of international targets include diplomats, tourists, foreign business people, and the locations they inhabit or visit. By bringing their struggle into the international arena, a low-budget, spectacular attack, on a symbolic or well-known target, can reap enormous, immediate publicity. Such political violence has been termed **asymmetrical warfare**, "a term used to describe tactics, organizational configurations, and methods of conflict that do not use previously accepted or predictable rules of engagement" (Martin, 2003: 216). This means that terrorists, who have an arsenal that is no match to, say, a superpower, can be successful when they apply imaginative, well-planned, low-budget, surprise attacks against a much stronger enemy. This occurred with the **9/11 attacks** on September 11, 2001, when al-Qaida suicide operatives, using box cutters, commandeered airliners and transformed them into missiles that totally destroyed the Twin Towers in New York City and damaged the Pentagon, while killing almost 3,000 people. Many countries, especially the United States, were forced to re-think their war fighting and counterterrorism capabilities because, clearly, the "wake-up call" is that conventional methods of deterrence, security, and war are no match against creative, determined terrorists.

In past decades, international terrorists targeted U.S. citizens and interests overseas. The most memorable attacks include the suicide truck bombings of the U.S. Marine barracks in Beirut, in 1983, killing 241 military service personnel; the 1988 bombing of Pan American Flight 103 over Lockerbie, Scotland, which killed 189 Americans; the 1996 bombing of Al-Khobar Towers in Dhahran, Saudi Arabia, resulting in the deaths of 19 U.S. military personnel; the 1998 bombing of the U.S. embassies in Kenya and Tanzania, which killed 12 Americans and many others; and the attack on the USS Cole in the port of Aden, Yemen, in 2000, killing 17 U.S. Navy members.

Many countries besides the United States have suffered from terrorists. For instance, Spain suffered what many called their 9/11 in 2004 when terrorists left backpacks filled with explosives on passenger trains converging on three train stations. The detonators on each bomb were wired to cell phones that were called to detonate the bombs as the trains reached the stations. About 200 people were killed and over 1,500 were injured. In 2004, Chechen rebels seized an elementary school in Beslan and during the Russian response 338 deaths resulted, mostly children. In 2005, London was the site of terrorist bombings of three subway trains and one bus resulting in 55 deaths and 700 wounded. Mumbai, India was the site of a 2010 terrorist attack by heavily armed men who killed almost 200 people and wounded about 300 others.

The true number of terrorist organizations is difficult to gauge because these groups are dynamic, splinter groups form, and groups change their names. The U.S. Department of State (2011), in *Country Reports on Terrorism 2010*, lists 47 foreign terrorist organizations and four state sponsors of terrorism: Cuba, Iran, Sudan, and Syria. The annual *Country Reports* contain a variety of information, including descriptions of terrorist groups, countries containing terrorist groups, and U.S. counterterrorism strategies.

Al-Qaida

Al-Qaida is the most significant terrorist threat to the United States and it has received much attention since its involvement in the 9/11 terrorist attacks. What follows here are descriptions of al-Qaida from the U.S. Department of State, the Federal Bureau of Investigation, and the insurance industry.

Country Reports on Terrorism (U.S. Department of State, 2011 and 2006) notes that there are various spellings of this group's name and several translations of the meaning of the name (a.k.a. International Front for Fighting Jews and Crusaders). This group was designated as a foreign terrorist organization in 1988. Al-Qaida was established by Osama bin Laden in 1988 with Arabs who fought in Afghanistan against the Soviet Union. The group helped finance, recruit, transport, and train Sunni Islamic extremists for the Afghan resistance. The goals of al-Qaida are to unite Muslims to fight the United States as a means of defeating Israel, overthrowing regimes it deems "non-Islamic," and expelling Westerners and non-Muslims from Muslim countries. Its eventual goal would be establishment of a pan-Islamic caliphate throughout the world. Al-Qaida leaders issued a statement in February 1998 under the banner of "The World Islamic Front for Jihad Against the Jews and Crusaders" saying it was the duty of all Muslims to kill U.S. citizens, civilian and military, and their allies everywhere.

Al-Qaida has strong ties to its affiliates in the Middle East, North Africa, Europe, and other locations. The group's planning and high-profile attacks continue. Besides the 9/11 attacks, other examples of al-Qaida terrorist attacks include the suicide attack on the USS Cole and the bombings of the U.S. embassies in Kenya and Tanzania. An example of a more recent plot (that failed) was a plan to attack the New York City subway system. In 2010 Najibullah Zazi, an Afghan immigrant, entered a guilty plea to terrorism charges in New York.

Al-Qaida's organizational strength is difficult to determine in the aftermath of extensive counterterrorist efforts since 9/11. The arrests and killings of mid-level and senior al-Qaida operatives have disrupted some communication, financial, and facilitation nodes and disrupted some terrorist plots. Additionally, supporters and associates worldwide who are inspired by the group's ideology may be operating without direction from al-Qaida's central leadership. *It is important to note that al-Qaida serves as a focal point of inspiration or imitation for a worldwide network of extremists.*

The group was based in Afghanistan until coalition forces removed the Taliban from power in late 2001. Since then, they have resided in Pakistan.

Al-Qaida primarily depends on donations from like-minded supporters and individuals who believe that their money is supporting a humanitarian or other cause. Some funds are diverted from Islamic charitable organizations. Additionally, parts of the organization raise funds through criminal activities; for example, al-Qaida raises funds through hostage-taking for ransom, and members in Europe have engaged in credit card fraud. United States and international efforts to block al-Qaida funding have hampered the group's ability to raise money.

A perspective not connected to the U.S. Department of State is by Shane (2011), a *New York Times* reporter. He writes that, although al-Qaida criticized Arab dictators as heretics and "puppets of the West," and called for their demise, al-Qaida played no role in the success of the 2011 "Arab Spring" that toppled regimes to facilitate Western-style democracies. In fact, these pro-democracy movements rejected al-Qaida teachings of brutal violence and religious fanaticism. Al-Qaida is seeking to survive and justify its relevance while becoming ever more decentralized and subject to internal struggles.

Lauren O'Brien (2011: 3–10), an FBI intelligence analyst, writes that the terrorist threat environment is increasingly complex and diverse. She points to several locations around the world, besides Pakistan, that harbor al-Qaida affiliates capable of launching global attacks. These include Yemen, Somalia, and Iraq. **Al-Qaida in the Arabian Peninsula** (AQAP), based primarily in Yemen, poses an especially dangerous global threat as illustrated in 2010 in the attempted detonation of explosives concealed in ink cartridges mailed to synagogues in Chicago and carried on two cargo planes. Another example is the November 2009 attack by U.S. Army Major Nidal Hasan who shot and killed 13 and wounded 30 at Fort Hood, Texas. He became radicalized while communicating with a cleric linked to AQAP. One month later an AQAP plot failed when a terrorist was unable to detonate an "underwear" bomb aboard a Northwest Airlines flight with 289 people on board as it approached Detroit.

O'Brien writes that a study of disrupted al-Qaida plots since 9/11 shows that this group aims for political, economic, symbolic, infrastructure and especially aviation targets. In addition, this group is involved in diversified attacks—long-term planning for spectacular attacks and short-term planning for small-scale attacks. Another threat noted by O'Brien is the "homegrown" terrorist (discussed later).

The insurance industry offers the following perspectives on al-Qaida and terrorism (Risk Management Solutions, 2011: 2–28):

- Despite being the most hunted terrorist group in the world today, Al-Qaeda and its affiliates continue to dictate the tempo of global terrorism activity.
- Although Western security and counterterrorism intelligence efforts have thwarted many attacks in the United States and Europe, other countries have been less fortunate. Since 9/11, more than 25,000 people have died, more than 45,000 have been injured, and 2,400 macro attacks (i.e., minimum impact of a car bomb) occurred from Islamic terrorist violence worldwide. Over 60 percent of the victims resided in Iraq and Afghanistan.
- Since 9/11, religious terrorism has increased and more than 92 percent of macro attacks have a link to religion and the belief that the violence is sanctioned by divine power.
- Terrorists often focus on soft targets because they lack the capabilities to attack hardened targets. An upsurge in attacks on hotels, transportation systems, and houses of worship has occurred in recent years.
- To reduce the terrorist threat, key leaders must be killed and the ideology must be discredited.
- The threat of terrorism in the United States has yet to improve. From 2002 to 2008 the number of plots declined. However, from 2008 onward the number of attacks linked to al-Qaida has increased.
- The terrorism threat in the United States will remain largely an urban challenge; over 90 percent of al-Qaida plots since 2001 have been focused on large cities.

The Killing of Osama bin Laden

In May of 2011, **Osama bin Laden**, head of al-Qaida, was killed by a U.S. joint operation of the CIA and Navy Seals who used helicopters to storm his compound in Abbottabad, Pakistan. His body was taken from the compound and buried at sea. Interestingly, the compound is less than one mile from the Pakistani equivalent of the U.S. Military Academy at West Point and in an area containing a major Pakistani military base. The raid resulted in tension and accusations between both nations.

On the killing of Osama bin Laden, Hoffman (2012b: 744–746) writes that when such a leader is eliminated, terrorist groups often lash out in revenge, work to survive, and seek to prove their continued power and influence. Hoffman argues that al-Qaida is not defeated, other terrorist threats remain, and we must not become complacent.

Domestic Terrorism

The FBI (n.d.) defines **domestic terrorism** as "groups or individuals who are based and operate entirely within the United States and Puerto Rico without foreign direction and whose acts are directed at elements of the U.S. government or population."

Domestic terrorism, also called "homegrown terrorism," is difficult to define, categorize, and research. White (2003: 207–208) offers an analysis of the problems of conceptualizing domestic terrorism. He notes that law enforcement in the United States has historically labeled terrorism or political violence under various crime labels (e.g., arson, homicide). White also writes that the United States has had a long history of political violence, but only recently have more than a few scholars studied the problem under the label of "terrorism."

White refers to three early authors who pioneered publications on domestic terrorism. He praised H. Cooper and coauthors who wrote the *Report of the Task Force on Disorders and Terrorism* in 1976. This publication focused on the civil disorders of the time, the political context of domestic terrorism, and the need for emergency planning. Two other pioneers are J. Bell and Ted Gurr (1979) who emphasized the historical context of terrorism and how it has been applied by the strong to control the weak, and by the weak to fight the strong. Examples in American history include the genocide against American Indians by Europeans, industrialists versus unions, and vigilantism (e.g., KKK).

Domestic Terrorists

To understand domestic terrorists and the history of terrorism in the United States we look to the work of Griset and Mahan (2003: 85–93) who divide domestic terrorists into five categories based on their ideology. These researchers state that their categories may overlap. The categories are described next with additional historical information from several sources. *The contemporary topics of "homegrown" and "lone wolf" terrorists are explained in subsequent pages.*

State-Sponsored

Government authorities, including Congress and the president, have passed laws and supported policies that have intimidated Americans and fostered violence. The Removal Act of 1830 required the forced march of Indian tribes from the east coast to Oklahoma; many Indians died during the march, and upon reaching reservations, the conditions were poor. In another example, during the late 1800s and early 1900s, many city and state police agencies were formed or expanded to serve as strikebreakers (Holden, 1986: 23).

Leftist Class Struggles

Identifying with Marxism, the 1960s and 1970s saw leftist groups embroiled in antiwar and civil rights struggles. The Students for a Democratic Society (SDS) began in 1960 to support the "liberal" policies of the Democratic Party. It called for an alliance of blacks, students, peace groups, and liberal organizations. As with many other leftist groups of the time, an "anti-establishment" theme prevailed (i.e., government and business were corrupt and change was needed to help the less fortunate). In 1965, the SDS organized the first anti-Vietnam War march in Washington, D.C., which resulted in increased popularity for the SDS among students. The SDS led many campus disturbances and sit-down demonstrations in buildings. Military draft resistance became a top priority. As the Vietnam War intensified, the SDS divided into competing groups, each supporting violence and terrorism. One of the most noted splinter groups was the Weather Underground Organization (WUO.) This group advocated revolution against capitalism. It was responsible for almost 40 bombings, including those inside the U.S. Capitol and the Pentagon. When the war ended, interest in the WUO waned (Poland, 1988: 91–92).

Violence against nonviolent civil rights workers sparked the Black Power movement during the 1960s. It advocated political, economic, and cultural awakening. One splinter group of this movement was the Black Panthers, formed in 1966, in Oakland, California. A tactic of the Black Panthers was to dispatch members to police stops to observe. The Panthers would arrive carrying law books and an open display of shotguns and rifles (legal at the time). This paramilitary movement grew and Panthers wore black berets and black leather jackets as they marched and chanted slogans. The FBI viewed the Panthers as a threat to domestic security; gun battles and arrests occurred, and by the late 1970s, the group began its decline (Martin, 2003: 311 and 318).

During the 1980s, the May 19 Communist Organization (M19CO) emerged from members of the SDS, Black Panthers, and other groups. The group's name came from the birthdays of Ho Chi Minh (North Vietnamese Communist leader) and Malcolm X (American Black Muslim leader). This group sought links to radical black, Hispanic, and women's movements. The overall goal was violent revolution. By the end of the decade, the group split apart and members were imprisoned for crimes such as bombings, murders, and robberies (Poland, 1988: 90).

Following the Spanish-American War in 1898, Puerto Rico was ceded to the United States by the Treaty of Paris. The resolve of Puerto Rican nationalists in seeking independence from the United States was illustrated in 1950 following an uprising that was quickly suppressed by government troops. On November 1, 1950, nationalists tried to assassinate President Truman while he was at Blair House, across from the White House. A gun battle erupted, one police officer was killed, Truman was unharmed, and the independence movement gained national attention. The next attack occurred on March 1, 1954, again in Washington, D.C., when several nationalists fired shots at legislators while Congress was in session, wounding five. In 1974, the Armed Forces for National Liberation (FALN) was formed. Its agenda for independence leaned to the far left. The FALN introductory attack consisted of bombing five banks in New York City. Over 200 bombings followed, targeting federal and local government buildings, the military (especially in Puerto Rico), and businesses. The FALN was also known for ambushing and shooting U.S. armed services personnel (Poland, 1988: 72–76).

Anarchists/Ecoterrorists

Anarchism, a political ideology that reached America from 19th century Europe, opposed centralized government and favored the poor and working class. Martin (2003: 38–39) argues that anarchists were among the first anti-establishment radicals who opposed both capitalism and Marxism and advocated revolution, although they never offered a plan for replacing a central government. Anarchism in the United States was linked to the labor movement and the advocacy of a bombing campaign against industry and government (Combs, 2003: 163).

The Ecoterror movement in the United States focuses on both the dangers of humans encroaching on nature and preserving wilderness. The FBI (2002) defines **ecoterrorism** as the use or threatened use of violence of a criminal nature against innocent victims or property by an environmentally oriented, subnational group for environmental-political reasons or aimed at an audience beyond the target, often of a symbolic nature.

Ted Kaczynski (A.K.A. the Unabomber) was a modern day ecoterrorist. He was an anti-industrialist who targeted scientists and engineers. In 1996 he was arrested after a 17-year investigation that involved 16 bombings, including three deaths and many injuries. Upon pleading guilty in federal court, he is now serving a life sentence without the possibility of release.

Two major domestic ecoterrorist groups are the **Animal Liberation Front (ALF)** and the **Earth Liberation Front (ELF)**. The Federal Bureau of Investigation (2011 and 2002) estimates that ALF and ELF have committed over 600 criminal acts in the United States, resulting in damage exceeding tens of millions of dollars in losses. Arson is the most destructive practice of ALF and ELF.

ALF is committed to ending the abuse and exploitation of animals. They seek to cause economic loss or destruction of the victim company's property. Victims include fur companies, mink farms, restaurants, and animal research laboratories.

ELF promotes "monkeywrenching," a euphemism for acts of sabotage and vandalism against companies that are perceived to be harming the environment. An example is "tree spiking," in which metal spikes are hammered into trees to damage logging saws.

Racial Supremacy

Continuing with the Griset and Mahan (2003: 86) typology of domestic terrorism, they view the racial supremacy and religious extremists classifications as right wing, but separate them based on their aims. Their research from multiple sources shows that "these two categories often overlap, but they have also developed separately—white supremacists without religion, and religious extremists without racism."

The KKK, as covered earlier, has advocated hate for over a century. Griset and Mahan add that there are also black separatist groups in the United States calling themselves the "Nation of Islam"; they are followers of the Messenger Elijah Muhammad.

White supremacists are often critical of the U.S. Constitution and our government. These right-wing groups are diversified and include Skinheads, neo-Nazi Aryan Nations, militant gun advocates, antitax protesters, and survivalists. The **sovereign citizen extremist movement** is viewed by the Federal Bureau of Investigation (2010) as a more recent threat. These anti-government extremists believe that they do not have to answer to courts, taxing agencies, police, and other government agencies. Crimes committed by this group range from financial scams to impersonating or threatening police. Since 2000, such extremists killed six law enforcement officers. The movement is fueled by the Internet (Federal Bureau of Investigation, Counterterrorism Analysis Section, 2011).

Religious Extremists

This is the fifth typology of domestic terrorism according to Griset and Mahan (2003: 91–92). As covered earlier in this chapter, global religious terrorism has had a significant impact on terrorism in many countries. Domestically, White (2004: 11) refers to the Christian Identity movement and their claims that Jews are descendants of the devil, nonwhite people evolved from animals, and Caucasians are created in the image of God. Hoffman (1998) claims that Christian Patriot influence played a role in the Oklahoma City bombing, and those religious extremists, not necessarily of the Islamic faith, are involved in terrorism.

Although a lot of attention and resources are focused on international terrorists, domestic terrorists are also a serious threat.

Ruby Ridge, Waco, and Oklahoma City

The decade of the 1990s provides an excellent example of how citizen confrontations with the federal government can foment hate and violence from extremists. Martin (2003: G-8) defines **extremism** as follows: "Political opinions that are intolerant toward opposing interests and divergent opinions. Extremism forms the ideological foundation for political violence."

The **Ruby Ridge** (Idaho) **incident** occurred in 1992 when U.S. Marshals tried to arrest Randy Weaver for failing to appear in court on charges of trying to sell illegal firearms to federal agents. Weaver was a white supremacist and believer in Christian Identity. Martin (2003: G-5) offers this definition for **Christian Identity**: "The American adaptation of Anglo-Israelism. A racial supremacist mystical belief that holds that Aryans are the chosen people of God, the United States is the Aryan 'Promised Land,' nonwhites are soulless beasts, and Jews are biologically descended from the devil." During the siege of Weaver's mountain cabin by federal agents, a U.S. Marshal was killed, as was Weaver's son. An FBI sniper fatally shot Weaver's pregnant wife as she stood in the cabin doorway prior to Randy Weaver's surrender. The Ruby Ridge incident became a symbol and inspiration for the struggle of right-wing extremists.

Another symbolic incident that inflamed right-wing extremists was the **Waco incident**, although it had nothing to do with right-wing extremism. On February 28, 1993, the Bureau of Alcohol, Tobacco, and Firearms (ATF), with numerous agents, tried to serve arrest and search warrants for illegal firearms on Vernon Howell, A.K.A. David Koresh. Koresh led the Branch Davidians at the Mount Carmel compound outside of Waco, Texas. He claimed that the end was near and that he was the second coming of Christ who would save the world. The ATF assault and gun battle involved thousands of rounds being fired from both sides. The media captured the fierce battle and showed, on television, ATF agents on an A-frame roof being hit by rounds fired through the wall from inside the compound as the agents attempted, but failed, to enter a window. Four ATF agents were killed and another 20 were injured, in addition to an unknown number of casualties on the Branch Davidian side.

Following the failed assault, the FBI was enlisted and a 51-day siege began. Continuous negotiations were unable to resolve the standoff as the Branch Davidians waited for a message from God. The FBI stopped electricity and water to the compound and bright lights and loud music were aimed at the compound at night to force members to leave. Some members did leave. Finally, on April 19, 1993, following unsuccessful negotiations, tear gas was pumped and fired into the compound. Fires began and then explosions from ammunition and combustibles created a huge fire. Seventy-five people, mostly women and children, perished; nine survived. The FBI, and especially the ATF, received an enormous amount of criticism, from many quarters, over the Waco incident (Combs, 2003: 178–179).

The **Oklahoma City bombing** was a major terrorist incident in the United States that occurred on April 19, 1995, when a powerful truck bomb exploded in front of the Alfred P. Murrah Federal Building (Figure 15-3). The attack killed 168 people and injured more than 500. The nine-story

FIGURE 15-3 Bombed Alfred P. Murrah Federal Building, Oklahoma City. *Courtesy: DOD photo by Sgt. Preston Chasteen.*

downtown building contained offices of the ATF, Secret Service, Drug Enforcement Administration, among other federal offices, and a day-care center. A rented Ryder truck served as the delivery weapon. It was converted into a mobile bomb containing ammonium nitrate, fuel oil, and Tovex (i.e., a high explosive that acted as a booster). The truck exploded near supporting columns at the front of the building, and then from the bottom up, floor slabs broke off in a "progressive collapse." The force of the explosion caused damage over several blocks in all directions (Poland, 2005: 184–186). Initially, it was thought that a Middle East terrorist group targeted the building, but this was not the case. A clean-cut white male named Timothy McVeigh was arrested by authorities and prosecuted, convicted, and sentenced to death. In 2001, he was executed by lethal injection at the U.S. penitentiary in Terre Haute, Indiana.

McVeigh was a member of the **Patriot movement** that developed in the early 1990s. This group believes in "true" American ideals of individualism, armed citizens, and minimum interference from government. They distrust government and view it as no longer reflecting the will of the people, intrusive, and violently oppressive (Martin, 2003: 313).

Besides McVeigh, Terry Nichols, a friend from their service in the U.S. Army, was also arrested and convicted for the bombing. Interestingly, April 19, 1995, the day of the Oklahoma City attack, was the second anniversary of the law enforcement disaster at Waco.

Homegrown and Lone Wolf Terrorists

As global efforts to thwart al-Qaida continue, many countries suffer from "homegrown" and "lone wolf" terrorists. **Homegrown terrorists** refer to individuals or groups radicalized within the country in which they reside for many years and then commit terrorist acts, whereby **lone wolf terrorists** commit terrorist acts solely by themselves and *without links to others*. An example of the former is the Oklahoma City bombing by Timothy McVeigh and Terry Nichols. Another example is the 2009 attack by U.S. Army Major Hasan at Fort Hood, Texas. He became radicalized while communicating with AQAP. An example of the lone wolf is Ted Kaczynski (A.K.A. the Unabomber).

We can argue here about what if a "homegrown" is also a "lone wolf" and goes overseas to commit a terrorist act. We can call this individual a "homegrown lone wolf transnational terrorist." More importantly, "homegrown" and "lone wolf" are sometimes used interchangeably and the definitions and explanations of each vary and are modified depending on the source and/or trends of terrorists.

What makes lone wolf terrorists especially difficult to detect is that they may use the Internet to become radicalized (e.g., study extremist literature) and learn of terrorist methods (e.g., bomb-making), without links to others (e.g., communications, meetings, training, and overseas trips), which makes investigations difficult. At the same time, poor training, supervision, and support can result in botched attacks.

Terrorist Methods

Dyson (2012: 6–12) explains four major advances of the 20th and 21st centuries that have facilitated terrorism: technology, communications, transportation, and weapons. Computer technology enables terrorists to communicate with each other; promote their ideologies; recruit; and learn about "the enemy," weapons, and targets. Besides e-mail, social network sites, and Internet chat rooms, communication includes the cellular telephone that permits conversation, access to the Internet, and the transmitting of text messages, videos, and photographs. At the same time, surveillance of communications systems has assisted authorities in eliminating terrorist threats. The Internet, radio, and television have enabled terrorists to gain a great deal of publicity for both their attacks and the justification for attacks, while instilling fear. Aviation is a great tool of terrorists who travel thousands of miles in hours to meet with others, attend training, and attack and escape.

The **cellular model of organization** has helped transnational and homegrown terrorist groups adapt to threats from law enforcement and the military. This model is decentralized, rather than centralized. A decentralized cell is characterized by more discretion; very little, if any, communication within a larger chain of command; more self-sufficiency; and greater pressure to be creative and resourceful. This approach has also been referred to as **leaderless resistance.** Because a cell consists of a few members, and they know each other well, infiltration by government agents is very difficult. Since communication within a centralized chain of command or between cells is unnecessary, the interception of telephone calls, e-mails, and other forms of communication are less of a risk. Examples of successful cells with leaderless resistance are the al-Qaida network and the McVeigh-Nichols group.

Prior to the 9/11 attacks, the Taliban in Afghanistan permitted Osama bin Laden to operate al-Qaida training camps until the Taliban fell to U.S. forces. Because of global concern about terrorism and the efforts of many nations in combating it, there has been a shift by many terrorists to "home schooling," online training, and on-the-job training.

The list that follows illustrates the range of methods terrorists employ and their cunning and creativity.

- Terrorists often falsify their identity, carry forged documents, and practice deception. They may change their appearance to deceive (e.g., shave a beard, wear a religious cross, carry a bible, and dress like the enemy).
- Terrorists, such as al-Qaida, are characterized by long-range planning, intense intelligence gathering, meticulous surveillance of potential targets, and the ability to conduct simultaneous attacks.
- They may communicate an anonymous threat to a potential target to monitor response and procedures.
- They may use the identification, uniforms, or vehicles of delivery, utility, emergency or other types of service to access a target. Or they may dress as females to lower the perceived threat.
- A terrorist may be planted in an organization as an employee to obtain information. The terrorist may work for a service or consulting firm to access many locations.
- Terrorists may attempt to detect surveillance by conducting dry runs of planned activities, using secondary roads and public transportation, employing neighborhood lookouts and tail vehicles, and establishing prearranged signals.

- Terrorists are keenly aware that counterterrorism analysts study terrorist groups, seek intelligence, and piece together information to anticipate terrorist plans and attacks. Studies of terrorist groups and their methods are available from many sources. Terrorists study these resources and, in a "cat and mouse game," seek to mislead analysts by releasing bogus information to disrupt the intelligence process. Terrorists maintain an advantage over analysts because terrorists plot and analysts must seek to uncover plots.

Terrorist Violence

Terrorists favor certain types of violent acts. These include bombings, suicide bombings, assassinations, assaults, kidnappings, and hijacking. Terrorists have tremendous advantages over authorities as they secretly plan their surprise attacks with discretion and flexibility over target location, victims, time of attack, weapons, and creative methods.

Crenshaw (2012: 165–182) writes of "old" versus **"new terrorism."** She argues that it is difficult to differentiate the two types and that the "new" contains characteristics of the "old." Citing several experts, Crenshaw explains that the "old" is characterized by terrorists being somewhat negotiable and hesitant to commit mass murder. "New terrorism" includes religion playing an increasing role, which facilitates fanaticism and suicide attacks, greater lethality, efforts to acquire weapons of mass destruction for a high "body count," and decentralization of operations. The U.S. Department of State (2011: 251), in *Country Reports on Terrorism 2010*, reported that bombings, including suicide attacks, were more lethal in 2010 than earlier, causing 70 percent of all deaths.

Terrorist Weapons

BOMBS

Bombs are very popular terrorist weapons because of the potential for high casualties and severe property damage. The terrorist can remain at a distance from a detonated bomb and avoid injury and initial capture. The **secondary explosion** is a horrible terrorist ploy: an initial explosion draws first responders, assets, bystanders, and the media; then, a second, larger bomb is detonated for more devastating losses.

A bomb can be placed in almost anything—a pipe, letter, book, package, cell phone, radio, vehicle, bicycle, podium, pillow, and so forth. Terrorist bombs are mostly improvised. In other words, they construct bombs from readily available materials or the misappropriation of military or commercial blasting supplies. **Improvised explosive device (IED)** is the name for such assembled bombs that kill or maim. The Internet, terrorist training, and manuals provide instructions. In an IED, the trigger activates the fuse that ignites the explosive. These three components are integrated in various ways in IEDs. Nails and other materials may be added to the IED to increase lethality. An IED may contain an anti-handling device that causes it to detonate when it is handled or moved. Coalition personnel in Iraq and Afghanistan have been killed when IEDs—hidden along roads in trash containers, light poles, dead people (including Americans) and animals, and other locations—were detonated while convoys and patrols passed nearby.

Risk Management Solutions (2011: 5) views counterterrorism efforts in the West as hindering terrorist capabilities, resulting in smaller IEDs. However, terrorists are often adept at targeting and timing with smaller bombs for maximum effect. Since 2001, about 2,000 vehicle bombs have been detonated by terrorists worldwide. As we know, terrorists are unpredictable and high-yield bombs, like a two-ton TNT bomb (about the yield of the Oklahoma City bomb), are inevitable.

Suicide bombings are especially effective, inexpensive, shock society and government, and guarantee media coverage. The terrorist can precisely deliver the bomb and there is no need for an escape plan or concern about interrogation by authorities. Ismayilov (2010) argues that suicide terrorism is such a complex social phenomenon of multiple causes that present theories to explain it are insufficient, which hinders countermeasures policies.

Johnson and Gorman (2011) report that surgically implanted explosives in suicide bombers present another threat, according to the U.S. Department of Homeland Security. Such bombs may be stitched into the buttocks or abdomen and detonated via a syringe. In 2009, an al-Qaida terrorist, claiming he wanted to leave the group, passed layers of security and met with the director of Saudi Arabia's counterterrorism operations. During the meeting, a second terrorist sent a text message to detonate a bomb that was hidden in the rectum of the terrorist at the meeting. The assassination attempt wounded the director and killed the suicide bomber. Drug smugglers have surgically implanted drugs for many years. Countermeasures are challenging; x-ray scanning results in health and privacy issues. Alternative strategies include profiling behavior and demeanor and use of canine teams.

OTHER WEAPONS

Four weapons used by terrorists are the AK-47, RPG-7, Stinger missile, and SA-7. These are explained next.

AK-47 (Soviet rifle). This weapon was invented by Mikhail Kalashnikov. It became the standard rifle for the Soviet Army in 1949 until it was succeeded by the AKM. During the Cold War, the AK-47 (Figure 15-4) was supplied by the former Soviet Union to anti-Western armed forces and insurgent terrorists. It became a symbol of left-wing revolution and it is still in the arsenal

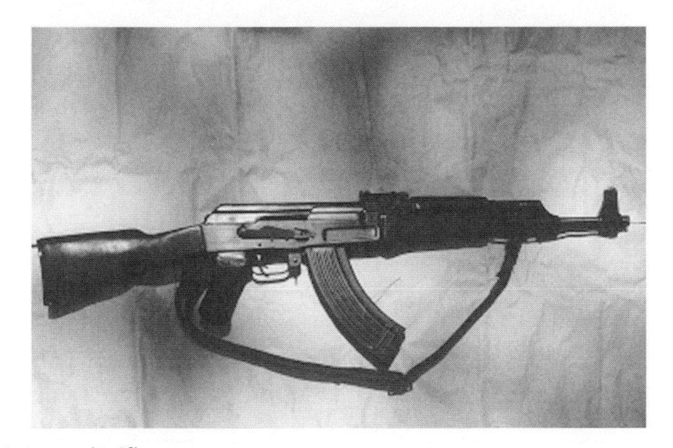

FIGURE 15-4 AK-47, Soviet assault rifle.

of armies, guerrillas, and terrorists. Between 30 and 50 million copies and variations of the AK-47 were distributed globally to make it the most widely used rifle in the world.

RPG-7 (Rocket Propelled Grenade). Issued by the former Soviet Union, China, and North Korea, this simple weapon is effective against a variety of targets within a range of about 500 meters for a fixed target (e.g., building) and about 300 meters when fired at a moving target. The RPG-7 (Figure 15-5) is a shoulder-fired, muzzle-loaded, antitank and antipersonnel grenade launcher. It can also bring down a helicopter. The RPG-7 fires a fin-stabilized, oversized grenade from a tube. The launcher weighs 15.9 pounds and has an optical sight. Upon firing, it leaves a telltale blue-gray smoke and flash signature that helps to identify the firer. U.S. forces in Vietnam used sandbags and chicken wire on vehicles to cause the RPG-7 grenade to explode before meeting the skin of vehicles. This technique continues to be applied, as in Iraq. Additional protection methods include avoiding the same route, pushing through ambush positions to avoid setting up a target, smoke grenades, and aerial surveillance (Mordica, 2003). The RPG is widely available in illegal international arms markets and is in the arsenal of many terrorists groups, such as those in the Middle East and Latin America.

Stinger. The U.S.-made stinger is a single person, portable, heat seeking, shoulder-fired surface-to-air missile (SAM). It proved its value when the Afghan Mujahedeen forced the Soviets out of Afghanistan during the late 1980s. The Stinger has targeted jets, helicopters, and commercial airliners.

SA-7. When the Soviet Union collapsed, thousands of earlier-model Soviet SAMs were sold and found their way into the arsenals of several terrorist and guerrilla groups. The Soviet SA-7, also called the Grail, uses an optical sight and tracking system with an infrared (i.e., heat-seeking) device to reach aircraft. Although very dangerous, the SA-7 and the Stinger are not readily available to terrorists.

FIGURE 15-5 Rocket propelled grenade and launcher. *Source: www.hamasonline.com*

Weapons of Mass Destruction

A **weapon of mass destruction (WMD)** is something capable of inflicting mass casualties and/or destroying or rendering high-value assets as useless. The CDC, DHS, and FEMA offer information on WMD. Later in this chapter, precautions relevant to WMD are explained.

Although chemical, biological, radiological, and nuclear (CBRN) weapons often serve as examples of WMD, many things can be used as a WMD. This became painfully true during the 9/11 attacks when passenger jets were used as missiles.

Risk Management Solutions (2011: 15–18), from an insurance industry perspective, estimates that a CBRN terrorist attack is most likely to target a major business center, causing significant business interruption resulting from casualties, evacuation, property damage, and decontamination. Al-Qaida continues to attempt to acquire CBRN weapons. A lot of relevant information is available online. This effort points to a case not of "if," but "when," such an attack will occur.

The term **weapon of mass effect (WME)** describes the human reactions and events surrounding the use of a WMD that may result in limited or no casualties or physical damage. The mass effect may be sensationalized media reporting, panic, and social and political change.

Chemical Weapons

Chemical weapons evolved more recently in history than did biological weapons. World War I became the first battleground noted for massive casualties from chemical weapons. Compared to other major WMD, chemical agents such as pulmonary, blood, and blister agents are relatively easy to manufacture. Nerve agents are more difficult to produce. These substances are also commercially available, examples being cleaning fluids and insecticides.

Al-Qaida plots employing chemicals have focused on chlorine and hydrogen cyanide. With so many chemicals in communities, an attack on a chemical plant, rail car, or truck could be devastating.

Chemical agents may work immediately or cause delayed reactions. Some are difficult to detect because they are odorless and tasteless. Most chemical agents are in a liquid form that, to be effective, must be aerosolized into droplets so they can be inhaled.

As with other WMD, dual use equipment and technology help terrorists disguise their intentions. Imagine the casualties from a chemical attack by a crop duster over a sports stadium, or by use of a Global Positioning System (GPS) to guide planes (manned or unmanned drones) to a target, or by backing up a truck to a building HVAC system.

Terrorists have employed chemical weapons in the past. The **Aum Shinrikyo cult**, established in 1987, aimed to take over Japan and then the world to purify everyone. On March 20, 1995, Aum members pierced bags of the chemical nerve agent sarin on five separate subway trains as they converged on central Tokyo. The attack killed twelve passengers and injured up to 6,000 as they fled while sickened and bleeding. The cult had previously tried, but failed, to use biological agents. Interestingly, Japanese police were scheduled to raid the cult's headquarters in Tokyo on March 22nd to seize WMD. However, the cult had infiltrated the Japanese police, were tipped off about the impending raid, and they hastily launched the attack. The casualties would have been worse, but the sarin was not pure and the method of release was primitive.

Biological Weapons

The use of biological weapons has a long history. In the 1346–1347 siege of Kaffa by the Mongols, soldiers catapulted diseased corpses over the walls of the city, causing deaths. In 1767, the British military gave smallpox-infested blankets to Native Americans causing an outbreak and deaths (Katz, 2013: 20). The United States, England, the former Soviet Union, and many other countries have developed biological weapons over many decades. Treaties attempt to eliminate production of such weapons and destroy current inventories.

Like chemical weapons, biological weapons may be perceived as a "poor person's atomic bomb" and developing countries may look to these WMD as an avenue to deter enemies. A biological weapons program is easy to disguise because the same biotechnology equipment is used in pharmaceutical manufacturing. Risk Management Solutions (2011: 16) reports that Western intelligence services are concerned about al-Qaida's recruitment of scientists and students for a long-term "jihadist biotech research project."

Biological weapons can be delivered via aerosols, animals, food and water, person-to-person, artillery shell, and missile. These weapons cause flu-like symptoms and then death. Detection and containment are difficult because the onset of symptoms and then the identification of the agent are delayed.

Anthrax is emphasized here because it was used to attack the United States in 2001. It is a deadly infectious disease caused by the bacterium called *Bacillus anthracis*. Its occurrence is rare in the United States, it is not contagious, and victims must be directly exposed to it. Infection occurs in three forms: cutaneous (i.e., from a cut on the skin), inhalation, and intestinal (i.e., eating tainted meat). Symptoms are as follows: for cutaneous anthrax, a skin ulcer; for inhalation anthrax, respiratory failure; and for intestinal anthrax, nausea, vomiting and diarrhea. Treatment includes antibiotics administered early on and a vaccine exists.

During the fall of 2001, following the 9/11 attacks, letters containing anthrax spores were mailed to news media personnel and congressional officials, leading to the first case of intentional release of anthrax in the Unites States (U.S. General Accounting Office, 2003). Several states and Washington, D.C., were affected. Eleven people were victimized by the cutaneous form and 11 from the inhalation form. Five people died, all from inhaling anthrax. The anthrax was in the form of white powder placed in envelopes. During mail processing at post offices, other government offices, and businesses, it was released. It contaminated air, equipment, buildings and people who touched and inhaled the spores. Besides the unfortunate casualties, the attacks caused enormous disruptions and expenses to organizations. At about the same time, other countries reported similar attacks. In 2008, as federal prosecutors in the United States were about to indict Dr. Bruce Ivins for the anthrax attacks, he committed suicide. Ivins was a government scientist who worked on vaccines for anthrax. Although prosecutors claimed to have a strong case, others viewed the evidence as weak.

Radiological Weapons

Radiological weapons, often called "dirty bombs," contain radioactive material that, although not explosive itself, is scattered by a conventional explosive. The technical name for this type of bomb is **radiological dispersion device** (RDD). It may contain radioactive material that is sealed in metal to prevent dispersal when handling it. Upon explosion, initial fatalities may be low, but climb over time due to victims developing cancer. However, the actual harm will depend on the grade of the radioactive material and the amount released. This weapon can make buildings and land unusable for many years.

Because terrorists are known to be creative, we can only surmise as to how they would construct a RDD and where it would be applied. It may become part of a vehicle bomb, detonated in a school, or introduced into food or water supplies.

Commercially available radiological materials are used in many industries to, for example, sterilize food, detect flaws in metal, and to treat cancer. Many sources are not secure. Terrorists are more likely to use a RDD than a nuclear weapon. In 1995, Chechen rebels placed a RDD, using dynamite and cesium 137, in Moscow's Izmailovo Park. It was not detonated. In 2002, suspected al-Qaida member Jose Padilla, trained in terrorist camps in Afghanistan, was arrested in Chicago for his role in a plot to explode a RDD in the United States.

Nuclear Weapons

To end World War II in the Pacific against the Japanese, and to avoid massive U.S. casualties from an invasion of the Japanese mainland, President Harry Truman approved the dropping of atomic bombs on Hiroshima and Nagasaki in 1945. Both cities were obliterated and sustained massive casualties. Then, an arms race began that continues today. The "nuclear club" has expanded from the United States to the United Kingdom, France, Russia, the People's Republic of China, India, Pakistan, Israel, South Africa, North Korea, and other countries that possess covert development programs (e.g., Iran).

According to the United Nations, Office on Drugs and Crime (n.d.), when the Cold War ended, a black market in Soviet-era weapons began to flourish. Former Soviet citizens participated in smuggling nuclear materials from poorly secured facilities to unknown buyers and transnational criminal organizations. The market for nuclear materials includes "rogue states," desperate national liberation movements, and suppressed ethnic groups. Al-Qaida has not only sought nuclear materials, its affiliates have targeted nuclear power plants in Canada, France, and Australia (Risk Management Solutions, 2011: 17).

The explosion of a nuclear weapon creates a large fireball. Everything in the fireball vaporizes, including soil and water, and the fireball goes upward, creating a large mushroom cloud. Radioactive material from the weapon mixes with the particles in the cloud and then the mixture falls to earth in what is called fallout. Because fallout is composed of particles, it may be carried by wind for long distances. Since the fallout is radioactive, it can contaminate anything it lands on, such as water resources and crops (Centers for Disease Control, n.d.).

Do you think terrorists will use WMD in the future? Why or why not?

Homeland Security

Homeland Security Defined

Homeland security is in a constant state of evolution and there are many perspectives on its definition. The U.S. Department of Homeland Security (DHS) is continuously developing its mission under changing circumstances and events. Numerous variables influence the DHS such as terrorist incidents, natural disasters, public opinion, politics, and funding.

The Office of Homeland Security (2002), formed prior to the DHS, provides the following definition: "**Homeland security** (HS) is a concerted national effort to prevent terrorist attacks within the United States, reduce America's vulnerability to terrorism, and minimize the damage and recover from attacks that do occur." Numerous publications help to define the problem of terrorism and the strategies of homeland security and national security (Figure 15-6).

To add to the question of how homeland security will evolve in the future, we can refer back to the Chapter 12 critical thinking perspective on resilience, which is defined as "the ability to adapt to changing conditions and withstand and rapidly recover from disruption due to emergencies" (U.S. Department of Homeland Security, 2011). The U.S. Government Accountability Office (2010) reported that the DHS, in its 2009 National Infrastructure Protection Plan (NIPP),

FIGURE 15-6 Numerous publications help to define the problem of terrorism and the strategies of homeland security and national security.

increased its encouragement of *resilience*, besides emphasizing the importance of *protection*. In addition, the National Security Strategy (White House, 2010), which faces global threats to U.S. interests, points to the need to "strengthen security and resilience at home."

National Security and Homeland Security

Here we look at the "big picture" of security for the United States before proceeding to specific topics on homeland security. The Office of Homeland Security (2002) discussed the link between the National Security Strategy of the United States (NSSUS) and the National Strategy for Homeland Security (NSHS). The Preamble to the U.S. Constitution defines the federal government's basic purpose as "to form a more perfect Union, establish justice, insure domestic Tranquility, provide for the common defense, promote the general Welfare, and secure the Blessings of Liberty to ourselves and our Posterity." The **National Security Strategy of the United States** (NSSUS) aims to ensure the sovereignty and independence of the United States, with our values and institutions intact. It includes the application of political, economic, and military power (Figure 15-7) to ensure the survival of our nation. The NSSUS and the NSHS complement each other.

Today, the military/intelligence complex and the domestic security/law enforcement complex strive to work together. This effort became especially important following the 9/11 attacks and subsequent criticism that agencies were not collaborating on threats.

The Office of Homeland Security (2002) released the **National Strategy for Homeland Security** (NSHS) as a comprehensive and shared vision of how best to protect America from terrorist attacks. This strategy was compiled with the assistance of many people in a variety of occupations in the public and private sectors. A major goal was to facilitate shared responsibility and cooperation.

FIGURE 15-7 U.S. Air Force F-16C fighter jet. Military power is an option to ensure the survival of the United States. However, other options, such as diplomacy, are available to settle differences among nations. *Courtesy: Senior Airman Sean Sides.*

The intent of the NSHS was to provide a strategic vision for the proposed DHS that was eventually supported by Congress through the Homeland Security Act of 2002. The NSHS concentrates on six critical mission areas:

- Intelligence and Warning
- Border and Transportation Security
- Domestic Counterterrorism
- Protecting Critical Infrastructure and Key Assets
- Defending Against Catastrophic Threats
- Emergency Preparedness and Response

Homeland security and national security are intertwined because of the global scope of terrorism and the interdependencies of the United States with other countries. Whether we are referring to aviation security, border security, security of containerized cargo entering ports, intelligence, or other aspects of security, international cooperation is essential to integrate domestic and international security strategies to avoid gaps that could be exploited by terrorists.

The Rand Corporation (Larson and Peters, 2001) provides additional insight into the relationship between national security and homeland security. When the prevention of terrorist attacks fail, government must respond through five missions:

- Assist civilians: First responders and local agencies and resources are the backbone of this effort.
- Ensure continuity of government: The 9/11 attacks illustrated the need for the protection of government leaders.
- Ensure continuity of military operations: Terrorist attacks could require domestic assistance from the military, besides an armed response to terrorists. Defense planners must protect military personnel and resources. The 9/11 attack on the Pentagon serves as an example.
- Border and coastal defense: The border of the United States includes thousands of miles along the Canadian border, the Mexican border, and the east and west coasts. Access controls and inspections are necessary for people, vehicles, and goods reaching the United States. This overwhelming workload falls on a variety of federal, state, and local agencies.
- National air defense: For many years, the Air Force operated the North American Air Defense Command (NORAD) to detect and provide warning of missile or bomber attacks, especially during the Cold War. Following the 9/11 attacks, the Department of Defense formed the **U.S. Northern Command (NORTH-COM)** to provide air and other types of defense for the United States.

In 2012, President Barack Obama's State of the Union address reiterated many of the goals of earlier NSHS. He emphasized counterterrorism, bio and nuclear security, improving intelligence and information sharing, protecting cyberspace, promoting resilience, securing borders, enhancing disaster management, and increasing the integration of the NSSUS and the NSHS (White House, 2012).

Besides the NSSUS and the NSHS, which together take precedence over all other national strategies, there are other, more specific, strategies of the United States that are subsumed within the twin concepts of national security and homeland security. They are:

- National Defense Strategy
- National Strategy for Counterterrorism
- National Strategy to Combat Weapons of Mass Destruction
- National Strategy to Secure Cyberspace
- National Drug Control Strategy

U. S. Government Action against Terrorism

The *National Strategy for Counterterrorism* (White House, 2011) builds on previous National Strategies and emphasizes a partnering approach of government departments and agencies, allies, the private sector, civilians, and others.

The core values of the Strategy include:

- Pursuing respect for human rights, civil liberties and civil rights
- Encouraging foreign governments to respond to the needs of their citizens so al-Qaida does not exploit grievances
- Seeking to balance security and openness in government
- Working to adhere to a durable legal framework for counterterrorism operations and bringing terrorists to justice

Key goals of The Strategy are to:

- Protect the American people and American interests around the globe
- Build a culture of resilience
- Disrupt, degrade, dismantle and defeat terrorist organizations through diplomatic, economic, information, law enforcement, military, financial, intelligence, and other instruments of power
- Prevent terrorist development, acquisition and use of WMD
- Eliminate safe havens where terrorists can train and plan
- Build counterterrorism partnerships and capabilities with other countries
- Reduce the sources of financing to terrorists

Legislative Action against Terrorism

Laws, executive orders, and presidential directives provided the legal foundation for the war against terrorism prior to and following the 9/11 attacks. Here is a list of relevant laws enacted by Congress.

The **Defense Against Weapons of Mass Destruction Act of 1996**, also called Nunn-Lugar, was influenced by three major terrorist attacks: the World Trade Center bombing in 1993, the Oklahoma City bombing in 1995, and the Tokyo subway sarin gas attack in 1995. This Act provided the impetus for increased federal government preparedness activities, and funding for training and equipment for first responders to respond to terrorism and WMD.

The **Antiterrorism and Effective Death Penalty Act of 1996** was designed to prevent terrorist acts, enhance counterterrorism methods, and increase punishments. It provided funding for antiterrorism measures by federal and state authorities and includes the death penalty under Federal law when a death results from a terrorist act.

The **USA Patriot Act of 2001** includes a broad array of measures to protect the United States. It focuses on counterterrorism funding, enhanced police surveillance, anti-money laundering programs, border protection, victim compensation, information sharing, strengthening criminal laws against terrorism, and improved intelligence.

The Act expanded police powers to investigate and apprehend terrorist suspects. It authorizes roving wiretaps. This means that police can obtain one warrant for multiple jurisdictions for wiretaps on any telephone used by a suspected terrorist. Prior to the Act, judicial authorization was required for each telephone. In addition, federal law enforcement officers can obtain search warrants that can be used nationwide and their subpoena powers were increased to obtain e-mail records of terrorists.

Another police power provided by the Act is referred to "sneak and peek." It authorizes police to enter a residence or business with a warrant when the owner is not present, search for evidence and seize it, and delay providing notice to the owner (possibly following arrest). This provision has been targeted for criticism because it typically requires police to break and enter secretly and is applied to other crimes, besides terrorism.

The Act relaxed restrictions on sharing information among U.S. law enforcement agencies and the intelligence community. The Treasury Department was provided with greater authority to force foreign banks dealing with U.S. banks to release information on suspected money laundering.

Under the Act, immigrant terrorist suspects can be held up to seven days for questioning without specific charges. Criminal penalties were enhanced for terrorist acts, financing and harboring terrorists, and possession of WMD.

An erosion of civil liberties became a major concern of those critical of the Patriot Act. To allay some of these fears, lawmakers included sunset provisions in this law so that expanded police powers would expire in four years and be subject to reauthorization.

The USA Patriot Improvement and Reauthorization Act of 2005 extended the Patriot Act, although it faced debate. Because of the continued threat of terrorism, provisions of the original Act have been repeatedly reauthorized by Congress, subject to slight modifications.

The **Homeland Security Act of 2002** began the largest single reorganization of the federal government since 1947 when President Truman created the Department of Defense. President George W. Bush signed it into law on November 25, 2002. The Act was created primarily because of criticism that the 9/11 attacks could have been prevented if federal agencies had an improved system of cooperating with each other and sharing intelligence. The Act:

- Established the **Department of Homeland Security (DHS)**, as an agency of the executive branch of government, with a DHS secretary who reports to the president. The DHS consolidated 22 agencies, with about 170,000 employees, under one unified organization with the aim of improving defenses against terrorism, coordinating intelligence, and minimizing the damage from attacks and natural disasters.

- Established within the office of the president the Homeland Security Council to advise the president.
- Detailed an organization chart and management structure for the DHS.
- Listed agencies and programs to be transferred to the DHS.
- Stated the responsibilities of the five directorates of the DHS: Information Analysis and Infrastructure Protection; Science and Technology; Border and Transportation Security; Emergency Preparedness and Response; and Management.

DHS does not have overarching authority for directing all aspects of the homeland security mission. As examples, the Department of Justice, the Department of Health and Human Services, and the Department of Defense are still major players in homeland security.

— Gilmore Commission, 2003

The 9/11 Commission Report

The 9/11 Commission Report was prepared by the National Commission on Terrorist Attacks Upon the United States (2004). This commission was an independent, bipartisan group created by congressional legislation and President George W. Bush in late 2002 to prepare an account of the 9/11 attacks, the U.S. response, and recommendations to guard against future attacks. This document of over 400 pages is summarized next from the *Executive Summary.*

Findings:

- U.S. air defense on 9/11/01 depended on close interaction between the Federal Aviation Administration (FAA) and the North American Aerospace Defense Command (NORAD). At the time of the attacks, protocols were unsuited for hijacked planes being used as WMD. NORAD planning focused on hijacked planes coming from overseas. Senior military and FAA leaders had no effective communication with each other. Even the president could not reach some senior officials.
- None of the measures adopted by the U.S. government from 1998 to 2001 disturbed or even delayed the progress of the 9/11 plot. The terrorists located and exploited a broad array of weaknesses in security. They studied publicly available materials on aviation security and used weapons with less metal content than a handgun. Airline personnel were trained to be nonconfrontational in the event of a hijacking.
- Operational failures included not sharing information linking individuals in the USS Cole attack to the 9/11 terrorists, not discovering fraud on visa applications and passports, and not expanding no fly lists to include names from terrorist watch lists.
- "The most important failure was one of imagination." Leaders did not understand the gravity of the threat from al-Qaida.
- Terrorism was not the overriding national security concern under the Clinton or pre-9/11 Bush administrations. The attention Congress gave terrorism was "episodic and splintered across several committees."

- The FBI was not able to link intelligence from agents in the field to national priorities.
- The U.S. government failed at pooling and efficiently using intelligence from the CIA, FBI, State Department, and military.
- U.S. diplomatic efforts were unsuccessful at persuading the Taliban regime in Afghanistan to stop offering sanctuary to al-Qaida and to expel bin Laden to a country where he could face justice.
- "Protecting borders was not a national security issue before 9/11."
- Following the attacks, first responders performed well and saved many lives. In New York, decision-making was hampered by problems in command and control. Radios were unable to assist multiple commands in responding in a unified fashion.

Recommendations:

- Because of action by our government, we are safer today than on September 11, 2001. "But we are not safe." Therefore, we make the following recommendations:
- "The enemy is not Islam, the great world faith, but a perversion of Islam."
- Post-9/11 efforts rightly included military action to topple the Taliban and pursue al-Qaida. A broad array of methods must be continued for success: diplomacy, intelligence, covert action, law enforcement, economic policy, foreign aid, and homeland defense.
- A strategy of three dimensions is proposed: attack terrorists and their organizations; prevent the continued growth of Islamist terrorism through coalition efforts, a vision for a better future, and support for public education and economic openness; and protect against and prepare for terrorist attacks by applying biometric identifiers, improving transportation security, sharing information, improving federal funding to high risk cities, enhancing first responder capabilities and communications, and cooperating with the private sector that controls 85 percent of the nation's infrastructure.
- Establish a National Counterterrorism Center (NCTC) to build unity of effort across the U.S. government, including the CIA, FBI, DOD, and Homeland Security. The NCTC would not make policy, which would be the work of the president and the National Security Council.
- Appoint a National Intelligence Director (NID) to oversee the NCTC. The NID should report to the president, yet be confirmed by the Senate. The system of "need to know" should be replaced by a system of "need to share."

The 9/11 Commission officially ended in August of 2004 after issuing its recommendations. However, it continued operating with private funds to monitor federal government progress on the recommendations. In December 2005, the Commission met for the last time to provide a "report card" on government action. The Commission gave the government several "Fs" and "Ds" and one "A." It characterized the failures as "shocking" and "scandalous" because of a lack of urgency by the government. The government received an "F" on such issues as allocating funds to cities based upon risk, improving radio communications for first responders, and pre-screening airline passengers. The only "A" was for controlling terrorist financing.

The National Security Preparedness Group (2011), the successor to the Commission, issued a *Tenth Anniversary Report Card.* It noted progress with agencies sharing information and airline passenger screening. Improvements were necessary with unity of command and interoperability (i.e., the ability of multiple agencies to communicate using technology) so all levels of government can respond to disasters without confusion. Also needing improvement was action on civil liberties, centralization of HS policies, strengthening the position of the Director of National Intelligence, screening of explosives at airports, an enhanced entry-exit border screening system, standards for sources of personal identification, and standards for the detention of terrorists.

Intelligence

Hughbank and Githens (2010: 31) refer to intelligence as "information that is analyzed and converted into a product to support a particular customer." For example, management in a company may use business intelligence to improve decisions. Here, we focus on intelligence in the context of security and counterterrorism.

Hughbank and Githens portray a realistic view of the work of intelligence analysts by explaining that these practitioners do not assassinate people, carry weapons, or wear a trench coat; very few are involved in clandestine intelligence gathering. Most of these practitioners sit in front of a computer terminal to complete tasks. However, Hughbank and Githens note that intelligence is crucial to identifying terrorists and preventing attacks.

Since the 9/11 attacks, the terms "intelligence" and "intelligence community" have been subject to attention and controversy. Most experts agree that there was a massive U.S. intelligence failure before the attacks. In the criticism over government incompetence, words echoed repeatedly were "stovepiping" and "connecting the dots." **"Stovepiping"** refers to organizations that operate in isolation from one another, closely protect internal information, avoid partnering with other organizations to improve overall efficiency and effectiveness, and do not recognize how their action or inaction impacts other organizations. **"Connecting the dots"** refers to clues that surfaced, prior to September 11, 2001, that were not pursued and shared by the CIA and the FBI to possibly prevent the attacks. Links between al-Qaida's previous plans and specific operatives pointed to indicators of an impending attack. Shultz and Vogt (2003) write that an FBI agent in Phoenix warned the agency in July 2001 that some Middle Eastern men training in American flight schools might be bin Laden agents; and, the CIA possessed information about two of the hijackers dating back to an al-Qaida meeting in Malaysia in January 2000.

Lowenthal (2012: 9) defines **intelligence** as follows: "Intelligence is the process by which specific types of information important to national security are requested, collected, analyzed, and provided to policymakers; the products of that process; the safeguarding of these processes and this information by counterintelligence activities; and the carrying out of operations as requested by lawful authorities." Examples of intelligence include the plans for a terrorist attack, the development of a new weapon by another country, and the ulterior motives of a rogue state.

Lowenthal (2012: 57–70) describes a seven step **intelligence process**: (1) identifying needs and priorities (e.g., rogue states, terrorists, weapons proliferation); (2) collecting intelligence (e.g., human intelligence such as espionage; satellites); (3) processing and exploiting collected information to produce intelligence (e.g., processing imagery; decoding); (4) analysis and production (e.g., analysts, who are experts in their fields, study information and prepare intelligence reports); (5) dissemination to policymakers (e.g., President's Daily Briefing presented by the CIA); (6) consumption (e.g., needs and preferences of policymakers); and (7) feedback (e.g., what intelligence has been useful and what intelligence requires increased emphasis).

The **intelligence community** of the United States consists of several government organizations that collect information. Since 1947, the Central Intelligence Agency (CIA) has brought together foreign intelligence for the president and the National Security Council. The Department of Defense (DOD) has enormous resources, beyond the CIA, to collect intelligence. DOD organizations that collect intelligence include the National Security Agency, Defense Intelligence Agency, National Imagery and Mapping Agency, Defense airborne reconnaissance programs, and the armed service intelligence units. The Departments of State, Treasury, and Energy also have intelligence units, although small. The FBI focuses on domestic intelligence, but it has expanded its work globally in recent years.

Since the 9/11 attacks, there have been many calls for reform of the intelligence community. The 9/11 Commission report proposed a restructuring of the U.S. intelligence community and the creation of an intelligence "czar" for better coordination and sharing of intelligence. President George W. Bush signed **The Intelligence Reform and Terrorism Prevention Act of 2004** to improve intelligence gathering and transform a system designed for the Cold War to a system more effective against terrorism. Although traditional intelligence gathering has focused on foreign countries and such topics as their policies, strategies, economies, technology, and military capabilities, the problem of terrorism requires an augmentation of intelligence gathering to include the methods, behaviors, strategies, tactics, and weapons of terrorists. The aim of the Act is to make the intelligence community more unified, coordinated, and effective.

The impact of the Act to improve the intelligence community will be an evolutionary process that will confront traditional resistance to changes, the protection of turf, and competition for resources. Success depends on several factors, such as the budgetary authority over intelligence of the Director of National Intelligence, leadership, and congressional oversight.

State and Local Governments

Security and loss prevention practitioners should understand the roles, capabilities, and limitations of federal, state, and local governments in HS. Such information serves as input for business continuity planning.

The Unites States has over 87,000 jurisdictions of overlapping federal, state, and local governments. All states and many local jurisdictions have established a homeland security office in the executive branch. Coordination is vital to produce the best possible multi-agency emergency response to serious events.

Every day, minor or major disasters strike in American and it is the local agencies that respond. Fire fighters, police, and emergency medical technicians are the **first responders** who rush to the scene to preserve life and to reduce property damage. Consequently, a terrorist attack or natural disaster becomes a *local event*, and the FBI, DHS, and FEMA do not arrive until later. A U.S. Conference of Mayors report noted in 2003, "When you dial 911, the phone doesn't ring in the White House. Those calls come in to your city's police, fire, and emergency medical personnel; our domestic troops." Today, we still have issues with HS at all levels of government. Funding is a leading issue.

National Terrorism Advisory System

In 2011, the DHS implemented the **National Terrorism Advisory System** (NTAS). It replaces the former color-coded alert system.

The earlier system was criticized and even ignored by the governments and citizens it was supposed to assist. It was labeled a "damned if you do, damned if you don't" system. When a heightened level of alert is declared and an attack does not occur, complaints surface about the cost, waste of resources, and unnecessarily frightening the public and desensitizing them to alerts. Too many alerts can cause "alert fatigue" and a nonchalant attitude during subsequent alerts. If a warning is not issued when intelligence indicates a possible attack, and an attack does occur, authorities face serious repercussions (Bannon, 2004).

The effectiveness of an advisory system is difficult to ascertain. An advisory may cause terrorists to abort an attack. Or, the intense vigilance of authorities and citizens may uncover and disrupt terrorist plans.

NTAS alerts will provide a summary of the potential threat, geographic region, mode of transportation, target of the threat, public safety action, and protection methods for communities. The threat will be defined in one of two ways:

1. Elevated Threat: warns of a credible terrorist threat against the United States.
2. Imminent Threat: warns of a credible, specific, and impending terrorist threat against the United States.

Responses to Weapons of Mass Destruction

The CDC, DHS, FEMA, and the American Red Cross websites offer a wealth of information on precautions for WMD attacks. Although the information here is aimed at family disaster planning and first responders, security practitioners in the private sector can benefit from these guidelines and recommendations.

Chemical Attack

The Federal Emergency Management Agency (2011) offers such information as the following:

- Build an emergency supply kit of such items as food, water, radio, flashlight, duct tape, and plastic to seal doors and windows for sheltering in place.
- Flee from the contaminated area, if possible. Otherwise, shelter in place and turn off HVAC.
- If exposed to chemical agents, remove clothes, wash with soap and water, and seek medical attention.

The U.S. Department of Homeland Security, Office for Domestic Preparedness (2004) distributed a video on guidelines for first responders facing WMD incidents. A summary of this program, which emphasizes a chemical attack, follows:

A WMD attack has potential to incapacitate first responders by direct or secondary contact (e.g., contaminated victims). The scene must be controlled and large numbers of victims may require decontamination. In the Tokyo sarin attack, some first responders and hospital staff were contaminated by victims' clothing, shoes, and skin. *It is important to remember that responders cannot help others if they become victims themselves.*

Steps for First Responders

Recognize the danger. Pay close attention to the dispatcher's report. For example, victims leaving a building are falling to the ground and coughing. Upon approaching the scene, be aware of wind direction, weather conditions, type of building, occupancy, and number of victims. The victims of a chemical attack may experience SLUDGE: salivation, lacrimation (tearing), urination, defecation, gastrointestinal distress, and emesis (vomiting).

Protect yourself. Do not rush into an area if you see multiple people down. Warn others about the situation and report the facts and dangers. Be aware that a secondary explosion or sniper may be next. Emergency vehicles should be at least 300 feet away from the victims or dangerous area. Anyone entering the "hot zone" should wear the highest level of protection (i.e., vapor-tight, fully-encapsulated suit with self-contained breathing apparatus).

Control the scene. Access controls should be maintained at 360 degrees around the scene and only authorized, properly trained, and equipped personnel should be admitted. When hazmat personnel arrive, they will identify the "hot zones" based upon readings from detection equipment. Public address systems can keep victims informed on procedures while responders maintain safe distances. At the same time, remember to preserve evidence, since the incident location is also a crime scene.

Rescue and decontaminate the victims. As soon as possible, victims should be brought to a shower area that can be established with fire trucks and related equipment.

Call for additional help. A variety of agencies may respond such as more local units, state units, FEMA, the FBI, the EPA, public health, and the DOD.

Biological Attack

Bioterrorism is difficult to detect. Indications of an attack are likely to surface when victims visit hospitals and a trend is discovered through centralized reporting to local and state health agencies. The CDC is the agency that monitors public health threats. The federal government has a "rapid response team" of medical personnel who have been vaccinated from a variety of biological threats and can enter infected locales. Vaccines do not exist for all biological threats. During an outbreak, an area (e.g., city) may require quarantine, backed by the military.

The Federal Emergency Management Agency (2011) offers such information as the following:

- Build an emergency supply kit.
- Use available technology to learn of the threat, symptoms, how it should be treated, and where to seek medical attention.

- Cover your mouth and nose with layers of fabric or a mask.
- If exposed to an agent, change clothes, bag the clothes, and wash yourself.
- Practice good hygiene to avoid spreading germs.

Radiation Attack

The Federal Emergency Management Agency (2011), Council on Foreign Relations (2004), and the Central Intelligence Agency (1998) offer such information as the following:

- Build an emergency supply kit.
- Use available technology to learn of the threat.
- Seek an internal room or public shelter for protection.
- If exposed to radiation, change clothes, bag the clothes, and wash yourself. These methods do not protect against "penetrating radiation" (e.g., gamma rays).
- Potassium iodide is recommended to prevent medical problems and cancer of the thyroid gland, but it offers no protection against other types of radiation problems.
- The onset of symptoms requires days to weeks.
- Specialized equipment is required to determine the size of the affected area and if the level of radioactivity presents long-term hazards. Radiological materials are not recognized by senses, and are colorless and odorless.
- The incident area would require evacuation and decontamination to remove radioactive material and keep radioactive dust from spreading. Contaminated buildings, roads, and soil may have to be removed.

Nuclear Attack

Although a nuclear weapon is difficult to construct, deliver, and detonate, worst case scenarios must be considered. Depending on the type of bomb, casualties could be enormous, along with an awful impact on physical and mental health, the economy, and society. If a warning occurs, people should take shelter in basements and subways, unless time is available for an evacuation of the area to several miles away.

First responders must be properly trained and equipped to perform their duties following a nuclear explosion. Enormous numbers of victims would need treatment and hospitals may be destroyed. Medical problems would include thermal radiation burns, wounds, fractures, and infection. The radioactive fallout from the blast requires people to be decontaminated. Refer to the above "radiation attack" discussion for additional information.

Precautions for Organizational Mail Systems

Organizations should take precautions for mail as it enters the premises. Here is a list of protection measures:

- Prepare emergency plans, policies, and procedures for suspect mail.
- Isolate the mail area and limit access. Isolate the ventilation system of the mail area. Consider handling mail at a separate, isolated building.

FIGURE 15-8 The September 11, 2001 terrorist attacks on the United States were followed by bioterrorism in which anthrax was spread through the mail system, necessitating precautions.

- If necessary, use gloves (Figure 15-8) and protective masks.
- Do not touch or smell anything suspicious.
- Consider technology: X-ray for bombs, letter and package tracking systems, automatic letter openers, and scan letters and send them electronically.
- Research the CDC and other websites for information on Anthrax and other threats.

Private Sector

In addition to government action against terrorism, there is a vast community of private sector businesses, organizations, associations, and volunteers that also play a major role in homeland security and the protection of American interests. Furthermore, joint public/private sector initiatives exist.

Public-Private Sector Partnerships

Reimer (2004: 10–12) sees the national system of preparedness for terrorism as a work in progress requiring a series of partnerships among the various levels of government, the public and private sectors, and the military. He notes that the strength of these partnerships will determine our level of preparedness. With the private sector owning 85 percent of U.S. critical infrastructure, Reimer emphasizes that private sector cooperation is essential.

Weidenbaum (2004: 189) argues that "the multifaceted national response to the September 11 terrorist attacks is altering the balance between the public and private sectors in the U.S." He describes the federal government effort against terrorism as a "mega priority" of programs and activities. This includes increased spending on the military and homeland security, special assistance and funding to companies heavily affected, and an expansion of government regulation of private activity, especially business. Weidenbaum notes that terrorism results in a hidden tax on businesses in the United States and overseas, in the form of added costs of operation.

To facilitate communications and to share information on homeland security with private industry, the federal government hosts conferences and training sessions. Another initiative is the **Information Sharing and Analysis Centers (ISACs)**. These are generally private sector networks of organizations that the federal government has helped to create to share information on threats to critical industries and coordinate efforts to identify and reduce vulnerabilities. For example, the North American Electric Reliability Council (NERC) shares threat indications and analyses to help participants in the electricity sector take protective action.

Fusion Centers, established by the federal government, enlist law enforcement and other first responders, intelligence, homeland security, and the private sector to share threat information. Located in states and major cities, fusion centers facilitate local understanding of national intelligence, provide interdisciplinary expertise and training, and offer connectivity to federal systems and grants.

Public police and private security partnerships vary. Both groups share information with each other, attend each other's conferences, and plan together for protection and emergencies. They even work side by side at certain sites such as downtown districts, government buildings, and special events. Additionally, many off-duty police officers work part-time in security and retire to assume a security position.

The **Overseas Security Advisory Council (OSAC)** promotes security cooperation between the U.S. Department of State and businesses with interests overseas. OSAC objectives include sharing information, promoting methods for security planning, and publishing a variety of helpful guides.

The Homeland Security Market

The homeland security market is a multi-billion dollar industry of products and services. Many companies have made handsome profits and have seen their stock prices rise considerably because of the need to enhance homeland security. Examples of products include luggage screening systems, WMD sensors, systems to track foreigners visiting the United States, and physical security. Services include the privatization of counterterrorism.

Business and Organizational Countermeasures against Terrorism

For a security practitioner considering countermeasures against terrorism, numerous factors come into play as a foundation for planning. These factors include employees and their families, facilities and assets, the type of products and/or services offered to customers, domestic and/or international operations, and terrorist groups that may target the entity and their methods and weapons. Businesses should plan for a variety of possible scenarios. Examples include casualties, the loss of manufacturing capacity or a major supplier, an extended power outage, the disruption of shipping at a port, or fear that causes employees and customers to stay home.

Planning to counter terrorism must be integrated with business continuity planning. Here is a list of suggestions to protect against terrorism:

- Terrorists have tremendous advantages over defenses because they search for vulnerabilities and strike almost anywhere with surprise. A British terrorism expert, Paul Wilkinson, stated, "Fighting terrorism is like a goalkeeper. You can make a hundred brilliant saves, but the only shot people remember is the one that gets past you" (Leader, 1997). Creative and astute security planning is essential.
- Maintain awareness of controversies that may increase exposure to terrorism.
- Reevaluate security, safety, and emergency plans. Seek cooperation with public safety agencies.
- Select a diverse committee for input for security and safety plans. Permit committee members to use their imaginations to anticipate weaknesses in protection.
- Increase employee awareness of terrorism and terrorist methods and tricks. Stifle and trick potential terrorists through creativity (e.g., avoid signs on the premises that identify departments, people, and job duties; periodically alter security methods).
- Know the businesses of tenants and whether they would be possible targets for violence.
- Consider the security and safety of childcare facilities used by employees.
- Intensify access controls by using a comprehensive approach covering people, vehicles, mail, deliveries, services, and any person or thing seeking access.
- Carefully screen employment applicants, vendors, and others seeking work.
- Secure and place under surveillance HVAC systems (e.g., air intake systems) and utilities.
- Once the best possible security and safety plans are implemented, remember that an offender (internal or external or both) may be studying your defenses to look for weaknesses to exploit.

Citizen Volunteers

The *National Strategy for Homeland Security* (Office of Homeland Security 2002: 12) makes the following statements concerning the importance of citizens becoming involved in homeland security.

> *All of us have a key role to play in America's war on terrorism. Terrorists may live and travel among us and attack our homes and our places of business, governance, and recreation. In order to defeat an enemy who uses our very way of life as a weapon—who takes advantage of our freedoms and liberties—every American must be willing to do his or her part to protect our homeland.*

Brian Jenkins, a noted authority on terrorism, points out that, although public and private sectors maintain contingency plans to put into effect during heightened states of alert and emergencies, citizens must "learn to take care of themselves, to be mentally tough and self-reliant. Those who expect the government to protect them from everything are in for a lot of disappointments." He writes that for homeland security to succeed, citizens must be involved in the defense of their communities, and with quality training, "the entire United States can be turned into a vast neighborhood watch—a difficult environment for terrorists." Jenkins goes on to note that there will never be enough police and firefighters to protect all citizens. And, the first people at the scene of an attack are the unfortunate people who are there at the time of the attack. They need to learn how to protect themselves and others until first responders arrive (Jenkins, 2003).

Volunteer organizations play a vital role of assistance in homeland security. Two noteworthy, national groups are the **American Red Cross** and the **Salvation Army**. Both groups have been involved in a variety of community-based programs for many years. These organizations partner with all levels of government to support the immediate, critical needs of disaster victims. These needs include food, shelter, and clothing.

Many other volunteer groups assist disaster victims. The **National Volunteer Organizations Against Disasters** (NVOAD) consists of several national groups that rely on NVOAD to coordinate disaster response for increased efficiency.

In 2002, the federal government launched **Citizen Corps** to coordinate and channel citizen volunteers to help with homeland security and disaster preparedness efforts. Citizen Corps is a part of **USA Freedom Corps**—a federal government program that facilitates a partnership of nonprofit, business, educational, faith-based, and other sectors to increase citizen involvement in their communities. USA Freedom Corps seeks to create a culture of service, citizenship, and responsibility, while increasing civic awareness and community involvement.

Search the Internet

Here is a list of websites relevant to this chapter:

American Red Cross: www.redcross
Center for Defense Information: www.cdi.org
Centers for Disease Control (CDC): www.cdc.gov
Center for the Study of Terrorism and Political Violence: www.st-andrews.ac.uk/intrel/research/cstpv
Central Intelligence Agency: www.cia.gov
Federal Bureau of Investigation: www.fbi.gov
Federal Emergency Management Agency: www.fema.gov
FEMA, Ready: www.ready.gov/terrorism
Government Accountability Office: www.gao.gov
International Policy Institute for Counterterrorism: www.ict.org.il
National Commission on Terrorist Attacks Upon the United States: www.9-11commission.gov
National Counterintelligence Center: www.ncix.gov
National Counterterrorism Center: www.nctc.gov
Office of the President: www.whitehouse.gov
Overseas Security Advisory Council (OSAC): www.ds-osac.org
Stockholm International Peace Research Institute: www.sipri.org

United Kingdom Security Service (MI5): www.mi5.gov.uk
United Nations: www.un.org
U.N. International Court of Justice: www.icj-cij.org
U.S. Department of Defense: www.defenselink.mil
U.S. Department of Homeland Security: www.dhs.gov
U.S. Department of State: www.state.gov
Animal Liberation Front: www.animalliberationfront.com
Earth Liberation Front: earth-liberation-front.org
People for the Ethical Treatment of Animals: www.peta.org
White Aryan Resistance: www.resist.com

Case Problems

15A. As a Chief Security Officer for a major international corporation, prepare a prioritized list of five general items about terrorism and homeland security that you would consider and apply in business continuity planning for domestic operations. In addition, prepare a prioritized list of five general items about terrorism and security that you would consider and apply in business continuity planning for international operations. Compare the lists and study the findings.

15B. As a security executive for a corporation based in the United States, your superior appoints you to a committee involved in strategic business planning. Business plans are being prepared for operations in Venezuela and Thailand. Your assignment is to research these countries and prepare reports on political stability, safety, security, and health issues. What resources do you use? What are your findings and recommendations?

15C. As a security executive for a manufacturer in the United States, what private and public sector groups would you seek partnerships with to enhance the survivability of your organization in case of a serious emergency?

15D. You are a security manager at a news media corporation with headquarters in the United States and offices throughout the world. An anthrax attack against news media organizations and politicians has recently begun. Your superior assigns to you the task of protecting the corporation and its employees. What are your plans and what actions do you take?

References

Bailey, G. (2003). Globalization. In A. del Carmen (Ed.), *Terrorism: An interdisciplinary perspective* (2nd ed.). Toronto: Thomas Learning, Inc.

Bandura, A. (1977). *Social learning theory*. NY: General Learning Press.

Bannon, A. (2004). Color-coding—or just color-confusing? *Homeland Protection Professional, 3* (September).

Barkan, S., & Snowden, L. (2001). *Collective violence*. Boston: Allyn and Bacon.

Bell, J., & Gurr, T. (1979). Terrorism and revolution in American. In H. Graham & T. Gurr (Eds.), *Violence in America*. Newbury Park, CA: Sage.

Centers for Disease Control. (n.d.). Frequently Asked Questions About a Nuclear Blast. <www.bt.cdc.gov/radiation> retrieved August 5, 2004.

Central Intelligence Agency. (1998). Chemical/Biological/Radiological Incident Handbook (October). <www.fas.org> retrieved June 4, 2004.

Chalmers, D. (1981). *Hooded Americanism: The history of the ku klux klan* (3rd ed.). Durham, NC: Duke University Press.

Clayton, C., et al. (2003). Terrorism as group violence. In H. Hall (Ed.), *Terrorism: Strategies for intervention.* Binghamton, NY: The Haworth Press.

Combs, C. (2003). *Terrorism in the twenty-first century* (3rd ed.). Upper Saddle River, NJ: Prentice-Hall.

Council on Foreign Relations. (2004). Responding to Radiation Attacks. <www.terrorismanswers.org> retrieved October 6, 2004.

Crenshaw, M. (2012). The debate over 'New' vs. 'Old' terrorism. In R. Howard & B. Hoffman (Eds.), *Terrorism and counterterrorism* (4th ed.). New York, NY: McGraw-Hill Pub.

Crenshaw, M. (1990). The logic of terrorism: Terrorist behavior as a product of strategic choice. In H. Hall (Ed.), *Terrorism: Strategies for intervention.* New York, NY: Routledge Pub.

Dyson, W. (2012). *Terrorism: An investigator's handbook* (4th ed.). Waltham, MA: Anderson Pub.

Federal Bureau of Investigation (n.d.). Counterterrorism. denver.fbi.gov/inteterr.htm, retrieved January 14, 2004.

Federal Bureau of Investigation (2010). Domestic Terrorism: The Sovereign Citizen Movement (April 13). <www.fbi.gov/page2/april10/sovereigncitizens_041310.html> retrieved April 14, 2010.

Federal Bureau of Investigation (2011). Fugitive Who Built Firebombs Linked to 2001 Arson of UW Center for Urban Horticulture Arrested Following Expulsion from China. (July 6, 2011). <www.fbi.gov/seattle/press-releases/2011> retrieved January14, 2011.

Federal Bureau of Investigation (2002). The Threat of Ecoterrorism. <www.fbi.gov/congress/congress02/jarboe021202.htm> retrieved May 3, 2004.

Federal Bureau of Investigation, Counterterrorism Analysis Section, (2011). Sovereign citizens: A growing domestic threat to law enforcement. *FBI Law Enforcement Bulletin, 80* (September).

Federal Emergency Management Agency (2011). Terrorist Hazards. <www.ready.gov/terrorism> retrieved January 25, 2012.

Ford, P. (2001). Why do they hate us? *The Christian Science Monitor* (September 27).

Greenberg, D. (2001). Is Terrorism New? <Slate.com.>, <historynewsnetwork.org> retrieved March 15, 2004.

Griset, P., & Mahan, S. (2003). *Terrorism in perspective.* Thousand Oaks, CA: Sage Pub.

Grosskopf, K. (2006). Evaluating the societal response to antiterrorism measures. *Journal of Homeland Security and Emergency Management, 3*

Hamid, S. (2011). The rise of the islamists: How islamists will change politics, and vice versa. *Foreign Affairs, 90* (May–June).

Hoffman, B. (2012a). Defining terrorism. In R. Howard & B. Hoffman (Eds.), *Terrorism and counterterrorism* (4th ed.). New York, NY: McGraw-Hill Pub.

Hoffman, B. (2012b). Bin ladin's killing and its effect on Al-Qa'ida: What comes next?. In R. Howard & B. Hoffman (Eds.), *Terrorism and counterterrorism* (4th ed.). New York, NY: McGraw-Hill Pub.

Hoffman, B. (1998). *Inside terrorism.* NY: Colombia University Press.

Holden, R. (1986). *Modern police management.* Englewood Cliffs, NJ: Prentice-Hall.

Hudson, R. (1999). The sociology and psychology of terrorism: Who becomes a terrorist and why?. In H. Hall (Ed.), *Terrorism: Strategies for intervention.* Washington, D.C.: Federal Research Division, Library of Congress.

Hughbank, R., & Githens, D. (2010). Intelligence and its role in protecting against terrorism. *Journal of Strategic Security, 3* (Spring).

Huntington, S. (1996). *The clash of civilizations and the remaking of the world order*. NY: Simon & Schuster.

Ismayilov, M. (2010). Conceptualizing terrorist violence and suicide bombing. *Journal of Strategic Security, 3* (Fall).

Jenkins, B. (2003). All citizens now first responders. *USA Today* (March 23).

Johnson, K., & Gorman, S. (2011). Bomb implants emerge as airline terror threat. *The Wall Street Journal* (July 7).

Kegley, C. (2003). *The new global terrorism: Characteristics, causes, controls*. Upper Saddle River, NJ: Prentice Hall.

Katz, R. (2013). *Essentials of public health preparedness*. Burlington, MA: Jones & Bartlett Learning.

Laqueur, W. (1999). *The new terrorism: Fanaticism and the arms of mass destruction*. NY: Oxford University Press.

Larson, E., and Peters, J. (2001). Preparing the US army for homeland security: Concepts, issues, and options. Santa Monica, CA: RAND. In, Kettl, D. (2004). System under stress: Homeland security and American politics. Washington, D.C.: CQ Press.

Leader, S. (1997). The rise of terrorism. *Security Management, April.*

Lowenthal, M. (2012). *Intelligence: From secrets to policy* (5th ed.). Washington, DC: CQ Press.

Martin, G. (2003). *Understanding terrorism: Challenges, perspectives, and issues*. Thousand Oaks, CA: Sage Pub.

Mordica, G. (2003). Phase four operations in iraq and the RPG-7. *News from the Front, November–December.*

National Commission on Terrorist Attacks Upon the United States. (2004). The 9/11 Commission Report. <www.9-11commission.gov> retrieved July 26, 2004.

National Counterterrorism Center (2011). About the National Counterterrorism Center. <www.nctc.gov/about_us/about_nctc.html> retrieved January 13, 2012.

National Institute of Justice (2011). Terrorism. <nij.gov/topics/crime/terrorism/welcome.htm> retrieved December 22, 2011.

National Security Preparedness Group (2011). Tenth Anniversary Report Card: The Status of the 9/11 Commission Recommendations. <www.bipartisanpolicy.org> retrieved January 24, 2012.

O'Brien, L. (2011). The evolution of terrorism since 9/11. *FBI Law Enforcement Bulletin, 80* (September).

Office of Homeland Security, (2002). *National strategy for homeland security*. Washington, D.C.: The White House.

Poland, J. (2005). *Understanding terrorism: Groups, strategies, and responses* (2nd ed.). Englewood Cliffs, NJ: Prentice Hall.

Poland, J. (1988). *Understanding terrorism: Groups, strategies, and responses*. Englewood Cliffs, NJ: Prentice Hall.

Purpura, P. (2007). *Terrorism and homeland security: An introduction with applications*. Burlington, MA: Elsevier Butterworth-Heinemann Pub..

Queller, D. (1989). Crusades *World book* (Vol. 4). Chicago: World Book, Inc.

Reimer, D. (2004). The private sector must be a partner in homeland security. *Homeland Security, 1* (September).

Risk Management Solutions (2011). Terrorism Risk in the Post-9/11 Era: A 10-Year Retrospective. <www.rms.com> retrieved January 18, 2012.

Roach, J., et al. (2005). The conjunction of terrorist opportunity: A framework for diagnosing and preventing acts of terrorism. *Security Journal, 18*

Shane, S. (2011). As regimes fall in arab world, Al-Qaeda sees history fly by. *The New York Times*(February 27) <www.nytimes.com/2011/02/28/world/middleeast/23qaeda.html> retrieved February 28, 2011.

Shultz, R., & Vogt, A. (2003). It's war! fighting post-11 september global terrorism through a doctrine of preemption. *Terrorism and Political Violence, 15* (Spring).

Sibalukhulu, N. (2011). Will there be an arab summer? *The Sunday Independent*(November 28) <www.iol.co.za/sundayindependent/will-there-be-an-arab-summer-1.1187879> retrieved December 23, 2011.

Silke, A. (1998). Terrorism. *The Psychologist, 14.*

Simonsen, C., & Spindlove, J. (2004). *Terrorism today* (2nd ed.). Upper Saddle River, NJ: Prentice-Hall.

United Nations, Office on Drugs and Crime. (n.d.). *Terrorism and Weapons of Mass Destruction.* <www.unodc.org/unodc/terrorism_weapons_mass_destruction> retrieved April 21, 2004.

U.S. Department of Homeland Security (2011). *Presidential Policy Directive/PPD-8: National Preparedness* (March 30). <www.dhs.gov> retrieved October 31, 2011.

U.S. Department of Homeland Security, Office for Domestic Preparedness. (2004). *Weapons of Mass Destruction and the First Responder* (Video) (January).

U.S. Department of State. (2011). *Country Reports on Terrorism 2010.* <www.state.gov/documents/organization/170479.pdf> retrieved January 10, 2012.

U.S. Department of State (2006). *Country Reports on Terrorism 2005.* <www.state.gov/s/ct/rls/crt/c17689.htm> retrieved January 4, 2007.

U.S. Department of State, Office of Inspector General (2005). *Review of the Department of State's Country Reports on Terrorism-2005.* <oig.state.gov/documents/organization/58021.pdf>, retrieved March 15, 2006.

U.S. General Accounting Office. (2003). Public Health Response to Anthrax Incidents of 2001. <www.gao.gov> retrieved January 16, 2004.

U.S. Government Accountability Office. (2010). *Critical Infrastructure Protection: Update to National Infrastructure Protection Plan Includes Increased Emphasis on Risk Management and Resilience* (March). <www.gao.gov/new.items/d10296.pdf> GAO-10-296, retrieved October 15, 2011.

Van Dongen, T. (2011). Break it down: An alternative approach to measuring effectiveness in counterterrorism. *Journal of Applied Security Research, 6* (September).

Weidenbaum, M. (2004). Government, business, and the response to terrorism. In T. Badey (Ed.), *Homeland security.* Guilford, CT: McGraw-Hill/Dushkin.

Weinberg, L., & Davis, P. (1989). *Introduction to political terrorism.* NY: McGraw Hill.

Weinzierl, J. (2004). Terrorism: Its origin and history. In A. Nyatepe-Coo & D. Zeisler-Vralsted (Eds.), *Understanding terrorism: Threats in an uncertain world.* Upper Saddle River, NJ: Pearson Prentice Hall.

White, J. (2003). *Terrorism: An introduction* (4th ed.). Belmont, CA: Wadsworth/Thomson Learning.

White House (2011). National Strategy for Counterterrorism (June). <www.whitehouse.gov/sites/default/files/counterterrorism_strategy.pdf> retrieved January 22, 2012.

White House (2010). National Security Strategy (May). <www.whitehouse.gov> retrieved January 27, 2012.

White House (2012). State of the Union (January 24). <www.whitehouse.gov/issues/homeland-security> retrieved January 24, 2012.

Wilkinson, P. (2003). Why modern terrorism? Differentiating types and distinguishing ideological motivations. In C. Kegley (Ed.), *The new global terrorism: Characteristics, causes, controls.* Upper Saddle River, NJ: Prentice-Hall.

16

Protecting Critical Infrastructure

OBJECTIVES

After studying this chapter, the reader will be able to:

1. Define critical infrastructure and key resources
2. Explain the role of the federal government in protecting critical infrastructure and key resources
3. Describe at least ten recommendations from the National Infrastructure Protection Plan for the private sector
4. Name and describe vulnerabilities and protection measures for at least five infrastructure sectors
5. Explain the threats to cyberspace and information technology and protection measures
6. Name and describe vulnerabilities and protection measures for at least four transportation subsectors
7. Name and explain the roles of the federal government agencies involved in border and transportation security

KEY TERMS

- cascade effect
- critical infrastructure (CI)
- Homeland Security Presidential Directive-7
- critical infrastructure protection (CIP)
- National Infrastructure Protection Plan
- sector-specific agencies
- sector-specific plans
- key resources
- key assets
- bioterrorism
- agroterrorism
- Public Health Security and Bioterrorism Preparedness and Response Act of 2002
- Project BioShield Act of 2004
- penetration test
- Fukushima nuclear disaster
- Chernobyl
- Three-Mile Island
- Bhopal, India
- Texas City, Texas
- BP Oil Spill
- Exxon Valdez
- Emergency Planning and Community Right-to-Know Act of 1986
- Chemical Facility Anti-Terrorism Standards
- denial-of-service attack
- compromised system
- malware
- Trojan horse
- virus
- worm

- logic bomb
- dumpster diving
- social engineering
- spam
- phishing
- key logging programs
- spyware
- blended cyber threats
- blended cyber-physical attack
- botnet
- hacktivism
- cyberwarfare
- cybersecurity
- cloud computing
- encryption
- National Industrial Security Program
- General Services Administration (GSA)
- Federal Protective Service
- U.S. Marshals Service
- Transportation Security Administration (TSA)
- Aviation and Transportation Security Act of 2001
- Intelligence Reform and Terrorism Prevention Act of 2004
- Federal Aviation Administration (FAA)
- Federal Air Marshal Service
- Man-Portable Air Defense Systems (MANPADS)
- Maritime Transportation Security Act of 2002
- U.S. Customs and Border Protection
- Security and Accountability For Every (SAFE) Port Act of 2006
- supply chain
- transportation choke points
- anthrax attacks
- U.S. Visitor and Immigrant Status Indicator Technology (US-VISIT)
- Container Security Initiative
- Advanced Manifest Rule
- Customs-Trade Partnership Against Terrorism (C-TPAT)
- U.S. Immigration and Customs Enforcement (ICE)
- U.S. Citizenship and Immigration Services

Critical Infrastructure

Following the 9/11 attacks, concern increased over the vulnerability of essential services and products that sustain our nation and its business. If vital services and products are harmed by an attack or subject to a natural disaster, our economy and daily lives would be impaired. We all depend on such systems as water, food, energy, transportation, and communications. In addition, our modern society depends on a complex network of interdependencies among services and products. If one essential service or product is weakened, other services and products may be affected. This is known as the **cascade effect**. For instance, if a power failure occurs, many daily activities would be at a standstill. The creativity and patience of terrorists make the protection of critical infrastructure essential. Imagine the chaos in a locale from simultaneous bombings of a high-rise building, electrical power, the 911 emergency communications system, and the water supply for fighting fires.

The Foundations of Critical Infrastructure Protection

The Patriot Act, Section 1016(e), defines **critical infrastructure (CI)** as "the systems and assets, whether physical or virtual, so vital to the United States that the incapacity or destruction of such systems and assets would have a debilitating impact on security, national economic security, national public health or safety, or any combination of those matters." A generally accepted estimate is that 85 percent of the U.S. critical infrastructure is owned by the private sector.

A major mission area stated in the *National Strategy for Homeland Security* is protecting critical infrastructure. **Homeland Security Presidential Directive-7** (HSPD-7), released in 2003, established a national policy for the federal government to identify and prioritize CI and key resources and to protect them from terrorist attacks.

The *National Strategy for Homeland Security* (Office of Homeland Security, 2002: ix) identifies eight major **critical infrastructure protection (CIP)** initiatives:

- Unify America's infrastructure protection effort in the Department of Homeland Security (DHS).
- Build and maintain a complete and accurate assessment of America's critical infrastructure and key assets.
- Enable effective partnership with state and local governments and the private sector.
- Develop a national infrastructure protection plan.
- Secure cyberspace.
- Harness the best analytic and modeling tools to develop effective protective solutions.
- Guard America's critical infrastructure and key assets against "inside" threats.
- Partner with the international community to protect our transnational infrastructure.

The National Strategy for the Physical Protection of Critical Infrastructures and Key Assets (White House, 2003a: 9) identified the following CI sectors and key assets for protection. As CIP evolved, terminology changed (e.g., the terms "key resources" and "key assets" are explained in a subsequent paragraph) and the terms "CI," "key resources," and "sectors" are applied to our nation's possessions (U.S. Department of Homeland Security, 2010a).

Agriculture and Food: 2.1 million farms, 880,500 firms, and over one million facilities
Water: 1,800 federal reservoirs; 1,600 municipal wastewater facilities
Public Health: 5,800 registered hospitals
Emergency Services: 87,000 U.S. localities
Defense Industrial Base: 250,000 firms in 215 distinct industries
Telecommunications: two billion miles of cable
Energy
    Electricity: 2,800 power plants
    Oil and Natural Gas: 300,000 producing sites
Transportation (Figure 16-1)
    Aviation: 5,000 public airports
    Passenger Rail and Railroads: 120,000 miles of major railroads

FIGURE 16-1 The transportation system is one of several critical infrastructure sectors.

Highways, Trucking, and Busing: 590,000 highway bridges
Pipelines: two million miles of pipelines
Maritime: 300 inland/coastal ports
Mass Transit: 500 major urban public transit operators
Banking and Finance: 26,600 FDIC-insured institutions
Chemical Industry and Hazardous Materials: 66,000 chemical plants
Postal and Shipping: 137 million delivery sites
Key Assets
National Monuments and Icons: 5,800 historic buildings
Nuclear Power Plants: 104 commercial nuclear power plants
Dams: 80,000 dams
Government Facilities: 3,000 government-owned/operated facilities
Commercial Assets: 460 skyscrapers

The **National Infrastructure Protection Plan** (NIPP) (U.S. Department of Homeland Security, 2009) meets the requirements of HSPD-7 by identifying CI, prioritizing protection, and facilitating a single national effort to protect CI. As the NIPP advanced, information technology and critical manufacturing were added as CI sectors; education became a subsector of government facilities (U.S. Department of Homeland Security, 2010b). As of early 2012, the NIPP of 2009 was still in use and on the DHS website.

The Role of Government in Protecting Critical Infrastructure and Key Resources (CIKR)

HSPD-7 designates a federal sector-specific department or agency for each sector to coordinate and collaborate with relevant agencies, state and local government, and the private sector. The DHS is responsible for (1) developing a national CIP plan consistent with the Homeland Security Act of 2002, (2) recommending CIP measures in coordination with public and private sector partners, and (3) disseminating information.

The term **sector-specific agencies** (SSAs) has been used to identify federal departments and agencies with protection responsibilities for specific sectors of CI. SSAs are required to develop **sector-specific plans** (SSPs). SSAs are listed next (U.S. Department of Homeland Security, 2009: 3; U.S. Department of Homeland Security, 2005a: 3; U.S. Department of Homeland Security, 2006a: 20; U.S. Government Accountability Office 2005a: 75–76).

- *Department of Homeland Security (DHS):* Responsible for emergency services; government facilities; information and telecommunications; transportation systems (with Department of Transportation); chemicals; postal and shipping sectors; dams; commercial facilities; and nuclear reactors, materials, and waste (with the Department of Energy and the Nuclear Regulatory Commission). Examples of specific functions include protection of federal property throughout the country by the Federal Protective Service, the Secret Service's role in coordinating site security at designated special events, and the National Cyber Response Coordination Group's role in coordinating national cyber emergencies.
- *Department of Defense (DOD):* Responsible for defense industrial base and physical security of military installations, activities, and personnel.
- *Department of Energy (DOE):* Responsible for developing and implementing policies and procedures for safeguarding power plants (except for commercial nuclear power facilities), oil, gas, research labs, and weapons production facilities.
- *Department of Justice (DOJ):* Through its Criminal Division and the FBI, the DOJ works to prevent the exploitation of the Internet, computer systems, and networks.
- *Department of State (DOS):* Responsible for matters of international CIP, given its overseas mission.
- *Department of Health and Human Services (DHHS):* Responsible for public health, healthcare, and food (other than meat, poultry, and egg products).
- *Environmental Protection Agency (EPA):* Responsible for drinking water and wastewater treatment systems.
- *Department of Agriculture:* Responsible for agriculture and food (meat, poultry, and egg products).
- *Department of the Treasury:* Responsible for banking and finance.
- *Department of the Interior:* Responsible for national monuments and icons.

The term *critical infrastructure/key resources* has replaced the term *critical infrastructure/ key assets* in the NIPP. As defined in the Homeland Security Act of 2002, **key resources** are publicly or privately controlled resources essential to the minimal operations of the economy

FIGURE 16-2 The Statue of Liberty National Monument is a key asset of national pride requiring protection. *Courtesy: DOD photo by Petty Officer 3rd class Kelly Newlin, U.S. Coast Guard.*

and government. **Key assets** are individual targets (Figure 16-2) whose destruction could cause large-scale injury, death, or destruction of property, and/or profoundly damage national prestige and confidence (U.S. Department of Homeland Security, 2009; U.S. Department of Homeland Security, 2006b).

NIPP objectives include protecting critical infrastructure and key resources (CIKR); managing long-term risk; assessing vulnerability; prioritizing; maintaining a national inventory of CIKR; mapping interdependencies among assets; articulating security partner roles and responsibilities; identifying incentives for voluntary action by the private sector; facilitating best practices, information exchange, training, and metrics; strengthening linkages between physical and cyber efforts; and promoting CIKR cooperation among international partners.

HSPD-7 requires that the NIPP be integrated with other national plans. During incident response, NIPP information will be shared through the National Incident Management System (NIMS) established communications mechanisms.

NIPP Risk Management Framework

The cornerstone of the NIPP is risk management applicable to "all-hazards" (U.S. Department of Homeland Security, 2009: 27). The U.S. Government Accountability Office (2007), the audit and investigative arm of Congress, notes that "risk management, a strategy for helping policymakers make decisions about assessing risk, allocating resources, and taking actions under conditions of uncertainty, has been endorsed by Congress, the President, and the Secretary of

DHS as a way to strengthen the nation against possible terrorist attacks." The NIPP risk management framework includes the following activities (U.S. Department of Homeland Security, 2009: 28; U.S. Department of Homeland Security, 2006a: 29–30):

- *Set security goals:* Define specific outcomes, conditions, end points, or performance targets that collectively constitute an effective protective posture.
- *Identify assets, systems, networks, and functions:* Develop an inventory of the assets, systems, and networks, including those located outside the United States, that comprise the nation's CIKR.
- *Assess risks:* Determine risk by combining potential direct and indirect consequences of a terrorist attack or other hazard.
- *Prioritize:* Establish priorities based on risk, and determine protection and business continuity initiatives that provide the greatest mitigation of risk.
- *Implement protective programs:* Select sector-appropriate protective actions or programs to reduce or manage the risk identified; secure the resources needed to address priorities.
- *Measure effectiveness:* Use metrics and other evaluation procedures at the national and sector levels to measure progress and assess the effectiveness of the national CIKR protection program in improving protection, managing risk, and increasing resiliency.

Both output and outcome metrics are used by SSAs and the DHS to track progress on specific activities outlined in the SSPs. Output metrics include the number of vulnerability assessments performed by a certain date. Outcome metrics include a reduced number of facilities assessed as high risk, following the implementation of protective measures. The DHS notes that selecting outcome metrics for protection programs is challenging because risk reduction is not directly observable. For example, it is difficult to measure the prevention of a terrorist attack or the extent of mitigation from a potential attack.

The Critical Infrastructure Risk Management Enhancement Initiative is a DHS program to ensure that NIPP and resilience activities achieve successful outcomes. This program emphasizes the ability to clearly understand CI risks, use of metrics, and adjusting resources (i.e., budgeting) where they are most needed (Keil, 2011).

NIPP Recommendations for the Private Sector

Kochems (2005: 3) describes the responsibilities of the federal government and the private sector to protect against terrorism:

> *The federal government—not the private sector—is responsible for preventing terrorist acts through intelligence gathering, early warning, and domestic counterterrorism. The private sector is responsible for taking reasonable precautions, much as it is expected to take reasonable safety and environmental precautions. The federal government also has a role in defining what is "reasonable" as a performance-based metric and facilitating information sharing to enable the private sector to perform due diligence (e.g., protection, mitigation, and recovery) in an efficient, fair, and effective manner.*

Research by Hayes and Ebinger (2011) refer to tension between the private sector and government over the amount of protection required to protect CI. They write that, although many factors influence spending on protection by the private sector, such as the ability to assess risk and government-sponsored incentives to the private sector, the most direct factors are cost-benefit analysis and ROI. Hayes and Ebinger emphasize the importance of government in providing tax and other incentives and a more aggressive regulatory framework.

Koski (2011) argues that CIP has been a monumental effort by the federal government in organizing partnerships and promoting the sharing of information. He writes that participation is clustered in sectors resulting from Clinton-era CIP policy and benefits from DHS funding. Koski adds that CIP policies are broad and there is a need for policy "glue" (i.e., incentives) for partnering among all sectors.

The following recommended homeland security practices for the private sector are summarized next from the NIPP (U.S. Department of Homeland Security, 2009: 167–169). The recommendations are based on best practices in use by various sectors.

- Asset, System, and Network Identification:
  - Incorporate the NIPP framework for the assets, systems, and networks under their control.
  - Voluntarily share CIKR-related information with appropriate partners.
- Assessment, Monitoring, and Reduction of Risks/Vulnerabilities:
  - Conduct appropriate risk and vulnerability assessment activities.
  - Implement measures to reduce risk and mitigate deficiencies and vulnerabilities.
  - Maintain the tools to detect possible insider and external threats.
  - Develop and implement personnel screening programs to the extent feasible for personnel working in sensitive positions.
  - Manage the security of information systems.
- Information Sharing:
  - Connect with and participate in the appropriate national, state, regional, local, and sector information-sharing mechanisms.
  - Develop and maintain close working relationships with local (and, as appropriate, federal, state, territorial, and tribal) law enforcement and first-responder organizations.
- Planning and Awareness:
  - Develop and exercise appropriate emergency response, mitigation, and business continuity-of-operations plans.
  - Participate in federal, state, local, or company exercises and other activities to enhance individual, organization, and sector preparedness.
  - Enhance security awareness and capabilities through periodic training, drills, and guidance that involve all employees annually to some extent and, when appropriate, involve others such as emergency response agencies or neighboring facilities.
  - Encourage employee participation in community preparedness efforts.

Sharing Information on Critical Infrastructure

During 2012, the DHS continued to facilitate a variety of information-sharing programs. Congress promoted information sharing through the Intelligence Reform and Terrorism Protection Act of 2004, also known as the 9/11 Reform Bill. It requires the president to facilitate an "information sharing environment" (ISE) with appropriate levels of government and private entities.

The Homeland Security Information Network (HSIN), an Internet-based tool that uses encryption and a secure network, facilitates collaboration between government agencies and the private sector. HSIN is made up of various groups such as CI sectors, emergency management, law enforcement, and intelligence. HSIN capabilities include 24/7 availability, training, libraries, web conferencing, and incident reporting.

The DHS/Information Analysis and Infrastructure Protection Daily Open Source Report is a Monday through Friday e-mail report divided by CI sectors. It is intended to educate and inform personnel engaged in the protection of CI. The content includes a variety of news reports on CI from many media outlets and information bulletins.

Sector Coordinating Councils (SCCs) are groups within a sector that serve as a point of entry for government to partner with the private sector on issues. The Government Coordinating Council (GCC) is the government counterpart to the SCC for each sector to facilitate interagency coordination.

Federally funded "Centers of Excellence" at universities share information on research of protection issues, best practices, technologies, behavioral aspects of terrorism, and all-hazards. These sites also offer training and undergraduate and graduate degrees relevant to CI.

As covered in the previous chapter, Information Sharing and Analysis Centers (ISACs), established by DHS, promote the sharing of information and work among CI sectors. These are voluntary organizations among CI sectors, such as food, electricity, and financial services. Industry associations (e.g., American Chemistry Council) often take the lead in forming ISACs and arrange meetings, establish websites, provide technical assistance, and issue warnings. The level of activity of ISACs varies. Fusion Centers and OSAC, explained in the previous chapter, also promote the sharing of information.

The Infrastructure Security Partnership (TISP) is a private sector group organized to bring together a variety of public and private sector organizations to collaborate on all-hazards facing the nation's built environment. TISP facilitates dialogue among those in the design and construction industries. A major goal of TISP is to improve security in the nation's CI. It promotes the transfer of research to codes, standards, and public policy.

One private sector concern is the possibility of a leak of sensitive information from the information-sharing process. For example, if a company reports a cyberattack and the information is reported in the media, profits may suffer. Also, private sector proprietary information must be protected because it is valuable.

Do you think businesses are reluctant to share critical infrastructure information with government? Why or why not?

Critical Infrastructure Sectors

Agriculture and Food

The U.S. agriculture and food industry accounts for almost one-fifth of the gross domestic product (GDP) and a significant percentage of that figure contributes to the U.S. export economy. About one in six jobs are linked to agriculture in this trillion-dollar industry. National and international confidence in the safety of its products is important. The greatest threats to it are disease and contamination. This sector is dependent on the water, transportation, energy, chemical, and other sectors (U.S. Departments of Homeland Security and Agriculture and U.S. Food and Drug Administration, 2010; Olson, 2012: 1).

"The use of natural agents in attacks on agriculture or directly on people is commonly described as **bioterrorism**" (Congressional Budget Office, 2004: 39). Olson (2012: 1) defines **agroterrorism** as "the deliberate introduction of an animal or plant disease for the purpose of generating fear, causing economic losses, or undermining social stability." Terrorism, disease, and disasters in this sector can cause serious health problems and panic.

An attack on the food supply can take place at numerous points between the farm and human consumption. Examples are farms, processing plants, retail stores, and restaurants. Whether domestic or imported, food could be subject to biological or chemical agents. An illustration of a terrorist attack on food occurred in 1984 when members of an Oregon religious group, who followed an Indian-born guru named Bhagwan Shree Rajneesh, secretly applied salmonella bacteria to a restaurant salad bar to poison residents of a community to influence an election. Over 750 people became ill. Disgruntled employees have caused other incidents of food sabotage.

The Centers for Disease Control (2011) estimates that "each year one in six Americans (or 48 million people) get sick from and 3,000 die of foodborne diseases." In one case, a CDC "disease detective" focused on shopper cards (i.e., what is swiped at the grocery store) of people in 40 states who fell ill, to identify one grocery store chain and the salami that the people who got ill had bought and ate.

Although unintentional, the 2001 outbreak of foot-and-mouth disease in Britain showed the high expense from disease in the food chain. About four million cattle were destroyed, and the cost to the British economy (primarily agriculture and tourism) was about $48 billion (Gewin, 2003).

The primary federal regulatory agencies responsible for food safety are the Food and Drug Administration (FDA), the U.S. Department of Agriculture (USDA), and the Environmental Protection Agency (EPA). Inspectors from the DHS's Bureau of Customs and Border Protection (CBP) assist other agencies with inspecting imports. States also conduct inspections.

The FDA has regulatory authority over about 80 percent of the nation's food supply. The USDA's Food Safety and Inspection Service regulates the safety, quality, and labeling of most meats, as well as poultry and egg products. The EPA's role is to protect public health and the environment from pesticides (Congressional Budget Office, 2004: 42–43).

Laws that defend the agriculture and food system against terrorism, disasters, and emergencies include the Homeland Security Act of 2002. Another law, the **Public Health Security and Bioterrorism Preparedness and Response Act of 2002** (Bioterrorism Act), bolsters the FDA's ability to identify domestic and foreign facilities that provide food, inspect food imports, and notify food businesses that become involved in the contamination of food supplies. Regulations under the authority of this law require food businesses in the supply chain to maintain records to identify the immediate previous sources and subsequent recipients of food. HSPD-9 directs HHS, USDA, and EPA to increase their efforts in prevention, surveillance, emergency response, and recovery. The **Project BioShield Act of 2004** requires action to expand and expedite the distribution of vaccines and treatments to combat potential bioterrorism agents (Congressional Budget Office, 2004: 43).

Gardner (2011) argues that the Bioterrorism Act has good intentions, but lacks "teeth." For instance, more FDA inspections of imported foods are hindered by limited funding. Another challenge is communications among agencies; for example, the CBP is unaware of 68 percent of imported food refused by the FDA. He writes that, although multiple agencies inspect imported food, about 98 percent are uninspected and susceptible to tampering.

Protection issues and strategies for the agriculture and food industry include the following (U.S. Departments of Homeland Security and Agriculture and U.S. Food and Drug Administration,2010; U.S. Food and Drug Administration, 2012; U.S. Department of Agriculture, 2011; Chalk, 2003; Congressional Budget Office, 2004: 44):

- Apply risk management, protection, and resiliency methods.
- Establish more extensive labeling and tracking systems for animals and food products to help pinpoint sources of contamination.
- Establish full tracking of ownership of the most hazardous materials (e.g., nitrates) that can be used as weapons.
- Establish enhanced incentives and protections for reporting new incidents of food contamination, improper sales of hazardous materials, unsafe processing and handling procedures, and incomplete inspections.
- Modify vulnerable practices in the industry.
- Increase the number of inspectors in the food chain and strengthen the capabilities of forensic investigations for early detection and containment.
- Foster improved links among the agriculture, food, and intelligence communities.
- Continue research on vaccinations.
- The FDA established the Food Defense and Emergency Response that includes preventive strategies and training for the food industry.
- The USDA Food Safety and Inspection Service (FSIS) offers training and a variety of information including notices, regulations, emergency resources, and security guidelines. It also maintains a Food Defense and Emergency Response.
- The Grocery Manufacturers Association maintains a food safety education and training program.

Water

As we know, water is essential to our lives, and its protection is vital. The water sector has two major components: fresh water supply and wastewater collection and treatment. The United

States has 160,000 public water systems that depend on reservoirs, dams, wells, aquifers, treatment systems, pumping stations, and pipelines. Also in the United States are about 16,255 publicly owned wastewater treatment works. This sector also includes storm water systems that collect storm water runoff (U.S. Department of Homeland Security and U.S. Environmental Protection Agency, 2010; U.S. Environmental Protection Agency, 2011).

To prioritize protection, the water sector studied four primary areas of possible attack (White House, 2003a: 39):

1. Physical damage to assets and release of toxic chemicals used in water treatment
2. Contamination of the water supply
3. Cyberattack on water information systems
4. Interruption of service from another CI sector

The U.S. Government Accountability Office (2005a: 87–88) reported similar vulnerabilities. It noted the challenges of protection include a lack of redundancy in vital systems; this increases the probability that an attack will render a water system inoperable. The GAO recommended that federal funds for utilities focus on high-density areas and those areas serving critical assets (e.g., military bases). Recommendations also included technological upgrades in physical security and monitoring of water systems.

Threats to water systems are varied. In one community, security was increased when vandals victimized a water reservoir. The offenders broke into a control room, opened two valves, and began to drain the water out of the reservoir. Making matters worse, they opened another valve and flooded the control room. A sensor signaled the drop in the water level of the reservoir. Following this event, an intrusion alarm system was installed at access points. Another threat is the disgruntled employee. This may be a more serious threat than outsiders because employees are knowledgeable of system weaknesses and security measures. However, almost anybody has access to thousands of miles of pipe, and an offender can patch into a neighborhood line and introduce a poison after the water has been treated and tested for safety. This type of attack could sicken a neighborhood or people in an office building. Experts argue that to poison a reservoir or other large water source would require truckloads of poisons or biological agents. If an airplane containing anthrax crashed into a water source, this attack is likely to fail because of fire or the inability of the anthrax to diffuse effectively (Winter and Broad, 2001).

HSPD-7 confirmed EPA as the lead agency for coordinating the protection of drinking water and water treatment systems. Under HSPD-7, the Water Security Division of the EPA is assigned the responsibility of preparing a water sector plan for the NIPP. The plan must identify assets, prioritize vulnerabilities, develop protection programs, and measure effectiveness. Other HSPDs relevant to the EPA are HSPD-8 (strengthens preparedness), HSPD-9 (surveillance program and laboratory network for early warning of an attack), and HSPD-10 (biodefense).

The Bioterrorism Act of 2002 addresses, in Title IV, drinking water protection. It requires water systems serving over 3,300 persons to conduct an assessment of vulnerabilities, prepare emergency response measures, and send relevant documents to the EPA (U.S. Environmental Protection Agency, 2012). Other federal laws support safe drinking water.

Protection issues and strategies for the water sector include the following (U.S. Department of Homeland Security and U.S. Environmental Protection Agency, 2010; White House, 2003a: 39–40):

- Protect public health through programs and strategies that focus on the contamination threat and resilience.
- Several methodologies are available to conduct a vulnerability assessment of water utilities. These are available from the EPA, Sandia National Laboratories, the Association of Metropolitan Sewerage Agencies, and other groups.
- As with all sectors, improve protection of digital control systems and supervisory control and data acquisition systems.
- Consider new technology, such as small robots placed in pipes to detect tainted water.
- Secure openings (e.g., manhole covers) to prevent the dumping of substances into systems.
- Properly store and secure chemicals.
- Use redundancy to increase the reliability of systems.
- Increase training and awareness.
- The EPA provides helpful information on security at its website.
- The DHS and EPA work with the water ISAC to coordinate information on threats and other topics of interest to this sector.
- The American Water Works Association (AWWA) is the largest organization of water supply professionals in the world. It is an international scientific and educational society dedicated to improving drinking water. The group offers training and education opportunities, conferences, publications, and guidelines on security (American Water Works Association, 2012). Two other related groups are the National Rural Water Association (NRWA) and the Association of State Drinking Water Administrators (ASDWA).

Protecting Digital Control Systems and Supervisory Control and Data Acquisition Systems

Many industries in the United States have transformed the way in which they control and monitor equipment by using digital control systems (DCSs) and supervisory control and data acquisition systems (SCADAs). These systems are computer-based and are capable of remotely controlling sensitive processes that were once controlled only manually on-site. The Internet is used to transmit data, a change from the closed networks of the past. Many sectors use this technology, including water, chemical, transportation, energy, and manufacturing. Sabotage of such systems can result in safety and health problems and deaths and injuries.

In one case, an engineer used radio telemetry to access a waste management system to dump raw sewage into public waterways and the grounds of a hotel. The offender worked for the contractor who supplied the remote equipment to the waste management system. In another case, during a **penetration test** (i.e., an authorized test of defenses), service firm personnel were able to access the DCS/SCADA of a utility within minutes. Personnel drove to a remote substation, noticed a wireless network antenna, and without leaving their vehicle, they used their wireless radios and connected to the network in five minutes. Within 20minutes, they had mapped the network, including SCADA equipment, and accessed the business network and data.

Methods to protect these systems include computer access controls, encryption of communications, virus protection, data recovery capabilities, and manual overrides. The DHS Industrial Control Systems Cyber Emergency Response Team (ICS-CERT) (U.S. Department of Homeland Security, 2012a), in collaboration with other entities, responds to incidents, conducts vulnerability and forensic analysis, and issues alerts, among other activities.

Energy

Energy, another CI sector, supports our quality of life, economy, and national defense. Presidential Decision Directive 63 (PDD 63), HSPD-3, and HSPD-7, among others, call for cooperation between government and individual infrastructures for increased protection. The Department of Energy (DOE) is designated as the lead agency for the energy sector. The American Recovery and Reinvestment Act of 2009 granted DOE funds to, among other things, enhance electricity reliability and research alternative forms of energy.

The energy sector is often divided into three industries: electricity, petroleum, and natural gas. In addition, this sector facilitates international coordination since it relies on energy and technology from other countries. This sector is dependent on pipelines and several other sectors (e.g., transportation), and virtually all sectors depend on energy (U.S. Departments of Homeland Security and Energy, 2010).

Electricity

The physical system of electricity consists of three major parts. Generation facilities include fossil fuel plants, hydroelectric dams, and nuclear power plants. Transmission and distribution systems link the national electricity grid while controlling electricity as customers use it. Control and communication systems maintain and monitor critical components.

The electricity industry is highly regulated. Regulators include the Federal Energy Regulatory Commission (FERC) and state utility regulatory commissions. The Nuclear Regulatory Commission (NRC) regulates nuclear plants and related activities.

The DOE designated the North American Electric Reliability Council (NERC) as the sector coordinator for electricity. Following the serious power blackout in New York in 1965, the electric industry established NERC consisting of public and private utilities from the United States and Canada. The aim of this group is to develop guidelines and procedures to prevent disruptions of power. This group also coordinates guidelines on physical and cybersecurity.

Serious vulnerabilities exist with the equipment used in the production and transmission of electricity and electronic monitoring devices, because these assets are outdoors and have minimal protection. The National Research Council reported that an assault on any individual segment of a network would likely result in minimal local disruption, but a coordinated attack on key assets could cause a long-term, multistate blackout.

In December 2004, the DHS reported that power company employees in Nevada were on routine patrol when they found that nuts, bolts, and supporting cross members had been removed from five transmission line towers. The company noted that if one tower fell, eight more could be brought down at the same time; however, because of redundancy in the transmission system, loss of electricity would not occur.

An example of a broad, regional disruption is the Northeast–Midwest blackout of August 14, 2003. It affected about 50 million people and took five days to fully restore service, although most customers had power within 24hours. The DOE estimated the total cost of this blackout at $6 billion, mostly from lost income and earnings. This emergency illustrated the importance of preparation. For example, in New York City, many high-rise buildings had no backup generators, traffic lights did not work (causing gridlock), service stations were unable to sell gas (causing vehicles to stop and block traffic), and clean water was in jeopardy because no backup power existed at water facilities. These problems were repeated throughout the region of the blackout, and it has been suggested that codes should be established for emergency backup generators (U.S. Departments of Homeland Security and Energy, 2010: 69; Congressional Budget Office, 2004: 37).

Protection issues and strategies for the electricity sector include the following (U.S. Departments of Homeland Security and Energy, 2010: 3; Congressional Budget Office, 2004: 32–37; White House, 2003a: 50–53):

- The owners and operators of the electric system are a heterogeneous group and data are needed to conduct analyses of this sector's interdependencies.
- Performance metrics have been developed to determine and justify protection investments.
- To increase reliability, this sector is seeking to build a less vulnerable grid, while looking to redundancy in its transmission and distribution facilities.
- To improve recovery from equipment failure (e.g., transformers), utilities should maintain adequate inventories of spare parts, and the design of parts should be standardized.
- DHS and this sector have initiated intra-industry working groups to address security, cybersecurity, and resilience issues.

Powell (2011: 86–92) writes of the "smart grid"—a system to improve energy efficiency by, for example, two-way communication between utilities and consumer devices at homes to manage demand cycles. At the same time, such a system increases vulnerabilities to cyberattacks. Powell contends that although the application of policies and standards can mitigate risks, the challenge is to produce effective policies and standards. He also notes that voluntary action on cybersecurity within CI sectors has not worked and is "a recipe for disaster."

Oil and Natural Gas

Oil and natural gas businesses are closely integrated. The oil industry includes crude oil transport, pipelines, refining, storage, terminals, ports, ships, trucks, trains, and retail stores. The natural gas industry includes production, transmission, and distribution. The oil and natural gas industries face similar challenges as the electric industry. Examples are issues of reliability, redundancy, risk management, and security.

Besides the risks of terrorism, other criminal acts, natural disasters, and accidents, critical infrastructure sectors also face the risk of equipment failure that can disrupt vital services.

Dams

According to the *National Strategy for the Physical Protection of Critical Infrastructures and Key Assets* (White House, 2003a: 76), dams are a key asset and vital to our economy and communities. The Congressional Budget Office (2004: 30–31) reported that about 80,000 dams were listed in the National Inventory of Dams. These structures are major components of CI that provide not only electricity but also water to population and agricultural areas. Failure of a dam can cause loss of life, property damage, serious economic loss, and the lost value of services (e.g., power, water, irrigation, and recreation). Locations downstream are most vulnerable, as in the Johnstown Flood of 1889 that caused 2,200 deaths. In 1976, the failure of the newly built Teton Dam, in Idaho, resulted in 11 deaths, 20,000 homes evacuated, and about $800 million in damages (more costly than the dam itself).

Most dams are small and the federal government is responsible for roughly 10 percent of them. The remaining ones belong to state and local governments, utilities, and the private sector. Under current law, owners are responsible for safety and security.

In 2011, the DHS Office of Inspector General issued a report on the dams sector and noted that failure of certain dams could affect 100,000 people and have an economic impact exceeding $10 billion. The Office found that DHS lacks assurance that risk assessments were conducted and risks mitigated at the most critical dam assets. This challenge of DHS results because it does not have the authority to ensure partnering and participation in protection by the dams sector. DHS has not been able to generate cooperation from certain segments of this sector and it is favoring a legislative proposal for regulatory authority similar to the chemical sector (U.S. Department of Homeland Security, Office of Inspector General, 2011: 22).

Nuclear Power Plants

The *National Strategy for the Physical Protection of Critical Infrastructures and Key Assets* (White House, 2003a: 74–75) views nuclear power plants as key assets that represent U.S. economic power and technological advancement. A disruption of these facilities could have significant impact on the safety and health of citizens and harm the economy.

The Nuclear Regulatory Commission (NRC), an independent agency created by Congress through the Energy Reorganization Act of 1974, regulates and licenses nuclear plants. It also regulates security and safety at the plants (U.S. Nuclear Regulatory Commission, 2011). NRC responsibilities include storage, transportation, and disposal of high-level waste. The nuclear sector does not include DOD and DOE nuclear facilities.

Nuclear power supplies about 20 percent of U.S. electricity from 104 commercial nuclear reactors in 31 states. This source of power has created much controversy (Massachusetts Public Interest Research Group, 2011; Congressional Budget Office, 2004: 9–19; Hebert, 2005). It has been touted as an option to reduce dependence on fossil fuels (i.e., oil, natural gas, and coal) and Middle East oil. Some environmentalists have favored nuclear power to reduce "greenhouse" gases caused by fossil fuels. Opponents refer to nuclear plant disasters such as the horrific **Fukushima nuclear disaster** where three reactors experienced meltdowns, exploded, and released radiation. The Japanese government was not accurate on the extent of the disaster and losses are still being incurred. (See the Chapter 12 coverage of "The Eastern Japan Great Earthquake Disaster of 2011.") In 1986, the **Chernobyl** nuclear plant, in Ukraine, exploded due to design deficiencies and safety problems not present in U.S. reactors. This disaster spread radioactive substances across northern Europe. Residents of Chernobyl had to permanently relocate, and agricultural lands were permanently contaminated. Thirty-one people died in the fire and meltdown, and cancer rates climbed in the area. The 1979 **Three-Mile Island** partial nuclear meltdown in Pennsylvania resulted in some radioactive material escaping into the atmosphere before the reactor could be shut down. The accident was blamed on feed water pumps that stopped, followed by pressure that began to build, and a safety valve that became stuck. About 150,000 people were forced to evacuate. The biggest cost was from the loss of Three Mile Island's Unit 2 that cost ratepayers $700 million to build and $1 billion to defuel and decontaminate. This incident resulted in a halt to nuclear reactor construction in the United States, although recently plants have been under construction again.

Zeller (2011) writes that although the NRC has one of the highest densities of PhDs in government, he cites examples of plant mishaps and views the NRC as being weak with its regulatory authority and slow in responding to serious problems. He also sees the NRC as being too close to the industry it regulates and refers to the NRC as being like a "prep school" for those employees interested in higher-paying jobs in the private sector nuclear industry.

Another controversy over nuclear power resulted from a report from the public interest group Riverkeeper (Welch, 2006: 96). The report focused on the Indian Point nuclear plant located 24 miles north of New York City. The group asserts that a successful terrorist attack could kill 44,000 people, increase cancer deaths, and cost over $2 trillion. In addition, evacuation of 17–20 million people would be very difficult. The report asks why a nuclear plant is located so close to the city. The NRC disputed the contents of the report and accused the group of sensationalism, especially because of the report title, "Chernobyl on the Hudson?"

Terrorists find nuclear plants attractive targets because of the potential for mass casualties and long-term environmental damage. Attacks may involve a 9/11 type of strike using an airliner, internal sabotage by an employee, a truck or boat bomb, an assault by a team aiming to plant an explosive, or a combination of these methods. Alternatives to attacking a reactor consist of planting explosives at nuclear waste or fuel locations to spread radioactive materials. Assaults of nuclear plants have occurred in Spain by Basque separatists, in Russia by Chechens, in South Africa during the apartheid era, and a rocket attack targeting a French nuclear plant. Captured al-Qaida documents in Afghanistan contained diagrams of U.S. nuclear plants.

Debate continues over whether the concrete containment structure of a reactor could withstand a crash by a wide-bodied, fully fueled airliner. Reactors have walls 3½ to six feet thick.

The NRC, in conjunction with other federal agencies, develops physical security criteria, or standards, for nuclear operations. These standards involve defenses against both internal and external threats. A possible internal threat is someone on the inside, with or without cooperation from others, who commits sabotage or theft of special nuclear materials (SNM). SNM can be used to manufacture nuclear weapons. An external threat could be posed by several people who are dedicated to their objectives, are well trained in military skills, and possess weaponry and explosives. An insider may be part of this threat.

FIGURE 16-3 Security officers at a nuclear generating plant.*Courtesy: Wackenhut Corporation. Photo by Ed Burns.*

NRC security criteria stress *redundancy* and *in-depth protection*. Examples are having two barriers and two intrusion alarm systems. A backup is required for every sensor, transmitter, processor, and alarm so terrorists cannot simply cut a line and catch a facility off-guard. Separate monitoring systems operate at nuclear plants. Three specific monitoring systems are security, fire, and safety. The security system is similar to a proprietary central station. A recordkeeping function is essential, especially to comply with stringent NRC recordkeeping regulations.

Title 10, Code of Federal Regulations, Part 73 (U.S. Nuclear Regulatory Commission, 2012), explains elements of an acceptable nuclear security program. It includes physical security organization, supervision, training, physical barriers, intrusion detection, access requirements, communications, testing, maintenance, armed response requirements, and secure transportation of nuclear materials.

NRC regulations require an armed response force present at a nuclear facility at all times to counter an adversary (Figure 16-3). Criteria established by the NRC reinforce adequate selection, training, and weapons. This entails the use of deadly force in self-defense or in the defense of others. Local law enforcement assistance is to be called in the event of an attack. However, because of a response time delay from local authorities, the primary armed defense is expected from the on-site nuclear security force.

The Congressional Budget Office (2004: 16) stated: “Despite the many efforts by nuclear plants since September 11, emergency preparedness that would help to contain losses may not fully reflect the realities of the current terrorist threat.” The U.S. Government Accountability Office (2005a: 91–92) noted challenges for the NRC. For example, the NRC does not have a system for better utilizing information from security inspections to identify problems that may be common to all plants.

Do you think security at nuclear power plants is weakened when U.S. Government Accountability Office reports and other reports are available for public access on the Internet and describe general security vulnerabilities and protection methods, even though classified versions of reports have limited circulation? Explain your answer.

Chemical Industry

The chemical sector is an essential component of the U.S. economy that employs almost one million people and generating revenues of over $637 billion annually. This industry provides another illustration of interdependencies among infrastructures and the ways a disruption in one sector can have a cascading affect. Chemical products depend on raw materials, electricity, and transportation, among other needs. The products from this industry include fertilizer for agriculture, chlorine for water purification, plastics for many household and industrial products, and a variety of medicines and other healthcare items. This industry also includes oil and gas production. The chemical sector is the top exporter in the United States, totaling 10 cents out of every dollar. As with other sectors, timely delivery of products is essential for customers. Businesses with "just-in-time" delivery systems maintain a small inventory of chemicals. Municipal water systems often hold only a few days' supply of chlorine for disinfecting drinking water (U.S. Department of Homeland Security, 2010c; Congressional Budget Office, 2004: 21–28; White House, 2003a: 65–66).

Protection of the chemical industry is important not only because of our dependence on its products and for economic reasons, but also because of its flammable and toxic substances and the risks of fire, explosion, theft, and pollution. The EPA is the primary federal agency responsible for protecting the public and the environment from chemical accidents.

A chemical accident that caused widespread losses occurred in 1984 in **Bhopal, India**, where a fatal pesticide was released from a Union Carbide Corporation plant. Nearly 4,000 people were killed, and India's courts ordered the company to pay $470 million in compensation to more than 566,000 survivors and dependents, including thousands of victims who were permanently disabled. When the accident occurred, there was no emergency response, atmospheric conditions were at their worst, building standards were inadequate, and there was no zoning to limit housing near the plant. In the United States, the greatest loss of life from a chemical accident came in 1947 from a ship containing fertilizer that exploded in **Texas City, Texas**, and spread fire to other ships and to industrial facilities. About 600 lives were lost.

In 2010, the **British Petroleum Oil Spill** became the leading accidental maritime oil leak in the history of the petroleum industry. From below a destroyed oil platform, about 4.9 million barrels of crude oil gushed from the sea floor into the Gulf of Mexico for over three months until the well was capped. Eleven men were killed and 17 injured in the platform explosion. Extensive and widespread damage occurred to marine and wildlife habitats and to fishing and tourism businesses. The Gulf continues to be contaminated with oil. British Petroleum established a multi-billion dollar fund for claimants. Another U.S. disaster was the **Exxon Valdez** oil spill in 1989 in Alaska. When this ship ran aground, 240,000 barrels of crude oil, with a value of $25 million, spilled. Exxon paid about $3 billion, mostly for cleanup activities.

Safety in the chemical industry has been strengthened through federal laws. The **Emergency Planning and Community Right-to-Know Act of 1986** resulted from the Bhopal accident and was the first federal initiative to promote safety specifically in chemical facilities. It requires plant operators to participate in community emergency planning, provide information to local planners about chemicals on-site, and notify local authorities of a release.

The Clean Air Act Amendments of 1990 mandated a role for the EPA in risk management planning for the chemical industry. Operators must have programs to prevent, detect, and mitigate releases of chemicals. The Maritime Transportation Security Act of 2002 requires vulnerability assessments for chemical facilities along U.S. waterways. State and local governments are heavily involved in chemical safety through emergency preparedness, and FEMA has provided technical support and training.

Since 2007, the **Chemical Facility Anti-Terrorism Standards** (CFATS) have enabled DHS to enforce, for the first time ever, national chemical facility security standards. DHS identifies facilities that contain "chemicals of interest," classifies them into one of four categories relevant to risk level, requires facilities to conduct risk assessments subject to DHS approval, and then security must be implemented to mitigate vulnerabilities. DHS risk-based performance standards enable flexibility at chemical facilities to accommodate the unique characteristics of each facility. Hodgson (2011: 60) writes that this flexibility permits facilities to choose security systems while providing wider opportunities for security businesses to bid on security systems. The DHS will inspect, audit, and penalize to ensure compliance with CFATS (U.S. Department of Homeland Security, 2011a).

Communications

The importance of the communications sector is illustrated when we are unable to use our cell phones. The communications sector has consistently provided reliable and vital communications. It provides voice and data service to public and private customers through various networks. Components of these systems include nearly two billion miles of fiber-optic and copper cables, besides cellular, microwave, and satellite technologies.

This sector is linked to the energy sector that provides power to operate cellular towers, the Internet and other critical services; the information technology sector for critical Internet-linked controls; the banking sector for business transactions; the emergency services sector for 911 calls and other dispatches; and the postal sector for communications on shipments.

The DHS National Communications System (NCS) is the SSA for the Communications Sector. The SSP, a public-private collaboration, includes protection and resilience in an all-hazards environment with special consideration for cybersecurity. As in other sectors, government and business leaders may have different perspectives on risks, security, and reliability. Protection issues and strategies in this sector include the following (U.S. Department of Homeland Security, 2010d; White House, 2003a: 48–49):

- As with other sectors, because interdependencies could result in a cascading effect, major concerns are redundancy, diversity, and resilience.
- The varieties of risk include adverse weather, unintentional cable cuts, and crime.
- Most of the communications sector is privately owned, requiring DHS to work with the private sector and industry groups to understand vulnerabilities and develop countermeasures.
- The DHS works with other nations to consider innovative communications paths to strengthen the reliability of this sector globally.

Information Technology

Threats

The threats to IT systems must be a high priority for mitigation to ensure the survival of our nation. Snow (2011), of the FBI, notes that with enough time, motivation, and funding, a determined offender will probably penetrate any system connected to the Internet. In reference to the costs of cybercrime, Snow writes that we cannot accurately define and calculate losses. The estimates of losses in the last five years range between millions to hundreds of billions of dollars. Snow cites a 2010 study by the Ponemon Institute that showed the median annual cost of cybercrime to a single organization ranges between $1 million and $52 million. Henry (2011), also of the FBI, looked to the 2010 Norton Cybercrime Report that put global annual losses to cybercrime at nearly $400 billion with over one million victims every day. He referred to one company victimized by a cyber-intrusion estimating that it lost 10 years of research and development work valued at about $1 billion dollars in one night.

There are many types of threats to cyberspace and information technology, and an all-hazards approach is best for protection. We often think about hackers attacking computer systems; however, the threats and hazards are broad and include disgruntled employees, errors, natural disasters, and accidents. Earlier chapters cover topics such as internal threats, business continuity, and resilience. Here an emphasis is placed on cyberattacks.

Cyberattacks offer several advantages to offenders. They include no physical intrusion, safety for the offender, no significant funding required, possible profit, usually no state sponsor, immense challenges for IT specialists and investigators, and enormous potential harm to victims. Computer crimes can be divided into four categories according to Taylor et al. (2006: 9–15). These categories are as follows:

- *Computer as a target:* This includes the attacker who alters data. A business can be harmed if, say, decisions are made based on the altered data. Another example is denying use of the system by legitimate individuals. Among the variations of this example is the **denial-of-service attack**. This networking prank initiates many requests for information to clog the system, slow performance, and crash the site. It may be used to cover up another cybercrime. Defacement of websites is another example in this category, which may be referred to as malicious acts.
- *Computer as instrument of a crime:* In this category, the offender uses the computer to commit a crime. Examples are theft, fraud, and threats.
- *Computer as incidental to a crime:* This category involves the use of a computer to facilitate and enhance crimes. For example, computers that speed transactions aid money laundering. In addition, offenders use computers to maintain records of their illegal enterprises.
- *Crimes associated with the prevalence of computers:* This category includes crimes against the computer-related industry and its customers. An example is software piracy.

Common terms associated with cyber threats follow. Experts differ on the definitions of such terms. According to Smith (2013: 12), a **compromised system** is no longer trustworthy to use because security methods many be disabled, proprietary information may be at risk, and unauthorized software may be secretly installed for a variety of harmful purposes. **Malware** (malicious software) is a general term referring to programs that create annoying or harmful actions. Often masquerading as useful programs or embedded into useful programs so users activate them, malware includes Trojan horses, viruses, worms, and spyware. A **Trojan horse** is a program, unknown to the user, which contains instructions that exploit a known vulnerability in software. Smith (2013: 440) explains that **viruses** and **worms** copy themselves and spread. He adds that "while a virus may infect USB drives, diskettes, and application programs, worms infect host computers." A worm will spread itself across networks and the Internet. Defenses include behavior blockers that stop suspicious code based on behavior patterns rather than signatures and applications that quarantine viruses in shielded areas. A **logic bomb** contains instructions in a program that creates a malicious act at a predetermined time or if the offender (e.g., employee) is not able to deactivate it on a regular basis. Programs are available that monitor applications seeking to change other applications or files when such a bomb goes off.

Dumpster diving involves searching garbage for information, sometimes used to support **social engineering** (i.e., using human interaction or social skills to trick a person into revealing sensitive information). An example of social engineering is a person being convinced to open an e-mail attachment or visit a malicious website. Another example is a hacker telephoning a corporate employee and claiming to be a corporate IT technician needing an access code to repair the system.

The U.S. Government Accountability Office (2005c) studied spam, phishing, and spyware threats to federal IT systems, and its findings have relevance to state and local governments and the private sector. Explanations of each threat follow.

Spam is the distribution of unsolicited commercial e-mail. It has been a nuisance to individuals and organizations by inundating them with e-mail advertisements for services, products, and offensive subject matter. Spammers can forge an e-mail header so the message disguises the actual source. The spam problem is made worse because it is a profitable business. Sending spam is inexpensive and sales do result.

As with other security methods through history, adversaries constantly seek methods to circumvent defenses. This is an ongoing "cat and mouse" competition. Anti-spam measures have caused spammers to design techniques to bypass detection and filtration. Spammer techniques include using alternate spelling, disguising the addresses in e-mails, and inserting the text as an image so a filter cannot read it. Compromised systems are regularly being used to send spam, making it difficult to track the source of spam.

Phishing is a word coined from the analogy that offenders use e-mail bait to fish for personal information. The origin of the word is from 1996 when hackers were stealing America Online (AOL) accounts by scamming passwords from unsuspecting AOL customers. Hackers often replace *f* with *ph*, and thus, the name *phishing* developed.

Phishing often uses spam or pop-up messages that trick people into disclosing a variety of sensitive identification information (e.g., credit card and Social Security numbers and passwords). For example, one ploy is for a phisher to send e-mails appearing as a legitimate business to potential victims. The e-mail requests an "update" of ID information or even participation in "enhanced protection against hackers, spyware, etc." Phishing applies a combination of technical methods and social engineering.

Zeller (2006) reports that **key logging programs** copy the keystrokes of computer users and send the information to offenders. These programs exploit security weaknesses and examine the path that carries information from the keyboard to other parts of the computer. Sources of the key logging programs include web pages, software downloads, and e-mail attachments. Since these programs are often hidden inside other software and infect computers, they are under the category of Trojan horses.

Spyware lacks an accepted definition by experts and even proposed legislation (U.S. Government Accountability Office, 2005c: 31). The definitions vary depending on whether the user consented to the downloading of the software, the types of information the spyware collects, and the nature of the harm. Spyware is grouped into two major purposes: advertising and surveillance. Often in exchange for a free service (e.g., allegedly scanning for threats), spyware can deliver advertisements. It can collect web surfing history and online buying habits, among other information. Other types of software are used for surveillance and to steal information. Consumers find it difficult to distinguish between helpful and harmful spyware.

Spyware is difficult to detect, and users may not know their system contains it. Spyware typically does not have its own uninstall program, and users must remove it manually or use a separate tool. Some types of spyware install multiple copies, and they can disable antispyware and antivirus applications and firewalls.

The potential for financial gain has caused spammers, malware writers, and hackers to combine their methods into blended cyber threats. Security analysts are seeing an increase in blended threats and destructive payloads. **Blended cyber threats** combine the characteristics of different types of malicious code to bypass security controls. The U.S. Government Accountability Office (2005c) and the National Institute of Standards and Technology have advised agencies to *use a layered security (defense-in-depth) approach, including strong passwords, patches, antivirus software, firewalls, software security settings, backup files, vulnerability assessments, and intrusion detection systems.*

Not only could a blended cyberattack cause harm, but also a **blended cyber-physical attack** can aggravate damage and hamper recovery. An example of a blended cyber-physical attack is penetrating a DCS/SCADA to alter a manufacturing process to cause a fire and then physically attacking a fire brigade and their equipment.

Smith (2013: 440) explains that many malware packages produce and operate botnets for financial gain. Snow (2011) explains that cyber criminals are business savvy. In reference to botnets, criminal groups hire programmers who write the malicious software, salespeople who sell or lease it, and support personnel to service customers. A **botnet** is produced when a hacker infiltrates a host computer via a Trojan, worm, or virus and it is hidden while providing a "backdoor" to control a computer which is referred to as a bot. "Networks of bots" refers to personal computers (PCs) infected with malicious software that enables the hacker to control the PCs. "Internet bots" operate automated tasks over the Internet. Bots perform repetitive tasks at a much higher rate than humans perform and use automated script to fetch, analyze, and file information from web servers.

Acohido and Swartz (2006) write about hackers who use botnets, as described next. From his home in Downey, California, Jeanson Ancheta, a 19-year-old high school dropout, controlled thousands of compromised PCs, or bots, which earned him enough cash to purchase a BMW and spend hundreds of dollars a week on clothes and car parts. However, he was caught by authorities and pleaded guilty to federal charges of hijacking thousands of PCs and selling access to others to spread spam and attack websites. The threat is global, as evidenced by the arrest of Farid Essebar, an 18-year-old resident of Morocco, who was linked to botnets. Tim Cranton, director of Microsoft's Internet Safety Enforcement Team, refers to botnets as the tool of choice for those using the Internet to commit crimes. They assemble networks of infected PCs to acquire cash and are paid for each PC they attack with ads. Although neophytes are slack about covering their tracks, they provide authorities with insight about their methods. In contrast, more sophisticated offenders work with organized crime groups who are more difficult to apprehend. Millions of PCs connected to the Internet globally are controlled by thousands of botnets. Smith (2013: 441) recommends that PC owners maintain up-to-date software, try to avoid using infected USB drives, and exercise caution prior to opening e-mail and attachments.

The *McAfee Threat Report* (Bu et al., 2012) described growth in mobile-based malware and botnets, with an increase in successful prosecutions of cybercriminals. Not only did Android become the largest target for mobile malware, an app allows penetration of a PC from a phone or tablet. Bu et al. report that data breaches have risen rapidly via hacking, malware, fraud, and insiders. During the third quarter, McAfee Labs recorded an average of 6,500 bad websites per day. Cyberattacks on infrastructures during the quarter targeted the South Houston Water and Sewer Department, a hospital in Georgia that was forced to stop admitting patients, and an ambulance communications system in New Zealand.

Hacktivism refers to politically motivated electronic protesting. Groups such as Anonymous use cyberspace to protest and commit cybercrimes. This group does not have a leader, but relies on collective, individual action in cyberspace such as the Distributed Denial of Service attacks against the recording and motion picture industries and various businesses. In 2011, Anonymous hacked into a U.S. security company with government contracts, stole thousands of e-mails, and posted them online. This was in retaliation for the company identifying members of Anonymous (Snow, 2011). Hacktivist activity also includes publishing police officer personal and family information on the Internet. In 2012, the FBI arrested a core member of Anonymous and one of the world's most-wanted cybercriminals. Hector Xavier Monsegur, 28, a self-taught computer programmer and welfare recipient, resided in a public housing project in New York City. When arrested, he became an informant and helped build cases against other offenders in the United States and Europe who hacked into corporate and government websites (Associated Press, 2012).

DiLonardo (2011: 66) sees exploitation on sites like Twitter because of abbreviated URLs (i.e., Uniform Resource Locators that are the addresses of documents on the Internet) that make it easier for criminals to direct users to bad websites. Noting that many cell phones, TVs, DVDs, iPads, and other devices are also web browsers, DiLonardo writes that this technology provides more opportunities for hackers.

Cyberwarfare

Cyberwarfare refers to a nation penetrating another nation's IT systems for gain and superiority. It can take several forms such as espionage or sabotage, or to degrade armed forces in conventional warfare. These activities are crucial for national survival. The U.S. Cyber Command defends U.S. military IT systems while focusing on the systems of other countries. The U.S. National Security Agency, a huge spy agency, listens to communications and penetrates foreign IT systems. Other nations have similar organizations. The DHS protects government in general and works with the private sector to protect critical infrastructures. Dilanian (2011) reports that U.S. officials said that China has laced the U.S. power grid and other systems with hidden malware capable of creating a disaster. At the same time, there are examples of damage to U.S. adversaries, such as the computer worm that harmed Iran's nuclear program.

Cybersecurity

Although the public and private sectors have for many years emphasized the importance of protecting cyberspace, many experts have warned repeatedly that more needs to be done about this challenge. For instance, in 2010, a U.S. Senate committee heard from government and industry experts on the potential harm from cyberattacks. The experts warned that the United States would be defeated by an all-out cyberwar unless the government takes action to reduce vulnerabilities. In addition, it was noted that government action may not occur until a catastrophic event (Mills, 2010).

Cybersecurity focuses on methods to counter threats from a host of offenders and others who harm cyberspace and IT and create losses. Numerous authorities and national strategies support the IT sector and its mission to secure cyberspace. Examples include the Homeland Security Act of 2002 and HSPD-7. Executive Order 13231 (amended by EO 13286) authorizes a protection program for cyberspace and the physical assets that support such systems. This order was complemented by the Federal Information Security Management Act that requires each federal agency to develop, document, and implement an agency-wide information security program. National strategies include *The National Strategy for Homeland Security* and *The National Strategy to Secure Cyberspace* (U.S. Department of Homeland Security, 2010e; U.S. Government Accountability Office, 2006: 4; White House, 2003b). In 2011, DHS published *Blueprint for a Secure Cyber Future: The Cybersecurity Strategy for the Homeland Security Enterprise* (U.S. Department of Homeland Security, 2011b). This publication specifies numerous guiding principles to enhance cybersecurity against, for example, multiple types of threats, and that CI must protect against advanced and persistent breaches of IT systems.

The DHS established the National Cyber Security Division (NCSD) to address cybersecurity issues and to coordinate a cybersecurity strategy. The NCSD also serves to coordinate a public/private partnership supporting the U.S. Computer Emergency Response Team (US-CERT), ISAC partnerships, and the Cyber Warning Information Network, among other initiatives. The 2010 DHS IT Sector-Specific Plan focused on: (1) prevention and protection through risk management; (2) situational awareness; (3) response, recovery, and resilience of America's IT infrastructure; and (4) continuous improvement of IT sector planning and response (U.S. Department of Homeland Security, 2010e).

The DHS is not the only entity involved in securing cyberspace. Examples of others include the U.S. Department of Justice and its FBI (FBI Cyber Division), the U.S. Secret Service (within DHS), state and local law enforcement agencies, the intelligence community, the military, universities, and many other public and private sector organizations. Because cyber-attacks often have a link to other countries, U.S. law enforcement and other government personnel are stationed overseas and work with similar specialists from host countries.

Cybersecurity is complex and those who work in this field face enormous challenges. Simultaneously, professionals must have vision. Henry (2011), of the FBI, offers the following:

> *Computer security has become an endless game of defense, which is both costly and unsurvivable in the long term if the status quo remains. Going after the threat actor is an absolutely necessary part of the risk equation and one that can be made far more effective with alternate architectures.*
>
> *Under the current environment, victims are often focused on how to get malware off their systems and on finding out what was taken. But what they should be asking is, 'What was left behind? And did it change my data?' Most users have no idea whether their software, hardware, or data integrity has been altered. Our current networks were never designed to detect that type of deviation.*
>
> *Continue the discussion about whether there is a need and enough demand to develop alternate networked environments that rely less on playing defense, and rely more on discovering and capturing threat actors so they change their own risk calculus on whether cyber-crime pays.*

Chittister and Haimes (2006: 1–20) write that the emphasis in IT security is tilted toward short-term tactical measures—such as firewalls and patches—rather than long-term, strategic approaches. They cite the work of Wulf and Jones (2004): "The truth is that we don't know how to build secure information systems." They stress enhanced security through the following principles: (1) cybersecurity must be a high priority, and (2) a realization that risk must be assessed, knowing that hackers will exploit every weakness in the system.

Croy (2005) notes that more entities understand the business value of IT. He writes:

> *Executives are also becoming more directly involved because of government regulations such as Gramm-Leach-Bliley, Sarbanes-Oxley, and HIPAA that hold them personally and legally liable for business issues including access to critical information, financial controls, customer privacy, and physical security.*

Best practices for IT security are varied and helpful to enhance security. However, it is important to note that "best practices" are an ideal state, and reality means that such practices are not universally applicable; each entity is different, and management must make organization-specific decisions (Landoll, 2005: 110).

Among the sources for best practices is ISO/IEC 27002:2005 (International Organization for Standardization, 2011). It is an internationally recognized generic standard for IT security. ISO/IEC 27031:2011 focuses on IT business continuity. Topics in ISO/IEC 27002:2005 include the following:

- Security policy
- Organization of information security
- Asset management
- Human resources security
- Physical and environmental security
- Communications and operations management
- Access control
- Information systems acquisition, development and maintenance
- Information security incident management
- Business continuity management
- Compliance

Following is a list of strategies from a variety of sources to enhance IT security (Smith, 2013; U.S. Department of Homeland Security, 2011b; Computer Security Institute/FBI, 2006; Savage, 2006: 23–28; Sager, 2000: 40; and Thompson, 1997: 25–30):

1. Establish an *IT security committee* to plan and lead through a *risk management* and *resilience* approach. Consider a third-party evaluation of plans.
2. Conduct an *economic evaluation* of security expenditures (e.g., return on investment).
3. Invest in quality *training* because IT security changes rapidly and it is important to ensure that all employees are involved in security.
4. Monitor and track *alert bulletins* and *best practices.*
5. Consider efforts to establish generally accepted benchmarks for securing computer networks. Such standards define levels of security against which an organization can measure itself.
6. Use a *layered approach to security*. This creates multiple "roadblocks" for offenders.
7. Ensure that *IT security and information security* are well coordinated and integrated. Remember that a variety of devices can hold data, including flash storage devices, personal digital assistants (PDA), digital cameras, cell phones, iPods, iPads, Xboxes, and DVD players. In addition, ordinary items such as watches and pens can hold memory cards. And, because of wireless technology, a wireless hard drive can be hidden almost anywhere.
8. Use *policies and procedures* to strengthen security.
9. Provide *physical security* for computer facilities and include servers, desktops, laptops, and other devices and equipment.
10. If designing a computer facility, avoid glass walls or doors, single paths for power to communications lines, uncontrolled parking, underground locations (because of

flooding), multitenant buildings, signs describing the facility, and information about the facility on the Internet.

11. Automatic *access control systems* for a computer facility are popular in combination with limited entrances; the double-door entry concept; visual verification; badge identification systems; and access control according to time, place, and specific personnel. Access controls are required not only for a computer facility, but also for all computers. This includes protection against unauthorized remote access. Biometric access control systems enable identification by fingerprints and so forth. Access to sensitive data must be safeguarded on the premises and from remote locations, even by legitimate computer users.
12. *Passwords* or *codes* are identification procedures that permit access only after the proper code is entered into the computer. The code should be changed periodically. Alarms to signal attempts at unauthorized access should be incorporated into software.
13. *Firewalls* are software and hardware controls that permit system access only to users who are registered with a computer. A firewall sits between a company's internal computer network and outside communications. Firewall products offer a range of features such as file or virus checking, log and activity reports, encryption, security and authentication schemes, and monitoring and alarm mechanisms for suspicious events or network intruders.
14. Use *intrusion detection software* that is like a physical intrusion detection system, only for the network.
15. Be proactive and conduct searches for hacker programs that may be used in an attack. Hackers tend to brag about their successes to the hacker community, so check out sites that attract hackers.
16. Disable unused services. Most software programs include services that are installed by default. These unused services can be a path for hackers.
17. Update software for improved security. Quickly install *security patches*. Software firms develop "patches" for protection when hackers attack their programs. Thus, when a patch is offered to a client, it should be installed as quickly as possible.
18. Use decoy programs that trick hackers into attacking certain sites where they can be observed and tracked while the important sites remain secure.
19. Because wireless networks increase the range of wired networks (by using radio waves to transmit data to other wireless devices), "signal leakage" is a security risk. Protection methods include policies, firewalls, and encryption.
20. *Audit* IT security through frequent, rigorous *vulnerability testing*.
21. Because of the trend of companies concentrating on their core businesses and outsourcing everything else, "managed security services" is an option. Essentially, the work contracted to an outside firm can be as broad as surrendering "enterprise security" to the contractor. This includes security technologies, infrastructure, services, and management. Many questions evolve from this concept. Some examples are: What is to be retained, and what is to be outsourced? How will proprietary information be protected? Who is to be responsible for losses?
22. Keep certain proprietary information off IT systems.

Cloud Computing

Cloud computing is the outsourcing of the delivery of information technology and computer services such as software, data storage, and maintenance. Customers access cloud applications (e.g., data stored at a remote location) through the Internet. The contract provider seeks to deliver quality services at a lower cost than what the customer would pay if they provided the services in-house. Cloud computing is characterized by shared services. Customers have the advantage of flexibility and speed when IT needs change.

McCumber (2011: 70) writes that entities "no longer have to manage data centers, the buildings they occupy or all the specialized IT personal." (Many organizations actually use a hybrid approach and retain personnel and resources.) Since centralized, contract-provided cybersecurity and related resources can improve protection, at the same time, each customer may face a loss of control over data. McCumber notes that although with cloud computing data is no longer in-house at the company's servers and mainframes, and much security has been delegated to the off-site service provider, the in-house IT security executive is ultimately responsible for losses. Bates (2012: 24–26) adds that decisions must be made about who controls IT security. Will it be the user entity, the cloud contractor, a third party, or a combination of the three? The decisions are especially important in light of regulatory and compliance requirements and the user's overall Enterprise Security Risk Management program.

One avenue to enhance security is through the Cloud Security Alliance. This not-for-profit organization promotes best practices and education.

Encryption

Many practitioners in the computer field consider encryption to be the best protection measure for data within a computer or while it is being transmitted. Once the domain of government, encryption has traditionally been used to protect military or diplomatic secrets. During the 1970s, the private sector began marketing encryption products, and with the growth of computers and the Internet, encryption likewise grew.

Encryption consists of hardware or software that scrambles (encrypts) data, rendering it unintelligible to an unauthorized person intercepting it. The coding procedure involves rules or mathematical steps, called *algorithms,* that convert plain data into coded data. This transformation of data is accomplished through what is called a *key*, which is a sequence of numbers or characters or both. The key is used in both transmitting and receiving data. Key security is vital because it is loaded into both ends of the data link (Smith, 2013: 345–374). Encryption tools should be changed periodically because breaches have become something of a game. Rapidly evolving technology has shortened the life of promising encryption systems. Computer expert John M. Carroll (1996: 249) adds that when you become mesmerized by the wonder of some promising crypto device, ask yourself one question: "How much do I trust the person who sold me this gadget?" He extends this question to the international level by claiming that it would be unlikely for any country, even an ally, to provide an encryption system to another without retaining the keys.

Controversy has developed over whether the U.S. government should have the power to tap into every telephone, fax, and computer transmission by controlling keys. From the law enforcement perspective, such control is necessary to investigate criminals, terrorists, and spies. Opponents claim violations of privacy and damage to the ability of American businesses to compete internationally.

The growth of the Internet and business on the Internet has created the need for encryption to secure electronic interactions. Internet users are seeking privacy, confidentiality, and verification of individuals and businesses they are dealing with. The public key infrastructure (PKI) and its authentication and encryption capabilities have evolved as a solution to these needs. Whereas the handshake or handwritten agreement has been tradition for centuries, a modern trend is the digital handshake and signature through the PKI. The PKI addresses three primary security needs: authentication, nonrepudiation, and encryption. The first need verifies an individual's identity. The second need means that an individual cannot deny he or she has provided a digital signature for a document or transaction.

Today, encryption products are widely used to protect public and private information. Modern encryption technology can secure data on mobile devices and encrypt and decrypt e-mail. Such protection is available for data on laptops or other computers. In one case, a hospital IT security director was concerned that physicians were using PDAs containing patient information without proper safeguards to protect data if the device was lost or stolen. Encryption served as the solution to protect the hospital from embarrassment and legal liability. Financial services organizations are especially concerned about the protection of financial transactions as well as storage. The retail and hospitality sectors are concerned about point-of-sale and payment information security (Piazza, 2007: 66–73; Smith, 2006: 10–13).

Defense Industry Base

Private sector defense industries play a crucial role in supporting the U.S. military. This worldwide industrial complex includes thousands of companies and their subcontractors that design and manufacture a wide variety of weapons (e.g., fighter jets, tanks, ships, and small arms), support equipment, and supplies essential to national defense and military operations (Figure 16-4). Also, many companies provide important services (e.g., transportation, IT, maintenance, security, and meals) to the military. The Department of Defense (DOD) is the sector-specific agency for the defense industry base (DIB), and in cooperation with DHS, prepares sector-specific plans (U.S. Departments of Homeland Security and Defense, 2010; White House, 2003a: 45–46).

The DOD, the DHS, and the private sector work to identify DIB protection requirements. These groups, and the intelligence and law enforcement communities, prepare policies and mechanisms to improve the exchange of security-related information. The DOD is also working with defense industry contractors to address national emergency situation requirements, such as response time and supply and labor availability.

For several decades, the DOD has identified its own critical assets and those it depends on in the private sector. For example, private utilities service many military bases. The U.S. Government Accountability Office (2005a: 93) acknowledged that the DOD has taken steps to protect military installations. However, the GAO claims the DOD lacks a single

FIGURE 16-4 Defense industries are a component of critical infrastructure and supply weapons, vehicles, uniforms, and other products and services to support the U.S. military. *Courtesy: DOD photo by Petty Officer 2nd Class Erik A. Wehnes, U.S. Navy.*

organization with authority to manage and integrate installation preparedness and prepare a comprehensive plan.

The DOD operates the **National Industrial Security Program** (NISP). This program integrates information, personnel, and physical security to protect classified information entrusted to contractors. The goal is to ensure that contractors' security programs deter and detect espionage and counter the threat posed by adversaries seeking classified information. According to DOD's Defense Security Service (DSS), which administers the NISP, attempts by foreign agents to obtain information from contractors have increased over the last several years and are expected to increase further. The DSS employs 285 Industrial Security Representatives in the United States and it is responsible for over 13,000 cleared contractor facilities (Defense Security Service, 2012). The U.S. Government Accountability Office (GAO) has frequently "slammed" DSS. For instance, the GAO found that the DSS cannot provide adequate assurances to federal government agencies that its oversight of contractors reduces the risk of information compromise. Remedies include establishing results-oriented performance goals and measures to assess whether DSS is achieving its mission and identifying information that needs to be analyzed to spot trends regarding how contractors are protecting classified information (U.S. General Accounting Office, 2004: 2–4). (The GAO changed its name in 2004.)

The following issues are especially important for national defense. Outsourcing and complex mergers and acquisitions of domestic and foreign companies have produced challenges for the DOD in ensuring that its prime contractors' second-, third-, and fourth-tier subcontractors fulfill supply and security requirements in a national emergency and during peacetime (National Defense Industrial Association, 2011). The U.S. Government Accountability Office (2005b: 3) noted that the DSS oversight of contractors under foreign ownership, control, or influence (FOCI) depends on contractors self-reporting. DSS then verifies the extent of the foreign relationship and works with the contractor on protection measures. The GAO found that the DSS cannot ensure that its approach under FOCI is sufficient to reduce the risk of a foreign nation gaining unauthorized access to U.S. classified information. Violations of FOCI policies were found by the GAO. Remedies include central collection and analysis of information, use of counterintelligence data, research tools, training for DSS staff, and lower turnover.

The National Industrial Security Program (2011) Advisory Committee meeting minutes included the vision of the new DSS director, as described next. The DSS will improve collaboration and communication with its government and industry partners, transparency so partners can understand challenges, the industry security review system, and threat data from intelligence agencies. The DSS will then share the threat data with specific companies.

Career: Government/Industrial Security

Government/industrial security professionals protect a variety of special categories of classified information in accordance with the National Industrial Security Program (NISP). Personnel within this specialty must meet the requirements, restrictions, and other safeguards within the constraints of applicable law and the Code of Federal Regulations necessary to prevent unauthorized disclosure of classified information released by U.S. government departments and agencies to their contractors.

Entry-level management positions require an academic degree from an accredited institution and three to five years of experience in the government or an industrial security field, preferably in a supervisory role. Necessary specialty-specific training or experience includes successful application of security regulations and some exposure to security budgetary issues. Professional certifications or a documented record of involvement in professional security activities or training and education are a plus. Other prerequisites include a proven, positive track record in problem solving and customer relations.

Mid-level management positions require a baccalaureate or advanced degree from an accredited institution and five to ten years of experience in the government or industrial security field with supervisory or previous managerial experience. Necessary specialty-specific training or experience includes professional certification (CPP, PCI, PSP, ISP, CFE, etc.) preferred, or a documented record of professional security training from U.S. government-sponsored training sessions. Other preferred prerequisites include a proven track record in problem solving, policy development, customer relations, successful application of security regulations, and budgetary development.

Modified from source: Courtesy of ASIS International. (2005). "Career Opportunities in Security." www.asisonline.org.

Critical Manufacturing

The critical manufacturing sector is diverse and includes metal, machinery, electrical, and transportation equipment manufacturing that are essential to other sectors. Employing over a million people, this sector contributes over $600 trillion to the U.S. economy. Under HSPD-7, DHS is the sector-specific agency for this sector and coordinates security and resilience with other government agencies and the private sector. The critical manufacturing sector is subject to numerous authorities such as OSHA and CFATS (U.S. Department of Homeland Security, 2010f).

Government Facilities

The government facilities sector, also labeled as key assets, includes a wide variety of buildings owned and leased by all levels of government. These buildings are involved in business, commercial, and recreational activities open to the public. Other government buildings, closed to the public, contain sensitive information, materials, and equipment. Examples of government buildings are general-use office buildings, courthouses, military facilities, and embassies. Facing varied internal and external threats, these buildings require comprehensive protection.

Reese and Tong (2010) of the Congressional Research Service report that the federal government alone manages about 446,000 buildings with an area of more than three billion square feet and a replacement value of $772.8 billion, plus more than 650 million acres of land. This sector also covers buildings operated by over 87,000 local governments.

In the federal system, the **General Services Administration (GSA)** is a primary agency charged with managing federal facilities. The GSA operates most of the major departmental headquarters in Washington, D.C. (e.g., Departments of Justice and Interior), and most of the key multiagency federal office buildings in major cities. The GSA has developed security standards and works with other groups (e.g., DHS) to enhance protection. It must balance security and public access. Other government organizations are involved in protecting facilities, such as the DOD, CIA, and the State Department's Diplomatic Security Service.

The DHS **Federal Protective Service** (FPS) is the SSA for the government facilities sector. It coordinates protection among all levels of government, especially those facilities determined to be nationally critical. The FPS offers a broad range of services, such as patrol, security, investigative, canine, and WMD response. Reese and Tong (2010) write that federal facility security methods have been criticized by government auditors and security experts. The criticism has focused on the use of private security officers, the practices of the FPS, and coordination of federal facility security. Privately owned buildings that house federal tenants present unique challenges that may create friction between federal and nonfederal occupants because of federal laws and regulations.

The FPS and the **U.S. Marshals Service** (USMS) provide law enforcement and security services to federal buildings that house court functions. These groups work together to enhance protection.

The 1995 Oklahoma City bombing led former President Clinton to direct the U.S. Department of Justice (1995) to assess the vulnerability of federal office buildings in the United States. Prior to the study and publication, "Vulnerability Assessment of Federal Facilities," there were no government-wide standards for security at federal facilities and no central database of the security in place at such facilities. Because of its expertise in court security, the USMS coordinated the study that focused on the approximately 1,330 federal office buildings housing about 750,000 federal civilian employees. One major result of the study was the development of standards focusing on perimeter security, entry security, interior security, and security planning. Another result of the study was the division of federal buildings into five security levels, based on staffing size, use, and the need for public access. A Level I building has 10 or fewer federal employees and is a "storefront" type of operation, such as a military recruiting office. A Level V building is critical to national security, such as the Pentagon. Recommended minimum security standards apply to each security level (Reese and Tong, 2010).

The Oklahoma City bombing also led to Executive Order 12977, establishing the DHS Interagency Security Committee (ISC) to address the quality and effectiveness of physical security requirements for federal facilities. Based on HSPD-7, the ISC reviews federal agencies' physical security plans and issues standards, best practices, and threat analyses to be applied by federal security professionals to protect nonmilitary federal facilities.

The ISC, composed of 50 federal departments and agencies, also publishes *Security Specialist Competencies* that explain the range of core competencies federal security specialists can possess to enhance their performance in protecting federal facilities. These competencies establish a common baseline of knowledge and abilities for any agency. Topics in this publication include risk management, security assessments, personnel security, information security, CPTED, physical security, safety, and many others (U.S. Department of Homeland Security, 2012b).

State and local government institutions also are involved in security efforts to protect government buildings, especially at criminal justice facilities. Jails and prisons have traditionally maintained tight security for obvious reasons. Police agencies, especially following the unrest of the 1960s, have strengthened security in and around police buildings, including communications centers, evidence and weapons storage rooms, and crime labs. Following violence in the judicial system, courts have also increased security (Figure 16-5). Because of the 6th Amendment right of defendants to a public trial, court security is especially challenging. Each major component of the criminal justice system—police, courts, and corrections—is

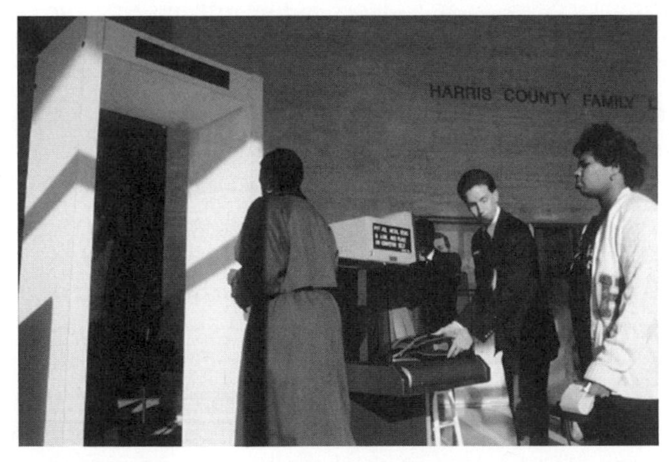

FIGURE 16-5 Government facilities, such as courts, are key assets requiring protection. Walk-through metal detector, left; X-ray scanner, right. *Courtesy: Wackenhut Corporation. Photo by Ed Burns.*

connected with organizations that promote either accreditation or performance measures containing security enhancements.

Transportation Systems

People and goods are moved through our vital transportation system consisting of aviation, maritime, mass transit and passenger rail, highway infrastructure and motor carrier, freight rail, and pipeline. These transportation subsectors are included in the DHS *Transportation System Sector-Specific Plan* (U.S. Department of Homeland Security, 2010g) and explained next.

Interdependencies exist between the transportation sector and nearly every other sector, and a disruption of this sector can have a cascading effect on other sectors. For example, the agriculture and food sector is dependent on the transportation sector to ship goods to customers. The transportation sector is vital to our economy and national security, including mobilization and deployment.

The **Transportation Security Administration (TSA)** and the United States Coast Guard (USCG) (both within DHS) are the sector-specific agencies for this sector in collaboration with the U.S. Department of Transportation (DOT) and numerous partners in state and local governments and the private sector (U.S. Department of Homeland Security, 2010g).

Numerous authorities influence the transportation sector. Examples are as follows. HSPD-7 focuses on CIKR identification, prioritization, and protection. The **Aviation & Transportation Security Act of 2001** established the TSA that has the responsibility of security for all modes of transportation. The **Intelligence Reform and Terrorism Prevention Act of 2004** directly affect the security of public transportation. It requires the DHS to develop a *National Strategy for Transportation Security*. This act contains provisions designed to address many of the transportation and border security vulnerabilities identified and recommendations made by the 9/11 Commission. It includes provisions designed to strengthen aviation security, information sharing, visa issuance, border security, and other areas.

International partnerships are especially important for transportation security. Not only must passengers be protected, but large volumes of products enter the United States every day through the global supply chain via several modes of transportation across oceans and our borders with Canada and Mexico.

Aviation

The aviation system is vast and made up of thousands of entry points. In the United States, it consists of two main parts: (1) airports, aircraft, and supporting personnel and assets; and (2) aviation command, control, communications, and IT systems to support and maintain safe use of airspace.

There are more than 19,800 general aviation and commercial aviation airports, 211,000 active aircraft, and 550,000 active pilots and instructors. General Aviation airports primarily involve recreational flying but also include such activities as medical services, aerial advertising, and aerial application of chemicals; this type of airport presents serious vulnerabilities.

FIGURE 16-6 Commercial airports maintain extensive security.

Here, the emphasis is on commercial airports (U.S. Department of Homeland Security, 2010g: 129; U.S. Government Accountability Office, 2004: 57).

The **Federal Aviation Administration (FAA)**, within the DOT, regulates civil aviation to promote safety and air traffic control. It issues airport operating certifications depending on the type of aircraft served. The FAA also issues certifications for aviation employees, aircraft, airlines, and other purposes.

Security varies at airports, depending on the types of airport and aircraft (Figures 16-6 and 16-7). Prior to the 9/11 attacks, security at airports was the responsibility of private carriers and state and local airport owners. Following the attack, Congress establishing the TSA as the responsible authority for aviation security. In other words, the TSA is charged with federalizing security at commercial airports in the United States and replacing private screeners with federal screeners. The TSA applies several layers of security to protect air transportation. For example, one layer that is implemented is screening passengers via a terrorist database. The GAO published several reports on aviation security and screening. It has repeatedly stated that challenges remain, such as hiring, training and deploying screeners and measuring performance of screeners and detection equipment.

Various technologies are applied to detect explosives, weapons, metals, and plastics. Examples include X-ray systems, metal detectors, biometrics, explosive trace detection (ETD) systems, and body scanning (Haas, 2010). Critics not only claim that privacy is violated during passenger screening, but radiation exposure is a health risk (Straw, 2011a).

FIGURE 16-7 Security varies at general aviation airports.

The screening of cargo transported on commercial passenger aircraft is another serious challenge. The TSA established procedures to require screening of 100 percent of cargo shipped on passenger aircraft originating in the United States, but for inbound flights from other countries this challenge requires diplomatic strategies. Screening of cargo on cargo planes is another problem.

A different program focuses on explosive detection through canine teams. On March 9, 1972, a Trans World Airlines jet took off from JFK International airport in New York bound for Los Angeles. Initially, the flight was normal, but then the airline received an anonymous telephone call warning of a bomb on the flight. The aircraft returned to JFK, passengers were evacuated, and a bomb-sniffing dog named Brandy conducted a search of the aircraft. Brandy found the bomb 12 minutes before it was set to explode. This successful conclusion of a dangerous situation resulted in the establishment of the *Explosives Detection Canine Team Program*, designed to place certified teams at strategic locations in the United States so aircraft with a possible bomb can be diverted quickly to an airport with such a program.

Airlines are required to provide basic security training for all aircrew members. The TSA is required by law to provide optional, hands-on self-defense training if crew members request it. Under the *Federal Flight Deck Officer Program*, a pilot, flight engineer, or navigator is authorized to use firearms to defend the aircraft. Crew members are trained in the use of force and firearms, defensive tactics, psychology of survival, and legal issues.

The **Federal Air Marshal Service** (FAMS) began in 1968 as the FAA Sky Marshal Program. This was during a time when hijackings of aircraft had intensified. The mission of the FAMS is to detect, deter, and defeat hostile acts targeting U.S. air carriers, passengers, and crews. The training requirements are stringent and include behavioral observation, intimidation tactics, close quarters self-defense, and a higher standard for handgun accuracy than officers of any other federal law enforcement agency. The number of Federal Air Marshals is classified information. Another TSA measure of protection is the *Alien Flight Student Program*. Specific aviation training entities are prohibited from offering flight training to aliens and other individuals in the operation of certain aircraft.

The 9/11 attacks resulted in improvements in physical security inside aircraft. For example, commercial airlines contain reinforced cockpit doors that remain locked during flight. CCTV cameras are installed directly outside the cockpit door in some aircraft.

Another concern, specifically against aircraft, is **Man-Portable Air Defense Systems (MANPADS)**. These handheld missile systems have been fired by terrorists toward commercial aircraft in the past. A counter-system faces the establishment of system requirements, the development of technology, cost-effectiveness, and funding.

Although the TSA and its partners work hard to secure aviation, the U.S. Department of Homeland Security, Office of Inspector General (2011), found through undercover audits that screening of both cargo transported on passenger aircraft and passenger carry-on items lack assurance of detection of contraband. Badging of employees and training also require improvements.

Maritime

The maritime shipping infrastructure is another vital component of the economy. This subsector consists of ports, ships that carry cargo and passengers, waterways, locks, dams, canals, and a network of transportation modes, such as railroads and pipelines that connect to ships in port. Ships carry over 80 percent of the world's trade by volume. About 7,000 commercial ships make approximately 60,000 U.S. port calls each year. Annually, The U.S. maritime industry manages about 2.3 billion tons of freight, three billions of tons of oil, more than 147 million ferry passengers, and about 13 million cruise ship travelers. The 361 seaports in the United States are diverse in size and characteristics. They are owned and operated by either private corporations or state and local governments. Ships are primarily owned and operated by the private sector. The DOD has designated certain seaports as strategic seaports for military purposes (U.S. Department of Homeland Security, 2010g: 171–178; White House, 2003a: 60).

The delivery and activation of a WMD on a ship near a port, in port, or in transit from a port would be a catastrophe. And, knowing the creativity and patience of terrorists, one possible scenario is the use of a fuel-laden supertanker as a WMD.

As with other transportation systems, maritime shipping faces the huge challenge of access controls. The inspection of all vessels and cargo that enter ports is impossible, and security concerns must be balanced with efficient access to ports by passengers and cargo. Compounding the need for improved security are international agreements and multinational authorities. The U.S. Department of State, the diplomatic arm of the U.S. government, is responsible for the negotiation of maritime practices with other countries.

The *International Ship and Port Facility Security Code* applies a risk management approach and seeks global standardization of security of port facilities and ships. Similar to U.S. Coast

Guard rules, the code requires minimal requirements, such as port security plans, a port security officer, ship security plans, a ship security officer, and training. Failure to comply with the code has resulted in ships being detained or denied entry by the U.S. Coast Guard.

The **Maritime Transportation Security Act of 2002** (MTSA) charged the Coast Guard with numerous homeland security responsibilities, including attention to waterways and coastal security, drug interdiction, migrant interdiction, and defense readiness to assist the U.S. Navy. Non-homeland security missions include marine safety, search and rescue, aids to navigation (e.g., buoys), marine resources (e.g., fishing treaties), environmental protection, and ice operations (Office of Inspector General, 2004). Critics argue that increased security tasks by the Coast Guard hinder their traditional duties such as boating safety and rescue.

The MTSA is a major law promoting maritime security. The Act mandates a National Maritime Transportation Security Plan and contingency plans in response to terrorist attacks, an increase in security personnel and screening equipment, a grant program to support security upgrades, security regulations to be developed through the DOT, training and certification, a maritime intelligence system, and the establishment of local port security committees to coordinate law enforcement agencies and private security. The MTSA established the Transportation Worker Identification Credential (TWIC) that is administered by the TSA and the Coast Guard. The TWIC is a biometric credential, subject to background screening, that is required of maritime employees for un-escorted access to maritime facilities.

The DOT works with the **U.S. Customs and Border Protection** (CBP) to ensure security in the shipping supply chain. Those shippers who are unable to comply with rules and regulations are subject to increased attention and delay when seeking entrance to U.S. ports.

In September 2005, the Departments of Homeland Security, Defense, and State collaborated to approve *The National Strategy for Maritime Security* (U.S. Department of Homeland Security, 2005b). This first-ever national maritime security plan, with contributors from over 20 government agencies and the private sector, contains supporting plans on awareness, intelligence, layered security, threat response, domestic and international outreach, and recovery. This national strategy considers threats from WMD, transnational crime and piracy, and environmental destruction.

Additional legislation that helps to protect ports and improve supply chain security is the **Security and Accountability For Every (SAFE) Port Act of 2006**. This Act establishes a Domestic Nuclear Detection Office within the DHS, appropriates funds toward the Coast Guard modernization program, and enhances port security training.

International Perspective: Supply Chain Risks and Security

Supply chain refers to a system of businesses, transportation modes, information technology, and employees that move a product or service from supplier to customer (Blasgen, 2011). Business executives planning to seek new markets and operate supply lines from around the globe should incorporate security risks and countermeasures in the planning process. Purtell and Rice (2006: 78–87) write that supply chain risk assessments may include the study of the manufacturing region and site, roads, trucking operations, freight forwarders, logistic warehouses, seaports, and airports. Corruption and anti-American or anti-Western sentiments are other factors to consider.

Numerous websites offer basic information on conditions in various countries. U.S. Department of State Travel Warnings on countries offers a variety of information. Another source is the Overseas Security Advisory Council (refer to the Chapter 10 box titled "International Perspective: Overseas Investigations"). Sources of information on supply chain security include the U.S. Customs and Border Protection.

Laden and Rogers (2006: 62–68) offer suggestions for supply chain security:

- Security, supply chain, and logistics practitioners should work together.
- The company should identify supply chain vulnerabilities prior to risk-based mitigation.
- The company must know its vendors and suppliers and communicate security expectations.
- The company should vet vendors and suppliers. The Internet and software tools can assist this process. Once a list and data are generated, the company should ensure that firms are not on the government's restricted lists, as published by the Departments of Commerce and Treasury and the U.S. Customs and Border Protection.
- Importers should consider restrictive language in orders so that they can control the use of subcontractors in manufacturing and shipping products.
- The company should use technology to improve supply chain visibility, so managers can see what is in the supply chain at all times. Examples of applicable technology are Radio Frequency ID (RFID) and Global Positioning Systems (GPS).

Mass Transit and Passenger Rail

The mass transit and passenger rail subsector (Figure 16-8) is very broad and includes buses, commuter rail, subways, trolleys, and long-distance rail (e.g., Amtrak). Millions of people in the United States ride these modes of transportation daily (U.S. Department of Homeland Security, 2010g: 216).

The U.S. Government Accountability Office (2005d) reports that the numerous stakeholders involved in securing rail transportation can lead to communication challenges, duplication of effort, and confusion of roles and responsibilities. Key federal stakeholders include the TSA, the Federal Transit Administration (FTA), and the Federal Railroad Administration (FRA). The FTA and the FRA are within the DOT. The FTA conducts safety and security activities, such as training, research, technical assistance, and awarding grants. The FRA has regulatory authority over rail safety for commuter rail lines and Amtrak. Other important stakeholders are state and local agencies and rail operators. Protection initiatives in this sector include improved screening and training for operators, security standards, and emergency and continuity of operations planning.

The U.S. Government Accountability Office (2005a: 58) notes that whereas the aviation system is housed in a closed and controlled system, mass transit systems are open so large numbers of people can be moved quickly. Security features that limit access, create delays, increase fares, or result in inconvenience, could make automobiles more attractive. At the same time, those

FIGURE 16-8 Mass transit systems maintain limited access controls and are a challenge to protect.

mass transit systems located in large urban areas or tourist spots make them attractive targets because of the potential for mass casualties and economic damage.

Jenkins (2004: 2) writes: "For those determined to kill in quantity and willing to kill indiscriminately, trains, subways and buses are ideal targets. They offer terrorists easy access and escape." He notes that an analysis of nearly 1,000 terrorist attacks on transportation found that the percentage of fatalities was much higher than the percentage for terrorist attacks in general.

In 2004, following the Madrid rail attacks, the TSA issued directives containing required security measures for passenger rail operators and Amtrak. The security measures produced controversy over limited dialogue to ensure industry "best practices" were included in the measures. Examples of the TSA security measures are as follows (U.S. Government Accountability Office, 2005d: 22):

- Designate coordinators to enhance security-related communication.
- Provide TSA with access to security assessments and plans.
- Reinforce employee watch programs.
- Ask passengers to report unattended property and suspicious behavior.
- Use clear plastic or bomb-resistant containers.
- Use canine explosive detection teams to screen passenger baggage, terminals, and trains.
- Allow TSA canine teams access to rail operations.
- Conduct frequent inspections of facilities and assets.

FIGURE 16-9 Buses are among other mass transit targets attractive to terrorists. Several security methods are applicable to all mass transit modes.

- Use surveillance systems.
- Ensure appropriate levels of policing and security that correlate with DHS threat levels.
- Lock all doors that allow access to train operators.

Additional measures from a variety of sources include security assessments and standards, public awareness, emergency planning and drills, sharing intelligence, visible and undercover patrols, behavioral observation skills of police and security, applying technology, emergency call boxes, research and testing of WMD detection equipment, ventilation systems that remove toxic smoke, and windows that blow out to reduce both the destructive force of a blast and smoke (U.S. Department of Homeland Security, 2010g: 223–235). Many security methods applied to trains are also applicable to buses (Figure 16-9).

The Reality of Mass Transit Security

Stoller (2010) reported that mass transit is much less secure than is aviation and the situation will likely remain that way—even though each weekday about 34 million people ride mass transit and rail and two million passengers fly. He writes that the TSA spends most of its budget on aviation with $30 billion spent over the decade since the 9/11 attacks, while $1.7 billion was spent on subway, bus, and rail during the same period. From a cost standpoint, Stoller cites terrorism expert Brian Jenkins, who estimates that the cost of screening a mass transit passenger is about $8 to $10, which would "destroy public transportation."

Stoller notes that 22 other nations have experienced terrorist attacks in the last five years on mass transit and rail and that the United States has foiled several plots, including those in Washington, DC, and New York City. Stoller refers to global statistics from the period 2005 to 2010 when 213 attacks were recorded on subways and trains and 197 on aviation. The former resulted in 700 killed and 3,262 wounded, while the latter resulted in 238 killed and 937 wounded.

Do you think it is inevitable that one day terrorists will succeed in attacking U.S. mass transit or rail? Why or why not?

Highway Infrastructure and Motor Carrier

The highway infrastructure and motor carrier subsector of the transportation sector has developed a sector-specific plan in cooperation with several government and private sector partners. This subsector includes nearly four million miles of roadway, almost 600,000 bridges, about 400 tunnels, trucks carrying assorted items and hazmat, motorcoaches, and school buses. Transporting people and goods in this network is vital for meeting the needs of citizens and businesses (U.S. Department of Homeland Security, 2010g: 253).

The trucking industry is highly fragmented, with thousands of trucking fleets and over 29 million trucks that haul more than 10 billion tons of freight annually. Just about every item consumed in the United States is transported on a truck at some point. The motorcoach industry has about 3,137 companies owning almost 30,000 buses and employing over 118,000 employees who transport about 750 million passengers annually. The school transportation industry transports approximately 23 million students daily on about 460,000 buses to over 80,000 schools (U.S. Department of Homeland Security, 2010g: 255–256).

As with other sectors and subsectors, the highway infrastructure and motor carrier subsector are subject to a variety of terrorist attacks, other threats, accidents, and natural disasters. Bridges, tunnels, trucks, and buses can be attacked directly, resulting in deaths, injuries, and property damage. In addition, a truck or other vehicle can be used to transport a bomb, as in the Oklahoma City bombing.

Transportation choke points (TCPs) are an important consideration in this subsector. These points include bridges, tunnels, highway interchanges, terminals, and border crossings that are crucial to transportation routes. TCPs are vital because they connect roads and modes of transportation. If subject to attack, accident or natural disaster, and destroyed, traffic could be blocked at TCPs, and the vulnerabilities of people, vehicles, and vehicle contents would be compounded.

Initiatives aimed at improving the protection of highways, trucking, and busing include cooperation among DHS, DOT, and stakeholders on

- Risk management
- Security and resilience
- Technological solutions for increased protection
- Information and intelligence sharing
- Development of criteria for identifying and mitigating TCPs
- National operator security education and awareness

Additional recommendations can be found through the DHS *Transportation System Sector-Specific Plan* (U.S. Department of Homeland Security, 2010g) and numerous organizations such as The American Trucking Association (ATA). The ATA supports the North American Transportation Management Institute's Certified Cargo Security Professional designation.

Freight Rail

Freight rail plays an important role in our economy by linking raw materials to manufacturers and carrying a wide variety of fuels and finished goods. In addition, numerous passenger rail systems operate at least partially over freight rail tracks. There are about 140,000 miles of railroad track in the United States used by carriers who employ about 180,000 employees. This business generates approximately $63 billion in revenues annually (U.S. Department of Homeland Security, 2010g: 283).

The freight rail subsector is complex because of differences in design, structure, and purpose. The differences result in disadvantages and advantages for security. For example, the size of the rail system makes responding to various threat scenarios difficult. At the same time, trains must follow specific routes, so if one were hijacked, for example, it could be diverted off a mainline by authorities. Similarly, if a bridge or tunnel were destroyed, the rail line would suffer disruption; however, national-level disruptions would be limited.

Numerous public and private sector groups partner to protect freight rail. These include TSA, FEMA, CBP and USCG—all within DHS, and the DOT, DOD, and DOJ. In the private sector, partners include the Association of American Railroads and several railway companies.

A serious risk in this subsector is the transportation of hazmat, especially through populated areas; therefore, industry and government coordination is necessary for decision making. Security and safety of containerized cargo are other challenges. Controversy has developed over the markings of containers to indicate the type of hazardous materials being transported. Although placards on rail cars provide helpful information to first responders during an emergency, the information can also assist terrorists planning an attack.

Nearly all locomotives and rail cars are tagged with automatic ID transponders that record and report location as it passes detectors. Such data is transmitted to centralized control systems. Consequently, cybersecurity is important.

The railroad industry maintains police and security forces for its individual railroad entities. The Association of American Railroads (n.d.) uses a multistage alert system and open lines of communications with government officials. This group's Railroad Security Task Force applied national intelligence community "best practices" to develop a security plan.

Freight rail protection initiatives include the following (U.S. Department of Homeland Security, 2010g: 291–300; White House, 2003a: 57):

- The rail mode works with the DOT to assess risk, security, and resilience.
- A surface transportation Information Sharing and Analysis Center (ISAC) exchanges information related to both cyber and physical threats.
- DHS and DOT partner with other federal agencies, state and local governments, and industry to improve the security and safety of hazmat.
- DHS and DOT work with the rail mode to identify and explore technologies and processes to efficiently screen rail passengers and baggage and to secure containers and detect threatening content.
- DHS and DOT work with industry to delineate infrastructure protection roles and responsibilities for surge requirements during emergencies.
- An emphasis on employee awareness, reporting suspicious activity, controlling sensitive information, and employee ID.

Are Rail Cars Containing Toxic Cargo "Sitting Ducks"?

Authorities face an enormous number of potential terrorist targets. Risks must be prioritized, since resources are limited. Toxic substances within rail cars that are in transit or at a standstill present attractive targets to terrorists and other criminals (Straw, 2011b). Unfortunately, as public safety agencies are doing their best to protect citizens from a variety of threats, inside their jurisdiction or nearby are toxic substances, necessary for our everyday lives, unprotected. In one case, a reporter and photographer gained easy access to an unguarded rail facility close to New York City and observed rail cars filled with deadly chlorine, ammonia, and other chemicals. They also spotted unlocked switching devices that could be used to cause an accident (Kocieniewski, 2006).

Imagine the carnage from a bomb placed on one of these rail cars, or a bomb placed on such a rail car sitting near fuel storage tanks or other target to facilitate a cascading effect. Although chemical plant security is improving, the shipments linking these plants to suppliers and customers must also be improved.

The U.S. Naval Research Lab reports that each year railroads transport 105,000 rail cars of toxic chemicals and 1.6 million rail cars of other hazardous materials. They estimate that an attack on such a rail car could kill 100,000 people. The TSA is working with railroads to reduce the time such rail cars are sitting unprotected. In early 2007, the TSA began monitoring rail cars containing toxic substances. Specifically, the TSA is recording how long these rail cars are stopped on tracks or sitting in unprotected storage yards in urban areas. Rerouting to reduce the danger to cities is an option; however, this would force transporting the toxic substances in trucks, which could be more dangerous (Frank, 2007). More recently, Straw (2011b) reported that DOT and TSA issued regulations to mitigate these vulnerabilities through risk-based hazmat routing, secure custody of stationary rail cars, and location tracking of cars when in transit.

Pipelines

Pipelines are considered part of the transportation sector. The pipeline industry moves a variety of substances such as crude oil, refined petroleum products, and natural gas within many hundreds of thousands of miles of pipelines that are mostly underground. This industry has a commendable record of safety, contingency plans, and the capabilities to repair or bypass localized disruptions. However, protection is an important issue because of the volatile nature of the products that pipelines deliver. Pipelines cross local, state, and international jurisdictions, and numerous businesses and other entities depend on a reliable flow of fuel. The problem of cascading is evident with pipelines, as it is with other CI sectors. DHS, DOT, and DOE, state and local governments, and the private sector collaborate to improve protection measures and emergency response plans The sector-specific plan for the pipeline subsector stresses security and resiliency, as well as the ability to detect physical and cyber intrusions, mitigate losses and restore service (U.S. Department of Homeland Security, 2010g: 315–317).

Postal and Shipping

The postal and shipping sector is essential to the U.S. economy, moving hundreds of millions of messages, products, and financial transactions each day. It handles small- and medium-size

FIGURE 16-10 The U.S. Postal Service is dependent on the transportation infrastructure, as are private carriers of mail and packages. Attacks on the mail system are extremely challenging for authorities.

packages by carriers who employ about 1.8 million people. Besides the U.S. Postal Service (USPS) (Figure 16-10), a few major private carriers comprise this sector. It depends on the transportation infrastructure and its trucks, aircraft, railroads, and ships to move items from millions of senders to millions of destinations (U.S. Department of Homeland Security, 2009; White House, 2003a: 67).

Several federal entities partner to protect the postal and shipping sector. These include TSA and CBP in DHS, the U.S. Postal Inspection Service, the Food and Drug Administration, and the Centers for Disease Control.

The networks of the USPS and private shipping companies present complex protection challenges. The Fall 2001 **anthrax attacks** through the USPS illustrate the impact of a serious disruption to this system. The attack resulted in five deaths and contamination of postal facilities and equipment, other government offices, and businesses. Numerous other risks face this sector such as robbery, theft, fraud, vandalism, transporting illegal substances, and hazmat. These risks necessitate comprehensive protection and resilience.

National Monuments and Icons

The National Monuments and Icons sector includes national monuments and symbols that have historical significance and represent U.S. traditions, values, and political power. Examples are the Washington Monument and the Statue of Liberty. The U.S. Department of the Interior (DOI), in cooperation with DHS, is the sector-specific agency responsible for the security and safety of public land that includes several national monuments. Protection challenges for the DOI are balancing security and public access, securing assets in remote areas, and addressing jurisdictional issues.

The DOI manages the third largest federal law enforcement force with 4,400 commissioned personnel. Safety and security is required for 70,000 employees, 200,000 volunteers, 1.3 million daily visitors, and over 507 million acres of public lands, including oil and gas production on federal and Indian trust lands. DOI is a partner in multiagency groups that share information and work on security and resilience (U.S. Departments of Homeland Security and Interior, 2010).

Border and Transportation Security

Border and transportation security are intertwined because various transportation modes and travelers continuously seek access through border entry and exit points. Government agencies, transportation systems, and travelers must comply with specific legal requirements to increase security and safety along borders and within borders. The legal issues are complex, considering the borders of the United States include a 5,525-mile border with Canada, a 1,989-mile border with Mexico, and 95,000 miles of shoreline and navigable waterways. All people and goods legally entering the United States must be processed through an air, land, or seaport point of entry. Hundreds of millions of people legally enter the United States each year and millions are noncitizens. In addition, many people enter the United States illegally each year, and if caught, are processed through various government agencies.

As we know, terrorists are elusive, crafty, and patient as they search for vulnerabilities in potential targets. Consequently, the global transportation system that crosses borders and has access to all communities must be protected against terrorists and other criminals.

A major federal government initiative of border security is **U.S. Visitor and Immigrant Status Indicator Technology (US-VISIT)**. This is a system for integrating data on the entry and exit of certain foreign nationals into and out of the United States The process begins overseas and continues through a visitor's arrival to and departure from the United States. The program involves entrance eligibility determinations by the Departments of Homeland Security and State. In developing this entry and exit data system, an emphasis has been placed on biometric technology (e.g., digital fingerprints and a photograph) and the development of tamper-resistant documents readable at ports of entry. It also requires that the system be able to interface with law enforcement databases to identify and detain individuals who pose a threat to national security. The ultimate goal of this challenging multibillion-dollar program is to check all people entering and departing the United States.

Government Agencies with Roles in Border and Transportation Security

U.S. Customs and Border Protection

U.S. Customs and Border Protection (CBP) is the unified border agency of the DHS. It contains the inspectional workforce and border authorities of legacy (i.e., predecessor agency) U.S. Customs, legacy Immigration and Naturalization Service (INS), legacy USDA Animal and Plant Health Inspection Service (APHIS), and the entire U.S. Border Patrol. This united force consists of over 40,000 employees.

The priority mission of CBP is homeland security. It is charged with managing, securing, and controlling the U.S. borders. CBP goals include facilitating the efficient movement of legitimate cargo and people while maintaining security and detecting and preventing contraband (i.e., anything illegal) from entering the country.

The CBP is involved in numerous initiatives, some of which are described next. The **Container Security Initiative** (CSI) extends the zone of security outward so that U.S. borders are the last line of defense, not the first. Maritime containers that present a risk for terrorism are identified and inspected at foreign ports before they are shipped to the US.

The CBP receives electronic information on over 95 percent of all U.S.-bound sea cargo before it arrives. Under the **Advanced Manifest Rule**, all sea carriers, with few exceptions (e.g., bulk carriers), must provide cargo descriptions and valid consignee addresses 24 hours before cargo is loaded at the foreign port for shipment to the United States. Under the Sea Automated Manifest System, the data includes cargo type, manufacturer, shipper, country of origin, routing, and the terms of payment. Using computer systems to analyze the data, the CBP assigns a numeric score to each shipment indicating its level of risk. High-risk containers are flagged to receive a security inspection.

Because CBP resources are limited, industry partnership programs bring the trade community into the process of homeland security to prevent terrorism and to focus on other threats such as money laundering and smuggling contraband. An example of an industry partnership is the **Customs-Trade Partnership Against Terrorism (C-TPAT)**. This initiative is part of the extended border strategy and entails CBP working with importers, carriers, brokers, and other businesses to improve supply chain security. C-TPAT has been criticized because it is a voluntary government-business initiative.

The CBP continues to rely on technology as an important component of border security. It employs X-ray and other detection technologies. A variety of technologies is used for surveillance. Examples are outdoor sensors, CCTV, night-vision equipment, and aircraft, including Unmanned Aerial Vehicles (UAVs).

U.S. Immigration and Customs Enforcement

The **U.S. Immigration and Customs Enforcement (ICE)** is the largest investigative arm of the DHS. It focuses on immigration and customs law enforcement *within* the United States. The Office of Detention and Removal is a division of ICE that removes unauthorized aliens from the United States.

ICE contains an Office of Investigations that concentrates on many types of violations, including those pertaining to national security, the transfer of WMD and arms, critical technology, commercial fraud, human trafficking, illegal drugs, child pornography and exploitation, and immigration fraud. In addition, ICE special agents conduct investigations involving the protection of critical infrastructure industries that may be victimized by attack, sabotage, exploitation, or fraud.

U.S. Citizenship and Immigration Services

The **U.S. Citizenship and Immigration Services**, within the DHS, grants immigration and citizenship benefits, promotes awareness and understanding of citizenship, ensures the integrity of the U.S. immigration system, and contributes to the security of the United States. This agency works with the U.S. Department of State, CBP, and ICE.

Search the Internet

Here is a list of websites relevant to this chapter:

Airports Council International: www.airports.org
American Association of Airport Executives: www.aaae.org
American Association of State Highway and Transportation Officials: www.transportation.org
American Bus Association: www.buses.org
American Chemistry Council: www.americanchemistry.com
American Public Transportation Association: www.apta.com
American Trucking Associations: www.truckline.com
American Water Works Association: www.awwa.org
ASIS International: www.asisonline.org
Association of American Railroads: www.aar.org
Cloud Computing Alliance: cloudsecurityalliance.org
Government website on food safety: www.foodsafety.gov
Grocery Manufacturers Association: www.gmaonline.org
Information Systems Audit Control Association: www.isaca.org
Information Systems Security Association: issa.org
International Information Systems Security Certifications Consortium: www.isc2.org
International Organization for Standardization (ISO): www.iso.org
Internet Crime Complaint Center (IC3): www.ic3.gov
IT Governance Institute: www.itgi.org
Office of the Metropolitan Transportation Authority Inspector General: mtaig.state.ny.us
National Infrastructure Protection Plan: www.dhs.gov/nipp
National Institute of Standards and Technology, Computer Security Resource Center: csrc.nist.gov
National Security Agency, Information Assurance Directorate: www.nsa.gov/ia
North American Transportation Management Institute: www.natmi.org
Nuclear Regulatory Commission: www.nrc.gov
Overseas Security Advisory Council (OSAC): www.osac.gov
The Infrastructure Security Partnership: www.tisp.org
United Motorcoach Association: www.uma.org
U.S. Citizenship and Immigration Services: www.uscis.gov
U.S. Coast Guard: www.uscg.mil
U.S. Computer Emergency Readiness Team (US-CERT): www.us-cert.gov
U.S. Customs and Border Protection: www.cbp.gov

U.S. Department of Agriculture, Food Safety and Inspection Service: www.fsis.usda.gov
U.S. Department of Defense: www.defenselink.mil
U.S. Department of Defense, Defense Security Service: www.dss.mil/isp
U.S. Department of Energy: energy.gov
U.S. Department of Homeland Security: www.dhs.gov
U.S. Department of the Interior: www.doi.gov
U.S. Department of Transportation: www.dot.gov
U.S. Department of Transportation, Maritime Administration: www.marad.dot.gov
U.S. Environmental Protection Agency: www.epa.gov
U.S. Food and Drug Administration, Food Defense & Emergency Response: www.fda.gov/Food/FoodDefense
U.S. Immigration and Customs Enforcement: www.ice.gov
U.S. Marshals Service: www.usmarshal.gov
U.S. Postal Inspection Service: postalinspectors.uspis.gov
U.S. Transportation Security Administration: www.tsa.gov

Case Problems

16A. Suppose you are a terrorist leader in the United States planning to attack the food industry. What is your plan?

16B. Suppose you are a disgruntled employee working in the chemical industry at a plant that produces chlorine. Your anger at your employer is so intense that you do not care about yourself or anyone else. All you can think about is your plan to get back at your employer. Besides what you know from your job, you conduct research on the Internet to find out everything you can about chlorine. What is your plan?

16C. As the Security and Safety Director of a company involved in the food industry, how would you prevent the attack planned in 16A? How would you respond if the plan were executed?

16D. As the Security and Safety Director of the chlorine plant in 16B, how would you prevent the crime(s) planned in 16B? How would you respond if the disgruntled employee executed the vindictive plan?

16E. You are the Chief Security Officer (CSO) of a manufacturing company with multiple plants in the United States. How do you protect the company from the cascade effect from all-hazards?

16F. As the security director for a city mass transit bus system, what are your top seven (prioritized) strategies for protecting passengers, employees, and assets associated with this system?

16G. The issue of security for cloud computing has surfaced in an in-house corporate meeting that you attended. Since you are a corporate security manager, your boss assigns you to an upcoming meeting on this issue. Research and prepare a report on this topic so you have information for the meeting.

References

Acohido, B., & Swartz, J. (2006). Malicious-Software spreaders get sneakier, more prevalent. *USA Today* (April 23) <www.usatoday.com/tech/news/computersecurity/infotheft/2006-04-23-bot-herders_x.htm> retrieved April 24, 2006.

American Water Works Association. (2012). Security. <www.awwa.org> retrieved February 11, 2012.

Associated Press. (2012). FBI busts hacker group after chief turns informant. (March 7).

Association of American Railroads (n.d.). Freight railroad security plan. <www.aar.org> retrieved October 23, 2005.

Bates, S. (2012). Managing risk in the cloud: The differing approaches in the three service models. *Security Technology Executive, 22* (January/February).

Blasgen, R. (2011). Supply chain management: The driving force behind the global economy. *The CIP Report, 10* (July). <cip.gum.edu> retrieved January 21, 2012.

Bu, Z. et al. (2012). McAfee Threats Report: Fourth Quarter 2011. <www.mcafee.com> retrieved February 25, 2012.

Carroll, J. (1996). *Computer security* (3rd ed.). Boston: Butterworth-Heinemann.

Centers for Disease Control. (2011). CDC and Food Safety. <www.cdc.gov> retrieved February 8, 2012.

Chalk, P. (2003). *Agroterrorism: What is the threat and what can be done about it?* Santa Monica, CA: Rand Corp.

Chittister, C., & Haimes, Y. (2006). Cybersecurity: From ad hoc patching to lifecycle of software engineering. *Journal of Homeland Security and Emergency Management, 3.*

Computer Security Institute/FBI. (2006). CSI/FBI Computer Crime and Security Survey. <www.GoCSI.com> retrieved July 14, 2006.

Congressional Budget Office, (2004). *Homeland security and the private sector* (December). Washington, D.C.: Congress of the United States.

Croy, M. (2005). Current and emerging trends in business continuity. *Disaster Recovery Journal,* 18 (Winter).

Defense Security Service. (2012). National Industrial Security Program.<www.dss.mil> retrieved February 28, 2012.

Dilanian, K. (2011). Virtual war a real threat. *Los Angeles Times* (March 28) <www.latimes.com> retrieved March 30, 2011.

DiLonardo, R. (2011). Emerging cyber-security threats. *Loss Prevention Magazine, 10* (March–April).

Frank, T. (2007). TSA to track rail shipments with toxic cargo. *USA Today* (January 21).

Gardner, C. (2011). Now serving 300 million: How the United States government plans to stop terrorists from using U.S. ports and imported food as instruments of terror. *Journal of Biosecurity, Biosafety and Biodefense Law, 1* <www.bepress.com/jbbbl/vol1/iss1/4> retrieved February 8, 2012.

Gewin, V. (2003). Agriculture shock. *Nature, January.*

Haas, D. (2010). Electronic security screening: Its origin with aviation security 1968–1973. *Journal of Applied Security Research, 5.*

Hayes, J., & Ebinger, C. (2011). The private sector and the role of risk and responsibility in securing the nation's infrastructure. *Journal of Homeland Security and Emergency Management, 8.*

Hebert, H. (2005). *Interest rises in nuclear plants for first time since disasters.* Associated Press. June 12.

Henry, S. (2011). Speech at the Information Systems Security Association International Conference, Baltimore, MD (October 20). <www.fbi.gov> retrieved February 24, 2012.

Hodgson, K. (2011). Why CFATS is good for the security industry. *Security Technology Executive, 21* (August).

International Organization for Standardization.(2011). ISO/IEC 27002:2005. <www.iso.org> retrieved February 26, 2012.

Jenkins, B. (2004). Terrorism and the Security of Public Surface Transportation." (April). Santa Monica, CA: RAND. Testimony presented to the Senate Committee on Judiciary on April 8, 2004.

Keil, T. (2011). The critical infrastructure risk management enhancement initiative. *The CIP Report, 10* (November). <cip.gum.edu> retrieved November 21, 2011.

Kochems, A. (2005). Who's on first? A strategy for protecting critical infrastructure. *Backgrounder, 1851* (May 9) <www.heritage.org> retrieved June 20, 2005.

Kocieniewski, D. (2006). Despite 9/11 effect, railyards are still vulnerable. *The New York Times* <www.nytimes.com/2006/03/27/nyregion> retrieved March 28, 2006.

Koski, C. (2011). Committed to protection? partnerships in critical infrastructure protection. *Journal of Homeland Security and Emergency Management, 8.*

Laden, M., & Rogers, K. (2006). Visualizing supply-chain security: Why LP should get on board. *Loss Prevention, 5* (May–June).

Landoll, D. (2005). Does IT security myth the point? *Security Management, 49* (January).

Massachusetts Public Interest Research Group (.2011). Unacceptable Risk: Two Decades of Close Calls, Leaks, and Other Problems at U.S. Nuclear Reactors (March). <www.masspirg.org> retrieved April 4, 2011.

McCumber, J. (2011). Lost in the cloud. *Security Technology Executive, 21* (March).

Mills, E. (2010). Experts warn of catastrophe from cyberattacks. *CNET News* (February 23). <news.cnet.com> retrieved February 21, 2012.

National Defense Industrial Association, (2011). National defense industrial association position paper on the defense industrial base. *The CIP Report, 9* (May). <cip.gum.edu> retrieved November 28, 2011.

National Industrial Security Program. (2011). Advisory Meeting Minutes (March 3). <www.archives.gov> retrieved February 29, 2012.

Office of Homeland Security. (2002). National Strategy for Homeland Security (July). <www.whitehouse.gov> retrieved September 14, 2004.

Office of Inspector General. (2004). *FY 2003 mission performance united states coast guard.* Washington, DC: DHS. (September).

Olson, D. (2012). Agroterrorism: The threat to America's economy and food supply. *FBI Law Enforcement Bulletin, 81* (February).

Piazza, P. (2007). Keys to encryption. *Security Management, 51* (January).

Powell, D. (2011). Protecting the smart grid. *Security Management, 55* (August).

Purtell, D., & Rice, J. (2006). Assessing cargo supply risk. *Security Management, 50* (November).

Reese, S., & Tong, L. (2010). Federal building and facility security. *Congressional Research Service* (April 27). [assets.opencrs.com, retrieved March 3, 2012.]

Sager, I. (2000). Cybercrime. *Business Week* (February 21).

Savage, M. (2006). Protect what's precious. *Information Security, 9* (December).

Smith, D. (2006). Encryption, once for the few, now an option for all. *IT Security, 2* (October).

Smith, R. (2013). *Elementary information security.* Burlington, MA: Jones & Bartlett Pub.

Snow, G. (2011). Statement Before the Senate Judiciary Committee, Subcommittee on Crime and Terrorism (April 12). <www.fbi.gov> retrieved February 24, 2012.

Stoller, G. (2010). Can trains, subways be protected from terrorists? *USA TODAY* (December 27) <www.usatoday.com> retrieved January 4, 2011.

Straw, J. (2011a). Backscatter safety questions persist. *Security Management, 55* (April).

Straw, J. (2011b). Is hazmat safety on track? *Security Management* (August).

Taylor, R., et al. (2006). Digital crime and digital terrorism: *Upper Saddle River, NJ.* Pearson: Prentice Hall.

Thompson, A. (1997). Smoking out the facts on firewalls. *Security Management* (January).

U.S. Department of Agriculture. (2011). Food Defense & Emergency Response. <www.fsis.usda.gov> February 8, 2012.

U.S. Department of Homeland Security. (2005a). Interim National Infrastructure Protection Plan (February). <www.dhs.gov> retrieved February 11, 2005.

U.S. Department of Homeland Security. (2005b). National Strategy for Maritime Security (September). <www.dhs.gov/interweb/assetlibrary/HSPD13_MaritimeSecurityStrategy.pdf> retrieved October 25, 2005.

U.S. Department of Homeland Security. (2006a). National Infrastructure Protection Plan. <www.dhs.gov/interweb/assetlibrary/NIPP_Plan.pdf> retrieved June 30, 2006.

U.S. Department of Homeland Security. (2006b). National Infrastructure Protection Plan (revised draft). E-mail, January 20, 2006.

U.S. Department of Homeland Security. (2009). National Infrastructure Protection Plan. <www.dhs.gov/xlibrary/assets/NIPP_Plan.pdf> retrieved January 4, 2012.

U.S. Department of Homeland Security. (2010a). Critical Infrastructure (November 30). <www.dhs.gov> retrieved February 8, 2012.

U.S. Department of Homeland Security. (2010b). More About the Office of Infrastructure Protection (December 27). <www.dhs.gov> retrieved February 5, 2012.

U.S. Department of Homeland Security. (2010c). Chemical Sector: Critical Infrastructure and Key Resources (June 21). <www.dhs.gov> retrieved February 18, 2012.

U.S. Department of Homeland Security (2010d). Communications Sector-Specific Plan: An Annex to the National Infrastructure Protection Plan. <www.dhs.gov> retrieved February 19, 2012.

U.S. Department of Homeland Security. (2010e). Information Technology Sector-Specific Plan: An Annex to the National Infrastructure Protection Plan. <www.dhs.gov> retrieved February 20, 2012.

U.S. Department of Homeland Security. (2010f). Critical Manufacturing Sector-Specific Plan: An Annex to the National Infrastructure Protection Plan. <www.dhs.gov> retrieved March 3, 2012.

U.S. Department of Homeland Security. (2010g). Transportation System Sector-Specific Plan: An Annex to the National Infrastructure Protection Plan. <www.dhs.gov> retrieved March 3, 2012.

U.S. Department of Homeland Security. (2011a). Chemical Facility Anti-Terrorism Standards (September 14). <www.dhs.gov> retrieved February 18, 2012.

U.S. Department of Homeland Security. (2011b). Blueprint for a Secure Cyber Future: The Cybersecurity Strategy for the Homeland Security Enterprise (November). <www.dhs.gov> retrieved February 21, 2012.

U.S. Department of Homeland Security. (2012a). Control Systems Security Program (February 11). <www.us-cert.gov/control_systems/ics-cert> retrieved February 11, 2012.

U.S. Department of Homeland Security. (2012b). Security Specialist Competencies: An Interagency Security Committee Guideline (January). <www.dhs.gov> retrieved February 12, 2012.

U.S. Departments of Homeland Security and Agriculture and U.S. Food and Drug Administration. (2010). Food and Agriculture Sector-Specific Plan, An Annex to the National Infrastructure Protection Plan. <www.dhs.gov> retrieved February 12, 2012.

U.S. Departments of Homeland Security and Defense. (2010). Defense Industrial Base Sector-Specific Plan, An Annex to the National Infrastructure Protection Plan. <www.dhs.gov> retrieved February 28, 2012.

U.S. Departments of Homeland Security and Energy. (2010). Energy Sector-Specific Plan, An Annex to the National Infrastructure Protection Plan. <www.dhs.gov> retrieved February 12, 2012.

U.S. Departments of Homeland Security and Interior. (2010). National Monuments and Icons Sector-Specific Plan: An Annex to the National Infrastructure Protection Plan. <www.dhs.gov> retrieved March 10, 2012.

U.S. Department of Homeland Security and U.S. Environmental Protection Agency. (2010). Water Sector-Specific Plan, An Annex to the National Infrastructure Protection Plan. <www.dhs.gov> retrieved February 14, 2012.

U.S. Department of Homeland Security, Office of Inspector General. (2011). Major Management Challenges Facing the Department of Homeland Security (November). <www.oig.dhs.gov/assets/Mgmt/OIG_12-8_Nov11.pdf> retrieved January 17, 2012.

U.S. Department of Justice, (1995). *Vulnerability assessment of federal facilities*. Washington, D.C.: U.S. Marshals Service.

U.S. Environmental Protection Agency. (2011). Basic Information about Water Security (November 9). water.epa.gov, retrieved February 11, 2012.

U.S. Environmental Protection Agency. (2012). Water Security Home (January 12). water.epa.gov, retrieved February 11, 2012.

U.S. Food and Drug Administration. (2012). Food Defense & Emergency Response. <www.fda.gov/Food/FoodDefense/default.htm> retrieved February 8, 2012.

U.S. General Accounting Office. (2004). Industrial Security: DOD Cannot Provide Adequate Assurances That Its Oversight Ensures the Protection of Classified Information (March). <www.gao.gov/cgi-bin/getrpt?GAO-04-332> retrieved July 20, 2005.

U.S. Government Accountability Office. (2004). General Aviation Security: Increased Federal Oversight Is Needed, but Continued Partnership with the Private Sector Is Critical to Long-Term Success (September). <www.gao.gov/cgi-bin/getrpt?GAO-05-144> retrieved December 13, 2004.

U.S. Government Accountability Office. (2005a). Homeland Security: Agency Plans, Implementation, and Challenges Regarding the National Strategy for Homeland Security (January). <www.gao.gov/cgi-bin/getrpt?GAO-05-213> retrieved February 15, 2005.

U.S. Government Accountability Office. (2005b). Industrial Security: DOD Cannot Ensure Its Oversight of Contractors under Foreign Influence Is Sufficient (July). <www.gao.gov/cgibin/getrpt?GAO-05-681> retrieved July 18, 2005.

U.S. Government Accountability Office. (2005c). Information Security: Emerging Cybersecurity Issues Threaten Federal Information Systems (May). <www.gao.gov/new.items.do5231.pdf> retrieved May 16, 2005.

U.S. Government Accountability Office. (2005d). Passenger Rail Security: Enhanced Federal Leadership Needed to Prioritize and Guide Security Efforts (October). <www.gao.gov/cgibin/getrpt?GAO-06-181T> retrieved October 21, 2005.

U.S. Government Accountability Office. (2006). Information Security: Agencies Need to Develop and Implement Adequate Policies for Periodic Testing (October). <www.gao.gov/new.items.d0765.pdf> retrieved November 21, 2006.

U.S. Government Accountability Office. (2007). Homeland Security: Applying Risk Management Principles to Guide Federal Investments (February 7). <www.gao.gov/cgi-bin/getrpt?CAO-07-386T> retrieved February 8, 2007.

U.S. Nuclear Regulatory Commission. (2011). Nuclear Security and Safeguards (July 21). <www.nrc.gov/security.html> retrieved February 15, 2012.

U.S. Nuclear Regulatory Commission. (2012). Part 73, Physical Protection of Plants and Materials (January 31). <www.nrc.gov/reading-rm/doc-collections/cfr/> retrieved February 18, 2012.

Welch, M. (2006). Seeking a safer society: America's anxiety in the war on terror. *Security Journal, 19*

White House. (2003a). The National Strategy for the Physical Protection of Critical Infrastructures and Key Assets (February). <www.whitehouse.gov> retrieved September 14, 2004.

White House. (2003b). The National Strategy to Secure Cyberspace (February). <www.whitehouse.gov> retrieved February 26, 2012.

Winter, G., & Broad, W. (2001). Added Security for Dams, Reservoirs and Aqueducts.<www.waterindustry.org> retrieved April 25, 2002.

Wulf, W., & Jones, A. (2004). A perspective on cybersecurity research in the United States: *In terrorism: Reducing vulnerabilities and improving response*. Washington, DC: National Research Council of the National Academies.

Zeller, T. (2006). Cyberthieves silently copy keystrokes. *News.Com* (February 27) news.com, retrieved February 28, 2006.

Zeller, T. (2011). Nuclear agency is criticized as too close to its industry. *The New York Times* (May 7) <www.nytimes.com> retrieved May 9, 2011.

17

Protecting Commercial and Institutional Critical Infrastructure

OBJECTIVES

After studying this chapter, the reader will be able to:

1. Name commercial and institutional critical infrastructure sectors and describe characteristics of each
2. Describe the problem of inventory shrinkage in retailing
3. Explain the threats and hazards facing retail businesses and countermeasures
4. Discuss the shoplifting problem and countermeasures
5. Describe robbery countermeasures for retail businesses
6. Explain the threats and hazards facing banks and financial businesses and countermeasures
7. Explain the threats and hazards facing educational institutions and countermeasures
8. Explain the threats and hazards facing healthcare institutions and countermeasures

KEY TERMS

- shrinkage
- shopping service
- checkout counters
- point-of-sale (POS) accounting systems
- exception reporting
- counterfeiting
- kleptomania
- organized retail theft
- electronic article surveillance
- source tagging
- benefit-denial technology
- civil recovery
- robbery
- burglary
- substitutability
- U.S. Department of the Treasury
- Bank Protection Act (BPA) of 1968
- savings and loan (S&L) scandal
- Bank Secrecy Act of 1986
- Anti-Drug Abuse Act of 1988
- The Antiterrorism and Effective Death Penalty Act of 1996
- Gramm-Leach-Bliley Act of 1999
- USA Patriot Act of 2001
- Sarbanes-Oxley (SOX) Act of 2002
- Regulation H, Code of Federal Regulations
- suspicious activity reports
- money laundering
- tear gas/dye packs
- Global Positioning System (GPS)
- skimming
- Safe and Drug-Free Schools and Communities Act
- Gun-Free Schools Act

- zero-tolerance policy
- Family Educational Rights and Privacy Act
- Gang Resistance Education and Training (GREAT)
- "soft" targets
- "hard" targets
- Columbine High School massacre
- Beslan Elementary School massacre
- Virginia Tech massacre
- Student-Right-to-Know and Campus Security Act of 1990
- Campus Sexual Assault Victims' Bill of Rights
- community policing
- Emergency Medical Treatment and Active Labor Act
- Health Insurance Portability and Accountability Act of 1996 (HIPAA)
- The Joint Commission
- National Center for Missing and Exploited Children
- Controlled Substances Act of 1970
- first receivers
- personal protective equipment (PPE)
- Public Health Security and Bioterrorism Preparedness and Response Act of 2002
- Centers for Disease Control and Prevention (CDC)

Introduction

This chapter focuses on protecting commercial and institutional sectors, including retail businesses, banks and financial businesses, educational institutions (a subsector of government facilities), and healthcare and public health institutions. Each of these topics begins with a broad perspective from U.S. Department of Homeland Security Sector-Specific Plans.

Commercial Facilities

The commercial facilities (CF) sector is diverse in the types of services and products offered to the public. CF subsectors are explained next from the *Commercial Facilities Sector-Specific Plan: An Annex to the National Infrastructure Protection Plan* (U.S. Department of Homeland Security 2010):

- *Entertainment and Media* includes television, radio, motion picture, and media (e.g., newspapers, magazines, and books) businesses.
- *Gaming Facilities* represents casinos and associated businesses such as hotels and restaurants.
- *Lodging* refers to hotels and motels.
- *Outdoor Events* are amusement parks, fairs, parks, and other outdoor venues.
- *Public Assembly* refers to conventions centers, stadiums, arenas, movie theaters, and other sites where many people gather.
- *Real Estate* represents office buildings, apartment buildings, condominiums, and other similar sites.
- *Sports Leagues* consist of the major sports leagues and federations.
- *Retail* includes shopping malls, strip malls, and freestanding retail sites.

A large number of people congregate at these locations to live, work, conduct business, shop, and enjoy restaurants and entertainment (Figure 17-1). The CF sector is dependent on other

FIGURE 17-1 Many people congregate at commercial facilities and protection is essential.

sectors such as energy, banking and finance, and transportation. The CF subsectors have similar and dissimilar protection needs, and since most of the CF sector is privately owned, the operators are responsible for assessing risks and implementing protection and resilience programs. Day-to-day protection is often the responsibility of private sector proprietary and contract security personnel with the assistance of physical security and fire protection technology. Public safety agencies provide emergency response.

HSPD-7 designated the U.S. Department of Homeland Security as the sector-specific agency for the CF sector. DHS provides assistance through advisories and alerts; efforts to bring groups together to enhance protection; site assistance; assessment tools; courses; and designation of an event (e.g., Superbowl) as a National Security Special Event, resulting in federal law enforcement participation and aid. Various guidelines, standards, codes, and regulations involving private sector organizations and government agencies also play an important role in protecting people and assets at CF. For example, FEMA published a series of risk management manuals providing guidance for designing buildings and mitigating threats and all-hazards.

Retail Subsector

The retail subsector is emphasized here. It contains many characteristics, vulnerabilities, threats, and hazards facing other CF subsectors. Examples are crimes associated with the acceptance of payment for products and services (e.g., internal theft and robbery), a constant flow of customers, access control and crowd control challenges, and the hiring of temporary and part-time employees. At the same time, the retail subsector is unique because store designs encourage people to enter. There is no payment to enter and very limited access controls. In addition, stores that "harden" security may discourage customer traffic and sales. For instance, customers often carry bags and other items that may contain contraband; searches would inconvenience customers.

The retail subsector generates over $4 trillion in annual sales, consists of 1.6 million U.S. establishments, and employs over 24 million (U.S. Department of Homeland Security 2010: 7). Collective interests of the retail industry are represented through the National Retail Federation and the Retail Industry Leaders Association. These professional groups enhance the retail industry through advocacy (i.e., communicating with legislators), research, professional development, education programs, loss prevention, protection of the global supply chain, and several other activities.

The retail subsector is regulated primarily at the state and local levels through building and fire codes that have an emphasis on safety. On the federal level, OSHA requirements promote safety and OSHA has increasingly encouraged workplace violence prevention programs as explained in the next chapter. Various public and private sector standard-setting organizations and government requirements also apply to retailing.

Retailers apply numerous security and loss prevention strategies as explained throughout this book. With a very small profit margin in many retail businesses, loss prevention is a necessity for survival. Of particular importance is screening applicants, training, business continuity planning, and coordination with public police and other public safety agencies.

Loss Prevention at Retail Businesses

Shrinkage

As defined in Chapter 11, **shrinkage** is the amount of merchandise that has disappeared through theft, has become useless because of breakage or spoilage, or is unaccounted for because of sloppy recording. It is often expressed as a percentage. In retail businesses, guidelines are often in place to hold managers of stores, departments, and loss prevention responsible for shrinkage. The reality of retailing is that those who fail to meet shrinkage goals may be dismissed.

Several organizations publish shrinkage figures and these organizations depend on reporting by retailers; there is no government requirement that retailers must report shrinkage figures. The Center for Retail Research publishes an annual Global Retail Theft Barometer (GRTB). For the twelve-month period ending June 2011, the GRTB reported shrinkage cost respondent retailers $119.1 billion, which equaled 1.45 percent of retail sales. The results were 6.6 percent higher than the previous year. The sources of shrinkage face continuous debate, especially in reference to shoplifting versus employee dishonesty. In reference to causation, the 2011 GRTB showed the following percentages of total shrinkage: 43.2 percent from shoplifting and organized retail theft, 35.0 percent from employee dishonesty, 16.2 percent from errors in pricing and accounting mistakes, and 5.6 percent from supplier/vendor fraud (DiLonardo, 2012: 60).

The University of Florida, noted for its *National Retail Security Survey* (NRSS), found that, in 2010, 140 responding retail chains reported an average shrinkage of 1.49 percent of total annual sales, a slight increase from the previous year. Employee theft resulted in 45 percent of shrinkage, whereas shoplifting resulted in 31 percent of shrinkage. These figures were almost identical to the previous year. This survey of U.S. retailers reported $35.28 billion of lost profits from shrinkage, $15.9 billion from employee theft, $10.94 billion from shoplifting, and the remainder from paperwork errors and vendor fraud (Hollinger, 2011: 24–26).

The Hayes International *Annual Retail Theft Survey* found that, in 2010, dishonest employees stole almost six times the amount stolen by shoplifters. On a per case average, employee thieves stole $640 and shoplifters stole $108 (Doyle, 2011: 10).

For career survival in retail loss prevention, an accurate inventory and precisely pin-pointing the cause of shrinkage are very important.

Human Resources Problems in Retailing

Because retail loss prevention is highly dependent on the efforts of all employees, it is important to discuss the realities of human resources problems in retailing. This includes many part-time and/or temporary employees, inexperienced workers, training challenges, employees dissatisfied with working conditions (e.g., low wages and long hours), and high turnover. Research by Hollinger (2011: 26) found the following: "When we looked at the most likely causes of inventory shrinkage, both sales associate turnover and heavy reliance on a part-time workforce are again the two most obvious correlates." Loss can become a by-product of each of these personnel factors. For instance, some part-time employees may be working during holiday seasons to make extra money and may steal to support gift expenses. A high rate of turnover creates additional training expenses and many inexperienced employees. These problems, coupled with poor performance, are especially troublesome to retailers because such employees are in direct contact with customers and this can have a negative impact on sales and customer loyalty.

Remedies for the above challenges include cost-effective screening, adequate socialization, auditing, and investigations. Another possible measure would be to assign part-time and temporary employees to be supervised by and work with permanent, experienced employees. Additionally, inexperienced employees could be barred from performing certain tasks and entering specific areas. Besides potential losses from part-time and temporary employees, loss prevention planning must also consider full-time regular employees.

Screening

The quality of job applicant screening is dependent on numerous factors. In a small store, the owner may interview the applicant; record pertinent information, such as address, telephone number, and social security number; and ask for a few references. In large multistore organizations, however, employment procedures are commonly more structured and controlled. Cambern (2010: 71) writes of vulnerabilities from retail employees obtained from staffing agencies. She writes that adverse results (e.g., criminal convictions) from background checks are higher for contract workers that for other employees. In addition, she notes that the most common alert is a bogus social security number, which signals false identity. Hollinger (2011: 26) reports that the "hottest" screening methods are criminal history checks, followed by honesty testing and computer-assisted interviews.

Socialization

Various training programs are applicable to adequately socialize an employee toward business objectives. Training can reduce employee mistakes and losses, raise productivity, create customer satisfaction, and reduce turnover.

Some retail businesses require employees to sign a statement that they understand the loss prevention program. This procedure reinforces loss prevention programming. Furthermore, as models of appropriate work attitudes and behavior, executives should *set a good example and practice what is preached.*

Numerous loss prevention awareness programs are applied by retailers. These include anonymous telephone hot lines, online reporting, discussions about shrinkage, bulletin board notices and posters, employee code of conduct, honesty incentives, and a variety of online and video training programs.

Internal Loss Prevention Strategies

The content of Chapter 7 on internal theft is pertinent to retailing. There are numerous targets for internal theft: merchandise, damaged items, cash, repair service, office supplies and tools, parts, time, samples for customers, food and beverages, and personal property. Research by Hollinger (2011) shows that apprehension and termination are common responses to employee theft. In addition, rapid detection of dishonest employees appears to keep losses down.

Although internal theft is a major part of the internal loss problem, there are other categories of internal losses. Examples are accidents, fire, unproductive employees, unintentional and intentional mistakes, and excessive absenteeism and lateness.

Besides screening applicants, training, and technological solutions, here are other strategies applied by companies to address internal losses:

- *Motivation, morale, and rewards:* Management must attempt to help employees feel as if they are an integral part of the business organization. Praise for an employee's accomplishments can go a long way in improving morale. Other methods of increasing morale and motivation are clean working conditions and participative management. Contest and reward programs are also helpful. The employee with the best loss prevention idea of the month could be rewarded with $50 and recognition in a company newspaper.
- *Employee discounts:* Employee discounts usually range between 15 percent and 25 percent. These discounts are an obvious benefit to employee morale even though employees sometimes abuse this discount by making purchases for relatives and friends.
- *Shopping service:* A **shopping service** is a business that assists retail loss prevention efforts by supplying investigators who pose as customers to test retail associates for honesty, accuracy, and demeanor. One common test of associates involves two shoppers who enter a store separately, acting as customers. One buys an item, pays for it with exact change, and leaves, while the second shopper, pretending to be a customer, observes whether the associate rings up the sale or pockets the money. Theft of cash is not the only source of loss; revenues are lost because of the curt or even abrasive behavior of some sales associates.

- *Undercover investigations:* Investigations can be used as a last resort when other controls fail. By penetrating employee informal organizations, investigators are able to obtain considerable information that may expose collusion and weaknesses in controls.

Preventing Losses at the Checkout Counters

Checkout counters are also called point-of-sale (POS) areas (Figure 17-2). These locations in a retail business accommodate customer payments, refunds, and service. Although most cashiers are honest, and scanning technology prevents losses, the following activities hinder profits:

1. Stealing money from the cash drawer
2. Overcharging customers, keeping a mental record, and stealing the money at a later time
3. Presenting the customer with the wrong change
4. Accepting bad checks, bad charge cards, and counterfeit money
5. Making pricing mistakes
6. Undercharging for relatives and friends
7. Failing to notice an altered price
8. Failing to notice shoplifted items secreted in legitimate purchases

FIGURE 17-2 Checkout counter, CCTV camera domes, and electric article surveillance (EAS) system (portals at exit). *Courtesy: Sensormatic, Inc.*

O'Donnell and Meehan (2012) report that retailers are increasingly adding self-checkouts for customer convenience. However, the technology creates opportunities for theft, such as leaving items in the cart, hiding items, or manipulating an item so it is not scanned.

Cashier Socialization of Procedures

The quality of training will have a direct bearing on accountability at checkout counters. Procedural training can include, but is not limited to, the following:

1. Assign each cashier to a particular cash drawer and register/computer.
2. Have each transaction recorded separately and the cash drawer closed afterward.
3. Establish a system for giving receipts to customers.
4. Train cashiers to count change carefully.
5. Show cashiers how to spot irregularities (e.g., altered prices).
6. Encourage cashiers to seek supervision when appropriate (e.g., question about customer credit).

Point-of-Sale Accounting Systems

Retailers apply **point-of-sale (POS) accounting systems** with bar code or Radio Frequency Identification (RFID) technologies to produce vital business information. (See Chapter 11 for the applications of RFID technology.) During inventory or at the checkout counters these systems collect information from the bar code or RFID attached to merchandise. This stored information provides a perpetual inventory and is helpful in ascertaining what is in stock and in ordering merchandise. In addition to shrinkage figures, POS systems produce a variety of loss prevention reports, exposing cashiers who repeatedly have cash shortages, bad checks, voids, and other problems. Voids are used to eradicate and record mistakes by cashiers at the checkouts. Theft occurs when, for instance, a cashier voids a legitimate no-mistake sale and pockets the money. Specialized software helps to pinpoint irregularities. Ratios are also helpful to spot losses; examples include cash to charge sales and sales to refunds. These processes and software within POS systems are designed to reduce and investigate irregularities and are referred to as **exception reporting**.

The Counterfeiting of Bar Codes

As technology improves retail operations, it also presents vulnerabilities. The counterfeiting of bar codes is a nuisance to retailers. The fraud involves using a computer to scan bar codes of inexpensive merchandise, printing copies of the bar codes, and placing the counterfeit bar codes onto expensive items. In addition, offenders are assisted by software available on the Internet. In one case, two married couples used counterfeit bar codes to purchase merchandise at Wal-Mart and two other retail chains, and then they returned the items for full value. Their con included entering stores at peak hours when employees were busy. Wal-Mart lost about $1.5 million from the scam. In another case, an offender used counterfeit bar codes to purchase Lego sets for a fraction of the actual value, and then he sold the sets on a website for toy collectors (Zimmerman, 2006; Beaulieu, 2005: 15).

CCTV and Video Analytics

POS systems are often linked to IP-based network cameras, video analytics, and network video recording (see Chapter 7). System capabilities enable retailers to view video of a cash register total, date, time, and other information as the merchandise is being sold. Furthermore, "exceptions" noted by the POS system can also trigger CCTV and a recording for later reference. Certain retailers are recording every transaction on every register. Today's technology permits the viewing of video images on personal digital assistants and cell phones.

The Internet has enhanced the capabilities of CCTV systems. At one retail store chain, the plan is to keep an eye on widely dispersed stores, avoid the expense of a central station, and reduce time and travel costs of loss prevention personnel. The installed system permits real-time video through the Internet from a PC, laptop, or other device. Video is accessed through a secure, password-protected website. The digital video is archived on a secure server at a remote location without the need for loss prevention personnel to maintain files. Optional features helped to sell the system to senior executives. Examples are customer traffic counting, time and attendance of employees, and the transmission of in-house commercials to store TV screens. CCTV cameras are placed at front and back doors, the POS, and the back hallway. The cameras at the doors are linked to contact alarms so that each time a door is opened, loss prevention personnel receive an e-mail. In one case, a manager was caught improperly opening a back door by not following strict procedures. The manager was called on the telephone immediately for corrective action (Anderson, 2001: 22–23).

Aubele (2011: 48–53) describes the combination of POS and video analytics as follows. Although these systems are improving, Aubele refers to the analysis of pixels of digitized video of customer and cashier behaviors to spot irregularities, such as when a cashier pretends to scan a bar code of an item so a friend receives the item for free. Such automation helps to mitigate the problem of fatigue from watching TV monitors. These systems are also capable of identifying an item left in or at the bottom of a shopping cart at the POS, reading the bar code, and adding it to the transaction. Retailers often test new technology at a few stores before installation company-wide.

Aubele (2011: 48–53) refers to the benefits of cloud computing (see Chapter 16) in that retailers often do not have the capital to invest in expanded IT, so it can be rented rather than purchased. Retailers can store POS and camera data off-site with the cloud vendor. Percoco (2011: 76–78) warns that the many POS systems with direct access to the Internet face malware, botnets, and other risks. He argues for strong cybersecurity to protect these systems.

As the effectiveness of facial recognition systems increase and as the cost drops, retailers will increasingly apply this technology to identify known shoplifters, banned customers, and others who committed fraud and other crimes against the retailer.

Electronic Payments

Electronic payments, such as the use of credit and debit cards, are more popular than payments by cash or check. Credit cards permit the extension of credit with delayed payment, whereas debit cards charge the account at the time of the transaction. Credit and debit cards have a magnetic stripe that contains information on the account, such as customer name and account number. The customer's card is "swiped" for electronic authorization to ensure that the account is valid and the purchase is within the account limit. Then the transaction is completed with the user's signature. Although technology assists retailers in speeding payment transactions and saving on fees, at the same time, technology brings vulnerabilities (Federal Reserve Bank of San Francisco, 2012).

In many locales throughout the world, a conversion is occurring from magnetic stripe to the "chip-and-PIN" card to speed transactions, especially for low-value payments (e.g., a cup of coffee), and to increase security. In the "chip-and-PIN" card, information is stored in a small computer chip. The customer places the card near a proximity reader, similar to an access card, and enters a personal identification number (PIN) instead of providing a signature. This newer type of card is less likely to be copied than the magnetic stripe card. A variation of this technology is referred to as "contactless payment systems," and a PIN is not required. Both the "chip-and-PIN" and "contactless" methods use a proximity reader to obtain information from the RFID computer chip imbedded in the card. Proponents claim that encryption protects the information. Critics argue that an offender nearby, armed with a specialized reader, might capture information on cards held by the customer (Federal Reserve Bank of San Francisco, 2012; DiLonardo, 2006: 116).

Another type of technology to save retailers money and provide quicker customer payment is known as "pay-by-finger." This biometric fingerprint technology requires customers to place a finger on a scanner to register the transaction with their bank account to debit their account without using a card, entering a PIN, or writing a check. Customers enroll at a kiosk by providing identification, bank information, and a finger scan. The information is encrypted and store employees do not have access to the information. Opponents of this technology claim that retailers should consider the operational impact on the business, data center issues, setup costs, and customer security (Gaur and Gilliland, 2006).

Card fraud is a huge problem costing billions of dollars. The three major groups involved in card usage are *card issuers* (banks, oil companies, retail businesses, travel, and entertainment groups), *acceptors* (merchants), and *users*. All of these groups are susceptible to losses due to fraud. Lost or stolen cards can cause monetary loss to users. Acceptors who are careless may become financially responsible for fraud under certain circumstances and could even be placed at a competitive disadvantage if no longer authorized to accept the issuer's card. Card issuers absorb billions of dollars in losses annually. Crime involving cards is varied and compounded because of identity theft. Vulnerabilities include theft of cards from the postal system and counterfeiting of cards.

The *Commercial Facilities Sector-Specific Plan: An Annex to the National Infrastructure Protection Plan* (U.S. Department of Homeland Security 2010: 160) refers to the Payment Card Industry Data Security Standard. It is a global information security standard designed to assist entities in preventing fraud when processing card payments. Businesses such as MasterCard, Visa, and American Express worked to prepare the standard. It covers topics such as network security and protecting cardholder data.

Although laws have been enacted to deter card fraud (e.g., Credit Card Fraud Act), the effectiveness of criminal laws is limited; we still have widespread card fraud. As with other risks, prevention is vital. Following are some loss prevention strategies (Federal Reserve Bank of San Francisco, 2012) for retailers, although, if a store is busy, employees may not have time to thoroughly check cards.

- Ask the customer for a photo ID containing their signature and compare the photo with the customer and the signature on the ID and card.
- Check for alterations in a card, such as letters and numbers not lining up.
- Look for the holograph that changes its color and image as the card is tilted.
- Search for fine-line printing, such as a repeated pattern of the card company name.
- If the card contains ultra-violet ink, it will be visible under ultra-violet light and display, for instance, the card company logo.

E-Business

E-business (using Internet-based technologies), phone order, and mail order are particularly vulnerable to fraud because the customer and the card are not present for the transaction. Offenders obtain card numbers from many sources (e.g., discarded receipts, stealing customer information, and hacking into a business), establish a mail drop, and then place fraudulent orders. Woodward (2012) reported that an *Internet Retailer* survey found that 72 percent of e-retailers have a fraud rate of less than 1.0 percent and 21 percent have a rate of between 1.0 and 3.0 percent.

In one case, the owner of a computer parts company in New Jersey became concerned when he received a $15,000 order from Bucharest, Romania. The owner tried to contact the credit card authorization company and the bank that works with the processing company to verify the credit cards. However, they were not able to assist, so the owner telephoned the customer and had him fax his Romanian driver's license and other documents to the owner. The owner then shipped the parts via UPS at about the same time the bank became suspicious and called the owner to say that all the cards used in the order were fraudulent. The owner was lucky when the shipment was intercepted in Bucharest.

Various tools are available to retailers to control fraud. Examples are IP geolocation products that analyze transaction risk based on geographic data, tools that use various metrics (e.g., dollar amounts and frequency) to monitor orders, and lists that show good customers. Barlas (2012) offers the following loss prevention strategies:

- Exercise caution (e.g., telephone the customer to confirm the order) if the shipping address differs from the billing address.
- Besides obtaining a card number and expiration date, request the three or four digit code on the card. The location of the code on the card varies with the issuer.
- Avoid processing an order if card and other necessary information are incomplete.
- Exercise caution with large orders, those from overseas, purchases with several cards, and orders from e-mail forwarding addresses.
- Employ card issuer services that authenticate customers.

Visa Security

Swartz (2012) describes security at a Visa Operations Center. Because hundreds of millions of daily card transactions are processed at Operations Center East (OCE), and at one other North American site, and because purchases on smartphones and tablets are expected to grow substantially, security is a high priority for Visa.

The main threats are profit-minded hackers and fraud. Visa employees work to keep hackers out of the network, address potential fraud, and ensure the network operates as designed. Visa has spent hundreds of millions of dollars on state-of-the-art technology and software (e.g., transaction risk scoring, transaction alerts, and encryption). Such efforts have reduced global fraud rates to 6 cents per $100 spent.

Security at OCE includes former military personnel, hundreds of cameras, bollards beneath the road, close turns to slow vehicles, and a drainage pond that serves as a moat. To access the huge facility, a guard station is the first layer of security, followed by a mobile security officer, and then a photo and fingerprint are required. Next, a "mantrap" is opened with the use of a badge and biometric image of the fingerprint.

The OCE command center looks like the one at NASA and includes a 40'×20' wall of screens to monitor the global network that processes about 2,500 transactions per second. The main walkway is approximately three football fields long and is connected to several pods, each containing 20,000 square feet for mainframes, storage, cables, and other items.

The facility can withstand earthquakes and hurricane winds of up to 170 mph. In the event of a power failure, diesel generators can supply electricity for nine days.

Data centers are increasing in number because of the enormous demand for digital data, the growth of cloud computing, and the use of smartphones and tablets. Google, Facebook, and Apple are among many companies that have built huge data centers, often in rural areas to save money on land and electricity.

Checks

Although the popularity of credit and debit cards has reduced the use of checks, retailers continue to sustain losses from checks. *A check is nothing more than a piece of paper until the money is collected.* It may be worthless. Characteristics of bad checks include an inappropriate date, written figures that differ from numeric figures on the same check, and smeared ink. Sources of bad checks are varied and include customers who write checks with no funds in their checking accounts. Offenders may write bad checks by using stolen checks or by writing checks from a nonexistent account. Counterfeit checks (e.g., bogus personal checks or bogus payroll checks) are another problem growing worse because of computers, desktop publishing, check-writing software, newer color copier machines, scanners, and laser printers.

Various strategies are applied to mitigate the problem of bad checks. A retailer may choose to hire a service that verifies the legitimacy of checks before accepting them. Retailer policies and procedures may include carefully examining checks, not accepting checks over a certain amount, seeking supervisory approval for checks written over a certain amount, prohibiting checks from out of state, never providing cash for a check, and obtaining a thumbprint of the customer. Retail employees should scrutinize customer identification for irregularities. Many employees look at identification cards but do not see irregularities. *Concentration is necessary to match (or not to match) the customer with the identification presented.*

A retailer's recovery from a bad check depends on the circumstances. Procedures depend on the state and retailers should query local police and prosecutors. A retailer may have to send the check writer a registered letter requiring payment (if the check writer has a legitimate address). If the letter is not effective, the retailer may be required to sign a warrant against the person who wrote the bad check.

In one jurisdiction, a "bad-check brigade" was formed. A magistrate coordinates the warrants, which are distributed to constables (part-time law enforcers) who are paid for each warrant served. Each week the brigade takes numerous offenders to jail. Retailers should check with public police about local practices. Some jurisdictions require collection through civil procedures. Another strategy, especially for large retailers, is to contract collection work to specialized firms.

If a retailer receives a check returned from a bank stating that there is "no account" or "account closed," then fraud may have been perpetuated. The police should be notified. An altered or forged U.S. government check should be reported to the U.S. Secret Service.

Refund Fraud

In loosely controlled businesses, employees have an opportunity to retain a customer receipt (or hope that the customer leaves one) and use it to substantiate a fraudulent customer refund (i.e., merchandise supposedly returned for money). With the receipt used to support the phony refund, the employee is "covered" and can pocket the cash.

Especially bold offenders might enter a store, pick up merchandise, and then go directly to the checkout counter and demand a cash refund without a receipt. Another ploy is to use a stolen credit card to purchase merchandise and then obtain a refund at another branch store. In other instances, collusion may take place between employees and customers.

Refund fraud is minimized through well-controlled supervision and accountability. Consider requesting photo identification and writing a check for refunds over a certain amount. A supervisor should account for returned merchandise before it is returned to the sales floor and sign and date the sales receipt after writing *refund*. Stelter (2011) reports that retailers are finding that issuing carefully worded warnings to suspect-customers reduces the problem. In addition, knowing that customer service is important and that a small percentage of offenders commit refund fraud, one retailer established a controlled environment for refunds whereby loss prevention professionals are notified if deviations from policy occur.

Gift Card Losses

Gift cards generate profits for retailers and the use of these cards is increasing. The methods of gift card fraud, committed mostly by employees, includes card swapping by cashiers at the POS, "skimming" gift cards to produce duplicates, and using fraudulent credit cards to purchase gift cards. Fraudulent gift cards are used by employees, traded, or sold to local residents or online. Countermeasures focus on inventories of cards, sales data, exception reports, and surveillance.

Counterfeiting

Counterfeiting is the unlawful duplication of something valuable in order to deceive. Counterfeit items can include money, coupons, credit or debit cards, clothes, and jewelry. Here, the emphasis is counterfeit money. The U.S. Secret Service investigates this federal offense.

Persons who recognize that they have counterfeit money will not be reimbursed when they give it to the Secret Service. Because of this potential loss, many people knowingly pass the bogus money to others. The extended chain of custody from the counterfeiter to authorities causes great difficulty during investigations.

Counterfeiting is a growing problem because of the newer color copier machines, scanners, computers, and laser printers that are in widespread use. The best method to reduce this type of loss is through the ability to recognize counterfeit money, and an excellent way to do this is to compare a suspect bill with a genuine bill. The U.S. government has countered counterfeiting through a variety of security features on bills. One should look for the red and blue fibers that are scattered throughout a genuine bill. These fibers are curved, about 1/4 inch long, hair thin, and difficult to produce on bogus bills. Other security features on genuine bills are a security thread embedded in the bill running vertically to the left of the Federal Reserve seal, microprinting on the rim of the portrait, a watermark (hold the bill up to light to see the faint image similar to the portrait), and color-shifting ink (tilt the bill repeatedly to see shifting ink color).

Watch for $1 bills that have counterfeit higher denomination numbers glued over the lower denomination numbers. Also, compare suspect coins with genuine coins.

Another technique is the bleaching of a real, lower denomination bill (to wash away the ink), followed by placing the blank bill on a printer to apply a higher denomination bill. This method of counterfeiting may defeat the use of counterfeit detector pens used by cashiers because the pens react to starch in paper by showing a black mark. However, genuine bills are actually cloth and devoid of starch; the pen will create a golden mark.

Shoplifting

Shoplifting is a multibillion-dollar problem as presented in the explanation of shrinkage. To reduce it, loss prevention practitioners should have an understanding of the types of shoplifters, motivational factors, shoplifting techniques, and countermeasures.

Types of Shoplifters and Motivational Factors

Amateur shoplifters, also referred to as *snitches*, represent the majority of shoplifters. These persons generally steal on impulse while often possessing the money to pay for the item. Individuals in this category represent numerous demographic variables (e.g., sex, age, social class, ethnicity, and race). The distinguishing difference between the amateur and the professional thief is that the former shoplifts for personal use, whereas the latter shoplifts to sell the goods for a profit.

Generally, *juveniles* take merchandise that they can use, such as clothing and recreational items. Frequently working in groups, their action is often motivated by peer pressure or a search for excitement.

Because of the nature of their work, *easy-access shoplifters* are less likely to be scrutinized than customers and they may be familiar with retail operations and loss prevention programs. Delivery personnel, repair technicians, and public inspectors are included in this category. Even public police and fire personnel have been known to represent this group, especially during emergencies. The motivation is obvious: to get something for nothing.

Fairly easy to detect, *drunk or vagrant shoplifters* usually steal liquor, food, and clothing for personal use or shoplift other merchandise to sell for cash. These persons often are under the influence of alcohol and have a previous alcohol-related arrest record.

Addict shoplifters are extremely dangerous because of their illegal drug dependence and accompanying desperation. These persons generally peddle stolen items for illegal drugs or cash. Addicts may "grab and run."

Kleptomania is a rare, persistent, neurotic impulse to steal. Kleptomaniacs usually shoplift without considering the value or personal use of the item, and seemingly want to be caught. This type of shoplifter usually has a criminal record from previous apprehensions and may have been caught several times at the same retail store.

Professionals or *"boosters"* account for a small percentage of those caught shoplifting. (This low figure could be the result of the professional's skill in avoiding apprehension.) The motive is profit or resale of shoplifted merchandise. Professionals may utilize a *booster box* (i.e., a box that looks wrapped and tied, but really contains a secret opening), hooks inside clothing, or extra long pockets. A criminal record is typical, as are ties to a fence and ORT.

Organized Retail Theft and Countermeasures

Organized retail theft (ORT) refers to gangs of offenders who victimize retail businesses by stealing and selling the merchandise for illegal profit. The Federal Bureau of Investigation (2011a) and the retail industry refer to losses from ORT at $30 billion annually; this figure includes other crimes against retailers such as card fraud. The FBI focuses primarily on significant cases involving the interstate transportation of stolen property. Its strategies include partnering with other police agencies and retailers, undercover operations, and recruiting offenders as informants (e.g., "letting a little fish off easy to prosecute a big fish"). Since police resources are limited, the retail industry is heavily involved in combating ORT through ORT investigative teams, use of data bases, training, and improving legislation. Cooperation is essential between retailers and law enforcement agencies, especially because vehicles may be stopped, search warrants executed, and subpoenas obtained for Internet records.

ORT gangs sell stolen merchandise to local residents, at flea markets and Internet auction sites, back to legitimate retailers, and/or work with a network of fences who sell the items to similar markets. The *e-fence* sells stolen items online.

Muscato and Pearson (2010: 58–66) refer to the "Booster Business Plan." They write that professional shoplifters and ORT gangs may set a daily monetary theft goal, work hard to reach the goal, and the results show on their "profit and loss statement" as tax-free profit. The business plan also considers risk management. For example, losses are mitigated by using multiple fences in case one is arrested and the booster's cash flow is interrupted.

The financial impact of just one booster gang in one targeted market is explained next from Muscato and Pearson. Before members of this ORT gang were arrested, they stole about $5,000 per day and worked about five days a week. Muscato and Pearson estimated a cumulative annual loss at retail of $1,300,000 and the gang's profit margin averaged 20 percent on the dollar, yielding a gross income of about $260,000. This example shows the importance of combatting ORT.

Muscato and Pearson offer the following remedies. Increase associate awareness and action, such as "Bob" and "Lisa." At the POS, the former refers to checking the bottom of the basket and the latter refers to looking inside always (i.e., merchandise hidden in merchandise). They also recommend using a greeter, educating police detectives who investigate property crimes, and more precise criminal laws that focus on those who try to defeat electronic article surveillance systems, create bar codes and receipts, and sell stolen items at flea markets.

Shoplifting Techniques

The following list presents only a few of the many shoplifting techniques:

1. Shoplifters may work alone or in a group.
2. A person may simply shoplift an item and conceal it in his or her clothing.
3. A person may "palm" a small item and conceal it in a glove.
4. Shoplifters often go into a fitting room with several garments and either conceal an item or wear it and leave the store.
5. An offender may ask to see more items than a clerk can control or send the clerk to the stockroom for other items; while the counter is unattended, items are stolen.
6. A self-service counter can provide an opportunity for a shoplifter to pull out and examine several items while returning only half of them.
7. Merchandise is often taken to a deserted location (e.g., restroom, elevator, stockroom, janitor's supply room, and so on) and then concealed.
8. A shoplifter may simply grab an item and quickly leave the store.
9. A shoplifter may drop an expensive piece of jewelry into a drink or food.
10. Shoplifters arrive at a store early or late to take advantage of any lax situation.
11. Disguised as a priest (or other professional), the shoplifter may have an advantage when stealing.
12. Some bold offenders have been known to impersonate salespeople while shoplifting and even to collect money from customers.
13. Price tags are often switched to allow merchandise to be bought at a lower price; sometimes the desired price will be written on the price tag of a sales item.
14. Shoplifters are aided by large shopping bags, lunch boxes, knitting bags, suitcases, flight bags, camera cases, musical instrument cases, and newspapers.
15. Dummy packages, bags, or boxes ("booster boxes") are used, which appear to be sealed and tied but contain false bottoms and openings to conceal items.
16. Shoplifters may use hollowed-out books.
17. Stolen merchandise is often concealed within legitimate purchases.
18. Expensive items are placed in inexpensive containers.

19. Shoplifters sometimes slide items off counters and into some type of container or clothing.
20. Sometimes shoplifters wear fake bandages or false plastic casts.
21. Professional shoplifters are known to carry store supplies (e.g., bag, box, stapler, or colored tape) to assist in stealing.
22. Baby carriages and wheelchairs have been utilized in various ways to steal.
23. Sometimes items are hidden in a store for subsequent pickup by an accomplice.
24. Various contrived diversions (e.g., dropping and breaking something, faking a medical problem, setting off a smoke bomb, or having someone call in a bomb threat) have been used to give an accomplice a chance to shoplift.
25. Teenagers sometimes converge on a particular retail department, cause a disturbance, and then shoplift.
26. Adult shoplifters have been known to use children to aid them.
27. Sometimes "blind" accomplices with "guide" dogs are used to distract and confuse sales personnel eager to assist the disadvantaged.
28. "Crotchwalking" is a method whereby a woman wearing a dress conceals an item between her legs and then departs.
29. Baillie (2012: 31) writes that the Internet, YouTube, and eBay contain information and products helpful to shoplifters in the form of "how to" articles and videos, plus devices for sale that circumvent alarms and EAS tags.

Prevention and Reduction of Shoplifting through People

The reality of retail loss prevention budgets is explained by Hollinger (2011: 26). He notes that in 2010, 0.46 percent of retail sales were directed to loss prevention, with most of this money spent on payroll expenses. In his articles over the years, Hollinger repeatedly points out, and correctly, the following: "With limited LP budgets and even less money for high-tech countermeasures, more of the day-to-day responsibility for loss prevention is being shifted to overworked store managers, untrained sales associates, and inexperienced LP personnel."

People are the first and primary asset for reducing shoplifting opportunities. The proper utilization of people is the test of success or failure in preventing and reducing shoplifting. Good training is essential and it can be applied in-store, online, or off-site.

- *Management:* Management has the responsibility for planning, implementing, and monitoring antishoplifting programs. The quality of leadership and the ability to motivate people are of paramount importance. The loss prevention manager and other retail executives must cooperate when formulating policies and procedures that do not hamper sales.
- *Salespeople:* One method to increase a shoplifter's anxiety and prevent theft is to require salespeople to approach all customers and say, for instance, "May I help you?" This approach informs the potential shoplifter that he or she has been noticed by salespeople and possibly by loss prevention personnel.
- *Loss Prevention (LP) Agents*: These specialists must have the ability to observe without being observed, remember precisely what happened during an incident, know criminal law and self-defense, effectively interview, testify in court, and recover stolen items. They

perform a variety of duties, such as observing the premises with a CCTV system, helping to reduce shrinkage, and conducting investigations of suspected internal theft and shoplifting. When they are on the sales floor and observing a suspect, they should blend in with the shopping crowd and look like shoppers. This can be done by dressing like average shoppers, carrying a package or two and even a bag of popcorn. The antiquated term "floor walker" should be reserved for historical discussions; this will help to professionalize the important LP position that has certification programs through the Loss Prevention Foundation.

- *Uniformed Officers:* Only a foolish shoplifter would steal in the presence of an officer who acts as a deterrent. A well-planned antishoplifting program must not lose sight of the systems approach to loss prevention, as described next (Altheide et al., 1978):

  The store had four security guards or personnel. These people were also looked upon as "jokes." The security personnel were primarily concerned with catching shoplifting customers. Most of their time was spent behind two-way mirrors with binoculars observing shoppers. Therefore, in reality while the security personnel caught a customer concealing a pair of pants in her purse, an employee was smuggling four pairs of Levi's out the front door. Security was concerned with shoppers on the floor while all employee thefts usually occurred in stockrooms.

- *Fitting-room personnel:* Employees who supervise merchandise passing in and out of fitting rooms play a vital role in reducing the shoplifting problem. Many stores place a limit on the number of items that can be brought into a fitting room.

Prevention and Reduction of Shoplifting through Physical Design and Physical Security

Physical design and physical security comprise a second major category of an effective antishoplifting program. In reference to physical security, these *systems are only as good as the people operating them.* A retail company can spend millions of dollars on antishoplifting systems; however, money will be wasted if employees are not properly trained and they are not knowledgeable of system capabilities and limitations.

PHYSICAL DESIGN

Physical design includes architectural design and store layout. Three objectives of physical design are to create an environment that stimulates sales, comply with the ADA, and create CPTED, as covered earlier in this book. The ADA requires the removal of barriers that hinder the disabled. An example of CPTED is increased visibility of customers by personnel (natural surveillance) by designing an employee lounge that is raised above the sales floor and has a large glass window. POS locations that are raised a few inches also increase visibility. Merchandise or other store features that obstruct the view of employees will aid shoplifters. A balance between attractiveness and loss prevention is essential.

Applying rational choice and situational crime prevention theories (see Chapter 3), Cardone and Hayes (2012: 22–58) interviewed shoplifters to identify store physical cues that influenced offenders' decisions to shoplift. Four primary cues influenced the decision: natural surveillance, guardianship levels, formal surveillance, and target accessibility.

Hayes (2006: 46) writes that a retailer can manipulate variables in a store to influence moods and behavior. As examples, he refers to number of entrances and exits, walkways, flooring, fixture types, aisle width, and employee workstations. The proper utilization of turnstiles, corrals, and other barriers can limit the circulation of shoplifters and funnel customer traffic to select locations (e.g., toward the POS).

Other methods to hinder theft include locking display cabinets, displaying only one of a pair, using dummy displays (e.g., empty cosmetic boxes), arranging displays neatly and in a particular pattern to allow for quick recognition of a disruption in their order, having hangers pointing in alternate directions on racks to prevent "grab and run" tactics, and placing small items closer to cash registers.

ELECTRONIC ARTICLE SURVEILLANCE

Electronic article surveillance (EAS) systems "watch" merchandise instead of people. Electronic tags are placed on merchandise and removed or deactivated by a salesperson at the POS. If a person leaves a designated area with tagged merchandise, a sensor at an exit activates an alarm.

These systems have been on the market since the late 1960s. Because these plastic tags tended to be large (about three or four inches) and difficult to attach to many goods, they were used on high-priced merchandise such as coats. However, improved technology permitted manufacturers to develop EAS tags the size of price tags, at a lower cost. Today, inexpensive EAS tags may remain on merchandise following a sale and being deactivated, whereas higher-cost large tags are removed and reused.

The *radio frequency* (RF) is a popular type of EAS system, and it contains a tiny circuit that can be hidden virtually anywhere on a product or in packaging. If a cashier does not remove or deactivate it (which can be done automatically as items are scanned for prices at the POS) and the customer walks near a sensor, an alarm may be triggered. Its weaknesses are that shoplifters can cover the labels with aluminum foil and that the tags cannot be used on metal objects. At the refund desk, a verifier can check for "live" tags that *may* indicate that an item was not purchased legitimately. A second type, *magnetic*, employs a metal strip (called an EM strip) that interferes with a magnetic field at an exit to activate an alarm. This system is often applied in libraries. Another type is the *acousto-magnetic system*, which is similar to magnetic tags. These tags are wider and have superior detection in comparison to magnetic tags. Another type is *microwave*. *Ink tags* are also applied by retailers; these tags contain ink that ruins clothes if the tag is broken during removal.

Each company contends that its system is best. None is without problems, nor is any foolproof. False alarms or failure of the cashier to deactivate or remove the tag has led to retailers being sued for false arrest. Excessive false alarms can result in limited responses by retail employees. Another problem results from employees who carry an EAS tag on their person and activate the system as a pretext to stop and search suspicious customers. Training, and politeness when approaching customers prevents litigation. Experts claim that professional shoplifters and employee thieves are not deterred by EAS. Others view the tags as too expensive to be justified for low-cost merchandise. Many tag only high-shrinkage items. Beck and Palmer (2011: 112) write: "This technology continues to generate an enormous amount of debate concerning how effective it really is and whether it provides a genuine return on investment."

Research by Read Hayes and Robert Blackwood (Gips, 2007: 26) showed that retail employees routinely ignored EAS alarms; of almost 4,000 alarms studied, employees requested receipts for merchandise in only 18 percent of cases. Prior to applying EAS, Hayes and Blackwood recommended strong evidence that a particular item is being shoplifted chainwide and EAS should be made obvious through signs.

Despite criticism of EAS, and initial costs and maintenance, these systems can be cost-effective. They are also popular. In fact, such systems are used in certain industries to "watch" inventories. Prisoners and patients also are being monitored with such technology.

Innovations have further enhanced EAS technology. One type of tag sends out an audible alarm when tampering occurs. Aubele (2011: 52) adds that although traditional EAS systems alert retailers that a theft may have occurred, these systems are unable to identify the merchandise. Modern RFID EAS systems, for instance, can inform the retailer that a belt selling for $59.99 passed through the door. Relevant data helps to plan loss prevention, such as items stolen frequently, from what department, on what days and times, who was on duty, and so forth.

Source tagging is popular; it involves the manufacturer placing a hidden EAS tag into products during manufacturing, to be deactivated at the POS. This saves the retailer the labor and cost of tagging and untagging merchandise. However, research by Beck and Palmer (2011: 110–122) showed that replacing large hard tags with less visible source tags resulted in a dramatic increase in shrinkage. When the choice was reversed, shrinkage dropped considerably. The research results stressed the importance of a visible deterrent and the use of large hard tags that are more difficult to defeat for both shoplifters and internal thieves. Some EAS manufacturers are addressing large hard tags as source tags.

Another loss prevention strategy is **benefit-denial-technology**. It is defined as technology that protects assets by rendering assets unusable to the thief unless purchased legitimately. Examples are ink tags and metal clamps that can ruin an item when removed without legitimate purchase. Other examples are electronic devices (e.g., cell phones and TVs) that do not work until activated following purchase. Benefit-denial technology has a great deal of protection potential for many products from the early stages of production and throughout the supply chain (Hayes, 2012: 39–46).

Alarms

In addition to EAS, various alarm sensors protect merchandise from theft. *Loop alarms* consist of a cable forming a closed electrical circuit that begins and ends at an alarm device. This cable is usually attached through merchandise handles and openings. When the electrical circuit is disrupted by someone cutting or breaking it, the alarm sounds. *Cable alarms* also use a cable that runs from the merchandise to an alarm unit. This alarm differs in that a pad attached to the end of the cable is placed on the merchandise. Each item has its own cable pad setup, which is connected to an alarm unit. Cable alarms are useful when merchandise does not have openings or handles. Retailers also use heavy nonalarmed cables that are also woven through expensive items such as leather coats. This cable usually has a locking device. *Wafer alarms* are sensing devices that react to negative pressure. This device, about the size of a large coin, is placed under the protected item; if a thief removes the item, an alarm sounds.

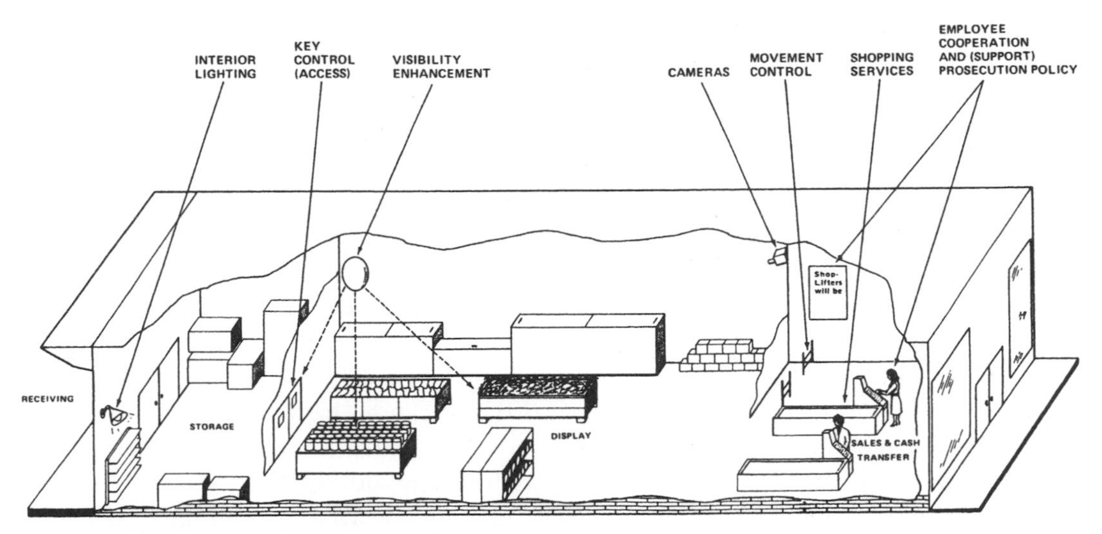

FIGURE 17- 3 Internal space protection.

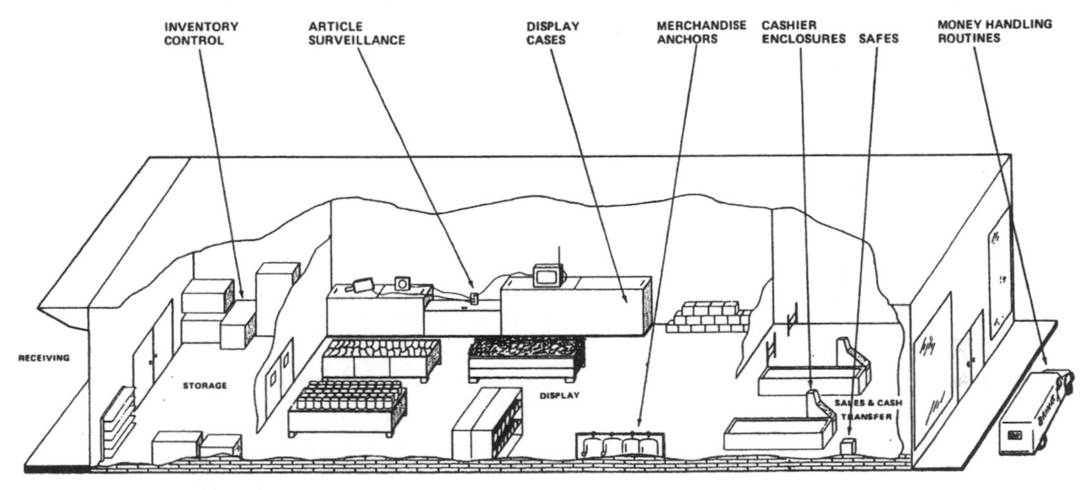

FIGURE 17-4 Point protection.

ADDITIONAL MEASURES

Numerous security measures are applicable to retail stores to thwart not only shoplifting but also other sources of loss. Figures 17-3 and 17-4 focus on internal space protection and point protection. Chapters 7 and 8 explain a variety of security measures, such as CCTV.

Two types of *mirrors* used to curb the shoplifting problem are see-through mirrors and wide-angle convex mirrors. These mirrors are relatively inexpensive and provide personnel with advantages that aid in the surveillance of shoplifters. See-through mirrors, also called two-way mirrors, are different than one-way mirrors. A one-way mirror is a regular mirror that reflects light back to the viewer. A see-through mirror enables the observer to see through the mirror; light strikes the surface and travels two ways: back to you as you see yourself and toward the observer on the other side. To detect this type of mirror, use a laser pointer or good flashlight to see the light go through (TwoWayMirrors.net, 2006). Observation booths are usually equipped with see-through mirrors; their location is above the sales floor at various places within a store. Personnel often use binoculars while watching customers (and employees) from these vantage points. A good communication system is necessary to summon aid because, when the observer leaves the booth, the shoplifter's actions are not being watched. If the shoplifter returns the item before being apprehended, the retail store may face a lawsuit. Wide-angle convex mirrors facilitate greater visibility and reduce blind spots around corners. Convex mirrors and regular mirrors can be located along walls, ceilings, at support columns, above merchandise displays, and at any point to create greater visibility and thus deter theft.

In one case, a juvenile was shoplifting recorded music items by sliding them into the bottom of a bag that had been stapled closed with a receipt attached. The incident occurred while the store was closing, and more than 20 retail employees attentively watched the shoplifter in action from the other side of a seven-foot-high partition that had a convex mirror placed above it. The situation was interesting because the music department had been cleared of retail employees, and the juvenile rushed around the department selecting items while nervously watching for people he did not know were watching him via the mirror.

Fake deterrents are debatable loss prevention methods. A dummy camera or a periodic, fake loudspeaker statement (e.g., "dispatch security to main floor") may prevent shoplifting, but the level of effectiveness is difficult to measure. Many stores combine fake and real methods.

Confronting the Suspected Shoplifter

Detection and Apprehension

Most suspected shoplifters exhibit the following activities and characteristics before, during, and after a theft: extreme sensitivity to those around them, surveillance of the sales floor, nervousness and anxiety, and walking repeatedly to certain areas. Other characteristics that invite suspicion are a group of juveniles or a person wearing excessive clothing on a warm day. Research by Hollinger (2011) showed that apprehension and civil recovery are the most common responses reported by retailers.

Two prerequisites to an apprehension are making sure to have seen the suspect conceal the store's merchandise and never losing sight of the suspect. The observer must be positive that he or she saw the item removed from a shelf because the customer could have brought the item into the store. If a customer wears a store item out of the store, or alters or switches a price tag, store personnel must make sure an eyewitness account is available. Some shoplifters panic on being observed and "ditch" the concealed merchandise or give it to an accomplice. Some people conceal merchandise, return it to the counter unnoticed, and hope to be apprehended in order to sue to collect damages. To help prove intent, store personnel should permit the suspect to pass the last cash register and begin to exit the store. If a register is outside (e.g., at a sidewalk sale), store personnel need to ensure the suspect passes it to strengthen the case.

Juveniles are handled differently than adults. Police and parents usually are called. A loss prevention practitioner should be familiar with local procedures regarding juveniles.

If a retail employee is not certain about a suspect, no apprehension should take place. In this situation, some practitioners recommend the "ghosting technique," which involves an employee carrying a duplicate of the item alleged to have been concealed as he or she walks close to the suspect. No words are spoken, but the message is obvious, if a theft has taken place. This technique has advantages for prevention, but it may permanently frighten customers from the store.

When retail employees positively witness a shoplifting incident, loss prevention personnel must be notified for apprehension purposes. The detector (e.g., sales associate) should assist as an eyewitness. *Although a reasonable degree of force can be used to control the subject, it is best to avoid a physical confrontation because the subject may be armed, people may be injured or killed, and the liability potential is significant. If the suspect flees, personnel should obtain a description and call police.*

Immediately upon approaching the subject, LP personnel should identify themselves by displaying identification or a badge. Personnel must never threaten the subject. *To reduce the problems associated with error, it is wise to not accuse the subject of stealing; the subject can be asked, "Would you mind answering a few questions about an item?" The next step is to quickly ask the subject to accompany personnel to the loss prevention office.* This is perhaps the most crucial point of the confrontation. At all times, careful observation of the subject is important because the merchandise may be "ditched" or escape may be attempted. If the subject escapes, again, personnel should avoid a physical confrontation and call police.

Shoplifter High on LSD Bites Security Officer

Deeth (2012) reports on one shoplifting case where a male teenager was observed ripping tags off of video games and secreting the games in his pants. When he was stopped by two security officers outside the door, a scuffle began and the officers tried to pin the subject to the ground. The subject kicked one officer in the head and then locked onto the other officer's arm with his teeth, necessitating hits to the subject's head as the subject bit a piece of flesh from the officer's arm. The officer had to endure weeks of antibiotic treatment to prevent disease, as well as a lasting injury. When public police arrived at the scene, the subject provided a false name, faked a seizure, and tried to escape. He was sentenced to almost seven months in jail.

Detention and Arrest

After entering the LP office, the subject is usually asked to produce some type of identification as well as the concealed merchandise. If the subject does not comply with these requests, the LP officer should call the public police.

After the merchandise is received, a receipt should be requested to reduce the possibility of error. Usually, a shoplifter will state some type of excuse in an attempt to cover up not having a receipt.

Many jurisdictions have adopted the retailer's privilege of detaining shoplifters. *Conditions for detention involve probable cause (reasonable grounds to justify legal action, such as an eyewitness account) and a reasonable time span for detention to accomplish questioning and documentation.* Such shoplifting statutes provide some protection from liability, provided legal action is conducted in a reasonable manner. The difference between an arrest and detention is that the former requires the arrester to turn the suspect over to the public police, whereas the latter does not.

Johnson and Carter (2009: 446–447) reported on studies of shoplifting arrests by retailers. The studies showed a range of 25 to 60 percent of shoplifting cases resulted in an arrest. They also noted that many shoplifters are not formally charged and that the arrest approach may not be an attractive means of controlling shoplifting because of the risk of injury, liability, and adverse publicity.

If an arrest is made, the crime charged by the retailer should be shoplifting (not larceny) to retain the right to detain under the shoplifting statute. Probable cause is a prerequisite for either detention or arrest. The detention of a person can evolve into an arrest. Many attorneys argue that any type of restricted movement placed on the suspect is equal to an arrest. The ultimate decision may rest with a jury in a lawsuit.

LP personnel—or any store personnel—should refrain from touching the subject. If force is used to exercise legal action, it must be reasonable. When a suspect is controlled after a struggle, it would be unreasonable to strike the suspect. Deadly force is restricted to life-threatening situations. *Unreasonable force can lead to prosecution difficulties, as well as criminal and civil action.*

Subsequent to proper detention or arrest, shoplifting statutes generally do not stipulate how merchandise is to be located and recovered from the suspect. Some states permit a search, whereas others forbid it. A shoplifter can be requested to empty his or her pockets and belongings to produce the stolen item(s). Public police should be called for obstinate suspects, and the police should conduct the search. If an arrest occurs, the stolen merchandise and the eyewitness to the shoplifting incident (plus a CCTV video recording, if available) will be the primary evidence that will aid in the prosecution.

If the subject complies voluntarily, an arrest has not been made unless specified. The LP officer should ask the offender to sign a civil release form, which is vital before releasing the subject. It provides some protection against civil liability and becomes a record of the incident. The form contains, in addition to basic information (e.g., name, date), a voluntary confession of the store's items stolen, the value, and a statement that retail personnel did not use force or coercion and that the subject's cooperation is voluntary. Although a release may state that the subject agrees not to sue, the retailer still can be sued. The release has psychological value in that the subject may believe that a suit is impossible and not seek legal advice, although a claim possibly can be made that the subject signed while under duress. If the subject is obstinate or violent, an arrest is appropriate; handcuffing may be necessary. With this situation, the public police must be summoned; they usually will act in an advisory capacity and transport the prisoner to jail. A decision to call the police and prosecute makes obtaining a signature on the form less important. A criminal conviction is perhaps the best protection against civil liability. If the retailer intends to release the subject without calling public police, then the civil release form is vital for some protection against litigation. Insurance companies write coverage protecting retailers against liability for false arrest, malicious prosecution, willful detention or imprisonment, and defamation.

The interviewer must not coerce the subject into a confession by prolonged questioning or tricky tactics. The U.S. Supreme Court has yet to require private security personnel to state to a suspect the *Miranda* rights prior to questioning, as public police are required to do. Courts have held that any involuntary confession, gained by public or private police, is inadmissible in court. To strengthen their case, many LP practitioners recite the *Miranda* rights anyway and request that subjects sign a waiver-of-rights form if willing to confess.

Throughout the confrontation, at least two retail employees should be present. One employee must be of the same sex as the subject. Problems inevitably develop if an item is forcibly removed from the subject. A video recording can prevent charges related to excessive force, coercion, a bribe, or sexual assault.

Racial profiling is another potential problem that can result in a lawsuit by a subject who claims that he or she was targeted for surveillance because of racial background. Since recordings of surveillance activities can be taken out of context by a plaintiff's attorney, security personnel should explain to the defense side how surveillance is conducted. Examples include monitoring any customer who continuously looks around to see who is watching or walks into a high-risk area. Also, it may be an employee who is under surveillance rather than the customer (Schmedlen, 2000: 70–73).

Prosecution

The deterrent effect of prosecution is debatable. Certain retailers may favor the prosecution of all shoplifters, believing that shoplifters will avoid a "tough" store. Retailers should institute policies that are cost effective. *Prevention appears to be less costly than a strict apprehension program.*

When a shoplifter is arrested and about to be prosecuted, the witness or LP practitioner will probably be asked to sign a complaint or warrant, the legal document containing the facts making up the essential elements of the crime. The LP practitioner should read the legal documents and safeguard all relevant records and forms. Public police, prosecutors, judges, and juries will be interested in facts. In most cases, the defendant pleads guilty in a lower court, pays a fine, and a trial is avoided. In other cases, a preliminary hearing may be necessary to give the judge an opportunity to review evidence and to decide on the necessity of a trial. Most defendants waive their rights to a hearing and a trial in exchange for a plea-bargaining opportunity. Some defendants choose to go to trial and retain an attorney. LP personnel facing a criminal trial must have an excellent case; otherwise, an acquittal is likely to lead to a lawsuit.

Prosecution evidence includes testimony of the eyewitness, CCTV recording, written confession, the unpaid merchandise in possession of the subject, and destroyed EAS tag and/or packaging. The responsibility for preserving physical evidence for a court appearance is part of the LP practitioner's job. Accurate records are necessary. The evidence should be properly labeled and secured in a box or plastic bag. It should be in a locked location with very limited access to maintain the chain of custody of evidence (see Chapter 10).

Two problems exist related to the preservation of physical evidence. First, millions of dollars of confiscated merchandise are not returned to the sales floor at hundreds of retail stores. Court congestion and delays can extend to months and years until the merchandise is out of date. Second, perishable items can deteriorate. A package of eight-month-old chicken breasts would present obvious resale problems when released after it was no longer needed as court evidence. A solution to these problems is photographing the stolen and recovered merchandise. Local requirements vary.

For each retailer, if the prosecution of shoplifters becomes expensive and time consuming, repeat offenders are not deterred, cases are not successful, and brand is harmed, then antishoplifting strategies must be changed.

Civil Recovery

Every state has a **civil recovery** law that holds shoplifters liable for paying damages to businesses and states periodically amended these laws to increase the recovery (because of inflation) from shoplifters (McIntyre, 2011: 63–64; Johnson and Carter, 2009: 445). Some states even extend these laws to employee theft. Civil recovery laws vary and may, for example, allow retailers to request three times the actual damages. These civil demand statutes aim to recover attorneys' fees, court costs, and even the cost of security. Essentially, the expense of theft and security are passed on to thieves.

To recover damages, a retailer sends a shoplifter a demand letter and a copy of the state's civil recovery law. Guidance can be obtained from a state retail association or an attorney. If there is no response, a second letter is sent stating that nonpayment may result in civil court action. With no response after a second letter, a retailer will have to decide whether to pursue the case in small claims court and whether a favorable judgment for damages can be collected.

LP personnel at some businesses may attempt civil recovery while the accused subject is still in custody because the chance for recovery diminishes once the subject leaves the store. The problem with this approach is that the shoplifting case may be based on faulty evidence. In addition, in some states, a guilty verdict may be necessary before civil action is sought. Other states permit simultaneous criminal and civil actions. In addition to the possibility of having a weak case against a suspected shoplifter, civil demand following an apprehension may result in legal action against the retailer for false imprisonment, extortion, and intentional or negligent infliction of emotion distress.

Retailers have the option of contracting civil recovery with an outside firm that usually charges 30 percent of any money collected. Outsourcing saves the retailer the time and expense of operating a civil recovery program.

Research is needed on the strategy of civil recovery. Topics for research can include ROI, effectiveness in reducing shoplifting, retailer frustration with the justice system, and comparisons to arrest and prosecution.

International Perspective: What Deters Shoplifters?

Gill, Bilby, and Turbin (1999: 29–39) interviewed 38 experienced shoplifters in England to assess the effectiveness of antishoplifting strategies. Their research is summarized here:

- *Retail staff:* While nine out of 38 respondents suggested that retail staff would sometimes deter them, 24 answered "never." They argued that salespeople were too busy or uninterested in shoplifting.

- *Security personnel:* Three-fourths of respondents said that they were able to spot plain-clothed store detectives. Only three claimed to be deterred always. The majority said that they would be deterred if a store detective actively observed them. Respondents claimed that one advantage of being followed by a store detective was that it would help an accomplice.
- *Signs:* Most of the sample claimed that signs stating "shoplifters will be prosecuted" would not deter them, especially because certain retailers do not follow what the sign states.
- *EAS and ink tags:* Twenty respondents stated that EAS tags would never deter them, and 14 claimed that they would remove the tags in the store, some with pliers and one with a bottle opener to pop it off. Some respondents avoided EAS tags, which illustrates some deterrent value. Respondents were evenly split on the deterrent effect of inktags. Claims were made that, by putting a condom over it or freezing it before removal, the merchandise would not be ruined.
- *CCTV:* Thirty-three out of 38 respondents claimed that CCTV would never deter them. Picture quality was perceived to be too poor for prosecution. [This is not as much of a problem today.] They claimed that CCTV helped them if they could see where cameras were aimed so they could steal away from the camera's view. Blind spots were also helpful. These research results show the value of dark camera domes so the offender cannot see the camera's direction.

This research demonstrated that offenders make rational choices and that situational crime prevention deters crime, but with the respondents in this study who were experienced shoplifters, a minority was deterred. A combination of security methods appears to have more of an impact than any one method. More research is needed to see how security methods impact amateurs.

Research That Examines the Empirical Relationship between LP Strategies and Levels of Shrinkage

Research of 103 retail firms by Langton and Hollinger (2005: 27–44) compared low- and high-shrinkage retailers within the theoretical framework of routine activities theory (see Chapters 3 and 8). Their research found that "in terms of blocking motivated offenders, there do seem to be significant differences between high- and low-shrinkage retailers in the number of pre-employment screening tests used for management and non-management, but not in the number of LP awareness programs implemented." They suggest that it is better to screen offenders out than try to prevent them from stealing once they are hired. Langton and Hollinger also found that background checks and drug tests were used, on average, by more low-shrinkage than high-shrinkage retailers. These researchers noted that they were unable to clearly identify which LP strategies are the most effective solutions to shrinkage, and many LP departments continue to apply the same strategies each year, using a "shot gun" approach, without researching the effectiveness of individual strategies in reducing shrinkage. Langton and Hollinger conclude that the most important difference between low- and high-shrinkage retailers appears to be the employment of capable personnel rather than physical security.

If you were a retailer, what strategies would you implement to prevent and reduce the shoplifting problem?

Table 17-1 Robbery/Burglary Matrix

	Robbery	Burglary
Harden the target	Yes[a]	Yes
Create time delay	No[b]	Yes
Reduce loot	Yes	Yes

[a]A retail business hardened for robbery may include an alarm and cameras; however, if the robber becomes trapped because of a metal gate that has blocked the only exit, then violence and a hostage situation may develop.
[b]Once a robbery is in progress, for safety's sake a time delay can be dangerous. The exception would be if there were no threatening situations (unlikely in a robbery); then a time delay might aid in immediate apprehension. Police agencies favor robber–police confrontation outside of the crime scene, away from innocent bystanders.

Robbery and Burglary

Robbery is the taking of something from an individual by force or threat of force. **Burglary** is unlawful entry into a structure to commit a felony or theft. For retailers, Hollinger (2011: 26) found that burglary causes greater average dollar loss and outnumber robbery cases each year. His research shows 3.56 burglaries versus 0.52 robberies per $100 million in sales.

LP specialists repeatedly call for retailers to prevent burglary by (1) hardening the target, (2) creating a time delay, and (3) reducing the loot. Hardening the target pertains to physical security such as locks and alarm systems. The reasoning behind creation of a time delay is to increase the time necessary to commit the crime and thereby frustrate the offender. The offender may abort the offense, or the delay may provide additional time for police apprehension. When the loot is reduced, losses are minimized. Some people may mistakenly favor all three of these measures for *robbery*. In Table 17-1, a robbery/burglary matrix illustrates some problems when all three strategies are incorrectly applied to robbery.

Burglary Countermeasures

Although most burglaries occur at residences, the problem is costly to retailers. Burglars often force open a door or window. Earlier chapters provide security methods to prevent burglary.

Robbery Countermeasures

Although most robberies occur on streets and highways, it is a dangerous and costly crime to retailers. Since employee socialization is a vital protection strategy, here is a list for guidance:

- *Opening:* The daily opening of a store is often referred to as the "opening routine." This can be a dangerous time because a "routine" can aid a robber who carefully studies a target. Therefore, a varied opening procedure reduces the chances for robbery. A typical procedure is for one employee to go inside the store while another waits outside; the first person returns within five minutes and signals that the "coast is clear." If the employee does not signal correctly or fails to appear, the police are called. Retailers in high-crime areas frequently open stores in the presence of security people or with a group of three to five employees. Many have permits for handguns.

- *Closing:* Closing procedures should include positioning a trusted store employee either inside or outside the store with a cell phone in hand. Various signaling procedures also are advisable.
- *Cash handling and transportation:* The POS should be located at the front of the store to enhance visibility to those passing by. A self-locking gate or Dutch door prevents access to the register by nonemployees. The cash drawer should require a key or code for access. Reducing the "loot" available will reduce the loss. When cash and other valuables (e.g., checks and card receipts) accumulate, a pickup system is necessary. A supervisor may walk to each register for collections while performing associated accountability procedures. Then the money is taken to the store money room. After the money is accounted for in the money room, a retail employee (or two) takes the money to a bank, or an armored car pickup takes place. Procedures vary. If an employee transports money to a bank, use of a moneybag is not advisable. It is better to use an innocuous paper bag.

Whatever procedures are employed to protect people and money, there is always the possibility that a present or former employee has provided such information to offenders. Many retail crimes have a connection to current or former employees. Therefore, *procedures and signals should be changed periodically.*

In the event of robbery, employees should be trained to act in the following ways:

1. Concentrate on *safety;* do not try to be a hero; accommodate the robber's requests as well as possible.
2. If safety permits, activate the alarm; give the robber bait money.
3. Concentrate on details: description of the robber(s); license tag; type of vehicle; and direction of travel.
4. After the robber leaves, telephone police immediately.

The following is a list of systems, devices, and services to protect against robbery:

- *Alarms:* A button or foot device signals a silent alarm to authorities. Another type of alarm is activated when money is removed from a money clip within the register drawer. The employee must try to prevent the robber from noticing the silent alarm activation by hand or foot movement.
- *CCTV:* A recording of a robbery is helpful in identifying and prosecuting robbers.
- *Safes: A drop safe* permits a deposit into the safe without opening it; only management can open the safe. Certain sections of a *time-delay safe* cannot be entered until preset times. *Dual-key locks* on safes require a key from two people, such as a retail employee and an armored car officer. Care must be exercised when designing safe access because a robber may become impatient and resort to violence; warning signs to deter robbers provide some protection.
- *Bait money:* Bait money, also known as *marked money*, has had the banks of issue, denominations, series years, and serial letters and numbers recorded by the retailer. This record should be kept in a safe place. Prosecution of the robber is strengthened if he or she is found with this money. Retail employees should carefully include the bait money with the loot.

- *Security officers:* Armed, rather than unarmed, officers should be used to prevent robbery at high-risk locations. Another method is to hire off-duty police officers.
- *Armored car service:* An armored car service can increase safety for retail employees while providing security for the transportation of money.

A retail executive from headquarters came to a chain store, displayed a company badge on his impressive suit, and systematically collected more than $2,000 from eight cash registers before leaving the store after the unauthorized collection. The purpose was to test security.

Shopping Mall Strategies

The size of a retail operation has a direct bearing on the extent of its LP program. A small store obviously will require a different program than a shopping mall. A mall's LP program would include not only countermeasures for crime, fire, accident, and other hazards, but also methods for numerous merchant needs, crowd control, parking lot problems (e.g., traffic, dead batteries, keys locked in autos), and lost people and merchandise. The size and complexity of malls compound loss problems.

A shopping mall's LP program should be centralized and headed by an executive who responds to overall needs. Monthly meetings with merchants will facilitate cooperation and the sharing of problems, ideas, and resources. These meetings can provide an opportunity for a short training program. Special sales and events require further preparation. For malls experiencing many crime incidents, the establishment of a police substation is helpful. Partnering with public safety agencies is important.

Since the 9/11 attacks, retail shopping malls have been referred to as "soft targets," as have many other locations (e.g., schools and houses of worship). Traditionally, merchant need for easy access by customers limits access controls at malls and retail stores. At this time, malls in the United States will not be protected like airports unless malls are attacked as they have been in other countries (e.g., Israel).

To improve the protection of malls, the International Council of Shopping Centers (ICSC), which serves the shopping center industry, has taken a lead role in developing training for mall security personnel. In cooperation with the U.S. Department of Homeland Security, the ICSC training includes a variety of topics such as terrorism, observational awareness, bombs, and emergencies.

Questions of concern are: What are the threats and hazards to U.S. malls today? Is it terrorism? Is it an assailant who enters a mall and shoots people? Is it internal theft and shoplifting? Is it gangs? A risk analysis can help to answer these questions.

Career: Retail Loss Prevention

The responsibilities of loss prevention agents and the role they play in the overall profitability of companies makes this one of the most critically important functions within retail organizations. Retail companies are plagued by theft and fraud from both internal and external sources. Entry-level

LP positions generally involve undercover shoplifting apprehensions, LP or safety auditing, and other functions. Part-time work as a loss prevention agent is an excellent job while one is in college.

LP managers have responsibility for overseeing the agents and addressing and resolving internal theft concerns. Depending on the size of the stores and the company, a LP manager may be responsible for the LP functions within a single store or responsible for overseeing LP at a number of locations within a chain.

Entry-level management positions generally require an undergraduate degree as well as three to five years of experience in LP. Mid-level management positions require expertise in multiple security disciplines, an undergraduate degree, and five to eight years of demonstrated success in the field.

Certifications from the Loss Prevention Foundation (2011) are especially helpful for applicants. Other helpful certifications are the Certified Protection Professional (CPP) and Certified Fraud Examiner (CFE) designations.

Modified from source: Courtesy of ASIS International. (2005). "Career Opportunities in Security." www.asisonline.org.

Banking and Finance

The banking and finance sector is the backbone for the world economy. It includes banks, insurance companies, mutual funds, pension funds, and other financial entities. This sector consists of a variety of physical structures, such as buildings and financial utilities. Physical structures to be protected house retail or wholesale banking operations, financial markets, regulatory institutions, and physical repositories for documents and financial assets. Today's financial utilities, such as payment systems, are primarily electronic, although some physical transfer of assets does still occur. Many financial services employees have highly specialized skills and are therefore considered essential elements of the industry's critical infrastructure.

The financial industry depends on continued public confidence and involvement to maintain normal operations. In times of crisis or disaster, maintaining public and global confidence demands that financial institutions, financial markets, and payment systems remain operational. Additionally, in times of stress the Secretary of the Treasury, the Chairman of the Federal Reserve, and the Securities and Exchange Commission proactively address public confidence issues, as was done following the 9/11 attacks. The Department of the Treasury and federal and state regulatory communities have developed emergency communications plans. With regard to retail financial services, physical assets are well distributed geographically throughout the industry. The sector's retail niche is characterized by a high degree of **substitutability**, which means that one type of payment mechanism or asset can be easily replaced with another during a short-term crisis. For example, in retail markets, consumers can make payments through cash, checks, or credit cards.

HSPD-7 established the **U.S. Department of the Treasury** as the sector-specific agency for this sector. It developed a sector-specific plan in collaboration with government agencies and the banking and finance industry. The security goals of this sector focus on prevention, detection, and correction. This is accomplished through resilience, risk management, and redundancy; addressing dependence on communications, IT, energy, and transportation sectors; and partnering with law enforcement, the private sector, and international partners (U.S. Department of Homeland Security, 2009; U.S. Departments of Homeland Security and Treasury, 2007).

Banking and Finance Sector Challenges

The potential for disruption of banking and finance sector systems from crime, power outages, disasters, and other events is an important concern. For example, the equity securities markets remained closed for four business days following the 9/11 attacks because the communications lines in lower Manhattan that connect key market participants were heavily damaged and could not be restored immediately. As a mitigation measure, financial institutions have made great strides to build resilience, redundancy, and backup into their systems and operations. Terrorist groups continue to view this sector as a high-value and symbolic target.

The banking and finance sector faces threats from robbery, burglary, larceny, fraud, embezzlement, and cybercrime. Money, securities, checks, and other liquid assets make this industry attractive to internal and external culprits. Services that present particular loss problems include automated teller machines (ATMs), electronic fund transfer systems (EFTSs), and online banking. Henry (2011), of the FBI, writes that an international network of hackers accessed a financial company's network, compromised its encryption, did reconnaissance inside the system for months, and then stole millions of dollars in less than 24hours.

Legal Responsibilities of the Banking and Finance Sector

The banking and finance sector is highly regulated with oversight and guidance from government regulators and through self-regulatory organizations. Industry professionals and government regulators regularly engage in identifying sector vulnerabilities and take appropriate protective measures, including sanctions for institutions that do not consistently meet standards. Numerous laws have been enacted that focus on examining the operations of financial institutions.

The **Bank Protection Act (BPA) of 1968** mandates minimum-security measures for protection against robbery, burglary, and larceny, primarily. These measures include a method of identifying suspects; CCTV systems are typically applied for this requirement. Other security measures are devices (e.g., a vault) to protect cash, an alarm system, and lighting, among other measures. The BPA was enacted prior to ATMs, online banking, and the use of mobile devices for transactions. In addition, a significant shortcoming of this legislation was that it did not establish standards for internal protection. For example, the **savings and loan (S&L) scandal** of the 1980s and 1990s, where bank executives approved risky loans to friends and relatives (i.e., "the fox was guarding the hen house"), cost U.S. taxpayers billions of dollars *(much more than embezzlement and bank robberies combined)*. Despite these drawbacks, the BPA was the first legislation reinforcing security for a large private commercial enterprise. It is impossible to ascertain the number of crimes that have been prevented because of the BPA.

The **Bank Secrecy Act of 1986** was implemented to establish reporting requirements for transactions of money to detect money laundering. The **Anti-Drug Abuse Act of 1988** requires banks to report any "suspicious transactions" that may be associated with illegal drug trafficking. **The Antiterrorism and Effective Death Penalty Act of 1996** makes it a criminal offense for persons in the United States, other than those exempted by the government, to engage in financial transactions with countries that condone or encourage terrorism. The **Gramm-Leach-Bliley Act of 1999**, which has broad application to a variety of businesses in this sector, requires safeguards on the privacy and security of customer information. The **USA PATRIOT Act of 2001** contains measures against money laundering and the financing of terrorism. The **Sarbanes-Oxley (SOX) Act of 2002**, which passed in the wake of major corporate scandals, requires public companies to assess internal controls over financial reporting to ensure accuracy.

As technology changes the banking and finance sector, security must also change. The Security Executive Council (2011: 24), a service group, explains some of these security changes as follows. The Federal Financial Institutions Examination Council (FFIEC), a formal interagency government body that prescribes standards for examinations of financial institutions, provides guidance on protecting customer information and authentication of customers during Internet banking to reduce fraud during transactions. The Federal Trade Commission established a "red flags rule" that requires financial institutions and creditors to implement a written identity theft prevention program designed to detect warning signs of identity theft. As explained earlier in the retail subsector, the Payment Card Industry Data Security Standard is a global information security standard designed to assist entities in preventing fraud when processing card payments.

Regulation H

Regulation H, Code of Federal Regulations, pertains to membership of state banking institutions in the Federal Reserve System. This regulation combines the BPA and provisions of the Bank Secrecy Act of 1986 into one document. Requirements include the following:

- Security procedures must be adopted to discourage robberies, burglaries, and larcenies, assist in the identification and prosecution of persons who commit such acts, and maintain records of crimes.
- The institution's board of directors must ensure that a written security program is developed and implemented.
- The board must designate a security officer to administer the security program and receive an annual report from that person.
- Program requirements include opening and closing procedures, warning signals, and safekeeping of currency and other valuables (e.g., vault controls).
- Policies, procedures, and training manuals must be developed, and initial and periodic training must be provided for employees.

A bank security officer's duties may include responsibility for Bank Secrecy Act requirements under Regulation H. This includes ensuring that **suspicious activity reports** (i.e., suspicious financial transactions over a specified amount) are filed with appropriate federal law enforcement agencies and the U.S. Department of the Treasury when violations of federal law are suspected. This includes insider abuse or money laundering.

Money laundering is a huge global problem with hundreds of billions of dollars laundered annually. Essentially, it is an attempt to make "dirty" money (i.e., obtained through illegal means, such as the drug trade) appear clean. Offenders often transfer "dirty" money, under disguised ownership, from one bank to another, globally, to increase the difficulty of tracing it. Leff (2012: 25) views money laundering as "taking criminal profits and moving them in a prohibited manner." He adds that although the 21st century has produced numerous technologically savvy professional money launderers, federal investigators have several sources of information available. Examples are mandatory currency transaction reports by businesses and Federal Reserve Bank fund transfer records.

Bank Robbery Countermeasures

A common feature of many financial locations is a warmer, more personal atmosphere gained through the elimination of security barriers. Although this may please customers, robbery—a crime characterized by a threat and the possible use of violence—becomes a serious danger to people. The Federal Bureau of Investigation (2011b) reported 5,546 bank robberies in the United States in 2010, plus 57 burglaries and five larcenies; total loot taken amounted to $43 million. For 2011, the Federal Bureau of Investigation (2012) reported 5,014 bank robberies in the United States, plus 60 burglaries and 12 larcenies; total loot taken amounted to $48 million. Not all bank crimes are reported to the FBI by state and local police.

When a robbery does take place at a bank, the teller is often the only one initially knowledgeable about the crime. The situation of a lone robber passing a holdup note to the teller is typical. In any robbery situation, the danger to life must not be taken lightly. The following suggestions can improve security and safety (Spadanuta, 2010: 50–57).

1. Apply carefully planned CPTED.
2. Institute a training program.
3. During a robbery, act cautiously and do not take any chances.
4. Activate an alarm, if possible, without the robber(s) noticing.
5. Provide the robber with bait money that has the serial numbers recorded. **Tear gas/dye packs** are also important. These devices look like packs of currency and emit tear gas and red smoke (which stains clothes and the money) when carried out of a bank. A radio transmitter activates the packet.
6. Study characteristics of the bandit, especially scars, shape of eyes, height, body structure, voice, speech patterns, and other permanent features.
7. Safely note the means of transportation, license number, and vehicle description.
8. Contact public police.

Uniformed security officers can offer comfort to customers and act as credible witnesses; however, their use is subject to controversy. They may deter an amateur robber, but not a professional. Tear gas/dye packs help to apprehend robbers and recover money. However, robbers have threatened violence if they receive one, and detonations have occurred inside banks. Banks are using the **Global Positioning System (GPS)**; a transmitting device is planted in the bait money to track the device via satellite signal, and tracking can be done by police and security through a secure website that shows a map on a computer screen. CCTV permits remote monitoring, and if a robbery, fraud, or other crime occurs, police (with appropriate equipment) can observe the crime while responding. Security personnel can e-mail a video record to police. Other protection methods include safes and vaults, bullet-resistive bandit barriers, metal detection portals, signage that describes security features, and height markers to help estimate robber height. A mantrap is another option. It consists of double doors capable of trapping a fleeing robber; however, careful planning and construction are essential to promote safety. In addition, the remote teller system can be used whereby business is conducted via CCTV and pneumatic tubes.

Kidnapping and Extortion

The financial executive and his or her family are potential victims of a kidnapper or extortionist. A bank teller is another potential victim. These crimes vary; two scenarios follow. An offender kidnaps a bank employee or family member while he or she is driving the family car or is at home. The kidnapped person is traded for cash. Another approach occurs when an extortionist calls a bank employee at work, claims to be holding a family member, and demands cash for the safe return of the kidnapped victim. The extortionist may even confront the bank employee personally. In these situations, violence is possible, even after cash is delivered.

A precise plan and employee training are essential prior to a confrontation. The following procedures should be observed by the person in contact with the kidnappers:

1. Try to remain calm during the ordeal. Contact public police as soon as possible.
2. If a hostage is involved, ask the caller to allow the hostage to come to the telephone to speak.
3. Repeat demands to the caller to double-check any directions and procedures.
4. Try to arrange for a person-to-person payoff and transfer of the hostage.
5. If prepared, trace the call.
6. After the call, record as much information as possible (e.g., date, time, words spoken, background noises). Management should provide a standard form.
7. If the hostage is brought to the bank or if the offender confronts a bank employee personally, notify other employees with a prearranged signal.
8. Include “bait” money in any payoff.

Automatic Teller Machines

Robbery and fraud have followed ATM self-service banking. The BPA and the Federal Electronic Funds Transfer Act focus on security of ATMs and fraudulent transactions, rather than the safety of ATM users. Several state and local government bodies have passed ATM safety laws. These bills typically require surveillance cameras, adequate lighting, mirrors, and low hedges to enhance visibility, and crime prevention tips for ATM cardholders. Although offenders watch for careless customers who do not protect their identities and PIN numbers, family members and friends commit a considerable amount of ATM fraud.

One low-tech crime victimizing ATMs is theft of the entire machine. A truck and chains or an inexpensive lift truck are used to pull the ATM from its foundation. Countermeasures include placing ATMs at locations that are busy 24 hours a day and using GPS to deter theft.

Skimming is another risk associated with ATMs. It is a risk to bank customers and banks, costing U.S. banks hundreds of millions of dollars annually. At an ATM, skimming involves the installation of a combination of the following devices: a small hidden camera (to record customer PINs) or a phony keypad over the real keypad (to record keystrokes for the PIN), and a phony card reader over the real one (to capture account information that is stored on a small, attached computer or cell phone or sent wirelessly to nearby offenders). The phony devices look real and are often installed with double-sided tape for just a few hours to obtain information from several customers. Then, customer information is encoded onto blank cards to be used to withdraw customer funds. The FBI and the U.S. Secret Service often partner on such cases. For example, two brothers from Bulgaria were charged in New York City with skimming to defraud two banks of over $1.8 million from 1,400 customers. Skimming is also applied at gas pumps and other POS sites (Federal Bureau of Investigation, 2011c).

Embezzlement, Fraud, and Online Risks

Banks have increased their exposure to crime because they are connected internally through an intranet and externally through online banking services. An employee may commit embezzlement by gaining unauthorized access to a bank's accounts payable system and create fictitious invoices and payments to the employee's post office box. Other internal crimes are to delete all records of a friend's loan, approve fictitious loans, or tap into dormant accounts.

Check swindles, which are a common externally perpetrated fraud, may involve the forgery of stolen checks or the manufacture of fictitious checks. Credit card fraud is another problem for banks; losses can result from a fraudulent application containing a stolen identity or from a stolen card.

Online banking risks are varied and include, for example, the hacking of bank websites to alter information or to steal customer information. Web spoofing seeks to create a look-alike website to trick customers into releasing confidential account information that is exploited or sold by the hacker. Financial institutions must balance offering new services to customers with secure online systems. A highly publicized bank loss or victimized customer could drive customers away.

Regulation E, Code of Federal Regulations, focuses on electronic funds transfers and limits a consumer's liability from unauthorized transfers. However, the limit does not apply to commercial bank accounts as described by Kroll (2012). She writes that guidance provided by the FFIEC for authentication and layered security in Internet banking can enhance protection of corporate bank accounts. The guidance is especially important because a corporation, rather than the bank, may incur the loss from a cybercriminal's theft from the corporate account. In one case, a cybercriminal accessed a company bank account and drained cash, resulting in financial difficulties for the company, reduced work hours and pay for employees, and the possibility of bankruptcy.

The FFIEC stresses the importance of dynamic security, whereby continuous risk assessments counter innovative techniques applied by cybercriminals. Another security measure includes "two-factor authentication" that goes beyond a user name and password. Examples include a customer choosing an image and phrase, or the use of a Universal Serial Bus (USB) token that must be plugged into a computer's USB port to access an online account. Banks also install software that monitors each online transaction for unusual patterns. Customer awareness is another measure, mentioned by the FFIEC (Spadanuta, 2011; Swartz, 2006).

What do you think is the most serious risk facing banks and financial institutions? What are your solutions?

Career

Banking and Financial Services Security

Careers in this field include those associated with banking (including retail banking, mortgage, credit/debit cards, Internet banking, commercial and consumer lending), stock brokerages, insurance companies, and other financial institutions. Financial institution security directors and managers must deal with a wide variety of concerns including theft, fraud, workplace violence, information security, investigations, executive protection, business continuity, and physical security in order to adequately protect their institution. Security managers must be effective leaders within their organization and able to successfully influence change.

Because of the increasing complexity of the financial services industry, which is heavily regulated, companies continue to seek the best and the brightest from law enforcement agencies, colleges and universities, and from other private sector companies, both within and outside the financial services sector.

Entry-level management positions in financial services generally require a degree in business, finance, or criminal justice from an accredited institution as well as three to five years of experience in either law enforcement or security.

Mid-level management positions, requiring expertise in multiple security disciplines, generally require a bachelor's degree in an appropriate discipline, as well as three to seven years of demonstrated success in the field. Professional certifications such as the Certified Financial Services Security Professional (CFSSP) (American Bankers Association, 2011), Certified Protection Professional (CPP), and Certified Fraud Examiner (CFE) are often desired as an indicator of professionalism and qualification.

Modified from: Courtesy of ASIS International. (2005). "Career Opportunities in Security." www.asisonline.org.

Educational Institutions

Two major types of educational systems are emphasized here: school districts and higher education. A key difference between school districts and college and university campuses is that, in the former, students normally go home at night and buildings are often empty. On campuses, students often live on the premises in dormitories. Exceptions are community

FIGURE 17-5 Protection must cater to the unique needs of institutions. *Courtesy: Talk-A-Phone.*

and technical colleges whose students typically commute. Both school districts and campuses schedule evening and weekend activities such as classes, sports events, and meetings. A major factor for those who plan and implement protection programs for these institutions is that the security and safety measures must cater to the needs and characteristics of the particular institution (Figure 17-5).

School districts and higher education are known as the education facilities subsector (EFS) of the government facilities sector under the National Infrastructure Protection Plan (NIPP). The U.S. Department of Education's (ED) Office of Safe and Drug-Free Schools (OSDFS) served as the sector-specific agency for the EFS. OSDFS promoted policy and recommendations on safety, health, security, emergency management, and resilience for the EFS. In 2011, the OSDFS became the Office of Safe and Healthy Students (U.S. Department of Education, 2011a). Under the Tenth Amendment ("powers…reserved to the States"), the general authority to administer the EFS is empowered to the states. The EFS is highly decentralized and protection is primarily a local responsibility (U.S. Departments of Homeland Security and Education, 2010).

Threats and Hazards at Educational Institutions

Educational institutions are subject to a host of threats and hazards similar to other segments of society. Examples are crimes of violence, bomb threats, illegal drugs, property crimes, cybercrimes, and vandalism. Additional concerns are fire, accident, disaster, gangs, terrorism, infectious disease outbreaks, food recalls, suicide, date rape, bullying, hazing, crowd control for sports and other events, traffic, parking, student activism, graffiti, the discharge of toxic substances (e.g., the release of pepper spray in a crowded hall), and the release of an animal in a school as a prank. Incidents may involve cross-sector protective efforts that assist in mitigation, for example the healthcare and public health, transportation (e.g., school buses), commercial facilities (e.g., stadiums and arenas), and IT communications sectors.

Protection for Educational Institutions

The following list describes measures for security and safety of school districts and institutions of higher education:

- Establish a security and safety committee and meet at least monthly.
- Ensure that all stakeholders and first responders are involved in security and safety planning and programs.
- Conduct risk analyses and prepare comprehensive, all-hazards protection (see earlier chapters).
- Offer counseling services and programs to assist students in crisis.
- Conduct substance abuse education and prevention programs.
- Typical security measures include access controls, emergency telephones, mass notification systems, intrusion alarms, patrols, CPTED, lighting, and CCTV.
- Use vandal-resistant construction materials.
- If graffiti is an issue, photograph it before removing it and show the photos to police.
- Carefully consider the use of unarmed students to supplement police and security forces; provide good training.
- Research online resources on protection, training programs, and grants (see the web addresses at the end of the chapter).

School Districts

More than 55 million students are enrolled in U.S. elementary through high schools (U.S. Census, 2011). From July 1, 2009 to June 30, 2010 there were 33 "school-associated violent deaths," defined as a homicide, suicide, or legal intervention (by police), in which the death occurred at an elementary or secondary school in the United States. In 2010, for students ages 12–18, 828,400 nonfatal victimizations occurred at school; this includes 358,600 victims of violence and 469,800 victims of theft. During this same year, for students ages 12–18, 652,500 nonfatal victimizations occurred away from school; this includes 281,200 victims of violence and 371,300 victims of theft. It is difficult to draw conclusions about crime victimization and safety "at school" versus "away from school" because of reporting practices of victims and officials. Besides the physical and psychological effects of victimization at schools, it can lead to increased dropout rates, decline in learning, early retirements, and increased fear (National Center for Education Statistics, 2012).

Legislation for School Districts

The **Safe and Drug-Free Schools and Communities Act** supports programs that prevent violence in and around schools, mitigates the illegal drug problem, facilitates parent and community involvement in school challenges, and appropriates funds to local schools and higher education facilities victimized by violence or a traumatic incident. As a requirement for grant funding, the Act necessitates a crisis management plan. In 2011, Congress eliminated programs implemented under this Act. At the same time, the Office of Safe and Healthy Students continues to promote drug and violence prevention programs (U.S. Department of Education, 2011a; U.S. Departments of Homeland Security and Education, 2010: 55).

In an attempt to reduce school violence, Congress enacted the **Gun-Free Schools Act,** which requires that each state receiving federal funds under the Elementary and Secondary Education Act must have a state law requiring local educational agencies to expel from school for a period of not less than one year a student who is determined to have brought a weapon to school. Each state's law also must allow the chief administering officer of the local educational agency to modify the expulsion requirement on a case-by-case basis (Anderson, 2012). This law resulted in many schools adopting a **zero-tolerance policy** for weapons being brought to school. In other words, any infraction results in full punishment. This same policy extends to alcohol and drugs. Meadows (2007: 170) writes that the intent of such a policy is both preventive and punitive. He notes the following: "Although zero-tolerance policies have a place in school security, there is the threat of overenforcement, which may undermine school-community relations and label students unfairly."

The **Family Educational Rights and Privacy Act** (FERPA) protects the privacy of students by limiting the types of information school officials can release. School records and personal information are protected. FERPA does not protect information disclosed during an emergency or obtained through observation or personal knowledge (e.g., a teacher hearing a threat from a student).

Teacher and staff victimization and intimidation in school districts are important topics often not given enough attention.

Protection for School Districts

A comprehensive school district loss prevention program must involve the community: students, teachers and administrators, parents, public safety agencies, civic groups, and businesses. The program can be divided into four components: special programs, personnel, physical security, and emergency management (Decker, 2000).

Special programs include character education to help students distinguish right from wrong, conflict resolution, diversity, prevention of bullying, anonymous tip lines, and programs that involve parents. Since gangs are a problem in many schools and communities, and they are often linked to violence and drugs, school administrators should be proactive to reduce this problem by, for example, meeting police gang specialists on a regular basis to exchange information and antigang strategies. One popular program is the **Gang Resistance Education and Training (GREAT)** program that provides students with tools to resist the lure and trap of gangs. Modeled after the Drug Abuse Resistance Education (DARE) program, the GREAT program seeks to prevent violence and introduces students to conflict resolution skills, cultural

FIGURE 17-6 Police using handheld metal detector at school access point. *Courtesy: Garrett Metal Detectors.*

sensitivity, and negative aspects of gang life. This program has spread to all 50 states and several other countries (Institute for Intergovernmental Research, 2012).

All employees at schools should be trained on early warning signs of inappropriate behavior or violence. These signs include feelings of isolation, rejection, and being persecuted, plus behaviors indicating anger or violence, such as threats. Intervention and counseling are vital in response to such signs.

A combined counseling and education approach might reduce student hostility and funnel student time into constructive activities. Traditional suspension from school often sends troublesome students to the streets, where more trouble is likely. On the other hand, if students remain at school in an appropriate program, improved results are probable.

Personnel consist of teachers, teacher aids, administrators, counselors, security officers, and School Resource Officers. SROs are police officers on duty at schools; they provide visibility, create rapport with students, and respond to incidents. Parents and volunteers play an important role in supplementing employees. All those who perform job duties at schools should undergo background screening, receive constant training, and clearly understand policies and procedures for day-to-day events, such as student discipline problems and how to care for people during emergencies.

Physical security examples are handheld (Figure 17-6) and walk-through metal detectors and duress alarms. Students must be safe without feeling as if they are in a prison. Students can swipe an ID card when they climb onto a bus and when they arrive and leave school; RFID tracks their movements. At the same time, CCTV cameras watch students on buses and on school premises. Visitors can be asked for their driver's licenses to check against a database of sex offenders. Whatever strategies are implemented, they should be subject to research and evaluation to produce the best possible solutions and utilization of resources.

Emergency management consists of plans to respond to violence, weapons, hostage situations, bombs/explosions, abused students, aggressive parents who are on the premises, and incidents involving parental rights. Many school districts distribute their crisis plans to public safety agencies that have a ready reference containing maps and building plans, utility shutoffs, staff and parent contact information, a yearbook with which to identify people, and a set of keys.

The National Center for Education Statistics (2012) surveyed public schools and reports a host of safety and security measures at schools, including the following:

- Controlled access and locked doors. Nearly all public schools require visitors to sign in.
- Restrictions on student access to certain websites on school computers.
- Prohibiting certain electronic devices (e.g., cell phones).
- Anonymous threat reporting.
- Drug testing of athletes and other students in extracurricular activities. Random dog sniffs to check for drugs.
- Other measures include picture ID, searches, student uniforms, and mass notification systems.

Research by Sobel (2012: 11–21) found that surveillance and searches at a high school were viewed in a positive light by male and female employees. Although students viewed surveillance positively, a majority of students viewed searches of lockers, backpacks and self negatively.

The *Guide for Preventing and Responding to School Violence*, 2nd edition (International Association of Chiefs of Police and Bureau of Justice Assistance, n.d.), offers numerous suggestions, including the topics above, plus the following: establish a climate for reporting threats and violence, create an anti-bullying program, use a student court for noncriminal offenses, and educate students about drugs, alcohol, suicide prevention, gangs, and other important topics.

What are the advantages and disadvantages of using unarmed students to supplement police and security forces at high schools?

Are Educational Institutions "Soft" Targets?

Terrorists search for "soft" targets containing limited security to increase the likelihood of a successful attack. Examples of **"soft" targets** are educational institutions, houses of worship, shopping malls, and theaters. **"Hard" targets** include military bases and fortified government buildings.

Emergency management became an important priority at many schools following the 1999 **Columbine High School massacre**. During this incident in Jefferson County, Colorado, two teenage students, armed with a variety of weapons and bombs, killed 12 students and a teacher, wounded 24, and then committed suicide. Prior to the attack, one of the killers showed warning signs of violence through a website, which contained violent threats, information on how to make bombs, and a log of mischief. The massacre fueled debate over gun violence, gun control, and the influence of the media on violence. In addition, the event reinforced attention to warning signs to prevent violence.

Another, more deadly, massacre, tied to terrorism, was the **Beslan Elementary School massacre** in 2004. This attack was part of the Chechen war for independence against Russia. Russia has fought an insurgency in the breakaway republic of Chechnya since the 1990s, a time when the former Soviet Union collapsed. The three-day hostage standoff at the elementary school in Beslan, where Chechen rebels rigged bombs around 1,200 hostages, ended in gunfire, explosions, and 338 deaths, mostly children. Each side blamed the other for the battle at the school that caused more controversy over Chechen terrorism and Russian responses.

The carnage began when 32 attackers—armed with AK-47s, a machine gun, grenade launchers, explosives, and two dogs (to protect against chemical attack)—drove to the crowded school and herded children, teachers, and parents into the school gym, which was quickly wired with explosives. Male hostages were forced to build barricades at doors and windows, and when the work was complete, they were executed and their bodies were thrown out of the building. Two female suicide bombers were among the attackers, which illustrated the resolve of the attackers and the possible outcome. The attackers demanded the release of Chechen rebels, the withdrawal of Russian troops from Chechnya, and Chechen independence. The Russian government's options were very limited, and giving in to terrorist demands fuels new hostage events. By the second day, the terrorists continued to refuse water, food, and medicine for the hostages. In their desperation, the hostages began to drink their own urine. On the third day, the terrorists permitted a crew to collect the bodies of 20 male hostages that had been thrown out of the building. At this point, a bomb exploded within the gym, supposedly by accident, a fire began, the roof collapsed, and as hostages escaped, the terrorists shot them. The terrorists were killed, and two who escaped were caught by a mob and lynched. As the disaster unfolded, "finger-pointing" began to explain what went wrong. Explanations were that Special Forces were not ordered to rescue the hostages and assault the building, no one wanted to take responsibility to order the rescue and assault, and armed volunteers interfered with the hostage situation and were not kept back. The disaster shocked the world, and the Russian government promised reform of the police and military (Abdullaev, 2004: 28–35).

On April 16, 2007, the **Virginia Tech massacre** became the worst school mass murder incident in U.S. history when a mentally disturbed student murdered 32 people and then committed suicide. Among the questions that followed the shootings were: Why did it take over two hours to release a notification that a killer was on the loose after two students were found shot earlier in another building? Was the police response quick enough to end the threat of an "active shooter"? Many factors must be considered when attempting to answer these questions. For instance, rushing into a building without a team and a plan is risky. Harwood (2007, 55–65) writes of strategies that require refinement: behavioral threat assessment teams to identify risky students, mass notification systems, and "active shooter" response. The federal government, including the U.S. Secret Service and the FBI, studied campus violence and concluded that although most (73%) violent acts on campus targeted a specific individual because of a triggering event (e.g., romantic breakup or academic failure), "understanding what leads an offender to exclusively target random individuals remains a complex and difficult challenge." The study also noted the importance of college and university threat assessment teams to evaluate "persons of concern" (Drysdale et al., 2010).

Colleges and Universities

Legislation for Colleges and Universities

In response to increasing crime on college campuses and the need for more accurate statistics, Congress passed the **Student-Right-to-Know and Campus Security Act of 1990**. This act is also known as the "Clery Act," named after Jeanne Clery, a college student who was raped and murdered in her dorm room. For institutions receiving federal student aid, this legislation requires crime awareness and prevention policies and an annual report of campus crime sent to the FBI Uniform Crime Reports program, while making these statistics available to students and the general public. Such data, available on the Internet, enables comparisons among colleges and universities. Amendments to this Act include the requirement that a campus community must be notified immediately of an emergency or threat to safety, disclosure of emergency plans, reporting on hate crimes, and reporting of relationship of campus police with state and local police. For campuses with housing, the amendments require procedures for missing students and methods of fire protection reporting (U.S. Department of Education, 2011b).

Congress enacted the **Campus Sexual Assault Victims' Bill of Rights** in 1992, requiring schools receiving federal student aid to afford sexual assault victims basic rights and to develop policies to deal with sexual assault on campus.

There is conjecture that some schools omit reporting acts of violence to protect recruiting efforts and their reputations (Jaeger, 2001: 6). FBI crime data has been criticized over the years because it represents crimes *reported* to police, and many crimes are never reported or recorded, as shown by victimization studies (Purpura, 1997: 32–36).

Protection for Colleges and Universities

Campus Security Guidelines were prepared by the Major Cities Chiefs Association and Bureau of Justice Assistance (2009). These guidelines focus on policies, coordination among local and campus police, interoperable communications, risk assessment, prevention, preparedness, emergency management, and external relations.

Numerous campuses have implemented the strategy of many public police agencies, namely, **community policing.** It aims to control crime through a partnership of police and citizens, and it strives to become a dominant philosophy throughout a police department. Community policing includes a proactive approach to problem solving, rather than responding repeatedly to the same problem. High priorities for protection on campuses are programs that focus on crime prevention, self-protection, and Neighborhood Watch.

In 2004, a National Summit on Campus Public Safety was held in Baltimore, Maryland. It was supported by the U.S. Department of Justice, Office of Community Oriented Policing Services (2005); the IACLEA (see below); the International Association of Chiefs of Police; the U.S. Department of Homeland Security; and the FBI. Some of the main points from this summit are as follows:

- There is a need for a national center for campus safety to support information sharing, standards, model practices, and research.

- Many campuses house sensitive materials and information and serve as contractors for the Department of Defense, Department of Justice, National Security Agency, other government bodies, and corporations.
- Securing chemical, biological, and radiological materials in an accessible environment, 24 hours a day, seven days a week, creates security challenges.
- Many campuses have a substantial number of international students who entered the country through student visas. In addition, many U.S. educational institutions maintain campuses overseas.
- Special events (e.g., sports, graduation, lectures) draw thousands of people to campuses and create vulnerabilities.
- U.S. educational institutions should be more closely linked to local and regional emergency management plans.

Research by Fisher and Sloan (1993: 67–77) produced the following points that should be considered when evaluating programs designed to reduce campus crime.

- A comprehensive approach includes security, faculty members, staff members, students, and public law enforcement personnel.
- Campus administrators should conduct surveys of the campus community to understand the nature and extent of crime and fear, perceptions of the effectiveness of security, and participation in crime prevention programs and whether participants adopted any of the preventive measures. Until evaluations become an integral part of responding to campus crime, administrators will continue to make poor decisions on security strategies.
- Research has confirmed that crime on campuses is influenced by poor lighting, excessive foliage, blocked views, and difficulty of escape by victims. (These issues relate to CPTED.)
- Location measures (e.g., proximity to urban areas with high unemployment) are predictors of high campus crime rates.

One particular group that has advanced the professionalism of campus safety and security is the International Association of Campus Law Enforcement Administrators (IACLEA). This group began in 1958 with 11 schools, and today it represents 1,200 colleges and universities in 20 countries. The group maintains a website of resources; holds an annual conference; offers training, professional development, and standards and accreditation; and publishes *Campus Law Enforcement Journal* (International Association of Campus Law Enforcement Administrators, 2011).

What do you view as the top three strategies to prevent violence at school districts? How would you answer this same question for colleges and universities?

Safety and Fire Protection at Educational Institutions

The NFPA 101 Life Safety Code (see Chapter 13) offers guidance to protect educational facilities from fire. Examples of fire hazards endangering life in places of public assembly are (1) overcrowding; (2) blocking, impairing, or locking exits; (3) storing combustibles in dangerous locations; and (4) using combustible decorations. Furthermore, hazards of educational buildings vary with construction characteristics and with the age group of students. Younger students, for example, require protection different from that for older students. The NFPA Life Safety Code specifies that kindergarten and first grade rooms should be on the floor of exit discharge so that stairs do not endanger these students. Because junior and senior high schools contain laboratories, shops, and home economics rooms, these facilities should have fire-resistant construction. School kitchens require similar protection. A fire alarm system is required for all educational buildings. Most schools conduct fire drills for pupils. The Life Safety Code and many good building codes provide numerous standards for increasing fire safety.

For colleges and universities, the Life Safety Code is applied depending on building characteristics and use. If buildings are windowless, the Life Safety Code requirements for special structures are applicable. This includes, for instance, venting systems for smoke. Because many campus buildings are multistory, specific safeguards are necessary. Fire drills and training are important for residence halls and academic buildings. If a campus contains "high-risk" chemicals, it may be subject to regulations of the Chemical Facilities Anti-Terrorism Standards (CFATS) (U.S. Departments of Homeland Security and Education, 2010: 24).

Healthcare and Public Health

The Healthcare and Public Health (HPH) sector is a significant portion of the U.S. economy. Over $2 trillion is spent on healthcare in the United States annually and this amount is increasing each year. Approximately 13 million people are employed in this sector. The U.S. Department of Health and Human Services (HHS) is responsible for public health, healthcare, and food (other than meat, poultry, and eggs, which is the responsibility of the U.S. Department of Agriculture). The sector-specific agency for the HPH sector is HHS and it is responsible for working with partners to develop a sector-specific plan (U.S. Departments of Homeland Security and Health & Human Services, 2010).

HHS publishes a National Health Security Strategy (U.S. Department of Health and Human Services, 2009). The goals of the Strategy are to increase community resilience and maintain the health and emergency response system. One criticism of the Strategy is that it requires local health departments to demonstrate preparedness capabilities without realistic support (Willman and Miller, 2010).

The public health sector is vast, diverse, and mostly privately owned. It includes federal, state, and local health departments and facilities, hospitals (Figure 17-7), health clinics, mental health facilities, nursing homes, laboratories, mortuaries, manufacturers of medical products, and pharmaceutical stockpiles. This sector is dependent on numerous other sectors and services, such as transportation to move patients, personnel, and supplies; communications and IT to support processes; energy to run facilities and equipment; emergency services for initial responses; chemical sector for medications; food; and water.

FIGURE 17-7 Healthcare institutions play a critical role in mitigating and recovering from the effects of natural disasters or deliberate attacks on the homeland. Physical damage to these facilities or disruption of their operations could prevent a full, effective response and exacerbate the outcome of an emergency.

Threats and Hazards of the Healthcare and Public Health Sector

Hospitals, nursing homes, and other healthcare institutions possess specific crime, fire, safety, and cyber weaknesses that require countermeasures. Violence is a recurring problem. A large inventory of consumable items is located within healthcare buildings, including food, medical supplies, drugs, and linens. Thousands of meals and prescriptions are served each day in many of these locations. Assorted crimes are a threat (e.g., assault, theft, kickbacks to purchasers, fraud). Drugs are susceptible not only to internal theft but also to robbery. Expensive medical and office equipment and patient and employee belongings are other tempting targets for offenders. Moreover, the safety of people is a high priority. There is a never-ending flow of employees (doctors, nurses, assorted specialists, nonprofessional support personnel, and volunteers), patients, visitors, salespeople, and repair technicians. A large number of female employees need protection, especially during nighttime shift changes. Patients are particularly vulnerable at all times because of their limited physical capabilities. What makes protection difficult is that these institutions remain open 24 hours a day. Cybersecurity and the protection of patient information are other major concerns.

Emergency plans and special equipment are necessary in case of fire, explosion, accident, natural disaster, bomb threat, strike, and other emergencies. Free access in hospitals makes it difficult to identify potential threats, prevent malicious entry, and detect potentially contaminated individuals from WMD that can have an impact on facility security and emergency operations. This sector must plan for possibilities such as hospital evacuation and fatality management.

The HPH sector plays a critical role in recovering from the effects of natural disasters or deliberate attacks on the homeland. Physical damage to HPH facilities or disruption of their operations could prevent a full, effective response and exacerbate the outcome of an emergency. In addition to established medical networks, the United States depends on several highly specialized laboratory facilities and assets, especially those related to disease control and vaccine development and storage, such as the HHS Centers for Disease Control and Prevention, the National Institutes of Health, and the National Strategic Stockpile (U.S. Departments of Homeland Security and Health & Human Services, 2010; White House, 2003: 41).

Legislative and Regulatory Authorities

Organizations in the HPH sector are bound by a wide range of legislative and regulatory authorities on the federal, state, and local levels. HSPD-7 establishes a national foundation for critical infrastructure protection that includes the HPH sector. The Pandemic and All-Hazards Preparedness Act promotes resilience through HHS and multiple medical programs. HSPD-21 establishes a plan to meet the medical needs of citizens in a catastrophic health event. The **Emergency Medical Treatment and Active Labor Act** requires hospitals to treat patients needing emergency care regardless of their insurance status. Disaster situations involving mass casualties tax the resources of critical facilities in terms of labor, medical supplies, and space.

Additional legal guidance for this sector is found through such laws as HIPAA, SOX, and the NLRA (see Chapter 4). The **Health Insurance Portability and Accountability Act of 1996 (HIPAA)** is a law designed to improve healthcare services delivery, lower costs by reducing paper records and claims, enhance electronic transmission of documents, secure medical data and patient information, prevent errors in the healthcare system, and transfer funds more securely. The regulations require that entities perform a risk analysis to identify risks to the confidentiality, integrity, and availability of the health information they control. Security measures must be documented. They include specific technologies (e.g., encryption) or specific procedures. Each breach of data security can cost healthcare entities millions of dollars for such expenses as notifying customers and paying legal fees and regulatory fines. In one case, HHS fined a healthcare entity $5.3 million when an employee left a laptop containing patient data on a train. Such incidents are mitigated through encryption, remotely wiping out data, and wiping out data after multiple failed password attempts. Besides HIPAA, state laws and regulations on patient data protection are becoming stricter (Wagley, 2011: 52–58).

HIPAA permits companies to release protected health information to a law enforcement agency under certain specified conditions. For example, a hospital must report certain types of wounds, such as gunshot wounds. Requested information must be disclosed in compliance with a court order, warrant, or judicially or grand-jury-issued subpoena or an administrative subpoena or summons.

The Joint Commission, previously known as the Joint Commission on Accreditation of Healthcare Organizations (JCAHO), is an independent, not-for-profit organization and dominant force in promoting standards in the healthcare field that affect funding from government. The Joint Commission addresses how hospitals should provide a secure environment for patients, staff, and visitors. Specific written plans for security are required for the "environment of care" (EC) standards. They include maintaining a security management program and addressing security concerns regarding protecting people and assets. Other standards focus on, for example, training and emergency management.

Examples of other organizations that oversee regulatory requirements include OSHA, government departments of health that perform a variety of duties to protect patient health and safety, Centers for Medicare and Medicaid Services whose duties focus on fraud mitigation, and local government building and fire departments that ensure code compliance. For those large medical centers/research facilities using radioactive materials, the Nuclear Regulatory Commission mandates strict security and safety requirements (Miehl, 2010: 52).

Although not a regulatory authority, the International Association for Healthcare Security and Safety (IAHSS), which was founded in 1968, developments standards for healthcare security practices, training, and certifications. This group enhances the professionalism of those who protect healthcare institutions. It is a not-for-profit organization for hospital security, law enforcement, and safety professionals. The IAHSS administers the Certified Healthcare Protection Administrator (CHPA) credential and publishes *The Journal of Healthcare Protection Management*. For the various certifications, students study IAHSS publications (International Association for Healthcare Security and Safety, 2012).

Healthcare and Public Health Sector Initiatives

HPH sector-specific plan accomplishments include improved information sharing, enhanced disaster planning for pandemics and other events, a cybersecurity working group, and an increase in the number of security site audits (U.S. Departments of Homeland Security and Health & Human Services,2010: i). In addition, this sector is focused on the following:

- In partnership with state health departments, HHS and DHS will identify and prioritize national-level critical medical centers and their services.
- Work with state and local health officials to develop isolation and quarantine standards and set priorities for the deployment of vaccination and prophylaxis resources during a public health crisis.
- HHS and DHS will work with this sector to protect stockpiles of medical supplies, distribution systems, and medical institutions, including basic surveillance capabilities necessary for tracking the spread of diseases and toxic agents. Additionally, HHS will identify providers of critical resources and ensure a ready stockpile of vital medicines for use in an emergency.

Strategies for Healthcare Institutions

Accountability and Inventory Control

Because much of the inventory in healthcare institutions can be used by employees at home or sold to others, accountability and inventory control can minimize shrinkage. For expensive items (e.g., medical equipment), an asset tracking system (electronic article surveillance) can be useful. Imaginative preventive techniques should also be applied where possible. For example, Russell L. Colling (2001: 459–460), in *Hospital and Healthcare Security*, explains that the loss of hospital scrub suits, which are a popular garment worn by a variety of people, can be reduced by issuing them to individuals, and a soiled suit can be exchanged for a clean one through either an issue window or an automatic uniform dispenser. The dispenser, similar to an ATM, credits a returned suit and issues a clean one; this is another use of an automated access card. Linen and other property permanently imprinted with the name of the institution will also aid shrinkage control. Disposable items (e.g., paper towels) are less expensive but also subject to pilferage. Soiled-linen chutes are convenient hiding places for stolen goods. Offenders often wrap stolen merchandise in dirty linen for later recovery. Daily inspection of soiled linens and trash collection and disposal systems ensures that these theft techniques are impeded.

Auditing

When accounting and other safeguards are checked for deviations, loss prevention programs are strengthened. How accurate are accounts receivable and accounts payable records? In the food service operation, how effective are the controls over the ordering of foods and food preparation and distribution? Major locations for theft are the food service area, receiving dock, pharmacy, and material storage rooms.

Applicant Screening

Both professional and nonprofessional staff members cause losses. Equipment and supplies may accompany doctors from a hospital to their private practice. Other employees (and patients and visitors) may steal. An enormous number of items are unguarded at healthcare facilities. Quality applicant screening prevents the hiring of thieves, quacks, and drug and sex offenders.

Access Controls

When entrances to a healthcare facility are limited, unauthorized entry is hindered. Of course, emergency exits are a necessity for safety and alarms on these doors will deter usage. "Smart card"/picture ID badges worn by employees assist in recognition by other employees and control access. Biometrics can be applied to access points at sensitive locations. Outside individuals on the premises (e.g., contractors or technicians) should also be issued access cards/picture IDs.

Koverman (2006: 30–35) writes that access controls in hospitals are especially challenging and require a "tricky balance between appropriate levels of security and appropriate levels of accessibility." The control of visitors must be handled with compassion and empathy. Visitors wanting to see an ill family member are often emotionally upset. Furthermore, the patient's recovery can be aided by visits from loved ones. These factors must be stressed in training programs. Many locations issue visitor passes to avoid overcrowding in patients' rooms and to inhibit a variety of problems such as the deviant who dresses like a doctor to "examine" female patients.

Options for access control include scanning driver's licenses of visitors and entering the information into a database prior to issuing visitor passes. Visitors without a driver's license can be asked for another form of ID. For children's hospitals, visitor names can be checked against sex registries (Miehl, 2010: 50–52).

Security Officers and CCTV

Surveillance of interior and exterior areas (e.g., parking lots and walkways), with patrols and CCTV, deters crime. Good lighting is an integral aspect of this effort. Large medical centers have a command center for communications and CCTV observation. More intense vigilance is usually needed at night, but assorted crimes are possible at any time. Offenders may break into automobiles in parking lots to obtain valuables. Doctors' automobiles are particularly vulnerable, especially if medical bags are left behind. Aggressive patrolling and surveillance are required during shift changes; women should be escorted to their vehicles to foil attempts at purse snatching, assault, and rape. During slow periods when traffic is limited, uniformed officers should resume patrolling while checking for safety hazards.

Some locations have chosen sports jackets or blazers for loss prevention personnel to create a "nonpolice image." Nevertheless, most facilities seem to favor uniforms to provide a "police image" to reinforce crime deterrence and to signify that the location is protected.

The arming of personnel at hospitals is controversial. Screening and training of officers are especially important. Training should include verbal de-escalation, weapon retention (in the event that a subject tries to wrestle away an officer's firearm), and judgment scenarios. Some medical institutions issue firearms to all officers, whereas other locations issue them only to supervisors or those assigned to external areas. Public police working security at hospitals are armed. Tasers are another choice (Spadanuta, 2012). A protective shield is an option for security personnel. It affords increased protection while decreasing injuries, such as from a violent person (Pilker, 2011).

Emergency Room

At least one officer must be stationed in the emergency room (ER) at all times. Access controls are important and include sealed off treatment areas, access cards, intercoms, door buzzers, and panic buttons. CCTV is useful with at least one monitor facing the waiting area so belligerent people can view their behavior and possibly control it. Depending on the crime rate, officers may have to be armed; wear bullet-resistant vests; and prepared to frisk people, confiscate weapons, and make arrests. Signs prohibiting weapons should be posted as well as notices of metal detectors. Disturbances occur regularly in the emergency rooms of busy hospitals. Verbal arguments, assaults, and destruction of property may be caused by patients or visitors who are intoxicated or by rival gangs. In one case, a grandmother/gang leader was shot five times and was being treated in an ER when numerous family and gang members entered and filled the ER. Hospital security was overwhelmed, necessitating many police officers from multiple jurisdictions to be dispatched to the scene. At another hospital, public police responded to so many calls each week that they established a police substation at the hospital.

Special rooms are an option for violent patients. Recurrent disturbances are the result of long waiting periods before treatment. Medical personnel can reduce a portion of disturbances if they adequately explain the reasons for delays.

Patient Property

A recurring puzzle sometimes accompanies patients prior to discharge: jewelry or other personal property is missing. In addition to theft, it is possible that the property is nonexistent, is at the patient's home, or has been misplaced. A few prevention measures are cost-effective. Forms prior to admission can contain a suggestion for the patient to keep valuables at home. The admitting form can contain a statement advising the patient to deposit valuables in a security envelope. These envelopes have the same serial numbers on both the envelope and the receipt. Valuables are inserted and the envelope is sealed in the presence of the patient. The patient and the clerk sign and date both the envelope and the receipt. An adequate safe is needed for these valuables.

Newborn Nursery

Infant abduction is a particularly disturbing problem for parents, employees, and police. The **National Center for Missing and Exploited Children** (NCMEC) created a database on infant abduction that led to the creation of guidelines for hospitals. The NCMEC website offers a variety of helpful information. Those who abduct infants from hospitals often impersonate medical staff and "case" the nursery prior to the crime. Countermeasures include taking footprints of infants, requiring photo ID cards of all employees and volunteers, applying access controls, restricting visitors, using CCTV, and not releasing birth information. Miehl (2010: 52) describes an infant protection system consisting of an electronic tag (RFID) attached to an infant upon birth, and if the tag passes a specific point, an alarm will result, CCTV will focus on the location, video will show on monitors at the nurses' stations and security control room, select doors and elevators will lock, and other systems will activate. Dubin (2011: 74) adds that one system based on RFID uses a tag containing skin sensing technology that activates an alarm if the tag is cut or removed. This added technology is important because, like GPS, the device is tracked and it may not be attached to what security or police think it is attached to.

If abduction occurs, hospital staff should immediately notify the police, the FBI, and the NCMEC; conduct a search; check exit points; and begin a thorough investigation. Such "target hardening" at hospitals can displace kidnapping to areas outside the hospital and to the home of the infant. Parents should protect the infant by, for example, not placing pretty bows or a sign outside the home to signify a birth.

Healthcare Violence and OSHA Guidelines

Certain locations at healthcare facilities have an increased chance of violence. They include emergency rooms, children's hospitals, psychiatric units, and home healthcare. To illustrate, a child in a pediatric unit may require protection following abuse or to prevent kidnapping. If the child's parents are estranged, family strife may spill into the hospital. Protection methods include providing crisis intervention training for healthcare employees, gathering information on the family at the preadmission stage, checking court papers showing parental rights or restraining orders, posting photos of persons to be barred from visiting the child, placing a false name on the nameplate of the patient's room, installing a CCTV camera in the hall near the child's room, and increasing patrols and access controls.

Because of violent crime at healthcare facilities, the federal government and certain states have developed safety standards to prevent violence in the workplace. On the federal level, the Occupational Safety and Health Administration (2004) published *Guidelines for Preventing Workplace Violence for Health Care and Social Service Workers*. According to OSHA, "All employers have a general duty to provide their employees with a workplace free from recognized hazards likely to cause death or serious physical harm" and "employers can be cited for violating the General Duty Clause if there is a recognized hazard of workplace violence in their establishments and they do nothing to prevent or abate it." The OSHA guidelines are advisory in nature and focus on management commitment, worksite analysis, hazard prevention and control (e.g., physical security), education and training, recordkeeping, and evaluation.

Pharmacy Protection

A small percentage of our population is addicted to drugs and will do anything to obtain them. Narcotics addiction among doctors, nurses and other healthcare personnel is a much more serious problem than recognized, and healthcare facilities are a major source of illegal drug traffic involving legal drugs (Hood, 2010; Colling, 2001: 443–450). There have been instances in which medical personnel have withheld drugs from patients for their own use or sale. One technique is to substitute flour for medication. At one hospital a surgical technician was addicted to narcotics while being a carrier of hepatitis C. She injected herself with syringes of a painkiller for post-operative patients, refilled the syringes with a saline solution, and caused unnecessary pain for patients, while infecting them with the disease. Hospital administrators are reluctant to report these acts for fear of lawsuits.

Another technique is to write a phony prescription or alter an existing one. In one case, a hospital pharmacist diverted thousands of dollars of drugs to his retail drugstore.

Longmore-Etheridge (2010: 30–32) writes of Milwaukee-area hospitals experiencing one case a month of a nurse stealing narcotics for personal use. Rather than prosecuting (unless a nurse is diverting for resale), nurses lose their jobs and are reported to the state for a license sanction—a penalty often greater than what would be received from the criminal justice system.

The **Controlled Substances Act of 1970**, amended several times, provides the legal foundation for the manufacture, importation, possession, and distribution of certain drugs. The Drug Enforcement Administration (DEA) enforces the Act. All states regulate pharmacies.

Pharmacy losses can be prevented through strict accountability and inventory controls. For the protection of the pharmacy, cashiering operations, and the business office, the following measures will deter burglary, robbery, and other crimes: (1) intrusion and holdup alarms integrated with CCTV, (2) bullet-resistant glass, (3) access controls, (4) patrols, and (5) consideration of removing signs identifying the location. To reduce employee theft of drugs, Longmore-Etheridge (2010: 30–32) suggests automated dispensing systems requiring a PIN, automated recordkeeping, auditing of pharmaceuticals, and employee awareness training.

Locker Rooms

Men and women who work in healthcare institutions are accustomed to using separate locker rooms before, during, and after shifts. Locker rooms are ordinarily located in basements or remote locations. CCTV and patrols can be applied to areas just outside locker rooms. Panic buttons that signal trouble are useful within locker rooms. Loss prevention personnel should conduct occasional locker inspections to deter assorted problems.

Mortuary

Mortuaries have been the sites of morbid crimes. People, including relatives of the deceased, have stolen jewelry directly from cadavers. Gold dental work has been extracted with pocketknives. The rare sexual perversion called necrophilia (i.e., sexual activity with a corpse) is a possibility in the mortuary. In addition, there are cases in which the wrong body was taken away for burial. A complete inventory should be conducted of the personal property of the deceased soon after death. Witnesses and appropriate paperwork must be a part of this procedure.

Fire and Other Disasters

A multitude of fire codes and standards for healthcare institutions emanate from local, state, and federal agencies. The NFPA 101 Life Safety Code refers to a "total concept approach" to the fire problem, consisting of construction, detection/suppression, and staff presence. NFPA 99 Healthcare Facilities Code establishes criteria to minimize hazards from fire, explosion, and electricity.

The Joint Commission has worked closely with the federal government to provide adequate patient care and safety. Legislation has stated that hospitals must meet federal requirements for health and safety if they are accredited by The Joint Commission. All healthcare locations are subject to OSHA, except those that are federal. At times, overlapping regulations produce confusion. A healthcare facility may require approval for life safety from multiple authorities having jurisdiction (AHJ). These authorities include The Joint Commission, OSHA, the state fire marshal, local building official, fire department, state healthcare licensing agency, HHS, and the facility's insurance carrier.

Fire protection at healthcare facilities must be a top priority, especially because of immobile patients and a great deal of disposables and rubbish that must be removed properly. The early detection and suppression of fire, plus fire-resistive construction, will play major roles if patients cannot be moved and must be "defended in place." Flammable substances require safe storage, use, and disposal. A full-time fire engineer or loss prevention manager should be in charge of fire protection. This individual coordinates training, evacuation plans, drills, equipment evaluation and purchasing, and other duties; in addition, he or she acts as liaison with local fire agencies.

If a WMD event should occur, healthcare workers would become **first receivers** who treat incoming victims. OSHA information assists hospitals in protecting healthcare workers and creating emergency plans for worst-case scenarios. It also offers suggestions for **personal protective equipment (PPE)**. Examples are gloves, mask, and protective eyewear, or a head-to-toes protective suit and respirator. In addition to healthcare workers, security officers at hospitals need personal protection and training so they do not become victims themselves. As first receivers treat victims, security officers will likely maintain order and control traffic and access by victims who may be in a state of panic and need to be decontaminated and quarantined. In addition, it is possible that the facility will become overwhelmed, civil disturbance may occur, and the hospital will become a disaster site.

The **Public Health Security and Bioterrorism Preparedness and Response Act of 2002** recognizes emergency room workers as major responders to the problem of terrorism, and it promotes a national curriculum of training to respond to biological agents. It also provides for a real-time surveillance system among emergency rooms, state public health departments, and the **Centers for Disease Control and Prevention (CDC)**—the lead agency, in HHS, if a communicable disease outbreak occurs.

What do you think is the most serious risk facing healthcare institutions? What are your solutions?

Search the Internet

Here is a list of websites relevant to this chapter:

American Bankers Association: www.aba.com
Centers for Disease Control and Prevention: www.cdc.gov
Center for the Prevention of School Violence: www.ncdjjdp.org/cpsv
International Association for Healthcare Security & Safety: www.iahss.org
International Association of Campus Law Enforcement Administrators: www.iaclea.org
International Council of Shopping Centers: www.icsc.org
Joint Commission: www.jointcommission.org
Loss Prevention Foundation (certifications): www.losspreventionfoundation.org
Loss Prevention Research Council: www.lpresearch.org
National Association of School Resource Officers: www.nasro.org
National Center for Missing and Exploited Children: www.missingkids.com
National Crime Prevention Council: www.ncpc.org
National Retail Federation: nrf.com
National Shoplifting Prevention Coalition: www.shopliftingprevention.org
Retail Industry Leaders Association: www.rila.org
U.S. Department of Education, Office of Safe and Healthy Schools: www.ed.gov/emergencyplan
U.S. Department of Education, REMS Technical Assistance Center: rems.ed.gov
U.S. Department of Health and Human Services: www.hhs.gov
U.S. Department of Homeland Security: www.dhs.gov
U.S. Secret Service: www.secretservice.gov

Case Problems

17A. A group of eight merchants who own retail stores at a small shopping center have hired you as a loss prevention consultant. These people are interested in reduced losses and increased profits. Their loss problems are employee theft, losses at the POS, shoplifting, robbery, and burglary. As a loss prevention consultant, what is your plan? Do not forget that you must earn your fee, satisfy the merchants to develop a good reputation, and produce effective loss prevention measures that will reduce losses and increase profits.

17B. As a regional loss prevention manager for a retail chain, you are growing increasingly suspicious of people employed by one of the stores in your region. The store manager and employees constantly state that the high shrinkage of four percent at the store is due to shoplifters. The manager argues that very little shrinkage results from damaged merchandise or employee theft. EAS and CCTV systems are functioning at the store, but no security personnel are employed; a variety of nonsecurity employees maintain and operate these systems. What action do you take?

17C. As a bank security director you are faced with the challenge of an upsurge of robberies of customers at automated teller machines and after-hours depositories (when retail managers deposit daily sales revenue). What do you do?

17D. You are a school district security director and it has been brought to your attention that students in the high school, in certain classes, routinely bring fast food, cell phones, and other electronic devices to their desks and intimidate teachers into permitting these items. The school district superintendent has asked you to work on a solution to this problem. What do you do?

17E. You are a hospital security director and your new challenge is to prevent the admittance of recidivist patients who fake illness to secure shelter and food. These individuals are often living on the street and use a different identity each time they seek admittance to the emergency room. The problem seems to increase during the winter months. What do you do?

17F. Of the four types of entities in this chapter, select two and describe why the ones you chose are unique in terms of loss problems and prevention strategies.

17G. Choose two types of entities from this chapter. For each, establish a priority list of the five most important measures to counter losses. Explain your reasoning behind each ranked list.

17H. Choose two types of entities from this chapter. (1) Refer to additional sources to gather further information, and (2) explain why you would prefer to work in one rather than the other. Maintain a bibliography.

17I. What do you think is the most demanding of the entities discussed in this chapter for a loss prevention manager? Justify your answer.

References

Abdullaev, N. (2004). Beslan, russia…terror in the schoolhouse!. *Homeland Defense Journal, 2* (September).

Altheide, D., et al. (1978). *The social meanings of employee theft crime at the top.* New York: J. B. Lippincott Co.

American Bankers Association. (2011). Certified Financial Services Security Professional (CFSSP). <www.aba.com/ICB/CFSSP.htm> April 11, 2012.

Anderson, J. (2012). Gun-Free schools act. *Education Law* (February 14) [lawhighereducation.com, retrieved April 15, 2012].

Anderson, T. (2001). Up and running. *Security Management, 45* (January).

Aubele, K. (2011). Checking out security solutions. *Security Management, 55* (December).

Baillie, C. (2012). The tactics of ORC: How shoplifters attempt to defeat security technology. *Security Technology Executive, 22* (May).

Barlas, D. (2012). How to reduce your risk of credit card fraud. *Small Business Information.* <sbinformation.about.com> retrieved March 28, 2012.

Beaulieu, E. (2005). Theft operation 'disturbs' retailers. *Security Director News, 2* (February).

Beck, A., & Palmer, W. (2011). The importance of visual situational cues and difficulty of removal in creating deterrence: The limitations of electronic article surveillance source tagging in the retail environment. *Journal of Applied Security Research, 6* (January–March).

Cambern, J. (2010). Screening contract employees: An important slice of retail security. *Loss Prevention, 2* (March–April).

Cardone, C., & Hayes, R. (2012). Shoplifter perceptions of store environments: An analysis of how physical cues in the retail interior shape shoplifter behavior. *Journal of Applied Security Research, 7* (January–March).

Colling, R. (2001). *Hospital and healthcare security.* Boston: Butterworth-Heinemann.

Decker, S. (2000). *Increasing school safety through juvenile accountability programs.* Washington, D.C.: U.S. Department of Justice.

Deeth, S. (2012). Shoplifter high on LSD bit chunk of zellers security guard's arm off. *The Peterborough Examiner* (April 26). <www.thepeterboroughexaminer.com> retrieved May 1, 2012.

DiLonardo, R. (2012). Global shrinkage rises 6.6 %. *Loss Prevention, 11* (January–February).

DiLonardo, R. (2006). Industry news. *Loss Prevention, 5* (May–June).

Doyle, M. (2011). Behind the statistics from the hayes international 23rd annual retail theft survey. *Security Technology Executive, 21* (September).

Drysdale, D., et al. (2010). *Campus attacks: Targeted violence affecting institutions of higher education.* Washington, DC: U.S. Secret Service, FBI, DHS, and U.S. Department of Education.

Dubin, C. (2011). Tag teams: Best-in-Class RFID applications. *Security Magazine, 48* (March).

Federal Bureau of Investigation. (2011a). Organized Retail Theft: A $30 Billion-a-Year Industry (January 1). <www.fbi.gov/news/stories/2011/january/retail_010311> retrieved March 30, 2012.

Federal Bureau of Investigation. (2011b). Bank Crime Statistics, 2010. <www.fbi.gov> retrieved April 25, 2012.

Federal Bureau of Investigation. (2011c). Taking a Trip to the ATM? Beware of Skimmers (July 14). <www.fbi.gov/news/stories/2011/july/atm_071411/atm_> retrieved April 10, 2012.

Federal Bureau of Investigation. (2012). Bank Crime Statistics, 2011. <www.fbi.gov> retrieved June 22, 2012.

Federal Reserve Bank of San Francisco. (2012). Plastic Fraud: Getting a Handle on Debit and Credit Cards. <www.frbsf.org> retrieved March 26, 2012.

Fisher, B., & Sloan, J. (1993). University response to the campus security act of 1990: Evaluating programs designed to reduce campus crime. *Journal of Security Administration, 16.*

Gaur, N., & Gilliland, W. (2006). New customer touch points. *Information Week* (June 1). <www.informationweek.com/> retrieved February, 13, 2007.

Gill, M., Bilby, C., & Turbin, V. (1999). Retail security: Understanding what deters shop thieves. *Journal of Security Administration, 22.*

Gips, M. (2007). EAS not a silver bullet. *Security Management, 51* (March).

Harwood, M. (2007). Preventing the next campus shooting. *Security Management, 51* (August).

Hayes, R. (2006). Store design and LP. *Loss Prevention, 5* (March–April).

Hayes, R. (2012). "You Can Steal It, But You Can't Use It: Moving Toward Benefit-Denial Technology." LP Magazine, 11.5 (September–October).

Henry, S. (2011). Speech at the Information Systems Security Association International Conference, Baltimore, MD. (October 20). <www.fbi.gov> retrieved February 24, 2012.

Hollinger, R. (2011). 2010 National retail security survey, executive summary. *Loss Prevention, 10* (November–December).

Hood, J. (2010). Healthcare professionals face unique addiction challenges. *Los Angeles Times* (December 24). <articles.latimes.com> retrieved April 24, 2012.

Institute for Intergovernmental Research. (2012). Welcome to the G.R.E.A.T. Web Site. <www.great-online.org> retrieved April 16, 2012.

International Association of Campus Law Enforcement Administrators. (2011). About IACLEA. <www.iaclea.org> retrieved April 17, 2012.

International Association of Chiefs of Police and Bureau of Justice Assistance. (n.d.). *Guide for Preventing and Responding to School Violence*, 2nd ed. <www.theiacp.org> retrieved April 16, 2012.

International Association for Healthcare Security and Safety. (2012). About Us. <www.iahss.org> retrieved April 22, 2012.

Jaeger, S. (2001). Crime concerns on campus. *Security Technology & Design, 12* (March).

Johnson, B., & Carter, T. (2009). Combatting the shoplifter: An examination of civil recovery laws. *Journal of Applied Security Research, 4*.

Koverman, R. (2006). Assessing and securing the hospital environment. *Security Technology & Design, 16* (May).

Kroll, K. (2012). With new bank-security guidance, how safe from cybercrime is your firm? *CFO World* (February 14) <www.cfoworld.com> retrieved February 24, 2012.

Langton, L., & Hollinger, R. (2005). Correlates of crime losses in the retail industry. *Security Journal, 18*.

Leff, D. (2012). Money laundering and asset forfeiture: Taking the profit out of crime. *FBI Law Enforcement Bulletin, 81* (April).

Longmore-Etheridge, A. (2010). Curing what ails hospitals. *Security Management, 54* (June).

Loss Prevention Foundation. (2011). Loss Prevention Foundation Certifications. <www.losspreventionfoundation.org/certification.html> retrieved April 11, 2012.

Major Cities Chiefs Association and Bureau of Justice Assistance. (2009). *Campus Security Guidelines: Recommended Operational Policies for Local and Campus Law Enforcement Agencies.* <www.majorcitieschiefs.com/pdf/MCC_CampusSecurity.pdf> retrieved April 19, 2012.

McIntyre, M. (2011). The need to modernize civil recovery statutes. *Loss Prevention, 10* (May–June).

Meadows, R. (2007). *Understanding violence and victimization* (4th ed.). Upper Saddle River, NJ: Pearson Prentice Hall.

Miehl, F. (2010). Hospital security strategies. *Security Technology Executive, 20* (June/July).

Muscato, F., & Pearson, J. (2010). Boosters versus fences: Which is the direct link to shrink? *Loss Prevention, 9* (March–April).

National Center for Education Statistics. (2012). *Indicators of School Crime and Safety: 2011.* nces.ed.gov retrieved April 15, 2012.

Occupational Safety and Health Administration. (2004). *Guidelines for Preventing Workplace Violence for Health Care and Social Service Workers.* <www.osha.gov/Publications/osha3148pdf> retrieved April 23, 2012.

O'Donnell, J., & Meehan, S. (2012). Self-checkout lanes boost convenience, theft risk. *USA Today* (April 9). <www.usatoday.com> retrieved April 9, 2012.

Percoco, N. (2011). The virtual 9/11: Point-of-Sale systems may be the ultimate IT security vulnerability. *Security Technology Executive, 21* (September).

Pilker, D. (2011). Security's successful operation. *Security Management, 55* (June).

Purpura, P. (1997). *Criminal justice: An introduction.* Boston: Butterworth-Heinemann.

Schmedlen, R. (2000). A picture of profiling. *Security Management, 44* (May).

Security Executive Council, (2011). Banking and financial regulations. *Security Technology Executive, 21* (January/February).

Sobel, R. (2012). Perception of violence on a high school campus. *Journal of Applied Security Research, 7* (January–March).

Spadanuta, L. (2012). Hospitals and guns. *Security Management, 56* (February).

Spadanuta, L. (2011). Minimizing the risks of online banking. *Security Management, 55* (November).

Spadanuta, L. (2010). Robbery risk reduction. *Security Management, 54* (January).

Stelter, L. (2011). As return fraud increases, retailers turn to data to identify anomalies. *Security Director News* (June 22).

Swartz, J. (2006). Banks pull out the big guns to guard online users. *USA Today* (November 20) <www.usatoday.com> retrieved November 21, 2006.

Swartz, J. (2012). Top secret visa data center banks on security, even has moat. *USA Today* (March 26) <www.usatoday.com> retrieved March 26, 2012.

TwoWayMirrors.net. (2006). Two Way Mirror FAQ. <www.twowaymirrors.net/mirrorfaq.htm> retrieved March 8, 2006.

U.S. Census. (2011). Back to School: 2011–2012. <www.census.gov> retrieved April 14, 2012.

U.S. Department of Education. (2011a). Office of Safe and Drug-Free Schools. <www.ed.gov> retrieved April 15, 2012.

U.S. Department of Education. (2011b). The Handbook for Campus Safety and Security Reporting. <www2.ed.gov/admins/lead/safety/handbook-2.pdf> retrieved March 10, 2011.

U.S. Department of Health and Human Services. (2009). *National Health Security Strategy of the United States of America.* <www.phe.gov/preparedness/planning/authority/nhss> retrieved March 25, 2012.

U.S. Department of Homeland Security. (2010). *Commercial Facilities Sector-Specific Plan: An Annex to the National Infrastructure Protection Plan.* <www.dhs.gov> retrieved March 27, 2012.

U.S. Department of Homeland Security.(2009). *National Infrastructure Protection Plan.* <www.dhs.gov/xlibrary/assets/NIPP_Plan.pdf> retrieved March 18, 2012.

U.S. Departments of Homeland Security and Education. (2010). *Education Facilities Sector-Specific Plan: An Annex to the Government Facilities Sector-Specific Plan.* <www.dhs.gov> retrieved April 14, 2012.

U.S. Departments of Homeland Security and Health & Human Services. (2010). *Healthcare and Public Health Sector-Specific Plan: An Annex to the National Infrastructure Protection Plan.* <www.dhs.gov> retrieved April 20, 2012.

U.S. Departments of Homeland Security and Treasury. (2007). *Banking and Finance: Critical Infrastructure and Key Resources Sector-Specific Plan as input to the National Infrastructure Protection Plan.* <www.dhs.gov> retrieved April 9, 2012.

U.S. Department of Justice, Office of Community Oriented Policing Services, (2005). *National summit on campus public safety.* Washington, DC: U.S. Department of Justice.

Wagley, J. (2011). Sick of data protection rules. *Security Management, 55* (October).

White House. (2003). *The National Strategy for the Physical Protection of Critical Infrastructures and Key Assets* (February). <www.whitehouse.gov> retrieved February 23, 2007.

Willman, A., & Miller, E. (2010). National health security strategy town hall. *National Association of County & City Health Officials* <www.hhs.gov/aspr/opsp/nhss/nhss0912.pdf> retrieved April 27, 2012.

Woodward, K. (2012). Most e-retailers say they have fraud under control. *Internet Retailer* (March 14). <www.internetretailer.com> retrieved March 22, 2012.

Zimmerman, A. (2006). As shoplifters use high-tech scams, retail losses rise. *The Wall Street Journal* (October 25). <online.wsj.com> retrieved October 30, 2006.

18 Topics of Concern

OBJECTIVES

After studying this chapter, the reader will be able to:

1. Discuss the problem of violence in the workplace and what can be done about it
2. List strategies of human resources protection
3. Describe the problems and remedies associated with substance abuse in the workplace
4. List the methods by which an adversary might obtain information assets and list strategies of information security
5. Explain the array of communications security
6. Describe electronic surveillance, wiretapping, and Technical Surveillance Countermeasures (TSCM)

KEY TERMS

- workplace violence
- OSHA's general duty clause
- workplace violence prevention program
- threat management team
- conflict resolution and nonviolent response
- active shooter
- human resources protection
- executive protection
- executive protection program
- country risk ratings
- creature of habit
- safe room
- identity theft
- phishing
- substance abuse
- anti-substance abuse program for the workplace
- Anti-Drug Abuse Act of 1988
- specimen validity testing
- trace detection
- alcoholic
- psychological dependence
- addiction
- tolerance
- withdrawal
- information security
- information assets
- espionage
- business espionage
- government espionage
- trade secret
- patent
- trademark
- copyright
- corporate intelligence
- competitive intelligence
- Economic Espionage Act of 1996
- reverse engineering

- "Four Faces of Business Espionage"
- pretext interview
- classified information
- information security program
- Operations Security (OPSEC)
- communications security
- TEMPEST
- electronic surveillance
- wiretapping
- communications tapping
- Technical Surveillance Countermeasures (TSCM)

Workplace Violence

The definition of workplace violence affects not only its volume but also its cost. As the definition expanded from "one employee attacks another" to "any threat or violence that occurs on the job," so too did the volume of workplace violence and its cost.

In explaining the range of workplace violence and publishing an excellent standard on workplace violence prevention, ASIS International (2011: 4) refers to it in broad terms and includes not only assaults but threats of violence, stalking, and aggressive behavior. ASIS considers perpetrators to be wide-ranging and to include a current or former employee, customer, vendor or contractor, family member or someone linked to an employee, or a person with an unusual fixation to an employee or the workplace. The location of problematic behavior is likewise diverse and ASIS refers to on-site, during work-related travel, or at a work-related event.

Perhaps in a subsequent ASIS published standard on workplace violence prevention, ASIS will place an emphasis on bullying, cyberbullying, and protecting employees required to work at home. These topics are covered in a subsequent page.

ASIS International (2011: 5) defines **workplace violence** as follows:

A spectrum of behaviors, including overt acts of violence, threats, and other conduct that generates a reasonable concern for safety from violence, where a nexus exists between the behavior and the physical safety of employees and others (such as customers, clients, and business associates), on-site, or off-site when related to the organization.

The Occupational Safety & Health Administration (2012) defines workplace violence as "any act or threat of physical violence, harassment, intimidation, or other threatening disruptive behavior that occurs at the work site."

Kenny (2005a: 45–56) emphasizes that while threats may not result in physical injury, they can leave victims traumatized and fearful. In addition, threats can harm productivity and "serve as the catalyst for progressively more dangerous behaviors." Kenny adds, "Those who ignore, misunderstand or fail to take threats seriously miss an opportunity to identify, diffuse, or resolve workplace problems." Research by Kenny (2005b: 55–66) indicated that women were more likely to be victimized. Tiesman et al. (2012: 277–284), found that intimate partner violence resulted in 142 homicides of women in the workplace in the United States between 2003 and 2008. This figure is 22 percent of the 648 women who were killed at work during this period.

OSHA reported 4,547 fatal workplace injuries in the United States in 2010; 506 were homicides. For homicides in the workplace in 2000, the number was 677, and in 1995, the number was 1,036. Homicide is the fourth-leading cause of death at work and the leading cause of death for women in the workplace. Almost two million American workers are victimized by workplace violence each year and many cases are unreported. Research has recognized factors that increase risk, including sites where money is exchanged with the public; working alone, late at night, in a high-crime area; or where alcohol is served. Other factors include locations where care is provided or where working with volatile and unstable people (Occupational Safety & Health Administration (2012)).

Workplace violence is costly to businesses. Losses reach into the billions of dollars each year and include medical and psychological care, lost wages, property damage, lost company production, impact on employee turnover and hiring, and litigation (Hughes, 2001: 69). An organization's brand can also be harmed, especially with today's technology of tweeting, Facebook, blogging, and other types of social media.

Obviously, incidents of workplace violence (e.g., homicides, assaults, rapes) occurred before increased attention focused on the problem in the early 1990s. What we have witnessed is a change in the way the problem is perceived and counted. This is beneficial for gauging increases and decreases in the problem and as a foundation for planning protection with scarce resources. The United States has been known to maintain good statistics on a number of problems and freely publicizes trends. From an international perspective, others may view the United States as the most violent society, when we may simply be the best at gathering data.

Do you think the United States is a violent society, or do we just maintain good data-gathering systems? Explain your viewpoint.

Legal Guidelines

There is no national law addressing violence in the workplace. Various states have enacted laws to prevent violent crime at work, especially for retail establishments and healthcare settings. Local ordinances may also require security for certain businesses (e.g., convenience stores). Labor contracts are another consideration. In addition, harassment, discrimination, and "hate crime" laws are an important source for planning to prevent hostile behavior from being directed at specific groups. Under the Family and Medical Leave Act and laws in certain states, employers are prohibited from discriminating against an employee victimized by an abusive relationship. Case law is another source on the issues of workplace violence.

OSHA has published voluntary guidelines for workers in late-night retail, healthcare, social services, and taxicab businesses. Although these guidelines are not legal requirements, OSHA has enhanced its efforts at reducing workplace violence. Under the **OSHA Act "general duty clause,"** employers are required to protect employees against recognized hazards (e.g., workplace violence) that are likely to cause serious injury or death. OSHA's strategies include targeting specific businesses and issuing citations for failure to safeguard employees. OSHA supports security strategies such as policies and procedures, training, reporting and investigating incidents, risk analyses, and physical security.

To boost safety, OSHA issued a federal program change titled "Enforcement Procedures for Investigating or Inspecting Workplace Violence." This change establishes policy guidance and procedures for field offices to apply when conducting inspections in response to incidents of workplace violence and emphasizes employer action to minimize the problem through precautions (Occupational Safety & Health Administration, 2011: 2).

Employers have a legal responsibility to prevent harm to people on the premises. Those who do not take measures to prevent violence in the workplace face exposure to lawsuits. Legal theories at the foundation of such lawsuits include premises liability, negligence, harassment (sexual and other forms), and respondeat superior (see Chapters 4 and 6). Workers' compensation may cover injured employees, but the exposure of employers is much greater.

Contradictions in the law make protection difficult: OSHA requires a safe working environment for employees, but the Americans with Disabilities Act (ADA) can create difficulties for an employer seeking to control an employee with a mental instability. Employees have successfully sued employers for defamation because instability was mentioned. The ADA restricts "profiling" of employees through the observation of traits thought to be potentially violent (Jaeger, 2001: 74).

Protection Methods

Kenny (2010: 162) writes that aggressors usually provide a warning prior to extreme violence. His pre-incident indicators include verbal attacks, intimidation, harassment, property destruction, simple assault, or bullying. Katz (2011: 34) refers to behaviors that *may* indicate subsequent violence: overreacts, belligerent, anger from constructive criticism, aggressive, reclusive, and making idle threats. Personal problems that *may* lead to violence include emotional instability, domestic, and substance abuse. Matsumoto et al. (2012: 1–11), report that "laypersons often do not recognize the important distinctions among emotions." "Anger, contempt, and disgust have different physiologies, mental states, and nonverbal expressions, implying different behaviors." Gaining an understanding of different emotions that serve to motivate people can help to predict violence. Meadows (2007: 121) views the worker who becomes violent as usually a white male, between 25 and 50 years old, with a history of interpersonal conflict and pathological blaming. He tends to be a loner and may have a mental health history of paranoia and depression. He also may have a fascination with weapons.

What follows is a list of strategies for a **workplace violence prevention program**.

- Establish a committee to assess risks and plan a violence prevention program. Include top management and specialists in security, risk management, human resources, psychology, law, and corporate communications. This committee can also serve as a **threat management team** to investigate and respond to workplace violence events (e.g., complaints, incidents, and emergencies).

- Establish policies (Figure 18-1) and procedures and communicate the problems of threats and violence to all employees. Include reporting requirements and procedures. Ensure a culture of reporting to prevent problems from escalating.
- Consider other policies linked to the organizational workplace violence prevention policy. These include anti-harassment, anti-discrimination, substance abuse, ethics, IT use (e.g., e-mail), searches, weapons on the premises, privacy, and others (ASIS International, 2011: 15).
- A survey by the Society for Human Resource Management found that 51 percent of participating companies had incidents of workplace bullying. It decreases morale and productivity, increases stress and depression, and can result in violence. Bullying can occur face to face or through cyberbullying—via e-mail, texts, tweets, and other ways. Company EEOC and harassment policies often cover bullying (Evans, 2012). Employers must take action and incorporate the problem into workplace violence prevention programs.
- Consider substance abuse testing as a strategy to prevent workplace violence. For years, Bureau of Justice Statistics data has shown a relationship between violent crime and substance abuse.
- Hershkowitz (2004: 83–88) recommends avoiding a strict zero tolerance policy that does not consider mitigating circumstances because legal problems may arise with cases involving harassment, labor contracts, disability, and discriminatory practices.

FIGURE 18-1 Policies are an essential component of a workplace violence prevention program.

- Ensure that a thorough and impartial investigation is conducted following a reported incident. Follow legal requirements, such as those pertaining to due process, privacy, and labor agreements. Consistently enforce policies.
- Although human behavior cannot be accurately predicted, screen employment applicants. The ADA limits certain questions; however, the following can be asked: "What was the most stressful situation you faced at work and how did you deal with it?" "What was the most serious incident you encountered in your work and how did you respond?"
- Managers and supervisors should be sensitive to disruptions in the workplace, such as terminations. Substance abuse and domestic and financial problems also can affect the workplace. An EAP is especially helpful for such problems.
- Train managers and supervisors to recognize employees with problems and report them to the human resources department. Include training in **conflict resolution and nonviolent response**. Train in active listening skills, such as repeating to the subject the message he or she is trying to communicate and asking open-ended questions to encourage the subject to talk about his or her problems. Listen and show that you are interested in helping to resolve the problem. Do not be pulled into a verbal confrontation; do not argue. Acknowledge and validate the anger by showing empathy, not sympathy. Speak softly and slowly. Ensure that a witness is present. Maintain a safe distance, without being obvious, to provide an extra margin of safety. If a threat is made or if a weapon is shown, call the police.
- Remember that outsiders (e.g., visitor, estranged spouse, or robber) may be a source of violence and protection programs must be comprehensive.
- If employees are required to work at home, employers should provide training for safety and security at home. This is enterprise security risk management and "extending security out." This approach can include a suggestion to contact public police crime prevention services or corporate security for a home security survey by a security specialist. Training topics can include protecting proprietary information and contacting social service and criminal justice agencies for domestic violence.
- If a violent incident occurs, a previously prepared crisis management plan becomes invaluable. (Chapter 13 explains emergencies and mass notification systems.) Without a plan, a committee should be formed immediately after emergency first responders complete their duties on the premises and affected employees and their families are assisted. At one major corporation, management was unprepared when the corporate security manager was shot. A committee was quickly formed to improve security and survey corporate plants. In addition to expenditures for physical security and training, an emphasis was placed on *awareness, access controls,* and *alerts* (Purpura, 1993: 150–157).

Active Shooter

"An **active shooter** is an individual, who may be accompanied by other shooters, who seeks to kill and wound as many random and/or specific people as possible by using firearms, explosives, and other weapons" (Purpura, 2011: 193). The Virginia Tech massacre serves as an example (see Chapter 17). Another example is the 2008 Mumbai attacks in India. Known as the financial capital of India and also a tourist destination, the city of Mumbai was attacked by ten terrorists who arrived in small boats, targeted major sites, and with automatic weapons and explosives, killed almost 200 people and wounded many over four days, until the terrorists were killed and one was captured and executed in 2013.

Active shooters possess various motivations (e.g., terrorism or personal problems) and apply a variety of methods to maximize their killing spree (e.g., multiple weapons, a lot of ammunition, and hindering the escape of people). Because the response time of authorities is unlikely to be instantaneous, potential victims must act to protect themselves. This includes escape or barricading doors and preparing makeshift weapons.

Since the Virginia Tech massacre and the Mumbai attacks, police have intensified training to confront active shooters immediately to reduce casualties (rather than wait for officers and equipment to respond and implement a plan). FEMA offers a free online course on the active shooter. The National Retail Federation updated its active shooter guidelines that coincide with U.S. Department of Homeland Security training.

"Dozens Sue over Lockheed Shootings"

The following account of workplace violence is from a news article in *The Clarion-Ledger*, Jackson, Mississippi (Hudson, 2004: 1A). Forty-seven employees and relatives of employees filed a federal lawsuit against Lockheed Martin claiming emotional distress following the shooting of 14 people when employees were allegedly forced back inside the plant for a "live head count." The shooting became the state's deadliest act of workplace violence when plant employee Doug Williams, 48, shot and killed six co-workers and wounded eight others prior to committing suicide. Lockheed Martin allegedly ordered employees to the canteen for a count, which is when employees supposedly walked near the victims' bodies. The lawsuit also claimed that Lockheed Martin failed to protect employees who complained that Williams had threatened to shoot black co-workers, and that the company denied employee requests for security officers prior to the shooting. A Lockheed Martin spokesperson stated: "Lockheed Martin has been cleared of responsibility for this incident by state and federal authorities and is confident that the same conclusion will be reached by the court."

One plaintiff, Henry Odom, a 35-year employee, stated in a court affidavit that he had complained to management about fights and auto thefts at the plant and asked for security officers for the premises. The affidavit stated that Williams shot him in his left arm twice; the second shot also entered his back and punctured a lung. The affidavit noted that the plant now has armed security officers on duty.

The lawsuit claimed that three weeks prior to the shooting, Williams placed a work-issued "bootie" on his head that appeared, according to some, to resemble a Ku Klux Klan hood. Management allegedly confronted him and he supposedly left the plant angry and did not return for about a week. According to the court papers, he was permitted to return to work, but required to attend an ethics course with black co-workers. He allegedly left a meeting and told workers he was angry and going to "take care of this." He allegedly returned to the meeting, fired on some in the room, and then went to the plant floor to shoot others. The lawsuit claimed that Lockheed Martin had sufficient time to stop him and to warn employees of the imminent danger he posed.

Workers' Comp Insider (Ryan, 2005) reported that a federal appeals court upheld workers' compensation as the exclusive remedy for the nine surviving victims and the families of the six workers who were killed in the Lockheed Martin shooting in Meridian. Ryan noted that this would limit damages to about $150,000. He reported the following: "Exclusive remedy is a strong concept that holds up under repeated legal challenges. Workers comp is no fault by its very nature, a quid pro quo arrangement in which employers agree to provide medical and wage replacement to injured workers, and in turn, this becomes the sole remedy. In all but the most unusual circumstances, employees lose the right to sue their employer for work-related injuries. Sometimes this seems unfair to a worker because benefits are paltry when stacked side by side with enormous awards from civil litigation. But when legal challenges succeed, they weaken the system's underpinnings. Workers comp is essentially a safety net, a system designed to provide the best for the most, not to provide individual redress for every wrong. When litigation is successful at piercing the exclusive remedy shield, it often involves employer misconduct that is highly egregious."…"Many states require proof of willful intent. It must be demonstrated that the employer had substantial certainty that an injury would occur. In this case, the shooting victims and their surviving families sued the company on the basis of having been deprived of civil rights, alleging that management knew of the threat and '… knew employee Doug Williams' racist views had created a volatile work environment but did too little to defuse the situation.'"

In 2008, a daughter of one of the victims reached a settlement following a lawsuit, not against Lockheed Martin, but against Lockheed's EAP (NEAS, Inc.). NEAS referred the offender to an affiliate without full background information on the offender. The affiliate cleared him following three counseling sessions (DiBianca, 2008).

Human Resources Protection

People are the most valued asset of an organization and, depending on the type of harm to them, the consequences can be devastating. At home and abroad, businesses have become the target of kidnappings, extortion, assassinations, bombings, and sabotage. Terrorists use these methods to obtain money for their cause, to alter business or government policies, or to change public opinion. Organized crime groups also are participants in such criminal acts, but in contrast to terrorists, their objective usually is money. It appears that successful criminal techniques employed in one country spread to other countries. This is likely to be one reason why companies are reluctant to release details of an incident or even to acknowledge it. Coca-Cola, B. F. Goodrich, and other companies have been victimized in the past. The problems of terrorism and political violence are growing globally, increasing risks for overseas businesses, and necessitating broader insurance coverage (Johnson, 2012).

Human resources protection (HRP) is a broad concept that focuses on methods to protect all employees and those linked to them from many types of harm. The links include family, customers, visitors, contractors, and others, depending on the business. This broad concept includes workplace violence prevention, and it can extend to safety, health, use of technology and social media, and personal and family protection, depending on organizational requirements. The broad concept of HRP is holistic and is part of enterprise security risk management (see Chapter 12). **Executive protection** concentrates on security methods to protect key management personnel who are high-value targets because of their positions of power and authority and their value to the business.

Following a risk analysis, management will have an improved understanding of the types of programs required to protect human resources. For example, a corporation may maintain a workplace violence prevention program, a protection program for all employees who travel overseas, an executive protection program for senior executives wherever they may be, and a broad program of protection for employees and their families. The following paragraphs begin with an emphasis on executive protection methods that are applicable to HRP.

Planning

A beginning point for an **executive protection program** is to develop a crisis management team and plan. An interdisciplinary group can greatly aid the program. The team may consist of top executives, the security executive, former federal agents, counter-terrorism experts, political analysts, insurance specialists, an experienced hostage negotiator, and an attorney. The composition and purpose of this team are different from the threat management team for workplace violence. The goals of the crisis management team are to reduce vulnerabilities and surprises and develop contingencies, especially with high-risk executives overseas. The crisis management plan can consist of threat assessments, countermeasures, policies, procedures, and lines of authority and responsibility. Contingency plans should include the possibility of safely extracting employees from areas subject to riots, revolution, attack, and other violence; there are firms that specialize in extraction of employees.

The early stages of the plan, if not the preplanning stages, would be devoted to convincing senior management that executive protection is necessary. To begin with, organizations have a legal responsibility to protect employees under "duty of care" laws, and alleged negligence may be an issue, as was described earlier with workplace violence prevention. Another issue is insurance coverage of employees overseas and premium reductions for protection programs. A risk manager can provide input on coverage (e.g., life, workers' compensation, medical, and kidnap). Executive protection can also be justified through a quality research report that focuses on risks, seeks to anticipate events, and answers the following questions: Which executives are possible targets? Where? When? Which individuals or groups may target the executive? What are their methods? What are the social and political conditions in the particular country? What role has the specific government played in past incidents? Were the police or military of the foreign country involved in past incidents? Such questions require research and intelligence gathering, as well as cooperative ties to government agencies of the United States and other countries. (Sources of assistance are found in the Chapter 10 box, "International Perspective: Overseas Investigations.") Interestingly, illnesses, vehicle accidents, and petty crimes are frequently cited in "country profiles" on the U.S. Department of State (2012) website. However, employees may also be at high risk for serious crimes, depending on the country.

Country risk ratings offered by private firms that conduct research have been used by international corporations for many years. These ratings help businesses gauge risk, decide on travel plans, educate travelers, and for insurers, set premiums. There is inconsistency in how these firms reach their conclusions. Consumers should inquire on the methodology used to prepare these ratings. Firms often acquire information from analysts located globally and from open sources (Elliott, 2006: 36–38).

The U.S. Secret Service completed a study of assassinations of public figures *in the United States* during the second half of the 20th century. The findings showed that threateners do not typically make good on their threats by attacking, and attackers do not usually issue threats to the target before striking. Although threats should not be ignored, this research showed that the most serious threats are unlikely to come from those who communicate threats. So if threats are not a major signal of an attack, what are the indicators? The research found that attackers planned the attack, spoke with others about the attack, followed the target, approached the target in a controlled and secure setting, and attempted repeatedly to contact the target and visit the target's home and a location regularly visited by the target. These latter behaviors signal a probing activity to test protection and attack strategies (Bowron, 2001: 93–97).

Education and Training

Depending on the extent of the executive protection program, many people can be brought into the education and training phase. Executives, their families, and security personnel (i.e., management, bodyguards, and uniformed officers) are top priorities. However, chauffeurs, servants, gardeners, and office workers also should be knowledgeable about terrorist and other criminal techniques, as well as countermeasures that include awareness, prevention strategies, personal security, recognizing and reporting suspicious occurrences, the proper response to bomb threats or postal bombs, and skills such as defensive driving. Most in-house security personnel are not experts in dealing with executive protection. Therefore, a consultant may have to be recruited.

Protection Strategies

Gips (2007: 52–60) writes that there are three essential components of executive protection. *Threat assessments* investigate potential harm to a principal and the likelihood of attack. *Advance procedures* focus on visiting the locations where the principal is expected to visit to coordinate comprehensive security. *Operations* involve protecting the principal in the field. This entails counter-surveillance (i.e., watching to see if anyone is observing the principal), assisting the principal with basic tasks to reduce exposure, defense, and rescue.

Principals should maintain a low profile and not broadcast their identity, affiliations, position, address, telephone number, e-mail address, net worth, or any information useful to enemies. Avoidance of publicity about future travel plans or social activities is wise. Those at risk should avoid social media and exercise care when communicating with others on the telephone, via e-mail or postal service, or in restaurants, and should dispose of sensitive information carefully.

Acohido (2011) reports how criminals use Internet services and mobile computer devices to gather research on victims. In one case, a senior executive of a large corporation traveled by plane to Monterey, Mexico. Prior to the executive landing, a driver waited at the airport holding a placard with the executive's name on it. Opportunistic kidnappers photographed the placard, used Google to research the victim and his employer, and then gave the driver $500 to leave the area and say he could not find the executive. Following the kidnapping, a ransom demand was not quickly answered by the employer and the executive was killed.

Avoidance of Predictable Patterns

The story of famous Italian politician Aldo Moro, murdered in 1978, is a classic case of a **creature of habit**. Moro was extremely predictable. He would leave his home in the morning to attend mass at a nearby church. Shortly after 9:00 A.M., he was en route to his office. The route was the same each morning, even though plans existed for alternatives. Although five armed men guarded Moro, he met an unfortunate fate. An attack characterized by military precision enabled terrorists to block Moro's vehicle and a following police car. Then, on the narrow street, four gunmen hiding behind a hedge opened fire. Eighty rounds hit the police car. Three police officers, Moro's driver, and a bodyguard were killed. Moro was dragged by his feet from the car. Almost two months later Moro was found dead in a car in Rome.

Recognition of Tricks

A terrorist group or a criminal may attempt to gain entry to an executive's residence or office under the pretext of repairing something or checking a utility meter. Repair people and government employees can be checked, before being admitted, by telephoning the employer. School authorities should be cautioned not to release an executive's child unless they telephone the executive's family to verify the caller.

Hiring bodyguards is a growth industry for the private sector. Bodyguards should be carefully vetted and trained. Other employees (e.g., servants or gardeners) surrounding an executive likewise should be vetted to hinder employment of those with evil motives. Employees surrounding the executive should be subject to integrity testing (e.g., an undercover agent offers an employee money to "look the other way").

An executive HR file should be stored in a secure location at the company's headquarters. If a kidnapping occurs, this data can be valuable to prevent deception by offenders, to aid the investigation, and to resolve the situation. Appropriate for the file are vitae for the executive, family members, and associated employees; full names, past and present addresses and telephone numbers; photographs, fingerprints, voice tapes, and handwriting samples; and copies of passports and other important documents.

Protection at Home

A survey of the executive's home will uncover physical security weaknesses. Deficiencies are corrected through investing in access controls, proper illumination, intrusion alarms, CCTV, protective dogs, and uniformed officers. Burglary-resistant locks, doors, and windows hinder offenders. Consideration should be given to the response time of reinforcements.

For a high-risk family, a **safe room** is an asset. This is a fortified room in the house that contains bullet- and explosive-resistant materials and a secure air supply. A first-aid kit, rations, and a bathroom are useful amenities. Multiple types of communication technology and a panic button connected to an external monitoring station will assist those seeking help. If weapons are stored in the room, proper training for their use is necessary. In a "perfect world," the whole family makes it to the safe room when criminals attack. However, a child could be playing in the backyard and a teenager could be in an upstairs bedroom listening to loud music during the attack. Consequently, consideration should be given to practical planning and ROI.

As with security in general, it is best when methods and expenditures for personnel protection remain secret to avoid providing information to an adversary. However, this may not always be possible. The public may have access to protection information through media reports. For example, in 2012, several media outlets published a *New York Times* report on tax breaks for executive protection, citing several companies and the names of executives (CNBC.COM, 2012). In another example, the Associated Press (2006) reported that The Charles Schwab Corporation's proxy filing with the Securities and Exchange Commission revealed that $2.68 million was spent over three years to protect the CEO, Charles R. Schwab, as part of business operations. The news report stated that the firm paid for a security system at the CEO's residence (based on recommendations of a consultant).

The following list provides some protection pointers for home and family:

1. Do not put a name on the mailbox or door of the home.
2. Have an unlisted telephone number.
3. Exercise caution when receiving unexpected packages.
4. Do not provide information to strangers.
5. Beware of unknown visitors or individuals loitering outside. Remain cautious with people who say they are police and show ID. Call for assistance.
6. Check windows for possible observation from outside by persons with or without binoculars. Install thick curtains; avoid blinds that may permit observation.
7. Make sure windows and doors are secure at all times.
8. Educate children and adults about protection.
9. Instruct children not to let strangers in the home or to supply information to outsiders.
10. When children leave the house, be sure to ascertain where they are going and who will be with them.
11. Keep a record of the names, addresses, and telephone numbers of children's playmates.
12. Tell children to refuse rides from strangers even if the stranger says that the parents know about the pickup.
13. Provide an escort for children if necessary.
14. Teach children how to seek assistance.

Protection at the Office

As with the residential setting, physical security is important at the executive's office. A survey may reveal that modifications will strengthen executive protection. The following list offers additional ideas:

1. Office windows should be curtained and contain bullet-resistant materials.
2. Equip the desk in the office with a hidden alarm button.

3. Establish policies and procedures for incoming mail and packages.
4. Beware of access by trickery.
5. Monitor access to the office by several controls.
6. Escort visitors.
7. Access during nonworking hours, by cleaning crew or maintenance people, should be monitored by uniformed officers and CCTV.
8. Educate and train employees.

Attacks While Traveling

History has shown that terrorists and other criminals have a tendency to strike when executives (and politicians) are traveling. Loss prevention practitioners should consider the following countermeasures:

1. Avoid using conspicuous limousines.
2. Maintain regular maintenance for vehicles.
3. Keep the gas tank at least half full at all times.
4. Use an armored vehicle and bullet-resistant clothing and vests.
5. Install an alarm that foils intrusion or tampering.
6. Maintain multiple types of communication technology in case of emergency.
7. A remote-controlled electronic car starter will enable starting the car from a distance. This will help to activate a bomb, if one has been planted, before the driver and the executive come into range.
8. Consider installing a bomb-scan device inside the auto.
9. Headlight delay devices automatically turn headlights off one minute after ignition is stopped.
10. High-intensity lights, mounted on the rear of the vehicle, will inhibit pursuers.
11. GPS can be used to track the executive's vehicle and the executive in case of kidnapping.
12. Protect auto parking areas with physical security.
13. Avoid using assigned parking spaces.
14. Keep doors, gas cap, hood, and trunk locked.
15. Practice vehicle key control.
16. Avoid a personalized license plate or company logo on vehicle.
17. Inspect outside and inside of vehicle before entering.
18. The chauffeur should have a duress signal if needed when picking up an executive.
19. Do not stop for hitchhikers, stranded motorists, accidents, or perhaps "police." It could be a trap. Summon aid, but keep moving.
20. Screen and train the chauffeur and bodyguards. Include the executive and the family in training.
21. Evasive driver training is vital.
22. Have weapons in vehicle ready to use.
23. Maintain the secrecy of travel itineraries.
24. Be unpredictable.

25. Know routes thoroughly as well as alternative routes.
26. Use safety belts.
27. If being followed, summon assistance, continuously sound the horn or alarm, and do not stop.
28. For air travel, use commercial airlines instead of company aircraft. Unless the company institutes numerous security safeguards, the commercial means of air transportation may be safer. Use carry-on luggage to avoid lost luggage or having to wait for workers to locate luggage.
29. Request the second or third floor at a hotel to improve chances of escape in case of fire or other emergency. Avoid rooms near roads in case of a vehicle bomb.
30. Ast (2010) recommends a "lifeline" program whereby each employee (or traveler) has a specific, trusted person who knows all the details of the employee's itinerary, in case of emergency.

Kidnap Insurance

As businesses become increasingly globalized, the risk of kidnapping also increases. Corporations obtain kidnap-ransom insurance policies for protection against the huge ransoms that they might be forced to pay in exchange for a kidnapped executive. Each year millions of dollars in premiums are paid to insurance companies for these policies. Of course, the insurance companies require certain protection standards to reduce premiums. Insurance companies are reluctant to admit writing these policies because terrorists and other criminals may be attracted to the insured company executive. Moreover, these policies often contain a cancellation clause if the insured company discloses the existence of the policy. Ransom payments are usually kept confidential. Consequently, statistics on global kidnapping are difficult to ascertain.

According to the Insurance Information Institute (2012), incidents of kidnapping for ransom money are rising. Kidnap and ransom insurance is sold as part of a comprehensive business insurance package, as a stand-alone policy for individuals, and from a few insurers as part of their homeowners insurance policy. Corporate policies generally cover most kidnapping-related expenses including hostage negotiation fees, lost wages, and the ransom amount. Policies for individuals pay for the expenses of dealing with a kidnapping but do not reimburse for ransom payments. (Chapter 12 also includes the topic of kidnap and ransom insurance.)

If Abduction Occurs

After abduction takes place, the value of planning and training becomes increasingly evident. Whoever receives notification from the kidnapper should express a willingness to cooperate. The recipient should ask to speak to the victim; this could provide an opportunity to detect a ruse. Asking questions about the hostage (e.g., birth date, mother's maiden name) to either the hostage or the kidnapper improves the chances of discovering a trick. Prearranged codes are effective. The recipient should notify appropriate authorities after the call. If a package or letter is received, the recipient should exercise caution, limit those who touch it, and contact authorities.

People who attempt to handle the kidnapping themselves can intensify the already dangerous situation. Loss prevention personnel and public law enforcement authorities (i.e., the FBI) are skilled in investigation, intelligence gathering, and negotiating. These professionals consider the safety of the hostage first and the capture of the offenders second, although the reverse may be true in certain foreign countries.

After abduction, the company's policies, in coordination with the insurer and authorities, should be instituted. These policies ordinarily answer such questions as who is to be notified, who is to inform the victim's family, what are the criteria for payment of the ransom, who will assemble the cash, and who will deliver it and how. Policies would further specify not disturbing the kidnapping site, whether or not to tap and record calls, how to ensure absolute secrecy to outsiders, and use of a code word with the kidnappers to impede any person or group who might enter the picture for profit.

The crisis management team should be authorized to coordinate the company's response to the kidnapping. Because this crime can occur at any time, the team members must be on call at all times.

Guidelines for the behavior of the hostage are as follows:

1. Do not struggle or become argumentative.
2. Try to remain calm.
3. Occupy your mind with all the incidents taking place.
4. Note direction of travel, length of time, speed, landmarks, noises, and odors.
5. Memorize the characteristics of the abductors (e.g., physical appearance, speech, and names).
6. Leave fingerprints, especially on glass.
7. Remember that an effort is being made to rescue you.
8. Do not attempt to escape unless the chances of success are in your favor.

Protection of Women

Whitzman et al. (2009) researched the effectiveness of women's safety audits that are applied in many regions of the world. These audits enable participants to identify safe and unsafe conditions and then recommend improved protection. The research found that the audit is adaptable to various locales, empowers women to take action to alert the public and authorities to partner to enhance the safety of women, and it results in situational crime prevention initiatives and environmental change (e.g., CPTED).

Avon Products, Inc. provides a superb illustration of an organization seeking to meet the protection needs of its employees. Initiating a global "Women and Security" program, Avon conducted extensive research on vulnerabilities of its female employees as they traveled the globe and faced greater risks than their male coworkers face. The Avon security team found that South Africa, the Philippines, Russia, and Latin America were high-risk areas for women. South Africa, for example, has one of the highest levels of sexual assault in the world. Latin America is noted for abductions at ATMs. The Avon program focused on brochures, self-defense training, and one-on-one evaluations. Brochures are country-specific and include tips such as not wearing expensive-looking jewelry because street robbers do not know a genuine from fake and avoiding public restrooms, if possible, because rapists sometimes disguise themselves as females. Employees expressed a need for self-defense training, which was provided. It aims to help women avoid and deter attack. The one-on-one evaluations consist of a security staff member who observes the daily routine of the employee working overseas to offer suggestions for improved protection. The "Women and Security" program has resulted in increased safety, less anxiety, and higher productivity (Shyman, 2000: 58–62).

Identity Theft

Human resources protection extends to the problem of **identity theft**. It is the illegal acquisition of another individual's personal identifying information to be used fraudulently for illegal gain. In addition, the more an adversary knows about a person or family to be harmed, the better the chances for successful victimization.

Identity theft is a multi-billion dollar crime, besides the challenges and time required by victims to correct their identities There are many ways in which an offender can steal personal and business information, and there are many types of crimes committed by using such information.

Traditional methods of obtaining personal and business information include looking over another's shoulder at a bank or ATM; searching trash (i.e., "dumpster diving"); impersonation over the telephone; stealing a wallet or mail; fraudulently ordering a copy of a victim's credit record; or going to a cemetery to locate a deceased person whose age, if living, would approximate the offender's age and securing documents to develop the identity. The Internet has expanded the opportunities to steal identities. Websites offer a wealth of personal information. In addition, hackers penetrate corporate databases and e-commerce websites and download a variety of personal information. **Phishing** is another threat; it is a word coined from the analogy that offenders use e-mail bait to fish for personal information. The technique often uses spam or pop-up messages that trick people into disclosing a variety of sensitive identification information.

An offender can use a victim's identity to secure credit cards, open bank and checking accounts, apply for a loan, purchase and sell a home or car, establish cellular service, file for bankruptcy, obtain a job, seek workers' compensation or a tax refund, file a lawsuit, and commit a crime in the victim's name. The possible offenses are endless, and victims often learn about the identity theft months after it occurred.

Congress passed the Identity Theft and Assumption Deterrence Act of 1998 to make identity theft a federal crime. Identity theft is investigated by the FBI, the U.S. Postal Service, the U.S. Secret Service, and state and local police. To increase consumer awareness of identity theft, the Federal Trade Commission created the Identity Theft Clearinghouse. It collects complaints from victims and offers resources to restore credit.

Personal protection strategies include the following:

- Protect personal and financial information in public, over the telephone, and online. Carry only necessary ID.
- Closely study bills and completely destroy unneeded papers containing personal information.
- When making online transactions, be sure they are made over secure, encrypted connections and verify the address and telephone numbers on websites.
- If victimized, contact creditors, the three major credit reporting agencies (Equifax, Experian, and TransUnion), and file a report (and obtain a copy) with police.

Company Pays Terrorists to "Protect" Workers

Apuzzo (2007) reported that Chiquita Brands International revealed that it had agreed to a $25 million fine after admitting it paid terrorists to protect its workers in a dangerous region of Colombia. The fine was part of a plea-bargaining agreement with the U.S. Department of Justice that investigated the company's payments to right-wing paramilitaries and leftist rebels the U.S. government classifies as terrorist groups. Federal prosecutors said company executives paid about $1.7 million between 1997 and 2004 to the United Self-Defense Forces of Colombia, the National Liberation Army, and the Revolutionary Armed Forces of Colombia. Prosecutors said that Chiquita disguised the payments in company accounting records. Colombia maintains one of the highest kidnapping rates among countries of the world. Companies are known to pay protection money; however, the amount paid is impossible to ascertain. Although companies supposedly have extensive security to protect employees, terrorist groups have fought intensely in Colombia's banana growing region.

What is your opinion of Chiquita Brands International paying money to terrorists to protect its workers in Colombia?

Maximize Technology

Here we cover technologies that assist in the protection of people (Coulombe, 2012: 17; Harwood, 2011: 42–50; Simovich, 2004: 73–80; Besse and Whitehead, 2000: 66–72). The Internet offers input for analyzing threats, planning HRP, and helping to convince senior management that protection is necessary. It contains a wealth of government information that is free. Private sources are also available, but for a fee. Companies that provide information on political violence, terrorism, and so forth, often make the information available through telecommunication devices such as GPS systems and satellite phones; however, some countries prohibit foreigners from entering with such equipment. The Internet offers opportunities to check people, businesses, trip routes, and many other subjects of inquiry. It also contains information of a negative nature from terrorists, activists, hate groups, and dissatisfied customers. "Sucks.com" sites, such as walmartsucks.com and wellsfargosucks.com, are used to vent at companies. A variety of intelligence can be gathered from such sites. (See Chapter 10 for government and private sector resources and websites.)

For advance planning, online maps are especially helpful. In addition, digital cameras can document travel routes, buildings, airports, etc., and images can be transmitted to planners for analysis. Portable, wireless alarm and CCTV systems offer protection at hotel rooms, vehicles, airplanes, and other locations. Pinhole lens cameras, built into almost anything, serve as witnesses to crimes and aid in identifying and prosecuting offenders. Thermal imagers, which detect heat rather than light, can be used in total darkness to detect intruders or for search and rescue. Another portable system to consider is the automatic external defibrillator (AED) that delivers electrical shocks to restore normal heart rate for those in cardiac arrest.

Global positioning systems (GPS) use a network of satellites that transmit data to ground receivers to navigate and map routes. In addition, GPS can track assets and an executive's vehicle, including speed, direction, and alarms transmitted from the vehicle. To send real-time position data from GPS technology, cellular, satellite radio, text, e-mail, and FM radio transmission can be applied. A GPS feature called "geo-fencing" permits a virtual perimeter using coordinates that can be used to determine if a person or asset breached the perimeter.

Another technology, geographic information systems (GIS), consists of hardware and software designed to present a variety of geographic data to interpret patterns in the form of maps, charts, and reports. GPS can provide location data to a GIS. Many images and videos posted on the Internet contain "geo-tags" (latitude and longitude data) that are automatically added by smartphones and other devices. This technology enables a user to track people or assets on a map. The vulnerability of such technology is that an adversary can learn about the location of a person's residence and office, plus travel information, if images are shared through social networking. Interesting sites are PleaseRobMe.com, Creepy (ilektrojohn.github.com/creepy), and ICANSTALKU.com.

Modern communications technology is essential during a crisis to notify personnel of danger, extraction plans, and other information. Smartphones are especially helpful to enable redundant, real time communications; for example, if cellphone lines are not working, text messages or e-mails are an alternative. A technology serving as a backup to a smartphone is satellite phone—also providing redundant paths of communication (i.e., voice, e-mail, and text). Satellites continue to orbit above a disaster area where communication infrastructure is destroyed. What is needed is a clear path to the sky and not dense urban or wilderness areas. We must not forget about the traditional telephone connected to a landline. Other, less attractive, alternatives for communication are Skype, Facebook, and Twitter. Users must ensure that they have charged devices and backup batteries.

For vehicles, cellular technology can be added to permit audio monitoring and remote start or kill of an engine. Remote systems are vulnerable to hacking, so defenses should include encryption. Although air bags in vehicles offer safety, if protection specialists ram their way out of an attack, or if an attacker backs into the protected vehicle, the activation of an air bag can hinder escape. One option for careful consideration is to disconnect the air bag on the driver's side.

Who has the advantage when a principal is targeted for attack, the principal and the protection team or the adversary? Explain your answer.

Substance Abuse in the Workplace

Substance abuse refers to human abuse of any substance that can cause harm to oneself, others, and/or organizations. The problem is pervasive. Millions of people abuse legal and illegal

FIGURE 18-2 Abuse of legal and illegal drugs and other substances is a problem in the workplace.

drugs and other substances (Figure 18-2). An employee substance abuser can cause harm in several ways such as losses in workplace productivity, an accident, selling drugs to others in the workplace, and stealing products or information assets to support a drug habit. Substance abuse is linked to tardiness, absenteeism, turnover, and violence. In addition, losses result from the costs of health care, criminal justice, and vehicle accidents.

Excessive alcohol consumption costs the United States hundreds of billions of dollars annually. The National Institute on Drug Abuse (2008) reports that "nearly 75 percent of all adult illicit drug users are employed, as are most binge and heavy alcohol users." Besides employees abusing alcohol and illicit drugs, Zezima and Goodnough (2010) write that the number of employees testing positive for prescription opiates is increasing. No occupation is immune to substance abuse. Those afflicted are from the ranks of blue-collar workers, white-collar workers, supervisors, managers, and professionals.

Countermeasures

Unenlightened managers ordinarily ignore substance abuse in the workplace. As with so many areas of loss prevention, when an unfortunate event occurs (e.g., drug-related crime, production decline, or accident due to substance abuse), these managers panic and react emotionally. Experienced people may be fired unnecessarily, arrests threatened, and litigation becomes a possibility. In contrast, action should begin before the first sign of abuse.

Here is a list of action for an **anti-substance abuse program for the workplace:**

1. Form a committee of specialists to pool ideas and resources.
2. Seek legal assistance from an employment law specialist.

3. Large corporations can afford to hire a substance abuse specialist. Outsourcing is another option. Also, contact the local government-supported alcohol and substance abuse agency.
4. Research online resources such as the U.S. Department of Labor "Drug-Free Workplace Policy Builder." It provides an opportunity to customize a *drug-free workplace policy statement*. However, the policy is one of five elements of a drug-free workplace program that also includes *supervisory training, employee education, employee assistance, and drug testing.*
5. Prepare policies that include input from a variety of employees. Policies should focus on the company's position on abuse of substances including alcohol, job performance and safety as it relates to substance abuse, drug deterrence such as urinalysis, the consequences of testing positive, the responsibility of employees to seek treatment for abuse problems, available assistance, and the importance of confidentiality.
6. Education and prevention programs can assist employees in understanding substance abuse, policies, and making informed decisions on life choices, health, and happiness. Use signs in the workplace and at entrances, and periodically distribute relevant educational materials.
7. Ensure that supervisors are properly trained to recognize and report substance abuse.
8. Consider an undercover investigation to ascertain drug usage in the workplace.

Employee Assistance Programs (EAPs)

Employee assistance can be traced to the origin of Alcoholics Anonymous (AA), founded in 1935. AA views alcoholism as a disease requiring long-term treatment. Employee assistance programs (EAPs) were first introduced in the 1940s, in U.S. corporations. Thousands of these programs exist today in the public and private sectors, where they incorporate a broad-based approach to such problems as substance abuse, depression, and marital and financial problems. These programs are characterized by voluntary participation by employees, referrals for serious cases, and confidentiality. The goal of EAPs is to help the employee so he or she can be retained, saving hiring and training costs. An organization may establish its own EAP or outsource the program. Initial EAP programs were characterized by "constructive confrontation" (i.e., correct the problem or leave). Today, the philosophy is that a company has no right to interfere in private matters, but it does have a right to impose rules of behavior and performance at work (Elliott and Shelley, 2005; Ivancevich, 2001: 464).

Today, EAPs vary from robust programs to web-only or phone delivery options. Research is needed on the effectiveness of EAPs, what problems it ameliorates, what problems it shows limited, if any, success for, and how it can be enhanced. Elliott and Shelley (2005) found that there were no differences in the accident rates of employees prior to and following EAP interventions. Research by EAP contractors show positive results and ROI, although the research methodology from such studies generate controversy (Sharar and Lennox, 2009).

Legal Guidelines

The federal **Anti-Drug Abuse Act of 1988** is an attempt to create a drug-free workplace. The law requires federal contractors and grantees to prepare and communicate policies banning illegal substances in the workplace and to create drug awareness programs and sanctions or rehabilitation for employees abusing substances. Federal contracts and grants are subject to suspension for noncompliance or excessive workplace drug convictions. Another form of regulation includes industries regulated by the U.S. Department of Transportation, such as airline, motor carrier, and rail, which are required to institute substance abuse programs, including drug testing. The Nuclear Regulatory Commission has similar regulations. Additional mandates that affect substance abuse programs in the workplace include Title VII of the Civil Rights Act of 1964, the Americans with Disabilities Act (ADA) of 1990, the Rehabilitation Act of 1973, U.S. Department of Defense regulations, state drug testing laws, and state workers' compensation laws.

Zezima and Goodnough (2010) write of the growing problem of employees being fired for consuming legal prescription medications that employers argue may cause impairment, accidents, and product defects—all liability issues. Employees contend that their privacy and the ADA are being violated when employers ask about prescription drugs, unless employees act in an unsafe manner. The EEOC is involved in litigation over cases where employees were fired after being required to disclose prescription drugs being consumed.

Drug Testing

Risk managers have seen a dramatic decrease in health insurance and workers' compensation claims following drug testing. In addition, insurance companies are encouraging companies to implement drug testing programs to reduce premiums. Furthermore, those companies that do not test become a magnet for those who are abusers. At the same time, laws pertaining to drug testing are complex and vary among the states (Zezima and Goodnough, 2010; Myshko, 2001: 44–46).

Here is a list of items to assist in planning a drug-testing program (Substance Abuse and Mental Health Services Administration, n.d.; Gips, 2006: 50–58; Smith, 2004):

- Drug testing must be well planned. Questions include the following: What type of test? Who will do the testing and what are the costs? Who will be tested? What circumstances will necessitate a test? What controls will prevent cheating and ensure accuracy? Do the laboratory and its personnel comply with state or federal licensing and certification requirements? Are all legal issues considered?
- Drug testing focuses on urine, hair, and blood (mostly urine tests). Hair analysis is the most expensive. Blood testing is used rarely, such as when an employee is unconscious from an accident and in an emergency room; for a person on dialysis; under a court order; or for a deceased person.
- Following legal research, especially of state law on drug testing and privacy, employers should consider an offer of employment to applicants contingent upon passing a drug test. Include this requirement in employment ads.
- Federal agencies must follow standardized testing procedures from the Substance Abuse and Mental Health Services Administration (SAMHSA). Private employers are on an improved legal foundation when following these procedures.

- Ensure drug testing is fair and applied equally.
- Randomly drug test (if legally permitted), test when an employee shows behavioral or physical indications of substance abuse, and test following an accident.
- Because of an industry of providing products on the Internet to adulterate or substitute for urine specimens, some companies are using multiple drug tests.
- Drug testing of hair is less likely to be tampered with when compared to urinalysis. Hair testing shows evidence of drugs much farther back in time than urinalysis.
- **Specimen validity testing** is gaining momentum from the federal government to ensure specimens are not adulterated or substituted.
- If a company shows that the percentage of drug tests that turn up positive is decreasing, caution is advised because employees may be getting better at subverting drug tests.

Trace detection should be used with caution. It consists of gathering minute particles of drugs in the workplace (e.g., at workstations or rest rooms) by using cloth swabs and then analyzing the swabs with a desktop instrument to detect a variety of drugs. This technique can possibly gauge the types of drugs in the workplace and serve as an aid to drug education and prevention. It presents difficulties if used for investigative and prosecutorial purposes.

What do you think are the most successful countermeasures against substance abuse in society and in the workplace?

Alcoholism

An **alcoholic** is defined as someone who cannot function on a daily basis without consuming an alcoholic beverage. Alcoholism is a disease characterized by craving (i.e., strong urge to drink), loss of control (i.e., unable to stop drinking once drinking begins), physical dependence (i.e., withdrawal symptoms such as nausea, sweating, shakiness, and anxiety), and tolerance (i.e., greater amounts are needed to get "high"). Alcohol is the most abused drug in America. In the United States, there are almost 18 million people who have alcohol-related problems (National Institute on Alcohol Abuse and Alcoholism, 2012). These figures do not include the millions who are on the fringe of alcoholism. It is often a hidden disease, whereby the alcoholic hides the problem from family, friends, physicians, and him- or herself. Some major indicators are heartburn, nausea, insomnia, tremor, high blood pressure, morning cough, and liver enlargement. The alcoholic often blames factors other than alcohol for these conditions.

Today, many businesses are no longer hiding the problem and rely on an EAP. In addition, Alcoholics Anonymous (AA), an organization for alcoholics and recovered alcoholics, run by people who have had a drinking problem, has had more success than most organizations.

An employee with a drinking problem affecting the workplace is advised of "helping agencies," in addition to internal and external policies and procedures and what is expected of him or her by the employer regarding steps for recovery. Health insurance benefits and company disability income are usually applicable. Unless the employee takes heed in seeking assistance, dismissal may occur because of poor job performance. The threat of job loss jolts many alcoholics into recognizing their serious situation and accepting treatment.

Types of Substances and Abuse

The explanation of four terms can assist the reader in understanding the human impact of various substance abuse categories.

- **Psychological dependence:** Users depend so much on the feeling of well-being from a substance that they feel compelled toward continued use. People can become psychologically dependent on a host of substances. Restlessness and irritability may result from deprivation of the desired substance.
- **Addiction:** Certain substances lead to physiological (or physical) addiction. This happens when the body has become so accustomed to a substance that the drugged state becomes "normal" to the body. Extreme physical discomfort results if the substance is not in the body.
- **Tolerance:** After repeated use of certain drugs, the body becomes so accustomed to the drug that increasing dosages are needed to reach the feeling of well-being afforded by earlier doses.
- **Withdrawal:** A person goes through physical and psychological upset as the body becomes used to the absence of the drug. Addicts ordinarily consume drugs to avoid pain, and possible death, from withdrawal. Symptoms vary from person to person and from substance to substance. An addict's life often revolves around obtaining the substance, by whatever means, to avoid withdrawal.

U.S. Department of Transportation drug and alcohol tests focus on alcohol, marijuana, cocaine, amphetamines, opiates, PCP, and ecstasy. Employers often test for several other substances such as the synthetic painkillers OxyContin and Vicodin and the anti-anxiety drug Xanax. Controversy surrounds the number and types of illicit and prescriptions drugs that should be tested. In addition, employers often wrongly assume that if an employee passes a drug test, the employee is drug-free (Zezima and Goodnough, 2010).

The website of the U. S. Drug Enforcement Administration contains a wealth of information on drugs. Topics include description/overview, control status, street names, short-term effects, long-term effects, trafficking trends, use/user population, arrests/sentencing, DEA drug seizures, legislation, treatment resources, photos, related news releases, and web links.

Information Security

The Office of the National Counterintelligence Executive (2011: 4) writes:

> *Estimates from academic literature on the losses from economic espionage range so widely as to be meaningless—from $2 billion to $400 billion or more a year—reflecting the scarcity of data and the variety of methods used to calculate losses.*

Information security protects an organization's information assets through a broad based, well-planned, and creative program of strategies that consider the widest possible risks from humans, technology, accidents, disasters, or any combination of factors. Information security

FIGURE 18-3 A balanced information security program avoids overemphasizing perimeter security since the greatest threat is from within.

is an extremely challenging undertaking because an organization can expend an enormous amount of resources on the protection of information assets and, as examples, one leak by one employee (e.g., on a telephone, in an e-mail, through another type of electronic device, or at a conference) or one intrusion by an outsider (e.g., electronic surveillance or cyber threat), can result in losses. A balanced information security program avoids overemphasizing perimeter security (Figure 18-3), since the greatest threat is from within. Information security professionals must avoid "tunnel vision." For instance, computer security professionals may only emphasize protection of data and information in IT systems and devices while overlooking such losses from overheard conversations or not shredding important documents.

Smith (2013: 5–6) provides information security principles from an IT perspective that are applicable to broad-based information security. He notes that information security challenges require a systematic process to work toward solutions. Smith recommends a six-phase security process familiar to practitioners in many security specializations:

1. Identify assets.
2. Analyze the risk of attack.
3. Establish security policy.
4. Implement defenses.
5. Monitor defenses.
6. Recover from attacks.

According to Smith, security practitioners should enhance the security process through the principle of *continuous improvement*, whereby the process never ends at the final step. In fact, any phase may reveal the need for change that will improve security.

Curtis and McBride (2005: 107–108) write that information security applies numerous strategies to ensure the availability, accuracy, authenticity, and confidentiality of information. Availability ensures that users can access information and not be blocked by, for example, a denial-of-service attack. Accuracy addresses issues of errors (e.g., by an employee) and the integrity of information (e.g., a hacker remotely accesses an IT system to manipulate data). Authenticity means that information and its sender are genuine. Phishing, for instance, can lead to identity theft and loss of sensitive information. Confidentially ensures that only authorized individuals access information.

Information assets contain a combination of economic value, property right, specialized knowledge, and competitive advantage that may be written, verbal, electronic, or in another form. The information is often extremely valuable (e.g., a secret formula) and may represent the lifeblood of an organization. Depending on the source, information assets may be termed "proprietary information" or "sensitive information." ASIS International (2007: 7–8) prepared an *Information Asset Protection Guideline* that offered the following terms and definitions:

> *Proprietary Information: As defined by the Federal Acquisition Regulation (48 CFR 27.402 Policy): A property right or other valid economic interest in data resulting from private investment. Protection of such data from unauthorized use and disclosure is necessary in order to prevent the compromise of such property right or economic interest.*
>
> *Sensitive Information: Information or knowledge that might result in loss of an advantage or level of security if disclosed to others.*

For simplicity, the term "information assets" is emphasized here. Subsequent pages explain corporate intelligence gathering, espionage, and countermeasures. **Espionage** is the act of spying, which is a criminal offense. **Business espionage** seeks a competitor's information assets through illegal means. **Government espionage** seeks not only a competitor's information assets, but also military and political secrets. Previous sections of this book elaborate on security, fire protection, and emergency management for assets, besides people. In addition, we must not forget that information pertaining to the privacy of individuals requires protection. This would include credit, medical, educational, and other records protected under various laws as covered earlier.

Common types of information assets that might be sought by a spy are the following: product design, financial reports, engineering data, tax records, secret formulas, marketing strategies, cost reduction methods, research data, client or customer information, trade secrets, human resources records, patent information, computer programs, oil or mineral exploration maps, mergers, and contract information. The Director of National Intelligence (Clapper, 2012: 7–8) reported that China and Russia, among other actors, "are responsible for extensive illicit intrusions into U.S. computer networks and theft of U.S. intellectual property." In addition, the insider threat is a major source of loss from espionage.

When we speak about information security we should maintain a broad perspective on the topic, consider the varieties of information that may be stolen or mishandled, adhere to the laws that aim to protect information, and apply an enterprise security risk management approach. Although these pages emphasize information that a spy may steal, a security professional may also be responsible for health information under HIPAA, financial information under FASTA, or, in the federal government, information security requirements under the Federal Information Security Management Act.

A **trade secret**, supposedly known only to certain individuals, is a secret process used to produce a salable product. It may involve a series of steps or special ingredients. A famous trade secret is the formula for Coca-Cola. The holder of a trade secret must take steps to maintain secrecy from competitors. If an employee were to reveal a trade secret to a competitor, the courts could issue an injunction, prohibiting the competitor from using the secret. Money damages might be awarded.

ASIS International (2007: 8) defines trade secret as follows: "All forms and types of financial, business, scientific, technical, economic, or engineering information, including patterns, plans, compilations, program devices, formulas, designs, prototypes, methods, techniques, processes, procedures, programs, or codes, whether tangible or intangible, and whether or how stored, compiled, or memorialized physically, electronically, graphically, photographically, or in writing if (a) the owner thereof has taken reasonable measures to keep such information secret; and (b) the information derives independent economic value, actual or potential, from not being generally known to, and not being readily ascertainable through proper means by, the public."

A **patent** provides protection for an invention or design. If a competitor duplicates the device, patent laws are likely to be violated and litigation would follow. Competitors often engineer around patents.

ASIS International (2007:7) defines patent as follows: "Information that has the government grant of a right, privilege, or authority to exclude others from making, using, marketing, selling, offering for sale, or importing an invention for a specified period (20 years from the date of filing) granted to the inventor if the device or process is novel, useful, and nonobvious."

A **trademark** includes words, symbols, logos, designs, or slogans that identify products or services as coming from a common source. McDonald's golden arches serve as an example.

A **copyright** provides protection for original works in any medium of expression by giving the creator or publisher exclusive rights to the work. This type of protection covers books, magazines, musical scores, movies, art, and computer software programs.

Corporate Intelligence Gathering: Putting it in Perspective

Corporate intelligence involves gathering information about competitors. It ranges from the illegal activity of business espionage to the acceptable, universally applied practice of utilizing salespeople to monitor public business practices of other companies. Corporate intelligence gathering, when done legally, makes good business sense, and this is why companies such as General Electric and Gillette have established formal intelligence programs. Because of unethical and illegal behavior by certain people and organizations when gathering intelligence, the whole specialization has earned a bad reputation. However, many avenues for gathering intelligence are legal. Let us first list the reasons for corporate intelligence gathering (Laczniak and Murphy, 1993: 1–4):

- Executives should take advantage of information that is publicly available to fulfill their fiduciary duty to shareholders. Because the Cordis Corporation, a pacemaker manufacturer, for example, was unsure of why its new line did not show improved sales, it asked its salespeople to check the tactics of the competition. The salespeople found that physicians were being offered cars and boats to stay with the competition. When Cordis increased educational support for doctors, added more salespeople, and matched the giveaways, sales increased.

- Competitive intelligence is a basis for strategic planning. One intelligence seminar director found a competitor using a "dirty trick" by enrolling in his course under an assumed name.
- It is necessary in order to be successful against global competitors. The Japanese have "deployed armies of engineers and marketing specialists" to other countries. Likewise, U.S.-based firms have set up offices abroad to gather information.
- It can be useful for the introduction of a new product. Coors did extensive chemical analysis on Gallo's wine coolers and found that it could not compete on price.

The Strategic and Competitive Intelligence Professionals (SCIP) (2012), formally the Society of Competitive Intelligence Professionals, is a world-wide nonprofit membership organization for practitioners involved in creating and managing business knowledge. This group offers networking and educational opportunities, a code of ethics, publications, and a certification program—Competitive Intelligence Professional (CIP). SCIP views its vocation as an honorable profession. The group defines **competitive intelligence** as "the legal and ethical collection and analysis of information regarding the capabilities, vulnerabilities, and intentions of business competitors." Furthermore, the group states:

> *While some decision makers may attempt to sail blindly through the global marketplace, it is the duty of the trained CI professional to show them alternative courses that will avoid potential dangers, and to take advantage of the tactics and strategies that lead to bottom-line success.*
>
> *CI is not spying. It is not necessary to use illegal or unethical methods in CI. In fact, doing so is a failure of CI, because almost everything decision makers need to know about the competitive environment can be discovered using legal, ethical means. The information that can't be found with research can be deduced with good analysis, which is just one of the ways CI adds value to an organization.*

Ethical sources of information include the following:

- Published material and public documents from government
- Purchasing access to business information databases (e.g., LexisNexis; Dun and Bradstreet)
- Disclosures made by competitors
- Market surveys and consultants' reports
- Financial reports and brokers' research reports
- Trade fairs, exhibits, and competitors' brochures
- Analysis of a competitor's products
- Legitimate employment interviews with people who worked for a competitor

A company wishing to detail prohibited activities in its policy should include the following (Horowitz, 2005; Ehrlich, 2002: 11–14):

- Bribery and theft
- Electronic eavesdropping or wiretapping
- Unethical and/or illegal cyber techniques (e.g., hacking)
- Trespassing
- Misrepresenting your identity or that of your company
- Inducing another to violate his duty of confidentiality to his current or former employer
- Accepting trade secret or proprietary information through a confidential relationship which you then violate
- Accepting trade secret or proprietary information from another knowing it was obtained through a violation of law

Economic Espionage Act of 1996

Because intellectual property assets are often more valuable to businesses than tangible assets, Congress passed the **Economic Espionage Act (EEA) of 1996**. This Act makes it a federal crime for any person to convert a trade secret to his or her own benefit or the benefit of others with the intent or knowledge that the conversion will injure the owner of the trade secret. The penalties for any person are up to 10 years of imprisonment and a fine up to $250,000. Corporations can be fined up to $5 million. If a foreign government benefits from such a crime, the penalties are even greater. Few of these complex cases are prosecuted by the federal government; in FY 2010, federal prosecutors and the FBI prosecuted ten trade secret cases and two economic espionage cases, most with a link to China (U.S. Department of Justice, 2010: 17). Interestingly, the Office of the National Counterintelligence Executive (2011: B-3) notes that Germany's espionage legislation has limited results when cases are initiated because of diplomatic immunity protections and attribution issues resulting from the global nature of cyberspace. Consequently, a proactive, preventive approach to information security is vital.

The EEA defines *trade secret* broadly as information that the owner has taken "reasonable measures" to keep secret because of the economic value from it. Case law has further defined the Act; the greater the protection and value of the information and the fewer people who know about the information, the more likely the courts will recognize its status as a protectable trade secret (Halligan, 2001: 53–58).

The Act raises two major concerns for management:

- *Protecting trade secrets:* This would include a comprehensive information security program.
- *Hiring employees from competitors:* Employers may violate the act if they hire employees from other firms who may bring with them trade secrets. Prevention includes a thorough interview of applicants, ascertaining whether the applicant signed contracts or agreements with others for the protection of sensitive information, and use of a company form that signifies that the new employee understands the act's legal requirements.

The Act also links the economic well-being of the nation to national security interests. In addition, it allows the FBI to investigate foreign intelligence services bent on acquiring sensitive information of U.S. companies.

At some point, a company may have to decide whether to report a violation of the Act to law enforcement authorities. The disadvantages are lost time and money, unwanted publicity, and the fact that the defendant's attorney may request secrets that could then be revealed in court. Although the Act offers some protection for information assets, this protection may depend on how a judge or attorneys in the case interpret the Act. Discovery proceedings may result in information loss greater than the original loss. Also, the case may be lost in criminal and civil courts. Therefore, management must carefully weigh decisions on legal action. Another point to consider is that the Act requires businesses to protect themselves from losses, which presents liability issues relevant to due diligence (Nolan, 1997: 54–57). *Prevention is seen here, as with many other vulnerabilities, as the key avenue for protection.*

Horowitz (1998: 6) wrote of the confusion and uneasiness for competitive intelligence professionals following the passage of the Economic Espionage Act of 1996. These specialists, and related contract firms and proprietary departments, were unsure of how they should conduct their business of gathering information. Horowitz wrote:

> *While the EEA makes trade secret law a federal criminal matter—this for the first time in U.S. history—the activities it criminalizes were prohibited under state law and/or unacceptable under SCIP's Code of Ethics. In other words, the rules are fundamentally the same, but the consequences of violating them are different. An activity that had always been a violation of state trade secret law can now result in not only state civil liability but federal criminal liability as well.*

Do Allies Commit Economic Espionage against Each Other?

The Office of the National Counterintelligence Executive (2011: B-2) reports the following:

> *According to a 2010 press report, the Germans view France and the United States as the primary perpetrators of economic espionage "among friends."*
>
> *France's Central Directorate for Domestic Intelligence has called China and the United States the leading "hackers" of French business, according to a 2011 press report.*

Espionage Techniques and the Vulnerabilities of Technology

The techniques used by adversaries to acquire information assets are so varied that defenders must not fall into the trap of emphasizing certain countermeasures while "leaving the back door open." For example, a company may spend hundreds of thousands of dollars defending against electronic surveillance and wiretapping while not realizing that most of the losses of information assets are from a few employees who are really spies for competitors.

Three patterns of illegally acquiring information assets are internal, external, and a conspiracy that combines the two. An *internal attack* can be perpetrated by an employee who sells a secret formula to a competitor, for example. An *external attack* occurs when an outsider gains unauthorized access to the premises physically or through cyberspace and steals product design data. The *combined conspiracy* is seen when an employee "just happens" to leave a secret mailing list on a desk and unlocks a rear door to aid an intruder or supplies access codes to a hacker who steals the same information from the comfort of their home. Furthermore, information assets are lost through legal means applied by competitors.

Spies use numerous techniques. A spy might assemble trash from a company and an executive's home to "piece together" information. Spies may claim they are students conducting surveys, as a "pretext" to acquire information. Several spies may each ask certain questions only, and then later, assemble the "big picture." Another method is tricking a key employee into being discovered in a compromising position (e.g., in bed with a prostitute), photographing the incident, and then blackmailing the employee to acquire information. A spy might attempt to seek employment, a contract, or joint research at a target company. Sometimes, proposals for a merger or acquisition are used as a cover to obtain information. Travelers who go overseas for business or other reasons may also be victimized through such methods as accessing computers, other devices, luggage, and hotel rooms.

Companies with inadequate information security programs can lose information assets in several easy ways, such as through company speeches, publications, media releases, trade meetings, disgruntled employees, consultants, and contractors. Information loss can occur at any location from conversations or phone calls. A spy might frequent a tavern or conference populated by engineers to listen to conversations. Another way in which businesses can lose information assets is when an overly eager salesperson supplies excessive information in an attempt to impress a customer.

Reverse engineering is a legal avenue to obtain a look at a competitor's product. The competitor simply purchases the product and dismantles it to understand the components. Patent applications, which are available to the public, can reveal valuable information. Some companies deliberately patent their failures to lead competitors astray.

Various devices are available to the spy. Wiretapping, electronic listening devices, and pinhole lens cameras are examples. A handheld document scanner, the size of a large pen, can capture numerous pages of text and graphics. A competitor could plant RFID readers to report on the movement of products to collect business data.

In our age of technological marvels, numerous devices we use daily can compromise information security. Ashley (2006: 84–85) describes such vulnerabilities and offers countermeasures. In order to attract customers, communications companies offer a variety of features beyond basic service. For instance, a ringtone is a sound file that is downloaded to a cell phone. If a virus has been loaded into the file, it infects the phone, so use only the ringtones that come with a phone. Bluetooth permits wireless communications over short distances (i.e., 30 feet or less), which facilitates the use of a cordless headset with a cell phone. Victimization may occur if the security feature of Bluetooth is turned off. An offender may be nearby, use a Bluetooth probe, transfer files into the victim's phone, and receive information from the phone. Bluetooth usage with other devices (e.g., laptop or PDA) creates additional vulnerabilities. Cell phones can also be compromised through instant and text messages, so avoid such messages from unknown sources. PDAs are subject to similar attacks.

Piazza (2005: 78–87) further explains the vulnerabilities of Bluetooth technology. This trade name is from a 10th century Danish king. It refers to a short-range wireless radio chip created in the 1990s that is in numerous devices. Bluetooth enables wireless communications in what is called a personal-area network (PAN). Attackers are developing technology that can access PANs from greater distances similar to traditional wireless networks. Several types of attacks can be applied to Bluetooth technology. One type can make an unauthorized connection to a cell phone and copy its contents, including the unique numerical identifier that an offender needs to clone a phone. In one scenario, Piazza describes how a tech-savvy individual was able to access the cell phone of a manager of a chain of coffee shops and retrieved door PIN codes, alarm codes, and safe combinations without the manager's knowledge. Other vulnerabilities include taking control of a victim's cell phone, making calls, sending and reading text messages, and performing other tasks. In another scenario, a tech-savvy individual focused on a group talking at a table in a bar with the cell phone of the victim sitting on the table. A connection was made to the targeted phone, it was signaled to dial the tech-savvy individual's voicemail, and it recorded the conversation without the knowledge of the victims.

Piazza (2007: 48) reports on the threats from USB flash drives. (USB are the initials for Universal Serial Bus, a standard that supports data transfer.) These storage devices can hold an enormous amount of data which can then be taken from the workplace. In addition, if a certain program infects a computer, the program will retrieve data from USB drives or other portable devices connected to the compromised computer. Another program takes the data from the USB drive and sends it out via an e-mail. Flash drives are capable of running software programs. One version allows an employee to circumvent security and visit prohibited websites via anonymous proxies.

Miller (2007: 22–30) asks us to think about all the ways we move and store data on mobile devices. She refers to USB ports that support a variety of portable storage devices, such as flash drives, portable hard drives, music and video players, and printers. Miller also refers to data storage on CDs and DVDs and the threat from unprotected WiFi and Bluetooth. WiFi means wireless fidelity. A WiFi-enabled device (e.g., laptop, cell phone, tablet, or PDA) can access the Internet when near an access point called a hotspot. WiFi allows networks to be deployed without cabling; however, WiFi networks can be vulnerable to monitoring and copying data, unless high-quality encryption is employed.

Nairn (2011) refers to espionage through social networks such as Facebook and Twitter and employees being duped into clicking on links that supposedly come from co-workers or social networking friends. In addition, all electronic tools such as instant messaging and Internet telephony have vulnerabilities. Nairn notes the risks of smartphones, tablets, and other devices in the workplace and the difficulty of preparing quality, up-to-date policies that are legal for information security without hindering productivity. He writes that hackers prepare information-stealing malware specifically designed for each new device that arrives on the market.

Although an adversary can travel to a targeted company to conduct surveillance, take photographs and video the location, the use of satellite images is increasing. These Internet services are free for basic services. More sophisticated services permit zoom, rotation, and 3D views of facilities and terrain. This vulnerability is challenging to counter. However, consideration should be given to facility design, landscape architecture, and transportation modes that disguise operations.

Mallery (2006: 76) writes of Internet-based methods of removing information assets from organizations. For example, an employee can import his or her client information into a variety of free e-mail programs.

A major point from these descriptions of vulnerabilities is that technology is "a blessing and a burden." Technology makes our lives easier; however, we face a never-ending "cat and mouse" cycle of new technology confronted by offenders seeking to exploit and profit from it as defenses follow.

Furthermore, a good spy does not get caught, and quite often, the victimized firm does not discover that it has been subjected to espionage. If the discovery is made, the company typically keeps it secret to avoid adverse publicity.

The Business Espionage Controls and Countermeasures Association (2012) states:

> *The purpose of the association is to research and exchange information about business espionage controls and countermeasures; to establish and encourage a code of ethics within the profession, and to promote our professional image within the business community through a Certified Confidentiality Officer (CCO) program.*
>
> *We identified four primary areas of risk as one of our first BECCA research projects. We called them the* **"Four Faces of Business Espionage,"** *a term now widely used by controls and countermeasures experts. These risk factors are Pretext Attacks [interviews], Computer Abuse, Technical Surveillance, and Undercover Attacks.*

Pretext interviews are disguised interviews or "surveys" that can take place in a variety of locations (e.g., over the telephone, at trade shows, in chat rooms, in bed). The people gathering information may not know the real reason behind the questions, and the victims may not know the identity of the interviewer. Technical surveillance includes planting a listening device (e.g., a bug).

Countermeasures

The first step in keeping information assets secure is to identify and classify it according to its value. Smith (2013: 767) refers to the important concept of **classified information** and defines it as "information explicitly protected by particular laws or regulations and marked to indicate its status." Governments typically classify information involving national security and intelligence activities. If a company has a Department of Defense (DOD) contract, then strict DOD criteria would apply. Each classification has rules for marking, handling, transmitting, storing, and access. The higher the classification the greater are the controls. Businesses without a DOD contract vary widely on information protection methods. Table 18-1 shows DOD and corporate classifications, explanations, and illustrations. ASIS International (2007: 37–39) offers the following classification system: unrestricted, internal use, restricted, and highly restricted.

Table 18-1 Classification Systems

	If unauthorized disclosure	Illustrations
Government classification*		
Top Secret	"Exceptionally grave damage" to national security	Vital national defense plans, new weapons, sensitive intelligence operations
Secret	"Serious damage" to national security	Significant military plans or intelligence operations
Confidential	"Identifiable damage" to national security	Strength of forces, munitions performance characteristics
Corporation classification		
Special controls	Survival at stake	New process or product; secret formula or recipe
Company confidential	Serious damage	Process, customer lists; depends on value to business
Private confidential	Identifiable damage, or could cause problems	Personnel data, price quote

*U.S. Department of Defense (2006). National Industrial Security Program Operating Manual (February 28). www.dss.mil/documents/odaa/nispom2006-5220.pdf, retrieved May 26, 2012. Also, Elsea, J. (2011). "The Protection of Classified Information: The Legal Framework." Congressional Research Service (January 10). www.fas.org/sgp/crs/secrecy/RS21900.pdf, retrieved May 26, 2012.

The list that follows here for an **information security program** is from multiple sources (Smith, 2013; Lowenthal, 2012; Office of the National Counterintelligence Executive, 2011). *The reader should refer to the Chapter 16 list on IT security to compare and contrast both lists for comprehensive security.*

1. *Prevention* is a key strategy to protect information assets, which can be stolen without anything being physically missing, and information assets often are not covered by insurance.
2. Establish formal policies and procedures for such activities as identifying and classifying information assets, handling, use, distribution, release of information on a "need-to-know" basis, storage, and disposal. Other examples are use of mobile devices, storage devices and devices owned by employees, security over passwords, and maintaining a "clean desk" policy so important items are not left in the open when they should be in a locked container.
3. Provide training and awareness programs for employees on all aspects of information security, including policies, methods used by spies, social engineering (see Chapter 16), reporting incidents, investigations, and auditing of the program.
4. Reinforce countermeasures through new employee orientation, the employee handbook, and performance evaluations.
5. Carefully screen employment applicants, maintain an insider threat program, and establish employee exit procedures (e.g., to remind them of protecting proprietary information).
6. Use employee nondisclosure agreements and employee noncompete agreements.

FIGURE 18-4 Sen Trac ID uses radio-frequency identification technology to provide hands-free access control and asset management to track people and products within a facility. *Courtesy: Sensormatic.*

7. Implement physical security and access controls for people and property entering, leaving, and circulating within a facility (Figure 18-4).
8. Secure information assets.
9. Review works written by employees prior to publication and their speeches, ensure protection during trade shows, and control media relations.
10. Control destruction of information assets.
11. Maintain state-of-the-art IT security. Refer to best practices such as ISO/IEC 27002:2005 (see Chapter 16). Use passwords, encrypt data, establish multi-factor authentication measures (e.g., biometrics, PINS, passwords, and knowledge-based questions), and create policies on mobile devices in the workplace and use of social media.
12. Be cautious when logging online in a wireless area. Ensure that your computer is not automatically connected to wireless access points that are unsecured.
13. Mark laptops with company name and telephone number and apply special software to increase the chances of recovery in case of theft.
14. Protect all forms of electronic communication—e-mail, network, fax, telephone, etc.
15. Establish controls over devices that contain a hard drive, electronic storage capacity, or embedded camera. Data are being stored in smaller spaces and so many ordinary items (e.g., pen, knife, watch) can contain a data storage device.
16. Control the variety of office machines (e.g., the combination copy machine, fax, scanner, and printer) that contain hard drives.
17. CDs and DVDs, rather than paper, are increasingly being used to store information. If a duplicator makes a copy of a master CD to its hard drive and then burns multiple copies,

the information is available to people who can access the duplicator, unless the data is purged.

18. Plan for resilience. Ensure that important data has a backup copy in case data are stolen, a disaster strikes, or IT fails.
19. Use technical surveillance countermeasures (TSCM).
20. To strengthen protection, conduct penetration testing and use internal and independent security audits

Operations Security

Operations Security (OPSEC) is defined by Isaacs (2004: 104) as follows: "OPSEC is a formal process for looking at the protection of critical information from the viewpoint of an adversary and then denying that adversary the information it needs." It is a government-developed approach to information security that began during the Vietnam War when it was discovered that lives were being lost, not only from espionage, but also from unclassified information that was being analyzed by the enemy. OPSEC is a way of thinking, rather than a series of steps. The components of OPSEC are: analyze the threat, identify critical information, examine vulnerabilities, assess risk, and apply countermeasures.

Smith (2013: 772) offers an illustration of OPSEC from the first Gulf War. As the U.S. military was about to strike Iraqi-held Kuwait, the pizza delivery business around the Pentagon spiked. OPSEC seeks to reduce the possibility of an enemy uncovering sensitive information from public activities by restricting public activities or through deceptive actions.

Destruction of Information Assets

Records, documents, computers, hard drives, cell phones, mobile devices, and other items that contain information assets should not simply be thrown into trash bins or discarded when no longer needed, because spies and other adversaries may retrieve the information. Total destruction affords better information security. Before pollution restrictions against burning, many firms placed unwanted records in incinerators. Today, *strip-cut shredders* (producing long strips of paper ¼ inch wide) are used by many organizations. However, security is limited. This became painfully evident in 1979, when Iranian militants stormed the U.S. Embassy in Tehran and pieced together top-secret documents that had been shredded by a strip-cut shredder. For increased security, *particle-cut shredders* (smaller pieces of paper) are the alternative (Figure 18-5). *Cross-cut shredders* offer higher security and *disintegrators* offer even more security. Other methods of destruction include chemical decomposition and composting. Vendors that sell high-security shredders seek to meet government national security standards.

Many companies outsource shredding to service firms that send a mobile shredding truck to the client to shred a variety of items besides paper. Examples are CDs, DVDs, hard drives, and credit cards. Security practitioners should exercise due diligence with shredding service firms and investigate the chain of custody of the shredded product. The National Association of Information Destruction (NAID) promotes professionalism and ethics of its member companies.

Shredding has increased in popularity because of privacy laws (e.g., HIPAA and FACTA), the problem of identity theft, and the U.S. Supreme Court case, *California v. Greenwood*,

FIGURE 18-5 A determined adversary might take the time to put small pieces of paper together for information.

which permits police warrantless search and seizure of garbage left on the street for collection (Wikipedia, 2012).

Unshredding is a growing specialization. Although unshredding can be done manually, computer technology speeds the process by scanning pieces on both sides and then the computer determines how the strips should be joined. In the Enron accounting case, many documents were fed through a shredder incorrectly, which made the pieces easier to put together. In reference to forensic identification, shredders contain device-specific characteristics that can be used to determine the specific device that shredded an item.

Comprehensive information security must consider that information is stored in many types of devices. Examples are servers, computers, handheld devices, phones, faxes, printers, copiers, cameras, access cards, readers, flash drives, mapping and navigation devices, and various media. When these devices are to be discarded, the information must be destroyed. If this is not accomplished in-house, vendors are available to contract this service. References on information destruction can be found through standards (e.g., NIST) and NAID.

Besides information destruction, we should remember that devices can be lost or stolen. Policies, security, software, and investigative methods are part of a comprehensive information security program. Information in a device that is "in the wrong hands" can be rendered inaccessible (e.g., passwords and encryption), destroyed (e.g., remotely), or recovered (e.g., GPS or other method).

Defenders against espionage must not fall into the trap of emphasizing certain counter-measures while "leaving the back door open."

Communications Security

Communications security involves defenses against the interception of communication transmissions. In the federal government, the National Security Agency (2000: 10) defines communications security (COMSEC) as follows:

> *Measures and controls taken to deny unauthorized persons information derived from telecommunications and to ensure the authenticity of such telecommunications. Communications security includes cryptosecurity [i.e., encryption or decryption], transmission security, emission security [i.e., intercept and analysis of emanations from equipment], and physical security of COMSEC material.*

In providing a comprehensive approach to protecting information assets, subfields of communications security are listed here (Carroll, 1996: 177–277):

- *Line security* protects communications lines of IT systems, such as a central computer and remote terminals. Line security is effective over lines an organization controls; a wiretap can occur in many locations of a line. Cryptographic security defeats wiretapping.
- *Transmission security* involves communications procedures that afford minimal advantage to an adversary bent on intercepting data communications from IT systems, telephones, radio, and other systems.
- *Emanation security* prevents undesired signal data emanations (e.g., from computer equipment) transmitted without wires (e.g., electromagnetic or acoustic) that could be intercepted by an adversary. **TEMPEST** is the code word used by the National Security Agency for the science of eliminating undesired signal data emanations. "Shielding," discussed soon, is one strategy to reduce data emanations.
- *Technical security,* also called *technical surveillance countermeasures,* provides defenses against the interception of data communications from microphones, transmitters, or wiretaps.

The above methods of attack can be used together, which is one reason why communications security is a highly complex field. What follows here is primarily technical security; however, *we must not lose sight of the importance of a comprehensive approach to protecting information assets.*

Electronic Surveillance and Wiretapping

Electronic surveillance utilizes electronic devices to covertly listen to conversations, whereas **wiretapping** pertains to the interception of telephone line communications. The term "wiretapping" is becoming antiquated and it should be replaced with the term "**communications tapping**" because so much electronic communications is wireless (e.g., cell phones). The private sector (e.g., private security, PIs, and citizens) are prohibited from applying these surveillance methods. Court and legislative restrictions and the actual use of electronic surveillance and wiretapping by federal, state and local police, the military, and the intelligence community vary. Government *criminal* investigations operate under higher legal standards (e.g., court order based upon probable cause) than investigations involving *spies, terrorists, or other national security threats* whereby the government operates under lower legal standards (e.g., National Security Letter issued by an FBI supervisor without court review). Because detection is so difficult, the exact extent of electronic surveillance and wiretapping is impossible to gauge, not only in criminal and intelligence investigations globally by all governments, but also by spies, PIs, and others. The prevalence of these activities applied illegally probably is greater than one would expect.

Winter (2012) reports that U.S. Magistrate Judge Stephen Smith, involved in approving secret warrants, estimates that 30,000 electronic surveillance orders are approved by federal judges each year. Such secret orders are authorized under the Electronic Communications Privacy Act (ECPA) of 1986 and provide law enforcement with access to telephone calls, e-mails, texts, websites visited, and other electronic communications. The ECPA permits electronic surveillance orders to be nonpublic through sealed court files, gag orders, and delayed-notice, although open to phone companies and other communication providers who execute the orders. Winter reports that "the balance between surveillance and privacy has shifted dramatically toward law enforcement."

In addition, we should also consider the extent of surveillance to curb national security threats under laws such as the Foreign Intelligence Surveillance Act, which was amended by the USA Patriot Act to include terrorism not supported by a foreign government. Another issue is the need to update these laws in light of changing technology.

Electronic surveillance and wiretapping technology are highly developed to the point where countermeasures have not kept up with the technology and methods. Consequently, only the most expertly trained and experienced specialist can counter this threat.

Surveillance equipment is easy to obtain. Transmitters are contained in toys and other items found in many homes. Retailers sell FM transmitters or microphones that transmit sound, without wires, to an ordinary FM radio after tuning to the correct frequency. These FM transmitters are advertised to be used by public speakers who favor wireless microphones so they can walk around as they talk without being hindered by wires; the voice is transmitted and then broadcast over large speakers. They are also advertised to listen in on a baby from another room. An electronically inclined person can simply enter a local electronics store or shop online and buy all the materials necessary to make a sophisticated bug. Pre-built models are also available. One type applies the same global system used by cell phones and it can be called from anywhere without it ringing for listening.

Miniaturization has greatly aided spying. With the advance of the microchip, transmitters are apt to be so small that these devices can be enmeshed in thick paper, as in a calendar, under a stamp, or within a nail in a wall. Bugs may be planted as a building is under construction, or a person may receive one hidden in a present or other item. Transmitters are capable of being operated by solar power (i.e., daylight) or local radio broadcast.

Bugging techniques are varied. Information from a hidden microphone can be transmitted via a radio transmitter or "wire run." Bugs are concealed in a variety of objects or carried on a person. Transmitting devices can be remotely controlled with a radio signal for turning them on and off. This makes detection difficult. A device known as a carrier current transmitter is placed in wall plugs, light switches, or other electrically operated components. It obtains its power from the AC wire to which it is attached. Sound systems with speakers serve as microphones.

Many spies use multiple systems. Multiple bugs are placed so they will be found which, in many instances, satisfies security and management. Other bugs are more cleverly concealed.

Gruber (2006: 280–283) notes that gun microphones are very effective. He writes that they can be aimed at a target from a significant distance; they are used with a headset and amplifier. Gun microphones can be seen at football games.

Traditional telephones use wires that enable calls to travel between stationary locations. Telephone lines are available in so many places that taps are difficult to detect. A tap can be direct or wireless. With a direct tap, as seen in Hollywood movies, a pair of wires is spliced to a telephone line and then connected to headphones or a recorder. There are several methods of modifying a telephone so it becomes a listening device, even when it is hung up. For a wireless tap, an FM transmitter, similar to a room bug, is employed. The transmitter is connected to the line and then a receiver (e.g., radio) picks up the signal. Wireless taps (and room bugs) are spotted by using special equipment. Direct taps are difficult to locate. A check of the entire line is necessary.

Today, many telephones are mobile and, because telephone traffic travels over space radio in several modes—for example, cellular, microwave, and satellite—the spy's job is made much easier and safer, since no on-premises tap is required. What is required is the proper equipment for each mode.

Diffie and Landau (2009) write:

Although big changes in telephony have given rise to equally big changes in wiretapping, the essentials remain the same. The interception and exploitation of communications has three basic components: accessing the signal, collecting the signal, and exfiltrating the signal. Access may come through alligator clips, a radio, or a computer program. Exfiltration is moving the results to where they can be used. Collection may be merged with exfiltration or may involve recording or listening. The tap can be in the phone itself, through introduction of a bug or malware that covertly exfiltrates the call, often by radio.

In one case a Mossad agent in Berne, Switzerland, was arrested after he tried to tap the telephone of a Hezbollah target. His technical system was a cellular telephone device that would be activated when the target telephone was put in use. The device would automatically call another cellular telephone where the target's telephone would be monitored (Business Espionage Controls and Countermeasures Association, 2007).

Consideration must be given to a host of methods and innovations that may be applied by a spy. These include infrared transmitters that use light frequencies below the visible frequency spectrum to transmit information. This can be defeated through physical shielding (e.g., closing the drapes). Another method, a laser listening device, "bounces" laser off a window to receive audio from the room. Inexpensive noise masking systems can defeat this technique (Jones, 2000: 1–17). Kaiser and Stokes (2006: 65) write: "Newer laser microphones are created by feeding two hair-thin strands of fiber-optic cable into the room being monitored. The microphone operates when a laser beam is sent down one of the fibers, where it bumps into a thin aluminum diaphragm and returns on the other fiber with the room conversation." A careful search is required to find this and other devices. Computer, e-mail, facsimile, and other transmissions are also subject to access by spies. A spy may conceal a recorder or pinhole lens camera on the premises, or wear a camera concealed in a jacket or tie. If drawings or designs are on walls or in sight through windows, a spy stationed in another skyscraper a few blocks away might use a telescope to obtain secret data, and a lip reader can enhance the information gathering. Or, a window washer might appear at a window for surveillance. Another method is a spy disguised as a janitor to be assigned to the particular site. All of these methods by no means exhaust the skills of spies as covered earlier under "espionage techniques."

Technical Surveillance Countermeasures

ASIS International (2007: 17) states the following:

> **Technical Surveillance Countermeasures (TSCM)** *refers to the use of services, equipment, and techniques designed to locate, identify, and neutralize the effectiveness of technical surveillance activities (electronic eavesdropping, wiretapping, bugging, etc.). Technical surveillance countermeasures should be a part of the overall protection strategy. Individuals within the organization responsible for physical security, facility security, information asset protection, telecommunications, meeting planning and information technology all have a stake in addressing these concerns.*

The physical characteristics of a building have a bearing on opportunities for surveillance. Some of these factors are poor access control designs, inadequate soundproofing, common or shared ducts, and space above false ceilings enabling access. The in-house security team can begin countermeasures by conducting a physical search for planted devices. If a decision is made to contact a specialist, *only the most expertly trained and experienced consultant should be recruited.*

The Countermeasures Consultant

Organizations often recruit a countermeasures consultant to perform contract work. As a consumer, ask for copies of certificates of TSCM courses completed and a copy of the insurance policy for errors and omissions for TSCM services. What equipment is used? What techniques are employed for the cost? Are sweeps and meticulous physical inspections conducted for the quoted price? Watch for scare tactics. Is the consultant really a vendor trying to sell surveillance detection devices, or a PI claiming to be a TSCM specialist? Will the consultant protect confidentiality? The interviewer should request a review of past reports to clients. Were names deleted to protect confidentiality? These questions help to avoid hiring an unqualified "expert." One practitioner offered clients debugging services and used an expensive piece of equipment to conduct sweeps. After hundreds of sweeps, he decided to have the equipment serviced. A service person discovered that the device was not working properly because it had no battery for one of its components. The surprised "expert" never realized a battery was required.

For a comprehensive countermeasures program, the competent consultant will be interested in sensitive information flow, storage, retrieval, and destruction. Extra cost will result from such an analysis, but it is often cost effective.

The employer should use a public telephone off the premises to contact the consultant in order not to alert a spy to impending countermeasures. An alerted spy may remove or turn off a bug or tap and the TSCM may be less effective.

Techniques and Equipment

Detection equipment is expensive and certain equipment is subject to puffing, but useless. A company should purchase its own equipment only if it retains a well-qualified TSCM technician, many sweeps are conducted, and the in-house TSCM program is cost-effective.

Equipment includes the nonlinear junction detector (NLJD), costing between $10,000 and $20,000. It is capable of detecting radio transmitters, microphones, infrared and ultrasonic transmitters, recorders, video cameras, cell phones, remote-controlled detonators, and other hidden electronic devices, even when they are not working. Gruber (2006: 284–285) offers the following on the NLJD. It transmits a microwave signal through its antenna and an internal receiver listens for a RF response that may mean a device is present. NLJDs are available in various power outputs to the restricted government version. The effectiveness of this equipment is poor in an area containing several electronic devices; in this case, a physical search is best.

The telephone analyzer is another tool designed for testing a variety of single and multiline telephones, answering machines, fax machines, and intercom systems. The spectrum analyzer is still another tool. Basically, it is a radio receiver with a visual display to detect airborne radio signals. Other types of specialized equipment are on the market. Buyer beware.

In one case, a TSCM specialist was conducting a sweep in a conference room of a major corporation when a harmless looking stapler sitting among other office supplies was found to contain a voice-activated recorder with memory. A pin-hole lense camera was then installed in the room and video showed an office worker exchanging the stapler every week for a similar looking one. When confronted and interviewed, the worker revealed who was behind the spying, that he was paid $500 for each stapler containing audio, and that he only transferred three staplers to the spy during his employment of five months. The worker was fired, police were not contacted, the media and stockholders never knew about possible leaks of information, and the spy was informed about the discovery and threatened with criminal and civil legal action.

Some security personnel or executives plant a bug for the sole purpose of determining if the equipment of the detection specialist is effective. This "test" can be construed as a criminal offense. Alternatives are specially designed test transmitters, commercially available, that have no microphone pickup and therefore can be used without liability. Another technique is to place a tape recorder with a microphone in a drawer.

A tool kit and standard forms are two additional aids for the countermeasures specialist. The tool kit consists of the common tools (e.g., screwdrivers, pliers, electrical tape) used by an electrician. Standard forms facilitate good recordkeeping and serve as a checklist. What was checked? What tests were performed? What were the readings? Where? When? Who performed the tests? Why were the tests conducted? Over a period, records can be used to make comparisons while helping to answer questions.

The following list offers topics of consideration for TSCM (Gruber, 2006: 277–304; Kaiser and Stokes, 2006: 60–68):

- Because a spy who learns of a TSCM search may turn off or remove his or her equipment, the TSCM specialist should be discreet by disguising vehicles, dress, and equipment. A top executive may choose to establish a cover story to avoid alerting anyone to the TSCM.
- An early step in TSCM is a physical search for devices, beginning from outside the building. The physical search, both outside and inside, is very important and time-consuming. On the outside, focus on items such as utilities, wires, ductwork, and openings (e.g., windows). A spy can tap into lines outside the building without needing to ever enter the building.
- Inside the building, the TSCM technician should check cabling and inside individual office equipment (e.g., telephones, faxes, and computers). Is there anything in the office equipment that appears odd?
- The technician should be knowledgeable about IT systems, computers, internal network or Local Area Network (LAN), and a connection to the outside or Wide Area Network (WAN). These systems can be bugged or tapped like telephone systems. For example, a LAN analyzer connected to a line can read all e-mail that travels through the line. The technician should have equipment to check what is attached to lines.
- Besides traditional cable, fiber optic cable can also be tapped. A tap on a fiber optic cable can be detected through an Optical Time Domain Reflectometer.
- Since devices may be hidden in walls, the technician can use an ultraviolet light to detect plaster repairs to walls. A NLJD or a portable x-ray machine can be used to detect devices in walls.
- Items in walls that should be checked are power outlets, phone jacks, and network jacks. Tools to check these items and inside walls are a flashlight, dental mirror, and a fiber optic camera.
- Plates at light switches, wall outlets, and HVAC vent covers should all be removed for the search and prior to the sweep.
- If a bug or tap is found, it should be documented and photographed. Caution is advised because the device could be booby-trapped. Although police could be contacted for assistance, their response and expertise will vary widely. Difficult questions surface as to whether the device should remain and whether to apply an OPSEC approach (e.g., feed false information). Seek legal assistance.
- The TSCM technician often finds nothing unusual. However, 100 percent protection is not possible. A spy may outfox the technician and the equipment. In addition, there are many ways to steal information. Security practitioners should be creative and think like a spy.

Another strategy to thwart listening devices is "shielding," also called *electronic soundproofing.* Basically, copper foil or screening and carbon filament are applied throughout a room to prevent acoustical or electromagnetic emanations from leaving. Although this method is very expensive, several organizations employ it to have at least one secure room or to protect information in computers.

Equipment is available on the market that *may* frustrate telephone taps and listening devices. Scramblers, attached to telephones, alter the voice as it travels through the line. However, no device or system is foolproof. Often, simple countermeasures are useful. For instance, an executive can wait until everybody is present for an important meeting, and then relocate it to a previously undisclosed location. Conversants can operate a radio at high volume during sensitive conversations, and exercise caution during telephone and other conversations.

Voice over Internet Protocol (VoIP) technology is popular with organizations and commercial telephony service providers because of lower costs and efficiency. VoIP enables voice to be transported digitally via a network using Internet Protocol standards. Such services may not even make contact with the traditional telephone network. One concern of VoIP technology relates to its inability to provide traditional location identification (i.e., Enhanced 911) for 911 emergency calls made to public safety agencies. Of particular interest for our discussion here is that traditional techniques for telephone intercepts and wiretaps are more difficult with VoIP, and end-to-end encryption compounds the challenges for the spy (National Institute of Justice, 2006).

As we know, information assets can be collected in many different ways besides with physical devices. Losses can occur through speeches and publications by employees, in company trash, and by unknowingly hiring a spy. Comprehensive, broad-based information security is necessary.

Who do you think has "the edge," those who seek information assets or those who protect them?

Search the Internet

Here is a list of websites relevant to this chapter:

ASIS, International: www.asisonline.org
Business Espionage Controls and Countermeasures Association: www.becca-online.org
Centers for Disease Control and Prevention: www.cdc.gov
Institute for a Drug-Free Workplace: www.drugfreeworkplace.org
National Association of Information Destruction, Inc.: www.naidonline.org
National Institute for Occupational Safety and Health (NIOSH): www.cdc.gov/niosh/homepage.html
Occupational Safety and Health Administration (OSHA): www.osha.gov
OSHA: www.osha.gov/SLTC/workplaceviolence/index.html
Strategic and Competitive Intelligence Professionals: www.scip.org
Substance Abuse and Mental Health Services Administration: www.samhsa.gov
U.S. Department of Labor: www.dol.gov/elaws/drugfree.htm
U.S. Department of State: www.state.gov
U.S. Drug Enforcement Administration: www.justice.gov/dea

Case Problems

18A. As a security manager, you just received an internal telephone call from a supervisor complaining about a subordinate who became angered by a work assignment. The employee told the supervisor that he knows where the supervisor lives and where his kids go to school. What do you do?

18B. You are a security manager at a plant. One day, a former employee shows up at the front gate and demands to see his estranged wife, an employee. In addition, he wants to talk with the human resources director about benefits. How do you handle this situation?

18C. John Smith, an employee who has just lost his job because of corporate downsizing, is in the office of the Director of Human Resources holding a pistol in the direction of the Director. As the security manager, you were summoned to the office earlier, not knowing that the pistol had been drawn. You enter the office and you stop upon seeing the pistol. John Smith states: "I've given 10 years of my life to this place. They had no right doing this to me. If I can't work, I can't support my family. It's management's fault and they are going to pay." As the security manager, what do you say and do? (This case problem was prepared with the assistance of Hasselt and Romano, 2004: 12–17).

18D. As the chief security officer for a corporation with plants in the United States and Europe, prepare a list of questions to answer as you plan a human resources protection program.

18E. As a security manager you hear through the grapevine that several employees smoke marijuana during lunch when they go to their vehicles. What do you do?

18F. As the new chief security officer for a corporation, you are reviewing the methods of information collection of the in-house competitive intelligence unit. The list includes using the Internet, public documents, public documents from government, private investigators, subscriptions to news services, purchasing securities to receive annual financial reports of competitors, collecting garbage from competitors, attending seminars and speeches of competitors, and purchasing competitor products for study. Do any of these methods necessitate closer attention? Explain and justify your answer.

18G. As the security director for a corporation engaged in research, you see the need for an information security consultant to improve protection. What criteria would you list to select such a specialist? What questions would you ask applicants during the selection process?

18H. Of the major topics in this chapter, which one would you select as a specialization and career? Why? How would you develop such a specialization and career?

References

Acohido, B. (2011). Kidnappers use google to select victims. *USA TODAY* (February 15). <content.usatoday.com> retrieved May 10, 2012.

Apuzzo, M. (2007). *Chiquita to pay $25M in terror case.* Associated Press. (March 14). <biz.yahoo.com/ap/070314/terrorism_bananas.html?.v=5> retrieved March 16, 2007.

Ashley, S. (2006). Cell phone vulnerabilities. *Law Officer Magazine, 2* (July).

ASIS International. (2007). *Information Asset Protection Draft Guideline.* <www.asisonline.org> retrieved February 9, 2007.

ASIS International. (2011). *Workplace Violence Prevention and Intervention, American National Standard.* <www.asisonline.org> retrieved October 10, 2011.

Associated Press, (2006). Schwab: $2m for CEO security. *Security Director News, 3* (May).

Ast, S. (2010). *Managing security overseas: Protecting employees and assets in volatile regions.* Boca Raton, FL: CRC Press.

Besse, W., & Whitehead, C. (2000). New tools of an old trade. *Security Management, 44* (June).

Bowron, E. (2001). All the world's a staging ground. *Security Management, 45* (April).

Business Espionage Controls and Countermeasures Association. (2007). About BECCA. <www.becca-online.org> retrieved March 18, 2007.

Business Espionage Controls and Countermeasures Association. (2012). BECCA History. <www.becca-online.org> retrieved May 24, 2012.

Carroll, J. (1996). *Computer security* (3rd ed.). Boston: Butterworth-Heinemann.

Centers for Disease Control and Prevention. (2011). Excessive Drinking Costs U.S. $223.5 Billion. <www.cdc.gov/Features/AlcoholConsumption> retrieved May 17, 2012.

Clapper, J. (2012). Unclassified Statement for the Record on the Worldwide Threat Assessment of the U.S. Intelligence Community for the Senate Select Committee on Intelligence (January 31). <www.dni.gov/testimonies/20120131_testimony_ata.pdf> retrieved May 22, 2012.

Cnbc.com. (2012). For Some Corporate Chiefs, Private Security is a Tax Break. <www.cnbc.com> retrieved April 16, 2012.

Coulombe, R. (2012). Positioning GPS. *Security Technology Executive, 22* (April).

Curtis, G., & McBride, R. (2005). *Proactive security administration.* Upper Saddle River, NJ: Pearson Prentice Hall.

DiBianca, M. (2008). Employee shooting results in unusual liability for workplace violence. *The Delaware Employment Law Blog* (May 8) <www.delawareemploymentlawblog.com> retrieved May 5, 2012.

Diffie, W., & Landau, S. (2009). Communications surveillance: Privacy and security at risk. *Privacy and Rights, 7* (September).

Ehrlich, C. (2002). Liar, liar: The legal perils of misrepresentation. *Competitive Intelligence Magazine, 5* (March–April).

Elliott, K., & Shelley, K. (2005). Impact of employee assistance programs on substance abusers and workplace safety. *Journal of Employment Counseling, 42* (September).

Elliott, R. (2006). What's behind country risk ratings? *Security Management, 50* (August).

Evans, J. (2012). Bullies aren't limited to school yards. *Coshocton Tribune* <www.coshoctontribune.com> retrieved March 27, 2012.

Gips, M. (2006). High on the Job. *Security Management, 50* (February).

Gips, M. (2007). My short life as an EP specialist. *Security Management, 51* (March).

Gruber, R. (2006). *Physical and technical security: An introduction.* Clifton Park, NY: Thomson Delmar Learning.

Halligan, R. (2001). Do your secrets pass the test? *Security Management, 45* (March).

Harwood, M. (2011). Planning for tumultuous times. *Security Management, 55* (June).

Hasselt, V., & Romano, S. (2004). Role-Playing: A vital tool in crisis negotiation skills training. *FBI Law Enforcement Bulletin, 73* (February).

Hershkowitz, R. (2004). Zero tolerance equals trouble. *Security Management, 48* (October).

Horowitz, R. (2005). A Comment on Drafting Corporate Competitive Intelligence Policies. <www.rhesq.com/CI/Comment%20on%20CI%20Policies.html> retrieved March 18, 2007.

Horowitz, R. (1998). The economic espionage act: The rules have not changed. *Competitive Intelligence Review, 9* <www.scip.org/pdf/9(3)horowitz.pdf> retrieved March 22, 2007.

Hudson, J. (2004). Dozens sue over lockheed shootings. *The Clarion-Ledger* (July 3).

Hughes, S. (2001). Violence in the workplace: Identifying costs and preventive solutions. *Security Journal, 14.*

Insurance Information Institute. (2012). What does kidnap and ransom insurance cover? <www.iii.org/articles/what-does-kidnap-and-ransom-ins> retrieved May 10, 2012.

Isaacs, R. (2004). How not to tell all. *Security Management, 48* (May).

Ivancevich, J. (2001). Human resource management (8th ed.). Boston: McGraw-Hill Irwin.

Jaeger, S. (2001). The age of rage. *Security Industry & Design, 11* (February).

Johnson, S. (2012). Doing business overseas gets riskier. *CFO* (April 24). <www3.cfo.com> retrieved April 25, 2012.

Jones, T. (2000). *Surveillance countermeasures in the business world.* Cookeville, TN: Research Electronics International.

Kaiser, M., & Stokes, R. (2006). Who's listening? *Security Management, 50* (February).

Katz, D. (2011). Preventing workplace violence. *Disaster Recovery Journal, 24* (Winter).

Kenny, J. (2010). Risk assessment and management teams: A comprehensive approach to early intervention in workplace violence. *Journal of Applied Security Research, 5*(2)

Kenny, J. (2005a). Threats in the workplace: The thunder before the storm? *Security Journal, 18*(3)

Kenny, J. (2005b). Workplace violence and the hidden land mines: A comparison of gender victimization. *Security Journal, 18*(1)

Laczniak, G., & Murphy, P. (1993). The ethics of corporate spying. *Ethics Journal* (Fall).

Lowenthal, M. (2012). *Intelligence: From secrets to policy* (5th ed.). Thousand Oaks, CA: Sage Pub.

Mallery, J. (2006). The hidden data thieves. *Security Technology & Design, 16* (March).

Matsumoto, D., et al. (2012). The role of emotion in predicting violence. *FBI Law Enforcement Bulletin, 81* (January).

Meadows, R. (2007). *Understanding violence and victimization* (4th ed.). Upper Saddle River, NJ: Pearson Prentice Hall.

Miller, S. (2007). Gone in a flash. *Information security, 10* (March).

Myshko, D. (2001). Just say yes to drug testing. *Risk and Insurance, 12* (April 16).

Nairn, G. (2011). Your wall has ears. *The Wall Street Journal* (October 19). <online.wsj.com> retrieved October 20, 2011.

National Institute of Justice. (2006). Telephony Implications of Voice over Internet Protocol. <www.ncjrs.gov/pdffiles1/nij/212976.pdf> retrieved June 18, 2007.

National Institute on Alcohol Abuse and Alcoholism. (2012). FAQs for the General Public. <www.niaaa.nih.gov> retrieved May 17, 2012.

National Institute on Drug Abuse. (2008). Drug Facts: Workplace Resources. <www.drugabuse.gov> retrieved May 16, 2012.

National Security Agency. (2000). National Information Systems Security (INFOSEC) Glossary. (UNCLASSIFIED). <security.isu.edu/pdf/4009.pdf> retrieved March 22, 2007.

Nolan, J. (1997). Economic espionage, proprietary information protection: Difficult times ahead. *Security Technology and Design* (January–February).

Occupational Safety & Health Administration. (2011). Enforcement Procedures for Investigating or Inspecting Workplace Violence. <www.osha.gov/OshDoc/Directive_pdf/CPL_02-01-052.pdf> retrieved May 4, 2012.

Occupational Safety & Health Administration. (2012). Workplace Violence. <www.osha.gov/SLTC/workplaceviolence/index.html> retrieved May 2, 2012.

Office of the National Counterintelligence Executive. (2011). Foreign Spies Stealing U.S. Economic Secrets in Cyberspace (October). <www.ncix.gov/publications/reports/fecie_all/Foreign_Economic_Collection_2011.pdf> retrieved May 23, 2012.

Piazza, P. (2005). From bluetooth to redfang. *Security Management, 49* (March).

Piazza, P. (2007). The ABCs of USB. *Security Management, 51* (January).

Purpura, P. (2011). *Security: An introduction*. Boca Raton, FL: CRC Press, Taylor & Francis Group.

Purpura, P. (1993). When the security manager gets shot: A corporate response. *Security Journal, 4*, 151 (July).

Ryan, L. (2005). Exclusive remedy upheld in lockheed martin shooting case. *Workers' Comp Insider* (July 21). <www.workerscompinsider.com/archives/2005_07.html> retrieved March 10, 2007.

Sharar, D. and Lennox, R. (2009). A New Measure of EAP Success. <www.shrm.org> retrieved May 16, 2012.

Shyman, R. (2000). Women at work. *Security Management, 44* (February).

Simovich, C. (2004). To serve and protect. *Security Management, 48* (October).

Smith, R. (2013). *Elementary information security*. Burlington, MA: Jones & Bartlett Pub.

Smith, S. (2004). What every employer should know about drug testing in the workplace. *Occupational Hazards, 66* (August).

Strategic and Competitive Intelligence Professionals. (2012). About SCIP. <www.scip.org> retrieved May 22, 2012.

Substance Abuse and Mental Health Services Administration. (n.d.). Drug Testing. <www.workplace.samhsa.gov/Dtesting.html> retrieved May 17, 2012.

Tiesman, H., et al. (2012). Workplace homicides among U.S. women: The role of intimate partner violence. *Annals of Epidemiology, 22*(4) (April).

U.S. Department of Justice. (2010). PRO IP ACT Annual Report FY 2010 (December 17). <www.justice.gov> retrieved May 23, 2012.

U.S. Department of State. (2012). Country Profiles. <www.state.gov> retrieved May 10, 2012.

Whitzman, C., et al. (2009). The effectiveness of women's safety audits. *Security Journal, 22*

Wikipedia. (2012). Paper Shredder. en.wikipedia.org/wiki/Paper_shredder, retrieved May 26, 2012.

Winter, M. (2012). Judge estimates 30K secret spying orders approved yearly. *USA Today* (June 5). <content.usatoday.com> retrieved June 6, 2012.

Zezima, K., & Goodnough, A. (2010). Drug testing poses quandary for employers. *The New York Times* (October 24). <www.nytimes.com> retrieved October 25, 2010.

19 Your Future in Security and Loss Prevention

OBJECTIVES

After studying this chapter, the reader will be able to:

1. Discuss security and loss prevention in the future
2. List at least 10 trends affecting security and loss prevention
3. Discuss security and loss prevention education, research, and training
4. Explain how to seek employment opportunities in the security and loss prevention profession

KEY TERMS

- smart security
- automated manufacturing
- cycle of protection
- zero-day exploits
- criminal entrepreneurs
- multinational and multicultural workforce
- five wars of globalization
- market forces
- decentralized networks
- security operations
- strategic security
- tradecraft
- clinical experience model
- evidence-based model
- DHS Centers of Excellence
- programmed, self-paced instruction
- distance learning online

Introduction

Future Shock, a world-famous book written by futurist Alvin Toffler in 1970, explained the impact on people from accelerated technological and social change in a short period of time. Toffler described society experiencing enormous structural change, information overload, and stress that overwhelms people. Forty years later Toffler Associates (2010) published *40 For the Next 40: A Sampling of the Drivers of Change That Will Shape Our World between Now and 2050*. This publication begins with the following paragraph:

> *We are in the midst of an accelerating, revolutionary transformation. Change is happening everywhere—in technology, business, government, economics, organizational structures, values and norms—and consequently affects how we live, work and play. As industry and*

government leaders, we must acknowledge that this change demands new ways of governing and of running our organizations. The ways in which we communicate and interact with each other will be different. The methods through which we gain and process information will be different. The means by which we earn and spend money will be different. Through the culmination of these and other changes, organizations will be radically transformed.

Seeking to accurately predict the future is difficult and risky. A professional will consider many variables when making educated guesses about the future. What follows here are possibilities for the future to which the reader can apply critical thinking skills.

Periodically, we hear older, experienced folks speak about how much simpler life was years ago. However, this view is subject to debate because technology, for example, has made life and work easier in many ways. At the same time, security and loss prevention practitioners face numerous challenges in their work today, brought about by many factors. These factors include the needs of businesses as they seek to survive and return profits, the needs of customers, globalization, an array of threats and hazards, limited security budgets, government laws and regulations, professional standards, and litigation. Our world is likely to continue to be increasingly complicated, and as difficult as it may be at times, problems should be perceived as challenges that have solutions and/or present opportunities.

Security and Loss Prevention in the Future

D'Addario et al. (2011: 44–48), write about security in 2020 and elements that security leaders should infuse into security planning to optimize the security function. The first element is to focus on risks of interest to the corporate board, such as financial, brand, resilience, human capital, IT, legal, regulation compliance, and new markets. Study which risks have security components and enhance security value in those areas. A second element is to establish cross-functional groups with whom to partner and share information and oversight on risk issues. A third element is all-hazards, enterprise-wide risk mitigation. A fourth element is training and mentoring for the next generation of security professionals. A fifth element seeks technology that shows value and ROI through efforts such as connecting integrators, technology providers, and customer needs.

As we know, technology has improved security in many ways. We have "smart cards" and "smart cameras," and we will see many more systems and devices that contain **"smart security,"** which means that never-ending enhanced technology will be built into systems and devices. We will see increasingly "smart" fences, contraband detection systems, WMD sensors, robot guards, and cybersecurity. Artificial intelligence will continue to surpass human capabilities. For instance, rather than security officers watching numerous CCTV monitors and becoming fatigued, the artificial intelligence that we have today will continue to help them to spot events deserving attention and response. Facial recognition systems will improve and eventually be capable of identifying offenders wherever they may be while they are moving. A vital question to consider in planning security as technology evolves is how it impacts humans, including customers of security and security practitioners.

Marc Goodman, a futurist who focuses on technology's impact on crime, offers the following points (Harwood, 2011: 30–34):

- Spoofing (i.e., pretending to be someone else) will increase. One example is when a sender of an e-mail alters the heading to make the e-mail appear as if it were sent by someone else. Another example is someone planting digital evidence to make a person appear as if they were at a crime scene based on location data from their smart phone.
- Location-based crime will increase. This involves gathering public information on people (e.g., through the Internet), plotting it on a map, and then victimizing them.
- A type of location-based crime called SWATing uses software in a smart phone to permit any phone number (i.e., spoofing) to be presented on the phone's outbound caller ID.
- Fake personas (e.g., photo and profile) in cyberspace dupe people into thinking they are communicating with the person described.
- Robotic crime will increase. An example is drug traffickers who use unmanned aerial vehicles (UAVs) and submersibles.
- Hacking into implanted medical devices (e.g., pacemaker) linked to the Internet and causing harm presents challenges for victims, manufacturers, investigators, and coroners.

With Goodman's points in mind, we can surmise how these methods can create challenges for security professionals. To illustrate, an investigation may become more complex if a defendant claims that they were "digitally set up" through spoofing. Location-based crime can involve a stalker or kidnapper victimizing an employee. SWATing enables an offender to mislead authorities who respond to the address linked to the phone number they receive. Police can possibly be duped into responding to a bogus emergency call (e.g., a shooting or bomb) at a certain address, such as a business. Security professionals are challenged to ensure employees are aware of fake personas and many other risks in cyberspace. Robotic crime can be applied to transport stole assets from a business or conduct surveillance. People with implanted medical devices can possibly be harmed or killed by a "hacker-assassin."

Zalud and Ritchey (2010: 30–38) predict increasing applications of UAVs, not only by the military and at borders, but also by public police and security. As UAVs become smaller and smaller, they will become another layer of security, patrolling and responding to incidents inside and outside of buildings while sending real-time video . UAVs are also helpful in observing disaster sites. In addition, Zalud and Ritchey write about speech recognition becoming the new "touch" for mobile devices. They also report on research to develop a helmet linked to a monitoring system that enables personnel wearing the helmet to take action through eye movements, as well as on liquid crystal displays (LCDs) that can be sewn into a sleeve, where personnel wearing the piece of clothing can read text messages.

Future integrated security systems will perform an array of loss prevention activities beyond what is accomplished today. For example, if an intruder enters a building, not only will a system pinpoint the entry location via a series of sensors and activate CCTV, it will also simultaneously dispatch a robot to apprehend and positively identify the intruder and uncover his or her background.

Access control systems will no longer use cards. An individual will stand in front of sophisticated sensors, and positive identification will be made by the analysis of a number of biometric characteristics: bone structure, teeth, and body odor, to name a few. It would be especially difficult for an offender to duplicate several of these characteristics. The same sensors could be placed at many locations to monitor personnel, such as parking lots, building entrances, elevators, high-security locations, copying machines, lunchrooms, and restrooms. Obviously, with such a system, it would be possible to know exactly where all employees were at every minute of every workday. If a fire developed or a crime occurred, the recorded location of everyone would aid loss prevention personnel. However, will employees welcome IT records revealing how many times and for how long they visit the restroom or other locations? Suppose a system had the capability to record every conversation, every day, within a building. This could give management the opportunity to "weed out" employees who are counterproductive to organizational goals. In addition, because this system would be capable of "recognizing" any conversations pertaining to losses, the loss prevention department could review these conversations and investigate vulnerabilities. With these possibilities in mind, one realizes the blessings and burdens of technology. Are such intense measures worth sacrificing privacy? Certainly not. *Countermeasures must strike a balance between preventing losses and protecting privacy. In the future, as today, the courts will be watching and ruling as technological innovations and loss prevention strategies are applied.*

As we progress in the era of **automated manufacturing** (AM), loss prevention methods will change to meet the new technology. AM operates machines that transform raw materials into finished products with limited human input. Robots with self-contained computers are an essential part of AM. Activities such as material handling, assembly, inspection, and quality control are automated. Human input comes from a computer control center. Cybersecurity is a major concern. Widespread use of "just-in-time production" also contributes to fewer employees in manufacturing. Production is based on the needs of retailers, which avoids costly inventory holding and producing items that do not sell well.

The future use of robots is promising. They operate 24hours a day without a coffee break or vacation. A salary and fringe benefits such as hospitalization and pensions are unnecessary for robots. Also, robots are immune to heat, cold, noise, radiation, and other hazards.

Imagine loss prevention for a plant operated by only three managers and six technicians. If internal pilfering occurred, the number of suspects would be narrowed to nine, excluding robots. Parking lots, frequently a source of crime and requiring traffic control, would be smaller. Fires and accidents, which are often caused by human error, would be reduced.

Can you describe an example of a clash between technology and civil liberties? What were the issues and solutions?

Trends Affecting Security and Loss Prevention

To begin with, the primary drivers of security and loss prevention practitioners, programs, and services are the needs and objectives of organizations and the changes within our society and world. Protection at a hotel will differ from protection at a nuclear power plant. The loss prevention needs of a retail chain in the United States will differ from the loss prevention needs of a multinational corporation with business interests in the Middle East. Although the protection needs of organizations differ, there are trends that affect all security and loss prevention programs.

The 21st century will continue to see no shortage of threats and hazards facing businesses and institutions. The list that follows presents trends and challenges that impact protection programs and the types of specialists required for security and loss prevention and, thus, employment opportunities:

- The challenges of crimes, fires, accidents, and disasters will remain but increase in complexity.
- Terrorism will become increasingly sinister and surprising because of the proliferation of WMD, mass casualties, severe economic harm, and its successes.
- Security and loss prevention metrics (see Chapter 2) will grow in importance to measure and enhance the success of protection programs. Campbell (2007: 60) writes: "Metrics are a tool used to facilitate influence, to demonstrate, argue, support, and convince."
- We often hear about security and loss prevention departments "doing more with less." Budget cuts are a fact of life requiring prioritization and creativity. The challenges can be enormous. However, "doing more with less" has its limitations, and a variety of outcomes is possible. Examples are victimization of people, litigation, financial losses, and ethical and legal issues.
- Risk analysis will become more challenging. Improved research methodologies for risk analyses must be sought for better decisions.
- Security managers are gaining new duties including risk management, safety, background investigations, and travel security.
- Many security departments are facing either placement under another department (e.g., facilities management), outsourcing, or even elimination.
- The market for security services and systems will continue to grow globally (see Chapter 2). Although many businesses and institutions will maintain proprietary security departments, we will see an increase in the outsourcing of partial and entire security departments. Service firms will continue to provide security officers, investigators, integrated systems, and cybersecurity, among other services.
- Numerous specializations will grow in importance. Examples are risk management, resilience, business continuity planning, emergency management, life safety, fire protection, homeland security, protection of critical infrastructures, border security, cybersecurity, fraud prevention and investigation, workplace violence prevention, human resources protection, and information security.
- The **cycle of protection** will remain: as new technology is developed, offenders will exploit it, and security specialists and offenders will remain in constant competition—one group striving to protect, the other striving to circumvent defenses. Both sides will win "battles," but neither will win the "war."
- Cybercrime will increase. The number of cybercrime cases reported to police will also increase and require additional resources and training for police, prosecutors, and judges.

"The largest computer crime problem affecting local law enforcement representing the largest number of victims and the largest monetary loss will be Internet fraud, including fraud via identity theft." Virtual crimes will require new criminal and civil laws and new methodologies for prevention and investigation. Organized crime and terrorist groups will increasingly become involved in cybercrime (Taylor et al., 2006: 354–383).

- Public police, especially on the local level, will continue to lag behind the technical expertise of cybercriminals. Consequently, the private sector will continue to fill the void.
- Information security will increase in intensity. Taylor et al., (2006: 374) write: "The character of espionage will continue to broaden into the arenas of information warfare, economic espionage, and theft of intellectual property."
- Chief Security Officers and Chief Information Security Officers will become increasingly involved in risk management and the protection of information. The drivers of these trends are legal compliance requirements to protect information and the general threat landscape (Griffin 2012).
- The market for **zero-day exploits** will continue to grow. These are previously unknown vulnerabilities in computer software that are exploited on the first day of being discovered and exploited until defenders fix the problem. The "good guys" may not know (and never know) the vulnerability exists, so the zero-day exploit can become a valuable commodity sold to the highest bidder or retained. In addition, "good guys" (e.g., a company or government agency) may purchase or discover and retain a zero-day exploit, striving to keep it a secret and unpatched to collect information on the unsuspecting. The problem extends to programmers who design a flaw into software and then sell information on the flaw (Schneier, 2012).
- The U.S. federal government is promoting global Internet freedom, online access, mobile phone applications, and social media, especially across the developing world (U.S. Department of State, 2012a). Although such goals strengthen U.S. foreign policy, offenders will exploit technology.
- As we become a cashless society, offenders will be ready to gain illegally from system weaknesses.
- E-business will increase along with protection needs.
- Senior management in criminal organizations is becoming increasingly well educated and trained. Criminals increasingly will have college degrees and experience in IT, engineering, money management, investments, accounting, and law. In addition, offenders will use just about all the technology available to government and businesses.
- **Criminal entrepreneurs**, who seek illicit profits from a wide variety of enterprises and launder the proceeds, are becoming increasingly knowledgeable of the operations of financial institutions and related vulnerabilities. They will continue to infiltrate and manipulate financial institutions for their own purposes (e.g., money laundering), as this target becomes a top priority. Financial security professionals will continue to be challenged not only by external offenders but by internal ones as well.
- Counterfeiting will continue to be a huge business for organized criminals, especially of designer clothes, software, entertainment items, vehicle parts, and medical supplies. Counterfeiters have production facilities and distribution networks ready for new product lines.

- Twenty-first century crime groups will own more and more shares of multinational corporations and be involved in management decisions. This will create new challenges for security professionals.
- The U.S. Department of State (2012b) promotes "Citizen Security Partnerships" in other countries to curb insecurity and violent crime while promoting effective government, the rule of law, and economic growth. These programs recognize "transnational threats that blur the lines between crime, terrorism, and military confrontation." At the same time, transnational, local, and white-collar (e.g., extortion) crimes are interconnected and require comprehensive security beyond counternarcotics and counterterrorism efforts. Private sector security professionals should be aware of U.S. security programs overseas and how these programs impact corporate security programs.
- Satellites will continue to expand to assist security through instantaneous communications throughout the world. People and assets will be more easily tracked. Such technology will assist in the investigations of kidnappings and hijackings.
- Corporate and institutional changes have had an impact on employee morale. The objectives of generating profit and improving quality while downsizing, and other workplace issues, have taken a toll on loyalty among workers. Loyalty has been an asset to protection programs. Its deterrent value today must be questioned and studied. New innovative strategies are required.
- Women, the elderly, and the disadvantaged will make up a greater portion of the workforce. Such groups present new challenges for security. Women and the elderly require protection not only in the workplace, but also at home and while traveling. Cases of domestic violence, sexual harassment, and stalking are likely to occupy more of the security professional's time. Women and the elderly will be involved in more workplace crime. If increasing numbers of disadvantaged workers are employed, they may bring with them problems of gangs and illegal drugs ("Here Comes the 21st Century: What Does It Hold for Security?" 1997: 4–7).
- More research is needed on issues of minority group members and women in security positions. The majority of women in one study of women security managers felt they experienced relatively high levels of sex discrimination, sexual harassment, and on-the-job stress; they also thought they were not paid the same as men for the same work. Despite these issues in this male-dominated vocation, the study showed three-fourths of the women surveyed were satisfied with their careers ("Survey Studies Women in Security," 1996: 87–88).
- Security must adapt to a **multinational and multicultural workforce**. As the workforce changes, security professionals must be aware of diversity and use it to the advantage of the business. Good communication can improve security and business, and even help develop new markets.
- Global markets require security to be aware of each culture and related risks.
- Twenty-first century police will spend most of their resources and time curbing violent crimes, while their efforts against property crimes will take a lower priority. Consequently, the private sector will fill the void.

- More and more citizens and business people are realizing that the police have limited ability and resources to curb crime. Police are primarily "reactive"; that is, they respond to calls for service and investigate. They are often under great pressure to solve serious cases. Because of limited resources, police are not able to be more "proactive." Citizen groups and the private sector will fill the void.
- A panel of law enforcement specialists predicted that, in 2035, private security agencies will perform more than 50 percent of all law enforcement responsibilities (Tafoya, 1991: 4).
- Ritter (2006: 8) writes: "There's no question that terrorism, the growth of multicultural populations, massive migration, upheavals in age-composition demographics, technological developments, and globalization over the next three or more decades will affect the world's criminal justice systems."
- First responders (i.e., police, firefighters, and EMS) will continue to have limited resources. Businesses and institutions must prepare for emergencies.
- As the world's population increases, competition for natural resources will intensify.
- Three key factors to assist protection professionals today and in the future are a broad-based education (e.g., business, security, IT), the skill to show that protection strategies have a ROI, and the flexibility to deal with rapid change.

The Director of National Intelligence offered the following assessments of global threats (Clapper, 2012):

- Although al-Qaida remains a threat to U.S. interests, it has been weakened significantly.
- A chemical, biological, radiological or nuclear (CBRN) limited attack in the United States or against U.S. interests overseas is of concern because of such a goal by some foreign groups (e.g., al-Qaida).
- Homegrown violent extremists will be characterized by lone actors or small groups.
- The proliferation of weapons of mass destruction (WMD) is a major threat, especially from Iran and North Korea.
- Transnational organized crime (TOC) is a serious threat and includes efforts at proliferation of nuclear materials and facilitating terrorism. TOC networks are increasingly linked to state leaders and foreign intelligence service personnel.
- Mexican drug cartels are responsible for high levels of violence and corruption while contributing to instability in Central America.
- Health threats and natural disasters continue to kill and injure while having the potential to destabilize governments.
- Water problems will contribute to government instability.
- Cyber threats are a critical national and economic security concern.

The following remarks from the former Director of the CIA (Petraeus, 2012) provide direction for security professionals protecting people, assets, and information. The CIA is focused on two major objectives: a war against al-Qaida and its affiliates and the need for global intelligence coverage of topics as described above by the Director of National Intelligence. Three major challenges of our era are "the utter transparency of the digital world, the enormous task of processing so-called Big Data [i.e., large and complex data sets], and the ever-greater need for speed." These challenges require a re-thinking of identity and secrecy. "Data is created constantly, often unknowingly and without permission." This leads to "information about location, habits, and, by extrapolation, intent and probable behavior." Thus, there are enormous intelligence opportunities and huge counterintelligence challenges. Two specific challenges for CIA officers are how to protect their identity when they have a digital footprint from birth through social media, for example, that can be accessed for decades and how to create a digital footprint for new identities for some officers. Petraeus goes on to say that we are moving from the "Internet of PCs" to the "Internet of Things"—all types of devices. Whereas machines in the 19th century learned to do and those in the 20th century learned to think, the machines in the 21st century sense and respond. Such progress is driven by devices that tag, sensors and wireless sensor networks, embedded systems that think and evaluate, and nanotechnology that produces devices small enough to operate almost anywhere.

The United States Commission on National Security/21st Century (1999: 1–8), commonly known as the Hart-Rudman Task Force on Homeland Security, produced a series of reports to meet the challenges of 2025. The following list is from the commission and focuses on trends in the future.

- Institutions designed for another age may not be appropriate for the future.
- Authoritarian regimes will increasingly collapse as they try to insulate their populations from free-flowing information and new economic opportunities.
- An economically strong United States is likely to remain a primary political, military, and cultural force in the world through 2025.
- Adversaries will resort to forms and levels of violence shocking to our sensibilities. Americans will likely die on American soil, possibly in large numbers.
- Emerging technologies, such as advances in biotechnology, will create new moral, cultural, and economic divisions and an anti-technology backlash.
- Energy, especially fossil fuel, will continue to have major strategic significance.
- Minorities will be less likely to tolerate prejudicial government. Consequently, new states, international protectorates, and zones of autonomy will be born in violence.
- Space will become a competitive military environment.
- Excellent U.S. intelligence will not prevent all surprises.

What trends do you anticipate in the security and loss prevention profession?

Five Wars of Globalization

As in the past, the private sector, as well as the public sector, will continue in the future to contend with what Naim (2003) calls the "**five wars of globalization**." These are the illegal trade in drugs, arms, intellectual property, people, and money. He emphasizes that religious zeal or political goals drive terrorists, whereas profit drives the other wars, and all of the wars result in murder, mayhem, and global insecurity. According to Naim, governments have been fighting the five wars for centuries and losing them. He does not include terrorism in this statement. Naim explains why governments cannot win the five wars of globalization:

- Criminal cartels can manipulate weak governments by corrupting politicians and police.
- International law, including embargoes, sanctions, and conventions, offer criminals opportunities to profit from illegal goods.
- Al-Qaida members are stateless and so are criminal networks involved in the five wars. Whereas terrorists and other criminals can seek refuge in and take advantage of porous borders, traditional notions of sovereignty frustrate governments.
- These wars pit governments against **market forces**. Thousands of independent, stateless organizations are motivated by large profits gained by exploiting international price differentials, unsatisfied demand, or the cost advantages resulting from theft (i.e., no cost to produce the product).
- These wars pit bureaucracies against networks. The same network that smuggles illegal drugs may be involved in counterfeit watches that are sold on the streets of Manhattan by illegal immigrants. Highly **decentralized networks** can act swiftly and flexibly, while often lacking a headquarters and central leadership to be targeted. Governments often meet the challenge by forming task forces or creating new bureaucracies.

Naim claims that governments may never be able to eliminate the international trade from the five wars, but they can and should do better. He offers four areas capable of producing ideas to meet the challenges from these wars:

- Negotiate more flexible notions of sovereignty. Since stateless networks regularly violate laws and cross borders to trade illegally, nations should develop agreements to "manage" sovereignty to combat criminal networks.
- Naim calls for stronger multilateral institutions (e.g., multinational police efforts such as Interpol). However, nations do not trust each other, some assume that criminals have infiltrated the police agencies of other countries, and today's allies may become tomorrow's enemies.
- The five wars render obsolete many institutions, legal frameworks, military doctrines, weapons systems, and law enforcement techniques that have been applied for numerous years. Rethinking and adaptation are needed to develop, for instance, new concepts of war "fronts" defined by geography and new functions for intelligence agents, soldiers, and enforcement officers.
- In all five wars, government agencies battle networks motivated by profits created by other government agencies that create an imbalance between demand and supply that makes prices rise and profits skyrocket. Since beating market forces is next to impossible, reality may force governments, in certain illegal markets, to change from repressing the markets to regulating them. In addition, creating market incentives may be better than creating bureaucracies to curb the excesses of markets. In certain instances, technology can possibly be used to replace government policies (e.g., encryption to protect software on CDs).

What can businesses and protection programs do to prevent and reduce losses from the "wars of globalization"?

In *40 For the Next 40*, Toffler Associates (2010) offers the following predictions:

- Private sector actors, NGOs [non-governmental organizations], religious groups, "hyper-empowered" individuals whose resources can exceed those of states, and a wide range of transnational networks—both licit and illicit—will create a radically different future.
- Open, collaborative partnering among entities and "problem-solver-networks" will create opportunities to grow innovation beyond internal research and development. [This will result in information security challenges.]
- Biotechnology and bio-implants will create enhanced human performance.
- Nanotechnology will have an enormous impact on our world and produce cheaper, smaller, and smarter surveillance devices that will result in unparalleled invasion of privacy.
- "Cyberdust" (i.e., data collected but not used) will become a more serious problem for intelligence agencies unless analysis of the enormous amounts of data (e.g., open-source information, video, and unclassified and classified information) can be automated rather than human-driven.
- Customers will be the main source of innovation within organizations.
- Organizations will find it impossible to hide improper activity because of the information explosion and readily available tools for information analysis. Consumer opinions of businesses will influence purchasing.
- Corporate global mobility will intensify. Businesses will become more adept at moving operations among countries.
- Energy resources will produce strength in nations during economic warfare.
- Information will continue to be a major asset, but it often has a limited shelf life due to the likelihood of it becoming obsolete because of accelerated change, resulting in costly degraded decisions.
- Climate change will impact many aspects of our world.
- Water filtration technology will advance, satisfy needs for drinkable water, and reduce conflict among nations.

Education

How relevant is a college education to the loss prevention careerist? A college degree will not guarantee a job or advancement opportunities. However, with a college degree, a person has improved chances of obtaining a favored position. If two equally experienced people are vying for the same professional loss prevention position and person A has a college degree while person B does not, person A will probably get the job. Of course, other characteristics of a person's background will improve job opportunities. Education and experience are top considerations. Training also is important. Two other characteristics are personality and common sense. As used here, *personality* pertains to one's ability to get along with others. Many consider this factor to be half of a person's job. *Common sense* is a subjective term that is used widely. It refers to an analysis of a situation that produces the "best" solution that most people would favor.

Security and Loss Prevention Education: Today and Tomorrow

Although business degree programs are an excellent location for security and loss prevention courses, the criminal justice (CRJ) degree programs at hundreds of campuses in the United States today are the greatest driving forces behind security and loss prevention courses and academic programs. Many institutions offer criminal justice degrees with courses in security, loss prevention, or a related discipline. In addition, several institutions maintain distinct departments focusing on security-related degree and certificate programs.

A check of the ASIS International (2010) website in 2012 under "Academic Institutions Offering Degrees and/or Courses in Security" shows many academic programs under a variety of names such as Private Security/Loss Prevention Management, Business/Organizational Security, Security Management, Protection Management, Global Security, Management of Personal Protection, Information Security, Security Technology and Intelligence, Security Studies, and Sport Security Management, among others. There are also courses such as Risk Management, Homeland Security, Emergency Management, and Safety.

Adolf (2011: 124–134) discusses the many challenges facing security studies. These include where to place security academic programs in institutions of higher education and three specific needs: more academic programs in security, more research, and accreditation of programs to maintain consistency and standards. Adolf supports his points by writing that businesses spend more on protection than do public police and that private security is now the primary protective resource for Americans. His research shows only four security academic programs within business programs and a small number of doctoral programs in security, which hinders expanded research. Adolf does write that security-related research emanates from students and educators from other fields such as criminal justice, business, and education.

During the late 1960s, many police science degree programs advanced to law enforcement and then to CRJ degree programs. As CRJ programs developed, so did the spectrum of course offerings: from primarily narrow police science courses to courses relating to the entire justice system (i.e., police, courts, corrections). CRJ programs became multidisciplinary because the answers to complex problems are more forthcoming from a broader spectrum of study. Likewise, security and loss prevention academic programs will evolve into broader-based, multidisciplinary programs and include the study of business, risk management, cybersecurity, terrorism, homeland security, infrastructure security, emergency management, resilience, intelligence, supply chain security, and green security.

ASIS International has sponsored an annual academic/practitioner symposium since 1997 to foster communications between educators and practitioners. These symposiums have produced core competencies, course outlines, undergraduate and graduate curriculum models, accreditation criteria, and directions for research, among other products. The ASIS website provides access to such publications.

Another source for competencies is the U.S. Department of Homeland Security (2012) publication titled "Security Specialist Competencies: An Interagency Security Committee Guideline." These competencies are designed for training federal government security specialists protecting nonmilitary federal facilities. Rather than each federal agency designing its own training program, this publication seeks uniformity and consistency. Topics include security assessments, physical security, CPTED, personnel security, information security, emergencies, and report writing.

Academic Programs Following the 9/11 Attacks

The September 11, 2001 terrorist attacks had a profound impact on many academic programs. Certain programs were rushed to market, proliferated, and/or morphed. As events unfolded, academic programs strived to meet the needs of a nation in shock.

Here we discuss the following academic programs: *homeland security, emergency management, security operations,* and *strategic security*. Each academic program, and others, including security and loss prevention, is continuously improving in quality while struggling with these questions: What should be taught? What are the needs of students, employers, and the profession? How do online courses compare to in-class courses in terms of quality of instruction, cheating, the knowledge and skill-sets acquired by students, employer hiring decisions, and meeting job requirements? The following discussion does not answer these questions for each academic program; the discussion here just scratches the surface. The issues are complex and there is a serious need to answer these questions and others through these activities: research, research, and more research!

Academic programs in emergency management (EM) have a longer history than do academic programs in homeland security (HS). Both are evolving with limited uniformity. Although a major source of direction for HS education and training is the National Strategy for Homeland Security, future catastrophic events, challenges, legislation, and government policy will influence curriculum design (Purpura, 2007: 452–454).

McCreight (2009) has a lot to say about the challenges of HS and EM academic programs. He writes of a general lack of consensus among educators and professionals concerning core curriculums, educational requirements, and education delivery methods. McCreight argues that DHS, FEMA, and professional associations must do more to produce standards for HS and EM academic programs. As a major issue, McCreight sees the need for "reconciliation" between HS and EM. He sees the former focused on preparedness and prevention to confront terrorism, while the latter addresses all-hazards. McCreight refers to "bridging these differences" and designing a "coherent curriculum."

Donahue et al. (2010), in a response to McCreight, argue that the "reconciliation" has already begun in practice; many state and local jurisdictions have centralized HS and EM into one authority. Donahue et al., see an issue in higher education of accommodating this integration.

The Federal Emergency Management Agency (2012) website lists many higher education programs in both HS and EM. The programs range from certificate to doctoral levels. *The future holds the challenge of whether educators can modify "their" educational programs to meet the needs of the homeland security and emergency management integration occurring in the field.*

The Office of the Director of National Intelligence (ODNI), which oversees sixteen government agencies involved in the collection of intelligence, in cooperation with the US Navy, has awarded grants to universities to develop bachelor degree programs in **security operations**. This effort seeks to produce graduates to fill dire workforce needs in the security and intelligence community. The ODNI Special Security Center (2010: 2) notes that "the security professional is an indispensable member of every government, contractor industry, commercial, and private organization." The ODNI curriculum guidelines emphasize a broad-based education focusing on problem solving and critical thinking skills in all courses. Besides security courses, and courses focusing on the government and private sector classified environment, the guidelines reflect multi-disciplinary courses in social science, communications, mathematics, statistics, and business. The guidelines bar online courses at this time. As these degree programs develop and proliferate, we will see a definition of security operations and model curricula. Eastern Kentucky University and Embry-Riddle Aeronautical University are developing such programs.

Dr. Sheldon Greaves (2008: 7–19), co-founder of Henley-Putnam University, explains the challenges of defining **strategic security** and its curriculum. He defines "security" as an activity and "ongoing behaviors designed to forestall reasonable or possible threats." The threats are viewed as very broad, including natural disasters, crime, social upheavals, and political corruption. The three major sub-disciplines of strategic security—intelligence, counterterrorism, and personal protection—are applied to "provide the right information to the right policy makers or decision makers so that they may act in our collective best interests based upon the facts as they are seen at that moment."

The strategic security curriculum is multi-disciplinary, drawing on history, science, humanities, language, politics, mathematics, and economics, among other fields. Greaves answers why strategic security, specifically intelligence, is a unique field of study. He refers to **tradecraft**, meaning skills developed in a clandestine vocation. Examples are methods used in espionage. Greaves adds that in the intelligence field, standard hiring practices involve the skill of "recruiting an asset" and, rather than preparing a research proposal, an "intelligence requirement" is created.

Greaves argues that the creation of the office of the Director of National Intelligence has increased the demand for intelligence specialists. In addition, a new generation of specialists is needed to replace retirees from the whole intelligence community.

Greaves notes multiple challenges in the development of strategic security curricula. These include the need for educational standards that currently face differences and gaps in training among agencies within the intelligence community. Another challenge is the requirement of secrecy in the intelligence vocation that inhibits academic discussion. Also, the lack of textbooks and standardized materials is a void that needs to be filled.

Dr. Robert Clark (2008: 1–6), a professor at Henley-Putnam University and a former analyst with the C.I.A., poses two key questions in reference to intelligence education and training: "What should we be teaching in universities?" "What should we leave to the intelligence community as training?" To begin with, Clark refers to the five core competencies for personnel to effectively function in the intelligence community, issued by the ODNI. These are critical thinking, communications, accountability for results, personal leadership and integrity, and engagement and collaboration. Academic institutions can provide the basics of these core competencies and many other subjects, according to Clark. He also notes that academic institutions offer introductions to counterintelligence, counterproliferation, counterterrorism, and counternarcotics. However, he emphasizes that case studies illustrating course content are largely classified, and, thus, learned in intelligence community training and on the job. He adds that the detail on cyber operations is, likewise, classified. Clark writes that an important contribution of academia is to inspire students to seek an intelligence career.

Academic programs in security operations and strategic security are needed and will likely proliferate. Although these programs show similarities and differences, it remains to be seen what content educators will design into courses and whether these programs will merge or become separate and distinct.

In reference to the academic programs explained above, we can see various specializations that meet unique needs. At the same time, security and loss prevention knowledge and skill sets likewise fulfill unique needs by protecting people and assets at businesses, institutions, and many other organizations as described in Chapter 2. For the reader, broad course selection and skill sets provide a foundation for improved and wider employment opportunities. What if, following graduation with a degree in security and loss prevention, no jobs are available for the applicant? Previous courses in HS, EM, and intelligence may provide opportunities. Likewise, if a degree in HS, EM, or a related field does not produce results in a job search in those fields, a transcript showing courses in security and loss prevention may enhance the job search.

Research

The criminal justice profession is one source for research direction for the security and loss prevention profession. The following perspectives on criminal justice research are verbatim from the *National Institute of Justice Journal* (Ritter, 2006: 8–11).

> *David Weisburd [University of Maryland] believes that the nature of criminal justice in 2040 will depend in large part on the primary research methodology. Is the criminal justice community better served by relying on the experiences and opinions of practitioners (the* **clinical experience model***) or by research that tests programs and measures outcomes (the* **evidence-based model***)?*
>
> *Currently, the clinical experience model is the research path most frequently followed. Policies and technologies are based primarily on reports from practitioners about what they have found to work or not work. Sharing approaches and programs that seemed to work in one community with another community allows for quick application of successful ideas. The downside of this model is that a program may be widely adopted before scientific research demonstrates its efficacy in more than one place or application. For example, in one youth program aimed at reducing delinquency, counselors and parents believed that the treatments were effective, based on initial measures of success. However, subsequent evaluation revealed that participation in the program actually increased the risk of delinquency.*

In the evidence-based model, a new program undergoes systematic research and evaluation before it is widely adopted. Now dominant in medicine—and becoming more popular in other areas such as education—the evidence-based model has been used successfully in criminal justice. For example, hot-spot policing (a policy adopted in the early 1990s that focuses police resources in high-crime areas) was preceded by studies that demonstrated its effectiveness.

But the evidence-based model also has shortcomings. Research requires a large investment of time and money, and many practitioners understandably would rather spend resources implementing an innovation than wait for confirming research. Time—always a precious commodity for policymakers and practitioners—can be a particularly frustrating component of the evidence-based model. Credible research requires time to adequately test an approach, often in more than one jurisdiction, before communities can adopt it on a large scale.

"Policymakers want to improve things while they have the power," Weisburd says. "They are under pressure to make an impact—so there is tension between the slowness of the evidence-based process and the pressure to move quickly."

Weisburd proposes making the evidence-based model "more realistic." He believes this can be done by

- Streamlining the process of developing evidence and conducting evaluations.
- Building an infrastructure to ensure that studies do not reinvent the wheel.
- Devising methods for getting studies off the ground faster, such as encouraging funders to help in the development of high-quality randomized experimental studies.
- Reinforcing a culture that emphasizes the exploration of which programs and practices do and do not work.

Weisburd also argues that federal investment in the scientific evaluation of new practices and programs must be increased. Researchers and practitioners must insist that "if you want us to make intelligent policy and not waste money by prematurely innovating in hundreds of departments, you must give us more money."

In what ways can the perspectives of Weisburd be applied to research in the security and loss prevention profession?

The most practical question for security and loss prevention researchers to address is, what strategies are best to prevent and reduce losses from crimes, fires, accidents, and other threats and vulnerabilities? Ongoing evaluative research is instrumental in strengthening successful strategies while eliminating those that are less useful.

Additional directions for research include protection against terrorism, WMD, protection of critical infrastructures and "soft targets," cybersecurity, security metrics, model training programs, determination of the most appropriate courses for relevant degree programs, effective job applicant screening, model statutes for licensing and regulation of the security industry, models for regulatory bodies, criteria for the selection of security services and systems, evaluation of services and systems, legal issues and liabilities, strategies to improve public–private sector cooperation, privatization, private justice system, a centralized data bank for compiling loss statistics, the feasibility of tax deductions for implementing loss prevention strategies by a variety of entities and residential settings, and the use and effect of robots and other technological innovations.

As well as college and university criminal justice and security programs sharing these research questions, other curricula are capable of providing valuable input. They include business, risk management, insurance, homeland security, emergency management, intelligence, safety, fire protection, public health and medicine, architecture, landscape architecture, and engineering degree programs. A multidisciplinary research effort is most beneficial. Without adequate research, loss prevention practitioners will be hindered in their decision-making roles.

Noted researchers who push a research agenda include Dr. Read Hayes and his team, who are affiliated with the University of Florida. They focus on evidence-based loss prevention strategies. Dr. Roger G. Johnston and his team from Argonne National Laboratory offer excellent articles in *The Journal of Physical Security*; these researchers report on vulnerabilities of physical security systems and how these systems operate in the field. Many other noted researchers and teams conduct studies. A portion of these studies are unable to be published because of such reasons as security or the information being proprietary.

ASIS International (2011), through its foundation, has been a source of research on many topics such as threats, best practices, and regulations. The research seeks "actionable knowledge for the security profession."

Universities have been actively fulfilling the research needs of homeland security and defense. Funding for university research for homeland security and defense is provided by federal departments (e.g., DHS, DOD) and agencies (e.g., CIA, EPA), the private sector, and state governments.

The U.S. Department of Homeland Security, Science and Technology Directorate (2012) is involved in numerous research initiatives. It partners with other government agencies, universities, and industries to find and develop innovative ideas. Major areas of focus are protection against catastrophic terrorism and response to disasters that could result in large-scale loss of life and major economic harm.

Homeland Security Presidential Directive-7 (HSPD-7) calls for a national critical infrastructure protection research and development (R&D) plan. *The National Plan for Research and Development in Support of Critical Infrastructure Protection* (Executive Office of the President, Office of Science and Technology Policy and The Department of Homeland Security Science and Technology Directorate, 2004) is essentially a roadmap for investment and protection that integrates cyber, physical, and human elements. This R&D plan was developed in coordination with the *National Infrastructure Protection Plan.*

Through the Homeland Security Act of 2002, Congress mandated the DHS to enhance U.S. leadership in science and technology aimed at homeland security issues. This goal is supported through DHS funding of undergraduate and graduate fellowships and scholarships, and establishing **DHS Centers of Excellence** (HS-Centers). HS-Centers (i.e., universities) are overseen by the Office of University Programs within the Science & Technology Directorate and are designed to create learning and research environments that bring together experts who focus on numerous topics including WMD, risk analysis related to the economic consequences of terrorism, infrastructure protection, biological threats and diseases, agro-security, and the behavioral aspects of terrorism and its consequences.

Larger sources of research on homeland security issues are the more than 700 federally funded research and development laboratories and centers throughout the United States. These sites include federal labs, college and university labs, and private industry labs. Each year billions of dollars support homeland security R&D.

What directions for research can you suggest for the security and loss prevention profession?

Training

The future direction of training largely depends on organizational needs and objectives, threats and vulnerabilities, technological innovations, and loss prevention strategies. Practitioners will need to know how to operate in an environment of new countermeasures and complex systems. But even though loss prevention will change, many key topics taught in the training programs of today will also be taught tomorrow. Strategies against crimes, fires, accidents, and disasters will still be at the heart of training programs.

Research by D'Addario et al. (2011: 48), showed "there is a wide gap in the transfer of valuable knowledge to new and advancing security leaders." They see the next generation of security leaders as having to begin anew, rather than building upon their predecessor's work. Thus, training and mentoring are necessary to maintain continuity of professional security operations.

Government involvement in training proprietary and contract service practitioners will probably increase for greater public safety. Uniformity in training will ensure that practitioners receive adequate information on basic topics such as laws of arrest, search and seizure, use of force, weapons, fire protection, safety, and emergency response. An interdisciplinary group would be the best choice to provide input for government-mandated training programs.

The training curriculum of tomorrow will continue to cater to the ever-increasing goal of professionalizing security and loss prevention personnel. The end product will be a thoroughly knowledgeable and skilled practitioner, able to provide a useful service to the community. Greater mutual respect will follow as public police and private-sector personnel share similar characteristics in education, training, salary, and professionalism.

Improved training helps to prevent costly liability suits caused by, for example, excessive force used by a uniformed officer. The quality of training, in terms of duration, intensity, and topics covered, is a prime consideration in such liability cases.

Programmed, self-paced instruction with the aid of computers will increase. This type of instruction is cost-effective, especially when an organization is faced with the challenge of turnover. Furthermore, new employees who learn quickly can move through the instructional program without having to wait for slower students. A varied training program may include not only programmed, self-paced instruction, but also lectures, audiovisual productions, role-playing, and demonstrations. **Distance learning online** will increase in popularity as students learn from remote locations.

The Internet shows numerous sources for training. The International Foundation for Protection Officers is an excellent source of training and certifications.

The Concept of the Security Institute

Every state should maintain a security institute, preferably at a college or university. Three goals of a security institute are as follows:

1. Develop and conduct college and continuing education courses based on the needs of customers.
2. Work with the state government agency that regulates the private security industry to develop training programs, conduct research, and professionalize the field.
3. Offer the institute's service area crime prevention initiatives involving security and criminal justice college students, who are an unrealized and underutilized national resource to control crime.

Employment

Many employment opportunities can be found in the security and loss prevention profession. Chapter 2 provides information on this industry and career specializations. The related profession of criminal justice, which includes police, courts, corrections, probation, and parole, plus the growth of privatization, provides opportunities and specializations. Here again, if the reader enrolls in a broad spectrum of courses (e.g., HS, EM, information security, cybersecurity, and intelligence), employment prospects are improved.

Entry-level security officer positions vary widely in pay, benefits, and training. Generally, these positions do not pay as well as public police. Conversely, depending on the locale, nuclear security and hospital security officers, for example, are paid comparable to public police. Entry-level, part-time security officer positions are available in many locales, the hours offer some flexibility, and the duties are less risky than public policing. Many college students gain valuable experience as security officers or loss prevention agents.

Supervisors in security have mastered the tasks of security officers, possess broader skills, and have good human relations qualities. Managers generally have more education, training, experience, and responsibilities than supervisors do. They are involved in planning, budgeting, organizing, marketing, recruiting, directing, and controlling.

Other careers in this profession include sales, self-employment (e.g., private investigations, consulting), and government security. The career boxes in previous chapters, courtesy of ASIS International, offer further information. In addition, refer to the *Occupational Outlook*

Handbook [search for "Protective Service"] published each year (online) by the U.S. Bureau of Labor Statistics (2012).

Sources of Employment Information

- *Online services:* The Internet is a major source of employment information. There are several advantages: offerings are broad, time is saved by reviewing opportunities open worldwide, it is more up-to-date than most publications, and you can respond electronically. Be aware that confidentiality is limited. Look for services that are free.
- *Social networking websites:* These sites offer an opportunity to let others know of your career interests and that you are seeking employment. Caution is advised when communicating with unfamiliar people.
- *Traditional networking:* This type of networking consists of communicating with people the applicant knows from past educational or employment experiences. It is a good idea for any professional to maintain contact, however slight, with peers. When employment-related challenges develop and solutions are difficult to obtain, networking may be an avenue for answers. Likewise, this mutual assistance is applicable to employment searches.
- *Periodicals:* Within these sources, trends in employment and employment opportunities are common topics.
- *Professional associations:* Professional organizations serve members through educational programs, publications, a variety of strategies aimed at increasing professionalism, and job listings.
- *Trade conferences:* People with a common interest attend trade conferences, which are advertised in trade publications. By attending these conferences, a practitioner or student can learn about a variety of topics. The latest technology and professional seminars are typical features. Trade conferences provide an opportunity to meet practitioners who may be knowledgeable about employment opportunities or are actively seeking qualified people.
- *Educational institutions:* Degree programs are another source of employment information. Usually, these programs have bulletin boards near faculty offices that contain career opportunities. College or university placement services and faculty members are other valuable sources.
- *Libraries:* A wealth of information for a career search is available at libraries. Examples include periodicals, newspapers, telephone books, directories, and books on career strategies.
- *Government buildings:* Public buildings often contain bulletin boards that specify employment news, especially near human resources offices.
- *Newspapers:* By looking under "security" in classified sections, one can find listings for entry-level positions. In large urban newspapers, more specialized positions are listed.
- *Public employment agencies:* These agency offices operate in conjunction with the U.S. Employment Service of the Department of Labor. Personnel will provide employment information without cost and actually contact recruiters or employers.

- *Telephone books:* As an alternative to online searches, the addresses and telephone numbers of both private and public entities are abundant in telephone books. By looking up "security," "guards," "investigators," and "government offices," one can develop a list of locations for possible employment opportunities.
- *Private employment agencies:* Almost all urban areas have private employment agencies that charge a fee. This source will probably be a last resort, and one should carefully study financial stipulations. At times, these agencies have fee-paid jobs, which means that the employer pays the fee.

Career Advice

1. Read at least two current books on searching for employment and careers; both the novice and the experienced person will learn or be reminded of many excellent tips that will "polish" the career search and instill confidence.
2. Begin your search by first focusing on yourself—your abilities and background, your likes and dislikes, and your personality and people skills.
3. If you are new to the security and loss prevention profession, aim to "get your foot in the door." As a college student, look for an intern position, work part-time, or do volunteer work. *Aim to graduate with experience.* If you are a retired government employee, or beginning a new career in security, market and transfer your accumulated skills and experience to this profession.
4. Most people will enter new careers several times in their lives.
5. Searching for a career opportunity requires planning, patience, and perseverance. Rejections are a typical part of every search. A positive attitude will make or break your career.
6. When planning elective courses in college or when training opportunities arise, consider that employers want people with skills (e.g., writing, speaking, interviewing, information technology).
7. Many students do not realize that a college education and some training programs teach the student *how to learn.* Although many bits of information studied and reproduced on examinations are forgotten months after being tested, the skills of how to study information, how to read a textbook or article, how to critically think and question, how to do research and solve problems, and how to prepare and present a report are skills for life that are repeated over and over in one's professional career. In our quickly changing information age, these skills are invaluable.
8. If you are fortunate enough to have a choice among positions, do not let salary be the only factor in your decision. Think about career potential and advancement, content of the work, free training, benefits, travel, and equipment.
9. Exercise due diligence on potential employers. For example, speak to present employees and check the organization's financial health.
10. Five key factors influencing an individual's chances for promotion are education, training, professional development and certification, experience, and personality.
11. Avoid quitting a job before you find another. Cultivate references, even in jobs you dislike; such jobs are learning experiences (Purpura, 1997: 366–384).

The employment situation in the security and loss prevention profession reflects a bright future. Good luck with your career!

Search the Internet

Here are websites relevant to this chapter:

ASIS International: www.asisonline.org
Central Intelligence Agency: www.cia.gov
CriminalJusticeJobsHelp.com: www.criminaljusticejobshelp.com
FEMA Emergency Management Institute: www.training.fema.gov/emiweb/edu
International Foundation for Protection Officers: www.ifpo.org
Law Enforcement Jobs: www.lawenforcementjobs.com
Loss Prevention Foundation: www.losspreventionfoundation.org
LPJobs.com: www.lpjobs.com
Nation Job Network: www.nationjob.com/security
U.S. Department of Homeland Security, Centers of Excellence: www.dhs.gov/files/programs/editorial_0498.shtm
U.S. Department of Homeland Security, Science and Technology Directorate: www.dhs.gov/xabout/structure/editorial_0530.shtm
U.S. Department of Labor, Occupational Outlook Handbook: www.bls.gov/oco
World Future Society: www.wfs.org

Case Problems

19A. As a contract security supervisor with a major security service firm, you are faced with a major career decision. You have a bachelor's degree and have been with your present employer for a total of seven years: two years part-time while in college, and five years full-time since graduating. These years have been spent at a hospital where you would like to advance to site security manager, but you face competition from one other supervisor, who also has a bachelor's degree, the same certifications and training that you possess, and about the same years of experience. The present contract site security manager is a former detective who wants to retire within the next year or two. You have repeatedly asked managers from the security service firm (i.e., your employer) about advancement opportunities, but nothing is available. Recently, you were offered the position of in-house security and safety training officer with a nearby urban school district. The pay is $800 less annually than what you are earning now. However, you would receive additional insurance and retirement benefits and work only day shifts, Monday through Friday. You would be the only college-educated officer on the school district security force, which is primarily composed of contract officers. The director of security is the only other in-house security officer, a retired police officer with no college degree. What career choice do you make?

19B. As a regional loss prevention manager with a major retailer, you have 12 years of retail loss prevention experience. You earned bachelor's and master's degrees and a CPP. The territory you work covers nearly 100 stores in six southeastern states. You are on the road each day responding to challenges and wish you could spend more time with your spouse and two children. After eight years with the same company, and no chance for advancement, you decide to consider an offer as loss prevention director with a major retailer in the New York City area. The pay is $10,000 more than what you are earning now, with similar benefits. You would have additional responsibilities, but you would travel much less. Your spouse and children do not want to move because of family and friends. What career choice do you make?

References

Adolf, D. (2011). Security studies and higher education. *Journal of Applied Security Research, 6* (January–March).

ASIS International. (2011). About the Foundation. <www.asisonline.org> retrieved June 15, 2012.

ASIS International. (2010). Academic Institutions Offering Degrees and/or Courses in Security. <www.asisonline.org> retrieved June 14, 2012.

Campbell, G. (2007). Security metrics in context. *Security Technology & Design, 17* (March).

Clapper, J. (2012). Unclassified Statement for the Record on the Worldwide Threat Assessment of the U.S. Intelligence Community for the Senate Select Committee on Intelligence (January 31). <www.dni.gov/testimonies/20120131_testimony_ata.pdf> retrieved May 22, 2012.

Clark, R. (2008). Dividing up intelligence education. *Journal of Strategic Security, 1* (November).

D'Addario, F., et al. (2011). What will security look like in 2020? *Security Technology Executive, 21* (October).

Donahue, D., et al. (2010). Meeting educational challenges in homeland security and emergency management. *Journal of Homeland Security and Emergency Management, 7*.

Executive Office of the President, Office of Science and Technology Policy and The Department of Homeland Security Science and Technology Directorate. (2004). *The National Plan for Research and Development in Support of Critical Infrastructure Protection*. <www.dhs.gov/xlibrary/assets/ST_2004_NCIP_RD_PlanFINALApr05.pdf> retrieved April 1, 2007.

Federal Emergency Management Agency. (2012). The College List. training.fema.gov, retrieved June 14, 2012.

Greaves, S. (2008). Strategic security as a new academic discipline. *Journal of Strategic Security, 1* (November).

Griffin, J. (2012). Roles of information security executives changing. *Security InfoWatch* (June 5) <www.securityinfowatch.com> June 11, 2012.

Harwood, M. (2011). Looking ahead: Technology and threats. *Security Management, 55* (September).

Here Comes the 21st Century: What Does It Hold for Security? (1997). *Security Management Bulletin* (March 25).

McCreight, R. (2009). Educational challenges in homeland security and emergency management. *Journal of Homeland Security and Emergency Management, 6*.

Naim, M. (2003). The five wars of globalization. *Foreign Policy* (January/February).

Office of the Director of National Intelligence, Special Security Center. (2010). Security Operations 2010: Curriculum and Academic Certification Guidelines for Undergraduate Degree Programs in Security Operations. e-commerce.sscno.nmci.navy.mil retrieved February 18, 2011.

Petraeus, D. (2012). Remarks by Director David H. Petraeus at In-Q-Tel CEO Summit (March 1). <www.cia.gov> retrieved June 5, 2012.

Purpura, P. (1997). *Criminal justice: An introduction*. Boston: Butterworth-Heinemann Pub.

Purpura, P. (2007). *Terrorism and homeland security: An introduction with applications*. Burlington, MA: Elsevier Butterworth-Heinemann.

Ritter, N. (2006). Preparing for the future: Criminal justice in 2040. *NIJ Journal* (November).

Schneier, B. (2012). The vulnerabilities market and the future of security. *Forbes* (May 5). <www.forbes.com> retrieved June 4, 2012.

Survey Studies Women in Security. (1996). *Security Management* (February).

Tafoya, W. (1991). The future of law enforcement? A chronology of events. *C J International, 7* (May–June).

Taylor, R., et al. (2006). *Digital crime and digital terrorism*. Upper Saddle River, NJ: Pearson Prentice Hall.

Toffler Associates. (2010). *40 For the Next 40: A Sampling of the Drivers of Change That Will Shape Our World between Now and 2050*. <www.toffler.com> retrieved June 6, 2012.

U.S. Bureau of Labor Statistics (2012). *Occupational Outlook Handbook, Protective Service Occupations* (March 29). <www.bls.gov/ooh/protective-service/home.htm> retrieved June 15, 2012.

United States Commission on National Security/21st Century. (1999). *New World Coming: American Security in the 21st Century* (September 15). <www.fas.org/man/docs/nwc/nwc.htm> retrieved March 31, 2007.

U.S. Department of Homeland Security. (2012). Security Specialist Competencies: An Interagency Security Committee Guideline. <www.dhs.gov> retrieved June 15, 2012.

U.S. Department of Homeland Security, Science and Technology Directorate. (2012). Science & Technology Directorate Mission, Goals, and Objectives. <www.dhs.gov/xabout/structure/st-mission.shtm> retrieved June 15, 2012.

U.S. Department of State. (2012a). Request for Statements of Interest: Internet Freedom Programs (May 4). <www.state.gov/j/drl/p/127829.htm> retrieved June 10, 2012.

U.S. Department of State. (2012b). The Central American Citizen Security Partnership (April 13). <www.state.gov/p/wha/rls/fs/2012/187121.htm> retrieved June 10, 2012.

Zalud, B., & Ritchey, D. (2010). Security innovations today and tomorrow: More personal, versatile and mobile. *Security, 47* (September).

Index

Note: Page numbers followed by "*f*", "*t*" and "*b*" refers to figures, tables and boxes respectively.

A

Abbreviated URLs, 498–499
ABC company, accountability, 309*f*
Abduction, executive protection program and, 602–603
Academic programs, 62–63, 649–651. *See also* Education; Training
Access controls. *See also* Lock-and-key systems
  automatic, 186–189
  biometric security systems, 189–190
  controlling employee traffic, 183–184
  defined, 183
  employee identification system, 185–186
  fingerprint-based, 190
  healthcare institutions, 578–579
  movement of packages and property, 185
  searching employees, 184
  visitors and, 184–185
Accidents. *See also* Workplace safety
  defined, 401
  external metrics for, 31
  Hardy Furniture Plant accident, 279–280
  human error and, 417–418
  human-made emergencies, 389
  investigations and, 417–418
  statistics and costs, 403
Accident proneness theory, 417
Accident reports, 402*f*, 408*f*
Accountability. *See also* Accounting; Auditing; Government Accountability Office
  ABC company, 309*f*
  defined, 306
  described, 307–312
  employee theft prevention and, 176
  healthcare institutions, 577–578
  importance of, 308
  inventory system, 178, 310–312
  purchasing and, 308–310
  RFID technology and, 311–312
Accounting. *See also* Accountability; Auditing
  American Institute of Certified Public Accountants, 313–315
  CPAs, 312–313
  described, 306
  employee theft prevention and, 176
  financial statements of two companies, 307*t*
  forensic, 313
  Generally Accepted Accounting Principles, 312
Acrylic windows, 232–233
"Acting under color of state law", 96
Active shooter, 594–596
Active vehicle barriers, 227
ADA. *See* Americans with Disabilities Act of 1990
Addict shoplifters, 543
Addiction, 611. *See also* Substance abuse in workplace
Adjusters, 340–341
Administrative agencies, 103
Administrative inspections, 104
Administrative law
  inspections, 103–104
  introduction, 103–106
  private security operations and, 104–105
  purpose of, 103
  searches, 103–104
Administrative searches, 104
ADT Security Services, Inc., 17
Advanced Manifest Rule, 522
Affirmative action, 137, 140
Age Discrimination in Employment Act of 1967, 137
Agent of socialization, 158
Agriculture and food sector, 484–485
Agroterrorism, 484
AHJ. *See* Authority having jurisdiction
AK-47, 449–450, 449*f*

Alarms. *See also* Burglar alarm systems; Intrusion detection systems
  alarm dogs, 254
  monitoring, 203, 243, 255*f*
  shoplifting reduction through, 548–550
  signaling systems, 241–243, 380
Alcoholism, 610. *See also* Substance abuse in workplace
ALF. *See* Animal Liberation Front
Alfred P. Murrah Federal Building bombing. *See* Oklahoma City bombing
Allbaugh, Joe, 350
All-hazards, 26
All-hazards preparedness concept, 347–348
al-Qaida, 431, 438–441. *See also* bin Laden; September 11, 2001 terrorist attacks
al-Qaida in the Arabian Peninsula (AQAP), 440
AM. *See* Automated manufacturing
Amateur shoplifters, 542
Amendments. *See specific Amendments*
American Institute of Certified Public Accountants, 313–315
American National Standards Institute (ANSI), 80
American Red Cross, 469
American Society for Industrial Security. *See* ASIS International
American Society for Testing and Materials (ASTM), 80, 191
American Water Works Association, 487
Americans with Disabilities Act of 1990 (ADA)
  described, 138
  entrances for handicapped, 192–193, 194*f*
  interview questions and, 147
  key-in-lever locks, 192, 193*f*
  medical examinations and, 149
  prescription drugs and, 609
  "profiling" of employees and, 592
  substance abuse programs and, 608–609
Analog cameras, 206
Analog technology, 205
Anarchism, 443
Ancient Egyptians, 10
Ancient Greece, 10
Ancient Rome, 10
Animal activism, 225
Animal barriers, 226
Animal Liberation Front (ALF), 443
Annealed glass, 231–232
Annual loss exposure, 68–70
Annunciator, 199
ANSI. *See* American National Standards Institute
Answers, to lawsuits, 102
Anthrax attacks, 19–20, 452, 466*f*, 486, 520
Anticlimb fences, 226, 227*f*
Anticrime lights, 248
Anti-Drug Abuse Act of 1988, 560–561, 608–609
Antiramming walls, 231
Anti-substance abuse program for the workplace, 607–608
Antiterrorism and Effective Death Penalty Act of 1996, 458, 560–561
Appellate jurisdiction, 91
Applicant background investigations, 277
Applicant screening. *See also* Employee socialization; Hiring
  applications and, 146–147
  defined, 136
  employee theft prevention and, 176
  employment law and, 137–144
  federal legislation and, 137–138
  healthcare institutions, 578
  hiring errors, 149
  infinity screening, 136, 176
  interviews and, 147–148
  methods, 145–154
  negligent hiring and, 145–146
  polygraph and, 153–154
  problems, in retailing, 533
  PSE and, 153
  purpose of, 136
  resumes and, 146–147
  retail industry and, 533
  shopping mall incident and, 136–137
  Supreme Court decisions and, 138–139
  tests and, 148–150
Apps, for public safety, 388–389, 388*b*–389*b*
AQAP. *See* al-Qaida in the Arabian Peninsula
Arab Spring, 430
Arafat, Yasser, 430
Archduke Ferdinand, 429
Area protection, intrusion detection systems, 240–241, 241*f*

Areas of refuge, 377
Armed Forces for National Liberation, 442–443
Armed security officers, unarmed *versus*, 251
Arraignment, 107
Arrests
  criminal justice procedures and, 106–112
  defined, 106
  of shoplifters, 551–553
Arson. *See* Fires
Artificial intelligence, investigations and, 293
Asia-Pacific governments, security risk assessment, 69–70
ASIS International
  American Society for Industrial Security, 18, 42, 48
  described, 48, 79
  diversity in security industry and, 141
  formation of, 18
  Private Security Officer Selection and Training, 45–46
  Security for Houses of Worship Project, 126
  "Women and Minorities in Security" conference, 141
ASIS International Standard: ASIS SPC 1-2009, Organizational Resilience: Security, Preparedness, and Continuity Management Systems-Requirements with Guidance for Use, 343–345
Assassinations study, 598
Assassins, 428
Assault, tort law, 95
Assaults in and Around Bars, 62
Assessment center, applicant testing method, 148
Assets. *See also* Information assets
  defined, 24–25
  key, 479–480
  National Strategy for the Physical Protection of Critical Infrastructures and Key Assets, 477–478, 490
Association of Certified Fraud Examiners
  Certified Fraud Examiner, 298, 559, 565
  defined, 313
  employee theft and, 169–171, 174–176
Associations, security, 64
ASTM. *See* American Society for Testing and Materials
Asymmetrical warfare, 437–438. *See also* September 11, 2001 terrorist attacks
Attack dogs, 254
Attacking locks, 193–196
Attacks, executive protection program and, 601–602
Attest function, 312
Attire, security and loss prevention, 131–132
Attorney/client privilege, 282–283
Audio detectors, passive, 202
Audio surveillance, 296
Auditing. *See also* Accountability; Accounting
  defined, 306
  described, 312–315
  employee theft prevention and, 176
  Generally Accepted Auditing Standards, 312
  healthcare institutions, 578
  internal control questionnaire, 313–315
Auditors, 312–313
Augustus, 10
Aum Shinrikyo cult, 451
Aura of security, 219–220
Authentication, access control and, 186
Authority, 83
Authority having jurisdiction (AHJ), 365
Authorization, access control and, 186
Autocratic style, 85
Automated manufacturing (AM), 640
Automatic access controls, 186–189
Automatic alarm signaling systems, 380
Automatic license plate recognition systems, 250
Automatic locking and unlocking devices, 198
Automatic monitoring systems, security officers, 252
Automatic telephone dialer systems, 243
Automatic teller machines, 563–564
AutoZone case, 287
Aviation and Transportation Security Act of 2001, 509
Aviation system, 509–512
Avoidance of predictable patterns, 599

B

Bacillus anthracis, 452. *See also* Anthrax attacks
Background investigations, 150–154, 277
Bail bond, 335
Bailment, 359

Bakke v. University of California, 139
Balanced capacitance, 239*t*
Balanced magnetic switch, 199
Bandwidth, CCTV and, 206
Bank Protection Act (BPA) of 1968, 560
Bank robbery countermeasures, 562–563
Bank Secrecy Act of 1986, 560–561
Banking and finance sector
  automatic teller machines, 563–564
  bank robbery countermeasures, 562–563
  career in, 565
  challenges, 560
  embezzlement, fraud, online risks, 564–565
  introduction, 559–565
  kidnapping and extortion, 563
  legal responsibilities of, 560–561
  Regulation H, Code of Federal Regulations, 561–562
Bar codes, 252, 536, 536*b*
Barbed tape, coiled, 226
Barbed wire fences, 226
Bar-coded cards, 188
Barr, Bob, 44
Barriers, 226–228
Barr-Martinez Bill, 44
Basic Protection Officer, 48
Basics of organization, 83–86
Battery, tort law, 95
Bearing Industries example, 381, 381*b*
Benchmark Survey, RIMS, 338–339, 338*b*–339*b*
Beslan Elementary School massacre, 571
Best Buy Service Company example, 378, 378*b*
Beyond reasonable doubt, 95*t*, 107–108
Bhopal chemical accident, 493
Biases, critical thinking and, 5–6
Bidding, competitive, 269
Bill of Rights, 93–96, 107–109
bin Laden, Osama, 439, 441, 447, 460–461
Bio Clinic Corp., 315
Biological attack. *See* Bioterrorism
Biological weapons, 452
Biometric security systems, 189–190
Bioterrorism
  anthrax attacks, 19–20, 452, 466*f*, 486, 520
  biological attack, 464–465
  defined, 484
  Project BioShield Act and, 484–485
  responding to, 464–465
Bioterrorism Act (Public Health Security and Bioterrorism Preparedness and Response Act of 2002), 484–486, 583
Black Hand, 429
Black September, 430
Black-collar crime, 92–93
Blast curtains, 232
Blast walls, 231
Blazers, traditional uniforms *versus*, 131–132, 131*f*
Blended cyber threats, 497
Blended cyber-physical attack, 497
Blizzards, 394
Bloodborne pathogens standard, OSHA, 409
Bollards, 227
Bombs, 389–392. *See also* Oklahoma City bombing; World Trade Center
  dirty, 452–453
  as human-made emergencies, 389–392
  suicide bombing, 253, 434, 448–449, 571
  terrorist, 448–449
  Unabomber, 443
Bomb threat form, 390*f*
Bona fide occupational qualification, 147
Bonds. *See also* Insurance. *specific bonds*
  claims, 341–342
  crime insurance and, 334–337
  defined, 334
  insurance and, 179
  types, 334–337
Booking, 106
Boosters, 543
Border and transportation security, 521–524
Botnets, 497–498
Bow Street Runners, 12–13
Brand, threats to, 27
Breach of contract, remedies for, 101
Brink, Washington Perry, 15–16
British Petroleum oil spill, 27, 493
Broder, James F., 67
"Broken Windows" theory, 221
Brokers, insurance, 326

Brown, Michael, 350
Budgeting
  employee training, 155
  management countermeasures for employee theft, 175
  security and loss prevention, 66
Building codes, 368, 370–371. *See also* Fire protection
Building designs. *See also* FEMA 426
  described, 368
  fire-resistive buildings, 385
  terrorism protection and, 228–233, 230*f*
Building inspections, fire protection and, 371
Building Owners and Managers Association Chicago Security Committee, 43
Bullet-resistant windows, 232
Burden of proof, 107
Burger King sexual harassment case, 143
Burglar alarm systems. *See also* Alarms; Intrusion detection systems
  citizen preventive activities, 120
  false alarm problem and, 47–49
  Tildesley and, 17
Burglar-resistant windows, 233
Burglary, robbery and
  countermeasures, 556–558
  defined, 556
  shopping mall strategies, 558
Burglary insurance, 335
Burglary-resistive safes, 210
Burke, Edmund, 428–429
Burns, William J., 15, 16*f*
Bush, George W., 19
Business as usual, critical thinking and, 4
Business at the Speed of Thought (Gates), 271
Business concepts, 64–66
Business continuity. *See also* Emergency management; PS-Prep; Resilience; Risk management
  CCTV failure and, 208
  defined, 342
  described, 342–345
  emergency management *versus*, 342
  guidance for, 342–345
  methodology for, 344–345, 344*b*–345*b*
  PS-Prep, 342–345, 344*b*
  risk management and, 325
Business countermeasures, against terrorism, 467–468
Business espionage, 613
Business ethics, 46–47
Business interruption insurance, 338
Business property insurance, 337–338
Business Standard Institution BS 25999-2:2007 Business Continuity Management, 343
Business-focused council of leaders, 8
Bypass, 238–240

C

Cameras. *See also* CCTV
  CCD, 206
  CCTV
    camera placement, 208
    dummy cameras, 207
    hidden cameras, 208
    IP-based network cameras, 205
  chip camera, 206
  digital *versus* analog, 206
  high definition security, 207
  Low-light-level, 244
  megapixel security, 207
  pinhole lens, 178
  thermal imaging, 244
Campus Sexual Assault Victims' Bill of Rights, 572
Canines. *See* Dogs
Capacitance sensors, 200, 201*f*
Capital budget, 66
Carbon monoxide detectors, 379
Card access systems
  biometric technology and, 190
  cards used in, 188
  lock-and-key systems *versus*, 198
  variance in, 188
Cardiac arrest, app for, 388
Carroll, John M., 503
Carry away method, safe attack, 212
Cascade effect, 476
Case law, 93
Casey, Kathleen, 151
Cash flow budget, 66

Castles, 11, 11*f*
Cause-of-loss form, 337–338
Cautious approach, employee suspect and, 179–180
CBP. *See* Customs and Border Protection
CCD. *See* Charged coupled device
CCTV (closed-circuit television). *See also* Intrusion detection systems
  cameras
    camera placement, 208
    dummy cameras, 207
    hidden cameras, 208
    IP-based network cameras, 205
  described, 204–208
  external threats and, 243–245
  green security and, 222
  healthcare institutions, 579
  history of, 204–205
  NLRB and, 106
  pharmaceutical company example, 225
  POS systems and, 537
  privacy issues and, 99–100
  retailing and, 537
  scientific method and, 81
  Tedson Manufacturing Corporation, 208
CDC. *See* Centers for Disease Control and Prevention
Cellular model of organization, 447
Centers for Disease Control and Prevention (CDC), 396–397, 403, 405–406, 576, 583
Centers of Excellence, 483, 653–654
Central Intelligence Agency (CIA), 284, 437, 441, 461–462, 465, 644–645
Central station alarm system, 242
Certifications, in security and loss prevention, 267–268
Certified Fraud Examiner, 298, 559, 565. *See also* Association of Certified Fraud Examiners
Certified Protection Officer, 48
Certified Protection Professional (CPP), 37, 48, 267–268, 559, 565
Certified public accountants (CPAs), 312–313
Certified Security Supervisor, 48
CF. *See* Commercial facilities
CFATs. *See* Chemical Facility Anti-Terrorism Standards
CFR. *See* Code of Federal Regulations
Chain of command, 83, 376
Chain of custody evidence, 283–284
Chain-link fences, 226
Challenges
  banking and finance sector, 560
  post-9/11 security, 19–20
  security industry, 40–49
Chandler Office Building example, 61
Chaos theory, 345
Charge the jury, 108
Charged coupled device (CCD), 206
Checkout counters, loss prevention at, 535–539
Checks, 540–541
"Checks and balances" system, 94
Chemical Facility Anti-Terrorism Standards (CFATs), 494, 574
Chemical industry sector, 493–494
Chemical weapons, 451, 463–464
Chernobyl nuclear disaster, 490–491
Chester Garment Company, 280–281
Chief security officer (CSO), 8
Chief Security Officer Organizational Standard, 8
Child pornography incident, 100–101
Chip camera, 206
"Chip-and-PIN" card, 538
Chiquita Brands International, terrorists and, 605
Chop method, safe attack, 210
Christian Identity, 444
CI. *See* Critical infrastructure
CIA. *See* Central Intelligence Agency
CIKR. *See* Critical infrastructure and key resources
CIP. *See* Critical infrastructure protection
Circumstances test, totality of the, 98
Circumstantial evidence, 283–284
Circumventing defenses, 9, 74, 496, 641
Citations, 106
Citizen Corps, 469–470
Citizen preventive activities, 120
Citizen volunteers, homeland security and, 468–470
Citizenship and Immigration Services, 523–524

Civil actions, tort law, 94–96
Civil Defense Programs, 348
Civil disturbances, 392
Civil investigations, 277
Civil justice procedures, 102–103
Civil law, 94
Civil recovery, shoplifting and, 554
Civil Rights Act of 1964, Title VII, 137
Civil Rights Act of 1991, 138
Civilian facilities protection, DHS and, 367, 367*b*
Claims. *See also* Insurance
  bonding, 341–342
  for crime losses, 341
  described, 339–342
Claims management, 339–340
Clark, Robert, 650–651
Classification systems, information assets, 620, 621*t*
Classified information, 620
Clauses and disclaimers, on applications, 146
Clean Air Act Amendments, 493–494
Clear zones, 224
Clearance rates, 278
Clery, Jeanne, 572
Clery Act, 572
Client/attorney privilege, 282–283
Clinical experience model, 651–652
Clinton, Bill, 507–508
Closed-circuit television. *See* CCTV
Close-ended questions, 286
Closing arguments, 103
Cloud computing
  convergence of IT and physical security, 183
  critical infrastructure and, 503, 503*b*
  data centers and, 503, 540
  defined, 183, 503
CMOS. *See* Complementary metal oxide semiconductor
Coaxial cable, ported, 239*t*
Codes, 78
  building, 368, 370–371
  fire, 370–371
  life safety and, 364
Code of ethics, 46–47
Code of Federal Regulations (CFR), 78, 364, 410, 492, 506, 561, 564
Code war, 77
Cognitive ability tests, 148
Coiled barbed tape, 226
Cold War, 18–19, 348, 449–450, 453, 456, 462
Collaborations. *See* Public-private sector partnerships
Colleges, universities and
  legislation for, 572
  protection for, 572–573
"Color of state law", 96
Color rendition, 248
Columbine High School massacre, 570–571
Combination detectors, 379
Combination locks, 197
Combination padlocks, 197
Comitatus, 11
Commerce, Department of, 284, 368
Commercial facilities (CF) sector, 591–592. *See also* Banking and finance sector; Educational institutions; Healthcare and Public Health sector; Retail subsector
Commercial package policy (CPP), 334, 338
Common law, 93
Common user provisioning, 187
Common wall, 228
Communications
  control center and, 254–256
  crisis communication plan, 127
  emergency communications systems, 393, 393*b*
  emergency management and, 352
  external-internal relations, 117
  interoperability and, 352
  risk communication theory, 326
  risk perception and communication theory, 346
Communications sector, 494–495
Communications security. *See also* Surveillance
  defined, 624–625
  introduction, 624–632
  subfields of, 625
Communications tapping, 625–626
Community, external relations with, 126
Community policing, 572
Community-based mitigation, 348–349
Company adjuster, 340
Competition, over standards, 77–78

Competitive bidding, 269
Competitive intelligence, 615
Complementary metal oxide semiconductor (CMOS), 206
Compliance, 317–319
Compression techniques, CCTV, 206
Compromised system, 495–496
Compulab Corporation, 166, 171*f*, 213, 218–219
Computers, investigative process and, 277, 293
Concertina fences, 226
Confined area entry, 410–411, 410*b*–411*b*
Conflict resolution and nonviolent response, 594
Confronting employee suspect, 179–180
Confronting suspected shoplifters, 550–555
Connecting the dots, 461
Conscious disregard, 98
Consensus standards, 76–77, 232
Constitution (U.S.), 93. *See also* Bill of Rights
Constitutional courts, 91–92
Constitutions, state, 93
Construct original viewpoint, 6
Consultants
  loss prevention programs and, 36*t*
  purchasing decisions and, 266–268
Contact memory buttons, 188
Contact switch sensors, 234, 235*f*
Contacting fire department, 379
Container Security Initiative (CSI), 522
Continuous improvement principle, 612
Contraband, 253
Contraband detection, 253
Contract construction bond, 334–335
Contract investigations, 277–278
Contract law, 101–102
Contract lifecycle management, 271–273
Contract security. *See also* Purchasing decisions
  basics of organization, 83–86
  defined, 32
  proprietary security *versus*, 32–35
  purchasing, 263–264
  shopping mall example and, 264–265
Contract undercover investigations, 266
Controls. *See also* Access controls
  land use, 229
  over private security, 94–97
  parking lot and vehicle controls, 249–250
  span of control, 83
Control center (security and loss prevention control center), 254–256
Control unit, 199
Controlled Substances Act of 1970, 581
Controllers, 61–62
Controlling employee traffic, 183–184
Convergence
  of IT and physical security, 180–183, 255
  organizational, 181–182
  technology, 181–182
Conway Excavation, 336–337, 336*b*–337*b*, 359
Copyright, 614
Corporate crimes, prosecuting, 130–131
Corporate intelligence, 614–617
Corporation restoration project, 252, 252*b*
Costs. *See also* Purchasing decisions
  accidents, 403
  cost/benefit analysis, 68
  of employee theft, 169–170
  of loss, risk analysis process, 68
  security officers, 251
Counter social media techniques, 279
Counterfeiting, 536, 536*b*, 542
Countermeasures. *See also* Access controls; CCTV; Employee theft; Intrusion detection systems; Locks; Safes; Technical Surveillance Countermeasures; Terrorism
  bank robbery, 562–563
  burglary and robbery, 556–558
  external threats, 219–256
    barriers, 226–228
    building designs and terrorism, 228–233, 230*f*
    CCTV, 243–245
    contraband detection, 253
    doors, 235–238
    environmental security design, 220–221
    lighting, 245–249
    parking lot and vehicle controls, 249–250
    perimeter security, 222–226, 223*f*
    protective dogs, 254
    security officers, 250–253
    sustainability and green security, 222
    window protection, 231–235

five “Ds”, 219–220
information assets protection, 620–624
management countermeasures for employee theft
accountability, accounting, auditing, 176
applicant screening, 176
budgeting, 175
confronting employee suspect, 179–180
employee socialization, 176
external and internal relations, 176
insurance and bonding, 179
inventory system, 178
investigations, 177
management support, 175
marking property, 178
metal detectors, 179
planning, 175
policy and procedural controls, 176
reporting losses, 177
signs, 176
physical security countermeasures for employee theft
convergence of IT and physical security, 180–183, 255
integrated systems, 180–183, 243
security officers and, 209
substance abuse in workplace, 607–610
Country Reports on Terrorism, 437–439, 448
Country risk ratings, 597–598
Courts. *See* Judicial systems
Courts of Appeals, U.S., 92
Courts of general jurisdiction, 91
Cover your ass (CYA), 307–308
CPAs. *See* Certified public accountants
CPP. *See* Certified Protection Professional; Commercial package policy
CPTED. *See* Crime Prevention Through Environmental Design
Cranton, Tim, 498
Creature of habit, 599
Credit cards, 537–539
Credit reporting
Fair and Accurate Credit Transaction Act of 2003, 150, 282
Fair Credit Reporting Act of 1971, 150, 282
Cressey, Donald R., 173
Cressey’s employee theft formula, 173
Crimes
criminal acts list, 26*t*
external metrics for, 30–31
insurance, bonds and, 334–337
pattern analysis, 28–29
prevention, citizen groupings and, 120
Crime Prevention and Community Safety: An International Journal, 63
Crime Prevention Through Environmental Design (CPTED), 59–61, 221
Criminal entrepreneurs, 642
Criminal justice, 56, 106–112
Criminal justice system, 38–40, 130
Criminal law, 94, 95*t*
Criminal offense investigations, 277
Criminology, 56
Criminology theories, 57–58
Crisis communication plan, 127
Critical infrastructure (CI). *See also* Transportation systems. specific *critical infrastructure sectors*
agriculture and food sector, 484–485
chemical industry sector, 493–494
cloud computing, 503, 503*b*
communications sector, 494–495
critical manufacturing sector, 506–507
cybersecurity, 499–503
dams and, 490
defined, 477
DIB and, 504–506
electricity sector, 488–489
energy sector, 488–493
facilities requiring security and loss prevention programs, 35, 36*t*
government facilities sector, 507–509
information sharing on, 483, 483*b*
internal threats, 168
introduction, 476–483
IT sector
cybersecurity, 499–503
cyberwarfare, 499, 499*b*
threats, 495–499
National Monuments and Icons sector, 520–521

Critical infrastructure (CI). *See also* Transportation systems. specific *critical infrastructure sectors* (*Continued*)
  nuclear power plants and, 490–492
  oil and natural gas sector, 489
  risk management and, 325
  sectors, 484–509
  water sector, 485–488
Critical infrastructure and key resources (CIKR), 387–388, 479–480
Critical infrastructure protection (CIP), 477–478
Critical Infrastructure Risk Management Enhancement Initiative, 481
Critical manufacturing sector, 506–507
Critical thinking
  biases and, 5–6
  construct original viewpoint, 6
  definitions for, 4–5
  evaluate viewpoints, 6
  four-step strategy for, 5–6
  how to think critically, 5–6
  objectivity and, 5, 7
  PS-Prep and, 344, 344*b*
  reasons for using, 4–5
  resilience movement and, 322–324
  security models and, 5, 75
  for security planning, 74–75
  *See*k other views, 5–6
  September 11 attacks and, 4
  understanding viewpoints, 5
Cross-examination tactics, 107–108, 300*t*–301*t*
Crusades, 431
Cryptography, 186–187
CSI. *See* Container Security Initiative
CSO. *See* Chief security officer
C-TPAT. *See* Customs-Trade Partnership Against Terrorism
Customer-driven security and loss prevention programs, 118
Customs and Border Protection (CBP), 513, 521–522
Customs-Trade Partnership Against Terrorism (C-TPAT), 522
CYA (cover your ass), 307–308
Cyber insurance, 336
Cybersecurity, 499–503
Cyberwarfare, 499, 499*b*
Cycle of protection, 641
Cylinders, 192, 192*f*

D

Dallas Law Enforcement and Private Security Program, 43
Dams, 490
Dark Ages, 11
Data centers, cloud computing and, 503, 540
Data mining, 293
Data transmission systems, alarms and, 242
Database information, investigations and, 290
Daubert challenge, 76–77
Day combination, safe attack and, 212
DCSs. *See* Digital control systems
Deadbolt, 191–192
Deadly force, 110
Debit cards, 537–539
Decentralized networks, 646
Deception, on resumes, 146
Decision theory, 346
Deductibles, 327
Defalcation, 168–169. *See also* Employee theft
Defamation, 96
Defendants, 102–103, 106–112
Defenses, circumventing, 9, 74, 496, 641
Defense Against Weapons of Mass Destruction Act of 1996, 457–458
Defense industry base (DIB), 504–506
Defensible space, 59
Delay, five “Ds”, 220
Delayed-egress locks, 366
Delegation of authority, 83
Delphi approach, 329
Democratic style, 85
Denial-of-service attacks, 495
Deny, five “Ds”, 220
Department of Commerce, 284, 368
Department of Defense, 479
Department of Energy, 479
Department of Health and Human Services, 104, 284, 349–350, 396–397, 405–406, 459, 479, 574

Department of Homeland Security. *See* Homeland Security Department
Department of Homeland Security's Approach to Risk Analysis report, 69
Department of Justice, 479
Department of State. *See also Country Reports on Terrorism*
al-Qaida and, 438–441
OSAC and, 285
overseas investigations and, 284
Department of the Interior, 479
Department of the Treasury, 479, 559–560
Depositions, 299
Designated spokesperson, 126–127
Destroy, five "Ds", 220
Destruction, of information assets, 623–624
Detect, five "Ds", 220
Detecting smoke and fire, 379–380. *See also* Fire suppression strategies
Detectives, 38
Detention, of shoplifters, 551–553
Deter, five "Ds", 219–220
Deterrence. *See also* Loss prevention; Security and loss prevention
defined, 38–39, 57
employee theft and, 173
prevention *versus*, 316
DHS. *See* Homeland Security Department
DHS/Information Analysis and Infrastructure Protection Daily Open Source Report, 483
Dial Corporation sex discrimination case, 139–140
DIB. *See* Defense industry base
Differential association, 172
Digital cameras, 206
Digital certificate systems, 187
Digital control systems (DCSs), 487–488, 487*b*–488*b*
Digital evidence, 293–296
Digital forensics, 294
Digital investigations, 293–296
Digital photos, 295
Digital technology, 205
Digital video recorders (DVRs), 205
Diploma mills, 146
Direct evidence, 283–284
Direct losses, 29
Directive system, 84
Dirty bombs, 452–453
Disabilities. *See* Americans with Disabilities Act
Disasters. *See* Emergency management; Human-made emergencies; Natural disasters
Disc tumbler locks, 197
Disclaimers, on applications, 146
Discovery, 102
Discrimination
Age Discrimination in Employment Act of 1967, 137
Dial Corporation sex discrimination case, 139–140
Genetic Information Nondiscrimination Act of 2008, 138
Pregnancy Discrimination Act of 1978, 138
reverse, 138, 140
sex discrimination cases, 139–140
unintentional, 138
WalMart sex discrimination case, 140
Disparate impact, 138
Disparate treatment, 138
Distance learning online, 654–655
Distributed Denial of Service attacks, 498
District Courts, U.S., 92
Diversity, 141
Division of work, 83
DNB, 290
Dogs
as barriers, 226
9/11 attacks and, 254
protective, 254
Domestic terrorism. *See also* Oklahoma City bombing
anarchism, 443
defined, 441
ecoterrorism, 443
homegrown terrorism, 446
leftist class struggles, 442–443
lone wolf terrorists, 446
racial supremacy, 443–444
religious extremists, 444
state-sponsored, 430, 442
Doors. *See also* Locks
countermeasures for external threats, 235–238
fire doors, 382
lock attacks and, 195

Doppler Effect, 201–202
Double-cylinder locks, 192
Double-hung window, 233–234, 234*f*
Drill method, safe attack, 210
Drug tests, 149, 609–610. *See also* Substance abuse in workplace
Drunk shoplifters, 543
Dry-pipe automatic sprinkler system, 382
Dual court system, 90
Dual sensor technologies, 202–203, 203*f*, 238–240
Due diligence, 271–273, 277
Duke of Normandy, William, 12
Dummy cameras, CCTV, 207
Dump truck accident report, 402*f*
Dumpster diving, 496
DVRs. *See* Digital video recorders
Dye packs/tear gas, 562

E

EAPs. *See* Employee assistance programs
Early America security. *See* Security
Early civilizations, 9–12
Earth Liberation Front (ELF), 443
Earthquakes
  all-hazards preparedness concept and, 347
  described, 394–395
  Eastern Japan Great Earthquake Disaster of 2011, 20, 323, 356–357, 356*b*–357*b*, 395, 490–491
  Northridge, 349
EAS. *See* Electronic article surveillance
Easy-access shoplifters, 543
E-business, 539–542
Economic espionage, 617
Economic Espionage Act of 1996, 616–617
Ecoterrorism, 443
eDiscovery amendments, 295–296
Education. *See also* Future in security and loss prevention; Training
  executive protection program, 598
  future in loss prevention and security, 647–651
Educational institutions
  colleges and universities
    legislation for, 572
    protection for, 572–573
  fire protection at, 573–574
  protection for, 567
  school districts
    introduction, 567–572
    legislation for, 568
    protection for, 568–570
  as soft targets, 570–571
  threats and hazards at, 566–567
  types, 565–574
Edward I (king), 12
EEO. *See* Equal employment opportunity
EEOC. *See* Equal Employment Opportunity Commission
Egypt, Ancient, 10
8th Amendment, 109
Electric field sensor, 239*t*
Electricity sector, 488–489
Electrocution, accident report, 408*f*
Electromechanical locks, 191
Electronic article surveillance (EAS), 547–548
Electronic lock, safe with, 209*f*
Electronic payments, 537–539
Electronic protection, for windows, 234–235
Electronic security glazing, 233
Electronic surveillance, 625–628. *See also* Technical Surveillance Countermeasures
ELF. *See* Earth Liberation Front
E-mail, internal relations and, 123
Embezzlement, 168–169, 564–565. *See also* Employee theft
Emergencies. *See also* Human-made emergencies; Life safety; Natural disasters
  life safety and, 363–364
  1910.38 Emergency Action Plans, 364
  overview, 389–398
  planning and training strategies, 389
  technological, 393
Emergency communications systems, 393, 393*b*
Emergency management. *See also* Business continuity; Katrina disaster; Resilience; Risk management
  business continuity *versus*, 342
  communications discipline, 352
  defined, 345
  described, 345–353
  disciplines, 351–353
  generic, 347–348

history of, 348–350
military and, 353–358
mitigation discipline, 351
NIMS and, 353
NPG and, 353
NRF and, 353
preparedness discipline, 352
recovery discipline, 352
response discipline, 351–352
risk management and, 325, 345–346
specialized, 347–348
Emergency medical services (EMS), 387–388
Emergency Medical Treatment and Active Labor Act, 576
Emergency Planning and Community Right-to-Know Act of 1986, 493–494
Emergency room, healthcare institutions and, 579
Emergency Services Sector (ESS), 387–388
Emergency Support Function Annex, 369
Emotional stress, infliction of, 95
Employees. *See also* Employee socialization; Employee theft; Workplace
groups, Speed's research and, 173–174
identification system, 185–186
needs, 158–160
personal problems, theft and, 170–171
suspect, confronting, 179–180
venting, 144–145
Employee assistance programs (EAPs), 608
Employee Polygraph Protection Act of 1988 (EPPA), 154, 180, 288–289
Employee selection. *See* Hiring
Employee socialization. *See also* Applicant screening; Hiring; Training
defined, 136
employee needs and, 158–160
employee theft prevention and, 176
example setting and, 158
orientation and, 154–155
overview, 154–160
Employee theft (internal theft). *See also* Countermeasures
causes of, 170–171
cost of, 169–170
defined, 168–169
deterrence and, 173
differential association and, 172
embezzlement, 168–169
environment and, 171
indicators of, 175
methods, 174–175
occupational fraud, 169–170
personal problems and, 170–171
pilferage, 168–169
prosecution and, 180
rational choice theory and, 173
rationalizations and, 173
reasons for, 170–174
routine activity theory and, 173
seriousness of, 168–175
situational crime prevention and, 173
Smith's lumberyard example, 172
Speed's research on, 173–174
terms for, 168–169
Employee theft formula, 173
Employee training. *See* Training
Employment law, 137–144
Employment opportunities, future in security and loss prevention, 655–658
Employment titles, 8
EMS. See Emergency medical services
Encryption, 186–187, 503–504
Energy barriers, 226
Energy sector, 488–493
England
common law, 93
security system, 12–13
contemporary times, 12–13
frankpledge system, 12
Magna Carta, 12
Peel's reforms, 13
tithing, 12
William I and, 12
workplace safety legislation, 403–404
Enron Corporation, 315–317
Enterprise risk management (ERM), 330–331
Enterprise security risk management (ESRM), 330–331. *See also* Risk management
Enterprises, operations of, 24–25
Entry Level Protection Officer, 48
Environment, employee theft and, 171

Environmental criminology, 59. *See also* Crime Prevention Through Environmental Design; Situational crime prevention
Environmental Protection Agency (EPA), 103, 409
Environmental security design, 220–221. *See also* Crime Prevention Through Environmental Design
EPA. *See* Environmental Protection Agency
EPPA. *See* Employee Polygraph Protection Act of 1988
Equal employment opportunity (EEO), 140
Equal Employment Opportunity Act of 1972, 137–138
Equal Employment Opportunity Commission (EEOC)
  description of, 103, 139
  Hi-Mark Home case, 141–142
  sex discrimination cases, 139–140
  sexual harassment, 142–144
Equal Pay Act of 1963, 137
ERM. *See* Enterprise risk management
Errors, in hiring, 149
Espionage
  business, 613
  defined, 613
  economic, 617
  Economic Espionage Act of 1996, 616–617
  Four Faces of Business Espionage, 620
  government, 613
  techniques, 617–620
ESRM. *See* Enterprise security risk management
ESS. *See* Emergency Services Sector
Essebar, Farid, 498
Esteem and status needs, 159
Ethics, 46–47
Evacuation plans, fire prevention and, 375–377
Evaluation
  of employee training, 156
  evaluate treatment options (risk management process), 328
  of security and loss prevention programs, 80–83
  of viewpoints, 6
Evening precautions, 120
Event risk management, 331
Evidence, investigations and, 283–285, 293–296
  evidence-based model, 651–652
Evidence-based research, 82–83
Example setting, poor, 158
Exception reporting, 536
Exclusionary clause, fidelity bond policy and, 341–342
Exclusionary rule, searches and, 110
Executive Order 12977, 508
Executive protection, 596–597
Executive protection program
  abduction and, 602–603
  attacks while traveling, 601–602
  avoidance of predictable patterns, 599
  education and training, 598
  kidnap insurance, 602
  planning for, 597–598
  protection at home, 599–600
  protection at office, 600–601
  protection strategies, 598–603
  recognition of tricks, 599
Experimental control group design, 81
Explosive blasts, 231
External loss prevention. *See also* Countermeasures
  defined, 218
  internal loss prevention *versus*, 218
  IT perspective and, 218
  unauthorized entry, 218–219
External metrics, 29–31
  for accidents, 31
  for crimes, 30–31
  defined, 29–30
  for fires, 31
  methodological problems and, 30
External relations. *See also* Internal relations
  benefits of, 117–118
  chart of, 117, 117*f*
  with community, 126
  defined, 115
  employee theft prevention and, 176
  with law enforcement agencies, 124–125
  with peers, 128
  with public safety agencies, 125–126
External threats, 167–168. *See also* Countermeasures; External loss prevention; Internal threats; Terrorism

Extinguishers, portable, 380–381, 381*b*. *See also* Fire suppression strategies
Extortion, 563, 602
Extortion insurance, 335
Extremism, 444
Extremists, 444–446
Exxon Valdez oil spill, 493

F

FAA. *See* Federal Aviation Administration
Face recognition, 189–190
Facebook
  benefits, 99
  espionage and, 619
  internal relations and, 123–124
Facility managers, 182–183
Facility planning, fire department protection efforts, 369–370
Factory Mutual Global, 369
Fair and Accurate Credit Transaction Act of 2003, 150, 282
Fair Credit Reporting Act of 1971 (FCRA), 150, 282
Fake deterrents, 550
False alarm problem, 47–49, 238–240
False imprisonment, tort law, 95
Family and Medical Leave Act of 1993, 138
Family Educational Rights and Privacy Act (FERPA), 568
FAMS. *See* Federal Air Marshal Service
Fargo, William, 15
Fastow, Andrew, 315–316
FBI (Federal Bureau of Investigation)
  al-Qaida and, 438–441
  Burns and, 15, 16*f*
  fraud and, 315
  IAFIS, 44–45
  National Crime Information Center, 151
  website, 48
FCRA. *See* Fair Credit Reporting Act of 1971
Federal Advisory Committee, OSAC and, 285. *See also* Overseas Security Advisory Council
Federal Air Marshal Service (FAMS), 511–512
Federal and state constitutions. *See* Constitution; State constitutions
Federal Aviation Administration (FAA), 510
Federal Bureau of Investigation. *See* FBI
Federal Civil Defense Administration, 348
Federal Contract Compliance Procedures, Office of, 138
Federal court system, 91–92
Federal Crime Insurance Program, 335
Federal Emergency Management Agency. *See* FEMA
Federal Energy Regulatory Commission, 488
Federal Financial Institutions Examination Council (FFIEC), 561, 564–565
Federal legislation, employment law, 137–138
Federal Protective Service (FPS), 479, 507
Federal Register, 103
Federal Rules of Civil Procedure, eDiscovery amendments and, 295–296
Federal Rules of Evidence, 295
Feedback
  customer-driven security and loss prevention programs and, 118
  on policies and procedures, 84
  systems perspective on loss prevention, 73–75
Felonies, 94
FEMA (Federal Emergency Management Agency), 349–350
FEMA 426 (*Reference Manual to Mitigate Potential Terrorist Attacks against Buildings*), 228–231, 230*f*, 250, 254–256. *See also* Terrorism
Fence-mounted detection, 239*t*
Fences, 226, 227*f*. *See also* Barriers
Ferdinand, Archduke, 429
FERPA. *See* Family Educational Rights and Privacy Act
Feudalism, 11
FFIEC. *See* Federal Financial Institutions Examination Council
Fiber optics, 202, 238, 243
Fidelity bond, 179, 334, 341–342
Fiduciary bond, 335
Fielding, Henry, 12–13
5th Amendment, 107–109, 111, 155
File cabinets, 213. *See also* Safes
Finance sector. *See* Banking and finance sector
Financial investigations, 277
Financial Modernization Act of 1999, 105. *See also* Gramm-Leach-Bliley Act

Financial risk management, 331
Financial statements, of two companies, 307*t*. *See also* Accounting
Financing, risk, 328
Finch Brothers Supermarket Company, 237
Fingerprint-based access control, 190
Fingerprint-based background checks, 152
Fires. *See also* Fire protection; National Fire Protection Association; U.S. Fire Administration
  Bearing Industries incident, 381, 381*b*
  external metrics for, 31
  investigations, 277
  problems of, 368
Fire brigades, 385–386
Fire codes, 370–371
Fire departments
  contacting, 379
  protection efforts, 369–370
Fire detectors, 379–380
Fire insurance, 337
Fire marshals, 365, 371–372, 582
Fire prevention strategies
  defined, 372
  evacuation plans, 375–377
  good housekeeping, 374
  hazmat incidents, 375
  inspections, 371–373
  list, 372, 372*t*
  medical services, 375–377
  planning, 373
  safety strategies, 373
  training, 377–378
Fire protection
  Ancient Romans, 10
  Best Buy Service Company, 378, 378*b*
  at educational institutions, 573–574
  healthcare institutions, 582–583
  performance-based, 385–386, 385*b*–386*b*
  private organizations in, 369
  problems of fire, 368
Fire suppression strategies
  access control, 384
  defined, 372
  fire walls and doors, 382
  fire-resistive buildings, 385
  integrated fire suppression systems, 378–379
  integrated systems, 378–379
  IPDACT, 379
  list, 372, 372*t*
  portable extinguishers, 380–381, 381*b*
  smoke and fire detection, 379–380
  sprinkler systems, 381–382, 383*f*
  stairwells, 384
  standpipe and hose systems, 382, 383*f*
  success of, 378
  training and, 385–386
Fire triangle, 372, 373*f*
Fire walls and doors, 382
Firefighters, hazards and, 385, 385*b*
Fire-resistive buildings, 385
Fire-resistive safes, 209–210
First party risks, 322
First receivers, 582–583
First responders, 462–463
First wave societies, 19
Five "Ds", 219–220
Five wars of globalization, 645–646
Flame detectors, 379
Floodlights, 248–249
Floods, 394
Floor wardens, 376
Fluorescent lamps, 247–248, 252
Fluorescent substances, marking property with, 178
Food sector. *See* Agriculture and food sector
Foot patrols, 250
Foot-candle, 246
Force, 110
Forced entry, 219
Ford, Henry, 18
Ford Motor Company, 18
Forensic accounting, 313
Forensics, digital, 294
Fore*See*ability, 97
Formal accountability, 307–308
Formal organization, 84
Fortress security, 120
Four Faces of Business Espionage, 620
Four-step strategy, critical thinking, 5–6
14th Amendment, 109, 139
4th Amendment, 104, 108–111, 411
FPS. *See* Federal Protective Service

Fragment retention film, 232
Frankpledge system, 12
Fraud. *See also* Association of Certified Fraud Examiners; Sarbanes-Oxley Act
  banks and, 564–565
  credit card fraud, 538–539
  described, 315–317
  IT directors and, 315
  refund, 541
  techniques of, 314
Freedom of Information Law, 283, 291
Freedonia Group, 32
Freight rail subsector, 518–519, 519*b*
French Revolution, 428–429
Frequency of loss, risk analysis process, 68
Fresnel lights, 248–249
Front de Liberation Nationale, 430
Fukushima nuclear disaster, 490–491
Fully thermally tempered glass, 231–232
Fundamentalism, 432–433
Fusion Centers, 467
Future in security and loss prevention
  challenges, 641–644
  education, 647–651
  employment opportunities, 655–658
  five wars of globalization, 645–646
  introduction, 637–638
  research, 651–654
  security institute concept, 655
  training, 654–655
  trends, 641–644
Future Shock (Toffler), 637–638

G

Gale, Stephen, 69–70
Game theory, 329
Gang Resistance Education and Training (GREAT), 568–569
GAO. *See* Government Accountability Office
Gas detectors, 379
Gates, 227
Gates, Bill, 271
General duty clause, OSHA's, 591–592
General jurisdiction courts, 91
"General Security Risk Assessment Guideline", 68–70
General Services Administration (GSA), 103, 232, 507
Generally Accepted Accounting Principles, 312
Generally Accepted Auditing Standards, 312
Generic emergency management, 347–348
Genetic Information Nondiscrimination Act of 2008 (GINA), 138
GHS. *See* Globally Harmonized System of Classification and Labeling of Chemicals
Gift card losses, 541
GINA. *See* Genetic Information Nondiscrimination Act of 2008
Glass, explosive blasts and, 231–233
Glass-breakage sensors, 234
Glazing, 231–233
Global perspective. *See* International perspective
Global Positioning System. *See* GPS
Globalization, 435–436, 645–646
Globally Harmonized System of Classification and Labeling of Chemicals (GHS), 409
Good housekeeping, fire prevention strategy, 374
Goodman, Marc, 639
Governance, risk management, and compliance (GRC), 317–319
Government
  CIKR protection by, 479–480
  risk management in, 346–347
Government Accountability Office (GAO)
  blended cyber threats and, 497
  corporate crimes and, 130–131
  cybersecurity strategies and, 499
  DSS and, 505–506
  Katrina disaster and, 355–356
  mass transit systems and, 514–515
  NRC and, 492
  rail transportation security and, 514
  resiliency and, 323–324, 454–455
  risk management and, 347, 480–481
  security companies and military installations, 34
  studies on spam, phishing, spyware, 496
  TSA security measures and, 515–516
  water vulnerabilities and, 486
Government espionage, 613
Government facilities sector, 507–509
Government insurance programs, 333
Government/industrial security, 506, 506*b*

GPS (Global Positioning System)
  bank robbery countermeasures and, 562–563
  central station alarm system and, 242
  defined, 296
  investigative process and, 296
  privacy issues and, 99
Gramm-Leach-Bliley Act, 105, 500, 560–561
Graph-based anomaly detection, 177
GRC. *See* Governance, risk management, and compliance
GREAT. *See* Gang Resistance Education and Training
Great Chicago Fire, 371
Great Depression, 348
Great Wall of China, 9
Great wars, 18
Greaves, Sheldon, 650
Greece, Ancient, 10
Green building standards, 222
Green security, 222
Greenpeace protest, 227
Grid wire sensors, 200
Griggs v. Duke Power, 139
Group 4 Falck, 16–17
Group 4 Securicor, 16–17
Groups, employee, 173–174
GSA. *See* General Services Administration
Guard dogs, 254
Guerilla warfare, 430
Guide for Preventing and Responding to School Violence, 570
Guilty plea, 107
Gun-Free Schools Act, 568

H

Hacking, into physical security systems, 245, 245*b*
Hacktivism, 498
Hallcrest Reports, 41–42, 44–45, 47, 263
Halogen lamps, 247–248
Hammurabi, 9
Hand geometry, biometric security systems, 189–190, 189*f*
Handicapped, entrances for, 192–193, 194*f*. *See also* Americans with Disabilities Act
Hard targets, 570
Hardy Furniture Plant accident, 279–280
Harmon Lorman Associates, 280–281
Harry Nash incident, 418, 418*b*
Hasan, Nidal, 440
Hasp, safety, 195, 196*f*
Hazards. *See also* Natural disasters; Workplace safety
  all-hazards, 26
  all-hazards preparedness concept, 347–348
  defined, 25
  at educational institutions, 566–567
  to HPH, 575–576
  list of, 25–26, 26*t*
Hazard Communication Standard (HCS), 407–409. *See also* Occupational Safety and Health Administration
Hazardous materials. *See* Hazmat incidents
Hazardous Waste Operations and Emergency Response Standard (HAZWOPER), 409
Hazmat (hazardous materials) incidents, 375. *See also* Fire prevention strategies
HAZWOPER. *See* Hazardous Waste Operations and Emergency Response Standard
HCS. *See* Hazard Communication Standard
Health and Human Services. *See* Department of Health and Human Services
Health and safety. *See* Occupational Safety and Health Administration; Safety; Workplace safety
Health Insurance Portability and Accountability Act of 1996 (HIPAA), 104, 282
Healthcare and Public Health (HPH) sector
  initiatives, 577
  introduction, 574–583
  legislative and regulatory authorities, 576–577
  pharmacy protection, 581
  threats and hazards to, 575–576
Healthcare institutions
  accountability and inventory control, 577–583
  applicant screening, 578
  auditing, 533–534
  CCTV at, 579
  emergency room and, 579
  fire protection, 582–583
  locker rooms and, 582

Healthcare violence, 580–581
Hearsay evidence, 283–284
Hedges, shrubbery and, 228
Henry II, 11
Hidden surveillance cameras, CCTV, 208
Hierarchy of human needs, 158–160, 159*f*
High definition security cameras, 207
High-pressure sodium lamps, 247–248
Highway infrastructure and motor carrier subsector, 517
Hi-Mark Home case, 141–142, 163
HIPAA. *See* Health Insurance Portability and Accountability Act of 1996
Hiring. *See also* Applicant screening; Employee socialization
  affirmative action and, 140
  applicant screening and, 136
  EEO and, 140
  errors in, 149
  negligent, 145–146
  quotas, 140
History of security and loss prevention, 8–9
  Ancient Greece, 10
  Ancient Rome, 10
  early civilizations, 9–12
  Middle Ages, 11–12
  in nutshell, 15
  reasons for studying, 8–9
Hitler, Adolf, 430
Hollow-core doors, 236
Holmes, Edwin, 17
Homegrown terrorism, 446. *See also* Domestic terrorism
Homeland security (HS). *See also* Terrorism
  big picture, 455–457
  citizen volunteers and, 468–470
  defined, 454–455
  intelligence and, 425
  legislative action against terrorism, 457–459
  NIPP recommendations to private sector, 481–483
  state and local governments, 462–463
  U.S. government action against terrorism, 457
Homeland Security Act of 2002, 458–459
Homeland Security Department (DHS)
  Centers of Excellence, 483, 653–654
  civilian facilities protection and, 367, 367*b*
  creation of, 19
  Department of Homeland Security's Approach to Risk Analysis report, 69
  ESS and, 387–388
  establishment of, 458–459
  hazards (definition), 25
  Interagency Security Committee, 46, 508
  PS-Prep and, 342–343
Homeland Security Information Network (HSIN), 483
Homeland Security Information Network-Critical Infrastructure (HSIN-CI), 44
Homeland security market, 467
Homeland Security Presidential Directives (HSPD), 353
  HSPD-3, 488
  HSPD-5, 353
  HSPD-7, 387–388, 477–480, 486, 488, 499, 506–509, 531, 559–560, 576, 653
  HSPD-8, 486
  HSPD-9, 484–486
  HSPD-10, 484–486
  HSPD-12, 182, 186–187
  HSPD-21, 576
Honesty tests, 149
Hoover, Herbert, 40
Hose and standpipe systems, 382, 383*f*. *See also* Fire suppression strategies
Hospital example, insider IT threats, 167–168
Hostile working environment, 142
Housekeeping strategies, fire prevention and, 374
"How Not to Waste Your Training Dollars", 157–158
HPH. *See* Healthcare and Public Health sector
HRP. *See* Human resources protection
HS. *See* Homeland security
HSIN. *See* Homeland Security Information Network
HSIN-CI. *See* Homeland Security Information Network-Critical Infrastructure
HSPD. *See* Homeland Security Presidential Directives
Human barriers, 226

Human error, accidents and, 417–418
Human relations, 120–121
Human resource problems, in retailing, 533–534
Human resources, 24–25
Human Resources Protection (HRP). *See also* Executive protection program
  Chiquita Brands example, 605
  defined, 596–597
  identity theft, 604–606
  introduction, 596–606
  protection of women, 603–604
  technologies for, 605–606
Human-made emergencies. *See also* Accidents; Bombs; Natural disasters; Weapons of mass destruction
  accidents, 389
  bomb threats, 389–392
  civil disturbances, 392
  strikes, 392
  technological, 393
  WMD, 392
Hurricanes. *See* Katrina disaster; Rita
Husick, Lawrence, 69–70
Hussein, Saddam, 431
Hype, from vendor, 262

I

IAFIS. *See* Integrated Automated Fingerprint Identification System
IAPSC (International Association of Professional Security Consultants), 268
ICC. *See* International Code Council
ICE. *See* Immigration and Customs Enforcement
ICS-CERT. *See* Industrial Control Systems Cyber Emergency Response Team
Identifying vulnerabilities, 68
Identi-Kit, 293
Identity management system (IDMS), 182
Identity theft, 604–606
Identity Theft and Assumption Deterrence Act of 1998, 604
IDMS. *See* Identity management system
IED. *See* Improved explosive device
IEEE. *See* Institute of Electrical and Electronics Engineers
IFPO. *See* International Foundation for Protection Officers
Illuminance, 246
Illuminating Engineering Society, 222, 246
Illumination, 246–247
Immigration and Customs Enforcement (ICE), 522
Implementation, risk management process, 328–329
Improved explosive device (IED), 448
Incandescent lamps, 247–248
Incapacitation, 57
Incentive programs, safety, 415–416, 416*b*
Incomplete protection plans, 72
Independent adjusters, 340
India, chemical accident in, 493
Indirect losses, 29
Industrial Control Systems Cyber Emergency Response Team (ICS-CERT), 488
Industrial Revolution, 12–13, 19, 403–404
Industrial/government security, 506, 506*b*
Infant abduction, 580
Infinity screening, 136, 176
Infliction of emotional stress, 95
Informal accountability, 307–308
Informal organization, 84
Informants, investigative process and, 292
Information. *See also* Investigations
  accuracy, 297
  information-gathering process, 297
  request for, 269
  sources, investigations, 289–296
Information assets
  classification of, 620, 621*t*
  defined, 613
  destruction of, 623–624
  protection countermeasures, 620–624
Information security. *See also* Espionage; Surveillance
  broad perspective, 613
  defined, 611–612
  introduction, 611–624
  program, 621–623
Information sharing, on critical infrastructure, 483, 483*b*
Information Sharing and Analysis Centers (ISACs), 467, 483

Information Systems Audit and Control Association, 8, 523
Information Systems Security Association, 8, 523
Information technology. *See* IT
InfraGard, 43
Infrared intrusion sensors, passive, 202
Infrared photoelectric beam sensors, 200–201, 201*f*, 239*t*
Infrastructure Security Partnership (TISP), 483
In-house research, 82
Initial appearance, 107
Injuries, definition, 401. *See also* Workplace safety
Insanity plea, 107
Insider IT threats, 167–168
Inspections
  as fire prevention strategy, 371–373
  OSHA, 411–413
  "search and seizure" *versus*, 110–111
  searches and, 103–104
  warrant, 104
Institute of Electrical and Electronics Engineers (IEEE), 199, 238
Insurance. *See also* Claims
  burglary, robbery, theft, 335
  business property, 337–338
  cyber, 336
  described, 332–339
  Federal Crime Insurance Program, 335
  fire, 337
  kidnapping and extortion, 335, 602
  liability, 338
  RIMS and, 338–339, 338*b*–339*b*
  terrorism, 336
  types, 333–334
  workers' compensation, 338, 404–405
Insurance agents, 340
Insurance brokers, 326
Insurance industry. *See also* Health Insurance Portability and Accountability Act of 1996; Risk management
  al-Qaida and, 438–441
  bonding and, 179
  investigations, 277
  loss prevention and, 15, 326
  research assistance by, 82–83
  risk management and, 326
Insurance Services Office (ISO), 333–334, 337
Integrated Automated Fingerprint Identification System (IAFIS), 44–45
Integrated fire suppression systems, 378–379
Integrated systems, 180–183, 243, 378–379. *See also* Countermeasures; Fire suppression strategies
Intelligence. *See also* Central Intelligence Agency
  community, 462
  competitive, 615
  corporate, 614–617
  described, 461–462
  process, 461–462
Intelligence Reform and Terrorism Prevention Act of 2004, 462, 483, 509
Intensive interview, 286–287
Interagency Security Committee, DHS, 46, 508
Interchangeable core locks, 198
Interior sensors, 199–209
Intermediate state appellate courts, 91
Internal control questionnaire, 313–315
Internal loss prevention. *See also* Employee theft; External loss prevention
  defined, 166
  external loss prevention *versus*, 218
  retailing, 534–535
Internal metrics, 28–29. *See also* Metrics
Internal relations. *See also* External relations
  benefits of, 117–118
  chart of, 116–117, 116*f*
  citizen preventive activities, 120
  customer-driven security and loss prevention programs, 118
  defined, 115
  e-mail and, 123
  employee theft prevention and, 176
  human relations, 120–121
  intranet and, 122–123
  involvement programs, 122
  loss prevention meetings, 122
  marketing concepts and, 118–120
  orientation and training programs, 121
  senior management support, 121
  social media and, 123–124
  surveys and, 119

Internal space protection, 549*f*
Internal surveys, 119
Internal theft. *See* Employee theft
Internal threats
  critical infrastructure, 168
  external threats *versus*, 167–168
  introduction, 166–167
  IT threats, 167–168
  spectrum of, 166–167
International Association of Chiefs of Police, 42, 570, 572–573
International Association of Fire Chiefs, 369
International Association of Fire Fighters, 369
International Association of Lighting Management Companies, 246
International Association of Professional Security Consultants (IAPSC), 268
International Code Council (ICC), 77, 368, 371
International Foundation for Protection Officers (IFPO), 48, 655
International Organization for Standardization (ISO), 80, 182
International perspective
  black-collar crime, 92–93
  crime prevention and citizen groupings, 120
  Eastern Japan Great Earthquake Disaster, 356–357, 356*b*–357*b*
  overseas investigations, 284–285
  pharmaceutical company security (examples), 225
  risk management in multinational business, 332, 332*b*
  Rockett's viewpoint on U.S. disaster paradigm, 395, 395*b*
  security impact assessments, 69–70
  shoplifter deterrence, 554–555, 554*b*–555*b*
  supply chain risks and security, 513–514, 513*b*–514*b*
International terrorism, 435–441. *See also* Terrorism
Internet. *See also* Social networks
  application soliciting through, 146–147
  ethics information and, 47
  lone wolf terrorists and, 446
Internet protocol (IP)-based network cameras, 205
Internet Protocol Digital Alarm Communicators/Transmitters (IPDACT), 379
Interoperability, 352
Interoperable facilities, 187
INTERPOL, 31, 646
Interrogation, 285–289. *See also* Investigations; Polygraph
Interviews
  applicant screening and, 147–148
  defined, 285
  intensive, 286–287
  investigations and, 285–289
  with media, suggestions for, 127
Intranet
  banking risks and, 564
  components of, 123
  convergence of IT and physical security, 181
  defined, 122–123
  internal relations and, 122–123
  loss reporting and, 177
  training programs and, 156
Intrusion detection systems. *See also* CCTV; Sensors
  alarm signaling systems, 241–243
  applications of, 240–241
  components of, 199
  defined, 199
  described, 238–245
  fiber optics and, 202, 238
  limitations of, 202
  operational zoning, 203
  overview, 199
  performance characteristics, 238–240
  standards, 199, 238
  types of, 239*t*
Invasion of privacy, 96
Inventory system, 178, 310–312, 577–583. *See also* Accountability
Investigations
  accidents and, 417–418
  background, 150–154, 277
  computers and, 277, 293
  database information and, 290
  digital, 293–296
  digital evidence and, 293–296

employee theft prevention and, 177
evidence, 283–285, 293–296
GPS and, 296
Hardy Furniture Plant accident, 279–280
important considerations, 281–282
informants and, 292
information sources, 289–296
as information-gathering process, 297
interrogation and, 285–289
interviews and, 285–289
introduction, 275–277
law and, 282–283
leads and, 291
loss scene and, 291–292
missing shirts example, 280–281
modus operandi and, 292–293
motive and, 292
overseas, 284–285
report writing and, 297–299
social networks and, 289–290
software and, 293
steps of, 276–277
surveillance and, 296–297
testimony and, 299, 300*t*–301*t*
types of, 277–282
undercover, 266, 278–281
victims and, 292
witnesses and, 292
Investigative leads, 291
Investigators, 38
Invoices, 308
Involvement programs, 122
Ionization detectors, 379
IP-based network cameras, 205
IP-based retrofit, 206
IPDACT (Internet Protocol Digital Alarm Communicators/Transmitters), 379
ISACs. *See* Information Sharing and Analysis Centers
Islam, 431
ISO. *See* Insurance Services Office; International Organization for Standardization
Israeli-Palestinian conflict, 432
IT (information technology)
convergence of IT and physical security, 180–183, 255
cybersecurity, 499–503
external loss prevention and, 218
fraud and, 315
insider threats, 167–168
IT sector. *See also* Cloud computing; Critical infrastructure
cybersecurity, 499–503
cyberwarfare, 499, 499*b*
threats, 495–499

J

Jamb, 193
Jamb peeling, 193–195
Japan
Eastern Japan Great Earthquake Disaster of 2011, 20, 323, 356–357, 395, 490–491
Fukushima nuclear disaster, 490–491
tsunami, 20, 356–357, 395
Jeffery, C. Ray, 60, 221. *See also* Crime Prevention Through Environmental Design
Jenkins, Brian, 468–469
Job applicant screening. *See* Applicant screening
Job reference immunity statutes, 151–152
Jobs. *See* Workplace
John (king), 12
John Warner National Defense Authorization Act of 2006, 351–352
The Joint Commission, 576–577, 582
Journal of Applied Security Research, 63
Journal of Physical Security, 63–64
Judges
civil justice procedures and, 102–103
criminal justice procedures and, 106–112
Judge-made law, 93
Judicial systems, 90–93
black-collar crime, 92–93
dual court system, 90
federal court system, 91–92
misconduct, 92–93
state court systems, 91
Jury
civil justice procedures and, 102–103
criminal justice procedures and, 106–112

K

Kaczynski, Ted, 443
Katrina disaster, 20, 323, 339, 346, 350–352, 350*f*, 354*f*, 355–358, 355*b*–356*b*
Keep out zones, 229
Kelling, George, 221
Kennedy, John F., 19, 348–349
Kennedy, Robert, 19
Key assets, 479–480
Key control, 198–199. *See also* Locks
Key logging programs, 497
Key performance indicators, 264
Key resources, 479–480. *See also* Critical infrastructure and key resources
Key systems, master, 198
Key-in-knob locks, 192
Key-in-lever locks, 192, 193*f*
Khan, Genghis, 9
Kickback, 309
Kidnapping, 563
Kidnapping insurance, 335, 602
King, Martin Luther, Jr., 19
King Edward I, 12
King Hammurabi, 9
King John, 12
Kiosks, 184–185, 185*f*
Kleptomania, 543
Klu Klux Klan, 429, 444
Korean War, 18–19
Koresh, David, 445

L

Labor law, 105–106
Labor matters, investigations of, 277
Labor unions, 17–18
Laminated glass, 232
Lamps, 247–248
Land use controls, 229
Las Vegas MGM Grand Hotel fire, 384
Latches, 191–192, 192*f*
Law. *See also specific laws*
  civil justice procedures, 102–103
  contract law, 101–102
  introduction, 90
  investigative process and, 282–283
  legal quiz, 94
  origins of, 93–94
  questioning and, 111–112
Law enforcement agencies, external relations with, 124–125
Law Enforcement and Private Security: Sources and Areas of Conflict, 41–42
Lawsuits
  answers to, 102
  negligence and, 77, 90, 97*f*, 98
Lay, Kenneth, 315–316
Layered security, 9, 220
Leaderless resistance, 447
Leadership in Energy and Environmental Design (LEED), 222
Leadership styles, 85
Leads, investigative, 291
Learning principles, training programs, 157
LEED (Leadership in Energy and Environmental Design), 222
Leftist class struggles, 442–443
Legal duty, 97
Legal guidelines
  substance abuse in workplace, 608–609
  workplace violence, 591–592
Legal implications, fire marshals and, 371–372
Legal quiz, 94
Legal responsibilities, of banking and finance sector, 560–561
Legal restrictions, information collection and, 290–291
Legal theory, of premises security claims, 97–101
Legislation. *See specific entries*
Legislative authorities, HPH sector, 576–577
Legislative courts, 91–92
Legislative law, 94
Lever tumbler locks, 197
LexisNexis, 290
Liability and negligence investigations, 277
Liability insurance, 338
Libel, 96
License plate recognition systems, 250
Life safety. *See also* Building designs; Emergency management; Fire protection
  defined, 363–364

introduction, 363–368
standards, regulations, codes, 364
Life Safety Code, NFPA 101, 77, 365–367. *See also* National Fire Protection Association
Lighting, 245–249
Lighting equipment, 248–249
Limited jurisdiction courts, 91
Lincoln, Abraham, 14–15, 14*f*
Line of sight, wireless video transmission, 205–206
Line personnel, 83
Link analysis, 293
Litigation bond, 335
Local alarm systems, 242
Local governments, homeland security and, 462–463
Locks. *See also* Doors
attacking, 193–196
automatic locking and unlocking devices, 198
classifying, 191
delayed-egress, 366
described, 190–199
double-cylinder, 192
electromechanical, 191
entrances for handicapped, 192–193, 194*f*
fingerprint-based access control and, 190
key control, 198–199
key-in-knob, 192
key-in-lever, 192, 193*f*
master key systems, 198
mechanical, 191
single-cylinder, 192
standards, 191
types, 191, 197–198
window, 233–234
Lock bumping, 195–196
Lock picking, 195–196
Lock-and-key systems
card access systems *versus*, 198
limitations, 187
purpose of, 190–191
trends with, 198
Locker rooms, healthcare institutions and, 582
Lockheed shootings, 595–596
Lockout/tagout, OSHA, 409–410
LOCOST retail company, 158
Logic bomb, 495–496
Logistics, 353
Lone wolf terrorists, 446
Losses
direct, 29
indirect, 29
total, 29
Loss prevention. *See also* Employee theft; External loss prevention; History of security and loss prevention; Security; Security and loss prevention
attire, 131–132
at checkout counters, 535–539
definition, 7
employment titles for, 8
insurance industry and, 326
meetings, internal relations and, 122
prevention theory, 57, 316
at retail businesses, 532–533
risk control, 328
security compared to, 7
Loss prevention planning. *See* Planning
Loss Prevention Retail Council, 82–83
Loss prevention survey, 67–68
Loss reporting, reward system for, 177
Loss scene, 291–292
Louis XIV (king), 6–7
Low-light-level cameras, 244
Low-pressure sodium lamps, 247
LSD, shoplifter and, 551
L-shaped iron plate, 195, 195*f*
Lumens, 246
Lux, 246
Lying. *See also* Polygraph
indicators of, 287–289
on resumes, 146

M

Madoff, Bernard L., 316
Madoff Ponzi scheme, 316
Magistrate's Courts, U.S., 92
Magna Carta, 12
Magnetic dot cards, 188
Magnetic field, 238, 253, 547. *See also* Intrusion detection systems

Magnetic stripe cards, 188
Mail alert, suspicious, 391*f*, 465–466. *See also* Anthrax attacks
Malcolm X, 442
Malicious prosecutions, tort law, 95
Malware, 495–496
Management countermeasures. *See* Countermeasures
Management theory, emergency management, 346
Manipulation method, safe attack, 212
MANPADS. *See* Man-Portable Air Defense Systems
Man-Portable Air Defense Systems (MANPADS), 512
Manuals, 85
Manufacturing sector, critical, 506–507
Manufacturing security, 37, 37*b*
Marijuana, 129–130, 611
Maritime shipping infrastructure, 512–514
Maritime Transportation Security Act of 2002, 493–494, 513
Market forces, 646
Market segmentation, 118
Marketing concepts, 118–120
Marketing strategies, 119–120
Marking property, 178
Marks, Ralph, 158
Martinez, Matthew, 44
Marxist theory, 346
Maslow, Abraham, 158–160, 159*f*
Mass notification systems, 393, 393*b*
Mass transit and passenger rail subsector, 514–517, 516*b*
Master budget, 66
Master key systems, 198
Material Safety Data Sheet, 407–409
Maxey Tool Company, 152–153
Maxims, of security, 74
May 19th Communist Organization, 442
McAfee Threat Report, 498
McClellan, John A., 14*f*
McCoy's case, internal surveys, 119
McVeigh, Timothy, 434–435, 445–447. *See also* Oklahoma City bombing
Measurements. *See* Metrics
Measuring terrorism, 436–437
Mechanical contact switches, 200
Mechanical locks, 191
Media. *See also* Social media
  designated spokesperson and, 126–127
  external relations with, 126–128
  interviews, guidelines for, 127
Medical examinations, 149
Medical services, fire prevention and, 375–377
Megapixel security cameras, 207
Megatrends (Naisbitt), 19
Mercenaries, 34
Mercury vapor lamps, 247–248
Metal detectors, 179
Metal halide lamps, 247–248
Methodological problems, metrics and, 29–30
Metrics. *See also* External metrics
  business continuity and, 344–345, 344*b*–345*b*
  crime pattern analysis and, 28–29
  defined, 27–28
  internal, 28–29
  methodological problems and, 29–30
  overview, 27–31
  public police performance, 82–83
Metropolitan Police Act, 13
Michigan Mining Company, 339, 339*b*, 359–360
Microdots, 178
Microwave data transmission systems, 243
Microwave motion detectors, 202
Microwave sensor, 239*t*
Middle Ages, 11–12
Middle East terrorism, 430
Military. *See also* Private military and private security companies
  DIB and, 504–506
  emergency management and, 353–358
  installations, security companies and, 34
Mines Act, 403–404
Minorities
  affirmative action, 137, 140
  diversity in workforce and, 141
  in security and loss prevention history, 6
  "Women and Minorities in Security" conference, 141

Miranda warnings, 107, 111, 388
Mirrors, shoplifting reduction and, 549–550
Misdemeanors, 94
Missing children, 580
Missing shirts example, 280–281
Mitigation, 348–349, 351. *See also* Emergency management
Models of security, 5, 75
Modus operandi, 292–293
Money laundering, 562
Money safe, 210
Monitoring and adjusting, risk management process, 329
Monitoring security officers, 251–252
Morgan Stanley, World Trade Center and, 367, 367*b*
Moro, Aldo, 599
Mortuaries, 582
Mosque, 432*f*
Motions, 102
  posttrial, 103
  pretrial, 107
Motivation+Opportunity+Rationalization= Theft, 173
Motive, investigative process and, 292
Movement of packages and property, 185
Movie theater sexual harassment case, 144
Moving surveillance, 296
Muhammad, Prophet, 431
Multicultural and multinational workforce, 643
Multinational and multicultural workforce, 643
Multinational business, risk management in, 332, 332*b*
Multiplex systems, video, 207
Multiplexing, alarm systems and, 243
Mumbai attacks, 438, 594–595
Murrah Federal Building bombing. *See* Oklahoma City bombing
Muslims, 431
Mussolini, Benito, 430
MySpace, 99, 123, 290. *See also* Facebook

N

Naisbitt, John, 19
National Advisory Committee on Criminal Justice Standards and Goals, 40–41
National Board of Fire Underwriters, 15, 326
National Center for Missing and Exploited Children, 580
National Commission on Terrorist Attacks Upon the United States, 459–460
National Communications System, 494–495
National Counterterrorism Center (NCTC), 437, 460
National Crime Information Center, 151
National Fire Data Center, 31
National Fire Protection Association (NFPA)
  code wars and, 77–78
  described, 31, 80
  NFPA 101 Life Safety Code, 77, 365–367
  NFPA 1600: 2007 and 2010 Standard on Disaster/Emergency Management and Business Continuity Programs, 343
  PS-Prep and, 342–343
National Guard, disaster response and, 351
National Incident Management System (NIMS), 353
National Industrial Security Program (NISP), 18, 505–506
National Infrastructure Protection Plan (NIPP)
  defined, 478
  recommendations for private sector, 481–483
  risk management framework, 480–481
National Institute for Occupational Safety and Health (NIOSH), 405–406. *See also* Occupational Safety and Health Administration
National Institute of Standards and Technology (NIST), 253, 368, 497, 624
National Intelligence Reform Act of 2004, 44–45
National Labor Relations Act (NLRA), 105–106, 576
National Labor Relations Board (NLRB), 103, 105–106, 144–145
National Lighting Bureau, 246–247
National Monuments and Icons sector, 520–521
National policy summit, 2004, 42–43
National Preparedness Goal (NPG), 353
National Research Council, 69, 488
National Response Framework (NRF), 353, 369

National Security Strategy of the United States (NSSUS), 455
National Sheriffs' Association, 42
National Strategy for Counterterrorism, 457
National Strategy for Homeland Security (NSHS), 455
National Strategy for the Physical Protection of Critical Infrastructures and Key Assets, 477–478, 490
National Technology Transfer and Advancement Act (NTTAA), 75–76, 78–79
National Terrorism Advisory System (NTAS), 463
National Volunteer Organizations Against Disasters (NVOAD), 469
Natural barriers, 226
Natural disasters. *See also* Earthquakes; Fires; Katrina disaster; Rita hurricane; Tsunamis
  blizzards, 394
  challenges, 20
  described, 393–397
  floods, 394
  list, 26*t*
  pandemics, 395–397
  windstorms, 393–394
Natural gas sector. *See* Oil and natural gas sector
Natural surveillance, 60
NCTC. *See* National Counterterrorism Center
Near field communications (NFC), 189
Needs
  employee, 158–160
  hierarchy of, 158–160, 159*f*
Negligence. *See also* Premises security claims
  defined, 96
  investigations, 277
  lawsuits and, 77, 90, 97*f*, 98
Negligent hiring, 145–146. *See also* Applicant screening
Neighborhood watch, 59, 120, 468–469, 572
Network video recorders (NVRs), 205
Neutralization techniques, 287
New terrorism, 448
Newborn nursery, 580
Newman, Oscar, 59–60, 221
NFC. *See* Near field communications
NFPA. *See* National Fire Protection Association
NFPA 101 Life Safety Code. *See* National Fire Protection Association
NFPA 1600: 2007 and 2010 Standard on Disaster/Emergency Management and Business Continuity Programs, 343
Nichols, Terry, 446–447. *See also* Oklahoma City bombing
Nighttime lighting, 246
NIMS. *See* National Incident Management System
9/11 attacks. *See* September 11, 2001 terrorist attacks
9/11 Commission, 342–343, 460, 627
9/11 Commission Report, 459–462
1910.38 Emergency Action Plans, 364
1993 bombing of World Trade Center. *See* World Trade Center
NIOSH. *See* National Institute for Occupational Safety and Health
NIPP. *See* National Infrastructure Protection Plan
NISP. *See* National Industrial Security Program
NIST. *See* National Institute of Standards and Technology
NLRA. *See* National Labor Relations Act
NLRB. *See* National Labor Relations Board
No contest plea, 107
NOGOs, 227
Nondelegable duty, 101
Nonviolent response and conflict resolution, 594
North American Electric Reliability Council, 488
North American interpretation, of security, 5
NORTH-COM (U.S. Northern Command), 353, 456
Northridge earthquake, 349
Not guilty by reason of insanity, 107
Not guilty plea, 107
NPG. *See* National Preparedness Goal
NRF. *See* National Response Framework
NSHS. *See* National Strategy for Homeland Security
NSSUS. *See* National Security Strategy of the United States
NTAS. *See* National Terrorism Advisory System
NTTAA. *See* National Technology Transfer and Advancement Act

Nuclear power plants, 490–492
Nuclear Regulatory Commission, 488
Nuclear weapons, 453, 465–466
Nuisance alarm rate, 238–240
NVOAD. *See* National Volunteer Organizations Against Disasters
NVRs. *See* Network video recorders

O

Object protection, intrusion detection systems, 240–241
Objectivity, critical thinking and, 5, 7
Observation, safe attack, 212
Observe and report, 38, 250, 265–266
Occupational fraud, 169–170
Occupational Outlook Handbook, 38, 655–656, 658
Occupational qualification, Bona fide, 147
Occupational Safety and Health Administration (OSHA)
  bloodborne pathogens standard, 409
  confined area entry, 410–411, 410*b*–411*b*
  criticism and controversy, 413–415
  defined, 78, 364
  development of, 405–406
  employer responsibilities, 411
  general duty clause, 591–592
  Hazard Communication Standard, 407–409
  HAZWOPER, 409
  inspections, 411–413
  lockout/tagout, 409–410
  1910.38 Emergency Action Plans, 364
  OSHA Act of 1970, 405–406
  poster, 412*f*
  recordkeeping and reporting, 410–411
  regulations, 364–365
  Small Business Handbook, 415
  standards, 407–410
  Strategic Plan, FY 2010–2016, 406
  Whistleblower Protection Program, 411
Office of Civil and Defense Mobilization, 348
Office of Defense Mobilization, 348
Office of Emergency Preparedness, 348–349
Office of Federal Contract Compliance Procedures, 138
Office search, safe attack, 212
Oil and natural gas sector, 489
Oklahoma City bombing (Alfred P. Murrah Federal Building bombing)
  blast walls and, 231
  described, 445–446, 445*f*
  emergency management and, 349–350
  Executive Order 12977 and, 508
  McVeigh and, 434–435, 445–447
  Nichols and, 446–447
  "Vulnerability Assessment of Federal Facilities" study and, 507–508
  warning signals and, 19
Online risks, banking, 564–565
Open-ended questions, 286
Opening statements, 102
Operation Partnership: Trends and Practices in Law Enforcement and Private Security Collaborations, 43
Operational zoning, 203
Operations of enterprises, 24–25
Operations security (OPSEC), 623
Opportunity, employee theft and, 173
OPSEC. *See* Operations security
Organization, basics of, 83–86
Organization charts, 84, 84*f*
Organizational behavior theory, 346
Organizational convergence, 181–182
Organizational countermeasures, against terrorism, 467–468
Organizational mail, suspicious, 391*f*, 465–466. *See also* Anthrax attacks
Organized retail theft (ORT), 543–544, 543*b*–544*b*
Orientation
  employee socialization and, 154–155
  internal relations and, 121
Original viewpoint, constructing, 6
ORT. *See* Organized retail theft
OSAC. *See* Overseas Security Advisory Council
OSHA. *See* Occupational Safety and Health Administration
OSHA's general duty clause, 591–592
Outdated technology, from vendor, 262
Outsourcing, 271–273. *See also* Purchasing decisions

Overseas investigations, 284–285
Overseas Security Advisory Council (OSAC), 43, 285, 467, 483, 514
Overt investigations, 278–281

P

Packages, movement of, 185
Padlocks, 197. *See also* Locks
Palestinian Liberation Organization, 430
Pandemics, 395–397
Paper trail, ABC company, 309*f*
Parking lot and vehicle controls, 249–250
Partnerships. *See* Public-private sector partnerships
Pass back, 187–188
Passenger rail subsector. *See* Mass transit and passenger rail subsector
Passive audio detectors, 202
Passive infrared intrusion sensors, 202
Passive vehicle barriers, 227
Past events, security and, 8–9. *See also* History of security and loss prevention
Patent, 614
Patriot Act of 2001, 19, 104, 282, 458, 477, 560–561, 626. *See also* September 11, 2001 terrorist attacks
Patriot Improvement and Reauthorization Act of 2005, 104, 458
Patriot movement, 446
Pay-by-finger, 538
PCAOB (Public Company Accounting Oversight Board), 317
PCI. *See* Professional Certified Investigator
Pearl Harbor, 18
Peculation, 168–169. *See also* Employee theft
Peel, Robert, 13
Peel or rip method, safe attack, 210
Peel's reforms, 13
Peer reviewed, 63
Peers, external relations with, 128
Penetration tests, 487–488
Pentagon, September 11 attack on, 4, 19, 392, 437–438
People, shoplifting reduction through, 545–546
Performance measures, 82–83
Performance-based fire protection, 385–386, 385*b*–386*b*
Perimeter protection, intrusion detection systems, 240–241, 241*f*
Perimeter security, 222–226, 223*f*
Periodic inventory system, 310–311
Periodicals, security, 63–64
Perpetual inventory system, 311
Personal protective equipment (PPE), 386, 410, 582–583
Personality inventories, 148
Pharmaceutical company security (examples), 225
Pharmacy protection, 581
Phishing, 496–497, 604
Physical ability tests, 148
Physical design, shoplifting reduction through, 546–548
Physical security. *See also* Countermeasures
  convergence of IT and physical security, 183, 255
  hacking into physical security systems, 245, 245*b*
  Journal of Physical Security, 63–64
Physical Security Professional (PSP), 48, 267–268, 272, 506
Physiological needs, 159
Picking locks, 195–196
Pilferage, 168–169. *See also* Employee theft
PIMall, 290
Pin tumbler locks, 197
Pinhole lens camera, 178
Pinkerton, Allan, 14–15, 14*f*
Pipeline industry, 519
Plaintiff, civil justice procedures and, 102–103
Planning
  critical thinking for, 74–75
  for emergencies, 389
  for employee training, 155–157
  executive protection program, 597–598
  as fire prevention strategy, 373
  incomplete protection plans, 72
  management countermeasures for employee theft, 175
  security and loss prevention, 66, 70–73
  from systems perspective, 73–75

Plastics, in windows, 232–233
Plate glass, 231–232
Plato, 10
Plea options, 107
Plea-bargaining, 107
Plinth wall, 227
PMPSCs. *See* Private military and private security companies
Point of view, critical thinking and, 5–6
Point protection, 240–241, 240*f*, 549*f*
Point-of-sale (POS) accounting systems, 536–537
Police. *See* Public police
Policies
  defined, 84, 176
  employee theft prevention and, 176
  policy statement, prosecution and, 129
Policing
  community, 572
  early U.S., 13–14
  Greek city-states and, 10
Polis, 10
Politics, terrorism and, 431–433
Polycarbonate windows, 232–233
Polygraph, 153–154, 288–289
Ponzi scheme, 316
Poor example setting, 158
Pope, Augustus, 17
Pornography, child, 100–101
Port Authority of New York and New Jersey, 146, 366
Portable fire extinguishers, 380–381, 381*b*
Ported coaxial cable, 239*t*
Posse comitatus, 11
Posse Comitatus Act, 11, 353–355
Post-9/11. *See* September 11, 2001 terrorist attacks
Postal and shipping sector, 519–520. *See also* Anthrax attacks
Posters, 412*f*, 418–419. *See also* Occupational Safety and Health Administration
Posttrial motions, 103
Power, 83
PPE. *See* Personal protective equipment
Practices, recommended, 78. *See also* Standards
Praetorian Guard, 10
Precedent cases, 93
Predictable patterns, executive protection program and, 599
Prefire planning, fire department protection efforts, 370
Pregnancy Discrimination Act of 1978, 138
Preliminary hearing, 107
Premises security claims, 97–101, 113
Preparedness discipline, 352. *See also* Emergency management
Pressure-sensitive mats, 200
Pretest-posttest design, 80–81
Pretext interviews, 620
Pretexting, 291
Pretrial conference, 102
Pretrial motions, 107
Prevention theory, 57, 316. *See also* Loss prevention; Security and loss prevention
PRIMA. *See* Public Risk Management Association
Printing plant example, 395, 395*b*
Prior similar incidents rule, 98
Privacy
  business justification *versus* employees' expectations of privacy, 99–101
  invasion of privacy (tort law), 96
Private consulting firms, 82
Private criminal justice system, 130
Private detectives, 38
Private insurance industry, 333
Private investigations, 278
Private military and private security companies (PMPSCs), 34–35
Private organizations, fire protection and, 369
Private sector, 24–25, 466–470, 481–483. *See also* Public-private sector partnerships
Private sector preparedness. *See* Business continuity
Private security. *See also* Security
  administrative legislation's impact on, 104–105
  controls over, 94–97
  public police *versus*, 39–40
  union contracts and, 96–97
Private Security and Police in America: The Hallcrest Report, 41. *See also Hallcrest Reports*

Private Security Officer Employment Authorization Act of 2004, 44–45, 151
Private Security Officer Selection and Training, 45–46
Private security officers
  Miranda warnings and, 111
  power of, 108
  training and, 45–46
Private Security Officers Quality Assurance Act, 44
Private Security Trends: 1970–2000, The Hallcrest Report II, 41
Private undercover investigations, 278–281
Private use, of public records, 152
Privatization, 33–35, 33*b*–35*b*
Privilege, 282–283
Proactive strategies, 40
Probabilities, risk analysis and, 67–68
Probability of detection, 238–240
Probable cause, 106
Procedural law, 94, 95*t*
Procedures
  defined, 84, 176
  employee theft prevention and, 176
Production budget, 66
Profession status, 56. *See also* Security and loss prevention
Professional Certified Investigator (PCI), 48, 267–268, 506
Programmed, self paced instruction, 654–655
Project BioShield Act, 484–485
Project budget, 66
Property
  insurance, 337–339
  marking, 178
  movement of, 185
Prophet Muhammad, 431
Proposal, request for, 269–270
Proprietary investigations, 277–278
Proprietary security
  basics of organization, 83–86
  contract security *versus*, 32–35
  defined, 32
Prosecution decisions, 128–131
  benefits of not prosecuting, 129
  benefits of prosecuting, 128–129
  employee theft and, 180
  marijuana incident, 129–130
  policy statement and, 129
  private criminal justice system, 130
  shoplifters, 553–554
Prosecution threshold, 130
Prosecutors, criminal justice procedures and, 106–112
Protected class, sexual orientation and, 137
Protecting critical infrastructure. *See* Critical infrastructure
Protection. *See also* Executive protection program; Fire protection; Human resources protection
  for colleges and universities, 572–573
  cycle of, 641
  for educational institutions, 567
  executive protection program, 598–603
  methods, workplace violence, 592–594
  pharmacy, 581
  for school districts, 568–570
  workplace violence and, 592–594
Protective dogs, 254
Proximate cause, 97
Proximity cards, 188. *See also* RFID
PSE. *See* Psychological stress evaluator
PSP. *See* Physical Security Professional
PS-Prep (Voluntary Private Sector Preparedness Accreditation and Certification Program), 342–345, 344*b*
Psychological causes, of terrorism, 434
Psychological dependence, 611
Psychological stress evaluator (PSE), 153. *See also* Polygraph
Psychopathic personality disorder, 434
PTZ, 204–205
Public adjusters, 340
Public Company Accounting Oversight Board (PCAOB), 317
Public education, fire department protection efforts, 370–372
Public Health sector. *See* Healthcare and Public Health sector
Public Health Security and Bioterrorism Preparedness and Response Act of 2002, 484–486, 583. *See also* Bioterrorism
Public investigations, 278

Public police
detectives and, 38
performance measures, 82–83
private security *versus*, 39–40
proactive strategies, 40
as public safety agency, 386–387
reactive strategies, 40
Public records, private use of, 152
Public Risk Management Association (PRIMA), 346–347
Public safety, 386
Public safety agencies. *See also* Public police
apps for public safety, 388–389, 388*b*–389*b*
EMS, 387–388
external relations with, 125–126
overview, 386–389
Public sector, 24–25
Public-private sector partnerships
challenges, 41–42
cooperation in, 41–44
false alarm problem and, 47–49
9/11 attacks and, 43–44
Operation Partnership study and, 43
terrorism and, 43–44, 466–470
Pueblo Indians, 9
Punch method, safe attack, 210
Purchase orders, 308
Purchase requisitions, 308
Purchasing. *See also* Accountability
accounting controls for, 310
described, 308–310
invoices and, 308
kickbacks and, 309
loss prevention strategies in, 310
Purchasing decisions for security services and systems
cardinal rules, 263
consultants and, 266–268
contract undercover investigations and, 266
introduction, 261–262
outsourcing, 271–273
pitfalls, 262–263
questions
security services, 263–264
security systems, 268–269
security services, 263–268
security systems, 268–271

Q
Qualified privilege, 282–283
Qualitative risk analysis, 68–70
Quality Loss Prevention Service financial statement, 307*t*
Quantitative risk analysis, 68–70, 328. *See also* Risk management
Quarantine enforcement, 397
Quartz halogen lamps, 247–248
Questioning. *See also* Interviews; Miranda warnings
interrogation and, 285–289
law and, 111–112
Questions. *See also* Investigations; Purchasing decisions
close-ended, 286
investigative, 276
open-ended, 286
purchasing decisions
security services, 263–264
security systems, 268–269
Quid pro quo harassment, 142
Quotas, 140
Quotation, request for, 269

R
Racial profiling, 553
Racial supremacy, 443–444
Radio frequency data transmission systems, 243
Radio Frequency Identification. *See* RFID
Radio transmissions, 9/11 transcripts, 366, 366*b*
Radiological dispersion device (RDD), 452–453
Radiological weapons, 452–453, 465
Rail cars. *See* Freight rail subsector
Railroads, 17
Rand Report, 40–41
Rape
Clery Act and, 572
one free rape rule, 98
Rational choice causes, of terrorism, 434–435
Rational choice theory, 58, 173
Rationalizations, employee theft and, 173
Razor ribbon, 226

RDD. *See* Radiological dispersion device
Reactive strategies, 40
Reasonable doubt, beyond, 95*t*, 107–108
Reasonable force, 110
'Reasonable person' standard, 97
Receiving report, 308–309
Recommended practices, 78. *See also* Standards
Record safe, 209–210
Recording capabilities, of CCTV systems, 207
Recordkeeping, OSHA, 410–411
Recovery discipline, 352. *See also* Emergency management
Recross-examination, 107–108
Redundant security, 9, 220
Reeve, 12
Reference Manual to Mitigate Potential Terrorist Attacks against Buildings. *See* FEMA 426
Refuge, areas of, 377
Refund fraud, 541
Regulations, 78. *See also* Code of Federal Regulations; Occupational Safety and Health Administration; Standards
  confusion over, 75
  defined, 78
  life safety and, 364
  OSHA, 364–365
  post-9/11, 78–79
  security industry, 44–45
Regulation H, Code of Federal Regulations, 561–562
Regulatory authorities, HPH sector, 576–577
Rehabilitation, 57
Rehabilitation Act of 1973, 138
"Reign of Terror", 428–429
Relations. *See* External relations; Internal relations
Reliability, research, 57
Religion, terrorism and. *See* Terrorism
Religious extremists, 444
Remedies for breach of contract, 101
Remote programming, 243
Remote workforce, 167–168. *See also* Telework
Reports. *See specific reports*
Report of the Task Force on Disorders and Terrorism, 441
Report of the Task Force on Private Security, 32, 40–41, 130, 263, 282
Reporting losses, 177
Request for information, 269
Request for proposal, 269–270
Request for quotation, 269
Research
  evidence-based, 82–83
  future, security and loss prevention, 651–654
  reliability, 57
  research assistance sources, 82–83
  on shrinkage, 555
  validity, 57
Resilience, 323. *See also* Business continuity; Emergency management; Risk management
  critical thinking perspective on, 322–324
  risk management and, 325
Respondeat superior, 101–102
Response discipline, 351–352. *See also* Emergency management
Responses to WMD. *See* Weapons of mass destruction
Responsibility, 83
Resumes, applicant screening and, 146–147
Retail loss prevention, 558–559
Retail subsector. *See also* Shoplifting
  CCTV and, 537
  checkout counters loss prevention, 535–539
  human resource problems, 533–534
  internal loss prevention strategies, 534–535
  introduction, 531–532
  loss prevention, 532–533
  socialization and, 534
Return on investment (ROI), 65, 121
Reverse discrimination, 138, 140
Reverse engineering, 618
Reward system, for loss reporting, 177
Reward/safety incentive programs, 416, 416*b*
RF online locking systems, 198
RFID (Radio Frequency Identification), 177–178, 183, 188–189, 311–312. *See also* Near field communications
Right-wing extremists, 444–446
RIMS. *See* Risk and Insurance Management Society
Rip or peel method, safe attack, 210

Risks
defined, 25, 325
first party, 322
shared, 332
third party, 322
Risk abatement, 330
Risk analysis
defined, 66–75
seven-step process, 68
three-step process, 67–68
Risk Analysis and the Security Survey (Broder), 67
Risk and Insurance Management Society (RIMS), 338–339, 338*b*–339*b*
Risk assumption, 330
Risk avoidance approach, 329
Risk communication theory, 326
Risk control, 328. *See also* Loss prevention
Risk financing, 328
Risk identification, risk management process, 328
Risk management. *See also* Business continuity; Emergency management; Resilience
critical infrastructure protection and, 325
defined, 318, 325
described, 325–332
emergency management and, 325, 345–346
ERM, 330–331
ESRM, 330–331
event, 331
financial, 331
in government, 346–347
GRC and, 317–319
in multinational business, 332, 332*b*
resilience and, 325
tools, 329–330
Risk Management: An International Journal, 63
Risk management framework, NIPP, 480–481
Risk management process, 328–329
Risk manager, 326–328
Risk modeling, 329
Risk perception theory, 326, 346
Risk spreading, 330
Risk transfer, 330
Risk treatment, 329–330
Rita hurricane, 20, 339, 356
Robbery. *See* Burglary
Robbery insurance, 335
Robert T. Stafford Disaster Relief and Emergency Assistance Act of 1988, 351–352
Robespierre, Maximilien, 428–429
Rockett's viewpoint on U.S. disaster paradigm, 395, 395*b*
ROI. *See* Return on investment
Rome, Ancient, 10
Roosevelt, Franklin D., 348
Routine activity theory, 59, 173
RPG-7, 450, 450*f*
RSS feeds, 127–128
Ruby Ridge incident, 444
Rucker, Edward A., 14–15

S

SA-7, 450
Safes
attacks, 210–212
burglary-resistive, 210
described, 209–210
with electronic lock, 209*f*
file cabinets, 213
fire-resistive, 209–210
fortification measures for, 212
UL testing of, 210, 211*t*
vaults, 212–213
Safe and Drug-Free Schools and Communities Act, 568
SAFE Port Act. *See* Security and Accountability For Every Port Act of 2006
Safe room, 599–600
Safety. *See also* Occupational Safety and Health Administration; Public safety agencies; Workplace safety
defined, 401
safety hasp, 195, 196*f*
security and safety needs, 159
Safety and health committee, 415
Safety incentive programs, 415–416, 416*b*
Safety performance assessment, 417–418
Safety posters/signs, 418–419
Safety strategies, fire prevention and, 374
Sales budget, 66

Salvation Army, 469
Sarbanes-Oxley (SOX) Act of 2002, 104–105, 150, 177, 282, 316–317, 328, 500, 560–561
SAS Institute, 347
Savings and loan (S&L) scandal, 560
Sawing the bolt, 193–195, 194*f*
SCADAs. *See* Supervisory control and data acquisition systems
SCCs. *See* Sector Coordinating Councils
Scene, of loss, 291–292
School districts
  introduction, 567–572
  legislation for, 568
  protection for, 568–570
Schwab, Charles R., 600
Scientific method, 57, 81
Screening. *See* Applicant screening
Screens, security, 233
"Search and seizure," inspections *versus*, 110–111
Searches
  employee, access controls and, 184
  inspections and, 103–104
  unreasonable, 110–111
SEC. *See* Securities and Exchange Commission
Second wave societies, 19
Secondary explosion, 448
Section 1983, 96
Sectors. *See* Critical infrastructure
Sector Coordinating Councils (SCCs), 483
Sector-specific agencies (SSAs), 479
Sector-specific plans (SSPs), 479
Securicor, 16–17
Securitas, 15
Securities and Exchange Commission (SEC), 104–105, 172, 315, 317, 559, 600
Securities fraud, 315
Security. *See also* Communications security; Contract security; History of security and loss prevention; Homeland Security Department; Information security; Loss prevention; Physical security; Planning; Private security
  academic programs, 62–63
  associations, 64
  attire, 131–132
  aura of, 219–220
  cybersecurity, 499–503
  definition, 7
  employment titles for, 8
  environmental security design, 220–221
  global operations and, 70
  green, 222
  layered, 9, 220
  loss prevention compared to, 7
  manufacturing, 37, 37*b*
  maxims of, 74
  models of, 5, 75
  negative connotations of, 7
  North American interpretation of, 5
  perimeter, 222–226, 223*f*
  periodicals, 63–64
  redundant, 9, 220
  safety and security needs, 159
  U.S.
    early history, 13–14
    security companies, 14–17
Security and Accountability For Every (SAFE) Port Act of 2006, 513–514
Security and loss prevention. *See also* Control center; Future in security and loss prevention; Planning
  certifications in, 267–268
  prevention theory, 57, 316
  profession
    profession status for, 56
    theoretical foundations, 56–59
  programs
    budgeting, 66
    consultants and, 36*t*
    customer-driven, 118
    evaluation of, 80–83
    facilities requiring, 35, 36*t*
    specialists and, 36*t*
Security business proposal, 66
Security companies, 14–17, 34
Security for Houses of Worship Project, 126
Security impact assessments (SIAs), 69–70
Security in depth, 220
Security industry
  challenges of, 40–49

diversity in, 141
ethics and, 46–47
growth factors for, 24
overview, 31–35
regulation of, 44–45
security and loss prevention products and services from, 36*t*
terms and definitions, 24–26
training and, 45–46
Security Industry Association (SIA), 80, 186–187, 242, 272
Security institute concept, 655
Security Journal, 63
Security Management, 48, 64
Security managers
duties, 85
facility managers *versus*, 182–183
new duties, 641
Security master plan, 66
Security officers
armed *versus* unarmed, 251
employee theft prevention and, 209
external threat countermeasures and, 250–253
job responsibilities, 38
monitoring, 251–252
observe and report, 38, 250, 265–266
Security operations, 650
Security risk assessment, Asia-Pacific governments, 69–70
Security sales, equipment, services (security specialty), 272
Security screens, 233
Security services, 261–262. *See also* Purchasing decisions
Security Supervisor, 48
Security systems, 261–262. *See also* Purchasing decisions
Security Systems International (SSI), 270, 273
*See*k other views, 5–6
Seismic sensor cable, 238, 239*t*
Self-actualization needs, 159
Self-defense classes, 120
Self-paced, programmed instruction, 654–655
Senior management support, internal relations, 121
Sensormatic Electronics Corp., 315
Sensors. *See also* Intrusion detection systems. *specific sensors*
defined, 199
dual sensor technologies, 202–203, 203*f*, 238–240
interior, 199–209
supervised wireless sensors, 202–203
window protection and, 234–235
Sentry dogs, 254
September 11, 2001 terrorist attacks. *See also* Patriot Act of 2001; Terrorism; World Trade Center
academic programs after, 649–651
asymmetrical warfare and, 437–438
critical thinking and, 4
9/11 Commission, 342–343, 460, 627
9/11 Commission Report, 459–462
Pentagon attack, 4, 19, 392, 437–438
post-9/11 security challenges, 19–20
post-9/11 standards and regulations, 78–79
release of transcripts, 366, 366*b*
successfulness of, 427
Sequence locking devices, 198
"Seven Deadly Sins of Training", 157–158
Sex discrimination cases, 139–140
Sexting, 144
Sexual harassment, 142–144
cases of, 143–144
defined, 142
hostile working environment, 142
quid pro quo, 142
social media and, 144–145
Sexual orientation, protected class and, 137
Shalloo, Jeremiah, 40
Shared risk, 332
Sharing information, on critical infrastructure, 483, 483*b*
Shatter-resistant film, 232
Sheltering-in-place, 377
Sheriffs, 12
Shiite, 431
Shipping sector. *See* Postal and shipping sector
Shire-reeve, 12
Shires, 12

Shoplifting
civil recovery and, 554
confronting suspect, 550–555
detention and arrest, 551–553
deterrence, international perspective, 554–555, 554*b*–555*b*
LSD user and, 551
motivational factors, 542–543
prevention and reduction
alarms, 548–550
EAS systems, 547–548
mirrors, 549–550
through people, 545–546
through physical design, 546–548
prosecution, 553–554
techniques, 544–545
types of, 542–543
Shopping malls
burglary prevention strategies, 558
examples
applicant screening, 136–137
contract security company, 264–265
WeGuardYou, 242
Shopping service, 534
Shrinkage
defined, 310
research on, 555
retail businesses and, 532–533
Shrubbery, hedges and, 228
SIA. *See* Security Industry Association
SIAs. *See* Security impact assessments
Sicarii, 427–428
Signs
employee theft prevention and, 176
safety, 418–419, 419*f*
Simpson, Carl, 270, 273
Simulation, risk management and, 329
Single-cylinder locks, 192
Situational crime prevention, 59–61, 173
6th Amendment, 109, 508–509
Skeleton key tumbler locks, 197
Skilling, Jeffrey, 315–316
Skimming, 564
Slander, 96
Small Business Handbook, OSHA, 415
Smart cards, 188
Smart grid, 489
Smart security, 638
Smash and grab attacks, 219
Smith Corporation bomb threat form, 390*f*
Smith Shirt manufacturing plant, 166, 170*f*, 213
Smith's lumberyard, employee theft, 172
Smoke detectors, 379–380
Snitches, 542
Sobriety instructions, app for, 388
Social constructionist theory, 346
Social engineering, 496
Social media. *See also* Social networks
counter social media techniques, 279
internal relations and, 123–124
sexual harassment and, 144–145
suggestions for using, 124
venting and, 144–145
Social networks
information source, investigative process and, 289–290
learning from mistakes of others, 99
super controllers and, 62
threats to brand and, 27
Socialization. *See also* Employee socialization
agent of, 158
retailing and, 534
safety strategies and, 415–416
Societal needs, 159
Soft targets, 392, 440, 558, 570–571, 652–653
Software, investigative process and, 293
Solid-core doors, 236
Source tagging, 548
Sources of research assistance, 82–83
Southeast Tool Company, 270, 273
Southeast Transportation Security Council, 43
Southern Manufacturing Plant marijuana incident, 129–130
Souza, Dawnmarie, 144–145
Sovereign citizen extremist movement, 444
SOX. *See* Sarbanes-Oxley Act of 2002
Spain, Norman M., 110
Spam, 496
Span of control, 83
Spartans, 10

Specialists, loss prevention programs and, 36*t*
Specialized emergency management, 347–348
Specimen validity testing, 610
Speed's research, on employee theft, 173–174
Spokesperson, designated, 126–127
Spoofing, 238–240
Spot protection, intrusion detection systems, 240–241
Springing the door, 193, 194*f*
Sprinkler systems, 381–382, 383*f*. *See also* Fire suppression strategies
Sprinkler water flow detectors, 379
Sputnik, 18–19
Spyware, 497
SSAs. *See* Sector-specific agencies
SSI (Security Systems International), 270, 273
SSPs. *See* Sector-specific plans
Staff personnel, 83
Stafford Act, 351–352
Stairwells, 384
Standards, 75–77. *See also* Occupational Safety and Health Administration; Regulations
  CCTV, 204
  competition over, 77–78
  confusion over, 75
  consensus standards, 76–77, 232
  doors, 235–236
  glazing, 232
  green building, 222
  intrusion detection systems, 199, 238
  life safety and, 364
  locks, 191
  post-9/11, 78–79
  PS-Prep, 342–345
Standard of care, 76
Standard-setting organizations, 79–80
Standoff distance, 231
Standpipe and hose systems, 382, 383*f*. *See also* Fire suppression strategies
Standpipes, 382
State constitutions, 93
State court systems, 91
State Department. *See* Department of State
State fire marshals. *See* Fire marshals
State governments, homeland security and, 462–463
State supreme courts, 91
State-sponsored terrorism, 430, 442
Stationary post, 250
Stationary surveillance, 296
Statistics, accident, 403
Status and esteem needs, 159
Statute of Westminster, 12
Stealing. *See* Employee theft
Stinger missile, 450
Stovepiping, 461
Strategic security, 650
Strikebreakers, 13–14, 17–18, 442
Strikes, 392
Structural barriers, 226
Structural causes, of terrorism, 435
Student-Right-to-Know and Campus Security Act of 1990, 572
Students for a Democratic Society, 442
Substance abuse in workplace. *See also* Workplace
  alcoholism, 610
  countermeasures, 607–610
  defined, 606–607
  drug testing, 609–610
  EAPs and, 608
  legal guidelines for, 608–609
  types of, 611
Substantive law, 94, 95*t*
Substitutability, 559
Subtlety, attire and, 131
Suicide bombing, 253, 434, 448–449, 571
Sumatran tsunami, 20, 395
Sunni, 431
Super controllers, 61–62
Supervised wireless sensors, 202–203
Supervisory control and data acquisition systems (SCADAs), 487–488, 487*b*–488*b*
Supply chains
  defined, 342–343
  risks and security, 513–514, 513*b*–514*b*
Supreme Court (U.S.), 92, 138–139
Surety bond, 334–335
Surreptitious entry, 219

Surveillance. *See also* CCTV; Communications security
- audio, 296
- EAS, 547–548
- electronic, 625–628
- hidden surveillance cameras, CCTV, 208
- moving, 296
- natural, 60
- stationary, 296
- Technical Surveillance Countermeasures, 623, 628–632

Surveys
- internal, 119
- loss prevention, 67–68

Suspicious activity reports, 561
Suspicious mail alert, 391*f*, 465–466. *See also* Anthrax attacks
Sustainability, 222
Sutherland, Edwin H., 92, 172
Switch sensors, 234, 235*f*
Switches, 199–200
Systems perspective, security planning and, 73–75
Systems theory, 345

T

Tablet PC, 252, 253*f*
Taco Bell sexual harassment case, 144
Tailgating, 187–188
Taliban, 439, 447, 460
Target hardening, 220
Target marketing, 118
Target-rich environment, 229
Taut wire, 238, 239*t*
TCPs. *See* Transportation choke points
Tear gas/dye packs, 562
Technical Surveillance Countermeasures (TSCM), 623, 628–632
Technological emergencies, 393
Technologies, HRP and, 605–606
Technology convergence, 181–182
Tecsonics protection plan, 72
Tedson Manufacturing Corporation, 208
Teledyne DALSA, 206
Telephone calls, 9/11 transcripts, 366, 366*b*
Telephone monitoring, 100
Telework, 167–168, 218, 397
Teller machines, automatic, 563–564
TEMPEST, 625
10th Amendment, 566
Territoriality, 59–60
Terrorism. *See also* al-Qaida; Bioterrorism; Department of State; Domestic terrorism; FEMA 426; Homeland security; September 11, 2001 terrorist attacks; Weapons of mass destruction
- Antiterrorism and Effective Death Penalty Act of 1996, 458, 560–561
- bombs, 448–449
- building designs and, 228–233, 230*f*
- business countermeasures against, 467–468
- causes of, 433–435
- Chiquita Brands International and, 605
- defined, 426–427
- explosive blasts and, 231
- fingerprint-based access control and, 190
- history of, 427–430
- insurance, 336
- international, 435–441
- legislative action against, 457–459
- measuring, 436–437
- methods, 446–454
- multinational business risk management and, 332, 332*b*
- National Commission on Terrorist Attacks Upon the United States, 459–460
- National Counterterrorism Center, 437, 460
- new, 448
- NTAS and, 463
- nuclear power plants and, 490–492
- organizational countermeasures against, 467–468
- parking lot design and, 250
- public-private sector partnerships and, 43–44, 466–470
- religion and, 427–428, 431–433
- state, 430
- U.S. government action against, 457
- violence, 448

Terrorism Risk Insurance Act of 2002, 336

Testimony, 299, 300*t*–301*t*. *See also* Investigations
Tests, applicant screening and, 148–150
Texas City chemical incident, 493
Theft formula, employee, 173
Theft insurance, 335
Theft of time, 166–167
Thermal detectors, 379
Thermal imaging cameras, 244
Thermally tempered glass, 231–232
Thinking critically. *See* Critical thinking
Third degree, 286–287. *See also* Interrogation
Third party risks, 322
Third wave societies, 19
Third-party administrators, claims and, 340
The Third Wave (Toffler), 19
Threats. *See also* Countermeasures; Employee theft; External threats; Internal threats
  to brand, 27
  defined, 25, 166
  at educational institutions, 566–567
  to HPH, 575–576
  to IT sector, 495–499
  list of, 25–26, 26*t*
  universal, 167–168
Threat assessment, 66–67
Threat management team, 592
Three-Mile Island nuclear meltdown, 490–491
Three-step risk analysis process, 67–68
Tiger traps, 227
Tildesley, burglar alarm and, 17
Timed lights, 120
TISP. *See* Infrastructure Security Partnership
Tithing, 12
Title VII, Civil Rights Act of 1964, 137
Toffler, Alvin, 19, 637–638
Tolerance, 611. *See also* Substance abuse in workplace
Topics of concern. *See* Communications security; Human resources protection; Information security; Substance abuse; Workplace violence
Torch method, safe attack, 210
Tort law
  defined, 94–95, 95*t*
  described, 94–97
  match the tort (quiz), 96
  sexual harassment and, 142
Tosti, Donald J., 157–158
Total cost of risk, *Benchmark Survey*, 338–339
Total losses, 29
Totality of the circumstances test, 98
Touch button technology, 252
Toxic cargo, freight rail and, 519, 519*b*
Trace detection, 610
Trade secret, 614
Tradecraft, 650
Trademark, 614
Traditional uniforms, blazers *versus*, 131–132, 131*f*
Traffic calming strategies, 250
Training
  for emergencies, 389
  executive protection program, 598
  fire prevention strategies and, 377–378
  fire suppression strategies and, 385–386
  future, security and loss prevention, 654–655
  learning principles and, 157
  orientation and, 121
  overview, 155
  planning, 155–157
  security industry and, 45–46
  wasted, 157–158
Transnational organized crime, 644
Transportation and border security, 521–524
Transportation choke points (TCPs), 517
Transportation Security Administration (TSA), 509–520
Transportation System Sector-Specific Plan, 509, 517
Transportation systems
  aviation system, 509–512
  freight rail subsector, 518–519, 519*b*
  highway infrastructure and motor carrier subsector, 517
  introduction, 509–520
  maritime shipping infrastructure, 512–514
  mass transit and passenger rail subsector, 514–517, 516*b*
  pipeline industry, 519
  postal and shipping sector, 519–520

Trespass to land, 95
Trespass to personal property, 95
Triage, 387–388
Trials
  civil justice procedures and, 102–103
  criminal justice procedures and, 106–112
  defined, 107–108
Tricks recognition, executive protection program, 599
Trico Corporation financial statement, 307*t*
Trip wire sensors, 200
Trojan horse, 495–496
True first responders, 386
Truman, Harry S., 442–443
TSA. *See* Transportation Security Administration
TSCM. *See* Technical Surveillance Countermeasures
Tsunamis
  Japan, 20, 356–357, 395
  Sumatra, 20, 395
Twin Towers. *See* World Trade Center
Twitter
  abbreviated URLs, 498–499
  benefits, 99
  espionage and, 619
  as information source, 289
  internal relations and, 124
  news releases and, 127–128
  super controllers and, 62
  threats to brand and, 27

U

UCR. *See* Uniform Crime Report
UL. *See* Underwriters Laboratories
Ultrasonic motion detectors, 201–202
Unabomber, 443
Unarmed security officers, armed *versus*, 251
Unauthorized entry, 218–219
"Under color of state law", 96
Undercover investigations
  contract, 266
  private, 278–281
Understanding Risky Facilities, 62
Understanding viewpoints, 5
Underwriters Laboratories (UL)
  described, 79
  testing, 210, 211*t*
Uniform Crime Report (UCR), 30–31, 572
"Uniform Guidelines on Employee Selection Procedures", 145
Uniformed Services Employment and Reemployment Rights Act of 1994, 138
Uniforms, blazers *versus*, 131–132, 131*f*
Unintentional discrimination, 138
Union contracts, private security and, 96–97
United Nations, 31
United States. *See* U.S.
Unity of command, 83
Universal threats, 167–168
Universities. *See* Colleges
University researchers, 82
Unlocking devices. *See* Locks
Unreasonable searches, 110–111
Unshielded twisted-pair cabling, CCTV and, 205
U.S. Citizenship and Immigration Services, 523–524
U.S. Constitution. *See* Constitution
U.S. Courts of Appeals, 92
U.S. Customs and Border Protection, 513, 521–522
U.S. District Courts, 92
U.S. Drug Enforcement Administration, 611
U.S. Fire Administration (USFA), 31, 368–369, 375–377
U.S. Immigration and Customs Enforcement (ICE), 522
U.S. Magistrate's Courts, 92
U.S. Marshals Service (USMS), 507–508
U.S. Northern Command (NORTH-COM), 353, 456
U.S. Secret Service Electronic Crimes Task Forces and Working Groups, 43
U.S. security. *See* Security
U.S. Supreme Court. *See* Supreme Court
U.S. Visitor and Immigrant Status Indicator Technology (US-VISIT), 521
USA Freedom Corps, 469–470
USA Patriot Act. *See* Patriot Act of 2001
USA Patriot Improvement and Reauthorization Act. *See* Patriot Improvement and Reauthorization Act of 2005

USC Section 1983, 96
USFA. *See* U.S. Fire Administration
USMS. *See* U.S. Marshals Service
USSearch, 290
US-VISIT. *See* U.S. Visitor and Immigrant Status Indicator Technology

V

Vagrant shoplifters, 543
Validity, research, 57
Value added, 64–65
Vaults, 212–213. *See also* Safes
VCRs, 204–205. *See also* CCTV
Vehicle and parking lot controls, 249–250
Vehicle barriers, 227
Vehicle patrols, 250
Vendors
  defined, 262
  security services and systems pitfalls from, 262–263
Venting, 144–145
Verdict, 103, 108
Vertical bolts, locks and, 198
Vibration sensors, 200, 200*f*, 234
Vicarious liability, 101–102
Victims, investigative process and, 292
Video
  analytics, 208
  DVRs, 205
  multiplex systems, 207
  NVRs, 205
  wireless video transmission, CCTV, 205–206
Video analytics, 208
Video motion detection (VMD), 207–208, 239*t*
Viewpoints, critical thinking and, 5–6
Vigiles, 10
Violence. *See also* Terrorism; Workplace violence
  Guide for Preventing and Responding to School Violence, 570
  healthcare, 580–581
  terrorist, 448
Violent Gangs and Terrorism Organization File, 44–45
Virginia Tech massacre, 571, 594–595
Viruses, 495–496
Visa security, 540, 540*b*
Visibility, attire and, 131
Visitors, access controls and, 184–185
VMD. *See* Video motion detection
Voice over Internet Protocol (VoIP) technology, 631
Voice stress analysis, 153
Voluntary Private Sector Preparedness Accreditation and Certification Program. *See* PS-Prep
Vulnerabilities
  assessment, 66–67
  to defeat, 238–240
  identifying, 68
"Vulnerability Assessment of Federal Facilities" study, 507–508

W

Wackenhut, George, 16–17
Wackenhut Corporation, 16–17
Waco incident, 445
Walls
  antiramming, 231
  as barriers, 228
  blast, 231
  fire walls, 382
WalMart sex discrimination case, 140
Warded locks, 197
Warfare, asymmetrical, 437–438
Warner Act, 351–352
Wars of globalization, 645–646
Wasted training, 157–158
Watch and ward, 12
Watch system, 13
Water sector, 485–488
Weak points, barriers, 228
Weapon of mass effect (WME), 451
Weapons of mass destruction (WMD)
  biological, 452
  chemical, 451, 463–464
  described, 392, 451
  nuclear, 453, 465–466
  radiological, 452–453, 465
Weather Underground Organization, 442
Weaver, Randy, 444

Weberian theory, 346
Websites, resumes on, 147
WeGuardYou, 242
Weigand cards, 188
Weingarten rights, 106
Weisburd, David, 651–652
Wells, Henry, 15
Wells Fargo, 16*f*
Westminster, Statute of, 12
Wet-pipe automatic sprinkler system, 382, 383*f*
Whistleblower Protection Program, OSHA, 411
White-collar crime, 92
Wickersham Commission, 40
William, Duke of Normandy, 12
William J. Burns Detective Agency, 15, 16*f*
William Steiger Occupational Safety and Health Act of 1970, 405. *See also* Occupational Safety and Health Administration
Wilson, James Q., 221
Window foil, 234
Window grating, 233
Window locks, 233–234
Window protection, 231–235
Windstorms, 393–394
Wireless locking systems, 198
Wireless sensors, supervised, 202–203
Wireless video transmission, CCTV, 205–206
Wire-reinforced glass, 231–232
Wiretapping, 625–628
Withdrawal, 611
Witnesses, investigative process and, 292
WMD. *See* Weapons of mass destruction
WME. *See* Weapon of mass effect
Women. *See also* Sexual harassment
  diversity in workforce and, 141
  HRP and, 603–604
  in security and loss prevention history, 6
  "Women and Minorities in Security" conference, 141
Woody's Lumber Company, 166, 169*f*, 213, 218–219
Work breakdown structure, 269–270
Work products, 65
Workers' compensation insurance, 338, 404–405
Workforce
  diversity in, 141
  remote, 167–168
  telework, 167–168, 218, 397
Workplace. *See also* Employees; Life safety; Occupational Safety and Health Administration; Substance abuse
  child pornography incident, 100–101
  harassment, 142–144
  hostile working environment, 142
  human relations and, 120–121
  investigations, 277
Workplace safety
  assistance with problems, 414–415, 414*b*–415*b*
  hazard examples, 401–403
  history of legislation, 403–406
  introduction, 415–419
Workplace violence
  active shooter and, 594–596
  defined, 590
  introduction, 590–596
  legal guidelines, 591–592
  Lockheed shootings and, 595–596
  prevention program, 592–594
  protection methods, 592–594
World Health Organization, 31
World Trade Center. *See also* September 11, 2001 terrorist attacks
  critical thinking and, 4
  Morgan Stanley and, 367, 367*b*
  9/11 attacks on, 4, 19, 352, 366–368, 366*b*, 433*f*
  1993 bombing, 19, 232, 349–350, 367, 457–458
World War I, 18
World War II, 18
WorldCom, 315–316
Worms, 495–496

X

X-ray equipment, safe attack and, 212
XYZ warehouse, undercover investigation, 279–281

Y

Yates, Ken, 98–99, 113
You Be the Judge

Conway Excavation, 336–337, 336*b*–337*b*, 359
Hi-Mark Home case, 141–142, 163
Michigan Mining Corporation, 339, 339*b*, 359–360
premises security claims, 98–99, 113
Southeast Tool Company, 270, 273
You Decide!
Chandler Office building, 61
corporation restoration project, 252, 252*b*
Maxey Tool Company, 152–153
printing plant example, 395, 395*b*
shopping mall and contract security company, 264–265
Tedson Manufacturing Corporation, 208
YouTube
as information source, 289
internal relations and, 124
lock picking and, 193
shoplifter information, 545
super controllers and, 62
threats to brand and, 27

Z

Zaleszny, Bronislav, 141–142, 163
Zealot-Sicarii, 427–428
Zero-day exploits, 642
Zero-tolerance policy, 568
Zoning, operational, 203
Zoonotic diseases, 395–396